Fundamentals of Digital Image Processing

PRENTICE HALL INFORMATION AND SYSTEM SCIENCES SERIES
Thomas Kailath, Editor

ANDERSON & MOORE	*Optimal Control: Linear Quadratic Methods*
ANDERSON & MOORE	*Optimal Filtering*
ASTROM & WITTENMARK	*Computer-Controlled Systems: Theory and Design, 2/E*
DICKINSON	*Systems: Analysis, Design and Computation*
GARDNER	*Statistical Spectral Analysis: A Nonprobabilistic Theory*
GOODWIN & SIN	*Adaptive Filtering, Prediction, and Control*
GRAY & DAVISSON	*Random Processes: A Mathematical Approach for Engineers*
HAYKIN	*Adaptive Filter Theory, 2/E*
JAIN	*Fundamentals of Digital Image Processing*
JOHNSON	*Lectures on Adaptive Parameter Estimation*
KAILATH	*Linear Systems*
KUMAR & VARAIYA	*Stochastic Systems: Estimation, Indentification, and Adaptive Control*
KUNG	*VLSI Array Processors*
KUNG, WHITEHOUSE, & KAILATH, EDS.	*VLSI and Modern Signal Processing*
KWAKERNAAK & SIVAN	*Modern Signals and Systems*
LANDAU	*System Identification and Control Design Using PIM+ Software*
LIUNG	*System Identification: Theory for the User*
MACOVSKI	*Medical Imaging Systems*
MIDDLETON & GOODWIN	*Digital Control and Estimation: A Unified Approach*
NARENDRA & ANNASWAMY	*Stable Adaptive Systems*
SASTRY & BODSON	*Adaptive Control: Stability, Convergence, and Robustness*
SOLIMAN & SRINATH	*Continuous and Discrete Signals and Systems*
SPILKER	*Digital Communications by Satellite*
WILLIAMS	*Designing Digital Filters*

Fundamentals of Digital Image Processing

ANIL K. JAIN
University of California, Davis

PRENTICE HALL, Englewood Cliffs, NJ 07632

Library of Congress Cataloging-in-Publication Data

JAIN, ANIL K.
Fundamentals of digital image processing.

Bibliography: p.
Includes index.
1. Image processing—Digital techniques. I. Title.
TA1632.J35 1989 621.36'7 88-12624
ISBN 0-13-336165-9

Editorial/production supervision: Colleen Brosnan
Manufacturing buyer: Mary Noonan
Page layout: Martin Behan
Cover design: Diane Saxe
Logo design: A.M. Bruckstein
Cover art: Halley's comet image by the author
reconstructed from data gathered by NASA's
Pioneer Venus Orbiter in 1986.

Printed in the United States of America

10 9 8 7

ISBN 0-13-336165-9

PRENTICE-HALL INTERNATIONAL (UK) LIMITED, *London*
PRENTICE-HALL OF AUSTRALIA PTY. LIMITED, *Sydney*
PRENTICE-HALL CANADA INC., *Toronto*
PRENTICE-HALL HISPANOAMERICANA, S.A., *Mexico*
PRENTICE-HALL OF INDIA PRIVATE LIMITED, *New Delhi*
PRENTICE-HALL OF JAPAN, INC., *Tokyo*
SIMON & SCHUSTER ASIA PTE. LTD., *Singapore*
EDITORA PRENTICE-HALL DO BRASIL, LTDA., *Rio de Janeiro*

Contents

PREFACE — ***xix***

ACKNOWLEDGMENTS — ***xxi***

1 INTRODUCTION — ***1***

1.1 Digital Image Processing: Problems and Applications 1

1.2 Image Representation and Modeling 4

1.3 Image Enhancement 6

1.4 Image Restoration 7

1.5 Image Analysis 7

1.6 Image Reconstruction from Projections 8

1.7 Image Data Compression 9

Bibliography 10

2 TWO-DIMENSIONAL SYSTEMS AND MATHEMATICAL PRELIMINARIES — ***11***

2.1 Introduction 11

2.2 Notation and Definitions 11

2.3 Linear Systems and Shift Invariance 13

2.4 The Fourier Transform 15
Properties of the Fourier Transform, 16
Fourier Transform of Sequences (Fourier Series), 18

2.5 The Z-Transform or Laurent Series 20
Causality and Stability, 21

2.6 Optical and Modulation Transfer Functions 21

2.7 Matrix Theory Results 22
Vectors and Matrices, 22
Row and Column Ordering, 23
Transposition and Conjugation Rules, 25
Toeplitz and Circulant Matrices, 25
Orthogonal and Unitary Matrices, 26
Positive Definiteness and Quadratic Forms, 27
Diagonal Forms, 27

2.8 Block Matrices and Kronecker Products 28
Block Matrices, 28
Kronecker Products, 30
Separable Operations, 31

2.9 Random Signals 31
Definitions, 31
Gaussian or Normal Distribution, 32
Gaussian Random Processes, 32
Stationary Processes, 32
Markov Processes, 33
Orthogonality and Independence, 34
The Karhunen Loève (KL) Transform, 34

2.10 Discrete Random Fields 35
Definitions, 35
Separable and Isotropic Covariance Functions, 36

2.11 The Spectral Density Function 37
Properties of the SDF, 38

2.12 Some Results from Estimation Theory 39
Mean Square Estimates, 40
The Orthogonality Principle, 40

2.13 Some Results from Information Theory 41
Information, 42
Entropy, 42
The Rate Distortion Function, 43

Problems 44

Bibliography 47

3 IMAGE PERCEPTION 49

3.1 Introduction 49

3.2 Light, Luminance, Brightness, and Contrast 49
Simultaneous Contrast, 51
Mach Bands, 53

3.3 MTF of the Visual System 54

3.4 The Visibility Function 55

3.5 Monochrome Vision Models 56

3.6 Image Fidelity Criteria 57

3.7 Color Representation 60

3.8 Color Matching and Reproduction 62
Laws of Color Matching, 63
Chromaticity Diagram , 65

3.9 Color Coordinate Systems 66

3.10 Color Difference Measures 71

3.11 Color Vision Model 73

3.12 Temporal Properties of Vision 75
Bloch's Law, 75
Critical Fusion Frequency (CFF), 75
Spatial versus Temporal Effects, 75

Problems 76

Bibliography 78

4 IMAGE SAMPLING AND QUANTIZATION 80

4.1 Introduction 80
Image Scanning, 80
Television Standards, 81
Image Display and Recording, 83

4.2 Two-Dimensional Sampling Theory 84
Bandlimited Images, 84
Sampling Versus Replication, 85
Reconstruction of the Image from Its Samples, 85
Nyquist Rate, Aliasing, and Foldover Frequencies, 87
Sampling Theorem, 88
Remarks, 89

4.3 Extensions of Sampling Theory 89
Sampling Random Fields, 90
Sampling Theorem for Random Fields, 90
Remarks, 90
Nonrectangular Grid Sampling and Interlacing, 91
Hexagonal Sampling, 92
Optimal Sampling, 92

4.4 Practical Limitations in Sampling and Reconstruction 93
Sampling Aperture, 93
Display Aperture/Interpolation Function, 94
Lagrange Interpolation, 98
Moire Effect and Flat Field Response, 99

4.5 Image Quantization 99

4.6 The Optimum Mean Square or Lloyd-Max Quantizer 101
The Uniform Optimal Quantizer, 103
Properties of the Optimum Mean Square Quantizer, 103
Proofs, 112

4.7 A Compandor Design 113
Remarks, 114

4.8 The Optimum Mean Square Uniform Quantizer for Nonuniform Densities 115

4.9 Examples, Comparison, and Practical Limitations 115

4.10 Analytic Models for Practical Quantizers 118

4.11 Quantization of Complex Gaussian Random Variables 119

4.12 Visual Quantization 119
Contrast Quantization, 120
Pseudorandom Noise Quantization, 120
Halftone Image Generation, 121
Color Quantization, 122

Problems 124

Bibliography 128

5 IMAGE TRANSFORMS 132

5.1 Introduction 132

5.2 Two-Dimensional Orthogonal and Unitary Transforms 134
Separable Unitary Transforms, 134
Basis Images, 135
Kronecker Products and Dimensionality, 137
Dimensionality of Image Transforms, 138
Transform Frequency, 138
Optimum Transform, 138

5.3 Properties of Unitary Transforms 138
Energy Conservation and Rotation, 138
Energy Compaction and Variances of Transform Coefficients, 139
Decorrelation, 140
Other Properties, 140

5.4 The One-Dimensional Discrete Fourier Transform (DFT) 141
Properties of the DFT/Unitary DFT, 141

5.5 The Two-Dimensional DFT 145
Properties of the Two-Dimensional DFT, 147

5.6 The Cosine Transform 150
Properties of the Cosine Transform, 151

5.7 The Sine Transform 154
Properties of the Sine Transform, 154

5.8 The Hadamard Transform 155
Properties of the Hadamard Transform, 157

5.9 The Haar Transform 159
Properties of the Haar Transform, 161

5.10 The Slant Transform 161
Properties of the Slant Transform, 162

5.11 The KL Transform 163
KL Transform of Images, 164
Properties of the KL Transform, 165

5.12 A Sinusoidal Family of Unitary Transforms 175
Approximation to the KL Transform, 176

5.13 Outer Product Expansion and Singular Value Decomposition 176
Properties of the SVD Transform, 177

5.14 Summary 180

Problems 180

Bibliography 185

6 *IMAGE REPRESENTATION BY STOCHASTIC MODELS* 189

6.1 Introduction 189
Covariance Models, 189
Linear System Models, 189

6.2 One-Dimensional Causal Models 190
Autoregressive (AR) Models, 190
Properties of AR Models, 191
Application of AR Models in Image Processing, 193
Moving Average (MA) Representations, 194
Autoregressive Moving Average (ARMA) Representations, 195
State Variable Models, 195
Image Scanning Models, 196

6.3 One-Dimensional Spectral Factorization 196
Rational SDFs, 197
Remarks, 198

6.4 AR Models, Spectral Factorization, and Levinson Algorithm 198
The Levinson-Durbin Algorithm, 198

6.5 Noncausal Representations 200
Remarks, 201
Noncausal MVRs for Autoregressive Sequences, 201
A Fast KL Transform, 202
Optimum Interpolation of Images, 204

6.6 Linear Prediction in Two Dimensions 204
Causal Prediction, 205
Semicausal Prediction, 206
Noncausal Prediction, 206
Minimum Variance Prediction, 206
Stochastic Representation of Random Fields, 207
Finite-Order MVRs, 208
Remarks, 209
Stability of Two-Dimensional Systems, 212

6.7 Two-Dimensional Spectral Factorization and Spectral Estimation Via Prediction Models 213
Separable Models, 214
Realization of Noncausal MVRs, 215
Realization of Causal and Semicausal MVRs, 216
Realization via Orthogonality Condition, 216

6.8 Spectral Factorization via the Wiener-Doob Homomorphic Transformation 219
Causal MVRs, 220
Semicausal WNDRs, 220
Semicausal MVRs, 222
Remarks and Examples, 222

6.9 Image Decomposition, Fast KL Transforms, and Stochastic Decoupling 223
Periodic Random Fields, 223
Noncausal Models and Fast KL Transforms, 224
Semicausal Models and Stochastic Decoupling, 225

6.10 Summary 226

Problems 227

Bibliography 230

7 *IMAGE ENHANCEMENT* 233

7.1 Introduction 233

7.2 Point Operations 235
Contrast Stretching, 235
Clipping and Thresholding, 235
Digital Negative, 238
Intensity Level Slicing, 238
Bit Extraction, 239
Range Compression, 240
Image Subtraction and Change Detection, 240

7.3 Histogram Modeling 241
Histogram Equalization, 241
Histogram Modification, 242
Histogram Specification, 243

7.4 Spatial Operations 244
Spatial Averaging and Spatial Low-pass Filtering, 244
Directional Smoothing, 245
Median Filtering, 246
Other Smoothing Techniques, 249
Unsharp Masking and Crispening, 249
Spatial Low-pass, High-pass and Band-pass Filtering, 250
Inverse Contrast Ratio Mapping and Statistical Scaling, 252
Magnification and Interpolation (Zooming), 253
Replication, 253
Linear Interpolation, 253

7.5 Transform Operations 256
Generalized Linear Filtering, 256
Root Filtering, 258
Generalized Cepstrum and Homomorphic Filtering, 259

7.6 Multispectral Image Enhancement 260
Intensity Ratios, 260
Log-Ratios, 261
Principal Components, 261

7.7 False Color and Pseudocolor 262

7.8 Color Image Enhancement 262

7.9 Summary 263

Problems 263

Bibliography 265

8 IMAGE FILTERING AND RESTORATION **267**

8.1 Introduction 267

8.2 Image Observation Models 268
Image Formation Models, 269
Detector and Recorder Models, 273
Noise Models, 273
Sampled Image Observation Models, 275

8.3 Inverse and Wiener Filtering 275
Inverse Filter, 275
Pseudoinverse Filter, 276
The Wiener Filter, 276
Remarks, 279

8.4 Finite Impulse Response (FIR) Wiener Filters 284
Filter Design, 284
Remarks, 285
Spatially Varying FIR Filters, 287

8.5 Other Fourier Domain Filters 290
Geometric Mean Filter, 291
Nonlinear Filters, 291

8.6 Filtering Using Image Transforms 292
Wiener Filtering, 292
Remarks, 293
Generalized Wiener Filtering, 293
Filtering by Fast Decompositions, 295

8.7 Smoothing Splines and Interpolation 295
Remarks, 297

8.8 Least Squares Filters 297
Constrained Least Squares Restoration, 297
Remarks, 298

8.9 Generalized Inverse, SVD, and Iterative Methods 299
The Pseudoinverse, 299
Minimum Norm Least Squares (MNLS) Solution and the Generalized Inverse, 300
One-step Gradient Methods, 301
Van Cittert Filter, 301
The Conjugate Gradient Method, 302
Separable Point Spread Functions, 303

8.10 Recursive Filtering For State Variable Systems 304
Kalman Filtering, 304
Remarks, 307

8.11 Causal Models and Recursive Filtering 307
A Vector Recursive Filter, 308
Stationary Models, 310
Steady-State Filter, 310
A Two-Stage Recursive Filter, 310
A Reduced Update Filter, 310
Remarks, 311

8.12 Semicausal Models and Semirecursive Filtering 311
Filter Formulation, 312

8.13 Digital Processing of Speckle Images 313
Speckle Representation, 313
Speckle Reduction: N-Look Method, 315
Spatial Averaging of Speckle, 315
Homomorphic Filtering, 315

8.14 Maximum Entropy Restoration 316
Distribution-Entropy Restoration, 317
Log-Entropy Restoration, 318

8.15 Bayesian Methods 319
Remarks, 320

8.16 Coordinate Transformation and Geometric Correction 320

8.17 Blind Deconvolution 322

8.18 Extrapolation of Bandlimited Signals 323
Analytic Continuation, 323
Super-resolution, 323
Extrapolation Via Prolate Spheroidal Wave Functions (PSWFs), 324
Extrapolation by Error Energy Reduction, 324
Extrapolation of Sampled Signals, 326
Minimum Norm Least Squares (MNLS) Extrapolation, 326
Iterative Algorithms, 327
Discrete Prolate Spheroidal Sequences (DPSS), 327
Mean Square Extrapolation, 328
Generalization to Two Dimensions, 328

8.19 Summary 330

Problems 331

Bibliography 335

9 IMAGE ANALYSIS AND COMPUTER VISION 342

9.1 Introduction 342

9.2 Spatial Feature Extraction 344
Amplitude Features, 344
Histogram Features, 344

9.3 Transform Features 346

9.4 Edge Detection 347
Gradient Operators, 348
Compass Operators, 350
Laplace Operators and Zero Crossings, 351
Stochastic Gradients, 353
Performance of Edge Detection Operators, 355
Line and Spot Detection, 356

9.5 Boundary Extraction 357
Connectivity, 357
Contour Following, 358
Edge Linking and Heuristic Graph Searching, 358
Dynamic Programming, 359
Hough Transform, 362

9.6 Boundary Representation 362
Chain Codes, 363
Fitting Line Segments, 364
B-Spline Representation, 364
Fourier Descriptors, 370
Autoregressive Models, 374

9.7 Region Representation 375
Run-length Codes, 375
Quad-Trees, 375
Projections, 376

9.8 Moment Representation 377
Definitions, 377
Moment Representation Theorem, 378
Moment Matching, 378
Orthogonal Moments, 379
Moment Invariants, 380
Applications of Moment Invariants, 381

9.9 Structure 381
Medial Axis Transform, 381
Morphological Processing, 384
Morphological Transforms, 387

9.10 Shape Features 390
Geometry Features, 391
Moment-Based Features, 392

9.11 Texture 394
Statistical Approaches, 394
Structural Approaches, 398
Other Approaches, 399

9.12 Scene Matching and Detection 400
Image Subtraction, 400
Template Matching and Area Correlation, 400
Matched Filtering, 403
Direct Search Methods, 404

9.13 Image Segmentation 407
Amplitude Thresholding or Window Slicing, 407
Component Labeling, 409
Boundary-based Approaches, 411
Region-based Approaches and Clustering, 412
Template Matching, 413
Texture Segmentation, 413

9.14 Classification Techniques 414
Supervised Learning, 414
Nonsupervised Learning or Clustering, 418

9.15 Image Understanding 421

Problems 422

Bibliography 425

10 IMAGE RECONSTRUCTION FROM PROJECTIONS 431

10.1 Introduction 431
Transmission Tomography, 431
Reflection Tomography, 432
Emission Tomography, 433
Magnetic Resonance Imaging, 434
Projection-based Image Processing, 434

10.2 The Radon Transform 434
Definition, 434
Notation, 436
Properties of the Radon Transform, 437

10.3 The Back-projection Operator 439
Definition, 439
Remarks, 440

10.4 The Projection Theorem 442
Remarks, 443

10.5 The Inverse Radon Transform 444
Remarks, 445
Convolution Back-projection Method, 446
Filter Back-projection Method, 446
Two-Dimensional Filtering via the Radon Transform, 447

10.6 Convolution/Filter Back-projection Algorithms: Digital Implementation 448
Sampling Considerations, 448
Choice of Filters, 448
Convolution Back-projection Algorithm, 450
Filter Back-projection Algorithm, 451
Reconstruction Using a Parallel Pipeline Processor, 452

10.7 Radon Transform of Random Fields 452
A Unitary Transform $\tilde{R}$, 452
Radon Transform Properties for Random Fields, 456
Projection Theorem for Random Fields, 457

10.8 Reconstruction from Blurred Noisy Projections 458
Measurement Model, 458
The Optimum Mean Square Filter, 458
Remarks, 458

10.9 Fourier Reconstruction 462
Algorithm, 462
Reconstruction of Magnetic Resonance Images, 463

10.10 Fan-Beam Reconstruction 464

10.11 Algebraic Methods 465
The Reconstruction Problem as a Set of Linear Equations, 465
Algebraic Reconstruction Techniques, 466

10.12 Three-Dimensional Tomography 468
Three-Dimensional Reconstruction Algorithms, 469

10.13 Summary 470

Problems 470

Bibliography 473

11 IMAGE DATA COMPRESSION 476

11.1 Introduction 476
Image Raw Data Rates, 476
Data Compression versus Bandwidth Compression, 477
Information Rates, 477
Subsampling, Coarse Quantization, Frame Repetition, and Interlacing, 479

11.2 Pixel Coding 479
PCM, 480
Entropy Coding, 480
Run-Length Coding, 481
Bit-Plane Encoding, 483

11.3 Predictive Techniques 483
Basic Principle, 483
Feedback versus Feedforward Prediction, 484
Distortionless Predictive Coding, 485
Performance Analysis of DPCM, 486
Delta Modulation, 488
Line-by-Line DPCM, 490
Two-Dimensional DPCM, 491
Performance Comparisons, 493
Remarks, 494
Adaptive Techniques, 495
Other Methods, 497

11.4 Transform Coding Theory 498
The Optimum Transform Coder, 498
Proofs, 499
Remarks, 501
Bit Allocation and Rate-Distortion Characteristics, 501

11.5 Transform Coding of Images 504
Two-Dimensional Coding Algorithm, 504
Transform Coding Performances Trade-offs and Examples, 507
Zonal versus Threshold Coding, 508
Fast KL Transform Coding, 510
Remarks, 512
Two-Source Coding, 513
Transform Coding under Visual Criterion, 515
Adaptive Transform Coding, 515
Summary of Transform Coding, 516

11.6 Hybrid Coding and Vector DPCM 518
Basic Idea, 518
Adaptive Hybrid Coding, 520
Hybrid Coding Conclusions, 521

11.7 Interframe Coding 521
Frame Repetition, 521
Resolution Exchange, 521
Conditional Replenishment, 522
Adaptive Predictive Coding, 522
Predictive Coding with Motion Compensation, 524
Interframe Hybrid Coding, 527
Three-Dimensional Transform Coding, 529

11.8 Image Coding in the Presence of Channel Errors 532
The Optimum Mean Square Decoder, 532
The Optimum Encoding Rule, 533
Optimization of PCM Transmission, 534
Channel Error Effects in DPCM, 536
Optimization of Transform Coding, 537

11.9 Coding of Two Tone Images 540
Run-length Coding, 540
White Block Skipping, 546
Prediction Differential Quantization, 547
Relative Address Coding, 547
CCITT Modified Relative Element Address Designate Coding, 548
Predictive Coding, 551
Adaptive Predictors, 552
Comparison of Algorithms, 553
Other Methods, 553

11.10 Color and Multispectral Image Coding 553

11.11 Summary 557

Problems 557

Bibliography 561

INDEX ***566***

Preface

Digital image processing is a rapidly evolving field with growing applications in science and engineering. Image processing holds the possibility of developing the ultimate machine that could perform the visual functions of all living beings. Many theoretical as well as technological breakthroughs are required before we could build such a machine. In the meantime, there is an abundance of image processing applications that can serve mankind with the available and anticipated technology in the near future.

This book addresses the fundamentals of the major topics of digital image processing: *representation, processing techniques,* and *communication.* Attention has been focused on mature topics with the hope that the level of discussion provided would enable an engineer or a scientist to design image processing systems or conduct research on advanced and newly emerging topics. Image representation includes tasks ranging from acquisition, digitization, and display to mathematical characterization of images for subsequent processing. Often, a proper representation is a prerequisite to an efficient processing technique such as enhancement, filtering and restoration, analysis, reconstruction from projections, and image communication. Image processing problems and techniques (Chapter 1) invoke concepts from diverse fields such as physical optics, digital signal processing, estimation theory, information theory, visual perception, stochastic processes, artificial intelligence, computer graphics, and so on. This book is intended to serve as a text for second and third quarter (or semester) graduate students in electrical engineering and computer science. It has evolved out of my class notes used for teaching introductory and advanced courses on image processing at the University of California at Davis.

The introductory course (Image Processing I) covers Chapter 1, Chapter 2 (Sections 2.1 to 2.8), much of Chapters 3 to 5, Chapter 7, and Sections 9.1 to 9.5. This material is supplemented by laboratory instruction that includes computer experiments. Students in this course are expected to have had prior exposure to one-dimensional digital signal processing topics such as sampling theorem, Fourier transform, linear systems, and some experience with matrix algebra. Typically, an entry level graduate course in digital signal processing is sufficient. Chapter 2 of the text includes much of the mathematical background that is needed in the rest of the book. A student who masters Chapter 2 should be able to handle most of the image processing problems discussed in the text and elsewhere in the image processing literature.

The advanced course (Image Processing II) covers Sections 2.9, 2.13, and selected topics from Chapters 6, 8, 9, 10, and 11. Both the courses are taught using visual aids such as overhead transparencies and slides to maximize discussion time and to minimize in-class writing time while maintaining a reasonable pace. In the advanced course, the prerequisites include Image Processing I and entry level graduate coursework in linear systems and random signals.

Chapters 3 to 6 cover the topic of image representation. Chapter 3 is devoted to low-level representation of visual information such as luminance, color, and spatial and temporal properties of vision. Chapter 4 deals with image digitization, an essential step for digital processing. In Chapter 5, images are represented as series expansion of orthogonal arrays or *basis images*. In Chapter 6, images are considered as random signals.

Chapters 7 through 11 are devoted to image processing techniques based on representations developed in the earlier chapters. Chapter 7 is devoted to image enhancement techniques, a topic of considerable importance in the practice of image processing. This is followed by a chapter on image restoration that deals with the theory and algorithms for removing degradations in images. Chapter 9 is concerned with the end goal of image processing, that is, image analysis. A special image restoration problem is image reconstruction from projections—a problem of immense importance in medical imaging and nondestructive testing of objects. The theory and techniques of image reconstruction are covered in Chapter 10. Chapter 11 is devoted to image data compression—a topic of fundamental importance in image communication and storage.

Each chapter concludes with a set of problems and annotated bibliography. The problems either go into the details or provide the extensions of results presented in the text. The problems marked with an asterisk (*) involve computer simulations. The problem sets give readers an opportunity to further their expertise on the relevant topics in image processing. The annotated bibliography provides a quick survey of the topics for the enthusiasts who wish to pursue the subject matter in greater depth.

Supplementary Course Materials

Forthcoming with this text is an instructors manual that contains solutions to selected problems from the text, a list of experimental laboratory projects, and course syllabus design suggestions for various situations.

Acknowledgments

I am deeply indebted to the many people who have contributed in making the completion of this book possible. Ralph Algazi, Mike Buonocore, Joe Goodman, Sarah Rajala, K. R. Rao, Jorge Sanz, S. Srinivasan, John Woods, and Yasuo Yoshida carefully read portions of the manuscript and provided important feedback. Many graduate students, especially Siamak Ansari, Steve Azevedo, Jon Brandt, Ahmed Darwish, Paul Farrelle, Jaswant Jain, Phil Kelly, David Paglieroni, S. Ranganath, John Sanders, S. H. Wang, and Wim Van Warmerdam provided valuable inputs through many examples and experimental results presented in the text. Ralph Algazi and his staff, especially Tom Arons and Jim Stewart, have contributed greatly through their assistance in implementing the computer experiments at the UCD Image Processing Facility and the Computer Vision Research Laboratory. Vivien Braly and Liz Fenner provided much help in typing and organizing several parts of the book. Colleen Brosnan and Tim Bozik of Prentice Hall provided the much-needed focus and guidance for my adherence to a schedule and editorial assistance that is required in the completion of a text to bring it to market. Thanks are also due to Tom Kailath for his enthusiasm for this work.

Finally, I would like to dedicate this book to my favorite image, Mohini, my children Mukul, Malini, and Ankit, and to my parents, all, for their constant support and encouragement.

1

Introduction

1.1 DIGITAL IMAGE PROCESSING: PROBLEMS AND APPLICATIONS

The term *digital image processing* generally refers to processing of a two-dimensional picture by a digital computer. In a broader context, it implies digital processing of any two-dimensional data. A digital image is an array of real or complex numbers represented by a finite number of bits. Figure 1.1 shows a computer laboratory (at the University of California, Davis) used for digital image processing. An image given in the form of a transparency, slide, photograph, or chart is first digitized and stored as a matrix of binary digits in computer memory. This digitized image can then be processed and/or displayed on a high-resolution television monitor. For display, the image is stored in a rapid-access buffer memory which refreshes the monitor at 30 frames/s to produce a visibly continuous display. Mini- or microcomputers are used to communicate and control all the digitization, storage, processing, and display operations via a computer network (such as the Ethernet). Program inputs to the computer are made through a terminal, and the outputs are available on a terminal, television monitor, or a printer/plotter. Figure 1.2 shows the steps in a typical image processing sequence.

Digital image processing has a broad spectrum of applications, such as remote sensing via satellites and other spacecrafts, image transmission and storage for business applications, medical processing, radar, sonar, and acoustic image processing, robotics, and automated inspection of industrial parts.

Images acquired by satellites are useful in tracking of earth resources; geographical mapping; prediction of agricultural crops, urban growth, and weather; flood and fire control; and many other environmental applications. Space image applications include recognition and analysis of objects contained in images obtained from deep space–probe missions. Image transmission and storage applica-

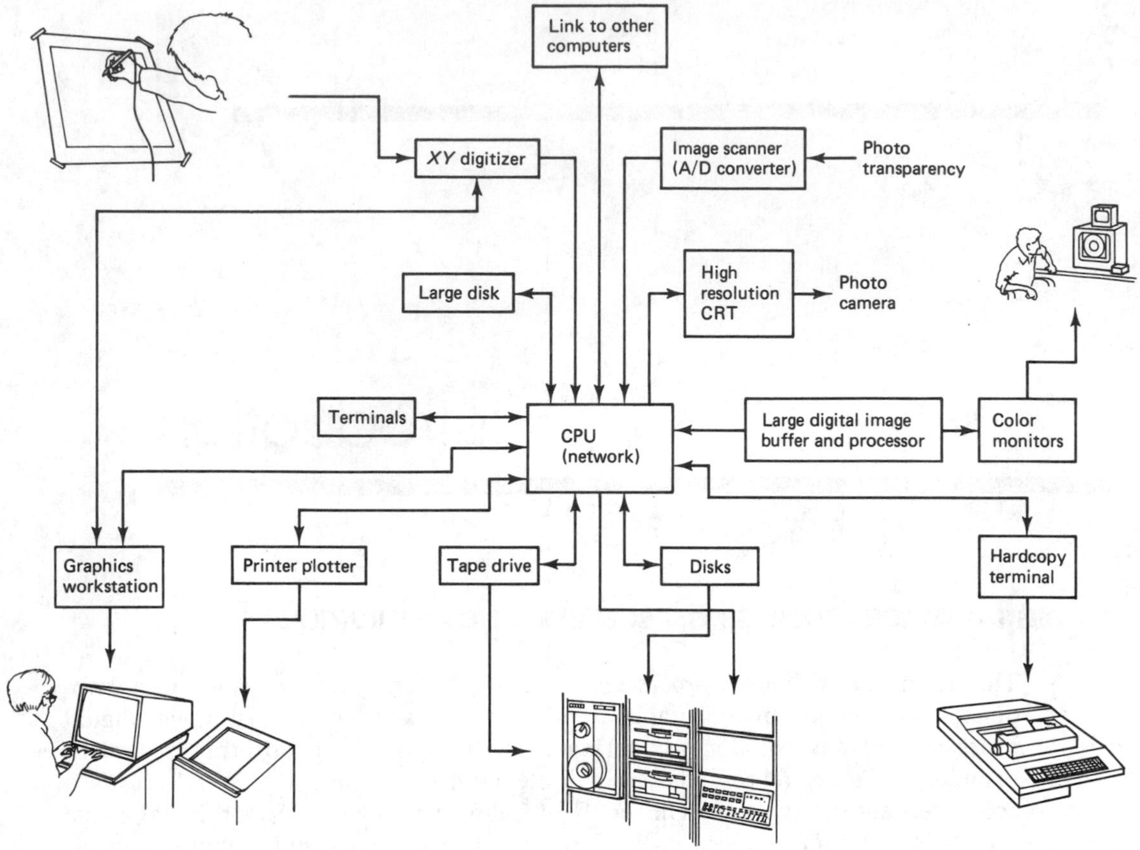

Figure 1.1 A digital image processing system (Signal and Image Processing Laboratory, University of California, Davis).

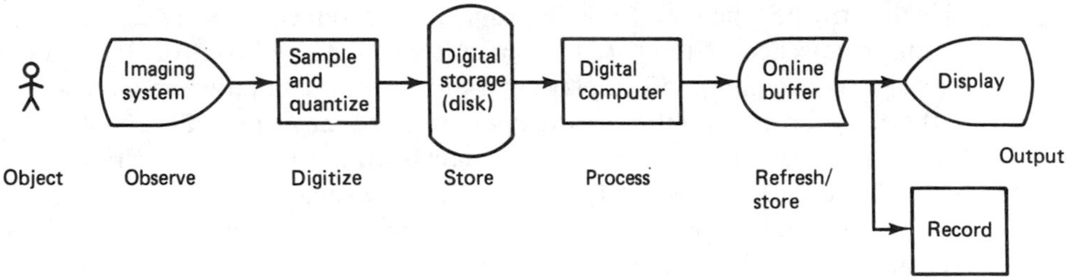

Figure 1.2 A typical digital image processing sequence.

tions occur in broadcast television, teleconferencing, transmission of facsimile images (printed documents and graphics) for office automation, communication over computer networks, closed-circuit television based security monitoring systems, and in military communications. In medical applications one is concerned with processing of chest X rays, cineangiograms, projection images of transaxial tomography, and other medical images that occur in radiology, nuclear magnetic resonance (NMR), and ultrasonic scanning. These images may be used for patient screening and monitoring or for detection of tumors or other disease in patients. Radar and sonar images are used for detection and recognition of various types of targets or in guidance and maneuvering of aircraft or missile systems. Figure 1.3 shows examples of several different types of images. There are many other applications ranging from robot vision for industrial automation to image synthesis for cartoon making or fashion design. In other words, whenever a human or a machine or any other entity receives data of two or more dimensions, an image is processed.

Although there are many image processing applications and problems, in this text we will consider the following basic classes of problems.

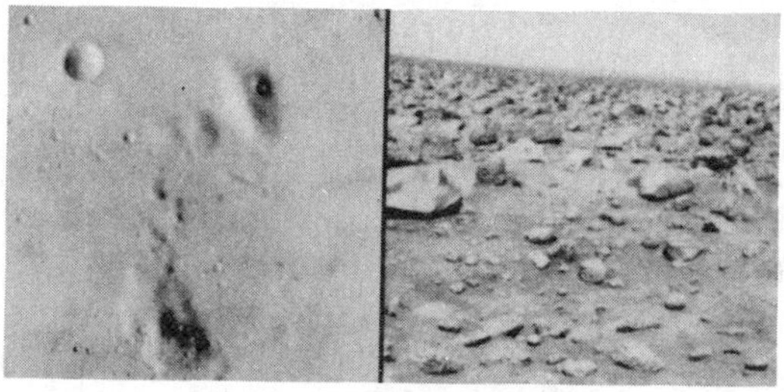

(a) Space probe images: Moon and Mars.

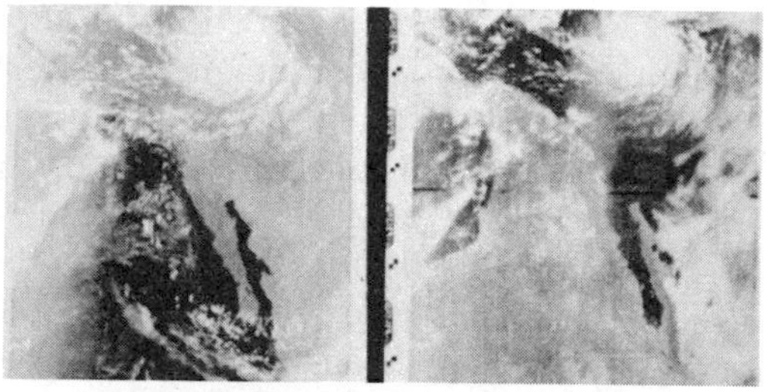

(b) multispectral images: visual and infrared.

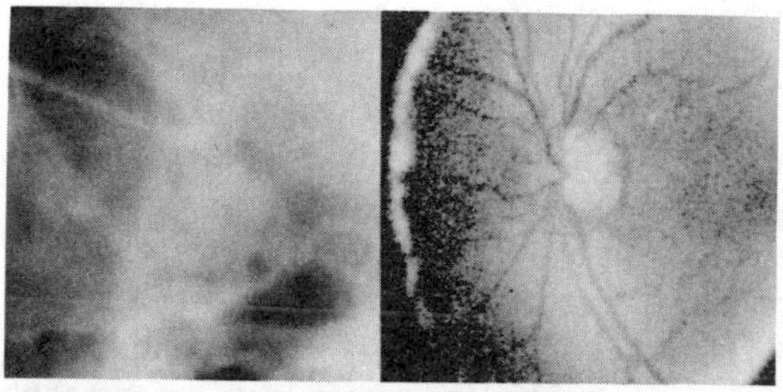

(c) medical images: Xray and eyeball.

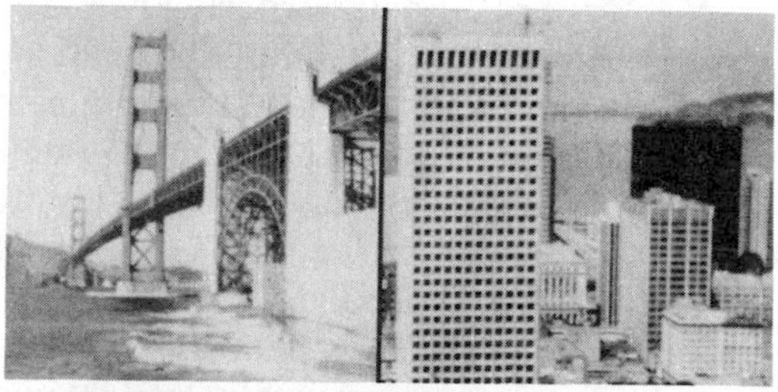

(d) optical camera images: Golden Gate and downtown San Francisco.

(e) television images: girl, couple, Linda and Cronkite.

Figure 1.3 Examples of digital images.

1. Image representation and modeling
2. Image enhancement
3. Image restoration
4. Image analysis
5. Image reconstruction
6. Image data compression

1.2 IMAGE REPRESENTATION AND MODELING

In image representation one is concerned with characterization of the quantity that each *picture-element* (also called *pixel* or *pel*) represents. An image could represent luminances of objects in a scene (such as pictures taken by ordinary camera), the absorption characteristics of the body tissue (X-ray imaging), the radar cross section of a target (radar imaging), the temperature profile of a region (infrared imaging), or the gravitational field in an area (in geophysical imaging). In general, any two-dimensional function that bears information can be considered an image. Image models give a logical or quantitative description of the properties of this function. Figure 1.4 lists several image representation and modeling problems.

An important consideration in image representation is the fidelity or intelligibility criteria for measuring the quality of an image or the performance of a processing technique. Specification of such measures requires models of perception of contrast, spatial frequencies, color, and so on, as discussed in Chapter 3. Knowledge of a fidelity criterion helps in designing the imaging sensor, because it tells us the variables that should be measured most accurately.

The fundamental requirement of digital processing is that images be sampled and quantized. The sampling rate (number of pixels per unit area) has to be large enough to preserve the useful information in an image. It is determined by the bandwidth of the image. For example, the bandwidth of raster scanned common television signal is about 4 MHz. From the sampling theorem, this requires a minimum sampling rate of 8 MHz. At 30 frames/s, this means each frame should contain approximately 266,000 pixels. Thus for a 512-line raster, this means each

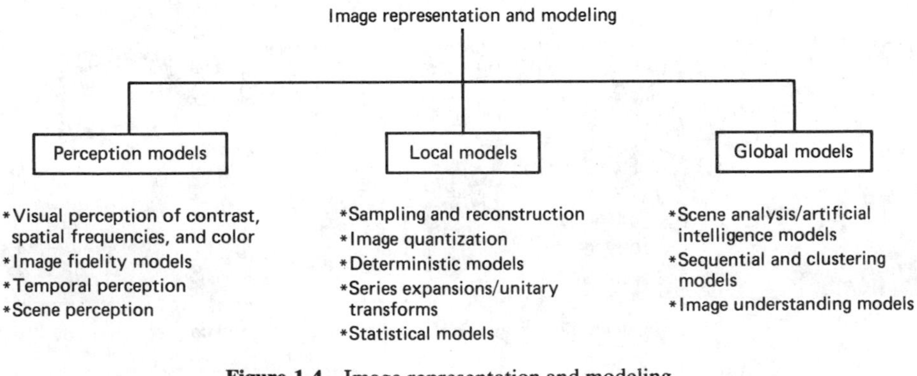

Figure 1.4 Image representation and modeling.

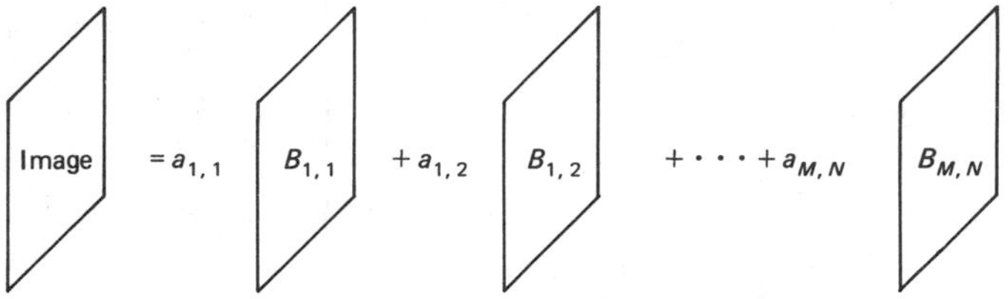

Figure 1.5 Image representation by orthogonal basis image series $B_{m,n}$.

image frame contains approximately 512×512 pixels. Image quantization is the analog to digital conversion of a sampled image to a finite number of gray levels. Image sampling and quantization methods are discussed in Chapter 4.

A classical method of signal representation is by an orthogonal series expansion, such as the Fourier series. For images, analogous representation is possible via two-dimensional orthogonal functions called *basis images.* For sampled images, the basis images can be determined from unitary matrices called *image transforms.* Any given image can be expressed as a weighted sum of the basis images (Fig. 1.5). Several characteristics of images, such as their spatial frequency content, bandwidth, power spectrum, and application in filter design, feature extraction, and so on, can be studied via such expansions. The theory and applications of image transforms are discussed in Chapter 5.

Statistical models describe an image as a member of an ensemble, often characterized by its mean and covariance functions. This permits development of algorithms that are useful for an entire class or an ensemble of images rather than for a single image. Often the ensemble is assumed to be stationary so that the mean and covariance functions can easily be estimated. Stationary models are useful in data compression problems such as transform coding, restoration problems such as Wiener filtering, and in other applications where global properties of the ensemble are sufficient. A more effective use of these models in image processing is to consider them to be spatially varying or piecewise spatially invariant.

To characterize short-term or local properties of the pixels, one alternative is to characterize each pixel by a relationship with its neighborhood pixels. For example, a linear system characterized by a (low-order) difference equation and forced by white noise or some other random field with known power spectrum density is a useful approach for representing the ensemble. Figure 1.6 shows three types of stochastic models where an image pixel is characterized in terms of its neighboring pixels. If the image were scanned top to bottom and then left to right, the model of Fig. 1.6a would be called a *causal model.* This is because the pixel A is characterized by pixels that lie in the "past." Extending this idea, the model of Fig. 1.6b is a noncausal model because the neighbors of A lie in the past as well as the "future" in both the directions. In Fig. 1.6c, we have a semicausal model because the neighbors of A are in the past in the j-direction and are in the past as well as future in the i-direction.

Such models are useful in developing algorithms that have different hardware realizations. For example, causal models can realize recursive filters, which require small memory while yielding an infinite impulse response (IIR). On the other hand,

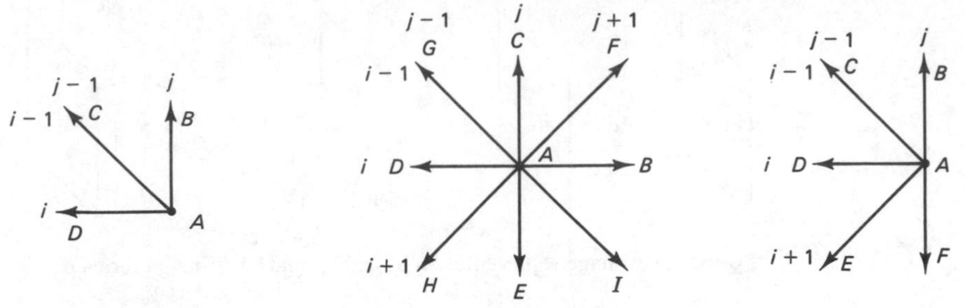

(a) Causal model. (b) Noncausal model. (c) Semicausal model.

Figure 1.6 Three canonical forms of stochastic models.

noncausal models can be used to design fast transform-based finite impulse response (FIR) filters. Semicausal models can yield two-dimensional algorithms, which are recursive in one dimension and nonrecursive in the other. Some of these stochastic models can be thought of as generalizations of one dimensional random processes represented by autoregressive (AR) and autoregressive moving average (ARMA) models. Details of these aspects are discussed in Chapter 6.

In global modeling, an image is considered as a composition of several objects. Various objects in the scene are detected (for example, by segmentation techniques), and the model gives the rules for defining the relationship among various objects. Such representations fall under the category of image understanding models, which are not a subject of study in this text.

1.3 IMAGE ENHANCEMENT

In *image enhancement,* the goal is to accentuate certain image features for subsequent analysis or for image display. Examples include contrast and edge enhancement, pseudocoloring, noise filtering, sharpening, and magnifying. Image enhancement is useful in feature extraction, image analysis, and visual information display. The enhancement process itself does not increase the inherent information content in the data. It simply emphasizes certain specified image characteristics. Enhancement algorithms are generally interactive and application-dependent.

Image enhancement techniques, such as contrast stretching, map each gray level into another gray level by a predetermined transformation. An example is the histogram equalization method, where the input gray levels are mapped so that the output gray level distribution is uniform. This has been found to be a powerful method of enhancement of low contrast images (see Fig. 7.14). Other enhancement techniques perform local neighborhood operations as in convolution, transform operations as in the discrete Fourier transform, and other operations as in pseudocoloring where a gray level image is mapped into a color image by assigning different colors to different features. Examples and details of these techniques are considered in Chapter 7.

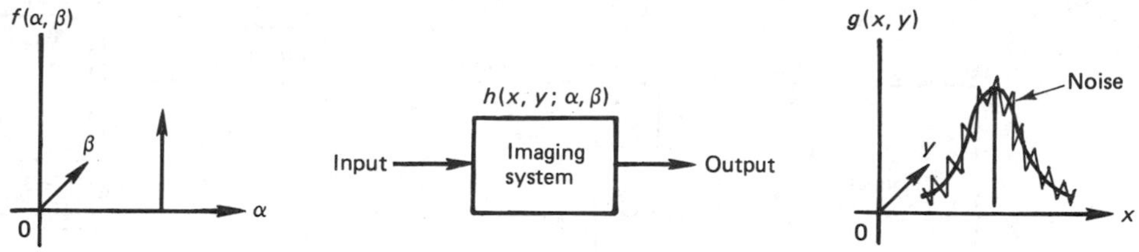

Figure 1.7 Blurring due to an imaging system. Given the noisy and blurred image the image restoration problem is to find an estimate of the input image $f(x, y)$.

1.4 IMAGE RESTORATION

Image restoration refers to removal or minimization of known degradations in an image. This includes deblurring of images degraded by the limitations of a sensor or its environment, noise filtering, and correction of geometric distortion or nonlinearities due to sensors. Figure 1.7 shows a typical situation in image restoration. The image of a point source is blurred and degraded due to noise by an imaging system. If the imaging sytem is linear, the image of an object can be expressed as

$$g(x, y) = \int_{-\infty}^{\infty}\int_{-\infty}^{\infty} h(x, y; \alpha, \beta) f(\alpha, \beta)\, d\alpha\, d\beta + \eta(x, y) \tag{1.1}$$

where $\eta(x, y)$ is the additive noise function, $f(\alpha, \beta)$ is the object, $g(x, y)$ is the image, and $h(x, y; \alpha, \beta)$ is called the *point spread function* (PSF). A typical image restoration problem is to find an estimate of $f(\alpha, \beta)$ given the PSF, the blurred image, and the statistical properties of the noise process.

A fundamental result in filtering theory used commonly for image restoration is called the *Wiener filter*. This filter gives the best linear mean square estimate of the object from the observations. It can be implemented in frequency domain via the fast unitary transforms, in spatial domain by two-dimensional recursive techniques similar to Kalman filtering, or by FIR nonrecursive filters (see Fig. 8.15). It can also be implemented as a semirecursive filter that employs a unitary transform in one of the dimensions and a recursive filter in the other.

Several other image restoration methods such as least squares, constrained least squares, and spline interpolation methods can be shown to belong to the class of Wiener filtering algorithms. Other methods such as maximum likelihood, maximum entropy, and maximum a posteriori are nonlinear techniques that require iterative solutions. These and other algorithms useful in image restoration are discussed in Chapter 8.

1.5 IMAGE ANALYSIS

Image analysis is concerned with making quantitative measurements from an image to produce a description of it. In the simplest form, this task could be reading a label on a grocery item, sorting different parts on an assembly line (Fig. 1.8), or

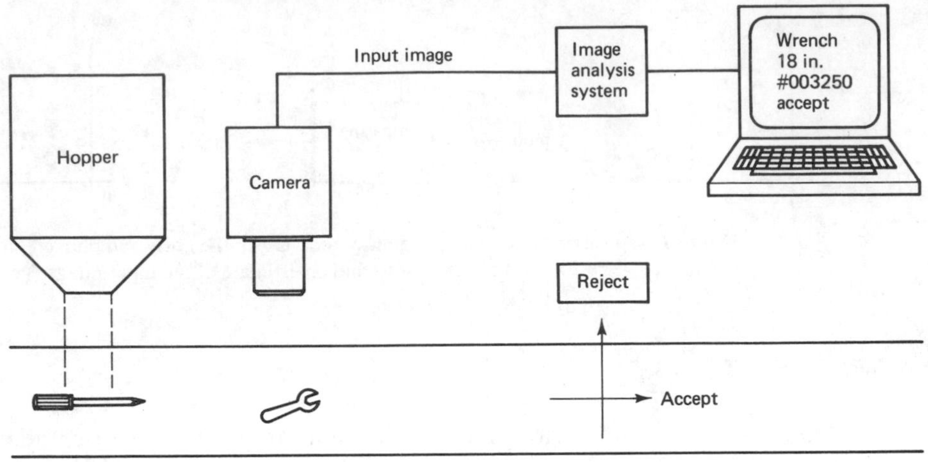

Figure 1.8 Parts inspection and sorting on an assembly line.

measuring the size and orientation of blood cells in a medical image. More advanced image analysis systems measure quantitative information and use it to make a sophisticated decision, such as controlling the arm of a robot to move an object after identifying it or navigating an aircraft with the aid of images acquired along its trajectory.

Image analysis techniques require extraction of certain features that aid in the identification of the object. Segmentation techniques are used to isolate the desired object from the scene so that measurements can be made on it subsequently. Quantitative measurements of object features allow classification and description of the image. These techniques are considered in Chapter 9.

1.6 IMAGE RECONSTRUCTION FROM PROJECTIONS

Image reconstruction from projections is a special class of image restoration problems where a two- (or higher) dimensional object is reconstructed from several one-dimensional projections. Each projection is obtained by projecting a parallel X ray (or other penetrating radiation) beam through the object (Fig. 1.9). Planar projections are thus obtained by viewing the object from many different angles. Reconstruction algorithms derive an image of a thin axial slice of the object, giving an inside view otherwise unobtainable without performing extensive surgery. Such techniques are important in medical imaging (CT scanners), astronomy, radar imaging, geological exploration, and nondestructive testing of assemblies.

Mathematically, image reconstruction problems can be set up in the framework of Radon transform theory. This theory leads to several useful reconstruction algorithms, details of which are discussed in Chapter 10.

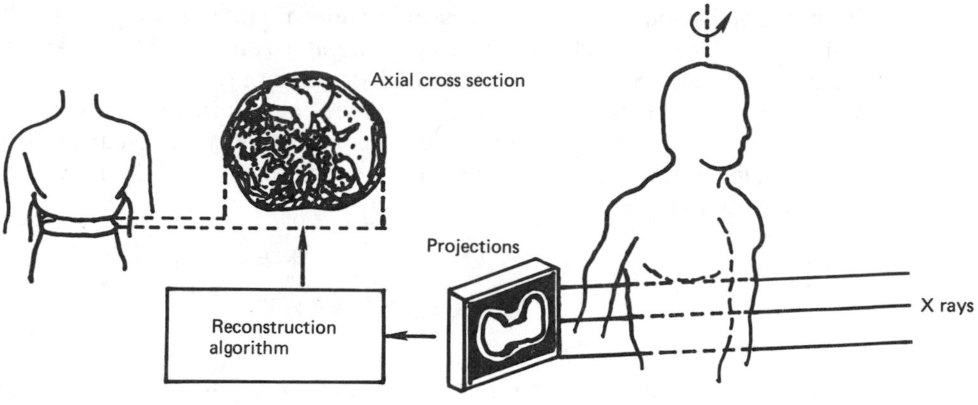

Figure 1.9 Image reconstruction using X-ray CT scanners.

1.7 IMAGE DATA COMPRESSION

The amount of data associated with visual information is so large (see Table 1.1a) that its storage would require enormous storage capacity. Although the capacities of several storage media (Table 1.1b) are substantial, their access speeds are usually inversely proportional to their capacity. Typical television images generate data rates exceeding 10 million bytes per second. There are other image sources that generate even higher data rates. Storage and/or transmission of such data require large capacity and/or bandwidth, which could be very expensive. *Image data compression techniques* are concerned with reduction of the number of bits required to store or transmit images without any appreciable loss of information. Image trans-

TABLE 1.1a Data Volumes of Image Sources (in Millions of Bytes)

National archives	12.5×10^9
1 h of color television	28×10^3
Encyclopeadia Britannica	12.5×10^3
Book (200 pages of text characters)	1.3
One page viewed as an image	.13

TABLE 1.1b Storage Capacities (in Millions of Bytes)

Human brain	125,000,000
Magnetic cartridge	250,000
Optical disc memory	12,500
Magnetic disc	760
2400-ft magnetic tape	200
Floppy disc	1.25
Solid-state memory modules	0.25

mission applications are in broadcast television; remote sensing via satellite, aircraft, radar, or sonar; teleconferencing; computer communications; and facsimile transmission. Image storage is required most commonly for educational and business documents, medical images used in patient monitoring systems, and the like. Because of their wide applications, data compression is of great importance in digital image processing. Various image data compression techniques and examples are discussed in Chapter 11.

BIBLIOGRAPHY

For books and special issues of journals devoted to digital imaging processing:

1. H. C. Andrews, with contribution by W. K. Pratt and K. Caspari. *Computer Techniques in Image Processing*. New York: Academic Press, 1970.
2. H. C. Andrews and B. R. Hunt. *Digital Image Restoration*. Englewood Cliffs, N.J.: Prentice-Hall, 1977.
3. K. R. Castleman. *Digital Image Processing*. Englewood Cliffs, N.J.: Prentice-Hall, 1979.
4. M.P. Ekstrom (Ed.). *Digital Image Processing Techniques*. New York: Academic Press, 1984.
5. R. C. Gonzalez and P. Wintz. *Digital Image Processing, 2nd ed.* Reading, Mass.: Addison-Wesley, 1987.
6.. E. L. Hall. *Computer Image Processing and Recognition*. New York: Academic Press, 1979.
7. T. S. Huang (ed.). *Topics in Applied Physics: Picture Processing and Digital Filtering*, Vol. 6. New York: Springer-Verlag, 1975. Also see, Vols. 42–43, 1981.
8. S. Lipkin and A. Rosenfeld. *Picture Processing and Psychopictorics*. New York: Academic Press, 1970.
9. W. K. Pratt. *Digital Image Processing*. New York: Wiley-Interscience, 1978.
10. A. Rosenfeld and A. C. Kak. *Digital Image Processing*, Vols. I and II. New York: Academic Press, 1982.
11. Special issues on image processing. *Proceedings IEEE*, 60, no. 7 (July 1972); *IEEE Computer*, 7, no. 5 (May 1974); *Proceedings IEEE*, 69, no. 5 (May 1981).

2

Two-Dimensional Systems and Mathematical Preliminaries

2.1 INTRODUCTION

In this chapter we define our notation and discuss some mathematical preliminaries that will be useful throughout the book. Because images are generally outputs of two-dimensional systems, mathematical concepts used in the study of such systems are needed. We start by defining our notation and then review the definitions and properties of linear systems and the Fourier and Z-transforms. This is followed by a review of several fundamental results from matrix theory that are important in digital image processing theory. Two-dimensional random fields and some important concepts from probability and estimation theory are then reviewed. The emphasis is on the final results and their applications in image processing. It is assumed that the reader has encountered most of these basic concepts earlier. The summary discussion provided here is intended to serve as an easy reference for subsequent chapters. The problems at the end of the chapter provide an opportunity to revise these concepts through special cases and examples.

2.2 NOTATION AND DEFINITIONS

A one-dimensional continuous signal will be represented as a function of one variable: $f(x)$, $u(x)$, $s(t)$, and so on. One-dimensional sampled signals will be written as single index sequences: u_n, $u(n)$, and the like.

A continuous image will be represented as a function of two independent variables: $u(x, y)$, $v(x, y)$, $f(x, y)$, and so forth. A sampled image will be represented as a two- (or higher) dimensional sequence of real numbers: $u_{m,n}$, $v(m, n)$, $u(i, j, k)$, and so on. Unless stated otherwise, the symbols $i, j, k, l, m, n, \ldots$ will be used to specify integer indices of arrays and vectors. The symbol roman j will represent $\sqrt{-1}$. The complex conjugate of a complex variable such as z, will be denoted by z^*. Certain symbols will be redefined at appropriate places in the text to keep the notation clear.

Table 2.1 lists several well-known one-dimensional functions that will be often encountered. Their two-dimensional versions are functions of the *separable form*

$$f(x, y) = f_1(x)f_2(y) \tag{2.1}$$

For example, the two-dimensional delta functions are defined as

$$\text{Dirac:} \quad \delta(x, y) = \delta(x)\delta(y) \tag{2.2a}$$

$$\text{Kronecker:} \quad \delta(m, n) = \delta(m)\delta(n) \tag{2.2b}$$

which satisfy the properties

$$\left.\begin{aligned} \int_{-\infty}^{\infty}\int_{-\infty}^{\infty} f(x', y')\delta(x - x', y - y')dx'\,dy' = f(x, y) \\ \lim_{\epsilon\to 0}\int_{-\epsilon}^{\epsilon}\int_{-\epsilon}^{\epsilon} \delta(x, y)dx\,dy = 1, \end{aligned}\right\} \tag{2.3}$$

$$\left.\begin{aligned} x(m, n) = \sum\sum_{m', n' = -\infty}^{\infty} x(m', n')\delta(m - m', n - n') \\ \sum\sum_{m, n = -\infty}^{\infty} \delta(m, n) = 1 \end{aligned}\right\} \tag{2.4}$$

The definitions and properties of the functions rect(x, y), sinc(x, y), and comb(x, y) can be defined in a similar manner.

TABLE 2.1 Some Special Functions

Function	Definition	Function	Definition
Dirac delta	$\delta(x) = 0, x \neq 0$	*Rectangle*	$\text{rect}(x) = \begin{cases} 1, & \lvert x\rvert \le 1/2 \\ 0, & \lvert x\rvert > 1/2 \end{cases}$
	$\lim_{\epsilon\to 0}\int_{-\epsilon}^{\epsilon} \delta(x)\,dx = 1$	*Signum*	$\text{sgn}(x) = \begin{cases} 1, & x > 0 \\ 0, & x = 0 \\ -1, & x < 0 \end{cases}$
Sifting property	$\int_{-\infty}^{\infty} f(x')\delta(x - x')\,dx' = f(x)$	*Sinc*	$\text{sinc}(x) = \dfrac{\sin \pi x}{\pi x}$
Scaling property	$\delta(ax) = \dfrac{\delta(x)}{\lvert a\rvert}$	*Comb*	$\text{comb}(x) = \sum_{n=-\infty}^{\infty} \delta(x - n)$
Kronecker delta	$\delta(n) = \begin{cases} 0, & n \neq 0 \\ 1, & n = 0 \end{cases}$	*Triangle*	$\text{tri}(x) = \begin{cases} 1 - \lvert x\rvert, & \lvert x\rvert \le 1 \\ 0, & \lvert x\rvert > 1 \end{cases}$
Sifting property	$\sum_{m=-\infty}^{\infty} f(m)\delta(n - m) = f(n)$		

2.3 LINEAR SYSTEMS AND SHIFT INVARIANCE

A large number of imaging systems can be modeled as two-dimensional linear systems. Let $x(m, n)$ and $y(m, n)$ represent the input and output sequences, respectively, of a two-dimensional system (Fig. 2.1), written as

$$y(m, n) = \mathcal{H}[x(m, n)] \tag{2.5}$$

This system is called *linear* if and only if any linear combination of two inputs $x_1(m, n)$ and $x_2(m, n)$ produces the same combination of their respective outputs $y_1(m, n)$ and $y_2(m, n)$, i.e., for arbitrary constants a_1 and a_2

$$\begin{aligned} \mathcal{H}[a_1 x_1(m, n) + a_2 x_2(m, n)] &= a_1 \mathcal{H}[x_1(m, n)] + a_2 \mathcal{H}[x_2(m, n)] \\ &= a_1 y_1(m, n) + a_2 y_2(m, n) \end{aligned} \tag{2.6}$$

This is called *linear superposition.* When the input is the two-dimensional Kronecker delta function at location (m', n'), the output at location (m, n) is defined as

$$h(m, n; m', n') \overset{\Delta}{=} \mathcal{H}[\delta(m - m', n - n')] \tag{2.7}$$

and is called the *impulse response* of the system. For an imaging system, it is the image in the output plane due to an ideal point source at location (m', n') in the input plane. In our notation, the semicolon (;) is employed to distinguish the input and output pairs of coordinates.

The impulse response is called the *point spread function* (PSF) when the inputs and outputs represent a positive quantity such as the intensity of light in imaging systems. The term *impulse response* is more general and is allowed to take negative as well as complex values. The *region of support* of an impulse response is the smallest closed region in the m, n plane outside which the impulse response is zero. A system is said to be a *finite impulse response* (FIR) or an *infinite impulse response* (IIR) system if its impulse response has finite or infinite regions of support, respectively.

The output of any linear system can be obtained from its impulse response and the input by applying the superposition rule of (2.6) to the representation of (2.4) as follows:

$$\begin{aligned} y(m, n) &= \mathcal{H}[x(m, n)] \\ &= \mathcal{H}\left[\sum_{m'}\sum_{n'} x(m', n')\delta(m - m', n - n')\right] \\ &= \sum_{m'}\sum_{n'} x(m', n')\mathcal{H}[\delta(m - m', n - n')] \end{aligned}$$

$$\Rightarrow \quad y(m, n) = \sum_{m'}\sum_{n'} x(m', n')h(m, n; m', n') \tag{2.8}$$

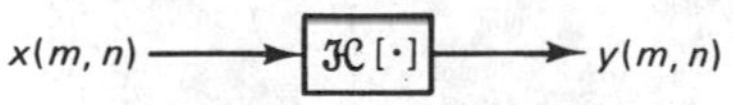

Figure 2.1 A system.

A system is called *spatially invariant* or *shift invariant* if a translation of the input causes a translation of the output. Following the definition of (2.7), if the impulse occurs at the origin we will have

$$\mathcal{H}[\delta(m, n)] = h(m, n; 0, 0)$$

Hence, it must be true for shift invariant systems that

$$\begin{aligned} h(m, n; m', n') &\triangleq \mathcal{H}[\delta(m - m', n - n')] \\ &= h(m - m', n - n'; 0, 0) \\ \Rightarrow \quad h(m, n; m', n') &= h(m - m', n - n') \end{aligned} \tag{2.9}$$

i.e., the impulse response is a function of the two displacement variables only. This means the shape of the impulse response does not change as the impulse moves about the *m, n* plane. A system is called *spatially varying* when (2.9) does not hold. Figure 2.2 shows examples of PSFs of imaging systems with separable or circularly symmetric impulse responses.

For shift invariant systems, the output becomes

$$y(m, n) = \sum\sum_{m', n' = -\infty}^{\infty} h(m - m', n - n')x(m', n') \tag{2.10}$$

which is called the *convolution* of the input with the *impulse response*. Figure 2.3 shows a graphical interpretation of this operation. The impulse response array is rotated about the origin by 180° and then shifted by (m, n) and overlayed on the array $x(m', n')$. The sum of the product of the arrays $\{x(\cdot,\cdot)\}$ and $\{h(\cdot,\cdot)\}$ in the overlapping regions gives the result at (m, n). We will use the symbol ⊛ to denote the convolution operation in both discrete and continuous cases, i.e.,

$$g(x, y) = h(x, y) \circledast f(x, y) \triangleq \int_{-\infty}^{\infty}\int_{-\infty}^{\infty} h(x - x', y - y')f(x', y')dx'\,dy'$$

$$y(m, n) = h(m, n) \circledast x(m, n) \triangleq \sum\sum_{m'n' = -\infty}^{\infty} h(m - m', n - n')x(m', n') \tag{2.11}$$

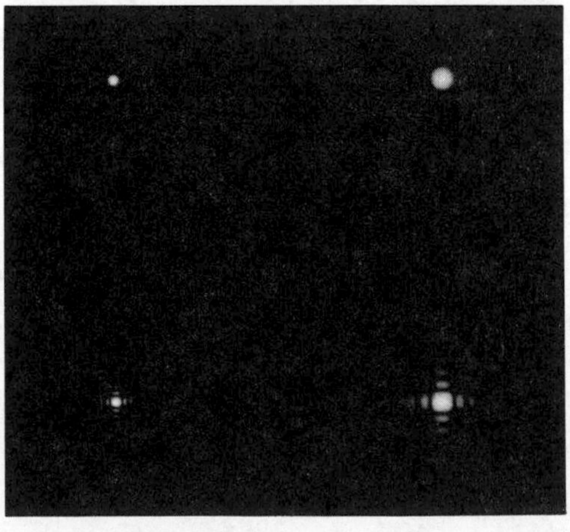

a	b
c	d

Figure 2.2 Examples of PSFs. (a) Circularly symmetric PSF of average atmospheric turbulence causing small blur; (b) atmospheric turbulence PSF causing large blur; (c) separable PSF of a diffraction limited system with square aperature; (d) same as (c) but with smaller aperture.

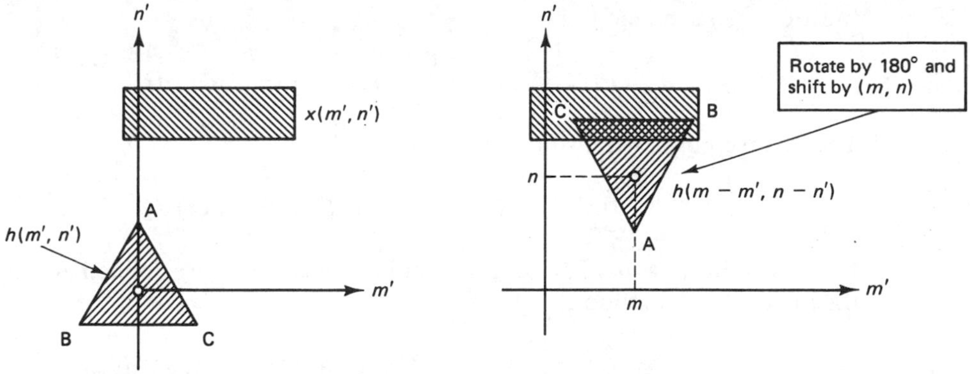

(a) impulse response

(b) output at location (m, n) is the sum of product of quantities in the area of overlap.

Figure 2.3 Discrete convolution in two dimensions

The convolution operation has several interesting properties, which are explored in Problems 2.2 and 2.3.

Example 2.1 (Discrete convolution)

Consider the 2×2 and 3×2 arrays $h(m, n)$ and $x(m, n)$ shown next, where the boxed element is at the origin. Also shown are the various steps for obtaining the convolution of these two arrays. The result $y(m, n)$ is a 4×3 array. In general, the convolution of two arrays of sizes $(M_1 \times N_1)$ and $(M_2 \times N_2)$ yields an array of size $[(M_1 + M_2 - 1) \times (N_1 + N_2 - 1)]$ (Problem 2.5).

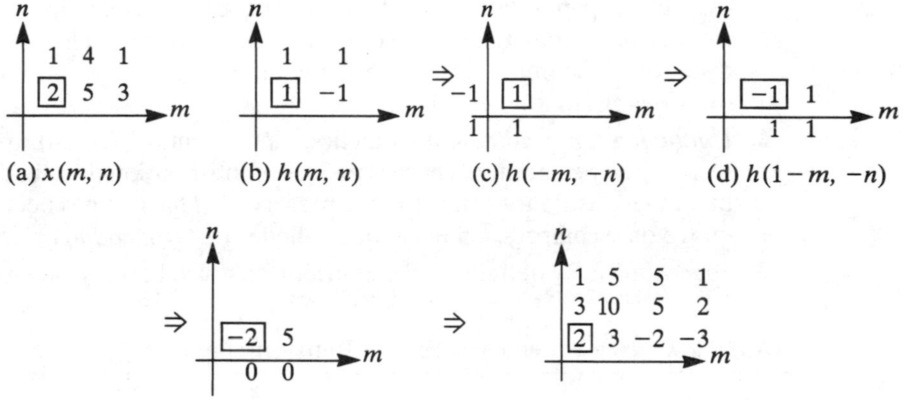

(a) $x(m, n)$ (b) $h(m, n)$ (c) $h(-m, -n)$ (d) $h(1-m, -n)$

(e) $y(1, 0) = -2 + 5 = 3$ (f) $y(m, n)$

2.4 THE FOURIER TRANSFORM

Two-dimensional transforms such as the Fourier transform and the Z-transform are of fundamental importance in digital image processing as will become evident in the subsequent chapters. In one dimension, the Fourier transform of a complex function $f(x)$ is defined as

$$F(\xi) \triangleq \mathcal{F}[f(x)] \triangleq \int_{-\infty}^{\infty} f(x) \exp(-j2\pi\xi x)\, dx \tag{2.12}$$

The inverse Fourier transform of $F(\xi)$ is

$$f(x) \triangleq \mathcal{F}^{-1}[F(\xi)] = \int_{-\infty}^{\infty} F(\xi) \exp(j2\pi\xi x)\, d\xi \tag{2.13}$$

Two-dimensional Fourier transform and its inverse are defined analogously by the linear transformations

$$F(\xi_1, \xi_2) = \int_{-\infty}^{\infty}\int_{-\infty}^{\infty} f(x, y) \exp[-j2\pi(x\xi_1 + y\xi_2)]\, dx\, dy \tag{2.14}$$

$$f(x, y) = \int_{-\infty}^{\infty}\int_{-\infty}^{\infty} F(\xi_1, \xi_2) \exp[j2\pi(x\xi_1 + y\xi_2)]\, d\xi_1\, d\xi_2 \tag{2.15}$$

Examples of some useful two-dimensional Fourier transforms are given in Table 2.2.

Properties of the Fourier Transform

Table 2.3 gives a summary of the properties of the two-dimensional Fourier transform. Some of these properties are discussed next.

1. *Spatial frequencies.* If $f(x, y)$ is luminance and x, y the spatial coordinates, then ξ_1, ξ_2 are the spatial frequencies that represent luminance changes with respect to spatial distances. The units of ξ_1 and ξ_2 are reciprocals of x and y, respectively. Sometimes the coordinates x, y are normalized by the viewing distance of the image $f(x, y)$. Then the units of ξ_1, ξ_2 are cycles per degree (of the viewing angle).
2. *Uniqueness.* For continuous functions, $f(x, y)$ and $F(\xi_1, \xi_2)$ are unique with respect to one another. There is no loss of information if instead of preserving the image, its Fourier transform is preserved. This fact has been utilized in an image data compression technique called *transform coding*.
3. *Separability.* By definition, the Fourier transform kernel is separable, so that it

TABLE 2.2 Two-Dimensional Fourier Transform Pairs

$f(x, y)$	$F(\xi_1, \xi_2)$
$\delta(x, y)$	1
$\delta(x \pm x_0, y \pm y_0)$	$\exp(\pm j2\pi x_0\xi_1)\exp(\pm j2\pi y_0\xi_2)$
$\exp(\pm j2\pi x\eta_1)\exp(\pm j2\pi y\eta_2)$	$\delta(\xi_1 \mp \eta_1, \xi_2 \mp \eta_2)$
$\exp[-\pi(x^2 + y^2)]$	$\exp[-\pi(\xi_1^2 + \xi_2^2)]$
$\text{rect}(x, y)$	$\text{sinc}(\xi_1, \xi_2)$
$\text{tri}(x, y)$	$\text{sinc}^2(\xi_1, \xi_2)$
$\text{comb}(x, y)$	$\text{comb}(\xi_1, \xi_2)$

TABLE 2.3 Properties of Two-Dimensional Fourier Transform

Property	Function $f(x, y)$	Fourier Transform $F(\xi_1, \xi_2)$
Rotation	$f(\pm x, \pm y)$	$F(\pm\xi_1, \pm\xi_2)$
Linearity	$a_1 f_1(x, y) + a_2 f_2(x, y)$	$a_1 F_1(\xi_1, \xi_2) + a_2 F_2(\xi_1, \xi_2)$
Conjugation	$f^*(x, y)$	$F^*(-\xi_1, -\xi_2)$
Separability	$f_1(x) f_2(y)$	$F_1(\xi_1) F_2(\xi_2)$
Scaling	$f(ax, by)$	$\dfrac{F(\xi_1/a, \xi_2/b)}{\lvert ab \rvert}$
Shifting	$f(x \pm x_0, y \pm y_0)$	$\exp[\pm j2\pi(x_0 \xi_1 + y_0 \xi_2)] F(\xi_1, \xi_2)$
Modulation	$\exp[\pm j2\pi(\eta_1 x + \eta_2 y)] f(x, y)$	$F(\xi_1 \mp \eta_1, \xi_2 \mp \eta_2)$
Convolution	$g(x, y) = h(x, y) \circledast f(x, y)$	$G(\xi_1, \xi_2) = H(\xi_1, \xi_2) F(\xi_1, \xi_2)$
Multiplication	$g(x, y) = h(x, y) f(x, y)$	$G(\xi_1, \xi_2) = H(\xi_1, \xi_2) \circledast F(\xi_1, \xi_2)$
Spatial correlation	$c(x, y) = h(x, y) \star f(x, y)$	$C(\xi_1, \xi_2) = H(-\xi_1, -\xi_2) F(\xi_1, \xi_2)$
Inner product	$I = \int_{-\infty}^{\infty}\int_{-\infty}^{\infty} f(x, y) h^*(x, y)\, dx\, dy$	$I = \int_{-\infty}^{\infty}\int_{-\infty}^{\infty} F(\xi_1, \xi_2) H^*(\xi_1, \xi_2)\, d\xi_1\, d\xi_2$

can be written as a separable transformation in x and y, i.e.,

$$F(\xi_1, \xi_2) = \int_{-\infty}^{\infty}\left[\int_{-\infty}^{\infty} f(x, y) \exp(-j2\pi x \xi_1)\, dx\right] \exp(-j2\pi y \xi_2)\, dy$$

This means the two-dimensional transformation can be realized by a succession of one-dimensional transformations along each of the spatial coordinates.

4. *Frequency response and eigenfunctions of shift invariant systems.* An eigenfunction of a system is defined as an input function that is reproduced at the output with a possible change only in its amplitude. A fundamental property of a linear shift invariant system is that its eigenfunctions are given by the complex exponential $\exp[j2\pi(\xi_1 x + \xi_2 y)]$. Thus in Fig. 2.4, for any fixed (ξ_1, ξ_2), the output of the linear shift invariant system would be

$$g(x, y) = \int_{-\infty}^{\infty}\int_{-\infty}^{\infty} h(x - x', y - y') \exp[j2\pi(\xi_1 x' + \xi_2 y')]dx'\, dy'$$

Performing the change of variables $\tilde{x} = x - x'$, $\tilde{y} = y - y'$ and simplifying the result, we get

$$g(x, y) = H(\xi_1, \xi_2) \exp[j2\pi(\xi_1 x + \xi_2 y)] \tag{2.16}$$

The function $H(\xi_1, \xi_2)$, which is the Fourier transform of the impulse response, is also called the *frequency response* of the system. It represents the (complex) amplitude of the system response at spatial frequency (ξ_1, ξ_2).

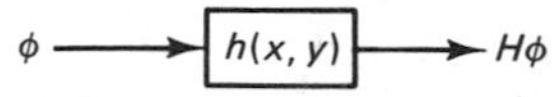

Figure 2.4 Eigenfunctions of a linear shift invariant system. $\phi \overset{\Delta}{=} \exp\{j2\pi(\xi_1 x + \xi_2 y)\}$, $H = H(\xi_1, \xi_2) \overset{\Delta}{=}$ Fourier transform of $h(x, y)$.

5. *Convolution theorem.* The Fourier transform of the convolution of two functions is the product of their Fourier transforms, i.e.,

$$g(x, y) = h(x, y) \circledast f(x, y) \Leftrightarrow G(\xi_1, \xi_2) = H(\xi_1, \xi_2)F(\xi_1, \xi_2) \qquad (2.17)$$

This theorem suggests that the convolution of two functions may be evaluated by inverse Fourier transforming the product of their Fourier transforms. The discrete version of this theorem yields a fast Fourier transform based convolution algorithm (see Chapter 5).

The converse of the convolution theorem is that the Fourier transform of the product of two functions is the convolution of their Fourier transforms.

The result of convolution theorem can also be extended to the *spatial correlation* between two real functions $h(x, y)$ and $f(x, y)$, which is defined as

$$c(x, y) = h(x, y) \star f(x, y) \triangleq \int_{-\infty}^{\infty}\int_{-\infty}^{\infty} h(x', y')f(x + x', y + y')\, dx'\, dy' \qquad (2.18)$$

A change of variables shows that $c(x, y)$ is also the convolution $h(-x, -y) \circledast f(x, y)$, which yields

$$C(\xi_1, \xi_2) = H(-\xi_1, -\xi_2)F(\xi_1, \xi_2) \qquad (2.19)$$

6. *Inner product preservation.* Another important property of the Fourier transform is that the inner product of two functions is equal to the inner product of their Fourier transforms, i.e.,

$$I \triangleq \int_{-\infty}^{\infty}\int_{-\infty}^{\infty} f(x, y)h^*(x, y)\, dx\, dy = \int_{-\infty}^{\infty}\int_{-\infty}^{\infty} F(\xi_1, \xi_2)H^*(\xi_1, \xi_2)\, d\xi_1\, d\xi_2 \qquad (2.20)$$

Setting $h = f$, we obtain the well-known *Parseval energy conservation formula*

$$\int_{-\infty}^{\infty}\int_{-\infty}^{\infty} |f(x, y)|^2\, dx\, dy = \int_{-\infty}^{\infty}\int_{-\infty}^{\infty} |F(\xi_1, \xi_2)|^2\, d\xi_1\, d\xi_2 \qquad (2.21)$$

i.e., the total energy in the function is the same as in its Fourier transform.

7. *Hankel transform.* The Fourier transform of a circularly symmetric function is also circularly symmetric and is given by what is called the *Hankel transform* (see Problem 2.10).

Fourier Transform of Sequences (Fourier Series)

For a *one-dimensional* sequence $x(n)$, real or complex, its Fourier transform is defined as the series

$$X(\omega) = \sum_{n=-\infty}^{\infty} x(n) \exp(-jn\omega), \qquad -\pi \le \omega < \pi \qquad (2.22)$$

The inverse transform is given by

$$x(n) = \frac{1}{2\pi}\int_{-\pi}^{\pi} X(\omega) \exp(jn\omega)\, d\omega \qquad (2.23)$$

Note that $X(\omega)$ is periodic with period 2π. Hence it is sufficient to specify it over one period.

The Fourier transform pair of a two-dimensional sequence $x(m, n)$ is defined as

$$X(\omega_1, \omega_2) \triangleq \sum\sum_{m,n=-\infty}^{\infty} x(m, n) \exp[-j(m\omega_1 + n\omega_2)], \qquad -\pi \le \omega_1, \omega_2 < \pi \tag{2.24}$$

$$x(m, n) = \frac{1}{4\pi^2}\int_{-\pi}^{\pi}\int_{-\pi}^{\pi} X(\omega_1, \omega_2) \exp[j(m\omega_1 + n\omega_2)]\, d\omega_1\, d\omega_2 \tag{2.25}$$

Now $X(\omega_1, \omega_2)$ is periodic with period 2π in each argument, i.e.,

$$X(\omega_1 \pm 2\pi, \omega_2 \pm 2\pi) = X(\omega_1 \pm 2\pi, \omega_2) = X(\omega_1, \omega_2 \pm 2\pi) = X(\omega_1, \omega_2) \quad (2.25)$$

Often, the sequence $x(m, n)$ in the series in (2.24) is absolutely summable, i.e.,

$$\sum\sum_{m,n=-\infty}^{\infty} |x(m, n)| < \infty \tag{2.26}$$

Analogous to the continuous case, $H(\omega_1, \omega_2)$, the Fourier transform of the shift invariant impulse response is called *frequency response*. The Fourier transform of sequences has many properties similar to the Fourier transform of continuous functions. These are summarized in Table 2.4.

TABLE 2.4 Properties and Examples of Fourier Transform of Two-Dimensional Sequences

Property	Sequence	Transform
	$x(m, n), y(m, n), h(m, n), \cdots$	$X(\omega_1, \omega_2), Y(\omega_1, \omega_2), H(\omega_1, \omega_2), \cdots$
Linearity	$a_1 x_1(m, n) + a_2 x_2(m, n)$	$a_1 X_1(\omega_1, \omega_2) + a_2 X_2(\omega_1, \omega_2)$
Conjugation	$x^*(m, n)$	$X^*(-\omega_1, -\omega_2)$
Separability	$x_1(m) x_2(n)$	$X_1(\omega_1) X_2(\omega_2)$
Shifting	$x(m \pm m_0, n \pm n_0)$	$\exp[\pm j(m_0\omega_1 + n_0\omega_2)] X(\omega_1, \omega_2)$
Modulation	$\exp[\pm j(\omega_{01} m + \omega_{02} n)] x(m, n)$	$X(\omega_1 \mp \omega_{01}, \omega_2 \mp \omega_{02})$
Convolution	$y(m, n) = h(m, n) \circledast x(m, n)$	$Y(\omega_1, \omega_2) = H(\omega_1, \omega_2) X(\omega_1, \omega_2)$
Multiplication	$h(m, n) x(m, n)$	$\left(\frac{1}{4\pi^2}\right) H(\omega_1, \omega_2) \circledast X(\omega_1, \omega_2)$
Spatial correlation	$c(m, n) = h(m, n) \star x(m, n)$	$C(\omega_1, \omega_2) = H(-\omega_1, -\omega_2) X(\omega_1, \omega_2)$
Inner product	$I = \sum\sum_{m,n=-\infty}^{\infty} x(m, n) y^*(m, n)$	$I = \frac{1}{4\pi^2}\int_{-\pi}^{\pi}\int_{-\pi}^{\pi} X(\omega_1, \omega_2) Y^*(\omega_1, \omega_2)\, d\omega_1\, d\omega_2$
Energy conservation	$\mathscr{E} = \sum\sum_{m,n=-\infty}^{\infty} \lvert x(m, n)\rvert^2$	$\mathscr{E} = \frac{1}{4\pi^2}\int_{-\pi}^{\pi}\int_{-\pi}^{\pi} \lvert X(\omega_1, \omega_2)\rvert^2\, d\omega_1\, d\omega_2$
	$\sum\sum_{m,n=-\infty}^{\infty} \exp[j(m\omega_{01} + n\omega_{02})]$	$4\pi^2 \delta(\omega_1 - \omega_{01}, \omega_2 - \omega_{02})$
	$\delta(m, n)$	$\frac{1}{4\pi^2}\int_{-\pi}^{\pi}\int_{-\pi}^{\pi} \exp[-j(\omega_1 m + \omega_2 n)]\, d\omega_1\, d\omega_2$

2.5 THE Z-TRANSFORM OR LAURENT SERIES

A useful generalization of the Fourier series is the Z-transform, which for a two-dimensional complex sequence $x(m, n)$ is defined as

$$X(z_1, z_2) = \sum_{m,n=-\infty}^{\infty}\sum x(m, n) z_1^{-m} z_2^{-n} \tag{2.27}$$

where z_1, z_2 are complex variables. The set of values of z_1, z_2 for which this series converges uniformly is called the *region of convergence*. The Z-transform of the impulse response of a linear shift invariant discrete system is called its *transfer function*. Applying the convolution theorem for Z-transforms (Table 2.5) we can transform (2.10) as

$$Y(z_1, z_2) = H(z_1, z_2) X(z_1, z_2)$$

$$\Rightarrow \quad H(z_1, z_2) = \frac{Y(z_1, z_2)}{X(z_1, z_2)}$$

i.e., the transfer function is also the ratio of the Z-transforms of the output and the input sequences. The inverse Z-transform is given by the double contour integral

$$x(m, n) = \frac{1}{(j2\pi)^2} \oint\oint X(z_1, z_2) z_1^{m-1} z_2^{n-1} \, dz_1 \, dz_2 \tag{2.28}$$

where the contours of integration are counterclockwise and lie in the region of convergence. When the region of convergence includes the unit circles $|z_1| = 1, |z_2| = 1$, then evaluation of $X(z_1, z_2)$ at $z_1 = \exp(j\omega_1)$, $z_2 = \exp(j\omega_2)$ yields the Fourier transform of $x(m, n)$. Sometimes $X(z_1, z_2)$ is available as a finite series (such as the transfer function of FIR filters). Then $x(m, n)$ can be obtained by inspection as the coefficient of the term $z_1^{-m} z_2^{-n}$.

TABLE 2.5 Properties of the Two-Dimensional Z-Transform

Property	Sequence	Z-Transform
	$x(m, n), y(m, n), h(m, n), \cdots$	$X(z_1, z_2), Y(z_1, z_2), H(z_1, z_2), \cdots$
Rotation	$x(-m, -n)$	$X(z_1^{-1}, z_2^{-1})$
Linearity	$a_1 x_1(m, n) + a_2 x_2(m, n)$	$a_1 X_1(z_1, z_2) + a_2 X_2(z_1, z_2)$
Conjugation	$X^*(m, n)$	$X^*(z_1^*, z_2^*)$
Separability	$x_1(m) x_2(n)$	$X_1(z_1) X_2(z_2)$
Shifting	$x(m \pm m_0, n \pm n_0)$	$z_1^{\pm m_0} z_2^{\pm n_0} X(z_1, z_2)$
Modulation	$a^m b^n x(m, n)$	$X\left(\frac{z_1}{a}, \frac{z_2}{b}\right)$
Convolution	$h(m, n) \circledast x(m, n)$	$H(z_1, z_2) X(z_1, z_2)$
Multiplication	$x(m, n) y(m, n)$	$\left(\frac{1}{2\pi j}\right)^2 \oint_{C_1}\oint_{C_2} X\left(\frac{z_1}{z_1'}, \frac{z_2}{z_2'}\right) Y(z_1', z_2') \frac{dz_1'}{z_1'} \frac{dz_2'}{z_2'}$

Causality and Stability

A one-dimensional shift invariant system is called causal if its output at any time is not affected by future inputs. This means its impulse response $h(n) = 0$ for $n < 0$ and its transfer function must have a one-sided Laurent series, i.e.,

$$H(z) = \sum_{n=0}^{\infty} h(n) z^{-n} \tag{2.29}$$

Extending this definition, any sequence $x(n)$ is called *causal* if $x(n) = 0, n < 0$; *anticausal* if $x(n) = 0, n \geq 0$, and *noncausal* if it is neither causal nor anticausal.

A system is called *stable* if its output remains uniformly bounded for any bounded input. For linear shift invariant systems, this condition requires that the impulse response should be absolutely summable (prove it!), i.e.,

$$\sum_{n=-\infty}^{\infty} |h(n)| < \infty \tag{2.30}$$

This means $H(z)$ cannot have any poles on the unit circle $|z| = 1$. If this system is to be *causal and stable*, then the convergence of (2.29) at $|z| = 1$ implies the series must converge for all $|z| \geq 1$, i.e., the poles of $H(z)$ must lie *inside* the unit circle.

In two dimensions, a linear shift invariant system is stable when

$$\sum_{m} \sum_{n} |h(m, n)| < \infty \tag{2.31}$$

which implies the region of convergence of $H(z_1, z_2)$ must include the unit circles, i.e., $|z_1| = 1, |z_2| = 1$.

2.6 OPTICAL AND MODULATION TRANSFER FUNCTIONS

For a spatially invariant imaging system, its *optical transfer function* (OTF) is defined as its normalized frequency response, i.e.,

$$\text{OTF} = \frac{H(\xi_1, \xi_2)}{H(0,0)} \tag{2.32}$$

The *modulation transfer function* (MTF) is defined as the magnitude of the OTF, i.e.,

$$\text{MTF} = |\text{OTF}| = \frac{|H(\xi_1, \xi_2)|}{|H(0,0)|} \tag{2.33}$$

Similar relations are valid for discrete systems. Figure 2.5 shows the MTFs of systems whose PSFs are displayed in Fig. 2.2. In practice, it is often the MTF that is measurable. The phase of the frequency response is estimated from physical considerations. For many optical systems, the OTF itself is positive.

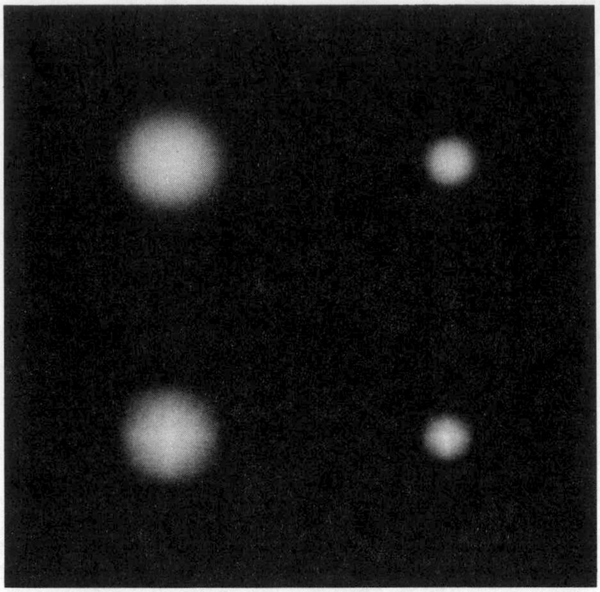

Figure 2.5 MTFs of systems whose PSFs are displayed in Figure 2.2.

Example 2.2

The impulse response of an imaging system is given as $h(x, y) = 2\sin^2[\pi(x - x_0)]/[\pi(x - x_0)]^2 \sin^2[\pi(y - y_0)]/[\pi(y - y_0)]^2$. Then its frequency response is $H(\xi_1, \xi_2) = 2\,\mathrm{tri}(\xi_1, \xi_2)\exp[-j2\pi(x_0\xi_1 + y_0\xi_2)]$, and $\mathrm{OTF} = \mathrm{tri}(\xi_1, \xi_2)\exp[-j2\pi(x_0\xi_1 + y_0\xi_2]$, $\mathrm{MTF} = \mathrm{tri}(\xi_1, \xi_2)$.

2.7 MATRIX THEORY RESULTS

Vectors and Matrices

Often one- and two-dimensional sequences will be represented by vectors and matrices, respectively. A column vector $\mathbf{u}$ containing N elements is denoted as

$$\mathbf{u} \triangleq \{u(n)\} = \begin{bmatrix} u(1) \\ u(2) \\ \cdot \\ \cdot \\ \cdot \\ u(N) \end{bmatrix} \tag{2.34}$$

The nth element of the vector $\mathbf{u}$ is denoted by $u(n)$, u_n, or $[\mathbf{u}]_n$. Unless specified otherwise, all vectors will be column vectors. A column vector of size N is also called an $N \times 1$ vector. Likewise, a row vector of size N is called a $1 \times N$ vector.

A matrix $\mathbf{A}$ of size $M \times N$ has M rows and N columns and is defined as

$$\mathbf{A} \stackrel{\Delta}{=} \{a(m, n)\} = \begin{bmatrix} a(1, 1) & a(1, 2) & \cdots a(1, N) \\ a(2, 1) & & \\ \cdot & \cdot & \cdot \\ \cdot & \cdot & \cdot \\ \cdot & \cdot & \cdot \\ a(M, 1) & a(M, 2) & \cdots a(M, N) \end{bmatrix} \tag{2.35}$$

The element in the mth row and nth column of matrix $\mathbf{A}$ is written as $[\mathbf{A}]_{m,n} \stackrel{\Delta}{=} a(m, n) \stackrel{\Delta}{=} a_{m,n}$. The nth column of $\mathbf{A}$ is denoted by $\mathbf{a}_n$, whose mth element is written as $a_n(m) = a(m, n)$. When the starting index of a matrix is not $(1, 1)$, it will be so indicated. For example,

$$\mathbf{A} = \{a(m, n) \qquad 0 \le m, n \le N - 1\}$$

represents an $N \times N$ matrix with starting index $(0, 0)$. Common definitions from matrix theory are summarized in Table 2.6.

In two dimensions it is often useful to visualize an image as a matrix. The matrix representation is simply a 90° clockwise rotation of the conventional two-dimensional Cartesian coordinate representation:

$$x(m, n) = \begin{array}{c|ccc} \overset{n}{\uparrow} & 2 & -1 & -3 \\ & 4 & 0 & 5 \\ & 1 & 2 & 3 \\ \hline & & & \xrightarrow{\;m\;} \end{array} \Rightarrow \mathbf{X} = \begin{array}{l} \xrightarrow{\qquad\qquad n} \\ \downarrow m \begin{bmatrix} 1 & 4 & 2 \\ 2 & 0 & -1 \\ 3 & 5 & -3 \end{bmatrix} \end{array}$$

Row and Column Ordering

Sometimes it is necessary to write a matrix in the form of a vector, for instance, when storing an image on a disk or a tape. Let

$$\mathscr{x} \stackrel{\Delta}{=} \mathcal{O}\{x(m, n)\}$$

be a one-to-one ordering of the elements of the array $\{x(m, n)\}$ into the vector $\mathscr{x}$. For an $M \times N$ matrix, a mapping used often is called the *lexicographic* or *dictionary ordering*. This is a *row-ordered vector* and is defined as

$$\begin{aligned} \mathscr{x}^T &= [x(1,1)x(1,2)\ldots x(1,N)x(2,1)\ldots x(2,N)\ldots x(M, 1)\ldots x(M, N)]^T \\ &\stackrel{\Delta}{=} \mathcal{O}_r\{x(m, n)\} \end{aligned} \tag{2.36a}$$

Thus $\mathscr{x}^T$ is the row vector obtained by stacking each row to the right of the previous row of $\mathbf{X}$. Another useful mapping is the column by column stacking, which gives a *column-ordered vector* as

$$\mathscr{x}^T = [x(1,1)x(2,1)\ldots x(M, 1)x(1,2)\ldots x(M, 2)\ldots x(1, M)\ldots x(M, N)]^T$$

TABLE 2.6 Matrix Theory Definitions

Item	Definition	Comments
Matrix	$\mathbf{A} = \{a(m, n)\}$	m = row index, n = column index
Transpose	$\mathbf{A}^T = \{a(n, m)\}$	Rows and columns are interchanged.
Complex conjugate	$\mathbf{A}^* = \{a^*(m, n)\}$	
Conjugate transpose	$\mathbf{A}^{*T} = \{a^*(n, m)\}$	
Identity matrix	$\mathbf{I} = \{\delta(m - n)\}$	A square matrix with unity along its diagonal.
Null matrix	$\mathbf{O} = \{0\}$	All elements are zero.
Matrix addition	$\mathbf{A} + \mathbf{B} = \{a(m, n) + b(m, n)\}$	$\mathbf{A}, \mathbf{B}$ have same dimensions.
Scalar multiplication	$\alpha\mathbf{A} = \{\alpha a(m, n)\}$	
Matrix multiplication	$c(m, n) \triangleq \sum_{k=1}^{K} a(m, k)\, b(k, n)$	$\mathbf{C} \triangleq \mathbf{AB}$, $\mathbf{A}$ is $M \times K$, $\mathbf{B}$ is $K \times N$, $\mathbf{C}$ is $M \times N$. $\mathbf{AB} \neq \mathbf{BA}$.
Commuting matrices	$\mathbf{AB} = \mathbf{BA}$	Not true in general.
Vector inner product	$\langle \mathbf{x}, \mathbf{y} \rangle \triangleq \mathbf{x}^{*T}\mathbf{y} = \sum_n x^*(n)\, y(n)$	Scalar quantity. If zero, $\mathbf{x}$ and $\mathbf{y}$ are called orthogonal.
Vector outer product	$\mathbf{x}\mathbf{y}^T = \{x(m)\, y(n)\}$	$\mathbf{x}$ is $M \times 1$, $\mathbf{y}$ is $N \times 1$, outerproduct is $M \times N$; is a rank 1 matrix.
Symmetric	$\mathbf{A} = \mathbf{A}^T$	
Hermitian	$\mathbf{A} = \mathbf{A}^{*T}$	A real symmetric matrix is Hermitian. All eigenvalues are real.
Determinant	$\lvert\mathbf{A}\rvert$	For square matrices only.
Rank [$\mathbf{A}$]	Number of linearly independent rows or columns.	
Inverse, $\mathbf{A}^{-1}$	$\mathbf{A}^{-1}\mathbf{A} = \mathbf{A}\mathbf{A}^{-1} = \mathbf{I}$	For square matrices only.
Singular	$\mathbf{A}^{-1}$ does not exist	$\lvert\mathbf{A}\rvert = 0$
Trace	$\mathrm{Tr}[\mathbf{A}] = \sum_n a(n, n)$	Sum of the diagonal elements.
Eigenvalues, λ_k	All roots $\lvert\mathbf{A} - \lambda_k \mathbf{I}\rvert = 0$	
Eigenvectors, $\boldsymbol{\phi}_k$	All solutions $\mathbf{A}\boldsymbol{\phi}_k = \lambda_k \boldsymbol{\phi}_k$, $\boldsymbol{\phi}_k \neq \mathbf{0}$	
ABCD lemma	$(\mathbf{A} - \mathbf{BCD})^{-1} = \mathbf{A}^{-1} + \mathbf{A}^{-1}\mathbf{B}(\mathbf{C}^{-1} - \mathbf{D}\mathbf{A}^{-1}\mathbf{B})^{-1}\mathbf{D}\mathbf{A}^{-1}$	$\mathbf{A}, \mathbf{C}$ are nonsingular.

$$= \begin{bmatrix} \mathbf{x}_1 \\ \mathbf{x}_2 \\ \cdot \\ \cdot \\ \cdot \\ \mathbf{x}_N \end{bmatrix} \triangleq \mathcal{O}_c\{x(m, n)\} \tag{2.36b}$$

where $\mathbf{x}_n$ is the nth column of $\mathbf{X}$.

Transposition and Conjugation Rules

1. $\mathbf{A}^{*T} = [\mathbf{A}^T]^*$
2. $[\mathbf{AB}]^T = \mathbf{B}^T\mathbf{A}^T$
3. $[\mathbf{A}^{-1}]^T = [\mathbf{A}^T]^{-1}$
4. $[\mathbf{AB}]^* = \mathbf{A}^*\mathbf{B}^*$

Note that the *conjugate transpose* is denoted by $\mathbf{A}^{*T}$. In matrix theory literature, a simplified notation $\mathbf{A}^*$ is often used to denote the conjugate transpose of $\mathbf{A}$. In the theory of image transforms (Chapter 5), we will have to distinguish between $\mathbf{A}$, $\mathbf{A}^*$, $\mathbf{A}^T$ and $\mathbf{A}^{*T}$ and hence the need for the notation.

Toeplitz and Circulant Matrices

A *Toeplitz* matrix $\mathbf{T}$ is a matrix that has constant elements along the main diagonal and the subdiagonals. This means the elements $t(m, n)$ depend only on the difference $m - n$, i.e., $t(m, n) = t_{m-n}$. Thus an $N \times N$ Toeplitz matrix is of the form

$$\mathbf{T} = \begin{bmatrix} t_0 & t_{-1} & & \cdots & t_{-N+1} \\ t_1 & t_0 & t_{-1} & & t_{-N+2} \\ t_2 & \ddots & \ddots & \ddots & \cdot \\ \cdot & & \ddots & \ddots & \cdot \\ \cdot & & & \ddots & \cdot \\ \cdot & & & & t_{-1} \\ t_{N-1} & \cdots & t_2 & t_1 & t_0 \end{bmatrix} \tag{2.37}$$

and is completely defined by the $(2N - 1)$ elements $\{t_k, -N + 1 \le k \le N - 1\}$. Toeplitz matrices describe the input-output transformations of one-dimensional linear shift invariant systems (see Example 2.3) and correlation matrices of stationary sequences.

A matrix $\mathbf{C}$ is called *circulant* if each of its rows (or columns) is a circular shift of the previous row (or column), i.e.,

$$\mathbf{C} = \begin{bmatrix} c_0 & c_1 & c_2 & \cdots & c_{N-1} \\ c_{N-1} & c_0 & c_1 & & c_{N-2} \\ \cdot & & \ddots & \ddots & \cdot \\ \cdot & & & \ddots & \cdot \\ \cdot & & & & \cdot \\ c_2 & & & & c_1 \\ c_1 & c_2 & \cdots & c_{N-1} & c_0 \end{bmatrix} \tag{2.38}$$

Note that **C** is also Toeplitz and

$$c(m, n) = c((m - n) \quad \text{modulo } N) \tag{2.39}$$

Circulant matrices describe the input-output behavior of one-dimensional linear periodic systems (see Example 2.4) and correlation matrices of periodic sequences.

Example 2.3 (Linear convolution as a Toeplitz matrix operation)

The output of a shift invariant system with impulse response $h(n) = n$, $-1 \le n \le 1$, and with input $x(n)$, which is zero outside $0 \le n \le 4$, is given by the convolution

$$y(n) = h(n) \circledast x(n) = \sum_{k=0}^{4} h(n-k)x(k)$$

Note that $y(n)$ will be zero outside the interval $-1 \le n \le 5$. In vector notation, this can be written as a 7×5 Toeplitz matrix operating on a 5×1 vector, namely,

$$\begin{bmatrix} y(-1) \\ y(0) \\ y(1) \\ y(2) \\ y(3) \\ y(4) \\ y(5) \end{bmatrix} = \begin{bmatrix} -1 & 0 & 0 & 0 & 0 \\ 0 & -1 & 0 & 0 & 0 \\ 1 & 0 & -1 & 0 & 0 \\ 0 & 1 & 0 & -1 & 0 \\ 0 & 0 & 1 & 0 & -1 \\ 0 & 0 & 0 & 1 & 0 \\ 0 & 0 & 0 & 0 & 1 \end{bmatrix} \begin{bmatrix} x(0) \\ x(1) \\ x(2) \\ x(3) \\ x(4) \end{bmatrix}$$

Example 2.4 (Circular convolution as a circulant matrix operation)

If two convolving sequences are periodic, then their convolution is also periodic and can be represented as

$$y(n) = \sum_{k=0}^{N-1} h(n-k)x(k), \qquad 0 \le n \le N-1$$

where $h(-n) = h(N-n)$ and N is the period. For example, let $N = 4$ and $h(n) = n + 3$ (modulo 4). In vector notation this gives

$$\begin{bmatrix} y(0) \\ y(1) \\ y(2) \\ y(3) \end{bmatrix} = \begin{bmatrix} 3 & 2 & 1 & 0 \\ 0 & 3 & 2 & 1 \\ 1 & 0 & 3 & 2 \\ 2 & 1 & 0 & 3 \end{bmatrix} \begin{bmatrix} x(0) \\ x(1) \\ x(2) \\ x(3) \end{bmatrix}$$

Thus the input-to-output transformation of a circular convolution is described by a circulant matrix.

Orthogonal and Unitary Matrices

An *orthogonal* matrix is such that its inverse is equal to its transpose, i.e., **A** is orthogonal if

$$\mathbf{A}^{-1} = \mathbf{A}^T$$

or

$$\mathbf{A}^T\mathbf{A} = \mathbf{A}\mathbf{A}^T = \mathbf{I} \tag{2.40}$$

A matrix is called unitary if its inverse is equal to its conjugate transpose, i.e.,

$$\mathbf{A}^{-1} = \mathbf{A}^{*T}$$

or

$$\mathbf{A}\mathbf{A}^{*T} = \mathbf{A}^{*T}\mathbf{A} = \mathbf{I} \tag{2.41}$$

A real orthogonal matrix is also unitary, but a unitary matrix need not be orthogonal. The preceding definitions imply that *the columns (or rows) of an $N \times N$ unitary matrix are orthogonal and form a complete set of basis vectors in an N-dimensional vector space.*

Example 2.5

Consider the matrices

$$\mathbf{A}_1 = \frac{1}{\sqrt{2}}\begin{bmatrix} 1 & 1 \\ 1 & -1 \end{bmatrix}, \qquad \mathbf{A}_2 = \begin{bmatrix} \sqrt{2} & j \\ -j & \sqrt{2} \end{bmatrix}, \qquad \mathbf{A}_3 = \frac{1}{\sqrt{2}}\begin{bmatrix} 1 & j \\ j & 1 \end{bmatrix}$$

It is easy to check that $\mathbf{A}_1$ is orthogonal and unitary. $\mathbf{A}_2$ is not unitary. $\mathbf{A}_3$ is unitary with orthogonal rows.

Positive Definiteness and Quadratic Forms

An $N \times N$ Hermitian matrix $\mathbf{A}$ is called *positive definite* or *positive semidefinite* if the quadratic form

$$Q \triangleq \mathbf{x}^{*T}\mathbf{A}\mathbf{x}, \qquad \forall \mathbf{x} \neq \mathbf{0} \tag{2.42}$$

is positive (>0) or nonnegative (≥ 0), respectively. Similarly, $\mathbf{A}$ is *negative definite* or *negative semidefinite* if $Q < 0$ or $Q \leq 0$, respectively. A matrix that does not satisfy any of the above is *indefinite.*

If $\mathbf{A}$ is a symmetric positive (nonnegative) definite matrix, then all its eigenvalues $\{\lambda_k\}$ are positive (nonnegative) and the determinant of $\mathbf{A}$ satisfies the inequality

$$|\mathbf{A}| = \prod_{k=1}^{N} \lambda_k \leq \prod_{k=1}^{N} a(k, k) \tag{2.43}$$

Diagonal Forms

For any Hermitian matrix $\mathbf{R}$ there exists a unitary matrix $\mathbf{\Phi}$ such that

$$\mathbf{\Phi}^{*T}\mathbf{R}\mathbf{\Phi} = \mathbf{\Lambda} \tag{2.44}$$

where $\mathbf{\Lambda}$ is a diagonal matrix containing the eigenvalues of $\mathbf{R}$. An alternate form of the above equation is

$$\mathbf{R}\mathbf{\Phi} = \mathbf{\Phi}\mathbf{\Lambda} \tag{2.45}$$

which is the set of eigenvalue equations

$$\mathbf{R}\boldsymbol{\phi}_k = \lambda_k \boldsymbol{\phi}_k \qquad k = 1, \ldots, N \tag{2.46}$$

where $\{\lambda_k\}$ and $\{\boldsymbol{\phi}_k\}$ are the eigenvalues and eigenvectors, respectively, of $\mathbf{R}$. For

Hermitian matrices, the eigenvectors corresponding to distinct eigenvalues are orthogonal. For repeated eigenvalues, their eigenvectors form a subspace that can be orthogonalized to yield a complete set of orthogonal eigenvectors. Normalization of these eigenvectors yields an orthonormal set, i.e., the unitary matrix $\mathbf{\Phi}$, whose columns are these eigenvectors. The matrix $\mathbf{\Phi}$ is also called the eigenmatrix of $\mathbf{R}$.

2.8 BLOCK MATRICES AND KRONECKER PRODUCTS

In image processing, the analysis of many problems can be simplified substantially by working with block matrices and the so-called Kronecker products. For example, the two-dimensional convolution can be expressed by simple block matrix operations.

Block Matrices

Any matrix $\mathcal{A}$ whose elements are matrices themselves is called a *block matrix*; for example,

$$\mathcal{A} = \begin{bmatrix} \mathbf{A}_{1,1} & \mathbf{A}_{1,2} & \cdots & \mathbf{A}_{1,n} \\ \mathbf{A}_{2,1} & \mathbf{A}_{2,2} & & \mathbf{A}_{2,n} \\ \cdot & \cdot & & \cdot \\ \cdot & \cdot & & \cdot \\ \cdot & \cdot & & \cdot \\ \mathbf{A}_{m,1} & \mathbf{A}_{m,2} & \cdots & \mathbf{A}_{m,n} \end{bmatrix} \tag{2.47}$$

is a block matrix where $\{\mathbf{A}_{i,j}\}$ are $p \times q$ matrices. The matrix $\mathcal{A}$ is called an $m \times n$ block matrix of basic dimension $p \times q$. If $\mathbf{A}_{i,j}$ are square matrices (say, $p \times p$), then we also call $\mathcal{A}$ to be an $m \times n$ block matrix of basic dimension p.

If the block structure is Toeplitz, $(\mathbf{A}_{i,j} = \mathbf{A}_{i-j})$ or circulant $(\mathbf{A}_{i,j} = \mathbf{A}_{((i-j) \text{modulo } n)}$, $m = n)$ then $\mathcal{A}$ is called *block Toeplitz* or *block circulant,* respectively. Additionally, if each block itself is Toeplitz (or circulant), then $\mathcal{A}$ is called *doubly block Toeplitz* (or doubly block circulant). Finally, if $\{\mathbf{A}_{i,j}\}$ are Toeplitz (or circulant) but $(\mathbf{A}_{i,j} \neq \mathbf{A}_{i-j})$ then $\mathcal{A}$ is called a *Toeplitz block* (or *circulant block*) matrix. Note that a doubly Toeplitz (or circulant) matrix need not be fully Toeplitz (or circulant), i.e., the scaler elements of $\mathcal{A}$ need not be constants along the subdiagonals.

Example 2.6

Consider the two-dimensional convolution

$$y(m, n) = \sum_{m'=0}^{2} \sum_{n'=0}^{1} h(m - m', n - n')x(m', n'), \qquad 0 \le m \le 3, \quad 0 \le n \le 2$$

where the $x(m, n)$ and $h(m, n)$ are defined in Example 2.1. We will examine the block structure of the matrices when the input and output arrays are mapped into column-ordered vectors. Let $\mathbf{x}_n$ and $\mathbf{y}_n$ be the column vectors. Then

$$\mathbf{y}_n = \sum_{n'=0}^{1} \mathbf{H}_{n-n'} \mathbf{x}_{n'}, \qquad \mathbf{H}_n = \{h(m - m', n), \qquad 0 \le m \le 3, \quad 0 \le m' \le 2\},$$

where

$$\mathbf{x}_0 = \begin{bmatrix} 2 \\ 5 \\ 3 \end{bmatrix}, \qquad \mathbf{x}_1 = \begin{bmatrix} 1 \\ 4 \\ 1 \end{bmatrix}, \qquad \mathbf{H}_0 = \begin{bmatrix} 1 & 0 & 0 \\ -1 & 1 & 0 \\ 0 & -1 & 1 \\ 0 & 0 & -1 \end{bmatrix}$$

$$\mathbf{H}_1 = \begin{bmatrix} 1 & 0 & 0 \\ 1 & 1 & 0 \\ 0 & 1 & 1 \\ 0 & 0 & 1 \end{bmatrix}, \qquad \mathbf{H}_{-1} = \mathbf{0}, \qquad \mathbf{H}_2 = \mathbf{0}$$

Defining $\mathscr{y}$ and $\mathscr{x}$ as column-ordered vectors, we get

$$\mathscr{y} = \begin{bmatrix} \mathbf{y}_0 \\ \mathbf{y}_1 \\ \mathbf{y}_2 \end{bmatrix} = \begin{bmatrix} \mathbf{H}_0 & \mathbf{0} \\ \mathbf{H}_1 & \mathbf{H}_0 \\ \mathbf{0} & \mathbf{H}_1 \end{bmatrix} \begin{bmatrix} \mathbf{x}_0 \\ \mathbf{x}_1 \end{bmatrix} \triangleq \mathscr{H}\mathscr{x}$$

where $\mathscr{H}$ is a doubly Toeplitz 3×2 block matrix of basic dimensions 4×3. However, the matrix $\mathscr{H}$ as a whole is not Toeplitz because $[\mathscr{H}]_{m,n} \neq [\mathscr{H}]_{m-n}$ (show it!). Hence the *one-dimensional system* $\mathscr{y} = \mathscr{H}\mathscr{x}$ is linear but not spatially invariant, even though the original two-dimensional system is. Alternatively, $\mathscr{y} = \mathscr{H}\mathscr{x}$ does not represent a one-dimensional convolution operation although it does represent a two-dimensional convolution.

Example 2.7

Block circulant matrices arise when the convolving arrays are periodic. For example, let

$$y(m, n) = \sum_{m'=0}^{2} \sum_{n'=0}^{3} h(m - m', n - n')x(m', n'), \qquad 0 \le m \le 2, \quad 0 \le n \le 3$$

where $h(m, n)$ is doubly periodic with periods $(3, 4)$, i.e., $h(m, n) = h(m + 3, n + 4)$, $\forall m, n$. The array $h(m, n)$ over one array period is shown next:

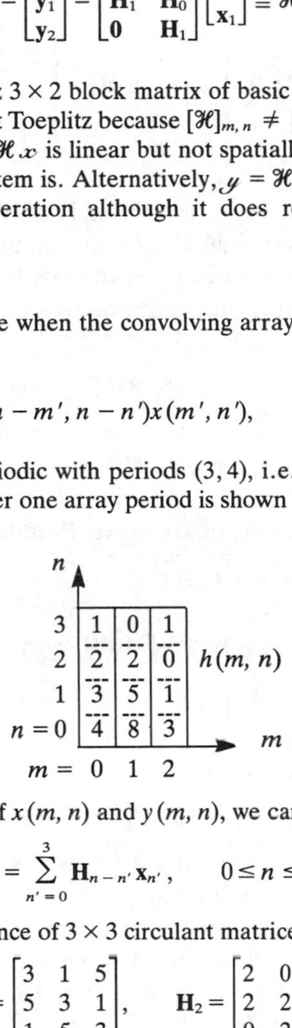

In terms of column vectors of $x(m, n)$ and $y(m, n)$, we can write

$$\mathbf{y}_n = \sum_{n'=0}^{3} \mathbf{H}_{n-n'} \mathbf{x}_{n'}, \qquad 0 \le n \le 3$$

where $\mathbf{H}_n$ is a periodic sequence of 3×3 circulant matrices with period 4, given by

$$\mathbf{H}_0 = \begin{bmatrix} 4 & 3 & 8 \\ 8 & 4 & 3 \\ 3 & 8 & 4 \end{bmatrix}, \quad \mathbf{H}_1 = \begin{bmatrix} 3 & 1 & 5 \\ 5 & 3 & 1 \\ 1 & 5 & 3 \end{bmatrix}, \quad \mathbf{H}_2 = \begin{bmatrix} 2 & 0 & 2 \\ 2 & 2 & 0 \\ 0 & 2 & 2 \end{bmatrix}, \quad \mathbf{H}_3 = \begin{bmatrix} 1 & 1 & 0 \\ 0 & 1 & 1 \\ 1 & 0 & 1 \end{bmatrix}$$

Written as a column-ordered vector equation, the output becomes

$$\mathcal{y} = \begin{bmatrix} \mathbf{y}_0 \\ \mathbf{y}_1 \\ \mathbf{y}_2 \\ \mathbf{y}_3 \end{bmatrix} = \begin{bmatrix} \mathbf{H}_0 & \mathbf{H}_3 & \mathbf{H}_2 & \mathbf{H}_1 \\ \mathbf{H}_1 & \mathbf{H}_0 & \mathbf{H}_3 & \mathbf{H}_2 \\ \mathbf{H}_2 & \mathbf{H}_1 & \mathbf{H}_0 & \mathbf{H}_3 \\ \mathbf{H}_3 & \mathbf{H}_2 & \mathbf{H}_1 & \mathbf{H}_0 \end{bmatrix} \begin{bmatrix} \mathbf{x}_0 \\ \mathbf{x}_1 \\ \mathbf{x}_2 \\ \mathbf{x}_3 \end{bmatrix} \stackrel{\Delta}{=} \mathcal{H}\mathcal{x}$$

where $\mathbf{H}_{-n} = \mathbf{H}_{4-n}$. Now $\mathcal{H}$ is a doubly circulant, 4×4 block matrix of basic dimension 3×3.

Kronecker Products

If $\mathbf{A}$ and $\mathbf{B}$ are $M_1 \times M_2$ and $N_1 \times N_2$ matrices, respectively, then their Kronecker product is defined as

$$\mathbf{A} \otimes \mathbf{B} \stackrel{\Delta}{=} \{a(m, n)\mathbf{B}\} = \begin{bmatrix} a(1, 1)\mathbf{B} \cdots a(1, M_2)\mathbf{B} \\ \vdots \qquad\qquad \vdots \\ a(M_1, 1)\mathbf{B} \cdots a(M_1, M_2)\mathbf{B} \end{bmatrix} \tag{2.48}$$

This is an $M_1 \times M_2$ block matrix of basic dimension $N_1 \times N_2$. Note that $\mathbf{A} \otimes \mathbf{B} \neq \mathbf{B} \otimes \mathbf{A}$. Kronecker products are useful in generating high-order matrices from low-order matrices, for example, the fast Hadamard transforms that will be studied in Chapter 5. Several properties of Kronecker products are listed in Table 2.7. A particularly useful result is the identity

$$(\mathbf{A} \otimes \mathbf{B})(\mathbf{C} \otimes \mathbf{D}) = (\mathbf{AC}) \otimes (\mathbf{BD}) \tag{2.49}$$

It expresses the matrix multiplication of two Kronecker products as a Kronecker product of two matrices. For $N \times N$ matrices, it will take $O(N^6) + O(N^4)$ oper-

TABLE 2.7 Properties of Kronecker Products

1. $(\mathbf{A} + \mathbf{B}) \otimes \mathbf{C} = \mathbf{A} \otimes \mathbf{C} + \mathbf{B} \otimes \mathbf{C}$
2. $(\mathbf{A} \otimes \mathbf{B}) \otimes \mathbf{C} = \mathbf{A} \otimes (\mathbf{B} \otimes \mathbf{C})$
3. $\alpha(\mathbf{A} \otimes \mathbf{B}) = (\alpha\mathbf{A}) \otimes \mathbf{B} = \mathbf{A} \otimes (\alpha\mathbf{B})$, where α is scalar.
4. $(\mathbf{A} \otimes \mathbf{B})^T = \mathbf{A}^T \otimes \mathbf{B}^T$
5. $(\mathbf{A} \otimes \mathbf{B})^{-1} = \mathbf{A}^{-1} \otimes \mathbf{B}^{-1}$
6. $(\mathbf{A} \otimes \mathbf{B})(\mathbf{C} \otimes \mathbf{D}) = (\mathbf{AC}) \otimes (\mathbf{BD})$
7. $\mathbf{A} \otimes \mathbf{B} = (\mathbf{A} \otimes \mathbf{I})(\mathbf{I} \otimes \mathbf{B})$
8. $\prod_{k=1}^{l} (\mathbf{A}_k \otimes \mathbf{B}_k) = \left(\prod_{k=1}^{l} \mathbf{A}_k\right) \otimes \left(\prod_{k=1}^{l} \mathbf{B}_k\right)$, where $\mathbf{A}_k$ and $\mathbf{B}_k$ are square matrices
9. $\det(\mathbf{A} \otimes \mathbf{B}) = (\det \mathbf{A})^m (\det \mathbf{B})^n$, where $\mathbf{A}$ is $m \times m$ and $\mathbf{B}$ is $n \times n$
10. $\mathrm{Tr}(\mathbf{A} \otimes \mathbf{B}) = [\mathrm{Tr}(\mathbf{A})][\mathrm{Tr}(\mathbf{B})]$
11. If $r(\mathbf{A})$ denotes the rank of a matrix $\mathbf{A}$, then $r(\mathbf{A} \otimes \mathbf{B}) = r(\mathbf{A})\, r(\mathbf{B})$.
12. If $\mathbf{A}$ and $\mathbf{B}$ are unitary, then $\mathbf{A} \otimes \mathbf{B}$ is also unitary.
13. If $\mathbf{C} = \mathbf{A} \otimes \mathbf{B}$, $\mathbf{C}\boldsymbol{\xi}_k = \gamma_k \boldsymbol{\xi}_k$, $\mathbf{A}\mathbf{x}_i = \lambda_i \mathbf{x}_i$, $\mathbf{B}\mathbf{y}_j = \mu_j \mathbf{y}_j$, then $\boldsymbol{\xi}_k = \mathbf{x}_i \otimes \mathbf{y}_j$, $\gamma_k = \lambda_i \mu_j$, $1 \le i \le m$, $1 \le j \le n$, $1 \le k \le mn$.

ations to compute the left side, whereas only $O(N^4)$ operations are required to compute the right side. This principle is useful in developing fast algorithms for multiplying matrices that can be expressed as Kronecker products.

Example 2.8

Let

$$\mathbf{A} = \begin{bmatrix} 1 & 1 \\ 1 & -1 \end{bmatrix}, \qquad \mathbf{B} = \begin{bmatrix} 1 & 2 \\ 3 & 4 \end{bmatrix}$$

Then

$$\mathbf{A} \otimes \mathbf{B} = \begin{bmatrix} 1 & 2 & 1 & 2 \\ 3 & 4 & 3 & 4 \\ 1 & 2 & -1 & -2 \\ 3 & 4 & -3 & -4 \end{bmatrix}, \qquad \mathbf{B} \otimes \mathbf{A} = \begin{bmatrix} 1 & 1 & 2 & 2 \\ 1 & -1 & 2 & -2 \\ 3 & 3 & 4 & 4 \\ 3 & -3 & 4 & -4 \end{bmatrix}$$

Note the two products are not equal.

Separable Operations

Consider the transformation (on an $N \times M$ image $\mathbf{U}$)

$$\mathbf{V} \triangleq \mathbf{AUB}^T$$

or

$$v(k, l) = \sum_m \sum_n a(k, m) u(m, n) b(l, n) \tag{2.50}$$

This defines a class of separable operations, where $\mathbf{A}$ operates on the columns of $\mathbf{U}$ and $\mathbf{B}$ operates on the rows of the result. If $\mathbf{v}_k$ and $\mathbf{u}_m$ denote the kth and mth row vectors of $\mathbf{V}$ and $\mathbf{U}$, respectively, then the preceding series becomes

$$\mathbf{v}_k^T = \sum_m a(k, m)[\mathbf{B}\mathbf{u}_m^T] = \sum_m [\mathbf{A} \otimes \mathbf{B}]_{k,m} \mathbf{u}_m^T$$

where $[\mathbf{A} \otimes \mathbf{B}]_{k,m}$ is the (k, m)th block of $\mathbf{A} \otimes \mathbf{B}$. Thus if $\mathbf{U}$ and $\mathbf{V}$ are *row-ordered* into vectors $\boldsymbol{u}$ and $\boldsymbol{v}$, respectively, then

$$\mathbf{V} = \mathbf{AUB}^T \Rightarrow \boldsymbol{v} = (\mathbf{A} \otimes \mathbf{B})\boldsymbol{u}$$

i.e., the separable transformation of (2.50) maps into a Kronecker product operating on a vector.

2.9 RANDOM SIGNALS

Definitions

A complex discrete random signal or a discrete random process is a sequence of random variables $u(n)$. For complex random sequences, we define

$$\text{Mean} \triangleq \mu_u(n) \triangleq \mu(n) = E[u(n)] \tag{2.51}$$

$$\text{Variance} \triangleq \sigma_u^2(n) = \sigma^2(n) = E[|u(n) - \mu(n)|^2] \tag{2.52}$$

$$\begin{aligned}\text{Covariance} &= \text{Cov}[u(n), u(n')] \triangleq r_u(n, n') \triangleq r(n, n') \\ &= E\{[u(n) - \mu(n)][u^*(n') - \mu^*(n')]\}\end{aligned} \tag{2.53}$$

$$\begin{aligned}\text{Cross covariance} &\triangleq \text{Cov}[u(n), v(n')] \triangleq r_{uv}(n, n') \\ &= E\{[u(n) - \mu_u(n)][v^*(n') - \mu_v^*(n')]\}\end{aligned} \tag{2.54}$$

$$\begin{aligned}\text{Autocorrelation} &\triangleq a_{uu}(n, n') \triangleq a(n, n') = E[u(n)u^*(n')] \\ &= r(n, n') + \mu(n)\mu^*(n')\end{aligned} \tag{2.55}$$

$$\text{Cross-correlation} = a_{uv}(n, n') = E[u(n)v^*(n')] = r_{uv}(n, n') + \mu_u(n)\mu_v^*(n') \tag{2.56}$$

The symbol E denotes the mathematical expectation operator. Whenever there is no confusion, we will drop the subscript u from the various functions. For an $N \times 1$ vector $\mathbf{u}$, its mean, covariance, and other properties are defined as

$$E[\mathbf{u}] = \boldsymbol{\mu} = \{\mu(n)\} \quad \text{is an } N \times 1 \text{ vector,} \tag{2.57}$$

$$\text{Cov}[\mathbf{u}] \triangleq E(\mathbf{u} - \boldsymbol{\mu})(\mathbf{u}^* - \boldsymbol{\mu}^*)^T \triangleq \mathbf{R}_u \triangleq \mathbf{R} = \{r(n, n')\} \quad \text{is an } N \times N \text{ matrix} \tag{2.58}$$

$$\text{Cov}[\mathbf{u}, \mathbf{v}] \triangleq E(\mathbf{u} - \boldsymbol{\mu}_u)(\mathbf{v}^* - \boldsymbol{\mu}_v^*)^T \triangleq \mathbf{R}_{uv} = \{r_{uv}(n, n')\} \quad \text{is an } N \times N \text{ matrix} \tag{2.59}$$

Now $\boldsymbol{\mu}$ and $\mathbf{R}$ represent the mean vector and the covariance matrix, respectively, of the vector $\mathbf{u}$.

Gaussian or Normal Distribution

The probability density function of a random variable u is denoted by $p_u(\mathscr{u})$. For a Gaussian random variable

$$p_u(\mathscr{u}) \triangleq \frac{1}{\sqrt{2\pi\sigma^2}} \exp\left\{\frac{-|\mathscr{u} - \mu|^2}{2\sigma^2}\right\}, \tag{2.60}$$

where μ and σ^2 are its mean variance and $\mathscr{u}$ denotes the value the random variable takes. For $\mu = 0$ and $\sigma^2 = 1$, this is called the *standard normal distribution.*

Gaussian Random Processes

A sequence, possibly infinite, is called a Gaussian (or normal) random process if the joint probability density of any finite sub-sequence is a Gaussian distribution. For example, for a Gaussian sequence $\{u(n), 1 \le n \le N\}$ the joint density would be

$$p_{\mathbf{u}}(\mathscr{u}) = p_{\mathbf{u}}(\mathscr{u}_1, \mathscr{u}_2, \ldots, \mathscr{u}_N) = [(2\pi)^{N/2}|\mathbf{R}|^{1/2}]^{-1} \exp\{-\tfrac{1}{2}(\mathscr{u} - \boldsymbol{\mu})^{*T}\mathbf{R}^{-1}(\mathscr{u} - \boldsymbol{\mu})\} \tag{2.61}$$

where $\mathbf{R}$ is the covariance matrix of $\mathbf{u}$ and is assumed to be nonsingular.

Stationary Processes

A random sequence $u(n)$ is said to be *strict-sense stationary* if the joint density of any partial sequence $\{u(l), 1 \le l \le k\}$ is the same as that of the shifted sequence

$\{u(l+m), 1 \le l \le k\}$, for any integer m and any length k. The sequence $u(n)$ is called *wide-sense stationary* if

$$E[u(n)] = \mu = \text{constant}$$
$$E[u(n)u^*(n')] = r(n-n') \tag{2.62}$$

This implies $r(n, n') = r(n - n')$, i.e., the covariance matrix of $\{u(n)\}$ is Toeplitz.

Unless stated otherwise, we will imply wide-sense stationarity whenever we call a random process stationary. Since a Gaussian process is completely specified by the mean and covariance functions, for such a process wide-sense stationarity is the same as strict-sense stationarity. In general, although strict-sense stationarity implies stationarity in the wide sense, the converse is not true.

We will denote the covariance function of a stationary process $u(n)$ by $r(n)$, the implication being

$$r(n) = \text{Cov}[u(n), u(0)] = \text{Cov}[u(n'+n), u(n')], \qquad \forall n', \forall n \tag{2.63}$$

Using the definitions of covariance and autocorrelation functions, it can be shown that the arrays $r(n, n')$ and $a(n, n')$ are conjugate symmetric and nonnegative definite, i.e.,

$$\textit{Symmetry:} \quad r(n, n') = r^*(n', n), \qquad \forall n, n' \tag{2.64}$$

$$\textit{Nonnegativity:} \quad \sum_n \sum_{n'} x(n) r(n, n') x^*(n') \ge 0, \qquad x(n) \ne 0, \forall n \tag{2.65}$$

This means the covariance and autocorrelation matrices are Hermitian and nonnegative definite.

Markov Processes

A random sequence $u(n)$ is called Markov-p, or pth-order Markov, if the conditional probability of $u(n)$ given the entire past is equal to the conditional probability of $u(n)$ given only $u(n-1), \ldots, u(n-p)$, i.e.,

$$\text{Prob}[u(n)|u(n-1), u(n-2), \ldots] = \text{Prob}[u(n)|u(n-1), \ldots, u(n-p)], \qquad \forall n \tag{2.66a}$$

A Markov-1 sequence is simply called Markov. A Markov-p scalar sequence can also be expressed as a $(p \times 1)$ Markov-1 vector sequence. Another interpretation of a pth-order Markov sequence is that if the "present," $\{u(j), n-p \le j \le n-1\}$, is known, then the "past," $\{u(j), j < n-p\}$, and the "future," $\{u(j), j \ge n\}$, are independent. This definition is useful in defining Markov random fields in two dimensions (see Chapter 6). For Gaussian Markov-p sequences it is sufficient that the conditional expectations satisfy the relation

$$E[u(n)|u(n-1), u(n-2), \ldots] = E[u(n)|u(n-1), \ldots, u(n-p)], \qquad \forall n \tag{2.66b}$$

Example 2.9 (Covariance matrix of stationary sequences)

The covariance function of a first-order stationary Markov sequence $u(n)$ is given as

$$r(n) = \rho^{|n|}, \qquad |\rho| < 1, \forall n \tag{2.67}$$

This is often used as the covariance model of a scan line of monochrome images. For an $N \times 1$ vector $\mathbf{u} = \{u(n), 1 \le n \le N\}$, its covariance matrix is $\{r(m-n)\}$, i.e.,

$$\mathbf{R} = \begin{bmatrix} 1 & \rho & \rho^2 & \cdots & \rho^{N-1} \\ \rho & \ddots & \ddots & \ddots & \vdots \\ \vdots & \ddots & \ddots & \ddots & \rho^2 \\ \vdots & & \ddots & \ddots & \rho \\ \rho^{N-1} & \cdots & & \rho & 1 \end{bmatrix} \tag{2.68}$$

which is Toeplitz. In fact the covariance and autocorrelation matrices of any stationary sequence are Toeplitz. Conversely, any sequence, finite or infinite, can be called stationary if its covariance and autocorrelation matrices are Toeplitz.

Orthogonality and Independence

Two random variables x and y are called *independent* if and only if their joint probability density function is a product of their marginal densities, i.e.,

$$p_{x,y}(\mathscr{x}, \mathscr{y}) = p_x(\mathscr{x})p_y(\mathscr{y}) \tag{2.69}$$

Two random sequences $x(n)$ and $y(n)$ are called *independent* if and only if for every n and n', the random variables $x(n)$ and $y(n')$ are independent.

The random variables x and y are said to be *orthogonal* if

$$E[xy^*] = 0 \tag{2.70}$$

and are called *uncorrelated* if

$$E[xy^*] = (E[x])(E[y^*])$$

or

$$E[(x - \mu_x)(y - \mu_y)^*] = 0 \tag{2.71}$$

Thus zero mean uncorrelated random variables are also orthogonal. Gaussian random variables which are uncorrelated are also independent.

The Karhunen-Loève (KL) Transform

Let $\{x(n), 1 \le n \le N\}$ be a complex random sequence whose autocorrelation matrix is $\mathbf{R}$. Let $\mathbf{\Phi}$ be an $N \times N$ unitary matrix, which reduces $\mathbf{R}$ to its diagonal form $\mathbf{\Lambda}$ [see (2.44)]. The transformed vector

$$\mathbf{y} = \mathbf{\Phi}^{*T}\mathbf{x} \tag{2.72}$$

is called the Karhunen-Loève (KL) transform of $\mathbf{x}$. It satisfies the property

$$E[\mathbf{y}\mathbf{y}^{*T}] = \mathbf{\Phi}^{*T}\{E[\mathbf{x}\mathbf{x}^{*T}]\}\mathbf{\Phi} = \mathbf{\Phi}^{*T}\mathbf{R}\mathbf{\Phi} = \mathbf{\Lambda}$$
$$\Rightarrow \quad E[y(k)y^*(l)] = \lambda_k \delta(k - l) \tag{2.73}$$

i.e., the elements of the transformed sequence $y(k)$ are *orthogonal.* If **R** represents the covariance matrix rather than the autocorrelation matrix of **x**, then the sequence $y(k)$ is *uncorrelated.* The unitary matrix $\mathbf{\Phi}^{*T}$ is called the KL transform matrix. *Its rows are the conjugate eigenvectors of* **R**, i.e., it is the conjugate transpose of the eigenmatrix of **R**. The KL transform is of fundamental importance in digital signal and image processing. Its applications and properties are considered in Chapter 5.

2.10 DISCRETE RANDOM FIELDS

In statistical representation of images, each pixel is considered as a random variable. Thus we think of a given image as a sample function of an ensemble of images. Such an ensemble would be adequately defined by a joint probability density of the array of random variables. For practical image sizes, the number of random variables is very large (262,144 for 512×512 images). Thus it is difficult to specify a realistic joint density function because it would be an enormous task to measure it. One possibility is to specify the ensemble by its first- and second-order moments only (mean and covariances). Even with this simplifying constraint, the task of determining realistic model parameters remains difficult. Various approaches for stochastic modeling are considered in Chapter 6. Here we consider some basic definitions that will be useful in the subsequent chapters.

Definitions

When each sample of a two-dimensional sequence is a random variable, we call it a *discrete random field.* When the random field represents an ensemble of images (such as television images or satellite images), we call it a *random image.* The term *random field* will apply to any two-dimensional random sequence.

The mean and covariance functions of a complex random field are defined as

$$E[u(m,n)] = \mu(m,n) \tag{2.74}$$

$$\begin{aligned}\text{Cov}[u(m,n), u(m',n')] &\triangleq E[(u(m,n) - \mu(m,n))(u^*(m',n') - \mu^*(m',n'))] \\ &= r_u(m,n;m',n') = r(m,n;m',n')\end{aligned} \tag{2.75}$$

Often, we will consider the stationary case where

$$\begin{aligned}\mu(m,n) &= \mu = \text{constant} \\ r_u(m,n;m',n') &= r_u(m-m', n-n') = r(m-m', n-n')\end{aligned} \tag{2.76}$$

As before, whenever there is no confusion, we will drop the subscript u from r_u. A random field satisfying (2.76) is also called *shift invariant, translational (or spatial) invariant, homogeneous, or wide-sense stationary.* Unless otherwise mentioned, stationarity will always be implied in the wide sense.

We will denote the covariance function of a stationary random field $u(m,n)$ by $r_u(m,n)$ or $r(m,n)$, implying that

$$r(m,n) = \text{Cov}[u(m,n), u(0,0)] = \text{Cov}[u(m'+m, n'+n), u(m',n')] \qquad \forall (m',n') \tag{2.77}$$

A random field $x(m, n)$ will be called a *white noise field* whenever any two different elements $x(m, n)$ and $x(m', n')$ are mutually uncorrelated, i.e., the field's covariance function is of the form

$$r_x(m, n; m', n') = \sigma_x^2(m, n)\delta(m - m', n - n') \tag{2.78}$$

A random field is called Gaussian if its every segment defined on an arbitrary finite grid is Gaussian. This means every finite segment of $u(m, n)$ when mapped into a vector will have a joint density of the form of (2.61).

Covariances and autocorrelations of two-dimensional fields have symmetry and nonnegativity properties similar to those of one-dimensional random processes:

$$\textit{Symmetry:} \quad r(m, n; m', n') = r^*(m', n'; m, n) \tag{2.79}$$

In general

$$r(m, n; m', n') \neq r(m', n; m, n') \neq r^*(m', n; m, n') \tag{2.80}$$

$$\textit{Nonnegativity:} \quad \sum_m \sum_n \sum_{m'} \sum_{n'} x(m, n) r(m, n; m', n') x^*(m', n') \geq 0, \quad x(m, n) \neq 0, \forall(m, n) \tag{2.81}$$

Separable and Isotropic Image Covariance Functions

The covariance function of a random field is called *separable* when it can be expressed as a product of covariance functions of one-dimensional sequences, i.e., if

$$r(m, n; m', n') = r_1(m, m')r_2(n, n') \qquad \text{(Nonstationary case)} \tag{2.82}$$

$$r(m, n) = r_1(m)r_2(n) \qquad \text{(Stationary case)} \tag{2.83}$$

A separable stationary covariance function often used in image processing is

$$r(m, n) = \sigma^2 \rho_1^{|m|} \rho_2^{|n|}, \qquad |\rho_1| < 1, \qquad |\rho_2| < 1 \tag{2.84}$$

Here σ^2 represents the variance of the random field and $\rho_1 = r(1, 0)/\sigma^2$, $\rho_2 = r(0, 1)/\sigma^2$ are the one-step correlations in the m and n directions, respectively.

Another covariance function often considered as more realistic for many images is the *nonseparable exponential* function

$$r(m, n) = \sigma^2 \exp\{-\sqrt{\alpha_1 m^2 + \alpha_2 n^2}\} \tag{2.85}$$

When $\alpha_1 = \alpha_2 = \alpha$, $r(m, n)$ becomes a function of the Euclidean distance $d \triangleq \sqrt{m^2 + n^2}$, i.e.,

$$r(m, n) = \sigma^2 \rho^d \tag{2.86}$$

where $\rho = \exp(-|\alpha|)$. Such a function is also called *isotropic* or *circularly symmetric*. Figure 2.6 shows a display of the separable and isotropic covariance functions. The parameters of the nonseparable exponential covariance function are related to the one-step correlations as $\alpha_1 = -\ln\rho_1$, $\alpha_2 = -\ln\rho_2$. Thus the covariance models (2.84) and (2.85) can be identified by measuring the variance and the one-step correlations of their zero mean random fields. In practice, these quantities are estimated from

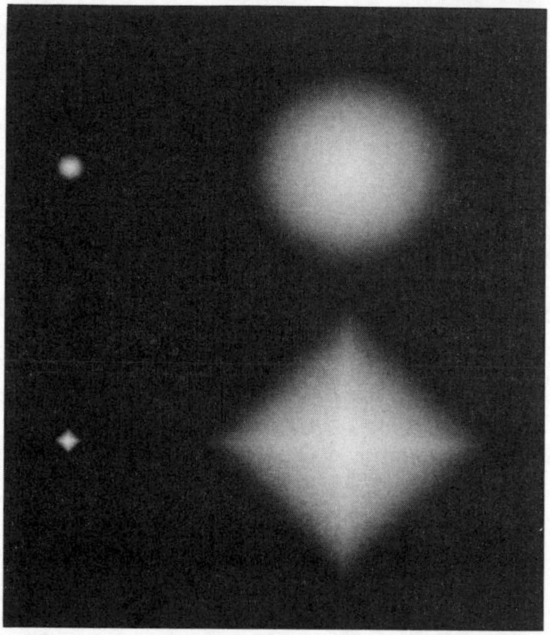

Figure 2.6 Two-dimensional covariance functions $\boxed{\frac{a}{b}}$ (a) Isotropic covariance and its log display; (b) separable covariance and its log display.

the given image data by replacing the ensemble averages by sample averages; for example, for an $M \times N$ image $u(m, n)$,

$$\mu \simeq \hat{\mu} = \frac{1}{MN} \sum_{m=1}^{M} \sum_{n=1}^{N} u(m, n) \tag{2.87}$$

$$r(m, n) \simeq \hat{r}(m, n) = \frac{1}{MN} \sum_{m'=1}^{M-m} \sum_{n'=1}^{N-n} [u(m', n') - \hat{\mu}][u(m + m', n + n') - \hat{\mu}] \tag{2.88}$$

For many image classes, ρ_1 and ρ_2 are found to be around 0.95.

Example 2.10 (Covariance matrices of random fields)

In Example 2.9 we saw that the covariance matrix of a one-dimensional stationary sequence is a symmetric Toeplitz matrix. Covariance matrices of stationary random fields mapped into vectors by row or column ordering are block Toeplitz matrices. For example, the covariance matrix of a segment of a stationary random field mapped into a vector by column (or row) ordering is a doubly Toeplitz block matrix. If the covariance function is separable, then this covariance block matrix is a Kronecker product of two matrices. For details see Problem 2.17.

2.11 THE SPECTRAL DENSITY FUNCTION

Let $u(n)$ be a stationary random sequence. Its *covariance generating function* (CGF) is defined as the Z-transform of its covariance function $r_u(n)$, i.e.,

$$\text{CGF}\{u(n)\} \triangleq S_u(z) \triangleq S(z) \triangleq \sum_{n=-\infty}^{\infty} r_u(n) z^{-n} \tag{2.89}$$

The *spectral density function* (SDF) is defined as the Fourier transform of $r_u(n)$, which is the CGF evaluated at $z = \exp(j\omega)$, i.e.,

$$\text{SDF}\{u(n)\} \triangleq S_u(\omega) \triangleq S(\omega) = \sum_{n=-\infty}^{\infty} r_u(n) \exp(-j\omega n) = S(z)|_{z=e^{j\omega}} \tag{2.90}$$

The covariance $r_u(n)$ is simply the inverse Fourier transform of the SDF, i.e.,

$$r_u(n) = \frac{1}{2\pi} \int_{-\pi}^{\pi} S_u(\omega) \exp(j\omega n)\, d\omega \tag{2.91}$$

In two dimensions the CGF and SDF have the analogous definitions

$$\text{CGF}\{u(m, n)\} = S_u(z_1, z_2) \triangleq S(z_1, z_2) \triangleq \sum\sum_{m,n=-\infty}^{\infty} r_u(m, n) z_1^{-m} z_2^{-n} \tag{2.92}$$

$$\text{SDF}\{u(m, n)\} = S_u(\omega_1, \omega_2) \triangleq S(\omega_1, \omega_2) \triangleq \sum\sum_{m,n=-\infty}^{\infty} r_u(m, n) \exp[-j(\omega_1 m + \omega_2 n)]$$

$$= S_u(z_1, z_2)|_{z_1=e^{j\omega_1}, z_2=e^{j\omega_2}} \tag{2.93}$$

$$r_u(m, n) = \frac{1}{4\pi^2} \iint_{-\pi}^{\pi} S_u(\omega_1, \omega_2) \exp[j(\omega_1 m + \omega_2 n)]\, d\omega_1\, d\omega_2 \tag{2.94}$$

This shows

$$\sigma_u^2 = E[|u(m, n) - \mu|^2] = r_u(0, 0) = \frac{1}{4\pi^2} \iint_{-\pi}^{\pi} S_u(\omega_1, \omega_2)\, d\omega_1\, d\omega_2 \tag{2.95}$$

i.e., the volume under $S_u(\omega_1, \omega_2)$ is equal to the average power in the random field $u(m, n)$. Therefore, physically $S_u(\omega_1, \omega_2)$ represents the power density in the image at spatial frequencies (ω_1, ω_2). Hence, the SDF is also known as the power spectrum density function or simply the power spectrum of the underlying random field. Often the power spectrum is defined as the Fourier transform of the autocorrelation sequence rather than the covariance sequence. Unless stated otherwise, we will continue to use the definitions based on covariances.

In the text whenever we refer to $S_u(z_1, z_2)$ as the SDF, it is implied that $z_1 = \exp(j\omega_1)$, $z_2 = \exp(j\omega_2)$. When a spectral density function can be expressed as a ratio of finite polynomials in z_1 and z_2, it is called a rational spectrum and it is of the form

$$S(z_1, z_2) = \frac{\sum_{k=-K}^{K} \sum_{l=-L}^{L} b(k, l) z_1^{-k} z_2^{-l}}{\sum_{m=-M}^{M} \sum_{n=-N}^{N} a(m, n) z_1^{-m} z_2^{-n}} \tag{2.96}$$

Such SDFs are realized by linear systems represented by finite-order difference equations.

Properties of the SDF

1. The SDF is real:

$$S(\omega_1, \omega_2) = S^*(\omega_1, \omega_2) \tag{2.97}$$

TABLE 2.8 Properties of SDF of Real Random Sequences

Property	One-Dimensional	Two-Dimensional
Fourier transform pair	$S(\omega) \leftrightarrow r(n)$	$S(\omega_1, \omega_2) \leftrightarrow r(m, n)$
Real	$S(\omega) = S^*(\omega)$	$S(\omega_1, \omega_2) = S^*(\omega_1, \omega_2)$
Even	$S(\omega) = S(-\omega)$	$S(\omega_1, \omega_2) = S(-\omega_1, -\omega_2)$
Nonnegative	$S(\omega) \geq 0, \quad \forall\omega$	$S(\omega_1, \omega_2) \geq 0, \quad \forall\omega_1, \omega_2$
Linear system output	$S_u(\omega) = \lvert H(\omega)\rvert^2 S_\epsilon(\omega)$	$S_u(\omega_1, \omega_2) = \lvert H(\omega_1, \omega_2)\rvert^2 S_\epsilon(\omega_1, \omega_2)$
Separability		$S(\omega_1,\omega_2) = S_1(\omega_1) S_1(\omega_2)$ if $r(m, n) = r_1(m) r_2(n)$

This follows by observing that the covariance function is conjugate symmetric, i.e., $r(m, n) = r^*(-m, -n)$. For real random fields the SDF is also even.

2. The SDF is nonnegative, i.e.,

$$S(\omega_1, \omega_2) \geq 0, \qquad \forall \omega_1, \omega_2 \tag{2.98}$$

Intuitively, this must be true because power cannot be negative. The formal proof can be obtained applying the nonnegativity property of (2.81) to the covariance functions of stationary random fields.

For a space-invariant system whose frequency response is $H(\omega_1, \omega_2)$ and whose input is a random field $\epsilon(m, n)$, the SDF of the output $u(m, n)$ is given by

$$S_u(\omega_1, \omega_2) = \lvert H(\omega_1, \omega_2)\rvert^2 S_\epsilon(\omega_1, \omega_2) \tag{2.99}$$

Table 2.8 summarizes the properties of the one- and two-dimensional SDFs. Similar definitions and properties hold for the SDFs of continuous random fields.

Example 2.11

The covariance function of a stationary white noise field is given as $r(m, n) = \sigma^2 \delta(m, n)$. The SDF is the constant σ^2 because

$$S(\omega_1, \omega_2) = \sigma^2 \sum_m \sum_n \delta(m, n) \exp[-j(\omega_1 m + \omega_2 n)] = \sigma^2$$

Example 2.12

Consider the separable covariance function defined in (2.84). Taking the Fourier transform, the SDF is found to be the separable function

$$S(\omega_1, \omega_2) = \frac{\sigma^2(1 - \rho_1^2)(1 - \rho_2^2)}{(1 + \rho_1^2 - 2\rho_1 \cos \omega_1)(1 + \rho_2^2 - 2\rho_2 \cos \omega_2)} \tag{2.100}$$

For the isotropic covariance function of (2.86), an analytic expression for the SDF is not available. Figure 2.7 shows displays of some SDFs.

2.12 SOME RESULTS FROM ESTIMATION THEORY

Here we state some important definitions and results from estimation theory that are useful in many image processing problems.

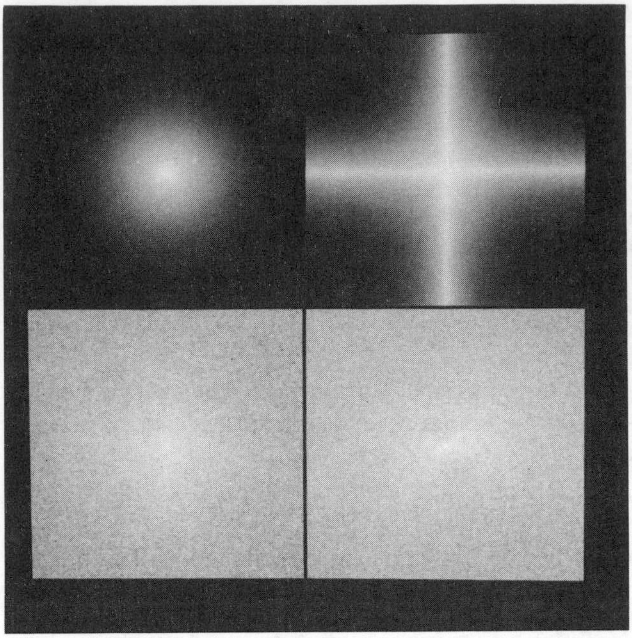

Figure 2.7 Spectral density functions

a	b
c	d

. SDF of (a) isotropic covariance function; (b) separable covariance function; (c) covariance function of the girl image; (d) covariance function of the moon image (see Figure 1.3).

Mean Square Estimates

Let $\{y(n), 1 \le n \le N\}$ be a real random sequence and x be any real random variable. It is desired to find $\hat{x}$, called the *optimum mean square* estimate of x, from an observation of the random sequence $y(n)$ such that the mean square error

$$\sigma_\epsilon^2 \triangleq E[(x - \hat{x})^2] \tag{2.101}$$

is minimized. It is simply the conditional mean of x given $y(n), 1 \le n \le N$ [9, 10]

$$\hat{x} = E(x|\mathbf{y}) \triangleq E[x|y(1), \ldots, y(N)] = \int_{-\infty}^{\infty} \xi p_{x|\mathbf{y}}(\xi)\, d\xi \tag{2.102}$$

where $p_{x|\mathbf{y}}(\xi)$ is the conditional probability density of x given the observation vector $\mathbf{y}$. If x and $y(n)$ are independent, then $\hat{x}$ is simply the mean value of x. Note that $\hat{x}$ is an unbiased estimate of x, because

$$E[\hat{x}] = E[E(x|y)] = E[x] \tag{2.103}$$

For zero mean Gaussian random variables, $\hat{x}$ turns out to be linear in $y(n)$, i.e.,

$$\hat{x} = \sum_{n=1}^{N} \alpha(n) y(n) \tag{2.104}$$

where the coefficients $\alpha(n)$ are determined by solving linear equations shown next.

The Orthogonality Principle

According to this principle the minimum mean square estimation error vector is orthogonal to every random variable functionally related to the observations, i.e., for any $g(\mathbf{y}) \triangleq g(y(1), y(2), \ldots, y(N))$,

$$E[(x - \hat{x})g(\mathbf{y})] = 0 \tag{2.105}$$

To prove this we write

$$E[\hat{x}g(\mathbf{y})] = E[E(x|\mathbf{y})g(\mathbf{y})] = E[E(xg(\mathbf{y})|\mathbf{y})] = E[xg(\mathbf{y})]$$

which implies (2.105). Since $\hat{x}$ is a function of $\mathbf{y}$, this also implies

$$E[(x - \hat{x})\hat{x}] = 0 \tag{2.106}$$

$$E[(x - \hat{x})g(\hat{x})] = 0 \tag{2.107}$$

i.e., the estimation error is orthogonal to every function of the estimate.

The orthogonality principle has been found to be very useful in linear estimation. In general, the conditional mean is a nonlinear function and is difficult to evaluate. Therefore, one often determines the optimum *linear* mean square estimate. For zero mean random variables this is done by writing x as a linear function of the observations, as in (2.104), and then finding the unknowns $\alpha(n)$ that minimize the mean square error. This minimization gives

$$\sum_{k=1}^{N} \alpha(k)E[y(k)y(n)] = E[xy(n)], \qquad n = 1, \ldots, N$$

In matrix notation this yields

$$\boldsymbol{\alpha} = \mathbf{R}_y^{-1}\mathbf{r}_{xy} \tag{2.108}$$

where $\boldsymbol{\alpha} = \{\alpha(n)\}$, $\mathbf{r}_{xy} = \{E[xy(n)]\}$ are $N \times 1$ vectors and $\mathbf{R}_y$ is the $N \times N$ covariance matrix of $\mathbf{y}$. The minimized mean square error is given by

$$\sigma_\epsilon^2 = \sigma_x^2 - \boldsymbol{\alpha}^T\mathbf{r}_{xy} \tag{2.109}$$

If x, $y(n)$ are nonzero mean random variables, then instead of (2.104), we write

$$\hat{x} - \mu_{\hat{x}} = \hat{x} - \mu_x = \sum_{n=1}^{N} \alpha(n)[y(n) - \mu_y(n)] \tag{2.110}$$

Once again $\boldsymbol{\alpha}$ is given by (2.108), where $\mathbf{R}_y$ and $\mathbf{r}_{xy}$ represent the covariance and cross-covariance arrays. If x, $y(n)$ are non-Gaussian, then (2.104) and (2.109) still give the best *linear* mean square estimate. However, this is not necessarily the conditional mean.

2.13 SOME RESULTS FROM INFORMATION THEORY

Information theory gives some important concepts that are useful in digital representation of images. Some of these concepts will be used in image quantization (Chapter 4), image transforms (Chapter 5), and image data compression (Chapter 11).

Information

Suppose there is a source (such as an image), which generates a discrete set of independent messages (such as gray levels) r_k, with probabilities $p_k, k = 1, \ldots, L$. Then the information associated with r_k is defined as

$$I_k = -\log_2 p_k \quad \text{bits} \tag{2.111}$$

Since

$$\sum_{k=1}^{L} p_k = 1 \tag{2.112}$$

each $p_k \leq 1$ and I_k is nonnegative. This definition implies that the information conveyed is large when an unlikely message is generated.

Entropy

The *entropy* of a source is defined as the average information generated by the source, i.e.,

$$\text{Entropy, } H = -\sum_{k=1}^{L} p_k \log_2 p_k \quad \text{bits/message} \tag{2.113}$$

For a digital image considered as a source of independent pixels, its entropy can be estimated from its histogram. For a given L, the entropy of a source is maximum for uniform distributions, i.e., $p_k = 1/L$, $k = 1, \ldots, L$. In that case

$$\max_{p_k} H = -\sum_{k=1}^{L} \frac{1}{L} \log_2 \frac{1}{L} = \log_2 L \quad \text{bits} \tag{2.114}$$

The entropy of a source gives the lower bound on the number of bits required to encode its output. In fact, according to Shannon's noiseless coding theorem [11, 12], it is possible to code without distortion a source of entropy H bits using an average of $H + \epsilon$ bits/message, where $\epsilon > 0$ is an arbitrarily small quantity. An alternate form of this theorem states that it is possible to code the source with H bits such that the distortion in the decoded message could be made arbitrarily small.

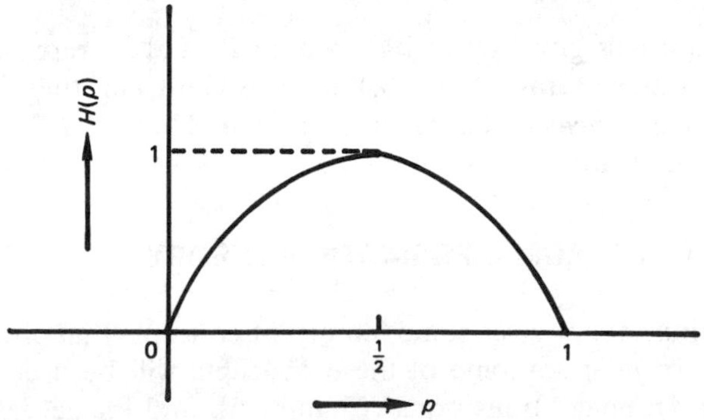

Figure 2.8 Entropy of a binary source.

Example 2.13

Let the source be binary, i.e., $L = 2$. Then, if $p_1 = p$, $p_2 = 1 - p$, $0 \leq p \leq 1$, the entropy is (Fig. 2.8)

$$H = H(p) = -p \log_2 p - (1-p) \log_2(1-p)$$

The maximum entropy is 1 bit, which occurs when both the messages are equally likely. Since the source is binary, it is always possible to code the output using 1 bit/message. However, if $p < \frac{1}{2}$, $p = \frac{1}{8}$ (say), then $H < 0.2$ bits, and Shannon's noiseless coding theorem says it is possible to find a noiseless coding scheme that requires only 0.2 bits/message.

The Rate Distortion Function

In analog-to-digital conversion of data, it is inevitable that the digitized data would have some error, however small, when compared to the analog sample. Rate distortion theory provides some useful results, which tell us the minimum number of bits required to encode the data, while admitting a certain level of distortion and vice versa.

The rate distortion function of a random variable x gives the minimum average rate R_D (in bits per sample) required to represent (or code) it while allowing a fixed distortion D in its reproduced value. If x is a Gaussian random variable of variance σ^2 and y is its reproduced value and if the distortion is measured by the mean square value of the difference $(x - y)$, i.e.,

$$D = E[(x - y)^2] \tag{2.115}$$

then the rate distortion function of x is defined as [11, 12]

$$\begin{aligned} R_D &= \begin{cases} (\tfrac{1}{2}) \log_2 (\sigma^2/D), & D \leq \sigma^2 \\ 0 \qquad\qquad\qquad\quad , & D > \sigma^2 \end{cases} \\ &= \max\left[0, (\tfrac{1}{2}) \log_2\left(\frac{\sigma^2}{D}\right)\right] \end{aligned} \tag{2.116}$$

Clearly the maximum value of D is equal to σ^2, the variance of x. Figure 2.9 shows the nature of this function.

Now if $\{x(0), x(1), \ldots, x(N-1)\}$ are Gaussian random variables encoded independently and if $\{y(0), \ldots, y(N-1)\}$ are their reproduced values, then the average mean square distortion is

$$D = \frac{1}{N} \sum_{k=0}^{N-1} E[|x(k) - y(k)|^2] \tag{2.117}$$

For a fixed average distortion D, the rate distortion function R_D of the vector $\mathbf{x}$ is given by

$$R_D = \frac{1}{N} \sum_{k=0}^{N-1} \max\left[0, \frac{1}{2} \log_2 \frac{\sigma_k^2}{\theta}\right] \tag{2.118}$$

where θ is determined by solving

$$D = \frac{1}{N} \sum_{k=0}^{N-1} \min(\theta, \sigma_k^2) \tag{2.119}$$

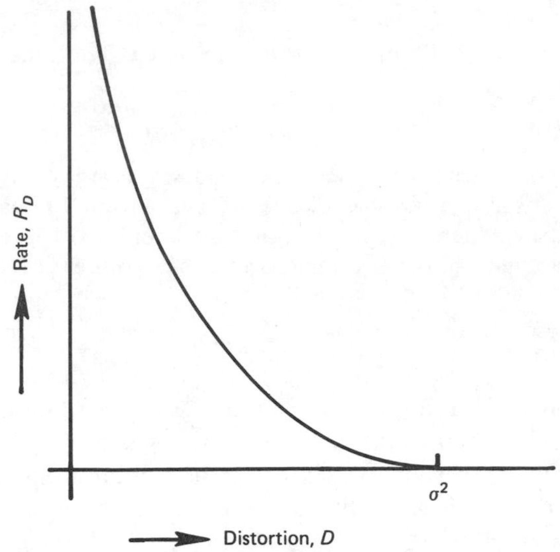

Figure 2.9 Rate distortion function for a Gaussian source.

Alternatively, if R_D is fixed, then (2.119) gives the minimum attainable distortion, where θ is obtained by solving (2.118). In general R_D is convex and a monotonically nonincreasing function of the distortion D.

PROBLEMS

2.1 **a.** Given a sequence $u(m, n) = (m + n)^3$. Evaluate $u(m, n)\delta(m-1, n-2)$ and $u(m,n) \circledast \delta(m-1, n-2)$.
b. Given a function $f(x, y) = (x + y)^3$. Evaluate $f(x, y)\delta(x-1, y-2)$ and $f(x, y) \circledast \delta(x-1, y-2)$.
c. Show that $\frac{1}{2\pi}\int_{-\pi}^{\pi} e^{\pm jn\theta}\, d\theta = \delta(n)$.

2.2 (*Properties of Discrete Convolution*) Prove the following:
a. $h(m, n) \circledast u(m, n) = u(m, n) \circledast h(m, n)$ (Commutative)
b. $h(m, n) \circledast [a_1 u_1(m, n) + a_2 u_2(m, n)] = a_1[h(m, n) \circledast u_1(m, n)] + a_2[h(m, n) \circledast u_2(m, n)]$ (Distributive)
c. $h(m, n) \circledast u(m - m_0, n - n_0 = h(m - m_0, n - n_0) \circledast u(m, n)$ (Shift invariance)
d. $h(m, n) \circledast [u_1(m, n) \circledast u_2(m, n)] = [h(m, n) \circledast u_1(m, n)] \circledast u_2(m, n)$ (Associative)
e. $h(m, n) \circledast \delta(m, n) = h(m, n)$ (Reproductive)
f. $\sum\sum_{m, n=-\infty}^{\infty} v(m, n) = \left[\sum\sum_{m, n=-\infty}^{\infty} h(m, n)\right]\left[\sum\sum_{m, n=-\infty}^{\infty} u(m, n)\right]$ (Volume conservation)
where $v(m, n) = h(m, n) \circledast u(m, n)$

2.3 Write and prove the properties of convolution analogous to those stated in Problem 2.2 for two-dimensional continuous systems.

2.4 In each of the following systems find the impulse response and determine whether or not the system is linear, shift invariant, FIR, or IIR.
a. $y(m, n) = 3x(m, n) + 9$
b. $y(m, n) = m^2 n^2 x(m, n)$

c. $y(m, n) = \sum_{m'=-\infty}^{m} \sum_{n'=-\infty}^{n} x(m', n')$

d. $y(m, n) = x(m - m_0, n - n_0)$

e. $y(m, n) = \exp\{-|x(m, n)|\}$

f. $y(m, n) = \sum_{m'=-1}^{1} \sum_{n'=-1}^{1} x(m', n')$

g. $y(m, n) = \sum_{m'=0}^{M-1} \sum_{n'=0}^{N-1} x(m', n') \exp\left\{-j\frac{2\pi mm'}{M}\right\} \exp\left\{-j\frac{2\pi nn'}{N}\right\}$

2.5 **a.** Determine the convolution of $x(m, n)$ of Example 2.1 with each of the following arrays, where the boxed element denotes the (0, 0) location.

i. $\begin{matrix} 0 & -1 & 1 \\ -1 & \boxed{4} & -1 \\ 0 & -1 & 0 \end{matrix}$ **ii.** $\begin{matrix} 1 & \boxed{2} & 3 \end{matrix}$ **iii.** $\begin{matrix} -2 \\ \boxed{3} \\ -1 \end{matrix}$

Verify your answers in each case via the *volume conservation property* discussed in Problem 2.2.

b. Show that in general the convolution of two arrays of sizes $(M_1 \times N_1)$ and $(M_2 \times N_2)$ yields an array of size $(M_1 + M_2 - 1) \times (N_1 + N_2 - 1)$.

2.6 Prove the convolution theorem i.e., (2.17) and from that prove (2.19).

2.7 Prove the Fourier transform relations of Table 2.2 and find the Fourier transforms of $\sin 2\pi x \eta_1 \cos 2\pi y \eta_2$ and $\cos[2\pi(x\eta_1 + y\eta_2)]$.

2.8. Prove the properties of the Fourier transform of two-dimensional sequences listed in Table 2.4.

2.9 For the optical imaging system shown in Fig. P2.9, show that the output image is a scaled and inverted replica of the object.

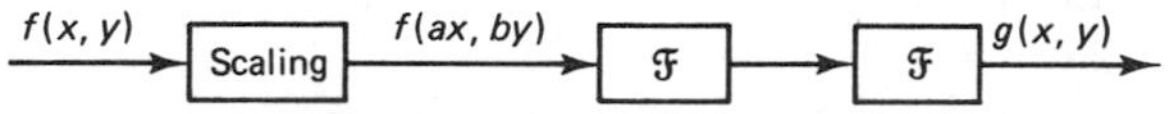

Figure P2.9

2.10 (*Hankel transform*) Show that in polar coordinates the two-dimensional Fourier transform becomes

$$F_p(\xi, \phi) \triangleq F(\xi \cos\phi, \xi \sin\phi) = \int_0^{2\pi} \int_0^{\infty} f_p(r, \theta) \exp[-j2\pi r\xi \cos(\theta - \phi)] r\, dr\, d\theta$$

where $f_p(r, \theta) = f(r\cos\theta, r\sin\theta)$. Hence, show that if $f(x, y)$ is circularly symmetric, then its Fourier transform is also circularly symmetric and is given by

$$F_p(\rho) = 2\pi \int_0^{\infty} r f_p(r) J_0(2\pi r\rho)\, dr, \quad J_0(x) \triangleq \frac{1}{2\pi} \int_0^{2\pi} \exp(-jx\cos\theta)\, d\theta$$

The pair f_p (r) and $F_p(\rho)$ is called the *Hankel transform pair of zero order.*

2.11 Prove the properties of the Z-transform listed in Table 2.5.

2.12 For each of the following linear systems, determine the transfer function, frequency response, OTF, and the MTF.

a. $y(m, n) - \rho_1 y(m-1, n) - \rho_2 y(m, n-1) = x(m, n)$

b. $y(m, n) - \rho_1 y(m-1, n) - \rho_2 y(m, n-1) + \rho_1\rho_2 y(m-1, n-1) = x(m, n)$

2.13 What is the impulse response of each filter?

a. Transfer function is $H_1(z_1, z_2) = 1 - a_1 z_1^{-1} - a_2 z_2^{-1} - a_3 z_1^{-1} z_2^{-1} - a_4 z_1 z_2^{-1}$.

b. Frequency response is $H(\omega_1, \omega_2) = 1 - 2\alpha\cos\omega_1 - 2\alpha\cos\omega_2$.

2.14 **a.** Write the convolution of two sequences $\{1, 2, 3, 4\}$ and $\{-1, 2, -1\}$ as a Toeplitz matrix operating on a 3×1 vector and then as a Toeplitz matrix operating on a 4×1 vector.

b. Write the convolution of two periodic sequences $\{1, 2, 3, 4, \ldots\}$ and $\{-1, 2, -1, 0, \ldots\}$, each of period 4, as a circulant matrix operating on a 4×1 vector that represents the first sequence.

2.15 [Matrix trace and related formulas].

a. Show that for square matrices $\mathbf{A}$ and $\mathbf{B}$, $\text{Tr}[\mathbf{A}] = \text{Tr}[\mathbf{A}^T] = \sum_{i=1}^{N} \lambda_i$, $\text{Tr}[\mathbf{A} + \mathbf{B}] = \text{Tr}[\mathbf{A}] + \text{Tr}[\mathbf{B}]$, and $\text{Tr}[\mathbf{AB}] = \text{Tr}[\mathbf{BA}]$ when λ_i are the eigenvalues of $\mathbf{A}$.

b. Define $D_{\mathbf{A}}(\mathbf{Y}) \overset{\Delta}{=} \dfrac{\partial}{\partial \mathbf{A}} \text{Tr}[\mathbf{Y}] \overset{\Delta}{=} \left\{ \dfrac{\partial}{\partial a(m, n)} \text{Tr}[\mathbf{Y}] \right\}$. Then show that $D_{\mathbf{A}}(\mathbf{AB}) = \mathbf{B}^T$, $D_{\mathbf{A}}(\mathbf{ABA}^T) = \mathbf{AB}^T + \mathbf{AB}$, and $D_{\mathbf{A}}(\mathbf{A}^{-1}\mathbf{BAC}) = -(\mathbf{A}^{-1}\mathbf{BACA}^{-1})^T + (\mathbf{CA}^{-1}\mathbf{B})^T$.

2.16 Express the two-dimensional convolutions of Problems 2.5(a) as a doubly Toeplitz block matrix operating on a 6×1 vector obtained by column ordering of the $x(m, n)$.

2.17 In the two-dimensional linear system of (2.8), the $x(m, n)$ and $y(m, n)$ are of size $M \times N$ and are mapped into column ordered vectors $\boldsymbol{x}$ and $\boldsymbol{y}$, respectively. Write this as a matrix equation

$$\boldsymbol{y} = \mathcal{H}\boldsymbol{x}$$

and show $\mathcal{H}$ is an $N \times N$ block matrix of basic dimension $M \times M$ that satisfies the properties listed in Table P2.17.

TABLE P2.17 Impulse Response (and Covariance) Sequences and Corresponding Block Matrix Structures

Sequence		Block matrix
Spatially varying	$h(m, n; m', n')$	$\mathcal{H}$, general
Spatially invariant in m;	$h(m - m', n; n')$	Toeplitz blocks
Spatially invariant in n;	$h(m, n - n'; m')$	Block Toeplitz
Spatially invariant in m, n;	$h(m - m', n - n')$	Doubly Toeplitz
Spatially invariant in m, n and periodic in m	$h(m \text{ modulo } M, n)$	Block Toeplitz with circulant blocks
Spatially invariant in m, n and periodic in n	$h(m, n \text{ modulo } N)$	Block circulant with Toeplitz blocks
Spatially invariant and periodic in m, n	$h(m \text{ modulo } M, n \text{ modulo } N)$	Doubly block circulant
Separable, spatially varying	$h_1(m, m')\, h_2(n, n')$	Kronecker product $\mathbf{H}_2 \otimes \mathbf{H}_1$
Separable, spatially invariant	$h_1(m - m')\, h_2(n - n')$	Toeplitz Kronecker product $\mathbf{H}_2 \otimes \mathbf{H}_1$, $\mathbf{H}_1$, $\mathbf{H}_2$ Toeplitz
Separable, spatially invariant, and periodic	$h_1(m)\, h_1(n)$ (m modulo M, n modulo N)	Circulant Kronecker product $\mathbf{H}_2 \otimes \mathbf{H}_1$, $\mathbf{H}_1$, $\mathbf{H}_2$ circulant

2.18 Show each of the following.

a. A circulant matrix is Toeplitz, but the converse is not true.

b. The product of two circulant (or block circulant) matrices is a circulant (or block circulant) matrix.
c. The product of two Toeplitz matrices need not be Toeplitz.

2.19 Show each of the following.
a. The covariance matrix of a sequence of uncorrelated random variables is diagonal.
b. The cross-covariance matrix of two mutually wide-sense stationary sequences is Toeplitz.
c. The covariance matrix of one period of a real stationary periodic random sequence is circulant.

2.20 In Table P2.17, if $h(m, n; m', n')$ represents the covariance function of an $M \times N$ segment of a random field $x(m, n)$, then show that the block matrix $\mathcal{H}$ represents the covariance matrix of column-ordered vector $\boldsymbol{x}$ for each of the cases listed in that table.

2.21 Prove properties (2.97) through (2.99) of SDFs. Show that (2.100) is the SDF of random fields whose covariance function is the separable function given by (2.84).

2.22 a.*Compute the entropies of several digital images from their histograms and compare them with the gray scale activity in the images. The gray scale activity may be represented by the variance of the image.
b. Show that for a given number of possible messages the entropy of a source is maximum if all the messages are equally likely.
c. Show that R_D given by (2.118) is a monotonically nonincreasing function of D.

BIBLIOGRAPHY

Sections 2.1–2.6

For fundamental concepts in linear systems, Fourier theory, Z-transforms and related topics:

1. T. Kailath. *Linear Systems.* Englewood Cliffs, N.J.: Prentice-Hall, 1980.
2. A. V. Oppenheim and R. W. Schafer. *Digital Signal Processing.* Englewood Cliffs, N.J.: Prentice-Hall, 1975.
3. A. Papoulis. *Systems and Transforms with Applications in Optics.* New York: McGraw-Hill, 1968.
4. J. W. Goodman. *Introduction to Fourier Optics.* New York: McGraw-Hill, 1968.
5. R. N. Bracewell. *The Fourier Transform and Its Applications.* New York: McGraw-Hill, 1965.
6. E. I. Jury. *Theory and Application of the Z-Transform Method.* New York: John Wiley, 1964.

Sections 2.7, 2.8

For matrix theory results and their proofs:

7. R. Bellman. *Introduction to Matrix Analysis,* 2d ed. New York: McGraw-Hill, 1970.
8. G. A. Graybill. *Introduction to Matrices with Applications in Statistics.* Belmont, Calif.: Wadsworth, 1969.

* Problems marked with an asterisk require computer simulation or other experiments.

Sections 2.9–2.13

For fundamentals of random processes, estimation theory, and information theory:

9. A. Papoulis. *Probability, Random Variables and Stochastic Processes.* New York: McGraw-Hill, 1965.
10. W. B. Davenport. *Probability and Random Processes.* New York: McGraw-Hill, 1970.
11. R. G. Gallager. *Information Theory and Reliable Communication.* New York: John Wiley, 1968.
12. C. E. Shannon and W. Weaver. *The Mathematical Theory of Communication.* Urbana: The University of Illinois Press, 1949.

3

Image Perception

3.1 INTRODUCTION

In presenting the output of an imaging system to a human observer, it is essential to consider how it is transformed into information by the viewer. Understanding of the visual perception process is important for developing measures of image fidelity, which aid in the design and evaluation of image processing algorithms and imaging systems. Visual image data itself represents spatial distribution of physical quantities such as luminance and spatial frequencies of an object. The perceived information may be represented by attributes such as brightness, color, and edges. Our primary goal here is to study how the perceptual information may be represented quantitatively.

3.2 LIGHT, LUMINANCE, BRIGHTNESS, AND CONTRAST

Light is the electromagnetic radiation that stimulates our visual response. It is expressed as a spectral energy distribution $L(\lambda)$, where λ is the wavelength that lies in the visible region, 350 nm to 780 nm, of the electromagnetic spectrum. Light received from an object can be written as

$$I(\lambda) = \rho(\lambda)L(\lambda) \tag{3.1}$$

where $\rho(\lambda)$ represents the reflectivity or transmissivity of the object and $L(\lambda)$ is the incident energy distribution. The illumination range over which the visual system can operate is roughly 1 to 10^{10}, or 10 orders of magnitude.

The retina of the human eye (Fig. 3.1) contains two types of photoreceptors called *rods* and *cones*. The rods, about 100 million in number, are relatively long and thin. They provide *scotopic* vision, which is the visual response at the lower several orders of magnitude of illumination. The cones, many fewer in number

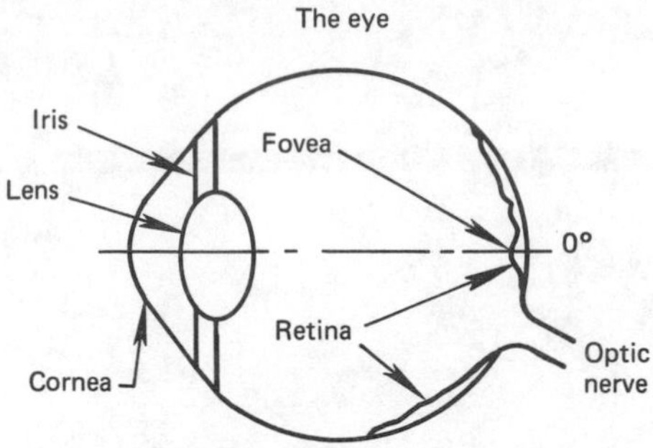

Figure 3.1 Cross section of the eye.

(about 6.5 million), are shorter and thicker and are less sensitive than the rods. They provide *photopic* vision, the visual response at the higher 5 to 6 orders of magnitude of illumination (for instance, in a well-lighted room or bright sunlight). In the intermediate region of illumination, both rods and cones are active and provide *mesopic* vision. We are primarily concerned with the photopic vision, since electronic image displays are well lighted.

The cones are also responsible for color vision. They are densely packed in the center of the retina (called *fovea*) at a density of about 120 cones per degree of arc subtended in the field of vision. This corresponds to a spacing of about 0.5 min of arc, or 2 μm. The density of cones falls off rapidly outside a circle of 1° radius from the fovea. The pupil of the eye acts as an aperture. In bright light it is about 2 mm in diameter and acts as a low-pass filter (for green light) with a passband of about 60 cycles per degree.

The cones are *laterally* connected by *horizontal cells* and have a forward connection with *bipolar cells.* The bipolar cells are connected to *ganglion cells,* which join to form the *optic nerve* that provides communication to the central nervous system.

The *luminance* or *intensity* of a spatially distributed object with light distribution $I(x, y, \lambda)$ is defined as

$$f(x, y) = \int_0^\infty I(x, y, \lambda)V(\lambda)\,d\lambda \tag{3.2}$$

where $V(\lambda)$ is called the *relative luminous efficiency function* of the visual system. For the human eye, $V(\lambda)$ is a bell-shaped curve (Fig. 3.2) whose characteristics

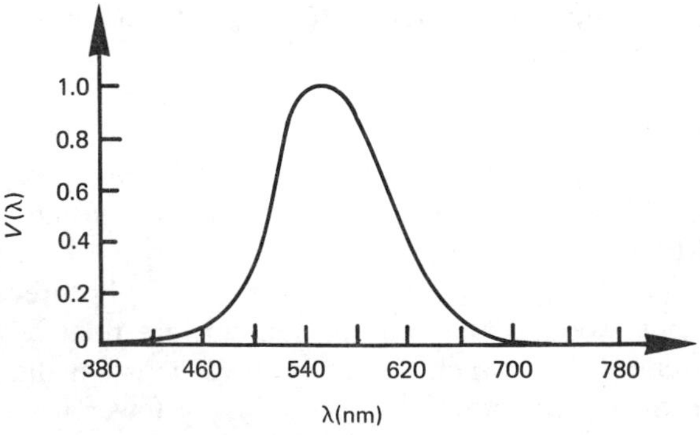

Figure 3.2 Typical relative luminous efficiency function.

depend on whether it is scotopic or photopic vision. The luminance of an object is independent of the luminances of the surrounding objects. The *brightness* (also called *apparent brightness*) of an object is the perceived luminance and depends on the luminance of the surround. *Two objects with different surroundings could have identical luminances but different brightnesses.* The following visual phenomena exemplify the differences between luminance and brightness.

Simultaneous Contrast

In Fig. 3.3a, the two smaller squares in the middle have equal luminance values, but the one on the left appears brighter. On the other hand in Fig. 3.3b, the two squares appear about equal in brightness although their luminances are quite different. The reason is that our perception is sensitive to luminance contrast rather than the absolute luminance values themselves.

According to *Weber's law* [2, 3], if the luminance f_0 of an object is just noticeably different from the luminance f_s of its surround, then their ratio is

$$\frac{|f_s - f_0|}{f_0} = \text{constant} \tag{3.3}$$

Writing $f_0 = f, f_s = f + \Delta f$ where Δf is small for just noticeably different luminances, (3.3) can be written as

$$\frac{\Delta f}{f} \simeq d(\log f) = \Delta c \qquad \text{(constant)} \tag{3.4}$$

The value of the constant has been found to be 0.02, which means that at least 50 levels are needed for the contrast on a scale of 0 to 1. Equation (3.4) says

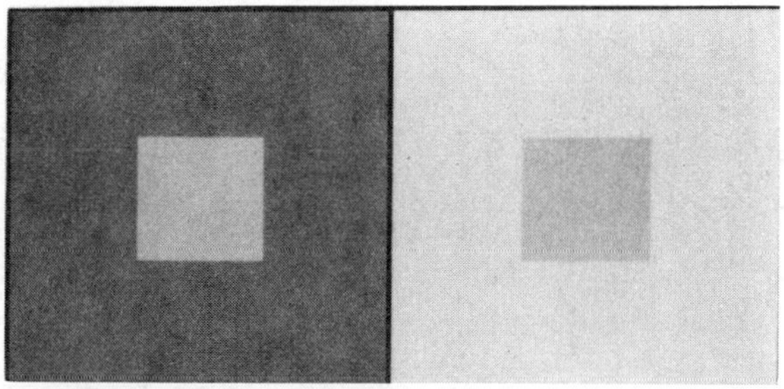

Figure 3.3 Simultaneous contrast: (a) small squares in the middle have equal luminances but do not appear equally bright;

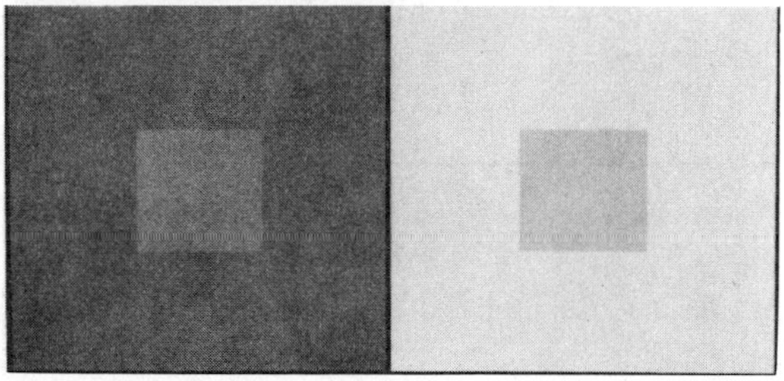

(b) small squares in the middle appear almost equally bright, but their luminances are different.

TABLE 3.1 Luminance to Contrast Models

1	Logarithmic law	$c = 50 \log_{10} f,\ 1 \le f \le 100$
2	Power law	$c = \alpha_n f^{1/n},\ n = 2, 3, \ldots.$ $\alpha_2 = 10,\ \alpha_3 = 21.9$
3	Background ratio	$c = \dfrac{f(f_B + 100)}{f_B + f}$ f_B = background luminance

The luminance f lies in the interval [0, 100] except in the logarithmic law. Contrast scale is over [0, 100].

equal increments in the log of the luminance should be perceived to be equally different, i.e., $\Delta(\log f)$ is proportional to Δc, the change in contrast. Accordingly, the quantity

$$c = a_1 + a_2 \log f \tag{3.5}$$

where a_1, a_2 are constants, is called the *contrast*. There are other *models* of contrast [see Table 3.1 and Fig. 3.4], one of which is the *root law*

$$c = f^{1/n} \tag{3.6}$$

The choice of $n = 3$ has been preferred over the logarithmic law in an image coding study [7]. However, the logarithmic law remains the most widely used choice.

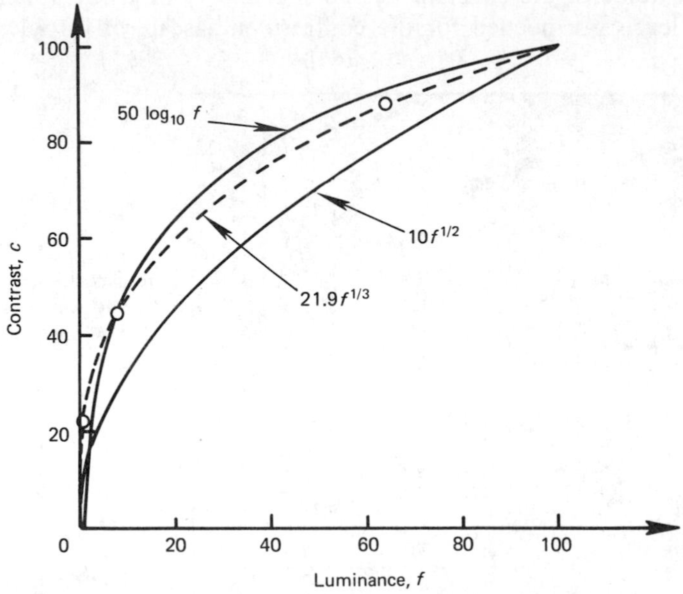

Figure 3.4 Contrast models.

Mach Bands

The spatial interaction of luminances from an object and its surround creates a phenomenon called the *Mach band* effect. This effect shows that brightness is not a monotonic function of luminance. Consider the gray level bar chart of Fig. 3.5a, where each bar has constant luminance. But the apparent brightness is not uniform along the width of the bar. Transitions at each bar appear brighter on the right side and darker on the left side. The dashed line in Fig. 3.5b represents the perceived brightness. The overshoots and undershoots illustrate the Mach band effect. Mach bands are also visible in Fig. 3.6a, which exhibits a dark and a bright line (marked D and B) near the transition regions of a smooth-intensity ramp. Measurement of the

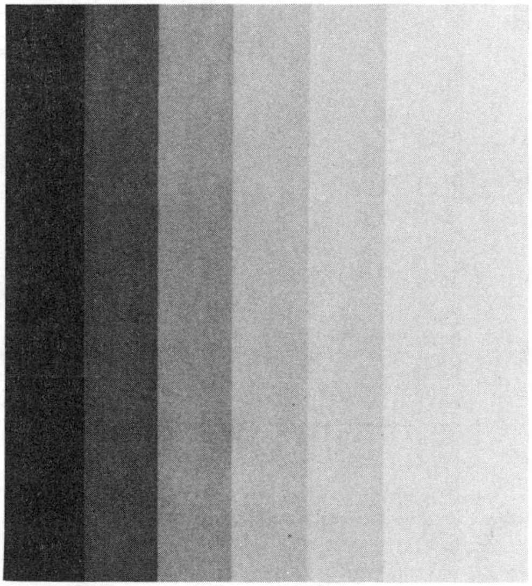

(a) Gray-level bar chart.

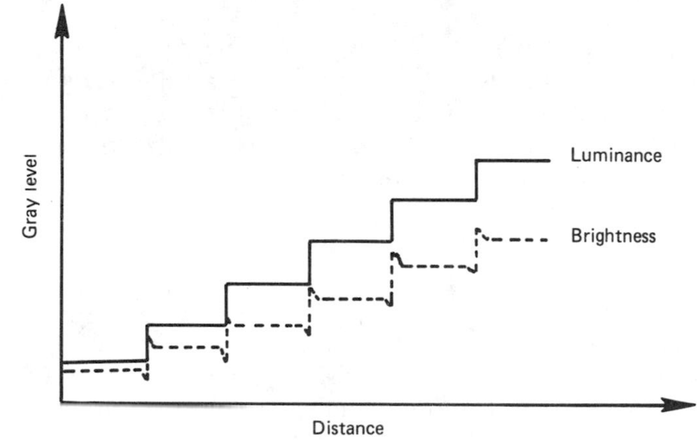

(b) Luminance versus brightness.

Figure 3.5 Mach band effect.

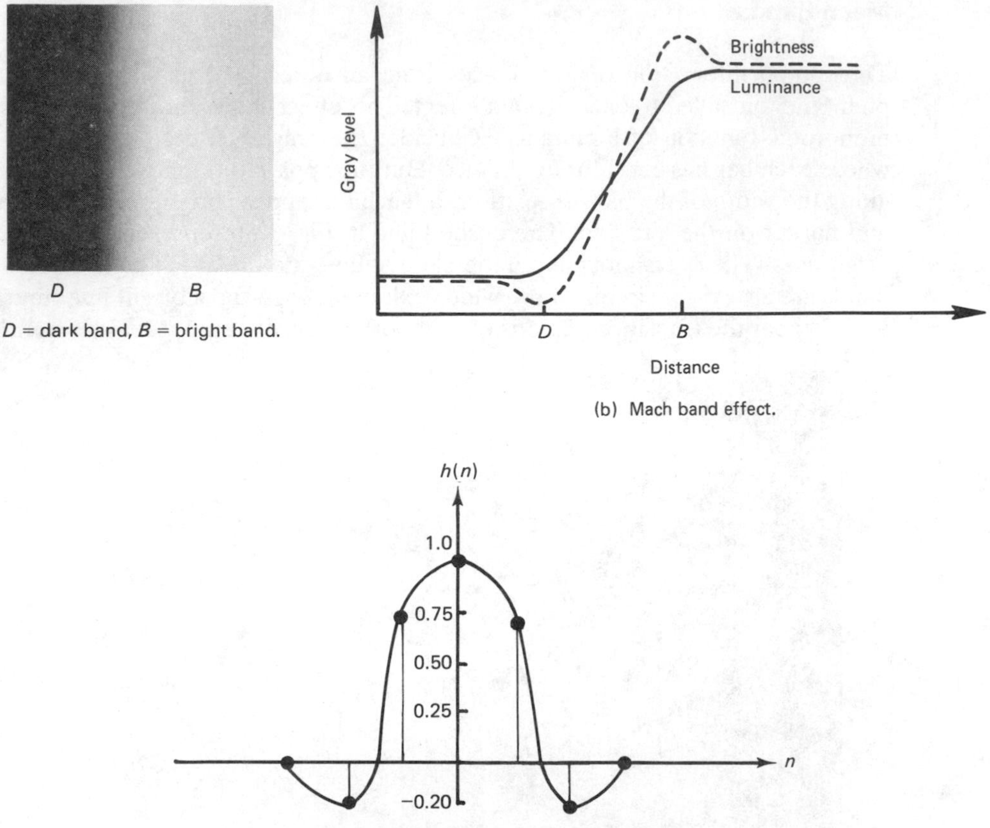

(a) D = dark band, B = bright band.

(b) Mach band effect.

(c) Nature of the visual system impulse response.

Figure 3.6 Mach bands.

Mach band effect can be used to estimate the impulse response of the visual system (see Problem 3.5).

Figure 3.6c shows the nature of this impulse response. The negative lobes manifest a visual phenomenon known as *lateral inhibition.* The impulse response values represent the relative spatial weighting (of the contrast) by the receptors, rods and cones. The negative lobes indicate that the *neural* (postretinal) signal at a given location has been inhibited by some of the laterally located receptors.

3.3 MTF OF THE VISUAL SYSTEM

The Mach band effect measures the response of the visual system in spatial coordinates. The Fourier transform of the impulse response gives the frequency response of the system from which its MTF can be determined. A direct measurement of the MTF is possible by considering a sinusoidal grating of varying contrast (ratio of the maximum to minimum intensity) and spatial frequency (Fig. 3.7a). Observation of this figure (at a distance of about 1 m) shows the thresholds of visibility at

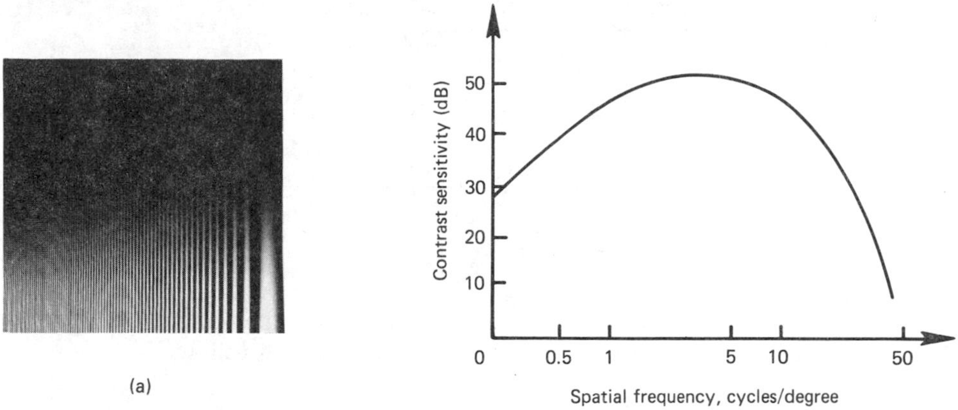

(a) (b)

Figure 3.7 MTF of the human visual system. (a) Contrast versus spatial frequency sinusoidal grating; (b) typical MTF plot.

various frequencies. The curve representing these thresholds is also the MTF, and it varies with the viewer as well as the viewing angle. Its typical shape is of the form shown in Fig. 3.7b. The curve actually observed from Fig. 3.7a is your own MTF (distorted by the printing process). The shape of the curve is similar to a band-pass filter and suggests that the human visual system is most sensitive to midfrequencies and least sensitive to high frequencies. The frequency at which the peak occurs varies with the viewer and generally lies between 3 and 10 cycles/degree. In practice, the contrast sensitivity also depends on the orientation of the grating, being maximum for horizontal and vertical gratings. However, the angular sensitivity variations are within 3 dB (maximum deviation is at 45°) and, to a first approximation, the MTF can be considered to be isotropic and the phase effects can be ignored. A curve fitting procedure [6] has yielded a formula for the frequency response of the visual system as

$$H(\xi_1, \xi_2) = H_p(\rho) = A\left[\alpha + \left(\frac{\rho}{\rho_0}\right)\right]\exp\left[-\left(\frac{\rho}{\rho_0}\right)^{\beta}\right]$$

$$\rho = \sqrt{\xi_1^2 + \xi_2^2} \text{ cycles/degree} \tag{3.7}$$

where A, α, β, and ρ_0 are constants. For $\alpha = 0$ and $\beta = 1$, ρ_0 is the frequency at which the peak occurs. For example, in an image coding application [6], the values $A = 2.6$, $\alpha = 0.0192$, $\rho_0 = (0.114)^{-1} = 8.772$, and $\beta = 1.1$ have been found useful. The peak frequency is 8 cycles/degree and the peak value is normalized to unity.

3.4 THE VISIBILITY FUNCTION

In many image processing systems—for instance, in image coding—the output image $u^{\cdot}(m, n)$ contains additive noise $q(m, n)$, which depends on $e(m, n)$, a function of the input image $u(m, n)$ [see Fig. 3.8]. The sequence $e(m, n)$ is sometimes called the *masking function*. A masking function is an image feature that is to be

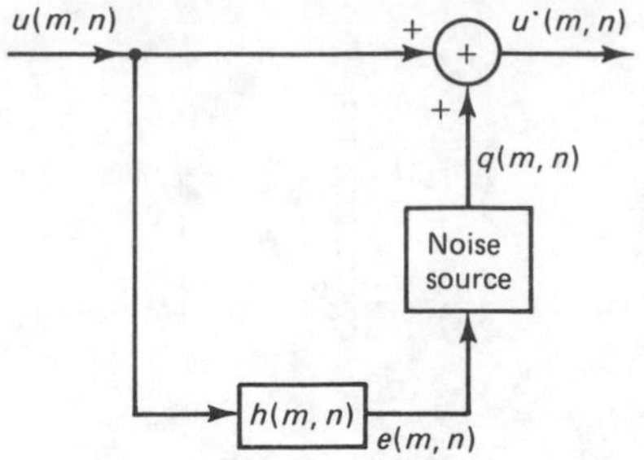

Figure 3.8 Visibility function noise source model. The filter impulse response $h(m, n)$ determines the masking function. Noise source output depends on the masking function amplitude $|e|$.

observed or processed in the given application. For example, $e(m, n) = u(m, n)$ in PCM transmission of images. Other examples are as follows:

1. $e(m, n) = u(m, n) - u(m-1, n)$
2. $e(m, n) = u(m, n) - a_1 u(m-1, n) - a_2 u(m, n-1) + a_3 u(m-1, n-1)$
3. $e(m, n) = u(m, n) - \alpha[u(m-1, n) + u(m+1, n) + u(m, n-1) + u(m, n+1)]$

The *visibility function* measures the subjective visibility in a scene containing this masking function dependent noise $q(m, n)$. It is measured as follows. For a suitably small Δx and a fixed interval $[x, x + \Delta x]$, add white noise of power P_e to all those pixels in the original image where masking function magnitude $|e|$ lies in this interval. Then obtain another image by adding white noise of power P_w to *all* the pixels such that the two images are subjectively equivalent based on a subjective scale rating, such as the one shown in Table 3.3. Then the visibility function $\nu(x)$ is defined as [4]

$$\nu(x) = \frac{-dV(x)}{dx} \tag{3.8}$$

where

$$V(x) = \frac{P_w}{P_e}$$

The visibility function therefore represents the subjective visibility in a scene of unit masking noise. This function varies with the scene. It is useful in defining a quantitative criterion for subjective evaluation of errors in an image (see Section 3.6).

3.5 MONOCHROME VISION MODELS

Based on the foregoing discussion, a simple overall model of *monochrome vision* can be obtained [5, 6] as shown in Fig. 3.9. Light enters the eye, whose optical characteristics are represented by a low-pass filter with frequency response $H_l(\xi_1, \xi_2)$. The spatial response of the eye, represented by the relative luminous efficiency function $V(\lambda)$, yields the luminance distribution $f(x, y)$ via (3.2). The nonlinear response of the rods and cones, represented by the point nonlinearity

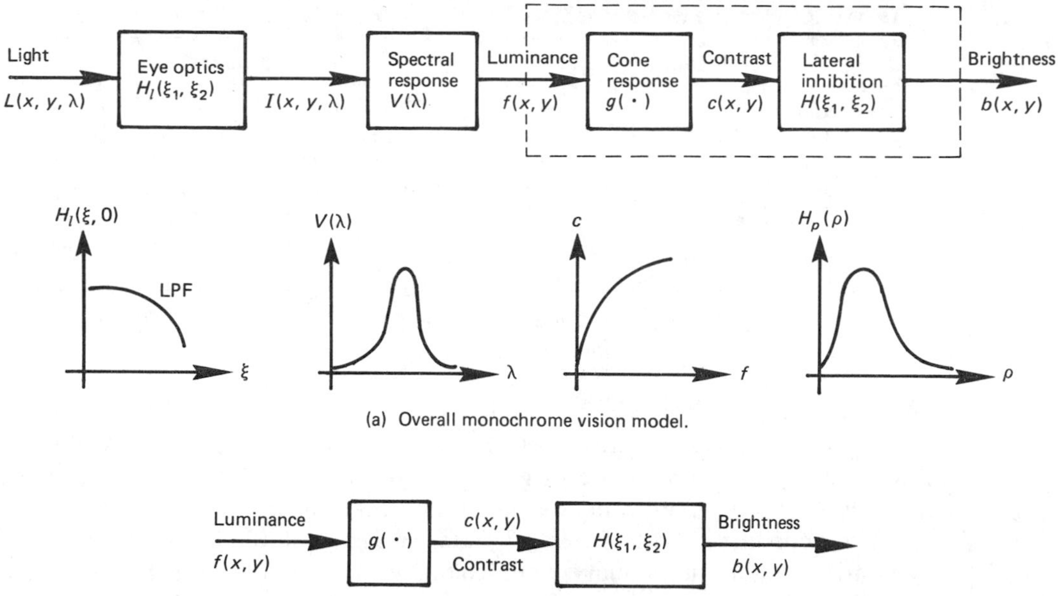

(a) Overall monochrome vision model.

(b) Simplified monochrome vision model.

Figure 3.9

$g(\cdot)$, yields the contrast $c(x, y)$. The lateral inhibition phenomenon is represented by a spatially invariant, isotropic, linear system whose frequency response is $H(\xi_1, \xi_2)$. Its output is the neural signal, which represents the apparent brightness $b(x, y)$. For an optically well-corrected eye, the low-pass filter has a much slower drop-off with increasing frequency than that of the lateral inhibition mechanism. Thus the optical effects of the eye could be ignored, and the simpler model showing the transformation between the luminance and the brightness suffices.

Results from experiments using sinusoidal gratings indicate that spatial frequency components, separated by about an octave, can be detected independently by observers. Thus, it has been proposed [7] that the visual system contains a number of independent spatial channels, each tuned to a different spatial frequency and orientation angle. This yields a refined model, which is useful in the analysis and evaluation of image processing systems that are far from the optimum and introduce large levels of distortions. For near-optimum systems, where the output image is only slightly degraded, the simplified model in Fig. 3.9 is adequate and is the one with which we shall mostly be concerned.

3.6 IMAGE FIDELITY CRITERIA

Image fidelity criteria are useful for measuring image quality and for rating the performance of a processing technique or a vision system. There are two types of criteria that are used for evaluation of image quality, subjective and quantitative. The subjective criteria use rating scales such as goodness scales and impairment

TABLE 3.2 Image Goodness Scales

Overall goodness scale		Group goodness scale	
Excellent	(5)	Best	(7)
Good	(4)	Well above average	(6)
Fair	(3)	Slightly above average	(5)
Poor	(2)	Average	(4)
Unsatisfactory	(1)	Slightly below average	(3)
		Well below average	(2)
		Worst	(1)

The numbers in parenthesis indicate a numerical weight attached to the rating.

scales. A goodness scale may be a global scale or a group scale (Table 3.2). The overall goodness criterion rates image quality on a scale ranging from excellent to unsatisfactory. A training set of images is used to calibrate such a scale. The group goodness scale is based on comparisons within a set of images.

The impairment scale (Table 3.3) rates an image on the basis of the level of degradation present in an image when compared with an *ideal* image. It is useful in applications such as image coding, where the encoding process introduces degradations in the output image.

Sometimes a method called *bubble sort* is used in rating images. Two images A and B from a group are compared and their order is determined (say it is AB). Then the third image is compared with B and the order ABC or ACB is established. If the order is ACB, then A and C are compared and the new order is established. In this way, the best image bubbles to the top if no ties are allowed. Numerical rating may be given after the images have been ranked.

If several observers are used in the evaluation process, then the mean rating is given by

$$R = \frac{\sum_{k=1}^{n} s_k n_k}{\sum_{k=1}^{n} n_k}$$

where s_k is the score associated with the kth rating, n_k is the number of observers with this rating, and n is the number of grades in the scale.

TABLE 3.3 Impairment Scale

Not noticeable	(1)
Just noticeable	(2)
Definitely noticeable but only slight impairment	(3)
Impairment not objectionable	(4)
Somewhat objectionable	(5)
Definitely objectionable	(6)
Extremely objectionable	(7)

Among the quantitative measures, a class of criteria used often is called *the mean square* criterion. It refers to some sort of average or sum (or integral) of squares of the error between two images. For $M \times N$ images $u(m, n)$ and $u'(m, n)$, (or $v(x, y)$ and $v'(x, y)$ in the continuous case), the quantity

$$\sigma_{ls}^2 \triangleq \frac{1}{MN} \sum_{m=1}^{M} \sum_{n=1}^{N} |u(m, n) - u'(m, n)|^2 \quad \text{or} \quad \iint_{\mathcal{R}} |v(x, y) - v'(x, y)|^2 \, dx \, dy \tag{3.9}$$

where $\mathcal{R}$ is the region over which the image is given, is called the *average least squares (or integral square) error.* The quantity

$$\sigma_{ms}^2 \triangleq E[|u(m, n) - u'(m, n)|^2] \quad \text{or} \quad E[|v(x, y) - v'(x, y)|^2] \tag{3.10}$$

is called the *mean square error,* where E represents the mathematical expectation. Often (3.9) is used as an estimate of (3.10) when ensembles for $u(m, n)$ and $u'(m, n)$ or $v(x, y)$ and $v'(x, y)$ are not available. Another quantity,

$$\sigma_a^2 = \frac{1}{MN} \sum_{m=1}^{M} \sum_{n=1}^{N} E[|u(m, n) - u'(m, n)|^2]$$

$$\text{or} \quad \iint_{\mathcal{R}} E[|v(x, y) - v'(x, y)|^2] dx \, dy \tag{3.11}$$

called the *average mean square* or *integral mean square error,* is also used many times. In many applications the (mean square) error is expressed in terms of a *signal-to-noise ratio* (SNR), which is defined in decibels (dB) as

$$\text{SNR} = 10 \log_{10} \frac{\sigma^2}{\sigma_e^2}, \qquad \sigma_e = \sigma_a, \sigma_{ms}, \text{ or } \sigma_{ls} \tag{3.12}$$

where σ^2 is the variance of the desired (or original) image.

Another definition of SNR, used commonly in image coding applications, is

$$\text{SNR}' = 10 \log_{10} \frac{(\text{peak-to-peak value of the reference image})^2}{\sigma_e^2} \tag{3.13}$$

This definition generally results in a value of SNR′ roughly 12 to 15 dB above the value of SNR (see Problem 3.6).

The sequence $u(m, n)$ (or the function $v(x, y)$) need not always represent the image luminance function. For example, in the monochrome image model of Fig. 3.9, $v(x, y) \triangleq b(x, y)$ would represent the brightness function. Then from (3.9) we may write for large images

$$\sigma_{ls}^2 = \iint_{-\infty}^{\infty} |b(x, y) - b'(x, y)|^2 \, dx \, dy \tag{3.14}$$

$$= \iint_{-\infty}^{\infty} |B(\xi_1, \xi_2) - B'(\xi_1, \xi_2)|^2 \, d\xi_1 \, d\xi_2 \tag{3.15}$$

where $B(\xi_1, \xi_2)$ is the Fourier transform of $b(x, y)$ and (3.15) follows by virtue of the Parseval theorem. From Fig. 3.9 we now obtain

$$\sigma_{ls}^2 = \iint_{-\infty}^{\infty} |C(\xi_1, \xi_2) - C'(\xi_1, \xi_2)|^2 \, |H(\xi_1, \xi_2)|^2 \, d\xi_1 \, d\xi_2 \tag{3.16}$$

which is a frequency weighted mean square criterion applied to the contrast function.

An alternate visual criteria is to define the expectation operator E with respect to the visibility (rather than the probability) function, for example, by

$$\sigma^2_{msse} \triangleq \int_{-\infty}^{\infty} |e|^2 \nu(e)\, de \tag{3.17}$$

where $e \triangleq u - u'$ is the value of the error at any pixel and $\nu(e)$ is its visibility. The quantity σ^2_{msse} then represents the *mean square subjective error*.

The mean square error criterion is not without limitations, especially when used as a global measure of image fidelity. The prime justification for its common use is the relative ease with which it can be handled mathematically for developing image processing algorithms. When used as a local measure, for instance, in adaptive techniques, it has proven to be much more effective.

3.7 COLOR REPRESENTATION

The study of color is important in the design and development of color vision systems. Use of color in image displays is not only more pleasing, but it also enables us to receive more visual information. While we can perceive only a few dozen gray levels, we have the ability to distinguish between thousands of colors. The perceptual attributes of color are *brightness, hue,* and *saturation.* Brightness represents the perceived luminance as mentioned before. The hue of a color refers to its "redness," "greenness," and so on. For monochromatic light sources, differences in

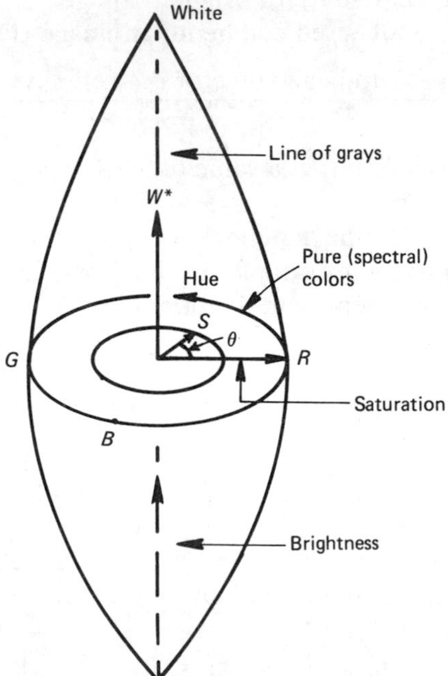

Figure 3.10 Perceptual representation of the color space. The brightness W^* varies along the vertical axis, hue θ varies along the circumference, and saturation S varies along the radius.

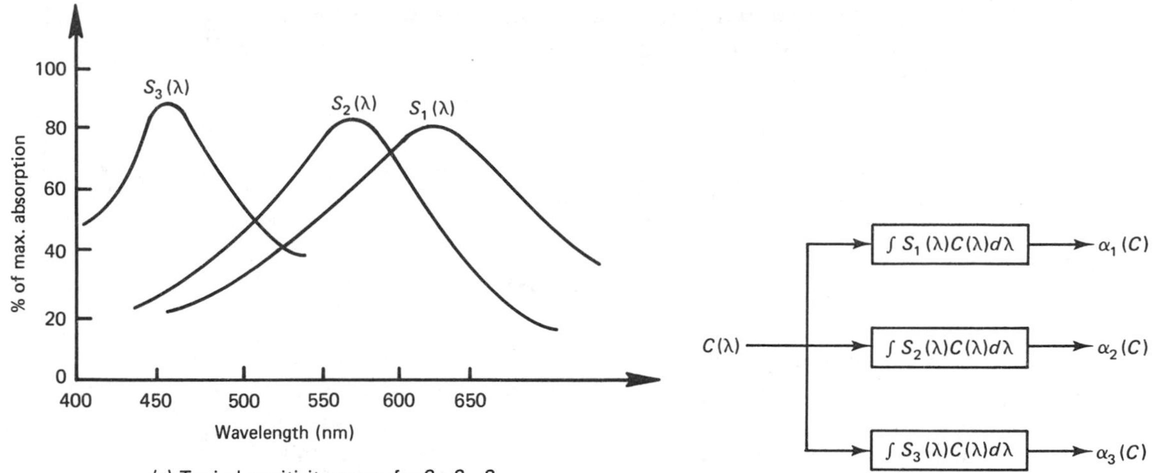

(a) Typical sensitivity curves for S_1, S_2, S_3 (not to scale).

(b) Three receptor model for color representation

Figure 3.11 (a) Typical absorption spectra of the three types of cones in the human retina; (b) three-receptor model for color representation.

hues are manifested by the differences in wavelengths. Saturation is that aspect of perception that varies most strongly as more and more white light is added to a monochromatic light. These definitions are somewhat imprecise because hue, saturation, and brightness all change when either the wavelength, the intensity, the hue, or the amount of white light in a color is changed. Figure 3.10 shows a perceptual representation of the color space. Brightness (W^*) varies along the vertical axis, hue (θ) varies along the circumference, and saturation (S) varies along the radial distance.† For a fixed brightness W^*, the symbols R, G, and B show the relative locations of the red, green, and blue spectral colors.

Color representation is based on the classical theory of Thomas Young (1802) [8], who stated that any color can be reproduced by mixing an appropriate set of three primary colors. Subsequent findings, starting from those of Maxwell [9] to more recent ones reported in [10, 11], have established that there are three different types of cones in the (normal) human retina with absorption spectra $S_1(\lambda)$, $S_2(\lambda)$, and $S_3(\lambda)$, where $\lambda_{min} \le \lambda \le \lambda_{max}$, $\lambda_{min} \simeq 380$ nm, $\lambda_{max} \simeq 780$ nm. These responses peak in the yellow-green, green, and blue regions, respectively, of the visible electromagnetic spectrum (Fig. 3.11a). Note that there is significant overlap between S_1 and S_2.

Based on the three-color theory, the spectral energy distribution of a "colored" light, $C(\lambda)$, will produce a color sensation that can be described by *spectral responses* (Fig. 3.11b) as

$$\alpha_i(C) = \int_{\lambda_{min}}^{\lambda_{max}} S_i(\lambda)C(\lambda)\,d\lambda, \qquad i = 1, 2, 3 \tag{3.18}$$

† The superscript * used for brightness should not be confused with the complex conjugate. The notation used here is to remain consistent with the commonly used symbols for color coordinates.

Equation (3.18) may be interpreted as an equation of color representation. If $C_1(\lambda)$ and $C_2(\lambda)$ are two spectral distributions that produce responses $\alpha_i(C_1)$ and $\alpha_i(C_2)$ such that

$$\alpha_i(C_1) = \alpha_i(C_2), \qquad i = 1, 2, 3 \tag{3.19}$$

then the colors C_1 and C_2 are perceived to be identical. Hence two colors that look identical could have different spectral distributions.

3.8 COLOR MATCHING AND REPRODUCTION

One of the basic problems in the study of color is the reproduction of color using a set of light sources. Generally, the number of sources is restricted to three which, due to the three-receptor model, is the minimum number required to match arbitrary colors. Consider three *primary sources* of light with spectral energy distributions $P_k(\lambda)$, $k = 1, 2, 3$. Let

$$\int P_k(\lambda)\, d\lambda = 1 \tag{3.20}$$

where the limits of integration are assumed to be $\lambda_{\min}$ and $\lambda_{\max}$ and the sources are linearly independent, i.e., a linear combination of any two sources cannot produce the third source. To match a color $C(\lambda)$, suppose the three primaries are mixed in proportions of β_k, $k = 1, 2, 3$ (Fig. 3.12). Then $\sum_{k=1}^{3} \beta_k P_k(\lambda)$ should be perceived as $C(\lambda)$, i.e.,

$$\alpha_i(C) = \int \left[\sum_{k=1}^{3} \beta_k P_k(\lambda)\right] S_i(\lambda)\, d\lambda = \sum_{k=1}^{3} \beta_k \int S_i(\lambda) P_k(\lambda)\, d\lambda \qquad i = 1, 2, 3 \tag{3.21}$$

Defining the ith *cone response* generated by one unit of the kth primary as

$$a_{i,k} \triangleq \alpha_i(P_k) = \int S_i(\lambda) P_k(\lambda)\, d\lambda, \qquad i, k = 1, 2, 3 \tag{3.22}$$

we get

$$\sum_{k=1}^{3} \beta_k a_{i,k} = \alpha_i(C) = \int S_i(\lambda) C(\lambda)\, d\lambda, \qquad i = 1, 2, 3 \tag{3.23}$$

These are the *color matching equations*. Given an arbitrary color spectral distribution $C(\lambda)$, the primary sources $P_k(\lambda)$, and the spectral sensitivity curves $S_i(\lambda)$, the quantities β_k, $k = 1, 2, 3$, can be found by solving these equations. In practice, the primary sources are calibrated against a reference white light source with known

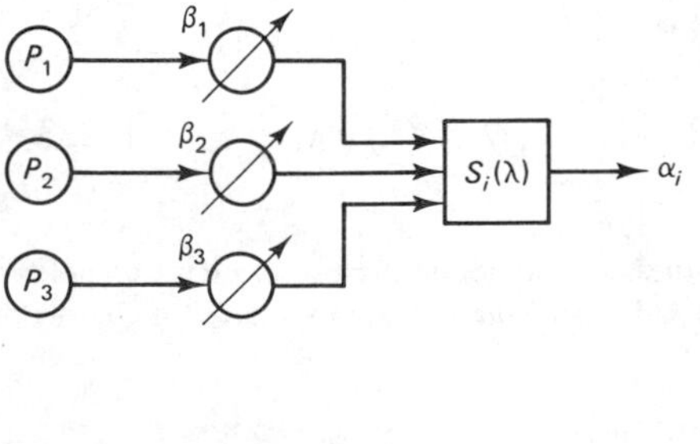

Figure 3.12 Color matching using three primary sources.

energy distribution $W(\lambda)$. Let w_k denote the amount of kth primary required to match the reference white. Then the quantities

$$T_k(C) = \frac{\beta_k}{w_k}, \qquad k = 1, 2, 3 \tag{3.24}$$

are called the *tristimulus values* of the color C. Clearly, the tristimulus values of the reference white are unity. The tristimulus values of a color give the relative amounts of primaries required to match that color. The tristimulus values, $T_k(\lambda)$, of unit energy spectral color at wavelength λ give what are called the *spectral matching curves*. These are obtained by setting $C(\lambda) = \delta(\lambda - \lambda')$ in (3.23), which together with (3.24) yield three simultaneous equations

$$\sum_{k=1}^{3} w_k a_{i,k} T_k(\lambda') = S_i(\lambda'), \qquad i = 1, 2, 3 \tag{3.25}$$

for each λ'. Given the spectral tristimulus values $T_k(\lambda)$, the tristimulus values of an arbitrary color $C(\lambda)$ can be calculated as (Problem 3.8)

$$T_k(C) = \int C(\lambda) T_k(\lambda)\, d\lambda, \qquad k = 1, 2, 3 \tag{3.26}$$

Example 3.1

The primary sources recommended by the CIE† as standard sources are three monochromatic sources

$$\begin{aligned} P_1(\lambda) &= \delta(\lambda - \lambda_1), & \lambda_1 &= 700 \text{ nm, red} \\ P_2(\lambda) &= \delta(\lambda - \lambda_2), & \lambda_2 &= 546.1 \text{ nm, green} \\ P_3(\lambda) &= \delta(\lambda - \lambda_3), & \lambda_3 &= 435.8 \text{ nm, blue} \end{aligned}$$

Using (3.22), we obtain $a_{i,k} = S_i(\lambda_k)$, $i, k = 1, 2, 3$. The standard CIE white source has a flat spectrum. Therefore, $\alpha_i(W) = \int S_i(\lambda) d\lambda$. Using these two relations in (3.23) for reference white, we can write

$$\sum_{k=1}^{3} w_k S_i(\lambda_k) = \int S_i(\lambda)\, d\lambda, \qquad i = 1, 2, 3 \tag{3.27}$$

which can be solved for w_k provided $\{S_i(\lambda_k), 1 \le i, k \le 3\}$ is a nonsingular matrix. Using the spectral sensitivity curves and w_k, one can solve (3.25) for the spectral tristimulus values $T_k(\lambda)$ and obtain their plot as in Fig. 3.13. Note that some of the tristimulus values are negative. This means that the source with negative tristimulus value, when mixed with the given color, will match an appropriate mixture of the other two sources. It is safe to say that any one set of three primary sources cannot match all the visible colors; although for any given color, a suitable set of three primary sources can be found. Hence, the primary sources for color reproduction should be chosen to maximize the number of colors that can be matched.

Laws of Color Matching

The preceding theory of colorimetry leads to a useful set of color matching rules [13], which are stated next.

† Commission Internationale de L'Eclairage, the international committee on color standards.

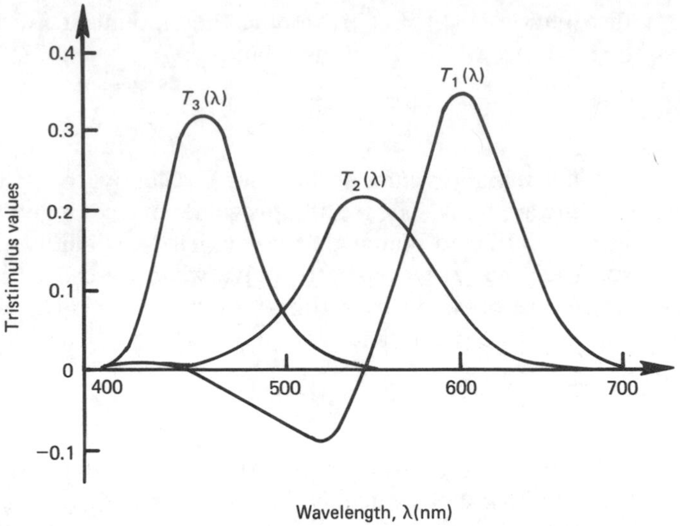

Figure 3.13 Spectral matching tristimulus curves for the CIE spectral primary system. The negative tristimulus values indicate that the colors at those wavelengths cannot be reproduced by the CIE primaries.

1. *Any color can be matched by mixing at most three colored lights.* This means we can always find three primary sources such that the matrix $\{a_{i,k}\}$ is nonsingular and (3.23) has a unique solution.
2. *The luminance of a color mixture is equal to the sum of the luminances of its components.* The luminance Y of a color light $C(\lambda)$ can be obtained via (3.2) as (here the dependence on x, y is suppressed)

$$Y = Y(C) = \int C(\lambda)V(\lambda)\,d\lambda \tag{3.28}$$

From this formula, the luminance of the kth primary source with tristimulus setting $\beta_k = w_k T_k$ (see Fig. 3.12) will be $T_k w_k \int P_k(\lambda)V(\lambda)\,d\lambda$. Hence the luminance of a color with tristimulus values T_k, $k = 1, 2, 3$ can also be written as

$$Y = \sum_{k=1}^{3} T_k \int w_k P_k(\lambda)V(\lambda)\,d\lambda \overset{\Delta}{=} \sum_{k=1}^{3} T_k l_k \tag{3.29}$$

where l_k is called the *luminosity coefficient* of the kth primary.

The reader should be cautioned that in general

$$C(\lambda) \neq \sum_{k=1}^{3} w_k T_k P_k(\lambda) \tag{3.30}$$

even though a color match has been achieved.

3. *The human eye cannot resolve the components of a color mixture.* This means that a monochromatic light source and its color are not unique with respect to each other, i.e., the eye cannot resolve the wavelengths from a color.
4. *A color match at one luminance level holds over a wide range of luminances.*

5. **Color addition:** *If a color C_1 matches color C_2 and a color C_1' matches color C_2', then the mixture of C_1 and C_1' matches the mixture of C_2 and C_2'.* Using the notation

$$[C_1] = [C_2] \Rightarrow \text{color } C_1 \text{ matches color } C_2$$
$$\alpha_1[C_1] + \alpha_2[C_2] \Rightarrow \text{a mixture containing an amount } \alpha_1 \text{ of } C_1 \text{ and an amount } \alpha_2 \text{ of } C_2$$

we can write the preceding law as follows. If

$$[C_1] = [C_1'] \quad \text{and} \quad [C_2] = [C_2']$$

then

$$\alpha_1[C_1] + \alpha_2[C_2] = \alpha_1[C_1'] + \alpha_2[C_2']$$

6. **Color subtraction:** *If a mixture of C_1 and C_2 matches a mixture of C_1' and C_2' and if C_2 matches C_2' then C_1 matches C_1'*, i.e., if

$$[C_1] + [C_2] = [C_1'] + [C_2']$$

and
$$[C_2] = [C_2']$$

then
$$[C_1] = [C_1']$$

7. **Transitive law:** *If C_1 matches C_2 and if C_2 matches C_3, then C_1 matches C_3*, i.e., if

$$[C_1] = [C_2] \quad \text{and} \quad [C_2] = [C_3]$$

then

$$[C_1] = [C_3]$$

8. **Color matches:** Three types of color matches are defined:
   a. $\alpha[C] = \alpha_1[C_1] + \alpha_2[C_2] + \alpha_3[C_3]$; i.e., α units of C are matched by a mixture of α_1 units of C_1, α_2 units of C_2, and α_3 units of C_3. This is a direct match. Indirect matches are defined by the following.
   b. $\alpha[C] + \alpha_1[C_1] = \alpha_2[C_2] + \alpha_3[C_3]$
   c. $\alpha[C] + \alpha_1[C_1] + \alpha_2[C_2] = \alpha_3[C_3]$

These are also called *Grassman's laws.* They hold except when the luminance levels are very high or very low. These are also useful in color reproduction *colorimetry,* the science of measuring color quantitatively.

Chromaticity Diagram

The chromaticities of a color are defined as

$$t_k \triangleq \frac{T_k}{T_1 + T_2 + T_3}, \qquad k = 1, 2, 3 \tag{3.31}$$

Clearly $t_1 + t_2 + t_3 = 1$. Hence, only two of the three chromaticity coordinates are independent. Therefore, the chromaticity coordinates project the three-dimensional color solid on a plane. The chromaticities t_1, t_2 jointly represent the

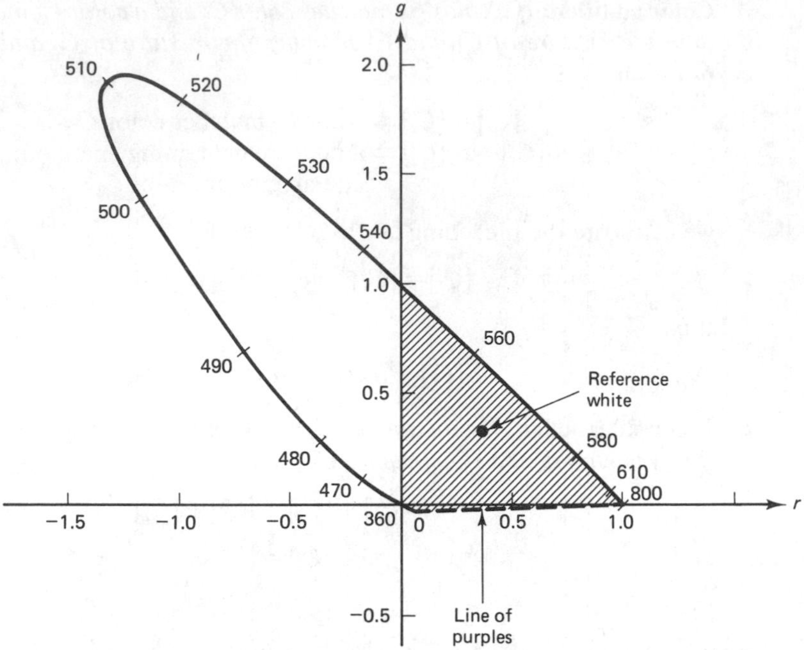

Figure 3.14 Chromaticity diagram for the CIE spectral primary system. Shaded area is the color gamut of this system.

chrominance components (i.e., hue and saturation) of the color. The entire color space can be represented by the coordinates (t_1, t_2, Y), in which any Y = constant is a chrominance plane. The chromaticity diagram represents the color subspace in the chrominance plane. Figure 3.14 shows the chromaticity diagram for the CIE spectral primary system. The chromaticity diagram has the following properties:

1. The locus of all the points representing spectral colors contains the region of all the visible colors.
2. The straight line joining the chromaticity coordinates of blue (360 nm) and red (780 nm) contains the purple colors and is called *the line of purples*.
3. The region bounded by the straight lines joining the coordinates $(0, 0)$, $(0, 1)$ and $(1, 0)$ (the *shaded region* of Fig. 3.14) contains all the colors reproducible by the primary sources. This region is called the *color gamut* of the primary sources.
4. The reference white of the CIE primary system has chromaticity coordinates $(\frac{1}{3}, \frac{1}{3})$. Colors lying close to this point are the less saturated colors. Colors located far from this point are the more saturated colors. Thus the spectral colors and the colors on the line of purples are maximally saturated.

3.9 COLOR COORDINATE SYSTEMS

There are several color coordinate systems (Table 3.4), which have come into existence for a variety of reasons.

TABLE 3.4 Color Coordinate Systems

Color coordinate system	Description
1. *C.I.E. spectral primary system: R, G, B*	Monochromatic primary sources P_1, red = 700 nm, P_2, green = 546.1 nm, P_3, blue = 435.8 nm. Reference white has flat spectrum and $R = G = B = 1$. See Figs. 3.13 and 3.14 for spectral matching curves and chromaticity diagram.
2. *C.I.E. X, Y, Z system* Y = luminance	$\begin{bmatrix} X \\ Y \\ Z \end{bmatrix} = \begin{bmatrix} 0.490 & 0.310 & 0.200 \\ 0.177 & 0.813 & 0.011 \\ 0.000 & 0.010 & 0.990 \end{bmatrix} \begin{bmatrix} R \\ G \\ B \end{bmatrix}$
3. *C.I.E. uniform chromaticity scale (UCS) system: u, v, Y*	$u = \dfrac{4X}{X + 15Y + 3Z} \equiv \dfrac{4x}{-2x + 12y + 3}$
u, v = chromaticities	$v = \dfrac{6Y}{X + 15Y + 3Z} \equiv \dfrac{6y}{-2x + 12y + 3}$
Y = luminance	$U = \dfrac{2X}{3},\ V = Y,\ W = \dfrac{-X + 3Y + Z}{2}$
U, V, W = tristimulus values corresponding to u, v, w	
4. U^*, V^*, W^* system (modified UCS system)	$U^* = 13W^*(u - u_0)$ $V^* = 13W^*(v - v_0)$ $W^* = 25(100Y)^{1/3} - 17,\ 1 \le 100Y \le 100$
Y = luminance [0.01, 1]	u_0, v_0 = chromaticities of reference white W^* = contrast or brightness
5. S, θ, W^* system: S = saturation θ = hue W^* = brightness	$S = [(U^*)^2 + (V^*)^2]^{1/2} = 13W^*[(u - u_0)^2 + (v - v_0)^2]^{1/2}$ $\theta = \tan^{-1}\left(\dfrac{V^*}{U^*}\right) = \tan^{-1}[(v - v_0)/(u - u_0)],\ 0 \le \theta \le 2\pi$
6. *NTSC receiver primary system* R_N, G_N, B_N	Linear transformation of X, Y, Z. Is based on television phosphor primaries. Reference white is illuminant C for which $R_N = G_N = B_N = 1$. $\begin{bmatrix} R_N \\ G_N \\ B_N \end{bmatrix} = \begin{bmatrix} 1.910 & -0.533 & -0.288 \\ -0.985 & 2.000 & -0.028 \\ 0.058 & -0.118 & 0.896 \end{bmatrix} \begin{bmatrix} X \\ Y \\ Z \end{bmatrix}$
7. *NTSC transmission system:* Y = luminance I, Q = chrominances	$Y = 0.299R_N + 0.587G_N + 0.114B_N$ $I = 0.596R_N - 0.274G_N - 0.322B_N$ $Q = 0.211R_N - 0.523G_N + 0.312B_N$
8. L^*, a^*, b^* *system:*	$L^* = 25\left(\dfrac{100Y}{Y_0}\right)^{1/3} - 16,\ 1 \le 100Y \le 100$
L^* = brightness	$a^* = 500\left[\left(\dfrac{X}{X_0}\right)^{1/3} - \left(\dfrac{Y}{Y_0}\right)^{1/3}\right]$
a^* = red-green content	$b^* = 200\left[\left(\dfrac{Y}{Y_0}\right)^{1/3} - \left(\dfrac{Z}{Z_0}\right)^{1/3}\right]$
b^* = yellow-blue content	X_0, Y_0, Z_0 = tristimulus values of the reference white

As mentioned before, the CIE spectral primary sources do not yield a full gamut of reproducible colors. In fact, no practical set of three primaries has been found that can reproduce all colors. This has led to the development of the CIE X, Y, Z system with hypothetical primary sources such that all the spectral tristimulus values are positive. Although the primary sources are physically unrealizable, this is a convenient coordinate system for colormetric calculations. In this system Y represents the luminance of the color. The X, Y, Z coordinates are related to the CIE R, G, B system via the linear transformation shown in Table 3.4. Figure 3.15 shows the chromaticity diagram for this system. The reference white for this system has a flat spectrum as in the R, G, B system. The tristimulus values for the reference white are $X = Y = Z = 1$.

Figure 3.15 also contains several ellipses of different sizes and orientations. These ellipses, also called *MacAdam ellipses* [10, 11], are such that colors that lie inside are indistinguishable. Any color lying just outside the ellipse is *just noticeably different* (JND) from the color at the center of the ellipse. The size, orientation, and eccentricity (ratio of major to minor area) of these ellipses vary throughout the color space. The *uniform chromaticity scale* (UCS) system u, v, Y transforms these elliptical contours with large eccentricity (up to 20 : 1) to near circles (eccentricity $\simeq$ 2 : 1) of almost equal size in the u, v plane. It is related to the X, Y, Z system via the transformation shown in Table 3.4. Note that x, y and u, v are the chromaticity coordinates and Y is the luminance. Figure 3.16 shows the chromaticity diagram of the UCS coordinate system. The tristimulus coordinates corresponding to u, v, and $w \triangleq 1 - u - v$ are labeled as U, V, and W respectively.

The U^*, V^*, W^* system is a modified UCS system whose origin (u_0, v_0) is shifted to the reference white in the u, v chromaticity plane. The coordinate W^* is a cube root transformation of the luminance and represents the contrast (or

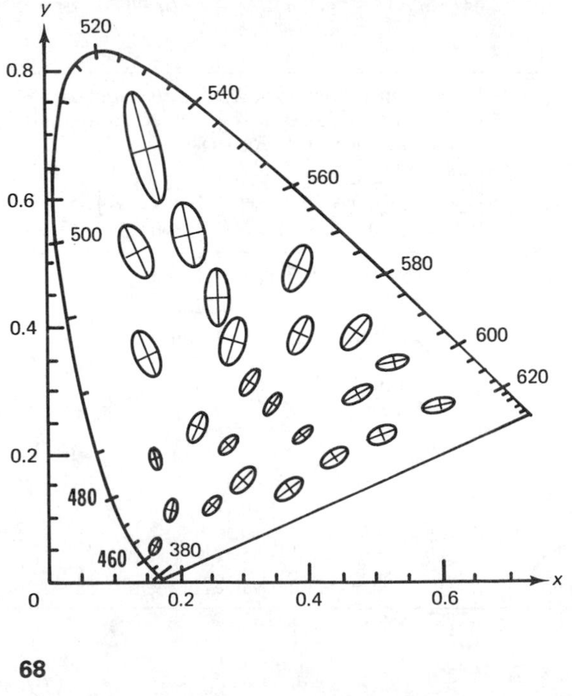

Figure 3.15 Chromaticity diagram for the CIE XYZ color coordinate system. The (MacAdam) ellipses are the just noticeable color difference ellipses.

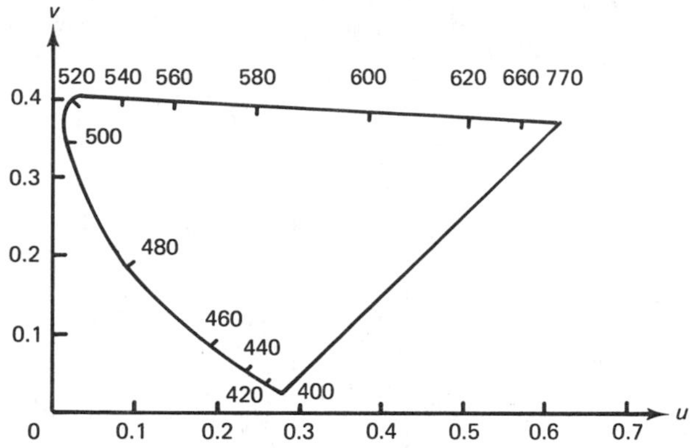

Figure 3.16 Chromaticity diagram for the CIE UCS color coordinate system.

brightness) of a uniform color patch. This coordinate system is useful for measuring color differences quantitatively. In this system, for unsaturated colors, i.e., for colors lying near the grays in the color solid, the difference between two colors is, to a good approximation, proportional to the length of the straight line joining them.

The S, θ, W^* system is simply the polar representation of the U^*, V^*, W^* system, where S and θ represent, respectively, the saturation and hue attributes of color (Fig. 3.10). Large values of S imply highly saturated colors.

The National Television Systems Committee (NTSC) *receiver primary system* (R_N, G_N, B_N) was developed as a standard for television receivers. The NTSC has adopted three phosphor primaries that glow in the red, green, and blue regions of the visible spectrum. The reference white was chosen as the illuminant C, for which the tristimulus values are $R_N = G_N = B_N = 1$. Table 3.5 gives the NTSC coordinates of some of the major colors. The color solid for this coordinate system is a cube (Fig. 3.17). The chromaticity diagram for this system is shown in Fig. 3.18. Note that the reference white for NTSC is different from that for the CIE system.

The NTSC *transmission system* (Y, I, Q) was developed to facilitate transmission of color images using the existing monochrome television channels without increasing the bandwidth requirement. The Y coordinate is the luminance (monochrome channel) of the color. The other two tristimulus signals, I and Q, jointly represent hue and saturation of the color and whose bandwidths are much smaller than that of the luminance signal. The I, Q components are transmitted on a subcarrier channel using quadrature modulation in such a way that the spatial

TABLE 3.5 Tristimulus and Chromaticity Values of Major Colors in the NTSC Receiver Primary System

	Red	Yellow	Green	Cyan	Blue	Magenta	White	Black
R_N	1.0	1.0	0.0	0.0	0.0	1.0	1.0	0.0
G_N	0.0	1.0	1.0	1.0	0.0	0.0	1.0	0.0
B_N	0.0	0.0	0.0	1.0	1.0	1.0	1.0	0.0
r_N	1.0	0.5	0.0	0.0	0.0	0.5	0.333	0.333
g_N	0.0	0.5	1.0	0.5	0.0	0.0	0.333	0.333
b_N	0.0	0.0	0.0	0.5	1.0	0.5	0.333	0.333

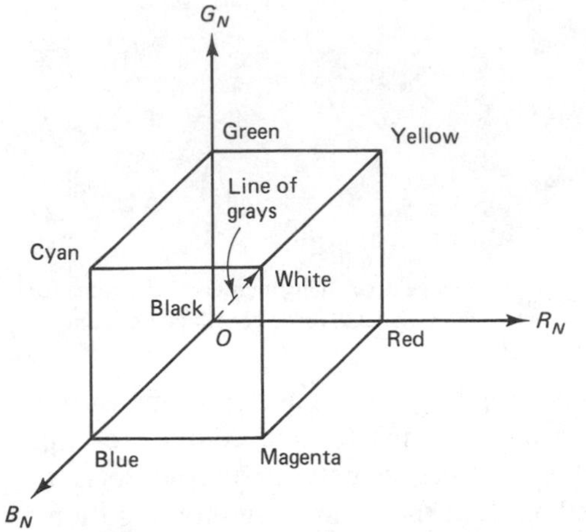

Figure 3.17 Tristimulus color solid for the NTSC receiver primary system.

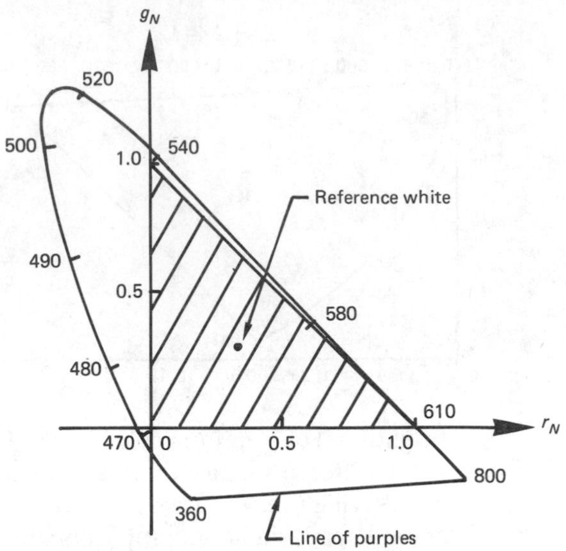

Figure 3.18 Chromaticity diagram for the NTSC receiver primary system.

spectra of I, Q do not overlap with that of Y and the overall bandwidth required for transmission remains unchanged (see Chapter 4). The Y, I, Q system is related to the R_N, G_N, B_N system via a linear transformation. This and some other transformations relating the different coordinate systems are given in Table 3.6.

The L^*, a^*, b^* system gives a quantitative expression for the Munsell system of color classification [12]. Like the U^*, V^*, W^* system, this also gives a useful color difference formula.

Example 3.2

We will find the representation of the NTSC receiver primary yellow in the various coordinate systems. From Table 3.5, we have $R_N = 1.0$, $G_N = 1.0$, $B_N = 0.0$.

Using Table 3.6, we obtain the CIE spectral primary system coordinates as $R = 1.167 - 0.146 - 0.0 = 1.021$, $G = 0.114 + 0.753 + 0.0 = 0.867$, $B = -0.001 + 0.59 + 0.0 = 0.058$.

The corresponding chromaticity values are

$$r = \frac{1.021}{1.021 + 0.867 + 0.058} = 0.525, \qquad g = \frac{0.867}{1.946} = 0.445, \qquad b = \frac{0.058}{1.946} = 0.030$$

Similarly, for the other coordinate systems we obtain:

$$X = 0.781, \quad Y = 0.886, \quad Z = 0.066; \quad x = 0.451, \quad y = 0.511, \quad z = 0.038$$

$$U = 0.521, \quad V = 0.886, \quad W = 0.972, \quad u = 0.219, \quad v = 0.373, \quad w = 0.408$$

$$Y = 0.886, \quad I = 0.322, \quad Q = -0.312$$

In the NTSC receiver primary system, the reference white is $R_N = G_N = B_N = 1$. This gives $X_0 = 0.982$, $Y_0 = 1.00$, $Z_0 = 1.183$. Note that X_0, Y_0, and Z_0 are

TABLE 3.6 Transformations from NTSC Receiver Primary to Different Coordinate Systems. Input Vector is $[R_N\ G_N\ B_N]^T$.

Output vector	Transformation matrix	Comments
$\begin{bmatrix} R \\ G \\ B \end{bmatrix}$	$\begin{pmatrix} 1.167 & -0.146 & -0.151 \\ 0.114 & 0.753 & 0.159 \\ -0.001 & 0.059 & 1.128 \end{pmatrix}$	CIE spectral primary system
$\begin{bmatrix} X \\ Y \\ Z \end{bmatrix}$	$\begin{pmatrix} 0.607 & 0.174 & 0.201 \\ 0.299 & 0.587 & 0.114 \\ 0.000 & 0.066 & 1.117 \end{pmatrix}$	CIE X, Y, Z system
$\begin{bmatrix} U \\ V \\ W \end{bmatrix}$	$\begin{pmatrix} 0.405 & 0.116 & 0.133 \\ 0.299 & 0.587 & 0.114 \\ 0.145 & 0.827 & 0.627 \end{pmatrix}$	CIE UCS tristimulus system
$\begin{bmatrix} Y \\ I \\ Q \end{bmatrix}$	$\begin{pmatrix} 0.299 & 0.587 & 0.114 \\ 0.596 & -0.274 & -0.322 \\ 0.211 & -0.523 & 0.312 \end{pmatrix}$	NTSC transmission system

not unity because the reference white for NTSC sources is different from that of the CIE. Using the definitions of u and v from Table 3.4, we obtain $u_0 = 0.201$, $v_0 = 0.307$ for the reference white.

Using the preceding results in the formulas for the remaining coordinate systems, we obtain

$$W^* = 25(88.6)^{1/3} - 17 = 94.45, \qquad U^* = 22.10, \qquad V^* = 81.04$$

$$S = 84.00, \qquad \theta = \tan^{-1}(3.67) = 1.30 \text{ rad}, \qquad W^* = 94.45$$

$$L^* = 25(88.6)^{1/3} - 16 = 95.45, \qquad a^* = 500\left[\left(\frac{0.781}{0.982}\right)^{1/3} - \left(\frac{0.886}{1}\right)^{1/3}\right] = -16.98$$

$$b^* = 200\left[(0.886)^{1/3} - \left(\frac{0.066}{1.183}\right)^{1/3}\right] = 115.67$$

3.10 COLOR DIFFERENCE MEASURES

Quantitative measures of color difference between any two arbitrary colors pose a problem of considerable interest in coding, enhancement, and analysis of color images. Experimental evidence suggests that the tristimulus color solid may be considered as a Riemannian space with a color distance metric [14, 15]

$$(ds)^2 = \sum_{i=1}^{3} \sum_{j=1}^{3} c_{i,j}\, dX_i\, dX_j \tag{3.32}$$

The distance ds represents the infinitesimal difference between two colors with coordinates X_i and $X_i + dX_i$ in the chosen color coordinate system. The coefficients

$c_{i,j}$ measure the average human perception sensitivity due to small differences in the ith and in the jth coordinates.

Small differences in color are described on observations of *just noticeable differences* (JNDs) in colors. A unit JND defined by

$$1 = \sum_{i=1}^{3} \sum_{j=1}^{3} c_{i,j}\, dX_i\, dX_j \tag{3.33}$$

is the describing equation for an ellipsoid. If the coefficients $c_{i,j}$ were constant throughout the color space, then the JND ellipsoids would be of uniform size in the color space. In that event, the color space could be reduced to a *Euclidean tristimulus space,* where the color difference between any two colors would become proportional to the length of the straight line joining them. Unfortunately, the $c_{i,j}$ exhibit large variations with tristimulus values, so that the sizes as well as the orientations of the JND ellipsoids vary considerably. Consequently, the distance between two arbitrary colors C_1 and C_2 is given by the minimal distance chain of ellipsoids lying along a curve $\mathcal{G}^*$ joining C_1 and C_2 such that the distance integral

$$d(C_1, C_2) \triangleq \int_{\mathcal{G}\;C_1(X_i)}^{C_2(X_i)} ds \tag{3.34}$$

is minimum when evaluated along this curve, i.e., for $\mathcal{G} = \mathcal{G}^*$. This curve is called the *geodesic* between C_1 and C_2. If $c_{i,j}$ are constant in the tristimulus space, then the geodesic is a straight line. Geodesics in color space can be determined by employing a suitable optimization technique such as dynamic programming or the calculus of

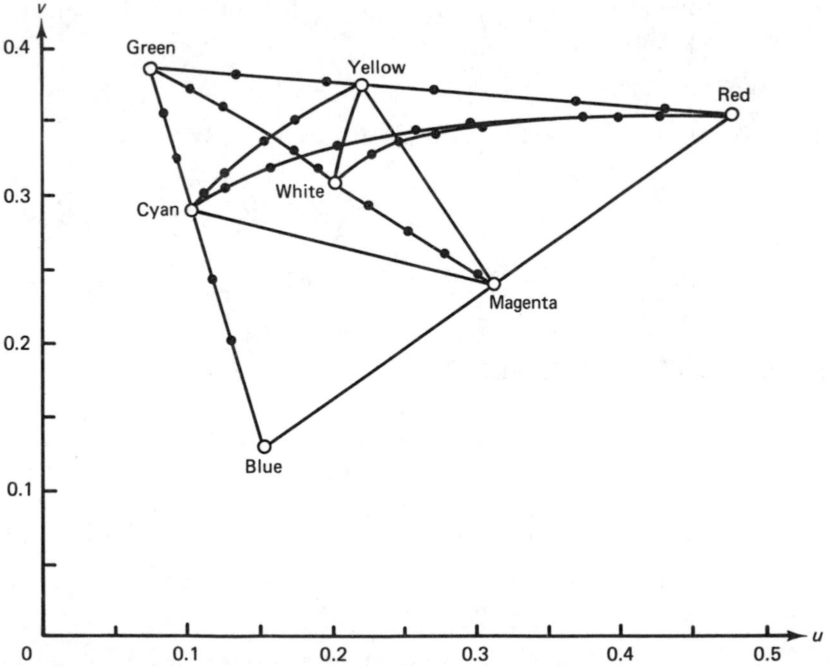

Figure 3.19 Geodesics in the (u, v) plane.

TABLE 3.7 CIE Color-Difference Formulas

Formula	Equation Number	Comments
$(\Delta s)^2 = (\Delta U^*)^2 + (\Delta V^*)^2 + (\Delta W^*)^2$	(3.35)	1964 CIE formula
$(\Delta s)^2 = (\Delta L^*)^2 + (\Delta u^*)^2 + (\Delta v^*)^2$ $L^* = 25\left(\frac{100Y}{Y_0}\right)^{1/3} - 16$ $u^* = 13L^*(u' - u_0)$ $v^* = 13L^*(v' - v_0)$ $u' = u$ $v' = 1.5v = \frac{9Y}{X + 15Y + 3Z}$	(3.36)	1976 CIE formula, modification of the u, v, Y space to u^*, v^*, L^* space. u_0, v_0, Y_0 refer to reference white.
$(\Delta s)^2 = (\Delta L^*)^2 + (\Delta a^*)^2 + (\Delta b^*)^2$	(3.37)	L^*, a^*, b^* color coordinate system.

variations [15]. Figure 3.19 shows the projections of several geodesic curves between the major NTSC colors on the UCS u, v chromaticity plane. The geodesics between the primary colors are nearly straight lines (in the chromaticity plane), but the geodesics between most other colors are generally curved.

Due to the large complexity of the foregoing procedure of determining color distance, simpler measures that can easily be used are desired. Several simple formulas that approximate the Riemannian color space by a Euclidean color space have been proposed by the CIE (Table 3.7). The first of these formulas [eq. (3.35)] was adopted by the CIE in 1964. The formula of (3.36), called the CIE 1976 L^*, u^*, v^* formula, is an improvement over the 1964 CIE U^*, V^*, W^* formula in regard to uniform spacing of colors that exhibit differences in sizes typical of those in the Munsell *book of color* [12].

The third formula, (3.37), is called the CIE 1976 L^*, a^*, b^* color-difference formula. It is intended to yield perceptually uniform spacing of colors that exhibit color differences greater than JND threshold but smaller than those in the Munsell book of color.

3.11 COLOR VISION MODEL

With color represented by a three-element vector, a color vision model containing three channels [16], each being similar to the simplified model of Fig. 3.9, is shown in Fig. 3.20. The color image is represented by the R_N, G_N, B_N coordinates at each pixel. The matrix $\mathbf{A}$ transforms the input into the three cone responses $\alpha_k(x, y, C)$, $k = 1, 2, 3$, where (x, y) are the spatial pixel coordinates and C refers to its color. In Fig. 3.20, we have represented the *normalized cone responses*

$$T_k \triangleq \frac{\alpha_k(x, y, C)}{\alpha_k(x, y, W)}, \qquad k = 1, 2, 3 \tag{3.38}$$

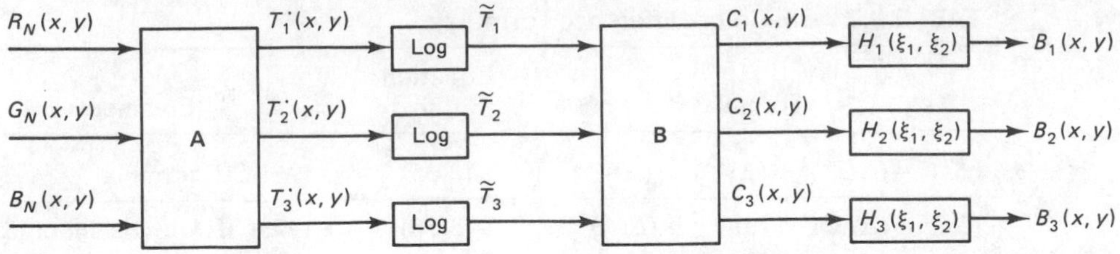

Figure 3.20 A color vision model.

In analogy with the definition of tristimulus values, T_k' are called the *retinal cone* tristimulus coordinates (see Problem 3.14). The cone responses undergo nonlinear point transformations to give three fields $\tilde{T}_k(x, y)$, $k = 1, 2, 3$. The 3×3 matrix **B** transforms the $\{\tilde{T}_k(x, y)\}$ into $\{C_k(x, y)\}$ such that $C_1(x, y)$ is the monochrome (achromatic) contrast field $c(x, y)$, as in Fig. 3.9, and $C_2(x, y)$ and $C_3(x, y)$ represent the corresponding chromatic fields. The spatial filters $H_k(\xi_1, \xi_2)$, $k = 1, 2, 3$, represent the frequency response of the visual system to luminance and chrominance contrast signals. Thus $H_1(\xi_1, \xi_2)$ is the same as $H(\xi_1, \xi_2)$ in Fig. 3.9 and is a bandpass filter that represents the lateral inhibition phenomenon. The visual frequency response to chrominance signals are not well established but are believed to have their passbands in the lower frequency region, as shown in Fig. 3.21. The 3×3 matrices **A** and **B** are given as follows:

$$\mathbf{A} = \begin{pmatrix} 0.299 & 0.587 & 0.114 \\ -0.127 & 0.724 & 0.175 \\ 0.000 & 0.066 & 1.117 \end{pmatrix}, \qquad \mathbf{B} = \begin{pmatrix} 21.5 & 0.0 & 0.00 \\ -41.0 & 41.0 & 0.00 \\ -6.27 & 0.0 & 6.27 \end{pmatrix} \tag{3.39}$$

From the model of Fig. 3.20, a criterion for color image fidelity can be defined. For example, for two color images $\{R_N, G_N, B_N\}$ and $\{R_N', G_N', B_N'\}$, their subjective mean square error could be defined by

$$e_{ls} = \frac{1}{A} \sum_{k=1}^{3} \iint_{\mathcal{R}} (B_k(x, y) - B_k'(x, y))^2 \, dx \, dy \tag{3.40}$$

where $\mathcal{R}$ is the region over which the image is defined (or available), A is its area, and $\{B_k(x, y)\}$ and $\{B_k'(x, y)\}$ are the outputs of the model for the two color images.

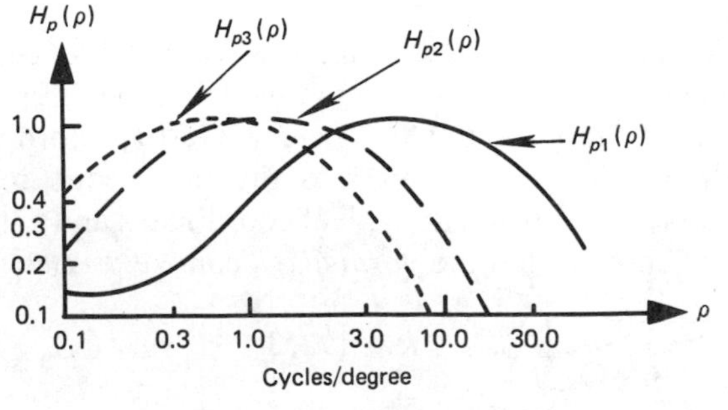

Figure 3.21 Frequency responses of the three color channels C_1, C_2, C_3 of the color vision model. Each filter is assumed to be isotropic so that $H_{pk}(\rho) \triangleq H_k(\xi_1, \xi_2)$, $\rho = \sqrt{\xi_1^2 + \xi_2^2}$, $k = 1, 2, 3$.

3.12 TEMPORAL PROPERTIES OF VISION

Temporal aspects of visual perception [1, 18] become important in the processing of motion images and in the design of image displays for stationary images. The main properties that will be relevant to our discussion are summarized here.

Bloch's Law

Light flashes of different durations but equal energy are indistinguishable below a critical duration. This critical duration is about 30 ms when the eye is adapted at moderate illumination level. The more the eye is adapted to the dark, the longer is the critical duration.

Critical Fusion Frequency (CFF)

When a slowly flashing light is observed, the individual flashes are distinguishable. At flashing rates above the *critical fusion frequency* (CFF), the flashes are indistinguishable from a steady light of the same average intensity. This frequency generally does not exceed 50 to 60 Hz. Figure 3.22 shows a typical temporal MTF.

This property is the basis of television raster scanning cameras and displays. Interlaced image fields are sampled and displayed at rates of 50 or 60 Hz. (The rate is chosen to coincide with the power-line frequency to avoid any interference.) For digital display of still images, modern display monitors are refreshed at a rate of 60 frames/s to avoid any flicker perception.

Spatial versus Temporal Effects

The eye is more sensitive to flickering of high spatial frequencies than low spatial frequencies. Figure 3.22 compares the temporal MTFs for flickering fields with different spatial frequencies. This fact has been found useful in coding of motion

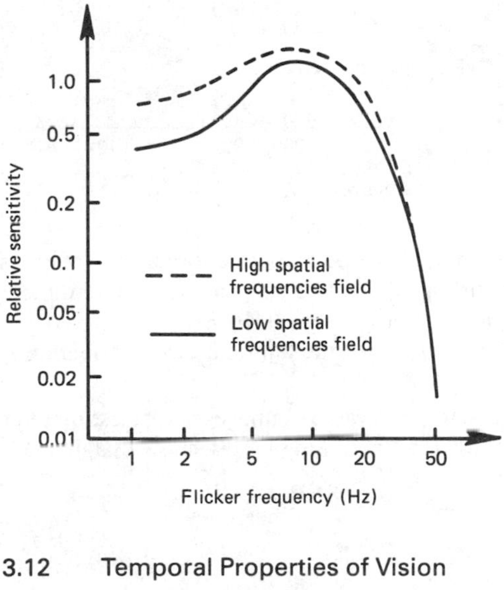

Figure 3.22 Temporal MTFs for flickering fields.

images by subsampling the moving areas everywhere except at the edges. For the same reason, image display monitors offering high spatial resolution display images at a noninterlaced 60-Hz refresh rate.

PROBLEMS

3.1 Generate two 256×256 8-bit images as in Fig. 3.3a, where the small squares have gray level values of 127 and the backgrounds have the values 63 and 223. Verify the result of Fig. 3.3a. Next change the gray level of one of the small squares until the result of Fig. 3.3b is verified.

3.2 Show that eqs. (3.5) and (3.6) are solutions of a modified Weber law: df/f^{γ} *is proportional to dc,* i.e., equal changes in contrast are induced by equal amounts of df/f^{γ}. Find γ.

3.3 Generate a digital bar chart as shown in Fig. 3.5a, where each bar is 64 pixels wide. Each image line is a staircase function, as shown in Fig. 3.5b. Plot the brightness function (approximately) as you perceive it.

3.4 Generate a 512×512 image, each row of which is a smooth ramp $r(n)$ as shown in Fig. P3.4. Display on a video monitor and locate the dark (D) and the bright (B) Mach bands.

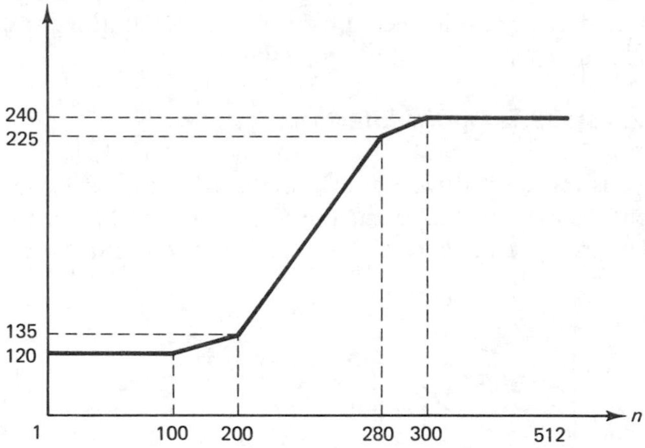

Figure P3.4

3.5 The Mach band phenomenon predicts the one-dimensional step response of the visual system, as shown by $s(n)$ in Fig. P3.5. The corresponding one-dimensional impulse response (or the vertical line response) is given by $h(n) = s(n) - s(n-1)$. Show that $h(n)$ has negative lobes (which manifest the lateral inhibition phenomenon) as shown in Fig. 3.6c.

3.6 As a rule of thumb, the peak-to-peak value of images can be estimated as $n\sigma$, where n varies between 4 to 6. Letting $n = 5$ and using (3.13), show that

$$\text{SNR}' \simeq \text{SNR} + 14\,dB$$

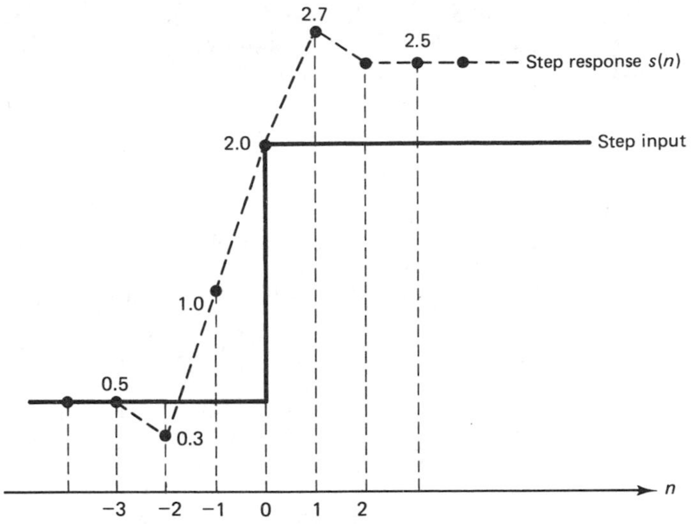

Figure P3.5

3.7 Can two monochromatic sources with different wavelengths be perceived to have the same color? Explain.

3.8 Using eqs. (3.23) through (3.25), show that (3.26) is valid.

3.9 In this problem we show that any two tristimulus coordinate systems based on different sets of primary sources are linearly related. Let $\{P_k(\lambda)\}$ and $\{P'_k(\lambda)\}$, $k = 1, 2, 3$, be two sets of primary sources with corresponding tristimulus coordinates $\{T_k\}$ and $\{T'_k\}$ and reference white sources $W(\lambda)$ and $W'(\lambda)$. If a color $C(\lambda)$ is matched by these sets of sources, then show that

$$\sum_{k=1}^{3} a_{i,k} w_k T_k(C) = \sum_{k=1}^{3} a'_{i,k} w'_k T'_k(C)$$

where the definitions of a's and w's follow from the text. Express this in matrix form and write the solution for $\{T'_k\}$.

3.10 Show that given the chromaticities t_1, t_2 and the luminance Y, the tristimulus values of a coordinate system can be obtained by

$$T_k = \frac{t_k Y}{\sum_{i=1}^{3} l_i t_i} \qquad k = 1, 2, 3$$

where l_i are the luminosity coefficients of the primary sources.

3.11* For all the major NTSC colors listed in Table 3.5, calculate their tristimulus values in the RGB, XYZ, UVW, YIQ, U^*V^*W, $L^*a^*b^*$, $S\theta W^*$, and $T_1^{\prime} T_2^{\prime} T_3^{\prime}$ coordinate systems. Calculate their chromaticity coordinates in the first three of these systems.

3.12 Among the major NTSC colors, except for white and black (see Table 3.5), which one (a) has the maximum luminance, (b) is most saturated, and (c) is least saturated?

3.13* Calculate the color differences between all pairs of the major NTSC colors listed in Table 3.5 according to the 1964 CIE formula given in Table 3.7. Which pair of colors is (a) maximally different, (b) minimally different? Repeat the calculations using the $L^*a^*b^*$ system formula given by (3.37).

3.14 [Retinal cone system; T_1^*, T_2^*, T_3^*] Let $P_k(\lambda), k = 1, 2, 3$ denote the primary sources that generate the retinal cone tristimulus values. Using (3.38), (3.24) and (3.23), show that this requires (for every x, y, C)

$$\sum_{k=1}^{3} \alpha_k(x, y, C) a_{i,k} = \alpha_i(x, y, C) \Rightarrow a_{i,k} = \delta(i - k) \tag{P3.14}$$

To determine $P_k(\lambda)$, write

$$P_k(\lambda) = \sum_{i=1}^{3} S_i(\lambda) b_{i,k}, \qquad k = 1, 2, 3$$

and show that (P3.14) implies $\mathbf{B} \triangleq \{b_{i,k}\} = \mathbf{\Sigma}^{-1}$, where $\mathbf{\Sigma} \triangleq \{\sigma_{i,j}\}$ and

$$\sigma_{i,j} = \int S_i(\lambda) S_j(\lambda)\, d\lambda$$

Is the set $\{P_k(\lambda)\}$ physically realizable? (*Hint:* Are $b_{i,k}$ nonnegative?)

BIBLIOGRAPHY

Sections 3.1–3.3

For further discussion on fundamental topics in visual perception:

1. T. N. Cornsweet. *Visual Perception.* New York: Academic Press, 1971.
2. E. C. Carterette and M. P. Friedman, eds. *Handbook of Perception,* vol. 5. New York: Academic Press, 1975.
3. S. Hecht. "The Visual Discrimination of Intensity and the Weber-Fechner Law." *J. Gen. Physiol.* 7, (1924): 241.

Section 3.4

For measurement and applications of the visibility function:

4. A. N. Netravali and B. Prasada. "Adaptive Quantization of Picture Signals Using Spatial Masking." *Proc. IEEE* 65, (April 1977): 536–548.

Sections 3.5–3.6

For a detailed development of the monochrome vision model and related image fidelity criteria:

5. C. F. Hall and E. L. Hall. "A Nonlinear Model for the Spatial Characteristics of the Human Visual System." *IEEE Trans. Syst. Man. Cybern.,* SMC-7, 3 (March 1977): 161–170.
6. J. L. Mannos and D. J. Sakrison. "The Effects of a Visual Fidelity Criterion on the Encoding of Images." *IEEE Trans. Info. Theory* IT-20, no. 4 (July 1974): 525–536.
7. D. J. Sakrison. "On the Role of Observer and a Distortion Measure in Image Transmission." *IEEE Trans. Communication* COM-25, (Nov. 1977): 1251–1267.

Sections 3.7–3.10

For introductory material on color perception, color representation and general reading on color:

8. T. Young. "On the Theory of Light and Colors." *Philosophical Transactions of the Royal Society of London,* 92, (1802): 20–71.
9. J. C. Maxwell. "On the Theory of Three Primary Colours." Lectures delivered in 1861. W. D. Nevin (ed.), *Sci. Papers* 1, Cambridge Univ. Press, London (1890): 445–450.
10. D. L. MacAdam. *Sources of Color Science.* Cambridge, Mass.: MIT Press, 1970.
11. G. W. Wyzecki and W. S. Stiles. *Color Science.* New York: John Wiley, 1967.
12. *Munsell Book of Color.* Munsell Color Co., 2441 North Calvert St., Baltimore, Md.
13. H. G. Grassman. "Theory of Compound Colours." *Philosophic Magazine* 4, no. 7 (1954): 254–264.

For color distances, geodesics, and color brightness:

14. C.I.E. "Colorimetry Proposal for Study of Color Spaces" (technical note). *J. Opt. Soc. Am.* 64, (June 1974): 896–897.
15. A. K. Jain. "Color Distance and Geodesics in Color 3 Space." *J. Opt. Soc. Am.* 62 (November 1972): 1287–1290. Also see *J. Opt. Soc. Am.* 63, (August 1973): 934–939.

Section 3.11

For the color vision model, their applications and related biography:

16. W. Frei and B. Baxter. "Rate Distortion Coding Simulation for Color Images." *IEEE Trans. Communications* COM-25, (November 1977): 1385–1392.
17. J. O. Limb, C. B. Rubinstein, and J. E. Thompson. "Digital Coding of Color Video Signals." *IEEE Trans. Communications* COM-25 (November 1977): 1349–1384.

Section 3.12

For further details on temporal visual perceptions, see [1] and:

18. D. H. Kelly. "Visual Responses to Time Dependent Stimuli. I. Amplitude Sensitivity Measurements." *J. Opt. Soc. Am.* 51, (1961): 422–429. Also see pp. 917–918 of this issue, and Vol. 59 (1969): 1361–1369.

4 Image Sampling and Quantization

4.1 INTRODUCTION

The most basic requirement for computer processing of images is that the images be available in digital form, that is, as arrays of finite length binary words. For digitization (Fig. 4.1), the given image is *sampled* on a discrete grid and each sample or pixel is *quantized* using a finite number of bits. The digitized image can then be processed by the computer. To display a digital image, it is first converted to an analog signal, which is scanned onto a display.

Image Scanning

A common method of image sampling is to scan the image row by row and sample each row. An example is the television camera with a vidicon camera tube or an image dissector tube. Figure 4.2 shows the operating principle. An object, film, or transparency is continuously illuminated to form an electron image on a photosensitive plate called the *target*. In a vidicon tube the target is photoconductive,

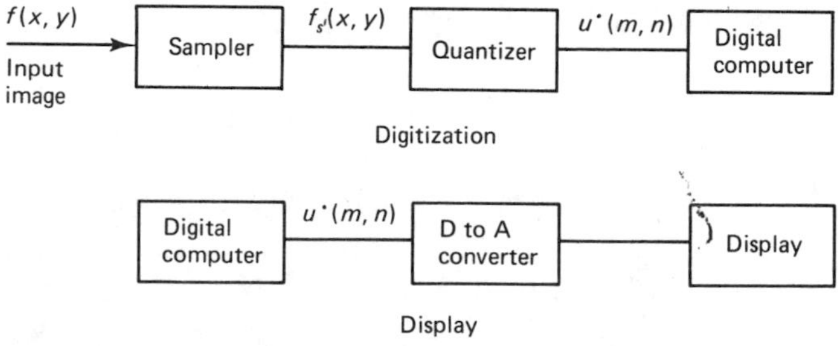

Figure 4.1 Sampling, quantization, and display of images.

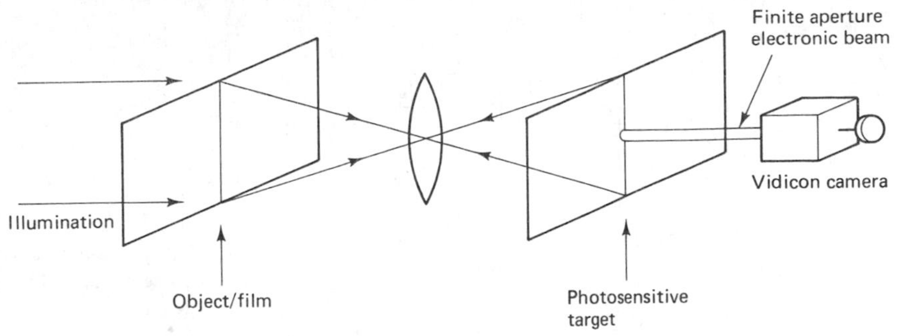

Figure 4.2 Scan-out method.

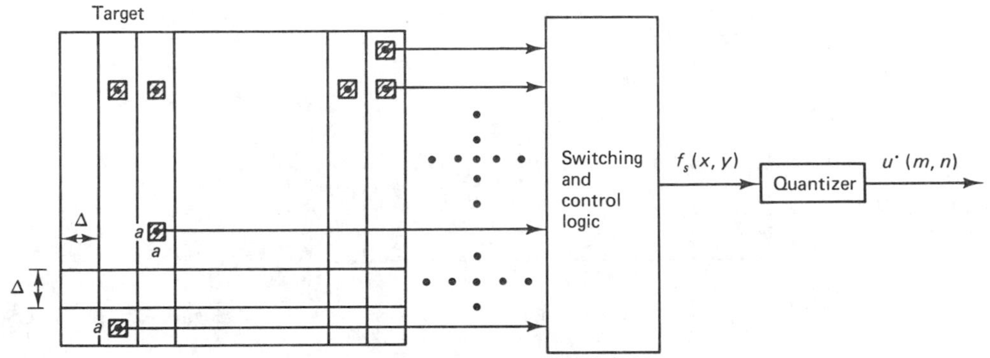

Figure 4.3 Self scanning array.

whereas in an image dissector tube it is photoemissive. A finite-aperture electron beam scans the target and generates current which is proportional to the light intensity falling on the target. A system with such scanning mechanism is called a *scan-out* digitizer. Some of the modern scanning devices, such as *charge-coupled device* (CCD) cameras, contain an array of photodetectors, a set of electronic switches, and control circuitry all on a single chip. By external clocking, the array can be scanned element by element in any desired manner (see Fig. 4.3). This is truly a two-dimensional sampling device and is sometimes called a *self-scanning* array.

In another technique, called the *scan-in* method, the object is scanned by a thin collimated light such as a laser beam, which illuminates only a small spot at a time. The transmitted light is imaged by a lens onto a photodetector (Fig. 4.4). Certain high-resolution flatbed scanners and rotating drum scanners use this technique for image digitization, display, or recording.

Television Standards

In the United States a standard scanning convention has been adopted by the RETMA†. Each complete scan of the target is called a *frame*, which contains 525

† Radio Electronics Television Manufacturers Association

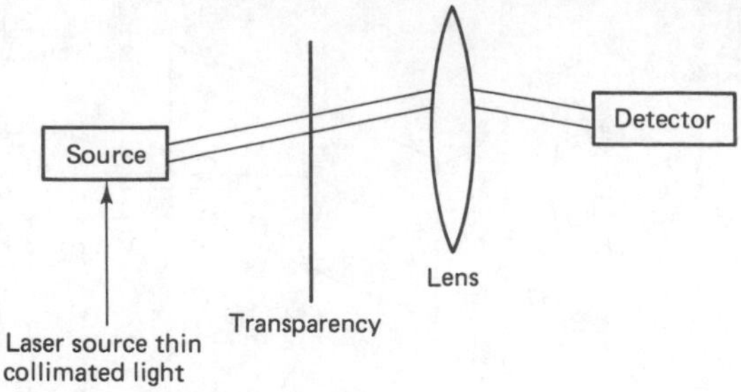

Figure 4.4 Scan-in method. Technique used by some high-resolution scanners.

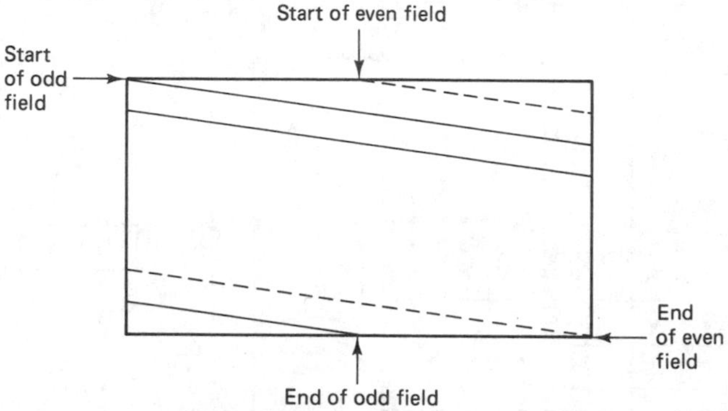

Figure 4.5 Interlaced scanning.

lines and is scanned at a rate of 30 frames per second. Each frame is composed of two interlaced fields, each consisting of 262.5 lines, as shown in Fig. 4.5. To eliminate flicker, alternate fields are sent at a rate of 60 fields per second. The scan lines have a tilt because of the slower vertical scan rate. The first field contains all the odd lines, and the second field contains the even lines. By keeping the field rate rather than the frame rate at 60 Hz, the bandwidth of the transmitted signal is reduced and is found to be about 4.0 MHz. At the end of the first field, the cathode-ray tube (CRT) beam retraces quickly upward to the top center of the target. The beam is biased off during the horizontal and vertical retrace periods so that its zigzag retrace is not visible. In each vertical retrace 21 lines are lost, so there are only 484 active lines per frame.

There are three color television standards, the NTSC, used in North America and Japan; the *Sequential Couleur à Mémoire* (SECAM, or sequential chrominance signal and memory), used in France, Eastern Europe, and the Soviet Union; and the *Phase Alternating Line* (PAL), used in West Germany, The United Kingdom, parts of Europe, South America, parts of Asia, and Africa.[‡]

The NTSC system uses 525 scan lines per frame, 30 frames/s, and two interlaced fields per frame. The color video signal can be written as a *composite signal*

[‡] Some television engineers have been known to refer to these standards as *Never Twice Same Color* (NTSC), *Something Essentially Contradictory to the American Method* (SECAM), and *Peace At Last* (PAL)!

$$u(t) = Y(t) + I(t)\cos(2\pi f_{sc}t + \phi) + Q(t)\sin(2\pi f_{sc}t + \phi) \tag{4.1}$$

where $\phi = 33°$ and f_{sc} is the subcarrier frequency. The quantities Y and (I, Q) are the *luminance and chrominance components*, respectively, which can be obtained by linearly transforming the R, G, and B signals (see Chapter 3). The half-power bandwidths of Y, I, and Q are approximately 4.2 MHz, 1.3 MHz, and 0.5 MHz, respectively. The color subcarrier frequency f_{sc} is 3.58 MHz, which is $455 f_l/2$, where f_l is the scan line frequency (i.e., 15.75 kHz for NTSC). Since f_{sc} is an odd multiple of $f_l/2$ as well as half the frame frequency, $f_r/2$, the phase of the subcarrier will change 180° from line to line and from frame to frame. Taking this into account, the NTSC composite video signal with 2 : 1 line interlace can be represented as

$$\begin{aligned} u(x, y, t) = Y(x, y, t) &+ I(x, y, t)\cos(2\pi f_{sc}x + \phi)\cos[\pi(f_r t - f_l y)] \\ &+ Q(x, y, t)\sin(2\pi f_{sc}x + \phi)\cos[\pi(f_r t - f_l y)] \end{aligned} \tag{4.2}$$

The SECAM system uses 625 lines at 25 frames/s with 2 : 1 line interlacing. Each scan line is composed of the luminance signal $Y(t)$ and one of the chrominance signals $\tilde{U} \triangleq (B - Y)/2.03$ or $\tilde{V} \triangleq (R - Y)/1.14$ alternating from line to line. These chrominances are related to the NTSC coordinates as

$$\begin{aligned} I &= \tilde{V}\cos 33° - \tilde{U}\sin 33° \\ Q &= \tilde{V}\sin 33° + \tilde{U}\cos 33° \end{aligned} \tag{4.3}$$

This avoids the quadrature demodulation and the corresponding chrominance shifts due to phase detection errors present in the NTSC receivers. The $\tilde{U}$ and $\tilde{V}$ subcarriers are at 4.25 and 4.41 MHz. SECAM also transmits a subcarrier for luminance, which increases the complexity of mixers for transmission.

The PAL system also transmits 625 lines at 25 frames/s with 2 : 1 line interlace. The composite signal is

$$u(t) = Y(t) + \tilde{U}\cos 2\pi f_c t + (-1)^m \tilde{V}\sin 2\pi f_c t$$

where m is the line number. Thus the phase of $\tilde{V}$ changes by 180° between successive lines in the same field. The cross talk between adjacent lines can be suppressed by averaging them. The $\tilde{U}$, $\tilde{V}$ are allowed the same bandwidths (1.3 MHz) with the carrier located at 4.43 MHz.

Image Display and Recording

An image display/recording system is conceptually a scanning system operating in the reverse direction. A common method is to scan the image samples, after digital to analog (D to A) conversion, onto a CRT, which displays an array of closely spaced small light spots whose intensities are proportional to the sample magnitudes. The image is viewed through a glass screen. The quality of the image depends on the spot size, both its shape and spacing. Basically, the viewed image should appear to be continuous. The required interpolation between the samples can be provided in a number of ways. One way is to blur the writing spot electrically, thereby creating an overlap between the spots. This requires control over the spot shape. Even then, one is not close to the "optimal solution," which, as we shall see, requires a perfect low-pass filter. In some displays a very small spot size can be achieved so that interpolation can be performed digitally to generate a larger array, which contains estimates of some of the missing samples in between the given samples. This idea is used in bit-mapped computer graphics displays.

The CRT display can be used for recording the image on a film by simply imaging the spot through a lens onto the film (basically the same as imaging with a camera with shutter open for at least one frame period). Other recorders, such as microdensitometers, project a rectangular aperture of size equal to that of the image pixel so that the image field is completely filled.

Another type of display/recorder is called a *halftone display*. Such a display can write only black or white dots. By making the dot size much smaller than the pixel size, white or black dots are dispersed pseudorandomly such that the average number of dots per pixel area is equal to the pixel gray level. Due to spatial integration performed by the eye, such a black and white display renders the perception of a gray-level image. Newspapers, magazines, several printer/plotters, graphic displays, and facsimile machines use the halftone method of display.

4.2 TWO-DIMENSIONAL SAMPLING THEORY

Bandlimited Images

The digitization process for images can be understood by modeling them as bandlimited signals. Although real-world images are rarely bandlimited, they can be approximated arbitrarily closely by bandlimited functions.

A function $f(x, y)$ is called *bandlimited* if its Fourier transform $F(\xi_1, \xi_2)$ is zero outside a bounded region in the frequency plane (Fig. 4.6); for instance,

$$F(\xi_1, \xi_2) = 0, \qquad |\xi_1| > \xi_{x0}, \qquad |\xi_2| > \xi_{y0} \tag{4.4}$$

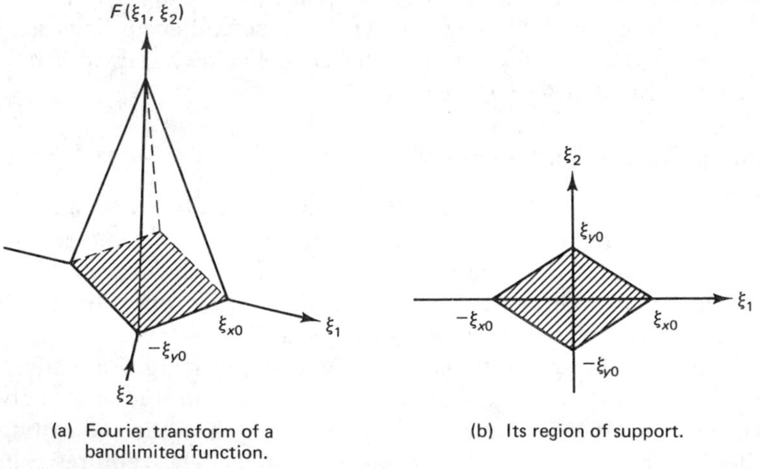

(a) Fourier transform of a bandlimited function.

(b) Its region of support.

Figure 4.6

The quantities ξ_{x0} and ξ_{y0} are called the x and y bandwidths of the image. If the spectrum is circularly symmetric, then the single spatial frequency $\xi_0 \triangleq \xi_{x0} = \xi_{y0}$ is called the bandwidth.

Sampling versus Replication

The sampling theory can be understood easily by remembering the fact that the Fourier transform of an arbitrary sampled function is a scaled, periodic replication of the Fourier transform of the original function. To see this, consider the ideal image sampling function, which is a two-dimensional infinite array of Dirac delta functions situated on a rectangular grid with spacing Δx, Δy (Fig. 4.7a), that is,

$$\text{comb}(x, y; \Delta x, \Delta y) \triangleq \sum_{m,n=-\infty}^{\infty}\sum \delta(x - m\Delta x, y - n\Delta y) \tag{4.5}$$

The sampled image is defined as

$$\begin{aligned} f_s(x, y) &= f(x, y)\,\text{comb}(x, y; \Delta x, \Delta y) \\ &= \sum_{m,n=-\infty}^{\infty}\sum f(m\Delta x, n\Delta y)\delta(x - m\Delta x, y - n\Delta y) \end{aligned} \tag{4.6}$$

The Fourier transform of a comb function with spacing Δx, Δy is another comb function with spacing $(1/\Delta x, 1/\Delta y)$, namely,

$$\begin{aligned} \text{COMB}(\xi_1, \xi_2) &= \mathscr{F}\{\text{comb}(x, y; \Delta x, \Delta y)\} \\ &= \xi_{xs}\xi_{ys}\sum_{k,l=-\infty}^{\infty}\sum \delta(\xi_1 - k\xi_{xs}, \xi_2 - l\xi_{ys}) \\ &= \xi_{xs}\xi_{ys}\,\text{comb}(\xi_1, \xi_2; 1/\Delta x, 1/\Delta y) \end{aligned} \tag{4.7}$$

where $\xi_{xs} \triangleq 1/\Delta x$, $\xi_{ys} \triangleq 1/\Delta y$. Applying the multiplication property of Table 2.3 to (4.6), the Fourier transform of the sampled image $f_s(x, y)$ is given by the convolution

$$\begin{aligned} F_s(\xi_1, \xi_2) &= F(\xi_1, \xi_2) \circledast \text{COMB}(\xi_1, \xi_2) \\ &= \xi_{xs}\xi_{ys}\sum_{k,l=-\infty}^{\infty}\sum F(\xi_1, \xi_2) \circledast \delta(\xi_1 - k\xi_{xs}, \xi_2 - l\xi_{ys}) \\ &= \xi_{xs}\xi_{ys}\sum_{k,l=-\infty}^{\infty}\sum F(\xi_1 - k\xi_{xs}, \xi_2 - l\xi_{ys}) \end{aligned} \tag{4.8}$$

From (4.8) the Fourier transform of the sampled image is, within a scale factor, a periodic replication of the Fourier transform of the input image on a grid whose spacing is (ξ_{xs}, ξ_{ys}) (Fig. 4.7b).

Reconstruction of the Image from Its Samples

From uniqueness of the Fourier tranform, we know that if the spectrum of the original image could be recovered somehow from the spectrum of the sampled image, then we would have the interpolated continuous image from the sampled

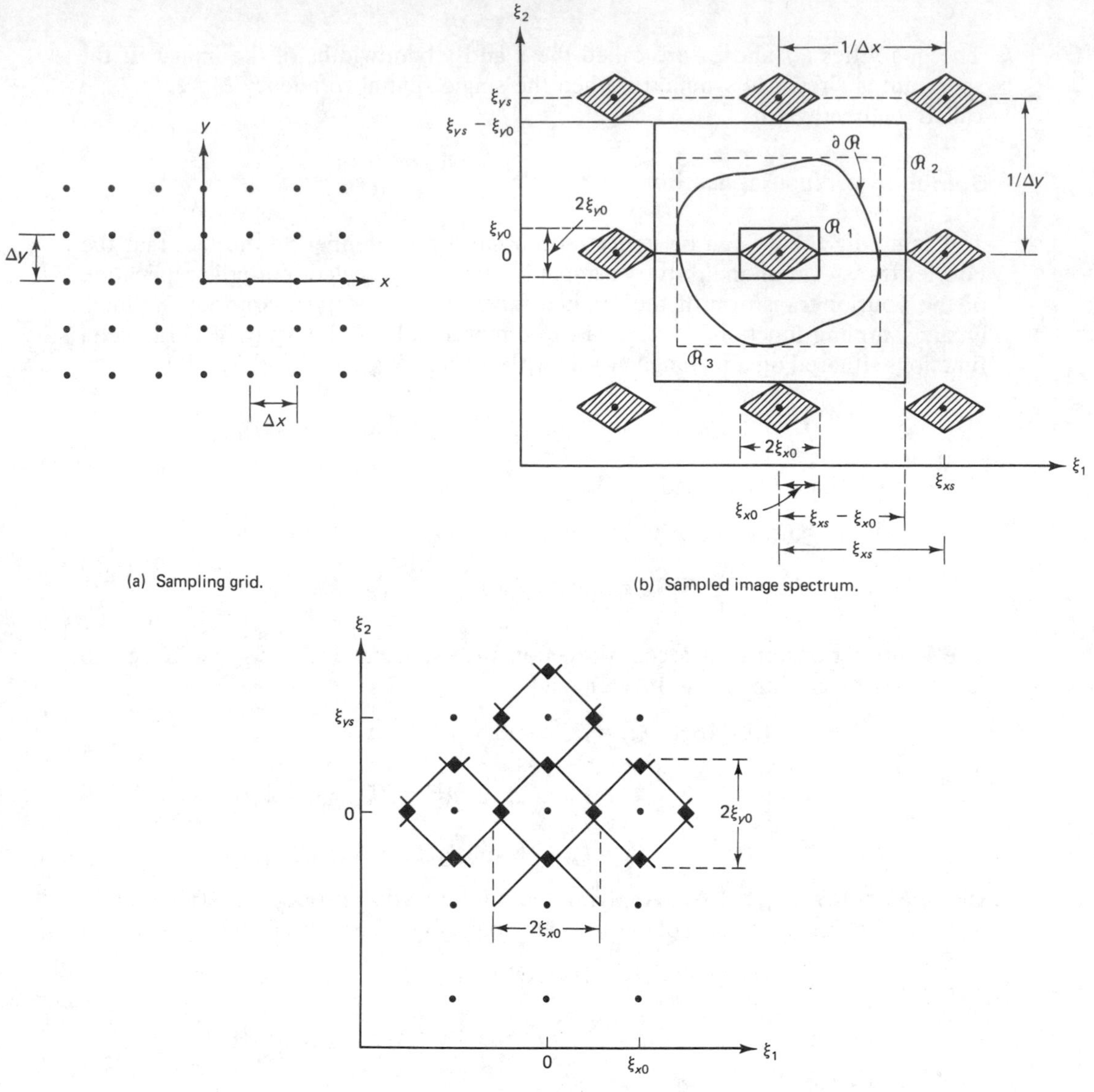

(a) Sampling grid.

(b) Sampled image spectrum.

(c) Aliasing and foldover frequencies (shaded areas).

Figure 4.7 Two-dimensional sampling.

image. If the x, y sampling frequencies are greater than twice the bandwidths, that is,

$$\xi_{xs} > 2\xi_{x0}, \qquad \xi_{ys} > 2\xi_{y0} \tag{4.9}$$

or, equivalently, if the sampling intervals are smaller than one-half of the reciprocal of bandwidths, namely,

$$\Delta x < \frac{1}{2\xi_{x0}}, \qquad \Delta y < \frac{1}{2\xi_{y0}} \tag{4.10}$$

then $F(\xi_1, \xi_2)$ can be recovered by a low-pass filter with frequency response

$$H(\xi_1, \xi_2) = \begin{cases} \dfrac{1}{(\xi_{xs}\xi_{ys})}, & (\xi_1, \xi_2) \in \mathcal{R} \\ 0, & \text{otherwise} \end{cases} \tag{4.11}$$

where $\mathcal{R}$ is any region whose boundary $\partial\mathcal{R}$ is contained within the annular ring between the rectangles $\mathcal{R}_1$ and $\mathcal{R}_2$ shown in Fig. 4.7b. This is seen by writing

$$\tilde{F}(\xi_1, \xi_2) \overset{\Delta}{=} H(\xi_1, \xi_2)F_s(\xi_1, \xi_2) = F(\xi_1, \xi_2) \tag{4.12}$$

that is, the original continuous image can be recovered exactly by low-pass filtering the sampled image.

Nyquist Rate, Aliasing, and Foldover Frequencies

The lower bounds on the sampling rates, that is, $2\xi_{x0}, 2\xi_{y0}$ in (4.9), are called the *Nyquist rates* or the *Nyquist frequencies*. Their reciprocals are called the *Nyquist intervals*. The sampling theory states that a bandlimited image sampled above its x and y Nyquist rates can be recovered without error by low-pass filtering the sampled image. However, if the sampling frequencies are below the Nyquist frequencies, that is, if

$$\xi_{xs} < 2\xi_{x0}, \qquad \xi_{ys} < 2\xi_{y0}$$

then the periodic replications of $F(\xi_1, \xi_2)$ will overlap (Fig. 4.7c), resulting in a distorted spectrum $F_s(\xi_1, \xi_2)$, from which $F(\xi_1, \xi_2)$ is irrevocably lost. The frequencies above half the sampling frequencies, that is, above $\xi_{xs}/2, \xi_{ys}/2$, are called the *foldover frequencies*. This overlapping of successive periods of the spectrum causes the foldover frequencies in the original image to appear as frequencies below $\xi_{xs}/2, \xi_{ys}/2$ in the sampled image. This phenomenon is called *aliasing*. Aliasing errors cannot be removed by subsequent filtering. Aliasing can be avoided by low-pass filtering the image *first* so that its bandwidth is less than one-half of the sampling frequency, that is, when (4.9) is satisfied.

Figures 4.8a and 4.8b show an image sampled above and below its Nyquist rate. Aliasing is visible near the high frequencies (about one-third distance from the center). Aliasing effects become invisible when the original image is low-pass filtered before subsampling (Figs. 4.8c and 4.8d).

If the region of support of the ideal low-pass filter in (4.11) is the rectangle

$$\mathcal{R} = [-\tfrac{1}{2}\xi_{xs}, \tfrac{1}{2}\xi_{xs}] \times [-\tfrac{1}{2}\xi_{ys}, \tfrac{1}{2}\xi_{ys}] \tag{4.13}$$

centered at the origin, then its impulse response is

$$h(x, y) = \text{sinc}(x\xi_{xs})\text{sinc}(y\xi_{ys}) \tag{4.14}$$

Inverse Fourier transforming (4.12) and using (4.14) and (4.6), the reconstructed image is obtained as

$$\tilde{f}(x, y) = \sum\sum_{m, n = -\infty}^{\infty} f(m\Delta x, n\Delta y)\text{sinc}(x\xi_{xs} - m)\text{sinc}(y\xi_{ys} - n) \tag{4.15}$$

(a) Sampled above Nyquist rate and reconstructed by ZOH;

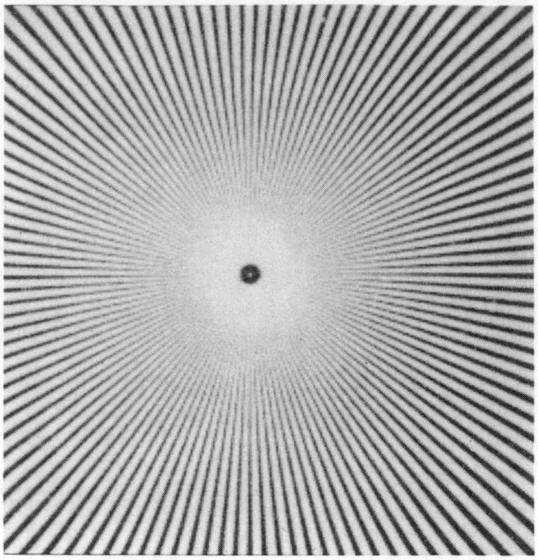

(b) sampled below Nyquist rate and reconstructed by ZOH;

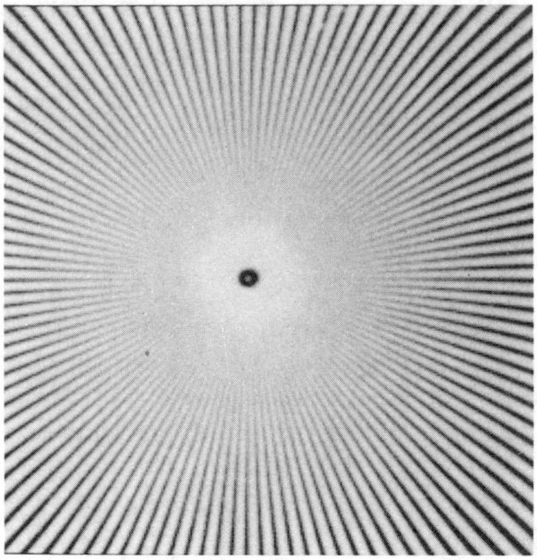

(c) lowpass filtered before subsampling and reconstructed by ZOH;

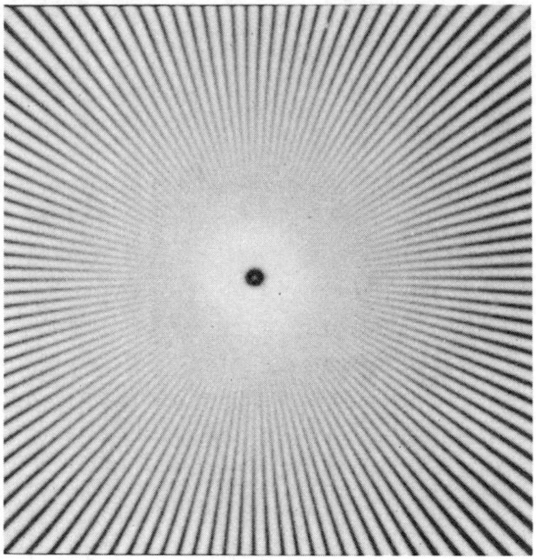

(d) lowpass filtered before subsampling and reconstructed by FOH.

Figure 4.8 Image sampling, aliasing, and reconstruction.

which is equal to $f(x, y)$ if Δx, Δy satisfy (4.10). We can summarize the preceding results by the following theorem.

Sampling Theorem

A bandlimited image $f(x, y)$ satisfying (4.4) and sampled uniformly on a rectangular grid with spacing Δx, Δy can be recovered without error from the sample values $f(m\Delta x, n\Delta y)$ provided the sampling rate is greater than the Nyquist rate, that is,

$$\frac{1}{\Delta x} = \xi_{xs} > 2\xi_{x0}, \qquad \frac{1}{\Delta y} = \xi_{ys} > 2\xi_{y0}$$

Moreover, the reconstructed image is given by the interpolation formula

$$f(x, y) = \sum\sum_{m,n=-\infty}^{\infty} f(m\Delta x, n\Delta y)\left(\frac{\sin(x\xi_{xs} - m)\pi}{(x\xi_{xs} - m)\pi}\right)\left(\frac{\sin(y\xi_{ys} - n)\pi}{(y\xi_{ys} - n)\pi}\right) \tag{4.16}$$

Remarks

1. Equation (4.16) shows that infinite order interpolation is required to reconstruct the continuous function $f(x, y)$ from its samples $f(m\Delta x, n\Delta y)$. In practice only finite-order interpolation is possible. However, sampling theory reduces the uncountably infinite number of samples of $f(x, y)$ over the area $\Delta x \Delta y$ to just one sample. This gives an infinite compression ratio per unit area.
2. The *aliasing energy* is the energy in the foldover frequencies and is equal to the energy of the image in the tails of its spectrum outside the rectangle $\mathcal{R}$ defined in (4.13).

Example 4.1

An image described by the function

$$f(x, y) = 2\cos 2\pi(3x + 4y)$$

is sampled such that $\Delta x = \Delta y = 0.2$. Clearly $f(x, y)$ is bandlimited, since

$$F(\xi_1, \xi_2) = \delta(\xi_1 - 3, \xi_2 - 4) + \delta(\xi_1 + 3, \xi_2 + 4)$$

is zero for $|\xi_1| > 3, |\xi_2| > 4$. Hence $\xi_{x0} = 3, \xi_{y0} = 4$. Also $\xi_{xs} = \xi_{ys} = 1/0.2 = 5$, which is less than the Nyquist frequencies $2\xi_{x0}$ and $2\xi_{y0}$. The sampled image spectrum is

$$F_s(\xi_1, \xi_2) = 25 \sum\sum_{k,l=-\infty}^{\infty} [\delta(\xi_1 - 3 - 5k, \xi_2 - 4 - 5l) + \delta(\xi_1 + 3 - 5k, \xi_2 + 4 - 5l)]$$

Let the low-pass filter have a rectangular region of support with cutoff frequencies at half the sampling frequencies, that is,

$$H(\xi_1, \xi_2) = \begin{cases} \frac{1}{25}, & -2.5 \le \xi_1 \le 2.5, \ -2.5 \le \xi_2 \le 2.5 \\ 0, & \text{otherwise} \end{cases}$$

Applying (4.12), we obtain

$$\tilde{F}(\xi_1, \xi_2) = \delta(\xi_1 - 2, \xi_2 - 1) + \delta(\xi_1 + 2, \xi_2 + 1)$$

which gives the reconstructed image as $\tilde{f}(x, y) = 2\cos 2\pi(2x + y)$. This shows that any frequency component in the input image that is above $(\xi_{xs}/2, \xi_{ys}/2)$ by $(\Delta\xi_x, \Delta\xi_y)$ is reproduced (or aliased) as a frequency component at $(\xi_{xs}/2 - \Delta\xi_x, \xi_{ys}/2 - \Delta\xi_y)$.

4.3 EXTENSIONS OF SAMPLING THEORY

There are several extensions of the two-dimensional sampling theory that are of interest in image processing.

Sampling Random Fields

In physical sampling environments, random noise is always present in the image, so it is important to consider sampling theory for random fields. A continuous stationary random field $f(x, y)$ is called bandlimited if its power spectral density function $S(\xi_1, \xi_2)$ is bandlimited, that is, if

$$S(\xi_1, \xi_2) = 0 \qquad \text{for } |\xi_1| > \xi_{x0}, |\xi_2| > \xi_{y0} \tag{4.17}$$

Sampling Theorem for Random Fields

If $f(x, y)$ is a stationary bandlimited random field, then

$$\tilde{f}(x, y) \triangleq \sum_{m, n = -\infty}^{\infty}\sum f(m\Delta x, n\Delta y)\operatorname{sinc}(x\xi_{xs} - m)\operatorname{sinc}(y\xi_{ys} - n) \tag{4.18}$$

converges to $f(x, y)$ in the mean square sense, that is,

$$E(|f(x, y) - \tilde{f}(x, y)|^2) = 0 \tag{4.19}$$

where $\xi_{xs} = 1/\Delta x,\ \xi_{ys} = 1/\Delta y,\ \xi_{xs} > 2\xi_{x0}, \xi_{ys} > 2\xi_{y0}$.

Remarks

This theorem states that if the random field $f(x, y)$ is sampled above its Nyquist rate, then a continuous random field $\tilde{f}(x, y)$ can be reconstructed from the sampled sequence such that $\tilde{f}$ converges to f in the mean square sense. It can be shown that the power spectral density function $S_s(\xi_1, \xi_2)$ of the sampled image $f_s(x, y)$ is a periodic extension of $S(\xi_1, \xi_2)$ and is given by

$$S_s(\xi_1, \xi_2) = \xi_{xs}\xi_{ys} \sum_{k, l = -\infty}^{\infty}\sum S(\xi_1 - k\xi_{xs}, \xi_2 - l\xi_{ys}) \tag{4.20}$$

When the image is reconstructed by an ideal low-pass filter with gain $1/(\xi_{xs}\xi_{ys})$, the reconstructed image power spectral density is given by

$$\begin{aligned}\tilde{S}(\xi_1, \xi_2) &= \left(\sum_{k, l = -\infty}^{\infty}\sum S(\xi_1 - k\xi_{xs}, \xi_2 - l\xi_{ys})\right) W(\xi_1, \xi_2) \\ &= S(\xi_1, \xi_2)\end{aligned} \tag{4.21}$$

where

$$W(\xi_1, \xi_2) = \begin{cases} 1, & (\xi_1, \xi_2) \in \mathcal{R} \\ 0, & \text{otherwise} \end{cases} \tag{4.22}$$

The aliasing power σ_a^2 is the power in the tails of the power spectrum outside $\mathcal{R}$, that is,

$$\sigma_a^2 = \iint_{\xi_1, \xi_2 \notin \mathcal{R}} S(\xi_1, \xi_2)\,d\xi_1\,d\xi_2 = \iint_{-\infty}^{\infty} [1 - W(\xi_1, \xi_2)]S(\xi_1, \xi_2)\,d\xi_1\,d\xi_2 \tag{4.23}$$

which is zero if $f(x, y)$ is bandlimited with $\xi_{x0} \le \xi_{xs}/2, \xi_{y0} \le \xi_{ys}/2$. This analysis is also useful when a bandlimited image containing wideband noise is sampled. Then the signal-to-noise ratio of the sampled image can deteriorate significantly unless it is low-pass filtered before sampling (see Problem 4.6).

Nonrectangular Grid Sampling and Interlacing

All of our previous discussion and most of the literature on two-dimensional sampling is devoted to rectangular sampling lattices. This is the desirable form of sampling grid if the spectrum $F(\xi_1, \xi_2)$ is limited over the rectangle $\mathscr{R}$ of (4.13). Other sampling grids may be more efficient in terms of sampling density (that is, samples/area) if the region of support of $F(\xi_1, \xi_2)$ is nonrectangular.

Consider, for example, the spectrum shown in Fig. 4.9, which can be tightly enclosed by a diamond-shaped region. On a rectangular sampling grid G_1, the

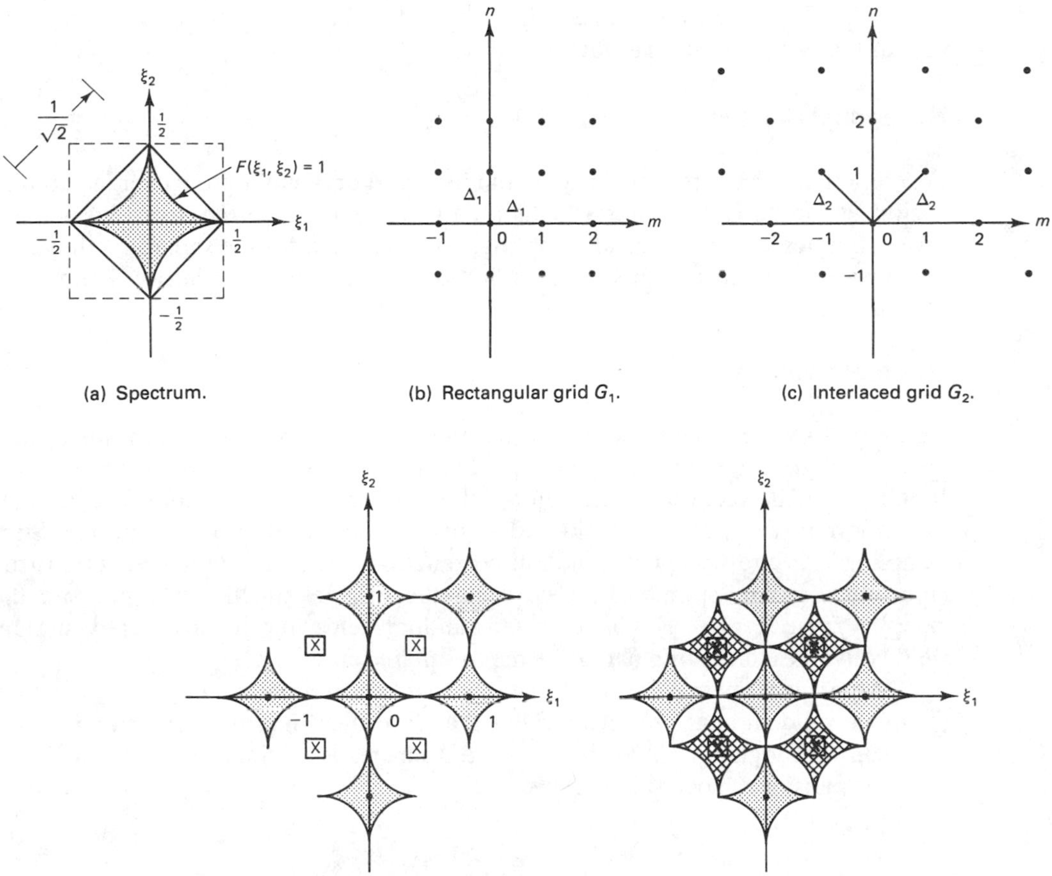

(a) Spectrum. (b) Rectangular grid G_1. (c) Interlaced grid G_2.

(d) Spectrum when sampled by G_1 (noninterlaced frames). (e) Spectrum of interlaced signal when sampled by G_2. Hatched area shows the new inserts.

Figure 4.9 Interlaced sampling.

Nyquist sampling intervals would be $\Delta x = \Delta y \triangleq \Delta_1 = 1$. If the sampling grid G_2 is chosen, which is a 45° rotation of G_1 but with intersample distance of Δ_2, the spectrum of the sampled image will repeat on a grid similar to G_2 (with spacing $1/\Delta_2$) (Fig. 4.9e). Therefore, if $\Delta_2 = \sqrt{2}$, there will be no aliasing, but the sampling density has been reduced by half. Thus if an image does not contain the high frequencies in both the dimensions simultaneously, then its sampling rate can be reduced by a factor of 2. This theory is used in line interlacing television signals because the human vision is insensitive to high spatial frequencies in areas of large motion (high temporal frequencies). The interlaced television signal can be considered as a three-dimensional signal $f(x, y, t)$ sampled in vertical (y) and temporal (t) dimensions. If ξ_1 and ξ_2 represent the temporal and vertical frequencies, respectively, then Fig. 4.9e represents the projection in (ξ_1, ξ_2) plane of the three-dimensional spectrum of the interlaced television signal.

In digital television, all three coordinates x, y, and t are sampled. The preceding interlacing concept can be extended to yield the *line quincunx* sampling pattern [10]. Here each field uses an interlaced grid as in Fig. 4.9c, which reduces the sampling rate by another factor of two.

Hexagonal Sampling

For functions that are circularly symmetric and/or bandlimited over a circular region, it can be shown that sampling on a hexagonal lattice requires 13.4 percent fewer samples than rectangular sampling. Alternatively, for the same sampling rate less aliasing is obtained on a hexagonal lattice than a rectangular lattice. Details are available in [14].

Optimal Sampling

Equation (4.16) provides the interpretation that the sampling process transforms a *continuous function* $f(x, y)$ into a *sequence* $f(m\Delta x, n\Delta y)$ from which the original function can be recovered. Therefore, the coefficients of any convergent series expansion of $f(x, y)$ can be considered to give a generalized form of sampling. Such sampling is not restricted to bandlimited functions. For bandlimited functions the sinc functions are optimal for recovering the original function $f(x, y)$ from the samples $f(m\Delta x, n\Delta y)$. For bandlimited random fields, the reconstructed random field converges to the original in the mean square sense.

More generally, there are functions that are optimal in the sense that they sample a random image to give a finite sequence such that the mean square error between the original and the reconstructed images is minimized. In particular, a series expansion of special interest is

$$f(x, y) = \sum_{m,n=0}^{\infty}\sum a_{m,n}\,\phi_{m,n}(x, y) \tag{4.24}$$

where $\{\phi_{m,n}(x, y)\}$ are the eigenfunctions of the autocorrelation function of the random field $f(x, y)$. This is called the *Karhunen-Loève (KL) series expansion* of the random field. This expansion is such that $a_{m,n}$ are orthogonal random variables, and, for a given number of terms, the mean square error in the reconstructed image is minimum among all possible sampling functions. This property is useful in developing data compression techniques for images.

The main difficulty in utilizing the preceding result for optimal sampling of practical (finite size) images is in generating the coefficients $a_{m,n}$. In conventional sampling (via the sinc functions), the coefficients $a_{m,n}$ are simply the values $f(m\Delta x, n\Delta y)$, which are easy to obtain. Nevertheless, the theory of KL expansion is useful in determining bounds on performance and serves as an important guide in the design of many image processing algorithms.

4.4 PRACTICAL LIMITATIONS IN SAMPLING AND RECONSTRUCTION

The foregoing sampling theory is based on several idealizations. Real-world images are not bandlimited, which means aliasing errors occur. These can be reduced by low-pass filtering the input image prior to sampling but at the cost of attenuating higher spatial frequencies. Such resolution loss, which results in blurring of the image, also occurs because practical scanners have finite apertures. Finally, the reconstruction system can never be the ideal low-pass filter required by the sampling theory. Its transfer function depends on the display aperture. Figure 4.10 represents the practical sampling/reconstruction systems.

Sampling Aperture

A practical sampling system gives an output $g_s(x, y)$, which can be modeled as (see Fig. 4.10)

$$\begin{aligned} g(x, y) &\triangleq p_s(x, y) \star f(x, y) = p_s(-x, -y) \circledast f(x, y) \\ &= \iint_{\mathcal{A}} p_s(x' - x, y' - y) f(x', y')\, dx'\, dy' \end{aligned} \tag{4.25}$$

$$g_s(x, y) = \text{comb}(x, y; \Delta x, \Delta y) g(x, y) \tag{4.26}$$

where $p_s(x, y)$ denotes the light distribution in the aperture and $\mathcal{A}$ denotes its shape. In practice the aperture is symmetric with respect to 180° rotation, that is, $p_s(x, y) = p_s(-x, -y)$. Equation (4.25) is the spatial correlation of $f(x, y)$ with

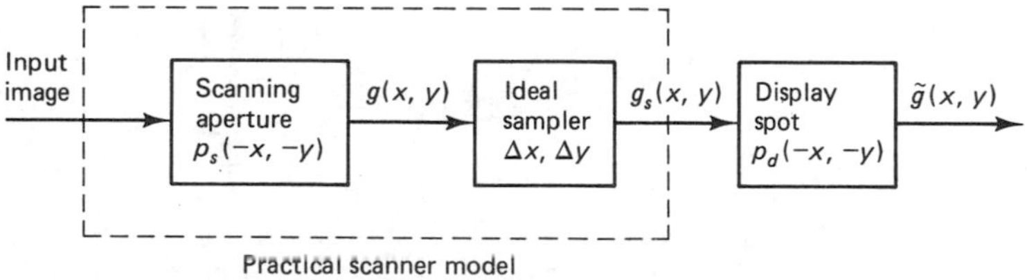

Figure 4.10 Practical sampling and reconstruction. In the ideal case $p_s(x, y) = p_d(x, y) = \delta(x, y)$.

$p_s(x, y)$ and represents the process of scanning through the aperture. Equation (4.26) represents the sampled output. For example, for an $L \times L$ square aperture with uniform distribution, we will have

$$g(x, y) = \int_{x-L/2}^{x+L/2} \int_{y-L/2}^{y+L/2} f(x', y')\, dx'\, dy' \tag{4.27}$$

which is simply the integral of the image over the scanner aperture at position (x, y). In general (4.25) represents a low-pass filtering operation whose transfer function is determined by the aperture function $p_s(x, y)$. The overall effect on the reconstructed image is a loss of resolution and a decrease in aliasing error (Fig. 4.11). This effect is also visible in the images of Fig. 4.12.

Display Aperture/Interpolation Function

Perfect image reconstruction requires an infinite-order interpolation between the samples $f(m\Delta x, n\Delta y)$. For a display system this means its display spot should have a light distribution given by the sinc function, which has infinite duration and negative lobes. This makes it impossible for an incoherent imaging system to perform near perfect interpolation.

Figure 4.13 lists several functions useful for interpolation. Two-dimensional interpolation can be performed by successive interpolation along rows and columns of the image. The zero-order- and first-order-hold filters give piecewise constant and linear interpolations, respectively, between the samples. Higher-order holds can give quadratic ($n = 2$) and cubic spline ($n = 3$) interpolations. With proper coordinate scaling of the interpolating function, the nth-order hold converges to the Gaussian function as $n \to \infty$. The display spot of a CRT is circular and can be modeled by a Gaussian function whose variance controls its spread. Figure 4.14 shows the effect of a practical interpolator on the reconstructed image. The resolution loss due to the reconstruction filter depends on the width of the main lobe. Since $|\text{sinc}(x)|^n < 1$ for every x, the main lobe of the nth-order-hold filter spectrum

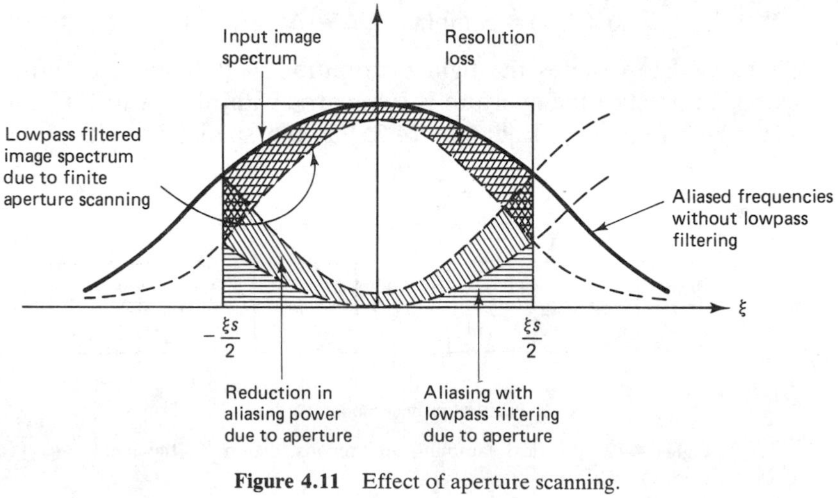

Figure 4.11 Effect of aperture scanning.

Figure 4.12 Comparison between zero- and first-order hold interpolators. Zero-order hold gives higher resolution and first-order hold gives greater smoothing.

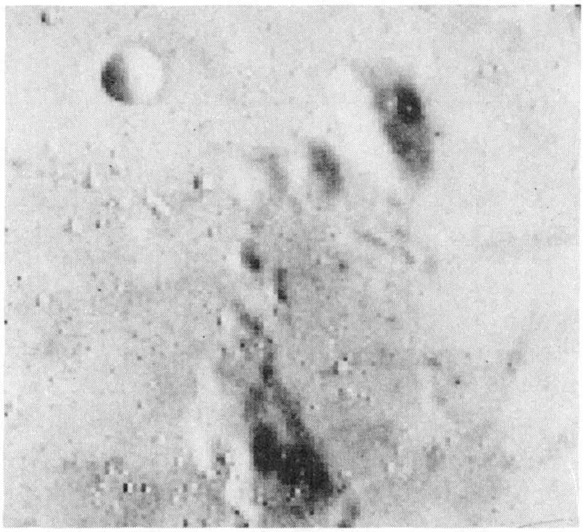

(a) 256 × 256 image interpolated to 512 × 512 by zero-order hold (ZOH).

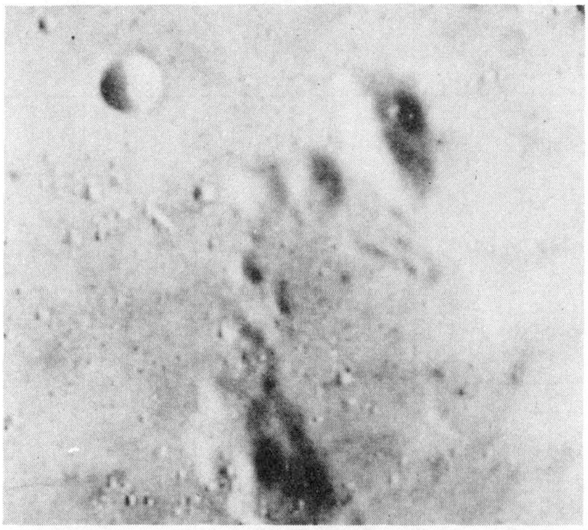

(b) 256 × 256 image interpolated to 512 × 512 by first-order hold (FOH).

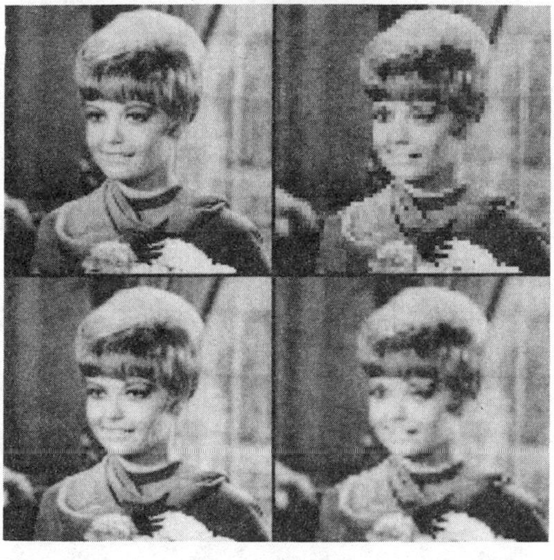

(c) 256 × 256 images after interpolation

(i)	(ii)
(iii)	(iv)

. (i) 128 × 128 image zoomed by ZOH; (ii) 64 × 64 image zoomed by ZOH; (iii) 128 × 128 image zoomed by FOH; (iv) 64 × 64 image zoomed by FOH.

One-dimensional interpolation function	Diagram	Definition $p(x)$	Two-dimensional interpolation function $p_d(x, y) = p(x)p(y)$	Frequency response $P_d(\xi_1, \xi_2)$	$P_d(\xi_1, 0)$
Rectangle (zero-order hold) ZOH $p_o(x)$	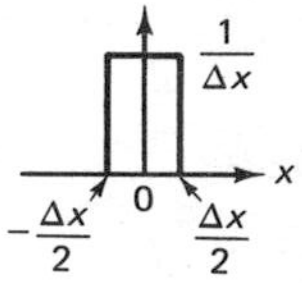	$\frac{1}{\Delta x} \text{rect}\left(\frac{x}{\Delta x}\right)$	$p_o(x)p_o(y)$	$\text{sinc}\left(\frac{\xi_1}{2\xi_{x0}}\right)\text{sinc}\left(\frac{\xi_2}{2\xi_{y0}}\right)$	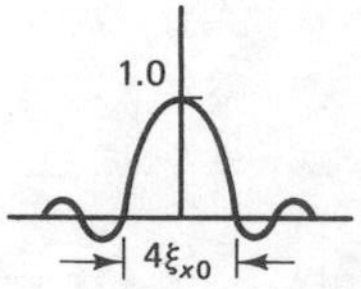
Triangle (first-order hold) FOH $p_1(x)$	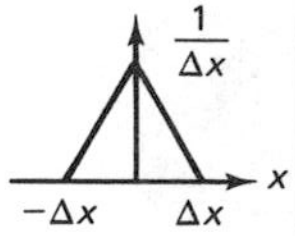	$\frac{1}{\Delta x} \text{tri}\left(\frac{x}{\Delta x}\right)$ $p_o(x) \circledast p_o(x)$	$p_1(x)p_1(y)$	$\left[\text{sinc}\left(\frac{\xi_1}{2\xi_{x0}}\right)\text{sinc}\left(\frac{\xi_2}{2\xi_{y0}}\right)\right]^2$	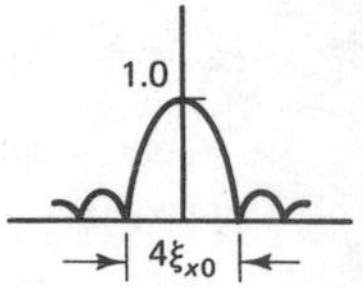
nth-order hold $n = 2$, quadratic $n = 3$, cubic splines $p_n(x)$	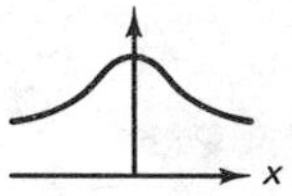	$p_o(x) \circledast \cdots \circledast p_o(x)$ n convolutions	$p_n(x)p_n(y)$	$\left[\text{sinc}\left(\frac{\xi_1}{\xi_{x0}}\right)\text{sinc}\left(\frac{\xi_2}{\xi_{y0}}\right)\right]^{n+1}$	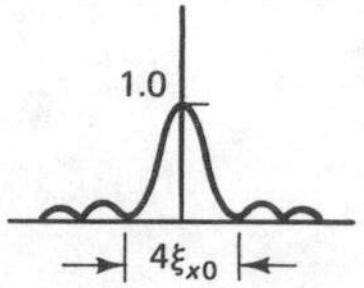
Gaussian $p_g(x)$	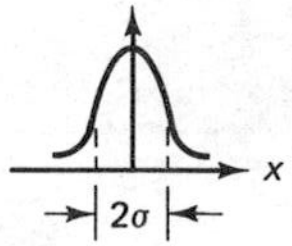	$\frac{1}{\sqrt{2\pi\sigma^2}} \exp\left[-\frac{x^2}{2\sigma^2}\right]$	$\frac{1}{2\pi\sigma^2} \exp\left[-\frac{(x^2 + y^2)}{2\sigma^2}\right]$	$\exp\left[-2\pi^2\sigma^2(\xi_1^2 + \xi_2^2)\right]$	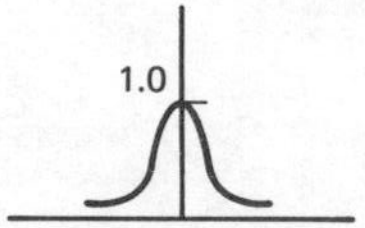
Sinc	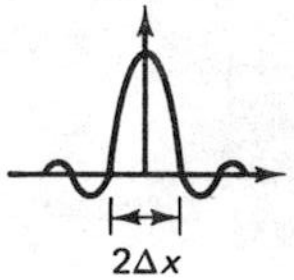	$\frac{1}{\Delta x} \text{sinc}\left(\frac{x}{\Delta x}\right)$	$\frac{1}{\Delta x \Delta y} \text{sinc}\left(\frac{x}{\Delta x}\right)\text{sinc}\left(\frac{x}{\Delta y}\right)$	$\text{rect}\left(\frac{\xi_1}{2\xi_{x0}}\right)\text{rect}\left(\frac{\xi_2}{2\xi_{y0}}\right)$	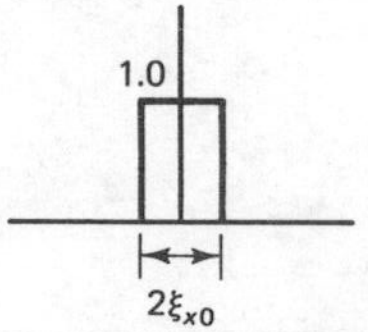

Figure 4.13 Image interpolation functions; $\xi_{x0} \stackrel{\Delta}{=} \frac{1}{2\Delta x}$, $\xi_{y0} \stackrel{\Delta}{=} \frac{1}{2\Delta y}$.

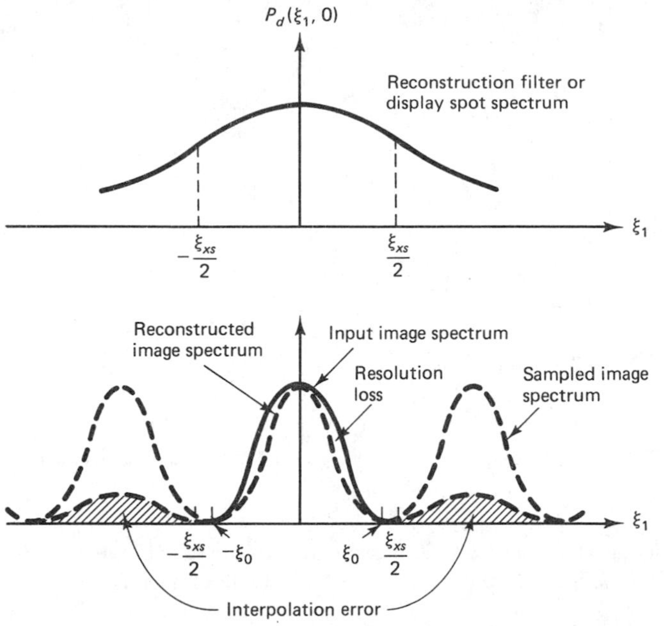

Figure 4.14 Effect of practical interpolation.

will become narrower as n increases. Therefore, among *the* n*th-order-hold interpolators of Figure 4.13, the zero-order-hold function will have minimum resolution loss and maximum interpolation error.* This effect is visible in Fig. 4.12, which contains images interpolated by zero- and first-order-hold functions. In practice bilinear interpolation (first-order hold) gives a reasonable trade-off between resolution loss and smoothing accuracy.

Example 4.2

A CCD camera contains a 256×256 array of identical photodetectors of size $a \times a$ with spacing $\Delta x = \Delta y = a \leq \Delta$ (see Fig. 4.3). The scanning electronics produces output pulses proportional to the response of each detector. The spatial response of a detector to a unit intensity impulse input at location (x, y) is $p_s(x, y) = p(x)p(y)$, where

$$p(x) = \begin{cases} \frac{2}{a}\left(1 - \frac{2|x|}{a}\right), & |x| \leq \frac{a}{2} \\ 0, & \text{otherwise} \end{cases}$$

Suppose an image $f(x, y) \overset{\Delta}{=} 2 \cos 2\pi(x/4a + y/8a)$, with $a = \Delta$, is scanned. Using (4.25) and taking the Fourier transform, we obtain

$$G(\xi_1, \xi_2) = \text{sinc}^2\left(\frac{a\xi_1}{2}\right)\text{sinc}^2\left(\frac{a\xi_2}{2}\right)F(\xi_1, \xi_2)$$

$$= \text{sinc}^2(\tfrac{1}{8})\text{sinc}^2(\tfrac{1}{16})F(\xi_1, \xi_2) \cong 0.94F(\xi_1, \xi_2)$$

where we have used $F(\xi_1, \xi_2) = \delta(\xi_1 - 1/4a, \ \xi_2 - 1/8a) + \delta(\xi_1 + 1/4a, \ \xi_2 + 1/8a)$. The scanner output signal can be written as

$$g_s(x, y) = g(x, y)w(x, y) \sum_{m, n = -\infty}^{\infty}\sum \delta(x - m\Delta, y - n\Delta)$$

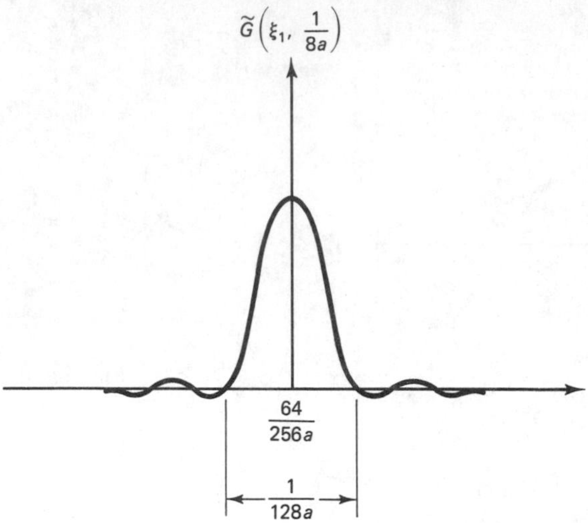

Figure 4.15 Array scanner frequency response.

where $w(x, y)$ is a rectangular window $[-L/2, L/2]$, which limits the field of view of the camera. With spacing Δ and $L = 256\Delta$, the spectrum of the scanned image is

$$G_s(\xi_1, \xi_2) = \xi_s^2 \sum_{m,n=-\infty}^{\infty}\sum \tilde{G}(\xi_1 - m\xi_s, \xi_2 - n\xi_s), \; \xi_s = 1/\Delta$$

where $\tilde{G}(\xi_1, \xi_2) \triangleq G(\xi_1, \xi_2) \circledast W(\xi_1, \xi_2)$, $W(\xi_1, \xi_2) \triangleq L^2 \operatorname{sinc}(\xi_1 L)\operatorname{sinc}(\xi_2 L)$. This gives

$$\tilde{G}(\xi_1, \xi_2) = 61{,}440a^2[\operatorname{sinc}(256a\xi_1 - 64)\operatorname{sinc}(256a\xi_2 - 32) + \operatorname{sinc}(256a\xi_1 + 64)\operatorname{sinc}(256a\xi_2 + 32)]$$

Figure 4.15 shows $\tilde{G}(\xi_1, \xi_2)$ at $\xi_2 = \frac{1}{8}a$, for $\xi_1 > 0$. Thus, instead of obtaining a delta function at $\xi_1 = \frac{1}{4}a$, a sinc function with main-lobe width of $\frac{1}{128}a$ is obtained. This degradation of $G(\xi_1, \xi_2)$ due to convolution with $W(\xi_1, \xi_2)$ is called *ripple*. The associated energy (in the frequency domain) leaked into the side lobes of the sinc functions due to this convolution is known as *leakage*.

Lagrange Interpolation

The zero- and first-order holds also belong to a class of polynomial interpolation functions called *Lagrange polynomials*. The Lagrange polynomial of order $(q - 1)$ is defined as

$$L_k^q(x) \triangleq \prod_{\substack{m=k_0 \\ m \neq k}}^{k_1} \left(\frac{x - m}{k - m}\right), \qquad k_0 \leq k \leq k_1, \quad q = 2, 3$$
$$L_k^1(x) \triangleq 1, \qquad \forall k \tag{4.28}$$

where $k_0 \triangleq -(q-1)/2$, $k_1 = (q-1)/2$ for q odd and $k_0 \triangleq -(q-2)/2$, $k_1 \triangleq q/2$ for q even. For a one-dimensional sampled sequence $f(m\Delta)$, with sampling interval Δ, the interpolated function between given samples is defined as

$$\hat{f}(x) = \hat{f}(m\Delta + \alpha\Delta) \triangleq \sum_{k=k_0}^{k_1} L_k^q(\alpha) f(m\Delta + k\Delta) \tag{4.29}$$

where $-\frac{1}{2} \le \alpha < \frac{1}{2}$ for q odd and $0 \le \alpha < 1$ for q even. This formula uses q samples, $f(\overline{m + k_0}\Delta), \ldots, f(\overline{m + k_1}\Delta)$, to interpolate at any location $(m + \alpha)\Delta$ between two samples. For $q = 1, 2, 3$, we get the formulas

$$q = 1 \Rightarrow \hat{f}(m\Delta + \alpha\Delta) = f(m\Delta), \qquad -\tfrac{1}{2} \le \alpha < \tfrac{1}{2}; \qquad \text{(zero-order hold)}$$

$$q = 2 \Rightarrow \hat{f}(m\Delta + \alpha\Delta) = (1 - \alpha)f(m\Delta) + \alpha f(\overline{m + 1}\Delta); \; 0 \le \alpha < 1; \; \text{(first-order hold)} \tag{4.30}$$

$$q = 3 \Rightarrow \hat{f}(m\Delta + \alpha\Delta) = (1 - \alpha)(1 + \alpha)f(m\Delta) + \frac{\alpha(1 + \alpha)}{2} f(\overline{m + 1}\Delta) - \frac{\alpha(1 - \alpha)}{2} f(\overline{m - 1}\Delta), \; -\tfrac{1}{2} \le \alpha < \tfrac{1}{2}$$

The Lagrange interpolation formula of (4.29) is useful because it converges to the sinc function interpolation as $q \to \infty$ [Problem 4.10]. In two dimensions the Lagrange interpolation formula becomes

$$\hat{f}(x, y) = \hat{f}(m\Delta x + \alpha\Delta x, n\Delta y + \beta\Delta y) \triangleq \sum_{k=k_0}^{k_1} \sum_{l=l_0}^{l_1} L_k^{q_1}(\alpha) L_l^{q_2}(\beta) f(\overline{m + k}\Delta x, \overline{n + l}\Delta y) \tag{4.31}$$

where q_1 and q_2 refer to the Lagrange polynomial orders in the x and y directions, respectively.

Moiré Effect and Flat Field Response [4, 40]

Another phenomenon that results from practical interpolation filters is called the *Moiré effect.* It appears in the form of beat patterns that arise if the image contains periodicities that are close to half the sampling frequencies. This effect occurs when the display spot size is small (compared to sampling distance) so that the reconstruction filter cutoff extends far beyond the ideal low-pass filter cutoff. Then a signal at frequency $\xi_x < \xi_{xs}/2$ will interfere with a companion signal at $\xi_{xs} - \xi_x$ to create a beat pattern, or the Moiré effect (see Problems 4.11 and 4.12). A special case of this situation occurs when the input image is a uniform gray field. Then, if the reconstruction filter does not have zero response at the sampling frequencies (ξ_{xs}, ξ_{ys}), scan lines will appear, and the displayed image will exhibit stripes and not a *flat field.*

4.5 IMAGE QUANTIZATION

The step subsequent to sampling in image digitization is quantization. A quantizer maps a continuous variable u into a discrete variable $u^{\cdot}$, which takes values from a finite set $\{r_1, \ldots, r_L\}$ of numbers. This mapping is generally a staircase function (Fig. 4.16) and the quantization rule is as follows: Define $\{t_k, k = 1, \ldots, L + 1\}$ as a set of increasing *transition* or *decision levels* with t_1 and t_{L+1} as the minimum and maximum values, respectively, of u. If u lies in interval $[t_k, t_{k+1})$, then it is mapped to r_k, the kth *reconstruction level.*

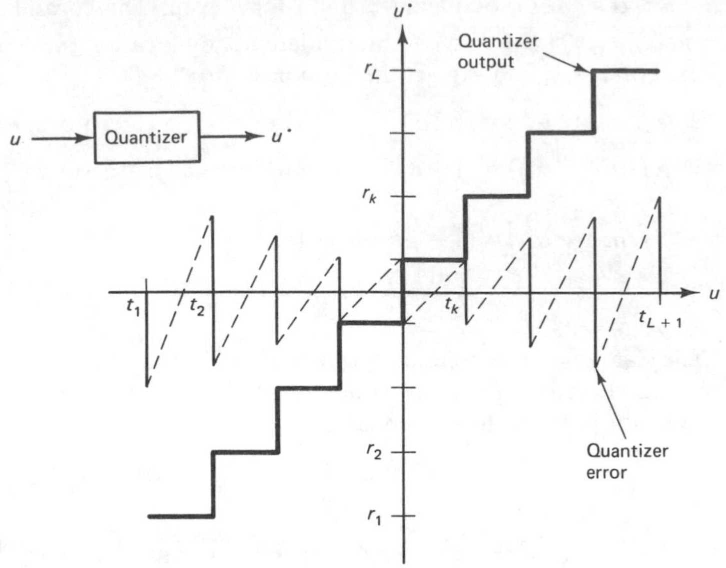

Figure 4.16 A quantizer.

Example 4.3

The simplest and most common quantizer is the uniform quantizer. Let the output of an image sensor take values between 0.0 to 10.0. If the samples are quantized uniformly to 256 levels, then the transition and reconstruction levels are

$$t_k = \frac{10(k-1)}{256}, \qquad k = 1, \ldots, 257$$

$$r_k = t_k + \frac{5}{256}, \qquad k = 1, \ldots, 256$$

The interval $q \triangleq t_k - t_{k-1} = r_k - r_{k-1}$ is constant for different values of k and is called the *quantization interval.*

In this chapter we will consider only *zero memory quantizers,* which operate on one input sample at a time, and the output value depends only on that input. Such quantizers are useful in image coding techniques such as pulse code modulation (PCM), differential PCM, transform coding, and so on. Note that the quantizer mapping is irreversible; that is, for a given quantizer output, the input value cannot be determined uniquely. Hence, a quantizer introduces distortion, which any reasonable design method must attempt to minimize. There are several quantizer designs available that offer various trade-offs between simplicity and performance. These are discussed next.

4.6 THE OPTIMUM MEAN SQUARE OR LLOYD-MAX QUANTIZER

This quantizer minimizes the mean square error for a given number of quantization levels. Let u be a real scalar random variable with a continuous probability density function $p_u(u)$. It is desired to find the decision levels t_k and the reconstruction levels r_k for an L-level quantizer such that the mean square error

$$\mathcal{E} = E[(u - u^{\cdot})^2] = \int_{t_1}^{t_{L+1}} (u - u^{\cdot})^2 p_u(u)\, du \tag{4.32}$$

is minimized. Rewriting this as

$$\mathcal{E} = \sum_{i=1}^{L} \int_{t_i}^{t_{i+1}} (u - r_i)^2 p_u(u)\, du \tag{4.33}$$

the necessary conditions for minimization of $\mathcal{E}$ are obtained by differentiating it with respect to t_k and r_k and equating the results to zero. This gives

$$\frac{\partial \mathcal{E}}{\partial t_k} = (t_k - r_{k-1})^2 p_u(t_k) - (t_k - r_k)^2 p_u(t_k) = 0$$

$$\frac{\partial \mathcal{E}}{\partial r_k} = 2\int_{t_k}^{t_{k+1}} (u - r_k) p_u(u)\, du = 0, \qquad 1 \le k \le L$$

Using the fact that $t_{k-1} \le t_k$, simplification of the preceding equations gives

$$t_k = \frac{(r_k + r_{k-1})}{2} \tag{4.34}$$

$$r_k = \frac{\displaystyle\int_{t_k}^{t_{k+1}} u\, p_u(u)\, du}{\displaystyle\int_{t_k}^{t_{k+1}} p_u(u)\, du} = E[u \mid u \in \mathcal{T}_k] \tag{4.35}$$

where $\mathcal{T}_k$ is the kth interval $[t_k, t_{k+1})$. These results state that the optimum transition levels lie halfway between the optimum reconstruction levels, which, in turn, lie at the center of mass of the probability density in between the transition levels. Together, (4.34) and (4.35) are nonlinear equations that have to be solved simultaneously given the boundary values t_1 and t_{L+1}. In practice, these equations can be solved by an iterative scheme such as the Newton method.

When the number of quantization levels is large, an approximate solution can be obtained by modeling the probability density $p_u(u)$ as a piecewise constant function as (see Fig. 4.17),

$$p_u(u) \simeq p_u(\hat{t}_j), \qquad \hat{t}_j \triangleq \tfrac{1}{2}(t_j + t_{j+1}), \qquad t_j \le u < t_{j+1} \tag{4.36}$$

Using this approximation in (4.33) and performing the required minimizations, an approximate solution for the decision levels is obtained as

$$t_{k+1} \simeq \frac{A\displaystyle\int_{t_1}^{z_k + t_1} [p_u(u)]^{-1/3}\, du}{\displaystyle\int_{t_1}^{t_{L+1}} [p_u(u)]^{-1/3}\, du} + t_1 \tag{4.37}$$

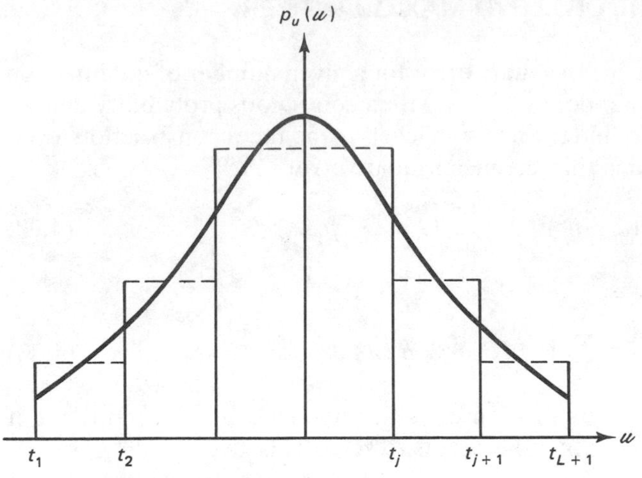

Figure 4.17 Piecewise constant approximation of $p_u(u)$.

where $A = t_{L+1} - t_1$ and $z_k = (k/L)A$, $k = 1, \ldots, L$. This method requires that the quantities t_1 and t_{L+1}, also called the *overload points,* be finite. These values, which determine the *dynamic range A* of the quantizer, have to be assumed prior to the placement of the decision and reconstruction levels. Once the transition levels $\{t_k\}$ have been determined, the reconstruction levels $\{r_k\}$ can be determined easily by averaging t_k and t_{k+1}. The quantizer mean square distortion is obtained as

$$\mathscr{E} \simeq \frac{1}{12L^2} \left\{ \int_{t_1}^{t_{L+1}} [p_u(u)]^{1/3} \, du \right\}^3 \tag{4.38}$$

This is a useful formula because it gives an *estimate* of quantizer error directly in terms of the probability density and the number of quantization levels. This result is exact for piecewise constant probability densities.

Two commonly used densities for quantization of image-related data are the Gaussian and the Laplacian densities, which are defined as follows.

Gaussian:

$$p_u(u) = \frac{1}{\sqrt{2\pi\sigma^2}} \exp\left(\frac{-(u - \mu)^2}{2\sigma^2}\right) \tag{4.39}$$

Laplacian:

$$p_u(u) = \frac{\alpha}{2} \exp(-\alpha|u - \mu|) \tag{4.40a}$$

where μ and σ^2 denote the mean and variance, respectively, of u. The variance of the Laplacian density is given by

$$\sigma^2 = \frac{2}{\alpha} \tag{4.40b}$$

Tables 4.1 and 4.2 (on pp. 104–111) list the design values for several Lloyd-Max quantizers for the preceding densities. For more extensive tables see [30].

The Uniform Optimal Quantizer

For uniform distributions, the Lloyd-Max quantizer equations become linear, giving equal intervals between the transition levels and the reconstruction levels. This is also called *the linear quantizer.* Let

$$p_u(u) = \begin{cases} \dfrac{1}{t_{L+1} - t_1}, & t_1 \le u \le t_{L+1} \\ 0, & \text{otherwise} \end{cases}$$

From (4.35) we obtain

$$r_k = \frac{(t_{k+1}^2 - t_k^2)}{2(t_{k+1} - t_k)} = \frac{t_{k+1} + t_k}{2} \tag{4.41}$$

Combining (4.34) and (4.41) we get

$$t_k = \frac{t_{k+1} + t_{k-1}}{2}$$

which gives

$$t_k - t_{k-1} = t_{k+1} - t_k = \text{constant} \triangleq q$$

Finally, we obtain

$$q = \frac{t_{L+1} - t_1}{L}, \qquad t_k = t_{k-1} + q, \qquad r_k = t_k + \frac{q}{2} \tag{4.42}$$

Thus all transition as well as reconstruction levels are equally spaced. The quantization error $e \triangleq u - u^{\cdot}$ is uniformly distributed over the interval $(-q/2, q/2)$. Hence, the mean square error is given by

$$\mathscr{E} = \frac{1}{q} \int_{-q/2}^{q/2} u^2 \, du = \frac{q^2}{12} \tag{4.43}$$

The variance σ_u^2 of a uniform random variable whose range is A is $A^2/12$. For a uniform quantizer having B bits, we have $q = A/2^B$. This gives

$$\frac{\mathscr{E}}{\sigma_u^2} = 2^{-2B} \quad \Rightarrow \quad \text{SNR} = 10 \log_{10} 2^{2B} = 6B \text{ dB} \tag{4.44}$$

Thus the signal-to-noise ratio achieved by the optimum mean square quantizer for uniform distributions is 6 dB per bit.

Properties of the Optimum Mean Square Quantizer

This quantizer has several interesting properties.

1. *The quantizer output is an unbiased estimate of the input,* that is,

$$E[u^{\cdot}] - E[u] \tag{4.45}$$

2. *The quantization error is orthogonal to the quantizer output,* that is,

$$E[(u - u^{\cdot})u^{\cdot}] = 0. \tag{4.46}$$

TABLE 4.1 Optimum mean square quantizers for Gaussian density with zero mean and unity standard deviation; $t_{-k} = -t_k, r_{-k} = -r_k, t_{L/2+1} \overset{\Delta}{=} \infty$.

Levels	2		3		4		5		6		7		8	
MSE	.3634		.1902		.1175		.0799		.0580		.0440		.0345	
SNR (dB)	4.3964		7.2085		9.3003		10.972		12.367		13.565		14.616	
Entropy	1.0000		1.5358		1.9111		2.2029		2.4428		2.6469		2.8248	
k	t_k	r_k	t_k	r_k	t_k	r_k	t_k	r_k	t_k	r_k	t_k	r_k	t_k	r_k
1	0.0000	.7979	.6120	0.0000	0.0000	.4528	.3823	0.0000	0.000	.3177	.2803	0.0000	0.0000	.2451
2				1.2240	.9816	1.5104	1.2444	.7646	.6589	1.0001	.8744	.5606	.5006	.7561
3								1.7242	1.4469	1.8936	1.6108	1.1882	1.0500	1.3440
4												2.0334	1.7480	2.1520

Levels	9		10		11		12		13		14		15	
MSE	.0279		.0229		.0192		.0163		.0141		.0122		.0107	
SNR (dB)	15.551		16.395		17.163		17.868		18.519		19.125		19.691	
Enthropy	2.9826		3.1245		3.2534		3.3716		3.4806		3.5819		3.6765	
k	t_k	r_k	t_k	r_k	t_k	r_k	t_k	r_k	t_k	r_k	t_k	r_k	t_k	r_k
1	.2218	0.0000	0.0000	.1996	.1838	0.0000	0.0000	.1685	.1569	0.0000	0.0000	.1457	.1370	0.0000
2	.6813	.4437	.4048	.6099	.5600	.3675	.3402	.5119	.4761	.3138	.2936	.4414	.4144	.2739
3	1.1977	.9189	.8339	1.0579	.9657	.7525	.6944	.8769	.8127	.6384	.5960	.7506	.7031	.5549
4	1.8656	1.4765	1.3247	1.5914	1.4359	1.1789	1.0814	1.2859	1.1843	.9871	.9182	1.0858	1.0132	.8513
5		2.2547	1.9683	2.3452	2.0593	1.6928	1.5345	1.7832	1.6231	1.3314	1.2768	1.4677	1.3607	1.1751
6						2.4259	2.1409	2.4986	2.2147	1.8647	1.7033	1.9388	1.7765	1.5463
7										2.5647	2.2820	2.6253	2.3439	2.0067
8														2.6811

Levels	16		17		18		19		20		21		22	
MSE	.0095		.0085		.0076		.0069		.0062		.0057		.0052	
SNR (dB)	20.222		20.723		21.196		21.644		22.071		22.477		22.865	
Entropy	3.7652		3.8486		3.9275		4.0023		4.0773		4.1410		4.2056	
k	t_k	r_k	t_k	r_k	t_k	r_k	t_k	r_k	t_k	r_k	t_k	r_k	t_k	r_k
1	0.0000	.1284	.1215	0.0000	0.0000	.1148	.1093	0.0000	0.0000	.1038	.0992	0.0000	0.0000	.9474
2	.2583	.3882	.3671	.2431	.2306	.3465	.3295	.2185	.2084	.3129	.2990	.1985	.1901	.2854
3	.5226	.6569	.6203	.4910	.4655	.5845	.5553	.4405	.4198	.5267	.5029	.3996	.3824	.4795
4	.7998	.9426	.8877	.7495	.7093	.8341	.7910	.6700	.6378	.7488	.7140	.6062	.5797	.6798
5	1.0995	1.2565	1.1785	1.0259	.9683	1.1024	1.0426	.9120	.8664	.9840	.9364	.8218	.7848	.8897
6	1.4374	1.6183	1.5080	1.3312	1.2513	1.4002	1.3187	1.1732	1.1114	1.2389	1.1756	1.0510	1.0016	1.1135
7	1.8438	2.0693	1.9060	1.6848	1.5733	1.7464	1.6340	1.4642	1.3814	1.5238	1.4399	1.3002	1.2355	1.3576
8	2.4011	2.7328	2.4542	2.1273	1.9638	2.1813	2.0177	1.8037	1.6906	1.8574	1.7437	1.5797	1.4949	1.6321
9				2.7810	2.5037	2.8261	2.5501	2.2317	2.0683	2.2791	2.1158	1.9078	1.7937	1.9553
10								2.8684	2.5937	2.9083	2.6349	2.3237	2.1606	2.3659
11												2.9460	2.6738	2.9817

Levels	23		24		25		26		27		28		29	
MSE	.0047		.0044		.0040		.0037		.0035		.0032		.0030	
SNR (dB)	23.237		23.593		23.935		24.264		24.581		24.887		25.182	
Entropy	4.2675		4.3267		4.3837		4.4384		4.4911		4.5420		4.5911	
k	t_k	r_k	t_k	r_k	t_k	r_k	t_k	r_k	t_k	r_k	t_k	r_k	t_k	r_k
1	.0909	0.0000	0.0000	.0871	.0839	0.0000	0.0000	.0807	.0778	0.0000	0.0000	.0751	.0726	0.0000
2	.2737	.1818	.1747	.2623	.2535	.1677	.1617	.2427	.2342	.1557	.1504	.2258	.2184	.1453
3	.4596	.3656	.3512	.4401	.4233	.3370	.3247	.4068	.3924	.3126	.3020	.3782	.3658	.2916
4	.6510	.5536	.5314	.6227	.5985	.5096	.4908	.5747	.5540	.4722	.4560	.5337	.5158	.4400
5	.8508	.7484	.7176	.8125	.7801	.6874	.6614	.7481	.7206	.6358	.6136	.6935	.6698	.5916
6	1.0626	.9531	.9126	1.0126	.9707	.8727	.8388	.9294	.8941	.8054	.7765	.8594	.8293	.7480
7	1.2918	1.1721	1.1199	1.2272	1.1739	1.0686	1.0254	1.1214	1.0772	.9829	.9465	1.0336	.9962	.9106
8	1.5466	1.4116	1.3448	1.4624	1.3949	1.2792	1.2249	1.3283	1.2732	1.1715	1.1263	1.2189	1.1729	1.0817
9	1.8408	1.6816	1.5954	1.7283	1.6416	1.5105	1.4423	1.5561	1.4872	1.3750	1.3191	1.4193	1.3628	1.2640
10	2.2029	2.0001	1.8854	2.0426	1.9277	1.7725	1.6854	1.8146	1.7270	1.5995	1.5300	1.6407	1.5708	1.4615
11	2.7107	2.4058	2.2431	2.4437	2.2813	2.0829	1.9679	2.1213	2.0062	1.8546	1.7668	1.8928	1.8047	1.6801
12		3.0156	2.7458	3.0479	2.7792	2.4797	2.3177	2.5141	2.3524	2.1578	2.0428	2.1928	2.0778	1.9293
13						3.0787	2.8111	3.1081	2.8416	2.5470	2.3856	2.5784	2.4174	2.2263
14										3.1363	2.8709	3.1634	2.8989	2.6085
15														3.1893

TABLE 4.1 Continued

Levels	30		31		32		33		34		35		36	
MSE	.0028		.0027		.0025		.0024		.0022		.0021		.0020	
SNR (dB)	25.468		25.744		26.012		26.272		26.525		26.770		27.009	
Entropy	4.6386		4.6846		4.7291		4.7723		4.8142		4.8550		4.8946	
k	t_k	r_k	t_k	r_k	t_k	r_k	t_k	r_k	t_k	r_k	t_k	r_k	t_k	r_k
1	0.0000	.0702	.0681	0.0000	0.0000	.0660	.0641	0.0000	0.0000	.0622	.0605	0.0000	0.0000	.0588
2	.1407	.2111	.2047	.1362	.1321	.1983	.1926	.1281	.1246	.1869	.1818	.1210	.1178	.1768
3	.2823	.3535	.3425	.2732	.2650	.3318	.3221	.2570	.2497	.3126	.3040	.2426	.2361	.2955
4	.4258	.4982	.4826	.4119	.3995	.4672	.4535	.3872	.3762	.4399	.4277	.3654	.3556	.4157
5	.5723	.6465	.6259	.5533	.5364	.6056	.5874	.5197	.5048	.5696	.5536	.4900	.4768	.5378
6	.7231	.7996	.7736	.6985	.6767	.7479	.7251	.6553	.6362	.7026	.6826	.6173	.6003	.6627
7	.8794	.9593	.9271	.8487	.8217	.8954	.8675	.7950	.7713	.8399	.8154	.7479	.7269	.7911
8	1.0434	1.1275	1.0884	1.0056	.9726	1.0497	1.0160	.9400	.9113	.9826	.9531	.8828	.8576	.9240
9	1.2173	1.3071	1.2596	1.1711	1.1313	1.2128	1.1723	1.0919	1.0574	1.1323	1.0972	1.0234	.9933	1.0625
10	1.4045	1.5019	1.4443	1.3481	1.3001	1.3874	1.3389	1.2527	1.2116	1.2909	1.2493	1.1710	1.1353	1.2081
11	1.6098	1.7177	1.6471	1.5404	1.4824	1.5773	1.5189	1.4250	1.3760	1.4611	1.4117	1.3276	1.2855	1.3628
12	1.8410	1.9642	1.8757	1.7538	1.6828	1.7883	1.7171	1.6127	1.5539	1.6467	1.5876	1.4958	1.4460	1.5292
13	2.1113	2.2584	2.1434	1.9977	1.9091	2.0298	1.9411	1.8215	1.7501	1.8534	1.7818	1.6794	1.6201	1.7110
14	2.4479	2.6375	2.4772	2.2892	2.1743	2.3188	2.2040	2.0607	1.9720	2.0905	2.0017	1.8842	1.8124	1.9138
15	2.9259	3.2143	2.9518	2.6653	2.5054	2.6912	2.5326	2.3473	2.2326	2.3748	2.2602	2.1192	2.0303	2.1469
16				3.2384	2.9768	3.2616	3.0010	2.7179	2.5588	2.7428	2.5841	2.4013	2.2869	2.4269
17								3.2840	3.0242	3.3056	3.0468	2.7669	2.6086	2.7902
18												3.3265	3.0685	3.3469

TABLE 4.1 Continued

Bit	6		7		Bits	6		7		Bits	7	
Levels	64		128		Levels	64		128		Levels	128	
MSE	.0006		.0002		MSE					MSE		
SNR (dB)	31.9094		37.8634		SNR (dB)					SNR (dB)		
Entropy	5.7074		6.6892		Entropy					Entropy		
k	t_k	r_k	t_k	r_k	k	t_k	r_k	t_k	r_k	k	t_k	r_k
1	.0000	.0336	0.0000	.0171	23	1.7173	1.7718	.7763	.7950	45	1.7472	1.7749
2	.0671	.1007	.0343	.0514	24	1.8300	1.8882	.8140	.8329	46	1.8035	1.8322
3	.1344	.1680	.0685	.0657	25	1.9510	2.0138	.8520	.8712	47	1.8618	1.8915
4	.2018	.2356	.1028	.1199	26	2.0824	2.1510	.8905	.9098	48	1.9223	1.9531
5	.2696	.3035	.1371	.1543	27	2.2270	2.3030	.9293	.9489	49	1.9851	2.0172
6	.3377	.3719	.1715	.1887	28	2.3892	2.4753	.9686	.9884	50	2.0507	2.0342
7	.4064	.4408	.2059	.2231	29	2.5757	2.6761	1.0084	1.0284	51	2.1193	2.1544
8	.4756	.5105	.2404	.2577	30	2.7986	2.9210	1.0487	1.0689	52	2.1913	2.2282
9	.5457	.5809	.2770	.2923	31	3.0824	3.2438	1.0895	1.1100	53	2.2673	2.3064
10	.6166	.6523	.3097	.3270	32	3.4955	3.7471	1.1309	1.1517	54	2.3479	2.3695
11	.6885	.7248	.3445	.3619	33			1.1729	1.1941	55	2.4339	2.4784
12	.7617	.7986	.3794	.3969	34			1.2156	1.2371	56	2.5264	2.5744
13	.8362	.8737	.4147	.4320	35			1.2590	1.2609	57	2.6266	2.6789
14	.9122	.9506	.4497	.4673	36			1.3032	1.3254	58	2.7366	2.7942
15	.9900	1.0294	.4851	.5028	37			1.3481	1.3708	59	2.8588	2.9234
16	1.0698	1.1103	.5207	.5387	38			1.3940	1.4172	60	2.9972	3.0711
17	1.1519	1.1936	.5564	.5744	39			1.4409	1.4645	61	3.1582	3.2453
18	1.2367	1.2798	.5924	.6105	40			1.4887	1.5129	62	3.3528	3.4602
19	1.3245	1.3692	.6285	.6468	41			1.5378	1.5625	63	3.6036	3.7470
20	1.4159	1.4625	.6651	.6834	42			1.5880	1.6134	64	3.9738	4.2006
21	1.5113	1.5601	.7019	.7203	43			1.6395	1.6657			
22	1.6115	1.6628	.7389	.7575	44			1.6926	1.7194			

TABLE 4.2 Optimum mean square quantizers for Laplacian density with zero mean and unity variance; $t_{-k} = -t_k$, $r_{-k} = -r_k$, $t_{L/2+1} \overset{\Delta}{=} \infty$.

Levels	2		3		4		5		6		7		8	
MSE	.5000		.2642		.1762		.1198		.0899		.0681		.0545	
SNR (dB)	3.0103		5.7800		7.5401		9.2152		10.464		11.669		12.638	
Entropy	1.100		1.3169		1.7282		1.9466		2.2071		2.3745		2.5654	
k	t_k	r_k	t_k	r_k	t_k	r_k	t_k	r_k	t_k	r_k	t_k	r_k	t_k	r_k
1	0.0000	.7071	.7071	0.0000	0.0000	.4198	.4198	0.0000	0.0000	.2998	.2998	0.0000	0.0000	.2334
2				.1414	1.1269	1.8340	1.5467	.8395	.7196	1.1393	1.0194	.5996	.5332	.8330
3								2.2538	1.8464	2.5535	2.1462	1.4391	1.2528	1.6725
4												2.8533	2.3797	3.0868

Levels	9		10		11		12		13		14		15	
MSE	.0439		.0365		.0306		.0262		.0225		.0197		.0173	
SNR (dB)	13.580		14.372		15.146		15.815		16.471		17.051		17.621	
Enthropy	2.7011		2.8519		2.9661		3.0907		3.1893		3.2955		3.3822	
k	t_k	r_k	t_k	r_k	t_k	r_k	t_k	r_k	t_k	r_k	t_k	r_k	t_k	r_k
1	.2334	0.0000	0.0000	.1912	.1912	0.0000	0.0000	.1619	.1619	0.0000	0.0000	.1404	.1405	0.0000
2	.7666	.4668	.4246	.6580	.6158	.3824	.3531	.5443	.5150	.3239	.3024	.4643	.4428	.2809
3	1.4862	1.0664	.9578	1.2576	1.1490	.8492	.7777	1.0111	.9396	.7062	.6555	.8467	.7959	.6047
4	2.6131	1.9060	1.6774	2.0971	1.8686	1.4488	1.3109	1.6107	1.4729	1.1731	1.0801	1.3135	1.2206	.9871
5		3.3202	2.8043	3.5114	2.9955	2.2883	2.0305	2.4503	2.1924	1.7727	1.6133	1.9131	1.7538	1.4540
6						3.7026	3.1574	3.8645	3.3193	2.6122	2.3329	2.7527	2.4733	2.0536
7										4.0264	3.4598	4.1669	3.6002	2.8931
8														4.3073

TABLE 4.2 Continued

Levels	16		17		18		19		20		21		22	
MSE	.0154		.0137		.0123		.0111		.0101		.0092		.0084	
SNR (dB)	18.133		18.636		19.094		19.545		19.959		20.368		20.746	
Entropy	3.4747		3.5521		3.6341		3.7040		3.7776		3.8413		3.9081	
k	t_k	r_k	t_k	r_k	t_k	r_k	t_k	r_k	t_k	r_k	t_k	r_k	t_k	r_k
1	0.0000	.1240	.1240	0.0000	0.0000	.1110	.1110	0.0000	0.0000	.1005	.1005	0.0000	0.0000	.0918
2	.2645	.4049	.3885	.2480	.2350	.3590	.3461	.2220	.2115	.3225	.3120	.2010	.1923	.2928
3	.5668	.7288	.6909	.5289	.4995	.6399	.6105	.4701	.4466	.5706	.5471	.4230	.4038	.5148
4	.9200	1.1111	1.0440	.8528	.8019	.9638	.9129	.7510	.7110	.8515	.8115	.6711	.6389	.7629
5	1.3446	1.5780	1.4686	1.2352	1.1550	1.3462	1.2660	1.0748	1.0134	1.1753	1.1139	.9520	.9033	1.0438
6	1.8778	2.1776	2.0018	1.7020	1.5796	1.8130	1.6907	1.4572	1.3665	1.5577	1.4671	1.2759	1.2057	1.3677
7	2.5974	3.0171	2.7214	2.3016	2.1129	2.4126	2.2239	1.9241	1.7912	2.0246	1.8917	1.6583	1.5589	1.7501
8	3.7243	4.4314	3.8483	3.1412	2.8324	3.2522	2.9435	2.5237	2.3244	2.6242	2.4249	2.1251	1.9835	2.2169
9				4.5554	3.9593	4.6664	4.0704	3.3632	3.0440	3.4637	3.1445	2.7247	2.5167	2.8165
10								4.7775	4.1709	4.8780	4.2714	3.5643	3.2363	3.6561
11												4.9785	4.3632	5.0703

Levels	23		24		25		26		27		28		29	
MSE	.0077		.0071		.0066		.0061		.0057		.0053		.0050	
SNR (dB)	21.120		21.467		21.811		22.133		22.452		22.751		23.048	
Entropy	3.9666		4.0277		4.0819		4.1382		4.1886		4.2408		4.2879	
k	t_k	r_k	t_k	r_k	t_k	r_k	t_k	r_k	t_k	r_k	t_k	r_k	t_k	r_k
1	.0918	0.0000	0.0000	.0845	.0845	0.0000	0.0000	.0783	.0783	0.0000	0.0000	.0729	.0729	0.0000
2	.2841	.1836	.1763	.2681	.2608	.1690	.1627	.2472	.2410	.1565	.1511	.2294	.2240	.1458
3	.4956	.3846	.3686	.4691	.4531	.3526	.3390	.4308	.4173	.3255	.3139	.3984	.3868	.3023
4	.7307	.6067	.5801	.6912	.6647	.5536	.5314	.6319	.6096	.5091	.4902	.5820	.5631	.4713
5	.9952	.8547	.8152	.9392	.8997	.7757	.7429	.8539	.8212	.7101	.6825	.7830	.7554	.6549
6	1.2976	1.1356	1.0797	1.2201	1.1642	1.0237	.9780	1.1020	1.0562	.9322	.8941	1.0051	.9670	.8559
7	1.6507	1.4595	1.3821	1.5440	1.4666	1.3046	1.2424	1.3829	1.3207	1.1803	1.1291	1.2532	1.2020	1.0780
8	2.0753	1.8419	1.7352	1.9264	1.8197	1.6285	1.5448	1.7068	1.6231	1.4612	1.3936	1.5341	1.4665	1.3261
9	2.6085	2.3087	2.1598	2.3932	2.2443	2.0109	1.8980	2.0892	1.9763	1.7851	1.6960	1.8580	1.7689	1.6070
10	3.3281	2.9083	2.6931	2.9929	2.7776	2.4778	2.3226	2.5560	2.4009	2.1675	2.0492	2.2404	2.1221	1.9309
11	4.4550	3.7479	3.4126	3.8324	3.4971	3.0774	2.8558	3.1556	2.9341	2.6343	2.4738	2.7072	2.5467	2.3133
12		5.1621	4.5395	5.2466	4.6240	3.9169	3.5754	3.9952	3.6537	3.2339	3.0070	3.3068	3.0799	2.7801
13						5.3311	4.7023	5.4094	4.7806	4.0735	3.7266	4.1464	3.7995	3.3797
14										5.4877	4.8535	5.5606	4.9264	4.2193
15														5.6335

TABLE 4.2 Continued

Levels	30		31		32		33		34		35		36	
MSE	.0046		.0044		.0041		.0039		.0036		.0034		.0033	
SNR (dB)	23.329		23.607		23.870		24.131		24.379		24.626		24.860	
Entropy	4.3366		4.3808		4.4264		4.4680		4.5109		4.5503		4.5907	
k	t_k	r_k	t_k	r_k	t_k	r_k	t_k	r_k	t_k	r_k	t_k	r_k	t_k	r_k
1	0.0000	.0682	.0682	0.0000	0.0000	.0641	.0641	0.0000	0.0000	.0604	.0604	0.0000	0.0000	.0572
2	.1411	.2140	.2093	.1364	.1323	.2005	.1964	.1282	.1245	.1886	.1850	.1209	.1176	.1781
3	.2922	.3705	.3605	.2822	.2734	.3463	.3375	.2646	.2568	.3250	.3173	.2491	.2422	.3063
4	.4550	.5395	.5232	.4387	.4245	.5028	.4886	.4104	.3979	.4708	.4584	.3855	.3745	.4427
5	.6313	.7231	.6995	.6077	.5873	.6718	.6514	.5669	.5491	.6274	.6096	.5313	.5156	.5884
6	.8236	.9242	.8919	.7914	.7636	.8555	.8277	.7359	.7119	.7964	.7723	.6878	.6668	.7450
7	1.0352	1.1462	1.1034	.9924	.9560	1.0565	1.0201	.9196	.8882	.9800	.9487	.8568	.8295	.9141
8	1.2703	1.3943	1.3385	1.2145	1.1675	1.2786	1.2316	1.1206	1.0805	1.1811	1.1410	1.0405	1.0059	1.0977
9	1.5348	1.6752	1.6030	1.4625	1.4026	1.5266	1.4667	1.3427	1.2921	1.4031	1.3526	1.2415	1.1982	1.2987
10	1.8372	1.9991	1.9054	1.7435	1.6671	1.8076	1.7312	1.5907	1.5272	1.6512	1.5877	1.4636	1.4098	1.5208
11	2.1903	2.3815	2.2585	2.0673	1.9695	2.1314	2.0336	1.8717	1.7917	1.9321	1.8522	1.7117	1.6449	1.7689
12	2.6149	2.8484	2.6832	2.4497	2.3227	2.5138	2.3868	2.1956	2.0941	2.2560	2.1546	1.9926	1.9094	2.0498
13	3.1482	3.4480	3.2164	2.9166	2.7473	2.9807	2.8114	2.5780	2.4472	2.6384	2.5077	2.3165	2.2118	2.3737
14	3.8677	4.2875	3.9360	3.5162	3.2805	3.5803	3.3446	3.0448	2.8719	3.1053	2.9323	2.6989	2.5649	2.7561
15	4.9945	5.7017	5.0629	4.3557	4.0001	4.4199	4.0642	3.6444	3.4051	3.7049	3.4656	3.1658	2.9896	3.2230
16				5.7700	5.1270	5.8341	5.1911	4.4840	4.1247	4.5444	4.1852	3.7654	3.5228	3.8226
17								5.8982	5.2516	5.9587	5.3120	4.6049	4.2424	4.6621
18												6.0191	5.3693	6.0764

TABLE 4.2 Continued

Bit	6		7		Bit	6		7		Bit	7	
Levels	64		128		Levels	64		128		Levels	128	
MSE	.0011		.0003									
SNR (dB)	29.7430		35.6880									
Entropy	5.4003		6.3826									
k	t_k	r_k	t_k	r_k	k	t_k	r_k	t_k	r_k	k	t_k	r_k
1	0.0000	.0326	0.0000	.0166	23	2.3936	2.4942	.8910	.9161	45	2.4416	2.4933
2	.0663	.1000	.0334	.0502	24	2.6052	2.7163	.9418	.9675	46	2.5477	2.6220
3	.1348	.1695	.0673	.0844	25	2.8403	2.9644	.9938	1.0202	47	2.6593	2.7165
4	.2055	.2415	.1018	.1191	26	3.1048	3.2453	1.0471	1.0741	48	2.7771	2.8376
5	.2787	.3159	.1368	.1544	27	3.4073	3.5692	1.1018	1.1295	49	2.9017	2.9659
6	.3545	.3930	.1724	.1903	28	3.7604	3.9516	1.1579	1.1863	50	3.0341	3.1024
7	.4330	.4730	.2086	.2269	29	4.1850	4.4185	1.2155	1.2447	51	3.1753	3.2483
8	.5146	.5563	.2455	.2640	30	4.7183	5.0181	1.2747	1.3047	52	3.3266	3.4049
9	.5995	.6427	.2829	.3019	31	5.4379	5.8576	1.3355	1.3664	53	3.4894	3.5740
10	.6879	.7330	.3211	.3404	32	6.5647	7.2719	1.3982	1.4299	54	3.6658	3.7577
11	.7801	.8272	.3600	.3796	33			1.4627	1.4964	55	3.8582	3.9587
12	.8764	.9257	.3995	.4195	34			1.5292	1.5629	56	4.0698	4.1809
13	.9774	1.0291	.4398	.4602	35			1.5978	1.6327	57	4.3049	4.4289
14	1.0834	1.1377	.4809	.5017	36			1.6687	1.7048	58	4.5694	4.7099
15	1.1949	1.2522	.5228	.5440	37			1.7420	1.7793	59	4.8718	5.0338
16	1.3127	1.3731	.5655	.5871	38			1.8180	1.8566	60	5.2250	5.4162
17	1.4373	1.5014	.6091	.6311	39			1.8967	1.9368	61	5.6496	5.6831
18	1.5697	1.6379	.6536	.6761	40			1.9764	2.0200	62	6.1829	6.4227
19	1.7108	1.7838	.6990	.7220	41			2.0634	2.1067	63	6.9024	7.3222
20	1.8621	1.9404	.7455	.7689	42			2.1518	2.1970	64	8.0293	8.7364
21	2.0249	2.1094	.7929	.8169	43			2.2441	2.2913			
22	2.2013	2.2931	.8414	.8659	44			2.3406	2.3899			

3. The variance of the quantizer output is reduced by the factor $1 - f(B)$, where $f(B)$ denotes the mean square distortion of the B-bit quantizer for unity variance inputs, that is,

$$\sigma_{u^\cdot}^2 = [1 - f(B)]\sigma_u^2 \tag{4.47}$$

4. It is sufficient to design mean square quantizers for zero mean unity variance distributions (see Problem 4.13).

Proofs

1. If p_k is the probability of r_k, that is,

$$p_k = \int_{t_k}^{t_{k+1}} p_u(u)\, du \tag{4.48}$$

then from (4.35),

$$E[u^\cdot] \triangleq \sum_{k=1}^{L} p_k E[u|u \in \mathcal{T}_k] = \sum_{k=1}^{L} \int_{t_k}^{t_{k+1}} u p_u(u)\, du = \int_{t_1}^{t_{L+1}} u p_u(u)\, du = E[u]$$

2. $$E[uu^\cdot] = E\{E[u^\cdot u|u \in \mathcal{T}_k]\} = \sum_{k=1}^{L} p_k\{r_k E[u|u \in \mathcal{T}_k]\} = \sum_{k=1}^{L} p_k r_k^2 = E[(u^\cdot)^2]$$

which proves (4.46). This gives an interesting model for the quantizer, as shown in Fig. 4.18. The *quantizer noise* η is uncorrelated with the *quantizer output,* and we can write

$$u = u^\cdot + \eta \tag{4.49}$$

$$\sigma_\eta^2 = E[(u - u^\cdot)^2] = E[u^2] - E[(u^\cdot)^2] \tag{4.50}$$

Since $\sigma_\eta^2 \geq 0$, (4.50) implies the average power of the quantizer output is reduced by the average power of quantizer noise. Also, the quantizer noise η *is dependent on the quantizer input u,* since

$$E[u\eta] = E[u(u - u^\cdot)] = E[\eta^2]$$

3. Since for any mean square quantizer $\sigma_\eta^2 = \sigma_u^2 f(B)$, (4.50) immediately yields (4.47).

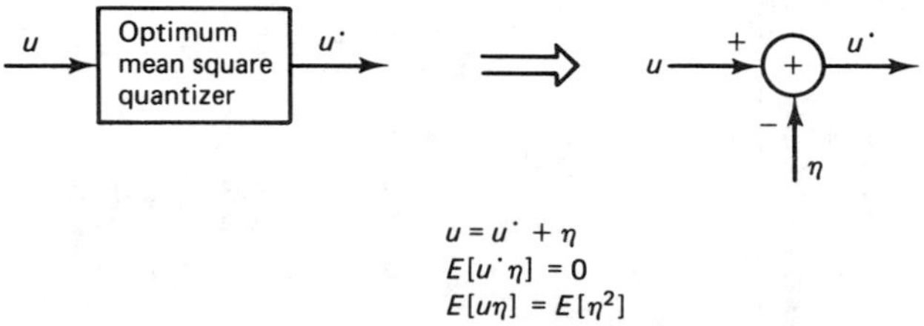

Figure 4.18 Optimum mean square quantizer and its signal dependent noise model.

4.7 A COMPANDOR DESIGN [20–23]

A *compandor* (compressor-expander) is a uniform quantizer preceded and succeeded by nonlinear transformations, as shown in Fig. 4.19. The input random variable u is first passed through a nonlinear memoryless transformation $f(\cdot)$ to yield another random variable w. This random variable is uniformly quantized to give $y \in \{y_i\}$, which is nonlinearly transformed by $g(\cdot)$ to give the output $u^{\cdot}$. The overall transformation from u to $u^{\cdot}$ is a nonuniform quantizer. The functions f and g are determined so that the overall system approximates the Lloyd-Max quantizer. The result is given by

$$g(x) = f^{-1}(x) \tag{4.51}$$

$$f(x) = 2a \left\{ \frac{\int_{t_1}^{x} [p_u(u)]^{1/3}\, du}{\int_{t_1}^{t_{L+1}} [p_u(u)]^{1/3}\, du} \right\} - a \tag{4.52}$$

where $[-a, a]$ is the range of w over which the uniform quantizer operates. If $p_u(u)$ is an even function, that is, $p_u(u) = p_u(-u)$, we get

$$f(x) = a \left\{ \frac{\int_{0}^{x} [p_u(u)]^{1/3}\, du}{\int_{0}^{t_{L+1}} [p_u(u)]^{1/3}\, du} \right\}, \qquad x \ge 0$$

$$f(x) = -f(-x), \qquad x < 0 \tag{4.53}$$

This gives the minimum and maximum values of $f(x)$ as $-a$ and a, respectively. However, the choice of a is arbitrary. As shown in Fig. 4.19, $f(x)$ and $g(x)$ turn out to be functions that compress and expand, respectively, their domains.

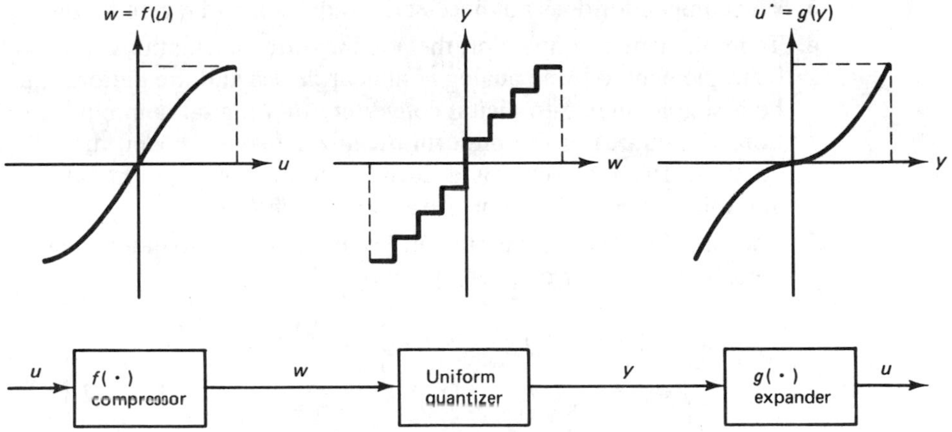

Figure 4.19 A compandor.

Example 4.4

Consider the truncated Laplacian density, which is often used as a probabilistic model for the prediction error signal in DPCM,

$$p_u(u) = ce^{-\alpha|u|}, \qquad -A \le u \le A \tag{4.54}$$

where $c = \frac{\alpha}{2}[1 - \exp(-\alpha A)]^{-1}$. Use of (4.53) gives

$$f(x) = \begin{cases} a\dfrac{[1 - \exp(-\alpha x/3)]}{[1 - \exp(-\alpha A/3)]}, & 0 \le x \le A \\ -f(-x), -A \le x < 0. \end{cases} \tag{4.55}$$

$$g(x) = \begin{cases} -\dfrac{3}{\alpha}\ln\left\{1 - \dfrac{x}{a}\left[1 - \exp\left(\dfrac{-\alpha A}{3}\right)\right]\right\}, & 0 \le x \le a \\ -g(-x), \qquad -a \le x < 0. \end{cases} \tag{4.56}$$

Transformations $f(x)$ and $g(x)$ for other probability densities are given in Problem 4.15.

Remarks

1. The compandor design does not cause the transformed random variable w to be uniformly distributed.
2. For large L, the mean square error estimate of the compandor can be approximated as

$$\mathscr{E} \cong \frac{1}{12L^2}\left(\int_{t_1}^{t_{L+1}} [p_u(u)]^{1/3}\, du\right)^3 \tag{4.57}$$

The actual performance characteristics of compandors are found to be quite close to those of the minimum mean square quantizers.

3. The compandor does not necessarily require t_1 and t_{L+1} to be finite.
4. To implement a compandor, the nonlinear transformations $f(\cdot)$ and $f^{-1}(\cdot)$ may be implemented by an analog nonlinear device and the uniform quantizer can be a simple analog-to-digital converter. In a digital communication application, the output of the uniform quantizer may be coded digitally for transmission. The receiver would decode, perform digital-to-analog conversion, and follow it by the nonlinear transformation $f^{-1}(\cdot)$.
5. The decision and reconstruction levels of the compandor viewed as a nonuniform quantizer are given by

$$\left.\begin{aligned} t_k &= g(kq), \qquad t_{-k} = -t_k,\ k = 0, \ldots, \frac{L}{2} \\ r_k &= g\left(\left(k - \frac{1}{2}\right)q\right), \qquad r_{-k} = -r_k,\ k = 1, \ldots, \frac{L}{2} \end{aligned}\right\} \tag{4.58}$$

for an even probability density function.

4.8 THE OPTIMUM MEAN SQUARE UNIFORM QUANTIZER FOR NONUNIFORM DENSITIES

Since a uniform quantizer can be easily implemented, it is of interest to know how to best quantize a nonuniformly distributed random varible by an L-level uniform quantizer. For simplicity, let $p_u(u)$ be an even function and let L be an even integer. For a fixed L, the optimum uniform quantizer is determined completely by the quantization step size q. Define

$$2a \triangleq Lq$$

where q has to be determined so that the mean square error is minimized. In terms of these parameters,

$$\mathcal{E} = \sum_{j=2}^{L-1} \int_{t_j}^{t_j+1} (u - r_j)^2 p_u(u)\, du + 2\int_{a-q}^{\infty} (u - r_L)^2 p_u(u)\, du\,.$$

Since $\{t_j, r_j\}$ come from a uniform quantizer, this simplifies to

$$\mathcal{E} = 2 \sum_{j=1}^{(L/2)-1} \int_{(j-1)q}^{jq} \left(u - \frac{(2j-1)q}{2}\right)^2 p_u(u)\, du + 2\int_{(L/2-1)q}^{\infty} \left(u - \frac{(L-1)q}{2}\right)^2 p_u(u)\, du$$

Now, the problem is simply to minimize $\mathcal{E}$ as a function of the single variable q. The result is obtained by iteratively solving the nonlinear equation (for $L > 2$)

$$\frac{d\mathcal{E}}{dq} = 0$$

For $L = 2$, the optimum uniform and the Lloyd-Max quantizers are identical, giving

$$a = 2\int_0^{\infty} u\, p_u(u)\, du\,.$$

which equals the mean value of the random variable $|u|$. Table 4.3 gives the optimum uniform quantizer design parameters for Gaussian and Laplacian probability densities.

4.9 EXAMPLES, COMPARISONS, AND PRACTICAL LIMITATIONS

Comparisons among various quantizers can be made in at least two different ways. First, suppose the quantizer output is to be coded by a fixed number of levels. This would be the case for a fixed-word-length analog-to-digital conversion. Then one would compare the quantizing error variance (or the signal-to-noise ratio, $\text{SNR} = 10 \log_{10} \sigma^2/\mathcal{E}$) as a function of number of quantization bits, B. Figure 4.20 shows these curves for the optimum mean square (Lloyd-Max), the compandor, and the optimum uniform quantizers for the Gaussian density function. As expected, the optimum mean square quantizers give the best performance. For Gaussian densities, the performance difference between the optimum mean square and the optimum uniform quantizers is about 2 dB for $B = 6$. In the case of the Laplacian

TABLE 4.3 Optimum Uniform Quantizers for Zero Mean, Unity Variance Gaussian, and Laplacian Densities

No. of	Gaussian			Laplacian		
Output levels	Step size	M.S.E.	Entropy	Step size	M.S.E.	Entropy
2	1.596	.363	1.00	1.414	.500	1.00
3	1.224	.190	1.54	1.414	.264	1.32
4	.996	.119	1.90	1.087	.196	1.75
5	.843	.082	2.18	1.025	.133	1.86
6	.733	.070	2.41	.871	.110	2.13
7	.651	.049	2.60	.822	.083	2.21
8	.586	.037	2.76	.731	.072	2.39
9	.534	.031	2.90	.694	.058	2.46
10	.491	.026	3.03	.633	.051	2.60
11	.455	.022	3.15	.605	.043	2.66
12	.424	.019	3.25	.561	.039	2.78
13	.397	.016	3.35	.539	.034	2.83
14	.374	.015	3.44	.506	.031	2.93
15	.353	.013	3.52	.487	.027	2.98
16	.335	.012	3.60	.461	.025	3.06
17	.319	.010	3.68	.445	.023	3.11
18	.304	.009	3.75	.424	.021	3.18
19	.291	.009	3.81	.411	.019	3.22
20	.279	.008	3.87	.394	.018	3.29
21	.268	.007	3.93	.383	.017	3.33
22	.258	.007	3.99	.368	.016	3.39
23	.248	.006	4.04	.358	.014	3.42
24	.240	.006	4.10	.346	.014	3.48
25	.232	.005	4.15	.337	.013	3.51
26	.224	.005	4.19	.326	.012	3.56
27	.217	.005	4.24	.318	.011	3.59
28	.211	.004	4.29	.309	.011	3.64
29	.204	.004	4.33	.302	.010	3.67
30	.199	.004	4.37	.294	.010	3.71
31	.193	.004	4.41	.287	.009	3.74
32	.188	.003	4.45	.280	.009	3.78
33	.183	.003	4.49	.274	.008	3.81
34	.179	.003	4.52	.268	.008	3.85
35	.174	.003	4.56	.263	.008	3.87
36	.170	.003	4.59	.257	.007	3.91
64	.104	.001	5.31	.166	.003	4.54
128	.057	.00030	6.18	.100	.001	5.32
256	.030	.00004	7.07	.055	.0003	6.15
512	.017	.00002	7.97	.031	.00009	6.96

density, this difference is about 4.3 dB. The compandor and optimum mean square quantizers are practically indistinguishable.

From rate distortion theory it is known that the minimum achievable rate, B_s, of a Gaussian random variable of variance σ^2 is given by the rate distortion function (see Section 2.13) as

$$B_s = \frac{1}{2} \log_2 \frac{\sigma^2}{D} \tag{4.59}$$

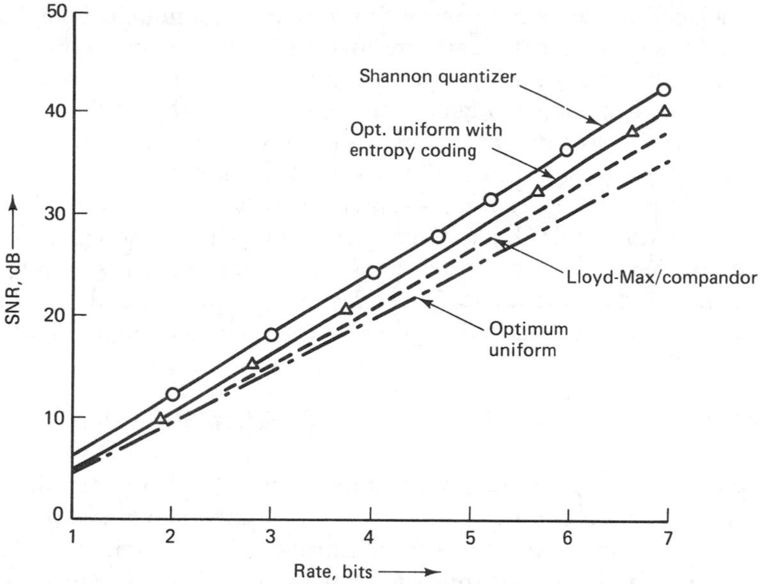

Figure 4.20 Performance of Gaussian density quantizers.

where $D < \sigma^2$, is the average mean square distortion per sample. This can also be written as

$$D = \sigma^2 2^{-2B_s}, \qquad B_s \geq 0 \tag{4.60}$$

and represents a lower bound on attainable distortion for any practical quantizer for Gaussian random variables. This is called the *Shannon lower bound.* The associated optimal encoder is hypothetical because it requires that a block of infinite observations of the random variable be quantized jointly. This is also called the *Shannon quantizer.* The various quantizers in Fig. 4.20 are also compared with this quantizer. Equation (4.60) also gives an *upper bound* on attainable distortion by the *Shannon quantizer,* for non-Gaussian distributed random variables. For a given rate B_s, zero memory quantizers, also called one-dimensional quantizers (block length is one), generally do not attain distortion below the values given by (4.60). The one-dimensional optimum mean square quantizer for a uniform random variable of variance σ^2 does achieve this distortion, however. Thus for any fixed distortion D, the rate of the Shannon quantizer may be considered to give *practically* the minimum achievable rate by a zero memory quantizer for most probability distributions of interest in image processing. (See Problems 4.17 and 4.18.)

The second comparison is based on the entropy of the quantizer output versus its distortion. If the quantized variables are entropy coded by a variable-length coding scheme such as Huffman coding (see Ch. 11, Section 11.2), then the average number of bits needed to code the output will often be less than $\log_2 L$. An optimum quantizer under this criterion would be the one that minimizes the distortion for a specified output entropy [27–29]. Entropy coding increases the complexity of the encoding-decoding algorithm and requires extra buffer storage at transmitters and receivers to maintain a constant bit rate over communication channels. From Fig. 4.20 it is seen that the uniform quantizer with entropy coding gives a better performance than the Lloyd-Max quantizer (without entropy coding). It has been found that the uniform quantizer is quite a good approximation of the "optimum quantizer" based on entropy versus mean square distortion criterion, if the quantization step size is optimized with respect to this criterion.

In practice the design of a quantizer boils down to the selection of number of quantization levels (L) and the dynamic range (A). For a given number of levels, a compromise has to be struck between the quantizer resolution ($t_j - t_{j-1}$) and the attainable dynamic range. These factors become particularly important when the input signal is nonstationary or has an unknown probability density.

4.10 ANALYTIC MODELS FOR PRACTICAL QUANTIZERS [30]

In image coding problems we will find it useful to have analytic expressions for the quantizer mean square error as a function of the number of bits. Table 4.4 lists the distortion function models for the Lloyd-Max and optimum uniform quantizers for Gaussian and Laplacian probability densities of unity variance. Note that the mean of the density functions can be arbitrary. These models have the general form $f(B) = a2^{-bB}$. If the input to the quantizer has a variance σ^2, then the output mean square error will be simply $\sigma^2 f(B)$. It is easy to check that the $f(B)$ models are monotonically decreasing, convex functions of B, which are properties required of distortion versus rate functions.

From Table 4.4 we see that for equal distortion, the number of bits, x, needed for the optimum mean square quantizer to match the performance of a B-bit Shannon quantizer is given by

$$2^{-2B} \simeq 2.26(2^{-1.963x})$$

for small distortions. Solving for x, we get

$$x \simeq B + 0.5$$

which means the zero memory (or a one-dimensional) optimum mean square quan-

TABLE 4.4 Quantizer Distortion Models, $f(B) = a2^{-bB}$

	$0 \le 2^B < 5$		$5 \le 2^B < 36$		$36 \le 2^B \le 512$	
Quantizer	a	b	a	b	a	b
Shannon	1	2	1	2	1	2
Mean square Guassian	1	1.5047	1.5253	1.8274	2.2573	1.9626
Mean square Laplacian	1	1.1711	2.0851	1.7645	3.6308	1.9572
Optimum uniform Gaussian	1	1.5012	1.2477	1.6883	1.5414	1.7562
Optimum uniform Laplacian	1	1.1619	1.4156	1.4518	2.1969	1.5944

tizer performs within about ½ bit of its lower bound achieved by an infinite-dimensional block encoder (for Gaussian distributions).

4.11 QUANTIZATION OF COMPLEX GAUSSIAN RANDOM VARIABLES

In many situations we want to quantize a complex random variable, such as

$$z = x + jy. \tag{4.61}$$

where x and y are independent, identically distributed Gaussian random variables. One method is to quantize x and y independently by their Lloyd-Max quantizers using B bits each. This would not be the minimum mean square quantizer for z. Now suppose we write

$$\begin{gathered} z = Ae^{j\theta} \\ A \triangleq \sqrt{x^2 + y^2}, \qquad \theta \triangleq \tan^{-1}\left(\frac{y}{x}\right) \end{gathered} \tag{4.62}$$

then A and θ are independent, where A has the Rayleigh density (see Problem 4.15b) and θ is uniformly distributed. It can be shown the minimum mean square quantizer for z requires that θ be uniformly quantized. Let L_1 and L_2 be the number of quantization levels for A and θ, respectively, such that $L_1 L_2 = L$ (given). Let $\{v_k\}$ and $\{w_k\}$ be the decision and reconstruction levels of A, respectively, if it were quantized independently by its own mean square quantizer. Then the decision levels $\{t_k\}$ and reconstruction levels $\{r_k\}$ of A for the optimum mean square reconstruction of z are given by [31]

$$\begin{gathered} t_k = v_k \\ r_k = w_k \operatorname{sinc}\left(\frac{1}{L_2}\right) \end{gathered} \tag{4.63}$$

If L_2 is large, then $\operatorname{sinc}(1/L_2) \to 1$, which means the amplitude and phase variables can be quantized independently. For a given L, the optimum allocation of L_1 and L_2 requires that for rates $\log_2 L \geq 4.6$ bits, the phase should be allocated approximately 1.37 bits more than the amplitude [32].

The performance of the joint amplitude-phase quantizer is found to be only marginally better than that of the independent mean square quantizers for x and y. However, the preceding results are useful when one is required to digitize the amplitude and phase variables, as in certain coherent imaging applications where amplitude and phase measurements are made directly.

4.12 VISUAL QUANTIZATION

The foregoing methods can be applied for gray scale quantization of monochrome images. If the number of quantization levels is not sufficient, a phenomenon called *contouring* becomes visible. When groups of neighboring pixels are quantized to the

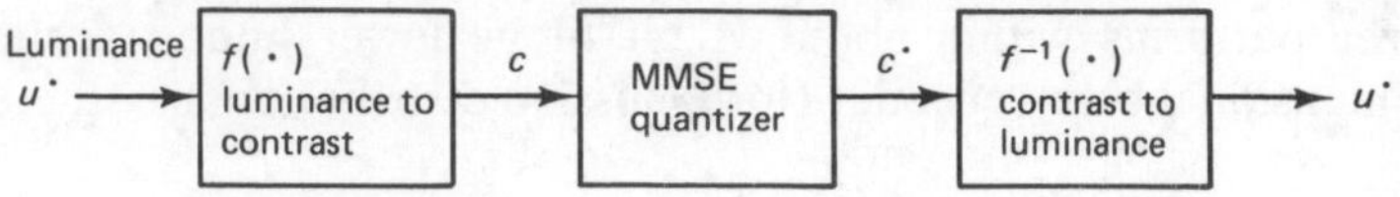

Figure 4.21 Contrast quantization.

same value, regions of constant gray levels are formed, whose boundaries are called *contours* (see Fig. 4.23a). Uniform quantization of common images, where the pixels represent the luminance function, requires about 256 gray levels, or 8 bits. Contouring effects start becoming visible at or below 6 bits/pixel. A mean square quantizer matched to the histogram of a given image may need only 5 to 6 bits/pixel without any visible contours. Since histograms of images vary quite drastically, optimum mean square quantizers for raw image data are rarely used. A uniform quantizer with 8 bits/pixel is usually used.

In evaluating quantized images, the eye seems to be quite sensitive to contours and errors that affect local structure. However, the contours do not contribute very much to the mean square error. Thus a visual quantization scheme should attempt to hold the quantization contours below the level of visibility over the range of luminances to be displayed. We consider two methods of achieving this (other than allocating the full 8 bits/pixel).

Contrast Quantization

Since visual sensitivity is nearly uniform to just noticeable changes in contrast, it is more appropriate to quantize the contrast function shown in Fig. 4.21. Two nonlinear transformations that have been used for representation of contrast c are [see Chapter 3]

$$c = \alpha \ln(1 + \beta u), \qquad 0 \le u \le 1 \tag{4.64}$$

$$c = \alpha u^{\beta} \tag{4.65}$$

where α and β are constants and u represents the luminance. For example, in (4.64) the values $\alpha = \beta/\ln(1 + \beta)$ for α lying between 6 and 18, and in (4.65) the values $\alpha = 1, \beta = 1/3$ have been suggested [34].

For the given contrast representation we simply use the minimum mean square error (MMSE) quantizer for the contrast field (see Fig. 4.21). To display (or reconstruct) the image, the quantized contrast is transformed back to the luminance value by the inverse transformation. Experimental studies indicate that a 2% change in contrast is just noticeable. Therefore, if uniformly quantized, the contrast scale needs 50 levels, or about 6 bits. However, with the optimum mean square quantizer, 4 to 5 bits/pixel could be sufficient.

Pseudorandom Noise Quantization

Another method of suppressing contouring effects [35] is to add a small amount of uniformly distributed pseudorandom noise to the luminance samples before quantization (see Fig. 4.22). This pseudorandom noise is also called *dither*. To display the

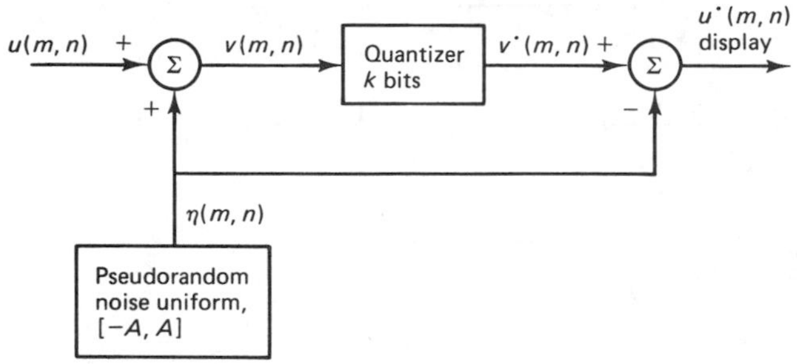

Figure 4.22 Pseudorandom noise quantization.

image, the same (or another) pseudorandom sequence is subtracted from the quantizer output. The effect is that in the regions of low-luminance gradients (which are the regions of contours), the input noise causes pixels to go above or below the original decision level, thereby breaking the contours. However, the average value of the quantized pixels is about the same with and without the additive noise. During display, the noise tends to fill in the regions of contours in such a way that the spatial average is unchanged (Fig. 4.23). The amount of dither added should be kept small enough to maintain the spatial resolution but large enough to allow the luminance values to vary randomly about the quantizer decision levels. The noise should usually affect the least significant bit of the quantizer. Reasonable image quality is achievable by a 3-bit quantizer.

Halftone Image Generation

The preceding method is closely related to the method of generating half-tone images from gray-level images. Halftone images are binary images that give a gray scale rendition. For example, most printed images, including all the images printed

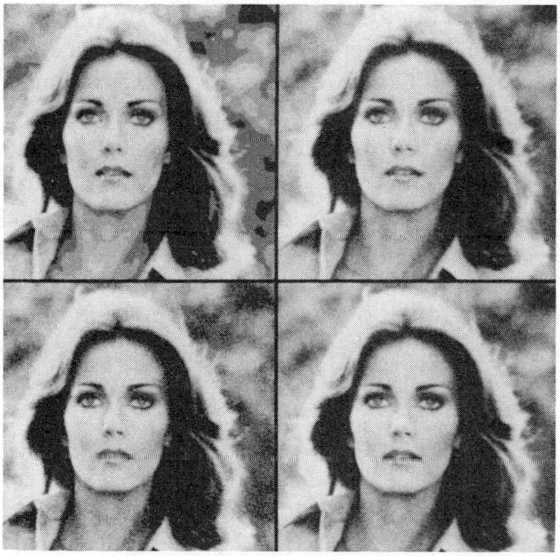

Figure 4.23 [a | b / c | d] (a) 3-bit image, contours are visible; (b) 8-bit image with pseudorandom noise uniform over $[-16, 16]$; (c) $v^{\cdot}(m,n)$, 3-bit quantized $v(m,n)$; (d) image after subtracting pseudorandom noise.

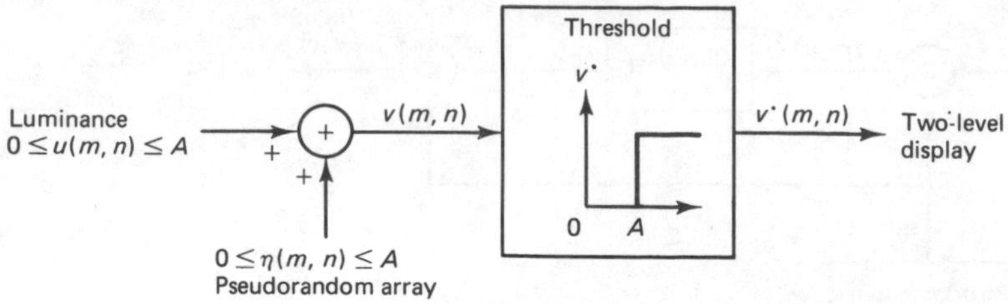

Figure 4.24 Digital halftone generation.

$$H_1 = \begin{bmatrix} 40 & 60 & 150 & 90 & 10 \\ 80 & 170 & 240 & 200 & 110 \\ 140 & 210 & 250 & 220 & 130 \\ 120 & 190 & 230 & 180 & 70 \\ 20 & 100 & 160 & 50 & 30 \end{bmatrix}$$

$$H_2 = \left[\begin{array}{cccc|cccc} 52 & 44 & 36 & 124 & 132 & 140 & 148 & 156 \\ 60 & 4 & 28 & 116 & 200 & 228 & 236 & 164 \\ 68 & 12 & 20 & 108 & 212 & 252 & 244 & 172 \\ 76 & 84 & 92 & 100 & 204 & 196 & 188 & 180 \\ \hline 132 & 140 & 148 & 156 & 52 & 44 & 36 & 124 \\ 200 & 228 & 236 & 164 & 60 & 4 & 28 & 116 \\ 212 & 252 & 244 & 172 & 68 & 12 & 20 & 108 \\ 204 & 196 & 188 & 180 & 76 & 84 & 92 & 100 \end{array}\right]$$

Figure 4.25 Two halftone patterns. Repeat periodically to obtain the full size array. H_2 is called 45° halftone screen because it repeats two 4×4 basic patterns at $\pm 45°$ angles.

in this text, are halftones. Figure 4.24 shows the basic concept of generating halftone images. The given image is oversampled (for instance, a 256×256 image may be printed on a 1024×1024 grid of black and white dots) to coincide with the number of dots available for the halftone image. To each image sample (representing a luminance value) a random number (halftone screen) is added, and the resulting signal is quantized by a 1-bit quantizer. The output (0 or 1) then represents a black or white dot. In practice the dither signal is a finite two-dimensional pseudorandom pattern that is repeated periodically to generate a halftone matrix of the same size as the image. Figure 4.25 shows two halftone patterns. The halftone image may exhibit Moiré patterns if the image pattern and the dither matrix have common or nearly common periodicities. Good halftoning algorithms are designed to minimize the Moiré effect. Figure 4.26 shows a 512×512 halftone image generated digitally from the original $512 \times 512 \times 8$-bit image. The gray level rendition in halftones is due to local spatial averaging performed by the eye. In general, the perceived gray level is equal to the number of black dots perceived in one resolution cell. One resolution cell corresponds to the area occupied by one pixel in the original image.

Color Quantization

Perceptual considerations become even more important in quantization of color signals. A pixel of a color image can be considered as a three-dimensional vector **C**,

Figure 4.26 (a) Original 8-bit/pixel image; (b) halftone screen H_2; (c) halftone image; (d) most significant 1-bit/pixel image.

its elements C_1, C_2, C_3 representing the three color primaries. From Chapter 3 we know the color gamut is a highly irregular solid in the three-dimensional space. Quantization of a color image requires allocating quantization cells to colors in the color solid in the chosen coordinate system. Even if all the colors were equally likely (uniform probability density), the quantization cells will be unequal in size because equal changes in color coordinates do not, in general, result in equal changes in perceived colors.

Figure 4.27 shows a desirable color quantization procedure. First a coordinate transformation is performed and the new coordinate variables T_k are independently quantized. The choice of transformation and the quantizer should be such that the perceptual color difference due to quantization is minimized. In the NTSC color coordinate R_N, G_N, B_N system, the reproducible color gamut is the cube $[0, 1] \times [0, 1] \times [0, 1]$. It has been shown that uniform quantization of each color coordinate in this system provides the best results as compared to uniform quantization in several other coordinate systems. Four bits per color have been found to be just adequate in this coordinate system.

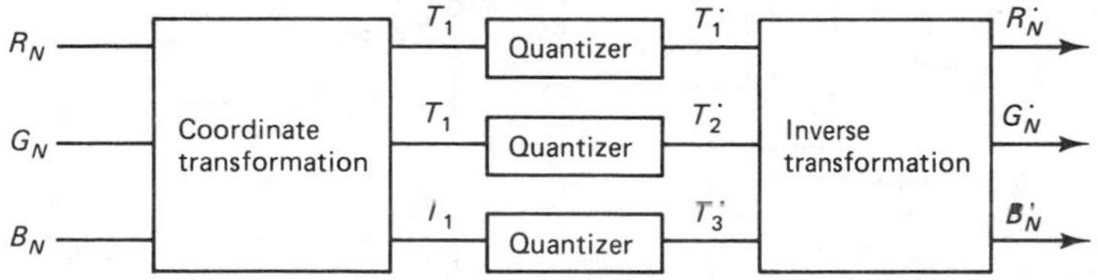

Figure 4.27 Color quantization.

PROBLEMS

4.1 In the RETMA scanning convention, 262.5 lines of each field are scanned in $\frac{1}{60}$ s. Show that the beam has a horizontal scan rate of 15.75 KHz and a slow downward motion at vertical scan rate of 60 Hz.

4.2 Show that a *bandlimited image cannot be space-limited, and vice-versa.*

4.3 The image $f(x, y) = 4 \cos 4\pi x \cos 6\pi y$ is sampled with $\Delta x = \Delta y = 0.5$ and $\Delta x = \Delta y = 0.2$. The reconstruction filter is an ideal low-pass filter with bandwidths $(\frac{1}{2}\Delta x, \frac{1}{2}\Delta y)$. What is the reconstructed image in each case?

4.4 The NTSC composite video signal of (4.2) is sampled such that $\Delta x = \frac{1}{2}\xi_{x0}$, $\Delta y = 1/f_l$. At any given frame, say at $t = 0$, what would be the spectrum of the sampled frame for the composite color signal spectrum shown in Fig. P4.4?

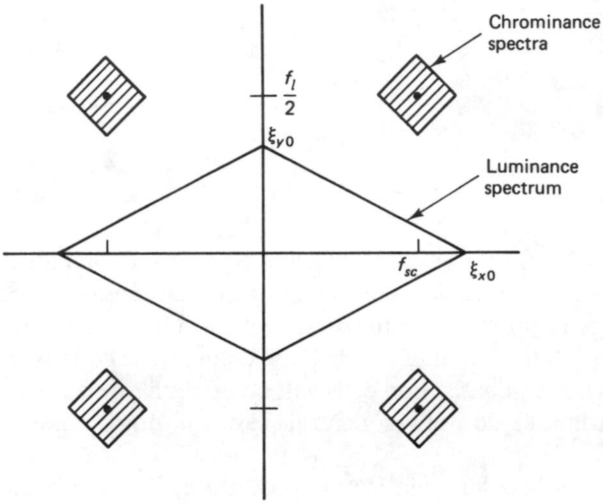

Figure P4.4 NTSC composite color signal spectrum.

4.5 Prove (4.19) for bandlimited stationary random fields.

4.6 *(Sampling Noisy Images)* A bandlimited image acquired by a practical sensor is observed as $g(x, y) = f(x, y) + n(x, y)$, where $\xi_{x0} = \xi_{y0} \stackrel{\Delta}{=} \xi_f$ and $n(x, y)$ is wideband noise whose spectral density function $S_n(\xi_1, \xi_2) = \eta/4$ is bandlimited to $-\xi_n \le \xi_1, \xi_2 \le \xi_n$, $\xi_n = 2\xi_f$. The random field $g(x, y)$ is sampled without prefiltering and with prefiltering at the Nyquist rate of the noiseless image and reconstructed by an ideal low-pass filter whose bandwidths are also ξ_{x0}, ξ_{y0} (Fig. P4.6). Show that (a) the SNR of the sensor output g over its bandwidth is $\sigma_f^2/(\eta\xi_f^2)$, where σ_f^2 is the image power, (b) the

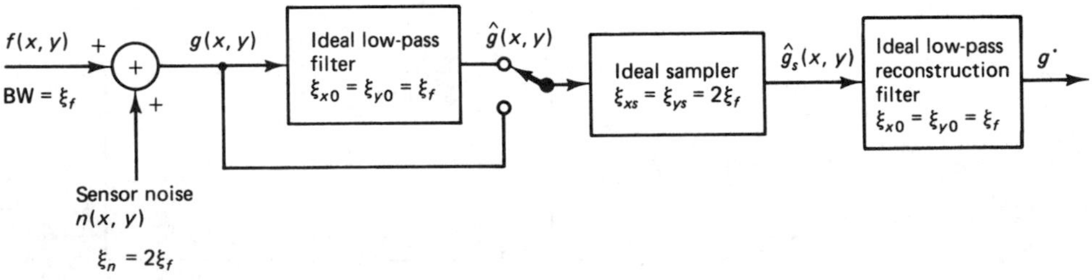

Figure P4.6 Sampling of noisy images.

SNRs of the reconstructed image with and without prefiltering are $\sigma_f^2/(\eta\xi_f^2)$ and $\sigma_f^2/(4\eta\xi_f^2)$, respectively. What would be the SNR of the reconstructed image if the sensor output were sampled at the Nyquist rate of the noise without any prefiltering? Compare the preceding sampling schemes and recommend the best way for sampling noisy images.

4.7 Show that (4.15) is an orthogonal expansion for a bandlimited function such that the least squares error

$$\sigma_{l.s.}^2 = \iint_{-\infty}^{\infty} | f(x, y) - \sum_{m,n}\sum a(m, n)\phi_m(x)\psi_n(y)|^2 \, dx \, dy$$

where $\phi_m(x) \triangleq \text{sinc}(x\xi_{xs} - m)$, $\psi_n(y) \triangleq \text{sinc}(y\xi_{ys} - n)$, is minimized to zero when $a(m, n) = f(m\Delta x, n\Delta y)$.

4.8 *(Optimal sampling)* A real random field $f(x,y)$, defined on a square $[-L, L] \times [-L, L]$, with autocorrelation function $R(x,y;x',y') \triangleq E[f(x,y)f(x',y')]$, $-L \le x, x', y, y' \le L$, is sampled by a set of orthogonal functions $\phi_{m,n}(x,y)$ to obtain the samples $a_{m,n}$ such that the reconstructed function $\tilde{f}_{M,N}(x,y) \triangleq \sum_{m=0}^{M-1}\sum_{n=0}^{N-1} a_{m,n}\phi_{m,n}(x,y)$ minimizes the mean square error $\sigma_{M,N}^2 \triangleq \iint_{-L}^{L} E[|f(x,y) - \tilde{f}_{M,N}(x,y)|^2] \, dx \, dy$. Let $\phi_{m,n}(x,y)$ be a set of complete orthonormal functions obtained by solving the eigenvalue integral equation

$$\iint_{-L}^{L} R(x,y;x',y')\phi_{m,n}(x',y') \, dx' \, dy' = \lambda_{m,n}\phi_{m,n}(x,y), \qquad -L \le x, y \le L$$

a. Show that $\{a_{m,n}\}$ are orthogonal random variables, that is,

$$E[a_{m,n}a_{m',n'}] = \lambda_{m,n}\delta(m - m', n - n').$$

b. Show that $\sigma_{M,N}^2$ is minimized when $\{\phi_{m,n}\}$ are chosen to correspond to the largest MN eigenvalues and the minimized error is $\sigma_{M,N}^2 = \sum_{m=M}^{\infty}\sum_{n=N}^{\infty} \lambda_{m,n}$

The preceding series representation for the random field $f(x, y)$ is called its KL series expansion.

4.9 *(Interlaced sampling)* The interlaced sampling grid G_2 of Fig. 4.9c can be written as a superposition of rectangular grids, that is,

$$g(x, y) = \sum_{m,n}\sum \delta(x - 2m, y - 2n) + \sum_{m,n}\sum \delta(x - 2m - 1, y - 2n - 1)$$

Verify Fig. 4.9e by showing the Fourier transform of this array is

$$G(\xi_1, \xi_2) = \tfrac{1}{2} \sum_{k+l=\text{even}}\sum \delta(\xi_1 - k\xi_0, \xi_2 - l\xi_0), \qquad \xi_0 \triangleq \tfrac{1}{2}$$

4.10 a. Show that the limiting Lagrange polynomial can be written as

$$\lim_{q\to\infty} L_k^q(x) = \prod_{\substack{m=-\infty \\ m\neq 0}}^{\infty} \left(\frac{x - k + m}{m}\right) = \prod_{m=1}^{\infty}\left[1 - \frac{(x-k)^2}{m^2}\right]$$

which is the well-known product expansion of $\text{sinc}(x - k)$.

b. Show that the Lagrange interpolation formula of (4.29) satisfies the properties $\hat{f}(m\Delta) = f(m\Delta)$—that is, no interpolation error at known samples—and $\int\hat{f}(x)\,dx = \Delta\sum_m f(m\Delta)$—that is, the area is preserved according to the trapezoidal rule of integration.

c. Write the two-dimensional interpolation formula for $q_1 = q_2 = 1, 2, 3$.

4.11 *(Moiré effect—one-dimensional)* A one-dimensional function $f(x) = 2 \cos \pi \xi_0 x$ is sampled at a rate ξ_s, which is just above ξ_0. Common reconstruction filters such as the zero- or first-order-hold circuits have a passband greater than $\pm\xi_s/2$ with the first zero crossings at $\pm\xi_s$, as shown in Fig. P4.11a.

a. Show that the reconstructed function is of the form

$$\tilde{f}(x) = 2(a + b \cos 2\pi\xi_s x) \cos \pi\xi_0 x + 2b \sin 2\pi\xi_s x \sin \pi\xi_0 x$$

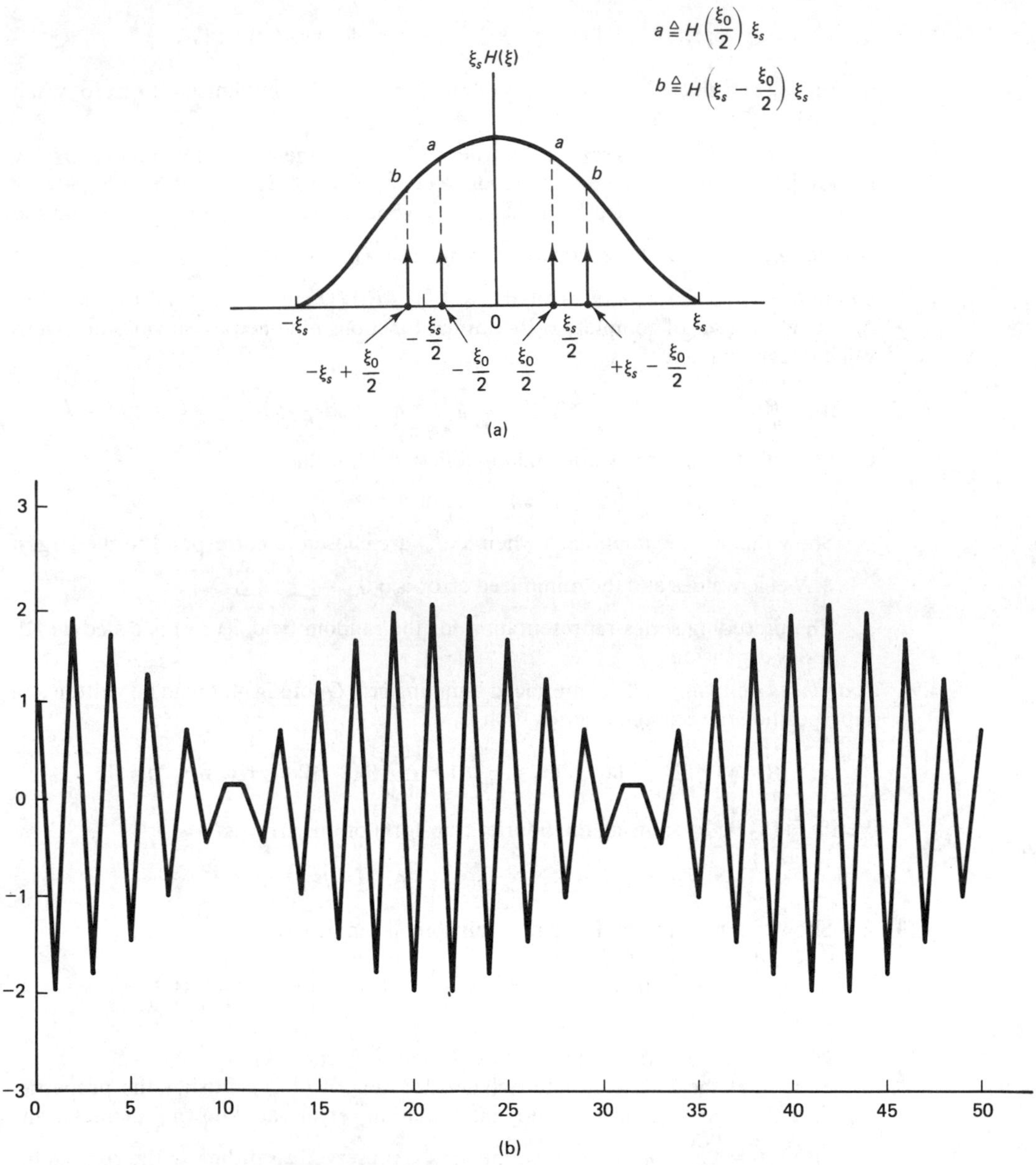

Figure P4.11 (a) Reconstruction filter frequency response. (b) A one-dimensional Moiré pattern.

which is a beat pattern between the signal frequency $\xi_0/2$ and one of its companion frequencies $\xi_s - (\xi_0/2)$ present in the sampled spectrum.

b. Show that if the sampling frequency is above twice the highest signal frequency, then the Moiré effect will be eliminated. Note that if $\xi_0 = 0$—that is, $f(x)=$ constant—then the reconstructed signal is also constant—that is, the sampling system has a *flat field* response.

c.* As an example, generate a sampled signal $f_s(k) = 2\cos(k\pi/1.05)$, which corresponds to $\xi_0 = 1$, $\xi_s = 1.05$. Now plot this sequence as a continuous signal on a line plotter (which generally performs a first-order hold). Fig. P4.11b shows the nature of the result, which looks like an amplitude-modulated sine wave. This is a Moiré pattern in one dimension.

4.12 *(Moiré effect—two dimensions)* An image, $f(x, y) = 4\cos 4\pi x \cos 4\pi y$, is sampled at a rate $\xi_{xs} = \xi_{ys} = 5$. The reconstruction filter has the frequency response of a square display spot of size 0.2×0.2 but is bandlimited to the region $[-5, 5] \times [-5, 5]$. Calculate the reconstructed image. If the input image is a constant gray instead, what would the displayed image look like? Would this display have a flat field response?

4.13 If t_k, r_k are the decision and reconstruction levels for a zero mean, unity variance random variable u, show that $\tilde{t}_k = \mu + \sigma t_k$, $\tilde{r}_k = \mu + \sigma r_k$, are the corresponding quantities for a random variable v having the same distribution but with mean μ and variance σ^2. Thus v may be quantized by first finding $u = (v - \mu)/\sigma$, then quantizing u by a zero mean, unity variance quantizer to obtain $u^\cdot$, and finally obtaining the quantized value of v as $v^\cdot = \mu + \sigma u^\cdot$.

4.14 Suppose the compandor transformations in Fig. 4.19 are $g(x) = f^{-1}(x)$ and

$$w \triangleq f(u) = \begin{cases} \int_0^u p_u(\alpha)\, d\alpha, & u > 0 \\ -f(-u), & u < 0 \end{cases}$$

where $p_u(\alpha) = p_u(-\alpha)$. This transformation (also called *histogram equalization*) causes w to be uniformly distributed over the interval $[-\frac{1}{2}, \frac{1}{2}]$. The uniform quantizer is now optimum for w. However, the overall quantizer need not be optimum for u.

a. Let $p_u(\alpha) = \begin{cases} 1 - |\alpha|, & -1 \le \alpha \le 1 \\ 0, & \text{otherwise} \end{cases}$

and let the number of quantizer levels be 4. What are the decision and reconstruction levels for the input u? Calculate the mean square error.

b. Show that this compandor is suboptimal compared to the one discussed in the text.

4.15 *(Compandor transformations)*

a. For zero mean Gaussian random variables, show that the compandor transformations are given by $f(x) = 2\,\text{erf}(x/\sqrt{6}\sigma)$, $x \ge 0$, and $g(y) = \sqrt{6}\sigma\, \text{erf}^{-1}(y/2)$, $y \ge 0$, where $\text{erf}(x) \triangleq (1/\sqrt{\pi}) \int_0^x \exp(-y^2)\, dy$.

b. For the Rayleigh density

$$p_u(\alpha) = \begin{cases} \dfrac{\alpha}{\sigma^2} \exp\left(-\dfrac{\alpha^2}{2\sigma^2}\right), & \alpha > 0 \\ 0, & \alpha < 0 \end{cases}$$

show that the transformation is

$$f(x) = c \int_0^x \alpha^{1/3} \exp\left(\frac{-\alpha^2}{6\sigma^2}\right) d\alpha - 1$$

where c is a normalization constant such that $f(\infty) = 1$.

4.16 Use the probability density function of Problem 4.14a.

a. Design the four-level optimum-uniform quantizer. Calculate the mean square error and the entropy of the output.

b. Design the four-level Lloyd-Max quantizer (or the compandor) and calculate the mean square error and the entropy of the output.

c. If the criterion of quantizer performance is the mean square error for a given entropy of the output, which of the preceding two quantizers is superior?

4.17 Show that the optimum mean square, zero-memory quantizer for a uniformly distributed random variable achieves the rate distortion characteristics of the Shannon quantizer for a Gaussian distribution having the same variance.

4.18 The differential entropy (in bits) of a continuous random variable u is defined as

$$H(u) = -\int_{-\infty}^{\infty} p_u(u)\log_2 p_u(u)\, du$$

a. Show that for a Gaussian random variable g whose variance is σ^2, $H(g) = \frac{1}{2}\log_2(2\pi e\sigma^2)$. Similarly, for a uniform random variable f whose variance is σ^2, $H(f) = \frac{1}{2}\log_2(12\sigma^2)$.

b. For an arbitrary random variable x, its *entropy power* Q_x is defined by the relation $H(x) = \frac{1}{2}\log_2(2\pi e Q_x)$. If we write $Q_x = \alpha_x\sigma^2$, then we have $Q_g = \sigma^2$, $Q_f = (6\sigma^2)/\pi e$, $\alpha_g = 1$, and $\alpha_f \simeq 0.702$. Show that among possible continuous density functions whose variance is fixed at σ^2, the Gaussian density function has the maximum entropy. Hence show that $\alpha_x \le 1$ for any random variable x.

c. For the random variable x with entropy power Q_x, the minimum achievable rate, $n_{\min}(x)$, for a mean square distortion D is given by $n_{\min}(x) = \frac{1}{2}\log_2(Q_x/D)$. For a uniform random variable f, its zero memory Lloyd-Max quantizer achieves the rate $n_f = \frac{1}{2}\log_2(\sigma^2/D)$. Show that

$$n_f \simeq n_{\min}(f) + \tfrac{1}{4}$$

that is, for uniform distributions, the one-dimensional optimum mean square quantizer is within $\frac{1}{4}$ bit/sample of its minimum achievable rate of its Shannon quantizer.

4.19* Take a 512×512 8-bit/pixel image and quantize it to 3 bits using a (a) uniform quantizer, (b) contrast quantizer via (4.65) with $\alpha = 1$, $\beta = 1/3$, and (c) pseudorandom noise quantizer of Fig. 4.22 with a suitable value of A (for instance, between 4 and 16). Compare the mean square errors and their visual qualities.

BIBLIOGRAPHY

Section 4.1

For scanning, display and other hardware engineering principles in image sampling and acquisition:

1. K. R. Castleman. *Digital Image Processing*. Englewood Cliffs, N.J.: Prentice Hall, 1979, pp. 14–51.
2. D. G. Fink, (ed.). *Television Engineering Handbook*. New York: McGraw-Hill, 1957.
3. H. R. Luxenberg and R. L. Kuehn (eds.). *Display Systems Engineering*. New York: McGraw-Hill, 1968.

4. P. Mertz and F. Grey. "A Theory of Scanning and its Relation to the Characteristics of the Transmitted Signal in Telephotography and Television." *Bell Sys. Tech. J.* 13 (1934): 464–515.

Section 4.2

The two-dimensional sampling theory presented here is a direct extension of the basic concepts in one dimension, which may be found in:

5. E. T. Whittaker. "On the Functions which are Represented by the Expansions of the Interpolation Theory." *Proc. Roy. Soc., Edinburgh, Section A* 35 (1915): 181–194.
6. C. E. Shannon. "Communications in the Presence of Noise," *Proc. IRE* 37 (January 1949): 10–21.

For extensions to two and higher dimensions:

7. J. W. Goodman. *Introduction to Fourier Optics.* New York: McGraw-Hill, 1968.
8. A. Papoulis. *Systems and Transforms with Applications in Optics.* New York: McGraw-Hill, 1966.
9. D. P. Peterson and D. Middleton. "Sampling and Reconstruction of Wave Number Limited Functions in *N*-dimensional Euclidean Spaces." *Inform. Contr.* 5 (1962): 279–323.
10. J. Sabatier and F. Kretz. "Sampling the Components of 625-Line Colour Television Signals." *Eur. Broadcast. Union Rev. Tech.* 171 (1978): 2.

Section 4.3

For extensions of sampling theory to random processes and random fields and for orthogonal function expansions for optimal sampling:

11. S. P. Lloyd. "A Sampling Theorem for Stationary (Wide Sense) Stochastic Processes." *Trans. Am. Math. Soc.* 92 (July 1959): 1–12.
12. J. L. Brown, Jr. "Bounds for Truncation Error in Sampling Expansions of Bandlimited Signals." *IEEE Trans. Inf. Theory* IT-15 (July 1969): 440–444.
13. A. Rosenfeld and A. C. Kak. *Digital Picture Processing.* New York: Academic Press, 1976, pp. 83–98.

For hexagonal sampling and related results:

14. R. M. Mersereau. "The Processing of Hexagonally Sampled Two Dimensional Signals." *Proc. IEEE* (July 1979): 930–949.

Section 4.4

For aliasing and other practical problems associated with sampling:

15. R. Legault. "The Aliasing Problems in Two Dimensional Sampled Imagery," in *Perception of Displayed Information,* L. M. Biberman (ed.). New York: Plenum Press, 1973.

Sections 4.5, 4.6

For comprehensive reviews of image quantization techniques and extended bibliography:

16. Special issue on Quantization. *IEEE Trans. Inform. Theory*. IT-28, no. 2 (March 1982).
17. A. K. Jain. "Image Data Compression: A Review." *Proc. IEEE* 69, no. 3 (March 1981): 349–389.

For mean square quantizer results:

18. S. P. Lloyd. "Least Squares Quantization in PCM." unpublished memorandum, Bell Laboratories, 1957. (Copy available by writing the author.)
19. J. Max. "Quantizing for Minimum Distortion," *IRE Trans. Inform. Theory* IT-6 (1960): 7–12.

Section 4.7

For results on companders:

20. P. F. Panter and W. Dite. "Quantizing Distortion in Pulse-Code Modulation with Nonuniform Spacing Levels." *Proc. IRE* 39 (1951): 44–48.
21. B. Smith. "Instantaneous Companding of Quantizing Signals." *Bell Syst. Tech. J.* 27 (1948): 446–472.
22. G. M. Roe. "Quantizing for Minimum Distortion." *IEEE Trans. Inform. Theory* IT-10 (1964): 384–385.
23. V. R. Algazi. "Useful Approximations to Optimum Quantization." *IEEE Trans. Commun. Tech.* COM-14 (1966): 297–301.

Sections 4.8, 4.9

For results related to optimum uniform quantizers and quantizer performance trade-offs:

24. T. J. Goblick and J. L. Holsinger. "Analog Source Digitization: A Comparison of Theory and Practice." *IEEE Trans. Inform. Theory* IT-13 (April 1967): 323–326.
25. H. Gish and J. N. Pierce. "Asymptotically Efficient Quantization." *IEEE Trans. Inform. Theory* IT-14 (1968): 676–681.
26. T. Berger. *Rate Distortion Theory*. Englewood Cliffs, N.J.: Prentice-Hall, 1971.

For optimum quantizers based on rate versus distortion characteristics:

27. T. Berger. "Optimum Quantizers and Permutation Codes." *IEEE Trans. Inform. Theory* IT-16 (November 1972): 759–765.
28. A. N. Netravali and R. Saigal. "An Algorithm for the Design of Optimum Quantizers." *Bell Syst. Tech. J.* 55 (November 1976): 1423–1435.
29. D. K. Sharma. "Design of Absolutely Optimal Quantizers for a Wide Class of Distortion Measures." *IEEE Trans. Inform. Theory* IT-24 (November 1978): 693–702.

Section 4.10

For analytic models of common quantizers:

30. S. H. Wang and A. K. Jain. "Application of Stochastic Models for Image Data Compression," Technical Report, Signal & Image Proc. Lab, Dept. of Electrical Engineering, University of California, Davis, September 1979.

Section 4.11

Here we follow:

31. N. C. Gallagher, Jr. "Quantizing Schemes for the Discrete Fourier Transform of a Random Time-Series." *IEEE Trans. Inform. Theory* IT-24 (March 1978): 156–163.
32. W. A. Pearlman. "Quantizing Error Bounds for Computer Generated Holograms," Tech. Rep. 6 503-1, Stanford University Information Systems Laboratory, Stanford, Calif., August 1974. Also see Pearlman and Gray, *IEEE Trans. Inform. Theory* IT-24 (November 1978): 683–692.

Section 4.12

For further details on visual quantization:

33. F. W. Scoville and T. S. Huang. "The Subjective Effect of Spatial and Brightness Quantization in PCM Picture Transmission." *NEREM Record* (1965): 234–235.
34. F. Kretz. "Subjectively Optimal Quantization of Pictures." *IEEE Trans. Comm.* COM-23, (November 1975): 1288–1292.
35. L. G. Roberts. "Picture Coding Using Pseudo-Random Noise." *IRE Trans. Infor. Theory* IT-8, no. 2 (February 1962): 145–154.
36. J. E. Thompson and J. J. Sparkes. "A Pseudo-Random Quantizer for Television Signals." *Proc. IEEE* 55, no. 3 (March 1967): 353–355.
37. J. O. Limb. "Design of Dither Waveforms for Quantized Visual Signals," *Bell Syst. Tech. J.*, 48, no. 7 (September 1969): 2555–2583.
38. B. Lippel, M. Kurland, and A. H. March. "Ordered Dither Patterns for Coarse Quantization of Pictures." *Proc. IEEE* 59, no. 3 (March 1971): 429–431. Also see IEEE Trans. Commun. Tech. COM-13, no. 6 (December 1971): 879–889.
39. C. N. Judice. "Digital Video: A Buffer-Controlled Dither Processor for Animated Images." *IEEE Trans. Comm.* COM-25, (November 1977): 1433–1440.
40. P. G. Roetling. "Halftone Method with Edge Enhancement and Moiré Suppression." *J. Opt. Soc. Am.* 66 (1976): 985–989.
41. A. K. Jain and W. K. Pratt. "Color Image Quantization." *National Telecomm. Conference 1972 Record,* IEEE Publication No. 72CH0601-S-NTC, Houston, Texas, December 1972.
42. J. O. Limb, C. B. Rubinstein, and J. E. Thompson. "Digital Coding of Color Video Signals—A Review." *IEEE Trans. Commun.* COM-25 (November 1977): 1349–1385.

5

Image Transforms

5.1 INTRODUCTION

The term *image transforms* usually refers to a class of unitary matrices used for representing images. Just as a one-dimensional signal can be represented by an orthogonal series of *basis functions,* an image can also be expanded in terms of a discrete set of basis arrays called *basis images.* These basis images can be generated by unitary matrices. Alternatively, a given $N \times N$ image can be viewed as an $N^2 \times 1$ vector. An image transform provides a set of coordinates or basis vectors for the vector space.

For continuous functions, orthogonal series expansions provide series coefficients which can be used for any further processing or analysis of the functions. For a one-dimensional sequence $\{u(n), 0 \le n \le N-1\}$, represented as a vector $\mathbf{u}$ of size N, a unitary transformation is written as

$$\mathbf{v} = \mathbf{A}\mathbf{u} \qquad \Rightarrow v(k) = \sum_{n=0}^{N-1} a(k,n)u(n), \qquad 0 \le k \le N-1 \tag{5.1}$$

where $\mathbf{A}^{-1} = \mathbf{A}^{*T}$ (*unitary*). This gives

$$\mathbf{u} = \mathbf{A}^{*T}\mathbf{v} \qquad \Rightarrow u(n) = \sum_{k=0}^{N-1} v(k)a^*(k,n) \qquad 0 \le n \le N-1 \tag{5.2}$$

Equation (5.2) can be viewed as a series representation of the sequence $u(n)$. The columns of $\mathbf{A}^{*T}$, that is, the vectors $\mathbf{a}_k^* \stackrel{\Delta}{=} \{a^*(k,n), 0 \le n \le N-1\}^T$ are called the *basis vectors* of $\mathbf{A}$. Figure 5.1 shows examples of basis vectors of several orthogonal transforms encountered in image processing. The series coefficients $v(k)$ give a representation of the original sequence $u(n)$ and are useful in filtering, data compression, feature extraction, and other analyses.

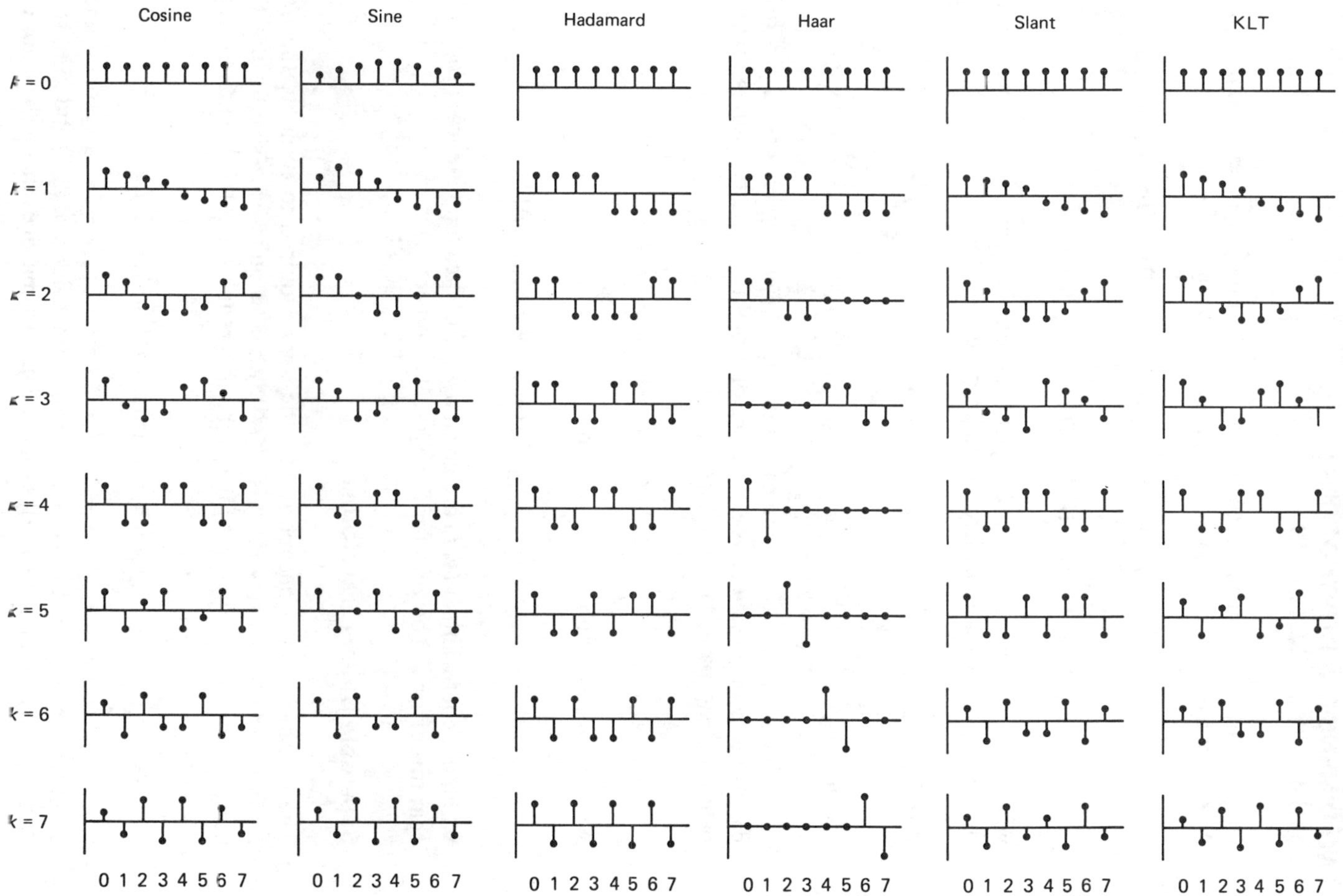

Figure 5.1 Basic vectors of the 8×8 transforms.

5.2 TWO-DIMENSIONAL ORTHOGONAL AND UNITARY TRANSFORMS

In the context of image processing a general orthogonal series expansion for an $N \times N$ image $u(m, n)$ is a pair of transformations of the form

$$v(k, l) = \sum_{m,n=0}^{N-1}\sum u(m, n)a_{k,l}(m, n), \qquad 0 \le k, l \le N-1 \tag{5.3}$$

$$u(m, n) = \sum_{k,l=0}^{N-1}\sum v(k, l)a^*_{k,l}(m, n), \qquad 0 \le m, n \le N-1 \tag{5.4}$$

where $\{a_{k,l}(m, n)\}$, called an *image transform,* is a set of complete orthonormal discrete basis functions satisfying the properties

$$\textit{Orthonormality:} \quad \sum_{m,n=0}^{N-1}\sum a_{k,l}(m, n)a^*_{k',l'}(m, n) = \delta(k - k', l - l') \tag{5.5}$$

$$\textit{Completeness:} \quad \sum_{k,l=0}^{N-1}\sum a_{k,l}(m, n)a^*_{k,l}(m', n') = \delta(m - m', n - n') \tag{5.6}$$

The elements $v(k, l)$ are called the *transform coefficients* and $\mathbf{V} \triangleq \{v(k, l)\}$ is called the *transformed image.* The orthonormality property assures that any truncated series expansion of the form

$$u_{P,Q}(m, n) \triangleq \sum_{k=0}^{P-1}\sum_{l=0}^{Q-1} v(k, l)a^*_{k,l}(m, n), \qquad P \le N, \quad Q \le N \tag{5.7}$$

will minimize the sum of squares error

$$\sigma_e^2 = \sum_{m,n=0}^{N-1}\sum [u(m, n) - u_{P,Q}(m, n)]^2 \tag{5.8}$$

when the coefficients $v(k, l)$ are given by (5.3). The completeness property assures that this error will be zero for $P = Q = N$ (Problem 5.1).

Separable Unitary Transforms

The number of multiplications and additions required to compute the transform coefficients $v(k, l)$ using (5.3) is $O(N^4)$, which is quite excessive for practical-size images. The dimensionality of the problem is reduced to $O(N^3)$ when the transform is restricted to be separable, that is,

$$a_{k,l}(m, n) = a_k(m)b_l(n) \triangleq a(k, m)b(l, n) \tag{5.9}$$

where $\{a_k(m), k = 0, \ldots, N-1\}$, $\{b_l(n), l = 0, \ldots, N-1\}$ are one-dimensional complete orthonormal sets of basis vectors. Imposition of (5.5) and (5.6) shows that $\mathbf{A} \triangleq \{a(k, m)\}$ and $\mathbf{B} \triangleq \{b(l, n)\}$ should be unitary matrices themselves, for example,

$$\mathbf{A}\mathbf{A}^{*T} = \mathbf{A}^T\mathbf{A}^* = \mathbf{I} \tag{5.10}$$

Often one chooses $\mathbf{B}$ to be the same as $\mathbf{A}$ so that (5.3) and (5.4) reduce to

$$v(k, l) = \sum_{m, n=0}^{N-1}\sum a(k, m)u(m, n)a(l, n) \leftrightarrow \mathbf{V} = \mathbf{AUA}^T \tag{5.11}$$

$$u(m, n) = \sum_{k, l=0}^{N-1}\sum a^*(k, m)v(k, l)a^*(l, n) \leftrightarrow \mathbf{U} = \mathbf{A}^{*T}\mathbf{VA}^* \tag{5.12}$$

For an $M \times N$ rectangular image, the transform pair is

$$\mathbf{V} = \mathbf{A}_M \mathbf{U} \mathbf{A}_N \tag{5.13}$$

$$\mathbf{U} = \mathbf{A}_M^{*T} \mathbf{V} \mathbf{A}_N^{*T} \tag{5.14}$$

where $\mathbf{A}_M$ and $\mathbf{A}_N$ are $M \times M$ and $N \times N$ unitary matrices, respectively. These are called two-dimensional separable transformations. Unless otherwise stated, we will always imply the preceding separability when we mention two-dimensional unitary transformations. Note that (5.11) can be written as

$$\mathbf{V}^T = \mathbf{A}[\mathbf{AU}]^T \tag{5.15}$$

which means (5.11) can be performed by first transforming each column of $\mathbf{U}$ and then transforming each row of the result to obtain the rows of $\mathbf{V}$.

Basis Images

Let $\mathbf{a}_k^*$ denote the kth column of $\mathbf{A}^{*T}$. Define the matrices

$$\mathbf{A}_{k,l}^* = \mathbf{a}_k^* \, \mathbf{a}_l^{*T} \tag{5.16}$$

and the matrix *inner product* of two $N \times N$ matrices $\mathbf{F}$ and $\mathbf{G}$ as

$$\langle \mathbf{F}, \mathbf{G} \rangle = \sum_{m=0}^{N-1} \sum_{n=0}^{N-1} f(m, n) g^*(m, n) \tag{5.17}$$

Then (5.4) and (5.3) give a *series representation for the image* as

$$\mathbf{U} = \sum_{k, l=0}^{N-1}\sum v(k, l) \mathbf{A}_{k,l}^* \tag{5.18}$$

$$v(k, l) = \langle \mathbf{U}, \mathbf{A}_{k,l}^* \rangle \tag{5.19}$$

Equation (5.18) expresses any image $\mathbf{U}$ as a linear combination of the N^2 matrices $\mathbf{A}_{k,l}^*$, $k, l = 0, \ldots, N-1$, which are called the *basis images*. Figure 5.2 shows 8×8 basis images for the same set of transforms in Fig. 5.1. The transform coefficient $v(k, l)$ is simply the inner product of the (k, l)th basis image with the given image. It is also called the projection of the image on the (k, l)th basis image. Therefore, any $N \times N$ image can be expanded in a series using a complete set of N^2 basis images. If $\mathbf{U}$ and $\mathbf{V}$ are mapped into vectors by row ordering, then (5.11), (5.12), and (5.16) yield (see Section 2.8, on Kronecker products)

$$\boldsymbol{\upsilon} = (\mathbf{A} \otimes \mathbf{A})\boldsymbol{u} \overset{\Delta}{=} \mathcal{A}\boldsymbol{u} \tag{5.20}$$

$$\boldsymbol{u} = (\mathbf{A} \otimes \mathbf{A})^{*T} \boldsymbol{\upsilon} = \mathcal{A}^{*T} \boldsymbol{\upsilon} \tag{5.21}$$

where

$$\mathcal{A} \overset{\Delta}{=} \mathbf{A} \otimes \mathbf{A} \tag{5.22}$$

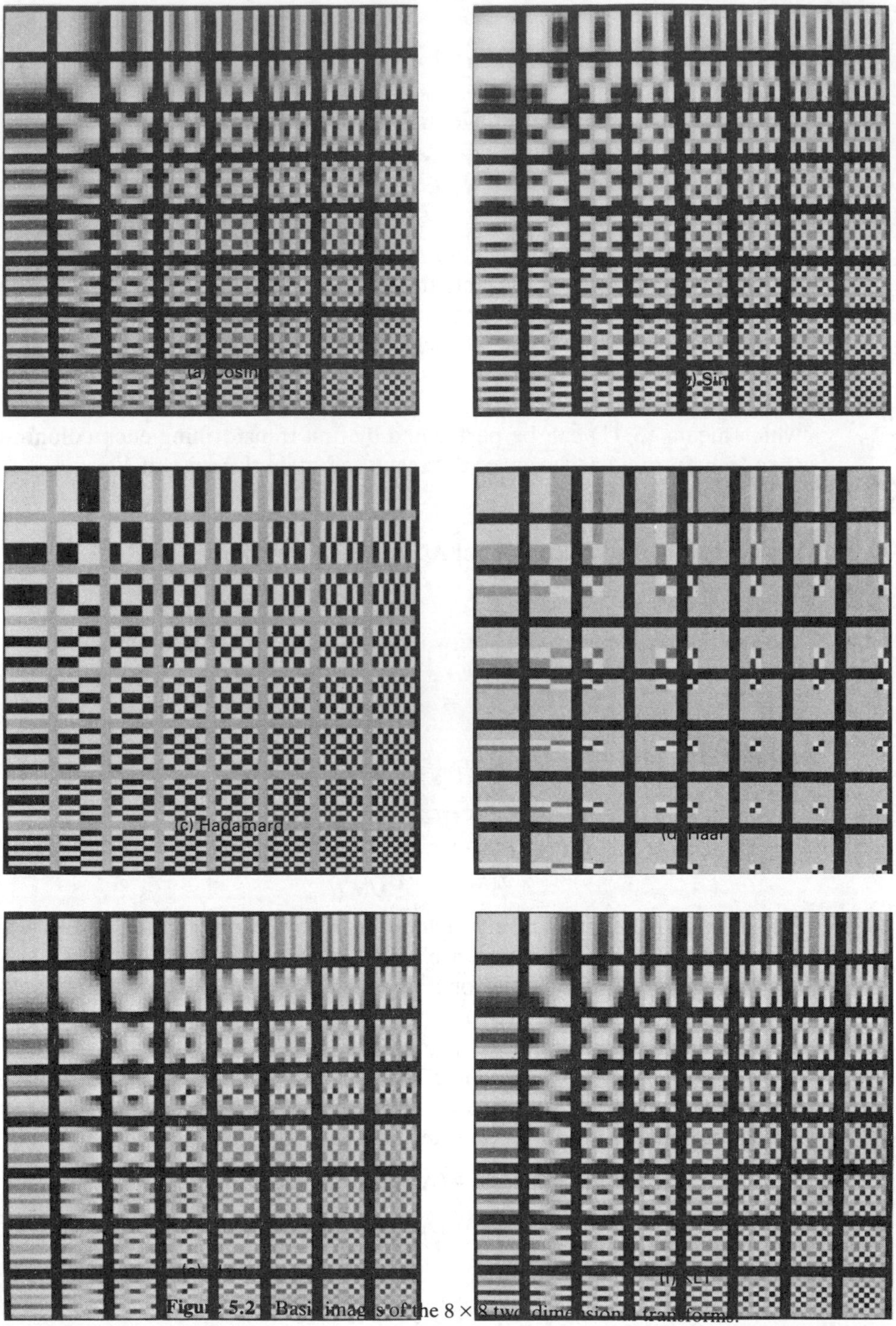

Figure 5.2 Basis images of the 8×8 two-dimensional transforms.

is a unitary matrix. Thus, given any unitary transform $\mathbf{A}$, a two-dimensional separable unitary transformation can be defined via (5.20) or (5.13).

Example 5.1

For the given orthogonal matrix $\mathbf{A}$ and image $\mathbf{U}$

$$\mathbf{A} = \frac{1}{\sqrt{2}}\begin{pmatrix} 1 & 1 \\ 1 & -1 \end{pmatrix}, \qquad \mathbf{U} = \begin{pmatrix} 1 & 2 \\ 3 & 4 \end{pmatrix}$$

the transformed image, obtained according to (5.11), is

$$\mathbf{V} = \tfrac{1}{2}\begin{pmatrix} 1 & 1 \\ 1 & -1 \end{pmatrix}\begin{pmatrix} 1 & 2 \\ 3 & 4 \end{pmatrix}\begin{pmatrix} 1 & 1 \\ 1 & -1 \end{pmatrix} = \tfrac{1}{2}\begin{pmatrix} 4 & 6 \\ -2 & -2 \end{pmatrix}\begin{pmatrix} 1 & 1 \\ 1 & -1 \end{pmatrix} = \begin{pmatrix} 5 & -1 \\ -2 & 0 \end{pmatrix}$$

To obtain the basis images, we find the *outer product of the columns* of $\mathbf{A}^{*T}$, which gives

$$\mathbf{A}^*_{0,0} = \tfrac{1}{2}\begin{pmatrix} 1 \\ 1 \end{pmatrix}(1 \quad 1) = \tfrac{1}{2}\begin{pmatrix} 1 & 1 \\ 1 & 1 \end{pmatrix}$$

and similarly

$$\mathbf{A}^*_{0,1} = \tfrac{1}{2}\begin{pmatrix} 1 & -1 \\ 1 & -1 \end{pmatrix} = \mathbf{A}^{*T}_{1,0}, \qquad \mathbf{A}^*_{1,1} = \tfrac{1}{2}\begin{pmatrix} 1 & -1 \\ -1 & 1 \end{pmatrix}$$

The inverse transformation gives

$$\mathbf{A}^{*T}\mathbf{V}\mathbf{A}^* = \tfrac{1}{2}\begin{pmatrix} 1 & 1 \\ 1 & -1 \end{pmatrix}\begin{pmatrix} 5 & -1 \\ -2 & 0 \end{pmatrix}\begin{pmatrix} 1 & 1 \\ 1 & -1 \end{pmatrix} = \tfrac{1}{2}\begin{pmatrix} 3 & -1 \\ 7 & -1 \end{pmatrix}\begin{pmatrix} 1 & 1 \\ 1 & -1 \end{pmatrix} = \begin{pmatrix} 1 & 2 \\ 3 & 4 \end{pmatrix}$$

which is $\mathbf{U}$, the original image.

Kronecker Products and Dimensionality

Dimensionality of image transforms can also be studied in terms of their Kronecker product separability. An arbitrary one-dimensional transformation

$$\mathscr{y} = \mathscr{A}\mathscr{x} \tag{5.23}$$

is called separable if

$$\mathscr{A} = \mathbf{A}_1 \otimes \mathbf{A}_2 \tag{5.24}$$

This is because (5.23) can be reduced to the separable two-dimensional transformation

$$\mathbf{Y} = \mathbf{A}_1 \mathbf{X} \mathbf{A}_2^T \tag{5.25}$$

where $\mathbf{X}$ and $\mathbf{Y}$ are matrices that map into vectors $\mathscr{x}$ and $\mathscr{y}$, respectively, by row ordering. If $\mathscr{A}$ is $N^2 \times N^2$ and $\mathbf{A}_1, \mathbf{A}_2$ are $N \times N$, then the number of operations required for implementing (5.25) reduces from N^4 to about $2N^3$. The number of operations can be reduced further if $\mathbf{A}_1$ and $\mathbf{A}_2$ are also separable. Image transforms such as discrete Fourier, sine, cosine, Hadamard, Haar, and Slant can be factored as Kronecker products of several smaller-sized matrices, which leads to fast algorithms for their implementation (see Problem 5.2). In the context of image processing such matrices are also called *fast image transforms*.

Dimensionality of Image Transforms

The $2N^3$ computations for $\mathbf{V}$ can also be reduced by restricting the choice of $\mathbf{A}$ to the *fast transforms,* whose matrix structure allows a factorization of the type

$$\mathbf{A} = \mathbf{A}_{(1)} \mathbf{A}_{(2)}, \ldots, \mathbf{A}_{(p)} \tag{5.26}$$

where $\mathbf{A}_{(i)}, i = 1, \ldots, p (p \ll N)$ are matrices with just a few nonzero entries (say r, with $r \ll N$). Thus, a multiplication of the type $\mathbf{y} = \mathbf{A}\mathbf{x}$ is accomplished in rpN operations. For Fourier, sine, cosine, Hadamard, Slant, and several other transforms, $p \simeq \log_2 N$, and the operations reduce to the order of $N \log_2 N$ (or $N^2 \log_2 N$ for $N \times N$ images). Depending on the actual transform, one operation can be defined as one multiplication and one addition or subtraction, as in the Fourier transform, or one addition or subtraction, as in the Hadamard transform.

Transform Frequency

For a one-dimensional signal $f(x)$, frequency is defined by the Fourier domain variable ξ. It is related to the number of zero crossings of the real or imaginary part of the basis function $\exp\{j2\pi\xi x\}$. This concept can be generalized to arbitrary unitary transforms. Let the rows of a unitary matrix $\mathbf{A}$ be arranged so that the number of zero crossings increases with the row number. Then in the transformation

$$\mathbf{y} = \mathbf{A}\mathbf{x}$$

the elements $y(k)$ are ordered according to increasing *wave number* or *transform frequency*. In the sequel any reference to frequency will imply the transform frequency, that is, discrete Fourier frequency, cosine frequency, and so on. The term *spatial frequency* generally refers to the continuous Fourier transform frequency and is not the same as the discrete Fourier frequency. In the case of Hadamard transform, a term called *sequency* is also used. It should be noted that this concept of frequency is useful only on a relative basis for a particular transform. A low-frequency term of one transform could contain the high-frequency harmonics of another transform.

The Optimum Transform

Another important consideration in selecting a transform is its performance in filtering and data compression of images based on the mean square criterion. The Karhunen-Loeve transform (KLT) is known to be optimum with respect to this criterion and is discussed in Section 5.11.

5.3 PROPERTIES OF UNITARY TRANSFORMS

Energy Conservation and Rotation

In the unitary transformation,

$$\mathbf{v} = \mathbf{A}\mathbf{u}$$

$$\|\mathbf{v}\|^2 = \|\mathbf{u}\|^2 \tag{5.27}$$

This is easily proven by noting that

$$\|\mathbf{v}\|^2 \triangleq \sum_{k=0}^{N-1} |v(k)|^2 = \mathbf{v}^{*T}\mathbf{v} = \mathbf{u}^{*T}\mathbf{A}^{*T}\mathbf{A}\mathbf{u} = \mathbf{u}^{*T}\mathbf{u} = \sum_{n=0}^{N-1} |u(n)|^2 \triangleq \|\mathbf{u}\|^2$$

Thus a unitary transformation preserves the signal energy or, equivalently, the length of the vector **u** in the N-dimensional vector space. This means every unitary transformation is simply a rotation of the vector **u** in the N-dimensional vector space. Alternatively, a unitary transformation is a rotation of the basis coordinates and the components of **v** are the projections of **u** on the new basis (see Problem 5.4). Similarly, for the two-dimensional unitary transformations such as (5.3), (5.4), and (5.11) to (5.14), it can be proven that

$$\sum_{m,n=0}^{N-1}\sum |u(m, n)|^2 = \sum_{k,l=0}^{N-1}\sum |v(k, l)|^2 \tag{5.28}$$

Energy Compaction and Variances of Transform Coefficients

Most unitary transforms have a tendency to pack a large fraction of the average energy of the image into a relatively few components of the transform coefficients. Since the total energy is preserved, this means many of the transform coefficients will contain very little energy. If $\boldsymbol{\mu}_u$ and $\mathbf{R}_u$ denote the mean and covariance of a vector **u**, then the corresponding quantities for the transformed vector **v** are given by

$$\boldsymbol{\mu}_v \triangleq E[\mathbf{v}] = E[\mathbf{A}\mathbf{u}] = \mathbf{A}E[\mathbf{u}] = \mathbf{A}\boldsymbol{\mu}_u \tag{5.29}$$

$$\begin{aligned}\mathbf{R}_v &= E[(\mathbf{v} - \boldsymbol{\mu}_v)(\mathbf{v} - \boldsymbol{\mu}_v)^{*T}] \\ &= \mathbf{A}(E[(\mathbf{u} - \boldsymbol{\mu}_u)(\mathbf{u} - \boldsymbol{\mu}_u)^{*T}])\mathbf{A}^{*T} = \mathbf{A}\mathbf{R}_u\mathbf{A}^{*T}\end{aligned} \tag{5.30}$$

The transform coefficient variances are given by the diagonal elements of $\mathbf{R}_v$, that is

$$\sigma_v^2(k) = [\mathbf{R}_v]_{k,k} = [\mathbf{A}\mathbf{R}_u\mathbf{A}^{*T}]_{k,k} \tag{5.31}$$

Since A is unitary, it follows that

$$\sum_{k=0}^{N-1} |\mu_v(k)|^2 = \boldsymbol{\mu}_v^{*T}\boldsymbol{\mu}_v = \boldsymbol{\mu}_u^{*T}\mathbf{A}^{*T}\mathbf{A}\boldsymbol{\mu}_u = \sum_{n=0}^{N-1} |\mu_u(n)|^2 \tag{5.32}$$

$$\sum_{k=0}^{N-1} \sigma_v^2(k) = \text{Tr}[\mathbf{A}\mathbf{R}_u\mathbf{A}^{*T}] = \text{Tr}[\mathbf{R}_u] = \sum_{n=0}^{N-1} \sigma_u^2(n) \tag{5.33}$$

$$\Rightarrow \quad \sum_{k=0}^{N-1} E[|v(k)|^2] = \sum_{n=0}^{N-1} E[|u(n)|^2] \tag{5.34}$$

The average energy $E[|v(k)|^2]$ of the transform coefficients $v(k)$ tends to be unevenly distributed, although it may be evenly distributed for the input sequence $u(n)$. For a two-dimensional random field $u(m, n)$ whose mean is $\mu_u(m, n)$ and covariance is $r(m, n; m', n')$, its transform coefficients $v(k, l)$ satisfy the properties

$$\mu_v(k, l) = \sum_m \sum_n a(k, m)a(l, n)\mu_u(m, n) \tag{5.35}$$

$$\sigma_v^2(k,l) = E[|v(k,l) - \mu_v(k,l)|^2]$$
$$= \sum_m \sum_n \sum_{m'} \sum_{n'} a(k,m)a(l,n)r(m,n;m',n')a^*(k,m')a^*(l,n') \tag{5.36}$$

If the covariance of $u(m,n)$ is separable, that is

$$r(m,n;m',n') = r_1(m,m')r_2(n,n') \tag{5.37}$$

then the variances of the transform coefficients can be written as a separable product

$$\sigma_v^2(k,l) = \sigma_1^2(k)\sigma_2^2(l)$$
$$\triangleq [\mathbf{A}\mathbf{R}_1\mathbf{A}^{*T}]_{k,k}[\mathbf{A}\mathbf{R}_2\mathbf{A}^{*T}]_{l,l} \tag{5.38}$$

where

$$\mathbf{R}_1 = \{r_1(m,m')\} \quad \text{and} \quad \mathbf{R}_2 = \{r_2(n,n')\}$$

Decorrelation

When the input vector elements are highly correlated, the transform coefficients tend to be uncorrelated. This means the off-diagonal terms of the covariance matrix $\mathbf{R}_v$ tend to become small compared to the diagonal elements.

With respect to the preceding two properties, the KL transform is optimum, that is, it packs the maximum average energy in a given number of transform coefficients while completely decorrelating them. These properties are presented in greater detail in Section 5.11.

Other Properties

Unitary transforms have other interesting properties. For example, the determinant and the eigenvalues of a unitary matrix have unity magnitude. Also, the entropy of a random vector is preserved under a unitary transformation. Since entropy is a measure of average information, this means information is preserved under a unitary transformation.

Example 5.2 (Energy compaction and decorrelation)

A 2×1 zero mean vector $\mathbf{u}$ is unitarily transformed as

$$\mathbf{v} = \tfrac{1}{2}\begin{pmatrix} \sqrt{3} & 1 \\ -1 & \sqrt{3} \end{pmatrix}\mathbf{u}, \quad \text{where} \quad \mathbf{R}_u \triangleq \begin{pmatrix} 1 & \rho \\ \rho & 1 \end{pmatrix}, \qquad 0 < \rho < 1$$

The parameter ρ measures the correlation between $u(0)$ and $u(1)$. The covariance of $\mathbf{v}$ is obtained as

$$\mathbf{R}_v = \begin{pmatrix} 1 + \sqrt{3}\rho/2 & \rho/2 \\ \rho/2 & 1 - \sqrt{3}\rho/2 \end{pmatrix}$$

From the expression for $\mathbf{R}_u$, $\sigma_u^2(0) = \sigma_u^2(1) = 1$, that is, the total average energy of 2 is distributed equally between $u(0)$ and $u(1)$. However, $\sigma_v^2(0) = 1 + \sqrt{3}\rho/2$ and $\sigma_v^2(1) = 1 - \sqrt{3}\rho/2$. The total average energy is still 2, but the average energy in $v(0)$ is greater than in $v(1)$. If $\rho = 0.95$, then 91.1% of the total average energy has been packed in the first sample. The correlation between $v(0)$ and $v(1)$ is given by

$$\rho_v(0,1) \triangleq \frac{E[v(0)v(1)]}{\sigma_v(0)\sigma_v(1)} = \frac{\rho}{2(1-\frac{3}{4}\rho^2)^{1/2}}$$

which is less in absolute value than $|\rho|$ for $|\rho| < 1$. For $\rho = 0.95$, we find $\rho_v(0,1) = 0.83$. Hence the correlation between the transform coefficients has been reduced. If the foregoing procedure is repeated for the 2×2 transform **A** of Example 5.1, then we find $\sigma_v^2(0) = 1 + \rho$, $\sigma_v^2(1) = 1 - \rho$, and $\rho_v(0,1) = 0$. For $\rho = 0.95$, now 97.5% of the energy is packed in $v(0)$. Moreover, $v(0)$ and $v(1)$ become uncorrelated.

5.4 THE ONE-DIMENSIONAL DISCRETE FOURIER TRANSFORM (DFT)

The discrete Fourier transform (DFT) of a sequence $\{u(n), n = 0, \ldots, N-1\}$ is defined as

$$v(k) = \sum_{n=0}^{N-1} u(n) W_N^{kn}, \qquad k = 0, 1, \ldots, N-1 \tag{5.39}$$

where

$$W_N \triangleq \exp\left\{\frac{-j2\pi}{N}\right\} \tag{5.40}$$

The inverse transform is given by

$$u(n) = \frac{1}{N} \sum_{k=0}^{N-1} v(k) W_N^{-kn}, \qquad n = 0, 1, \ldots, N-1 \tag{5.41}$$

The pair of equations (5.39) and (5.41) are not scaled properly to be unitary transformations. In image processing it is more convenient to consider the *unitary DFT,* which is defined as

$$v(k) = \frac{1}{\sqrt{N}} \sum_{n=0}^{N-1} u(n) W_N^{kn}, \qquad k = 0, \ldots, N-1 \tag{5.42}$$

$$u(n) = \frac{1}{\sqrt{N}} \sum_{k=0}^{N-1} v(k) W_N^{-kn}, \qquad n = 0, \ldots, N-1 \tag{5.43}$$

The $N \times N$ unitary DFT matrix **F** is given by

$$\mathbf{F} = \left\{\frac{1}{\sqrt{N}} W_N^{kn}\right\}, \qquad 0 \le k, n \le N-1 \tag{5.44}$$

Future references to DFT and unitary DFT will imply the definitions of (5.39) and (5.42), respectively. The DFT is one of the most important transforms in digital signal and image processing. It has several properties that make it attractive for image processing applications.

Properties of the DFT/Unitary DFT

Let $u(n)$ be an arbitrary sequence defined for $n = 0, 1, \ldots, N-1$. A circular shift of $u(n)$ by l, denoted by $u(n-l)_c$, is defined as $u[(n-l) \text{ modulo } N]$. See Fig. 5.3 for $l = 2, N = 5$.

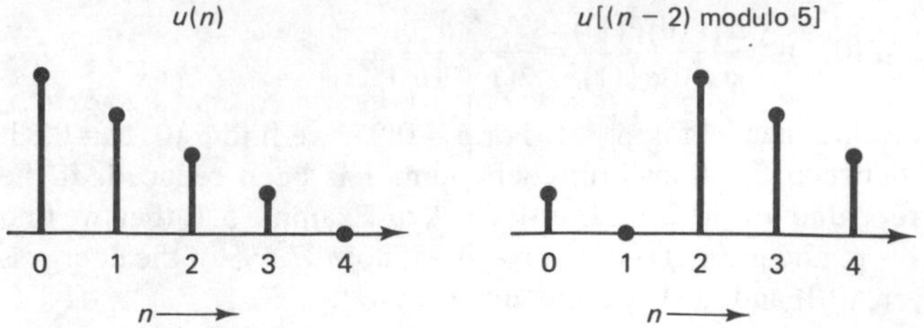

Figure 5.3 Circular shift of $u(n)$ by 2.

The DFT and Unitary DFT matrices are symmetric. By definition the matrix $\mathbf{F}$ is symmetric. Therefore,

$$\mathbf{F}^{-1} = \mathbf{F}^* \tag{5.45}$$

The extensions are periodic. The extensions of the DFT and unitary DFT of a sequence and their inverse transforms are periodic with period N. If for example, in the definition of (5.42) we let k take all integer values, then the sequence $v(k)$ turns out to be periodic, that is, $v(k) = v(k+N)$ for every k.

The DFT is the sampled spectrum of the finite sequence $u(n)$ extended by zeros outside the interval $[0, N-1]$. If we define a zero-extended sequence

$$\tilde{u}(n) \triangleq \begin{cases} u(n), & 0 \le n \le N-1 \\ 0, & \text{otherwise} \end{cases} \tag{5.46}$$

then its Fourier transform is

$$\tilde{U}(\omega) = \sum_{n=-\infty}^{\infty} \tilde{u}(n) \exp(-j\omega n) = \sum_{n=0}^{N-1} u(n) \exp(-j\omega n) \tag{5.47}$$

Comparing this with (5.39) we see that

$$v(k) = \tilde{U}\left(\frac{2\pi k}{N}\right) \tag{5.48}$$

Note that the unitary DFT of (5.42) would be $\tilde{U}(2\pi k/N)/\sqrt{N}$.

The DFT and unitary DFT of dimension N can be implemented by a fast algorithm in $O(N \log_2 N)$ operations. There exists a class of algorithms, called the *fast Fourier transform* (FFT), which requires $O(N \log_2 N)$ operations for implementing the DFT or unitary DFT, where one operation is a real multiplication and a real addition. The exact operation count depends on N as well as the particular choice of the algorithm in that class. Most common FFT algorithms require $N = 2^p$, where p is a positive integer.

The DFT or unitary DFT of a real sequence $\{x(n), n = 0, \ldots, N-1\}$ is conjugate symmetric about $N/2$. From (5.42) we obtain

$$v^*(N-k) = \sum_{n=0}^{N-1} u^*(n) W_N^{-(N-k)n} = \sum_{n=0}^{N-1} u(n) W_N^{kn} = v(k)$$

$$\Rightarrow \quad v\left(\frac{N}{2} - k\right) = v^*\left(\frac{N}{2} + k\right), \qquad k = 0, \ldots, \frac{N}{2} - 1 \tag{5.49}$$

$$\Rightarrow \quad \left|v\left(\frac{N}{2} - k\right)\right| = \left|v\left(\frac{N}{2} + k\right)\right| \tag{5.50}$$

Figure 5.4 shows a 256-sample scan line of an image. The magnitude of its DFT is shown in Fig. 5.5, which exhibits symmetry about the point 128. If we consider the periodic extension of $v(k)$, we see that

$$v(-k) = v(N - k)$$

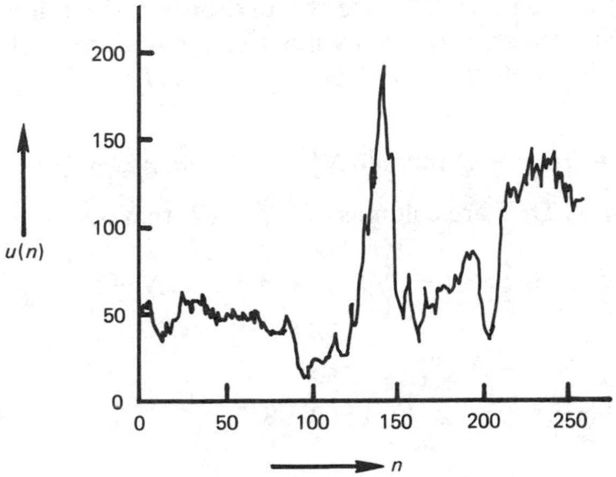

Figure 5.4 A 256-sample scan line of an image.

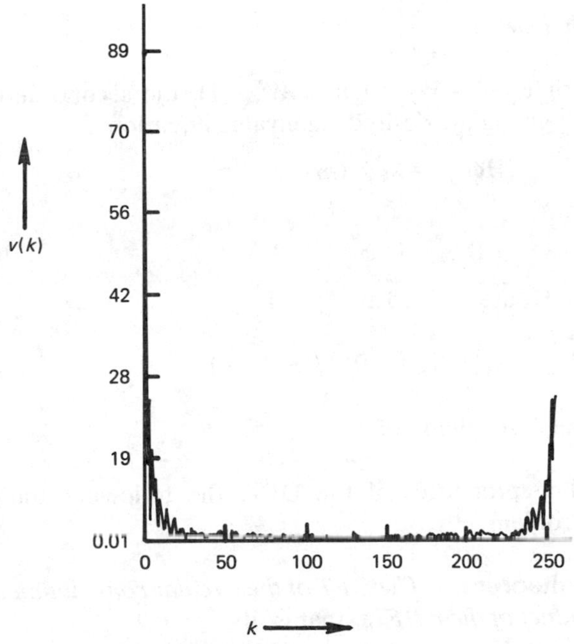

Figure 5.5 Unitary discrete Fourier transform of Fig. 5.4.

Hence the (unitary) DFT frequencies $N/2 + k$, $k = 0, \ldots, N/2 - 1$, are simply the negative frequencies at $\omega = (2\pi/N)(-N/2 + k)$ in the Fourier spectrum of the finite sequence $\{u(n), 0 \le n \le N - 1\}$. Also, from (5.39) and (5.49), we see that $v(0)$ and $v(N/2)$ are real, so that the $N \times 1$ real sequence

$$v(0), \left\{\text{Re}\{v(k)\}, k = 1, \ldots, \frac{N}{2} - 1\right\}, \left\{\text{Im}\{v(k)\}, k = 1, \ldots, \frac{N}{2} - 1\right\}, v\left(\frac{N}{2}\right) \tag{5.51}$$

completely defines the DFT of the real sequence $u(n)$. Therefore, it can be said that the DFT or unitary DFT of an $N \times 1$ real sequence has N degrees of freedom and requires the same storage capacity as the sequence itself.

The basis vectors of the unitary DFT are the orthonormal eigenvectors of any circulant matrix. Moreover, the eigenvalues of a circulant matrix are given by the DFT of its first column. Let $\mathbf{H}$ be an $N \times N$ circulant matrix. Therefore, its elements satisfy

$$[\mathbf{H}]_{m,n} = h(m - n) = h[(m - n) \text{ modulo } N], \qquad 0 \le m, n \le N - 1 \tag{5.52}$$

The basis vectors of the unitary DFT are columns of $\mathbf{F}^{*T} = \mathbf{F}^*$, that is,

$$\boldsymbol{\phi}_k = \left\{\frac{1}{\sqrt{N}} W_N^{-kn}, 0 \le n \le N - 1\right\}^T, \qquad k = 0, \ldots, N - 1 \tag{5.53}$$

Consider the expression

$$[\mathbf{H}\boldsymbol{\phi}_k]_m = \frac{1}{\sqrt{N}} \sum_{n=0}^{N-1} h(m - n) W_N^{-kn} \tag{5.54}$$

Writing $m - n = l$ and rearranging terms, we can write

$$[\mathbf{H}\boldsymbol{\phi}_k]_m = \frac{1}{\sqrt{N}} W_N^{-km} \left[\sum_{l=0}^{N-1} h(l) W_N^{kl} + \sum_{l=-N+m+1}^{-1} h(l) W_N^{kl} - \sum_{l=m+1}^{N-1} h(l) W_N^{kl}\right] \tag{5.55}$$

Using (5.52) and the fact that $W_N^{-l} = W_N^{N-l}$ (since $W_N^N = 1$), the second and third terms in the brackets cancel, giving the desired eigenvalue equation

$$[\mathbf{H}\boldsymbol{\phi}_k]_m = \lambda_k \boldsymbol{\phi}_k(m)$$

or

$$\mathbf{H}\boldsymbol{\phi}_k = \lambda_k \boldsymbol{\phi}_k \tag{5.56}$$

where λ_k, the eigenvalues of $\mathbf{H}$, are defined as

$$\lambda_k \triangleq \sum_{l=0}^{N-1} h(l) W_N^{kl}, \qquad 0 \le k \le N - 1 \tag{5.57}$$

This is simply the DFT of the first column of $\mathbf{H}$.

Based on the preceding properties of the DFT, the following additional properties can be proven (Problem 5.9).

Circular convolution theorem. *The DFT of the circular convolution of two sequences is equal to the product of their DFTs,* that is, if

$$x_2(n) = \sum_{k=0}^{N-1} h(n-k)_c x_1(k), \qquad 0 \le n \le N-1 \tag{5.58}$$

then

$$\text{DFT}\{x_2(n)\}_N = \text{DFT}\{h(n)\}_N \text{DFT}\{x_1(n)\}_N \tag{5.59}$$

where $\text{DFT}\{x(n)\}_N$ denotes the DFT of the sequence $x(n)$ of size N. This means we can calculate the circular convolution by first calculating the DFT of $x_2(n)$ via (5.59) and then taking its inverse DFT. Using the FFT this will take $O(N \log_2 N)$ operations, compared to N^2 operations required for direct evaluation of (5.58).

A linear convolution of two sequences can also be obtained via the FFT by imbedding it into a circular convolution. In general, the linear convolution of two sequences $\{h(n), n = 0, \ldots, N'-1\}$ and $\{x_1(n), n = 0, \ldots, N-1\}$ is a sequence $\{x_2(n), 0 \le n \le N' + N - 2\}$ and can be obtained by the following algorithm:

Step 1: Let $M \ge N' + N - 1$ be an integer for which an FFT algorithm is available.

Step 2: Define $\tilde{h}(n)$ and $\tilde{x}_1(n), 0 \le n \le M-1$, as zero extended sequences corresponding to $h(n)$ and $x_1(n)$, respectively.

Step 3: Let $\tilde{y}_1(k) = \text{DFT}\{\tilde{x}_1(n)\}_M, \lambda_k = \text{DFT}\{\tilde{h}(n)\}_M$. Define $\tilde{y}_2(k) = \lambda_k \tilde{y}_1(k)$, $k = 0, \ldots, M-1$.

Step 4: Take the inverse DFT of $\tilde{y}_2(k)$ to obtain $\tilde{x}_2(n)$. Then $x_2(n) = \tilde{x}_2(n)$ for $0 \le n \le N + N' - 2$.

Any circulant matrix can be diagonalized by the DFT/unitary DFT. That is,

$$\mathbf{F}\mathbf{H}\mathbf{F}^* = \mathbf{\Lambda} \tag{5.60}$$

where $\mathbf{\Lambda} = \text{Diag}\{\lambda_k, 0 \le k \le N-1\}$ and λ_k are given by (5.57). It follows that if $\mathbf{C}$, $\mathbf{C}_1$ and $\mathbf{C}_2$ are circulant matrices, then the following hold.

1. $\mathbf{C}_1\mathbf{C}_2 = \mathbf{C}_2\mathbf{C}_1$, that is, circulant matrices commute.
2. $\mathbf{C}^{-1}$ is a circulant matrix and can be computed in $O(N \log N)$ operations.
3. $\mathbf{C}^T$, $\mathbf{C}_1 + \mathbf{C}_2$, and $f(\mathbf{C})$ are all circulant matrices, where $f(x)$ is an arbitrary function of x.

5.5 THE TWO-DIMENSIONAL DFT

The two-dimensional DFT of an $N \times N$ image $\{u(m, n)\}$ is a separable transform defined as

$$v(k, l) = \sum_{m=0}^{N-1} \sum_{n=0}^{N-1} u(m, n) W_N^{km} W_N^{ln}, \qquad 0 \le k, l \le N-1 \tag{5.61}$$

and the inverse transform is

$$u(m, n) = \frac{1}{N^2} \sum_{k=0}^{N-1} \sum_{l=0}^{N-1} v(k, l) W_N^{-km} W_N^{-ln}, \qquad 0 \le m, n \le N-1 \tag{5.62}$$

The two-dimensional unitary DFT pair is defined as

$$v(k, l) = \frac{1}{N} \sum_{m=0}^{N-1} \sum_{n=0}^{N-1} u(m, n) W_N^{km} W_N^{ln}, \qquad 0 \le k, l \le N-1 \tag{5.63}$$

$$u(m, n) = \frac{1}{N} \sum_{k=0}^{N-1} \sum_{l=0}^{N-1} v(k, l) W_N^{-km} W_N^{-ln}, \qquad 0 \le m, n \le N-1 \tag{5.64}$$

In matrix notation this becomes

$$\mathbf{V} = \mathbf{FUF} \tag{5.65}$$

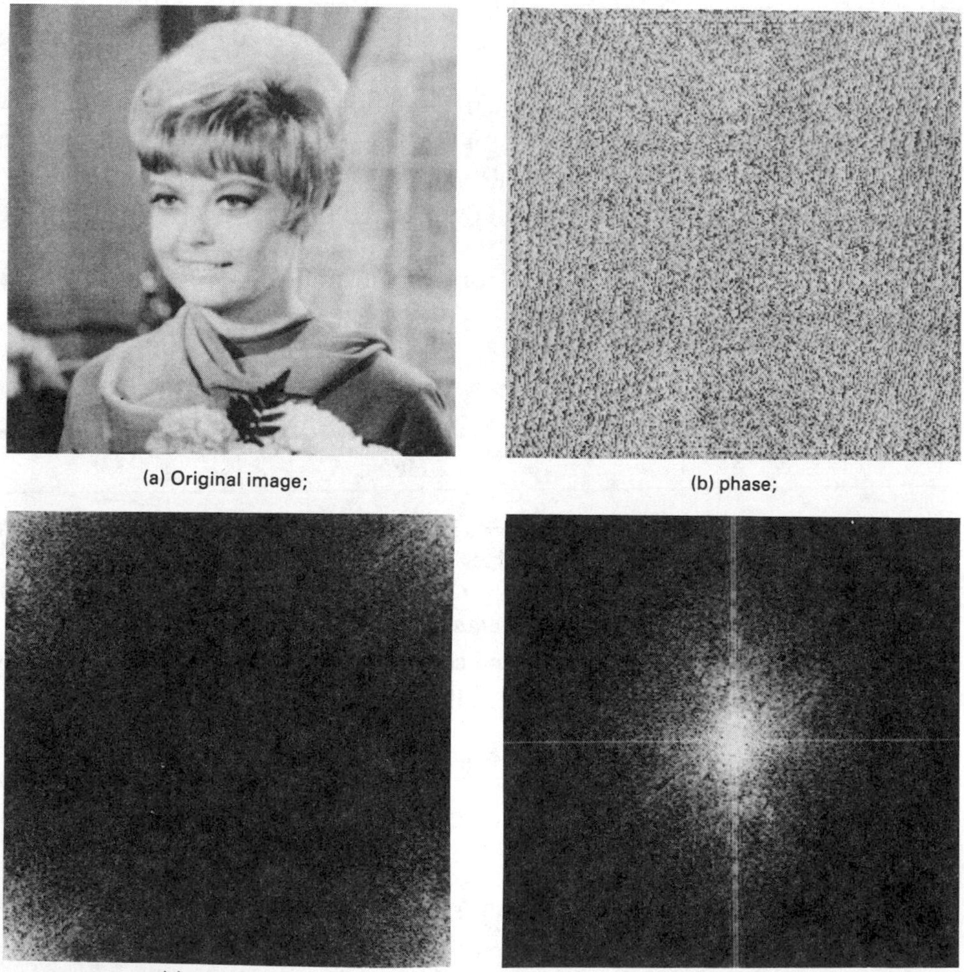

(a) Original image; (b) phase; (c) magnitude; (d) magnitude centered.

Figure 5.6 Two-dimensional unitary DFT of a 256 × 256 image.

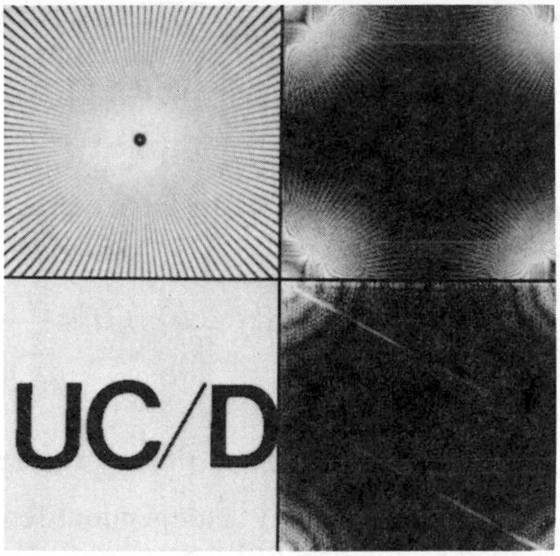

a	*b*
c	*d*

Figure 5.7 Unitary DFT of images

(a) Resolution chart;
(b) its DFT;
(c) binary image;
(d) its DFT. The two parallel lines are due to the '/' sign in the binary image.

$$\mathbf{U} = \mathbf{F}^*\mathbf{V}\mathbf{F}^* \tag{5.66}$$

If $\mathbf{U}$ and $\mathbf{V}$ are mapped into row-ordered vectors $\mathscr{u}$ and $\mathscr{v}$, respectively, then

$$\mathscr{v} = \mathscr{F}\mathscr{u}, \qquad \mathscr{u} = \mathscr{F}^*\mathscr{v} \tag{5.67}$$

$$\mathscr{F} = \mathbf{F} \otimes \mathbf{F} \tag{5.68}$$

The $N^2 \times N^2$ matrix $\mathscr{F}$ represents *the* $N \times N$ *two-dimensional unitary* DFT. Figure 5.6 shows an original image and the magnitude and phase components of its unitary DFT. Figure 5.7 shows magnitudes of the unitary DFTs of two other images.

Properties of the Two-Dimensional DFT

The properties of the two-dimensional unitary DFT are quite similar to the one-dimensional case and are summarized next.

Symmetric, unitary.

$$\mathscr{F}^T = \mathscr{F}, \qquad \mathscr{F}^{-1} = \mathscr{F}^* = \mathbf{F}^* \otimes \mathbf{F}^* \tag{5.69}$$

Periodic extensions.

$$\begin{aligned} v(k+N, l+N) &= v(k,l), \qquad \forall k, l \\ u(m+N, n+N) &= u(m,n), \qquad \forall m, n \end{aligned} \tag{5.70}$$

Sampled Fourier spectrum. If $\tilde{u}(m,n) = u(m,n), 0 \le m, n \le N-1$, and $\tilde{u}(m,n) = 0$ otherwise, then

$$\tilde{U}\left(\frac{2\pi k}{N}, \frac{2\pi l}{N}\right) = \text{DFT}\{u(m,n)\} - v(k,l) \tag{5.71}$$

where $\tilde{U}(\omega_1, \omega_2)$ is the Fourier transform of $\tilde{u}(m,n)$.

Fast transform. Since the two-dimensional DFT is separable, the transformation of (5.65) is equivalent to $2N$ one-dimensional unitary DFTs, each of which can be performed in $O(N \log_2 N)$ operations via the FFT. Hence the total number of operations is $O(N^2 \log_2 N)$.

Conjugate symmetry. The DFT and unitary DFT of *real images* exhibit conjugate symmetry, that is,

$$v\left(\frac{N}{2} \pm k, \frac{N}{2} \pm l\right) = v^*\left(\frac{N}{2} \mp k, \frac{N}{2} \mp l\right), \qquad 0 \le k, l \le \frac{N}{2} - 1 \qquad (5.72)$$

or

$$v(k, l) = v^*(N - k, N - l), \qquad 0 \le k, l \le N - 1 \qquad (5.73)$$

From this, it can be shown that $v(k, l)$ has only N^2 independent real elements. For example, the samples in the shaded region of Fig. 5.8 determine the complete DFT or unitary DFT (see problem 5.10).

Basis images. The basis images are given by definition [see (5.16) and (5.53)]:

$$\mathbf{A}^*_{k,l} = \boldsymbol{\phi}_k \boldsymbol{\phi}_l^T = \frac{1}{N}\{W_N^{-(km + ln)}, \qquad 0 \le m, n \le N - 1\}, \qquad 0 \le k, l \le N - 1 \qquad (5.74)$$

Two-dimensional circular convolution theorem. *The DFT of the two-dimensional circular convolution of two arrays is the product of their DFTs.*

Two-dimensional circular convolution of two $N \times N$ arrays $h(m, n)$ and $u_1(m, n)$ is defined as

$$u_2(m, n) = \sum_{m'=0}^{N-1} \sum_{n'=0}^{N-1} h(m - m', n - n')_c u_1(m', n'), \qquad 0 \le m, n \le N - 1 \qquad (5.75)$$

where

$$h(m, n)_c = h(m \text{ modulo } N, n \text{ modulo } N) \qquad (5.76)$$

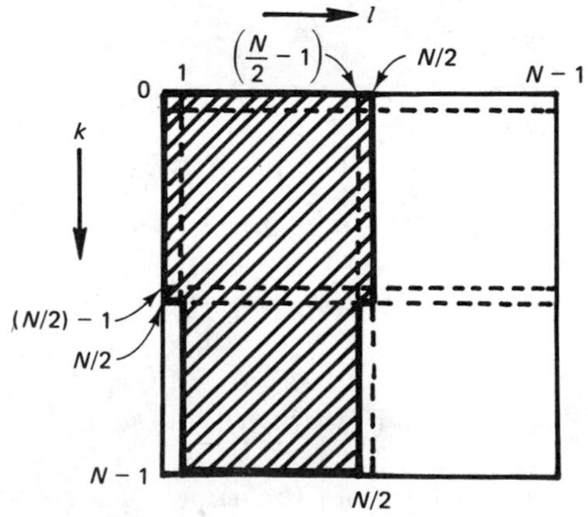

Figure 5.8 Discrete Fourier transform coefficients $v(k, l)$ in the shaded area determine the remaining coefficients.

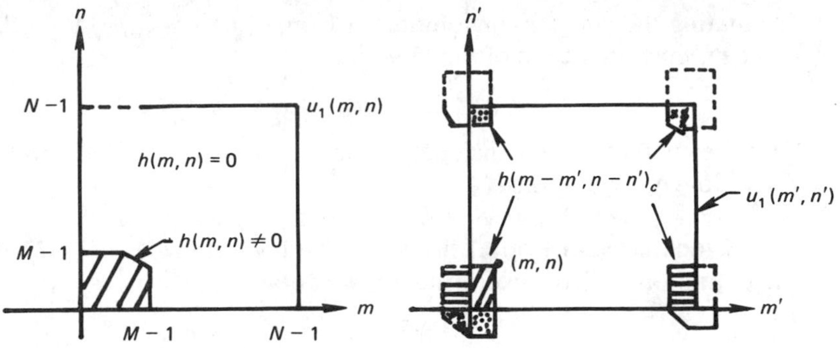

(a) Array $h(m, n)$.

(b) Circular convolution of $h(m, n)$ with $u_1(m, n)$ over $N \times N$ region.

Figure 5.9 Two-dimensional circular convolution.

Figure 5.9 shows the meaning of circular convolution. It is the same when a periodic extension of $h(m, n)$ is convolved over an $N \times N$ region with $u_1(m, n)$. The two-dimensional DFT of $h(m - m', n - n')_c$ for fixed m', n' is given by

$$\begin{aligned}\sum_{m=0}^{N-1}\sum_{n=0}^{N-1} h(m - m', n - n')_c W_N^{(mk+nl)} &= W_N^{(m'k+n'l)} \sum_{i=-m'}^{N-1-m'} \sum_{j=-n'}^{N-1-n'} h(i, j)_c W_N^{(ik+jl)} \\ &= W_N^{(m'k+n'l)} \sum_{m=0}^{N-1}\sum_{n=0}^{N-1} h(m, n) W_N^{(mk+nl)} \\ &= W_N^{(m'k+n'l)} \text{DFT}\{h(m, n)\}_N\end{aligned} \tag{5.77}$$

where we have used (5.76). Taking the DFT of both sides of (5.75) and using the preceding result, we obtain†

$$\text{DFT}\{u_2(m, n)\}_N = \text{DFT}\{h(m, n)\}_N \text{DFT}\{u_1(m, n)\}_N \tag{5.78}$$

From this and the fast transform property (page 142), it follows that an $N \times N$ circular convolution can be performed in $O(N^2 \log_2 N)$ operations. This property is also useful in calculating two-dimensional convolutions such as

$$x_3(m, n) = \sum_{m'=0}^{M-1}\sum_{n'=0}^{M-1} x_2(m - m', n - n') x_1(m', n') \tag{5.79}$$

where $x_1(m, n)$ and $x_2(m, n)$ are assumed to be zero for $m, n \notin [0, M - 1]$. The region of support for the result $x_3(m, n)$ is $\{0 \leq m, n \leq 2M - 2\}$. Let $N \geq 2M - 1$ and define $N \times N$ arrays

$$\tilde{h}(m, n) \triangleq \begin{cases} x_2(m, n), & 0 \leq m, n \leq M - 1 \\ 0, & \text{otherwise} \end{cases} \tag{5.80}$$

$$\tilde{u}_1(m, n) \triangleq \begin{cases} x_1(m, n), & 0 \leq m, n \leq M - 1 \\ 0, & \text{otherwise} \end{cases} \tag{5.81}$$

† We denote $\text{DFT}\{x(m, n)\}_N$ as the two-dimensional DFT of an $N \times N$ array $x(m, n), 0 \leq m, n \leq N - 1$.

Evaluating the circular convolution of $\tilde{h}(m, n)$ and $\tilde{u}_1(m, n)$ according to (5.75), it can be seen with the aid of Fig. 5.9 that

$$x_3(m, n) = u_2(m, n), \qquad 0 \le m, n \le 2M - 2 \tag{5.82}$$

This means the two-dimensional linear convolution of (5.79) can be performed in $O(N^2 \log_2 N)$ operations.

Block circulant operations. Dividing both sides of (5.77) by N and using the definition of Kronecker product, we obtain

$$(\mathbf{F} \otimes \mathbf{F})\mathcal{H} = \mathcal{D}(\mathbf{F} \otimes \mathbf{F}) \tag{5.83}$$

where $\mathcal{H}$ is doubly circulant and $\mathcal{D}$ is diagonal whose elements are given by

$$[\mathcal{D}]_{kN+l,\, kN+l} \triangleq d_{k,l} = \text{DFT}\{h(m, n)\}_N, \qquad 0 \le k, l \le N - 1 \tag{5.84}$$

Eqn. (5.83) can be written as

$$\mathcal{F}\mathcal{H} = \mathcal{D}\mathcal{F} \qquad \text{or} \qquad \mathcal{F}\mathcal{H}\mathcal{F}^* = \mathcal{D} \tag{5.85}$$

that is, a doubly block circulant matrix is diagonalized by the two-dimensional unitary DFT. From (5.84) and the fast transform property (page 142), we conclude that a doubly block circulant matrix can be diagonalized in $O(N^2 \log_2 N)$ operations. The eigenvalues of $\mathcal{H}$, given by the two-dimensional DFT of $h(m, n)$, are the same as operating $N\mathcal{F}$ on the first column of $\mathcal{H}$. This is because the elements of the first column of $\mathcal{H}$ are the elements $h(m, n)$ mapped by lexicographic ordering.

Block Toeplitz operations. Our discussion on linear convolution implies that any doubly block Toeplitz matrix operation can be imbedded into a double block circulant operation, which, in turn, can be implemented using the two-dimensional unitary DFT.

5.6 THE COSINE TRANSFORM

The $N \times N$ cosine transform matrix $\mathbf{C} = \{c(k, n)\}$, also called the *discrete cosine transform* (DCT), is defined as

$$c(k, n) = \begin{cases} \dfrac{1}{\sqrt{N}}, & k = 0,\ 0 \le n \le N - 1 \\ \sqrt{\dfrac{2}{N}} \cos \dfrac{\pi(2n + 1)k}{2N}, & 1 \le k \le N - 1,\ 0 \le n \le N - 1 \end{cases} \tag{5.86}$$

The one-dimensional DCT of a sequence $\{u(n), 0 \le n \le N - 1\}$ is defined as

$$v(k) = \alpha(k) \sum_{n=0}^{N-1} u(n) \cos\left[\frac{\pi(2n + 1)k}{2N}\right], \qquad 0 \le k \le N - 1 \tag{5.87}$$

where

$$\alpha(0) \triangleq \sqrt{\frac{1}{N}}, \qquad \alpha(k) \triangleq \sqrt{\frac{2}{N}} \quad \text{for} \quad 1 \le k \le N - 1 \tag{5.88}$$

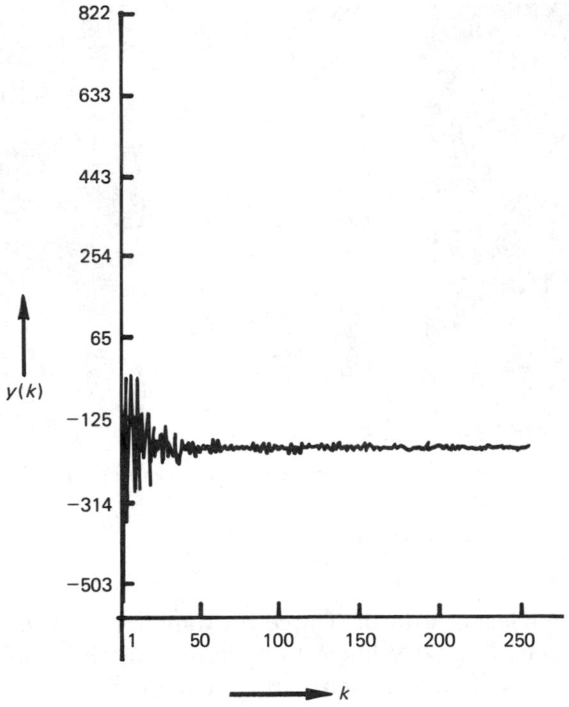

Figure 5.10 Cosine transform of the image scan line shown in Fig. 5.4.

The inverse transformation is given by

$$u(n) = \sum_{k=0}^{N-1} \alpha(k)v(k) \cos\left[\frac{\pi(2n+1)k}{2N}\right], \qquad 0 \le n \le N-1 \tag{5.89}$$

The basis vectors of the 8×8 DCT are shown in Fig. 5.1. Figure 5.10 shows the cosine transform of the image scan line shown in Fig. 5.4. Note that many transform coefficients are small, that is, most of the energy of the data is packed in a few transform coefficients.

The two-dimensional cosine transform pair is obtained by substituting $\mathbf{A} = \mathbf{A}^* = \mathbf{C}$ in (5.11) and (5.12). The basis images of the 8×8 two-dimensional cosine transform are shown in Fig. 5.2. Figure 5.11 shows examples of the cosine transform of different images.

Properties of the Cosine Transform

1. The cosine transform is real and orthogonal, that is,

$$\mathbf{C} = \mathbf{C}^* \quad \Rightarrow \mathbf{C}^{-1} = \mathbf{C}^T \tag{5.90}$$

2. The cosine transform is not the real part of the unitary DFT. This can be seen by inspection of $\mathbf{C}$ and the DFT matrix $\mathbf{F}$. (Also see Problem 5.13.) However, the cosine transform of a sequence is related to the DFT of its symmetric extension (see Problem 5.16).

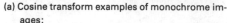

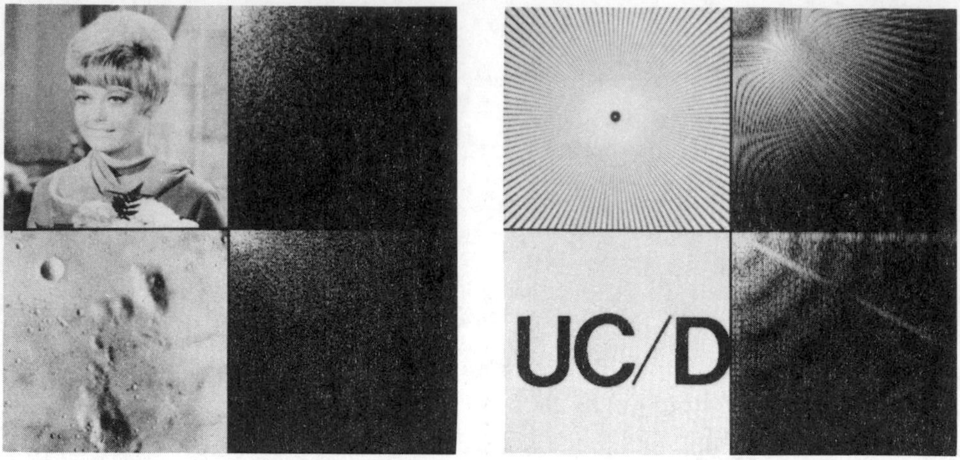

(a) Cosine transform examples of monochrome images; (b) Cosine transform examples of binary images.

Figure 5.11

3. The cosine transform is a fast transform. The cosine transform of a vector of N elements can be calculated in $O(N \log_2 N)$ operations via an N-point FFT [19]. To show this we define a new sequence $\tilde{u}(n)$ by reordering the even and odd elements of $u(n)$ as

$$\left.\begin{aligned} \tilde{u}(n) &= u(2n) \\ \tilde{u}(N-n-1) &= u(2n+1) \end{aligned}\right\}, \qquad 0 \le n \le \left(\frac{N}{2}\right) - 1 \tag{5.91}$$

Now, we split the summation term in (5.87) into even and odd terms and use (5.91) to obtain

$$\begin{aligned} v(k) &= \alpha(k) \left\{ \sum_{n=0}^{(N/2)-1} u(2n) \cos\left[\frac{\pi(4n+1)k}{2N}\right] \right. \\ &\quad \left. + \sum_{n=0}^{(N/2)-1} u(2n+1) \cos\left[\frac{\pi(4n+3)k}{2N}\right] \right\} \\ &= \alpha(k) \left\{ \left[\sum_{n=0}^{(N/2)-1} \tilde{u}(n) \cos\left[\frac{\pi(4n+1)k}{2N}\right] \right. \right. \\ &\quad \left. \left. + \sum_{n=0}^{(N/2)-1} \tilde{u}(N-n-1) \cos\left[\frac{\pi(4n+3)k}{2N}\right] \right] \right\} \end{aligned}$$

Changing the index of summation in the second term to $n' = N - n - 1$ and combining terms, we obtain

$$v(k) = \alpha(k) \sum_{n=0}^{N-1} \tilde{u}(n) \cos\left[\frac{\pi(4n+1)k}{2N}\right] \tag{5.92}$$

$$= \text{Re}\left[\alpha(k)e^{-j\pi k/2N} \sum_{n=0}^{N-1} \tilde{u}(n)e^{-j2\pi kn/N}\right] = \text{Re}[\alpha(k)W_{2N}^{k/2}\,\text{DFT}\{\tilde{u}(n)\}_N]$$

which proves the previously stated result. For inverse cosine transform we write (5.89) for even data points as

$$u(2n) = \hat{u}(2n) \triangleq \text{Re}\left[\sum_{k=0}^{N-1} [\alpha(k)v(k)e^{j\pi k/2N}]e^{j2\pi nk/N}\right], \qquad 0 \le n \le \left(\frac{N}{2}\right) - 1 \tag{5.93}$$

The odd data points are obtained by noting that

$$u(2n+1) = \hat{u}[2(N-1-n)], \qquad 0 \le n \le \left(\frac{N}{2}\right) - 1 \tag{5.94}$$

Therefore, if we calculate the N-point inverse FFT of the sequence $\alpha(k)v(k)\exp(j\pi k/2N)$, we can also obtain the inverse DCT in $O(N \log N)$ operations. Direct algorithms that do not require FFT as an intermediate step, so that complex arithmetic is avoided, are also possible [18]. The computational complexity of the direct as well as the FFT based methods is about the same.

4. The cosine transform has excellent energy compaction for highly correlated data. This is due to the following properties.
5. The basis vectors of the cosine transform (that is, rows of $\mathbf{C}$) are the eigenvectors of the symmetric tridiagonal matrix $\mathbf{Q}_c$, defined as

$$\mathbf{Q}_c = \begin{bmatrix} 1-\alpha & -\alpha & & \mathbf{0} \\ -\alpha & 1 & \ddots & \\ & \ddots & 1 & -\alpha \\ \mathbf{0} & & -\alpha & 1-\alpha \end{bmatrix} \tag{5.95}$$

The proof is left as an exercise.

6. The $N \times N$ cosine transform is very close to the KL transform of a first-order stationary Markov sequence of length N whose covariance matrix is given by (2.68) when the correlation parameter ρ is close to 1. The reason is that $\mathbf{R}^{-1}$ is a symmetric tridiagonal matrix, which for a scalar $\beta^2 \triangleq (1-\rho^2)/(1+\rho^2)$ and $\alpha \triangleq \rho/(1+\rho^2)$ satisfies the relation

$$\beta^2\mathbf{R}^{-1} = \begin{bmatrix} 1-\rho\alpha & -\alpha & & \mathbf{0} \\ -\alpha & 1 & \ddots & \\ & \ddots & 1 & -\alpha \\ \mathbf{0} & & -\alpha & 1-\rho\alpha \end{bmatrix} \tag{5.96}$$

This gives the approximation

$$\beta^2 R^{-1} \cong \mathbf{Q}_c \quad \text{for} \quad \rho \cong 1 \tag{5.97}$$

Hence the eigenvectors of **R** and the eigenvectors of $\mathbf{Q}_c$, that is, the cosine transform, will be quite close. These aspects are considered in greater depth in Section 5.12 on sinusoidal transforms.

This property of the cosine transform together with the fact that it is a fast transform has made it a useful substitute for the KL transform of highly correlated first-order Markov sequences.

5.7 THE SINE TRANSFORM

The $N \times N$ sine transform matrix $\boldsymbol{\Psi} = \{\psi(k, n)\}$, also called the *discrete sine transform* (DST), is defined as

$$\psi(k, n) = \sqrt{\frac{2}{N+1}} \sin \frac{\pi(k+1)(n+1)}{N+1}, \qquad 0 \le k, n \le N-1 \tag{5.98}$$

The sine transform pair of one-dimensional sequences is defined as

$$v(k) = \sqrt{\frac{2}{N+1}} \sum_{n=0}^{N-1} u(n) \sin \frac{\pi(k+1)(n+1)}{N+1}, \qquad 0 \le k \le N-1 \tag{5.99}$$

$$u(n) = \sqrt{\frac{2}{N+1}} \sum_{k=0}^{N-1} v(k) \sin \frac{\pi(k+1)(n+1)}{N+1}, \qquad 0 \le n \le N-1 \tag{5.100}$$

The two-dimensional sine transform pair for $N \times N$ images is obtained by substituting $\mathbf{A} = \mathbf{A}^* = \mathbf{A}^T = \boldsymbol{\Psi}$ in (5.11) and (5.12). The basis vectors and the basis images of the sine transform are shown in Figs. 5.1 and 5.2. Figure 5.12 shows the sine transform of a 255×255 image. Once again it is seen that a large fraction of the total energy is concentrated in a few transform coefficients.

Properties of the Sine Transform

1. The sine transform is real, symmetric, and orthogonal, that is,

$$\boldsymbol{\Psi}^* = \boldsymbol{\Psi} = \boldsymbol{\Psi}^T = \boldsymbol{\Psi}^{-1} \tag{5.101}$$

Thus, the forward and inverse sine transforms are identical.

Figure 5.12 Sine transform of a 255×255 portion of the 256×256 image shown in Fig. 5.6a.

2. The sine transform is not the imaginary part of the unitary DFT. The sine transform of a sequence is related to the DFT of its antisymmetric extension (see Problem 5.16).
3. The sine transform is a fast transform. The sine transform (or its inverse) of a vector of N elements can be calculated in $O(N \log_2 N)$ operations via a $2(N+1)$-point FFT.
   Typically this requires $N+1=2^p$, that is, the fast sine transform is usually defined for $N = 3, 7, 15, 31, 63, 255, \ldots$. Fast sine transform algorithms that do not require complex arithmetic (or the FFT) are also possible. In fact, these algorithms are somewhat faster than the FFT and the fast cosine transform algorithms [20].
4. The basis vectors of the sine transform are the eigenvectors of the symmetric tridiagonal Toeplitz matrix

$$\mathbf{Q} = \begin{bmatrix} 1 & -\alpha & & \mathbf{0} \\ -\alpha & \ddots & \ddots & \\ & \ddots & \ddots & -\alpha \\ \mathbf{0} & & -\alpha & 1 \end{bmatrix} \tag{5.102}$$

5. The sine transform is close to the KL transform of first order stationary Markov sequences, whose covariance matrix is given in (2.68), when the correlation parameter ρ lies in the interval $(-0.5, 0.5)$. In general it has very good to excellent energy compaction property for images.
6. The sine transform leads to a fast KL transform algorithm for Markov sequences, whose boundary values are given. This makes it useful in many image processing problems. Details are considered in greater depth in Chapter 6 (Sections 6.5 and 6.9) and Chapter 11 (Section 11.5).

5.8 THE HADAMARD TRANSFORM

Unlike the previously discussed transforms, the elements of the basis vectors of the Hadamard transform take only the binary values ± 1 and are, therefore, well suited for digital signal processing. The Hadamard transform matrices, $\mathbf{H}_n$, are $N \times N$ matrices, where $N \triangleq 2^n$, $n = 1, 2, 3$. These can be easily generated by the core matrix

$$\mathbf{H}_1 = \frac{1}{\sqrt{2}} \begin{pmatrix} 1 & 1 \\ 1 & -1 \end{pmatrix} \tag{5.103}$$

and the Kronecker product recursion

$$\mathbf{H}_n = \mathbf{H}_{n-1} \otimes \mathbf{H}_1 = \mathbf{H}_1 \otimes \mathbf{H}_{n-1} = \frac{1}{\sqrt{2}} \begin{pmatrix} \mathbf{H}_{n-1} & \mathbf{H}_{n-1} \\ \mathbf{H}_{n-1} & -\mathbf{H}_{n-1} \end{pmatrix} \tag{5.104}$$

As an example, for $n = 3$, the Hadamard matrix becomes

$$\mathbf{H}_3 = \mathbf{H}_1 \otimes \mathbf{H}_2 \tag{5.105}$$

$$\mathbf{H}_2 = \mathbf{H}_1 \otimes \mathbf{H}_1 \tag{5.106}$$

which gives

$$\mathbf{H}_3 = \frac{1}{\sqrt{8}} \left[\begin{array}{rrrr|rrrr} 1 & 1 & 1 & 1 & 1 & 1 & 1 & 1 \\ 1 & -1 & 1 & -1 & 1 & -1 & 1 & -1 \\ 1 & 1 & -1 & -1 & 1 & 1 & -1 & -1 \\ 1 & -1 & -1 & 1 & 1 & -1 & -1 & 1 \\ \hline 1 & 1 & 1 & 1 & -1 & -1 & -1 & -1 \\ 1 & -1 & 1 & -1 & -1 & 1 & -1 & 1 \\ 1 & 1 & -1 & -1 & -1 & -1 & 1 & 1 \\ 1 & -1 & -1 & 1 & -1 & 1 & 1 & -1 \end{array}\right] \quad \begin{array}{c} \text{Sequency} \\ 0 \\ 7 \\ 3 \\ 4 \\ 1 \\ 6 \\ 2 \\ 5 \end{array} \tag{5.107}$$

The basis vectors of the Hadamard transform can also be generated by sampling a class of functions called the *Walsh functions*. These functions also take only the binary values ± 1 and form a complete orthonormal basis for square integrable functions. For this reason the Hadamard transform just defined is also called the Walsh-Hadamard transform.

The number of zero crossings of a Walsh function or the number of transitions in a basis vector of the Hadamard transform is called its *sequency*. Recall that for sinusoidal signals, frequency can be defined in terms of the zero crossings. In the Hadamard matrix generated via (5.104), the row vectors are not sequency ordered. The existing sequency order of these vectors is called the *Hadamard order*. The Hadamard transform of an $N \times 1$ vector $\mathbf{u}$ is written as

$$\mathbf{v} = \mathbf{H}\mathbf{u} \tag{5.108}$$

and the inverse transform is given by

$$\mathbf{u} = \mathbf{H}\mathbf{v} \tag{5.109}$$

where $\mathbf{H} \triangleq \mathbf{H}_n$, $n = \log_2 N$. In series form the transform pair becomes

$$v(k) = \frac{1}{\sqrt{N}} \sum_{m=0}^{N-1} u(m)(-1)^{b(k,m)}, \qquad 0 \le k \le N-1 \tag{5.110}$$

$$u(m) = \frac{1}{\sqrt{N}} \sum_{k=0}^{N-1} v(k)(-1)^{b(k,m)}, \qquad 0 \le m \le N-1 \tag{5.111}$$

where

$$b(k, m) = \sum_{i=0}^{n-1} k_i m_i; \qquad k_i, m_i = 0, 1 \tag{5.112}$$

and $\{k_i\}$, $\{m_i\}$ are the binary representations of k and m, respectively, that is,

$$\left.\begin{aligned} k &= k_0 + 2k_1 + \cdots + 2^{n-1} k_{n-1} \\ m &= m_0 + 2m_1 + \cdots + 2^{n-1} m_{n-1} \end{aligned}\right\} \tag{5.113}$$

The two-dimensional Hadamard transform pair for $N \times N$ images is obtained by substituting $\mathbf{A} = \mathbf{A}^* = \mathbf{A}^T = \mathbf{H}$ in (5.11) and (5.12). The basis vectors and the basis

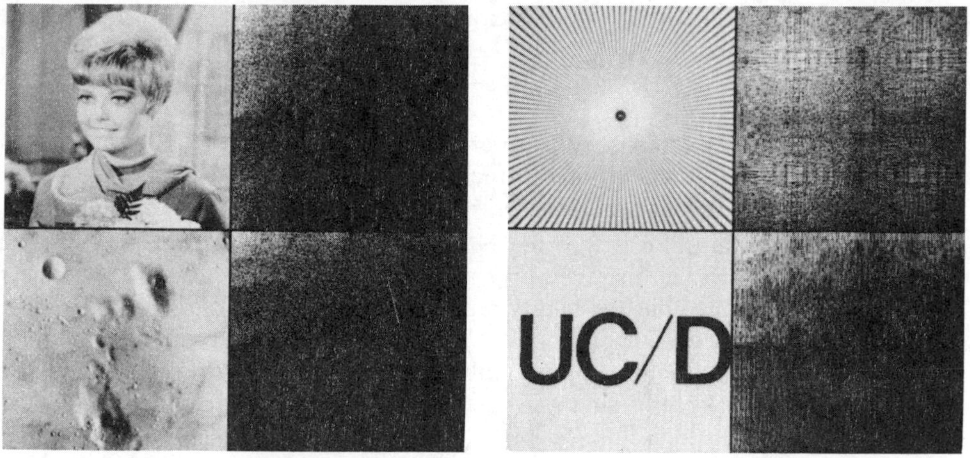

(a) Hadamard transforms of monochrome images. (b) Hadamard transforms of binary images.

Figure 5.13 Examples of Hadamard transforms.

images of the Hadamard transform are shown in Figs. 5.1 and 5.2. Examples of two-dimensional Hadamard transforms of images are shown in Fig. 5.13.

Properties of the Hadamard Transform

1. The Hadamard transform $\mathbf{H}$ is real, symmetric, and orthogonal, that is,

$$\mathbf{H} = \mathbf{H}^* = \mathbf{H}^T = \mathbf{H}^{-1} \tag{5.114}$$

2. The Hadamard transform is a fast transform. The one-dimensional transformation of (5.108) can be implemented in $O(N \log_2 N)$ additions and subtractions.

   Since the Hadamard transform contains only ± 1 values, no multiplications are required in the transform calculations. Moreover, the number of additions or subtractions required can be reduced from N^2 to about $N \log_2 N$. This is due to the fact that $\mathbf{H}_n$ can be written as a product of n sparse matrices, that is,

$$\mathbf{H} = \mathbf{H}_n = \tilde{\mathbf{H}}^n, \qquad n = \log_2 N \tag{5.115}$$

where

$$\tilde{\mathbf{H}} \triangleq \frac{1}{\sqrt{2}} \left[\begin{array}{cccccccc} 1 & 1 & 0 & 0 & & & & \\ 0 & 0 & 1 & 1 & 0 & 0 & & \\ \vdots & \vdots & \vdots & \vdots & & & & \\ 0 & 0 & & & & \cdots & 1 & 1 \\ \hline 1 & -1 & 0 & 0 & & & & \\ 0 & 0 & 1 & -1 & & & & \\ \vdots & \vdots & \vdots & \vdots & & & & \\ 0 & 0 & & & & \cdots & 1 & -1 \end{array}\right] \begin{array}{l} \uparrow \\ \frac{N}{2} \text{ rows} \\ \downarrow \\ \uparrow \\ \frac{N}{2} \text{ rows} \\ \downarrow \end{array} \tag{5.116}$$

Since $\tilde{\mathbf{H}}$ contains only two nonzero terms per row, the transformation

$$\mathbf{v} = \tilde{\mathbf{H}}_n^n \mathbf{u} = \underbrace{\tilde{\mathbf{H}}\tilde{\mathbf{H}} \dots \tilde{\mathbf{H}}}_{n \text{ terms}}\mathbf{u}, \qquad n = \log_2 N \tag{5.117}$$

can be accomplished by operating $\tilde{\mathbf{H}}$ n times on $\mathbf{u}$. Due to the structure of $\tilde{\mathbf{H}}$ only N additions or subtractions are required each time $\tilde{\mathbf{H}}$ operates on a vector, giving a total of $Nn = N \log_2 N$ additions or subtractions.

3. The natural order of the Hadamard transform coefficients turns out to be equal to the bit reversed gray code representation of its sequency s. If the sequency s has the binary representation $b_n b_{n-1} \dots b_1$ and if the corresponding gray code is $g_n g_{n-1} \dots g_1$, then the bit-reversed representation $g_1 g_2 \dots g_n$ gives the natural order. Table 5.1 shows the conversion of sequency s to natural order h, and vice versa, for $N = 8$. In general,

$$\begin{aligned} g_k &= b_k \oplus b_{k+1}, \qquad k = 1, \dots, n-1 \\ g_n &= b_n \\ h_k &= g_{n-k+1} \end{aligned} \tag{5.118}$$

and

$$\begin{aligned} g_k &= h_{n-k+1} \\ b_k &= g_k \oplus b_{k+1}, \qquad k = n-1, \dots, 1 \\ b_n &= g_n \end{aligned} \tag{5.119}$$

give the forward and reverse conversion formulas for the sequency and natural ordering.

4. The Hadamard transform has good to very good energy compaction for highly correlated images. Let $\{u(n), 0 \le n \le N-1\}$ be a stationary random

TABLE 5.1 Natural Ordering versus Sequency Ordering of Hadamard Transform Coefficients for $N = 8$

Natural order h	Binary representation $h_3 h_2 h_1$	Gray code of s or reverse binary representation $g_3 g_2 g_1 = h_1 h_2 h_3$	Sequency binary representation $b_3 b_2 b_1$	Sequency s
0	000	000	000	0
1	001	100	111	7
2	010	010	011	3
3	011	110	100	4
4	100	001	001	1
5	101	101	110	6
6	110	011	010	2
7	111	111	101	5

sequence with autocorrelation $r(n), 0 \le n \le N-1$. The fraction of the expected energy packed in the first $N/2^j$ sequency ordered Hadamard transform coefficients is given by [23]

$$\mathscr{E}\left(\frac{N}{2^j}\right) \triangleq \frac{\sum_{k=0}^{(N/2^j-1)} D'_k}{\sum_{k=0}^{N-1} D_k} = \frac{\left[1 + 2\sum_{k=1}^{2^j-1}\left(1 - \frac{k}{2^j}\right)\frac{r(k)}{r(0)}\right]}{2^j} \tag{5.120}$$

$$D_k \triangleq [\mathbf{HRH}]_{k,k}, \qquad \mathbf{R} \triangleq \{r(m-n)\} \tag{5.121}$$

where D'_k are the first $N/2^j$ sequency ordered elements D_k. Note that the D_k are simply the mean square values of the transform coefficients $[\mathbf{Hu}]_k$. The significance of this result is that $\mathscr{E}(N/2^j)$ depends on the first 2^j autocorrelations only. For $j = 1$, the fractional energy packed in the first $N/2$ sequency ordered coefficients will be $(1 + r(1)/r(0)/2$ and depends only upon the one-step correlation $\rho \triangleq r(1)/r(0)$. Thus for $\rho = 0.95$, 97.5% of the total energy is concentrated in half of the transform coefficients. The result of (5.120) is useful in calculating the energy compaction efficiency of the Hadamard transform.

Example 5.3

Consider the covariance matrix $\mathbf{R}$ of (2.68) for $N = 4$. Using the definition of $\mathbf{H}_2$ we obtain

$$\mathbf{D} = \text{diagonal}\,[\mathbf{H}_2\mathbf{R}\mathbf{H}_2] = \frac{1}{4}\begin{bmatrix} 4+6\rho+4\rho^2+2\rho^3 & & & \mathbf{0} \\ & 4-6\rho+4\rho^2-2\rho^3 & & \\ & & 4+2\rho-4\rho^2-2\rho^3 & \\ \mathbf{0} & & & 4-2\rho-4\rho^2+2\rho^3 \end{bmatrix} \begin{matrix} \text{Sequency} \\ 0 \\ 3 \\ 1 \\ 2 \end{matrix}$$

This gives $D'_0 = D_0$, $D'_1 = D_2$, $D'_2 = D_3$, $D'_3 = D_1$ and

$$\mathscr{E}\left(\frac{1}{2}\right) = \frac{1}{4}\sum_{k=0}^{1} D'_k = \frac{1}{16}(4+6\rho+4\rho^2+2\rho^3+4+2\rho-4\rho^2-2\rho^3) = \frac{(1+\rho)}{2}$$

as expected according to (5.120).

5.9 THE HAAR TRANSFORM

The Haar functions $h_k(x)$ are defined on a continuous interval, $x \in [0, 1]$, and for $k = 0, \ldots, N-1$, where $N = 2^n$. The integer k can be uniquely decomposed as

$$k = 2^p + q - 1 \tag{5.122}$$

where $0 \le p \le n-1$; $q = 0, 1$ for $p = 0$ and $1 \le q \le 2^p$ for $p \ne 0$. For example, when $N = 4$, we have

k	0	1	2	3
p	0	0	1	1
q	0	1	1	2

Representing k by (p, q), the Haar functions are defined as

$$h_0(x) \triangleq h_{0,0}(x) = \frac{1}{\sqrt{N}}, \qquad x \in [0, 1]. \tag{5.123a}$$

$$h_k(x) \triangleq h_{p,q}(x) = \frac{1}{\sqrt{N}} \begin{cases} 2^{p/2}, & \dfrac{q-1}{2^p} \le x < \dfrac{q-\frac{1}{2}}{2^p} \\ -2^{p/2}, & \dfrac{q-\frac{1}{2}}{2^p} \le x < \dfrac{q}{2^p} \\ 0, & \text{otherwise for } x \in [0, 1] \end{cases} \tag{5.123b}$$

The Haar transform is obtained by letting x take discrete values at m/N, $m = 0, 1, \ldots, N-1$. For $N = 8$, the Haar transform is given by

$$\mathbf{Hr} = \frac{1}{\sqrt{8}} \begin{bmatrix} 1 & 1 & 1 & 1 & 1 & 1 & 1 & 1 \\ 1 & 1 & 1 & 1 & -1 & -1 & -1 & -1 \\ \sqrt{2} & \sqrt{2} & -\sqrt{2} & -\sqrt{2} & 0 & 0 & 0 & 0 \\ 0 & 0 & 0 & 0 & \sqrt{2} & \sqrt{2} & -\sqrt{2} & -\sqrt{2} \\ 2 & -2 & 0 & 0 & 0 & 0 & 0 & 0 \\ 0 & 0 & 2 & -2 & 0 & 0 & 0 & 0 \\ 0 & 0 & 0 & 0 & 2 & -2 & 0 & 0 \\ 0 & 0 & 0 & 0 & 0 & 0 & 2 & -2 \end{bmatrix} \begin{matrix} \text{Sequency} \\ 0 \\ 1 \\ 2 \\ 2 \\ 2 \\ 2 \\ 2 \\ 2 \end{matrix} \tag{5.124}$$

The basis vectors and the basis images of the Haar transform are shown in Figs. 5.1 and 5.2. An example of the Haar transform of an image is shown in Fig. 5.14. From the structure of **Hr** [see (5.124)] we see that the Haar transform takes differences of the samples or differences of local averages of the samples of the input vector. Hence the two-dimensional Haar transform coefficients $y(k, l)$, except for $k = l = 0$, are the differences along rows and columns of the local averages of pixels in the image. These are manifested as several "edge extractions" of the original image, as is evident from Fig. 5.14.

Although some work has been done for using the Haar transform in image data compression problems, its full potential in feature extraction and image analysis problems has not been determined.

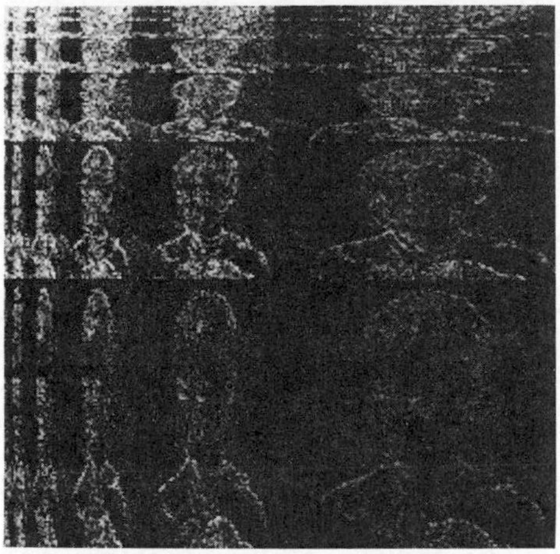

Figure 5.14 Haar transform of the 256 × 256 image shown in Fig. 5.6a.

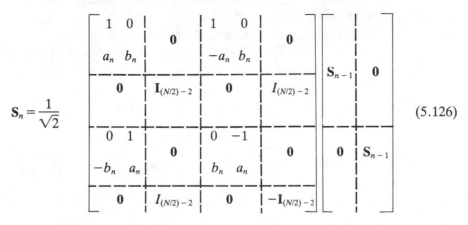

Figure 5.15 Slant transform of the 256 × 256 image shown in Fig. 5.6a.

Properties of the Haar Transform

1. The Haar transform is real and orthogonal. Therefore,

$$\mathbf{Hr} = \mathbf{Hr}^*$$
$$\mathbf{Hr}^{-1} = \mathbf{Hr}^T \tag{5.125}$$

2. The Haar transform is a very fast transform. On an $N \times 1$ vector it can be implemented in $O(N)$ operations.
3. The basis vectors of the Haar matrix *are* sequency ordered.
4. The Haar transform has poor energy compaction for images.

5.10 THE SLANT TRANSFORM

The $N \times N$ Slant transform matrices are defined by the recursion

$$\mathbf{S}_n = \frac{1}{\sqrt{2}} \left[\begin{array}{cc|c|cc|c} 1 & 0 & \mathbf{0} & 1 & 0 & \mathbf{0} \\ a_n & b_n & & -a_n & b_n & \\ \hline \multicolumn{2}{c|}{\mathbf{0}} & \mathbf{I}_{(N/2)-2} & \multicolumn{2}{c|}{\mathbf{0}} & I_{(N/2)-2} \\ \hline 0 & 1 & \mathbf{0} & 0 & -1 & \mathbf{0} \\ -b_n & a_n & & b_n & a_n & \\ \hline \multicolumn{2}{c|}{\mathbf{0}} & I_{(N/2)-2} & \multicolumn{2}{c|}{\mathbf{0}} & -\mathbf{I}_{(N/2)-2} \end{array}\right] \left[\begin{array}{c|c} \mathbf{S}_{n-1} & \mathbf{0} \\ \hline \mathbf{0} & \mathbf{S}_{n-1} \end{array}\right] \tag{5.126}$$

where $N = 2^n$, $\mathbf{I}_M$ denotes an $M \times M$ identity matrix, and

$$\mathbf{S}_1 = \frac{1}{\sqrt{2}}\begin{bmatrix} 1 & 1 \\ 1 & -1 \end{bmatrix} \tag{5.127}$$

The parameters a_n and b_n are defined by the recursions

$$\left.\begin{aligned} b_n &= (1 + 4a_{n-1}^2)^{-1/2}, \qquad a_1 = 1 \\ a_n &= 2b_n a_{n-1} \end{aligned}\right\} \tag{5.128}$$

which solve to give

$$a_{n+1} = \left(\frac{3N^2}{4N^2 - 1}\right)^{1/2}, \qquad b_{n+1} = \left(\frac{N^2 - 1}{4N^2 - 1}\right)^{1/2}, \qquad N = 2^n \tag{5.129}$$

Using these formulas, the 4×4 Slant transformation matrix is obtained as

$$\mathbf{S}_2 = \frac{1}{2}\begin{bmatrix} 1 & 1 & 1 & 1 \\ \frac{3}{\sqrt{5}} & \frac{1}{\sqrt{5}} & \frac{-1}{\sqrt{5}} & \frac{-3}{\sqrt{5}} \\ 1 & -1 & -1 & 1 \\ \frac{1}{\sqrt{5}} & \frac{-3}{\sqrt{5}} & \frac{3}{\sqrt{5}} & \frac{-1}{\sqrt{5}} \end{bmatrix} \begin{matrix} \text{Sequency} \\ 0 \\ 1 \\ 2 \\ 3 \end{matrix} \tag{5.130}$$

Figure 5.1 shows the basis vectors of the 8×8 Slant transform. Figure 5.2 shows the basis images of the 8×8 two dimensional Slant transform. Figure 5.15 shows the Slant transform of a 256×256 image.

Properties of the Slant Transform

1. The Slant transform is real and orthogonal. Therefore,

$$\mathbf{S} = \mathbf{S}^*, \qquad \mathbf{S}^{-1} = \mathbf{S}^T \tag{5.131}$$

2. The Slant transform is a fast transform, which can be implemented in $O(N \log_2 N)$ operations on an $N \times 1$ vector.
3. It has very good to excellent energy compaction for images.
4. The basis vectors of the Slant transform matrix $\mathbf{S}$ are not sequency ordered for $n \geq 3$. If $\mathbf{S}_{n-1}$ is sequency ordered, the ith row sequency of $\mathbf{S}_n$ is given as follows.

$$\begin{aligned} i &= 0, && \text{sequency} = 0 \\ i &= 1, && \text{sequency} = 1 \\ 2 \leq i &\leq \frac{N}{2} - 1, && \text{sequency} = \begin{cases} 2i, & i = \text{even} \\ 2i + 1, & i = \text{odd} \end{cases} \\ i &= \frac{N}{2}, && \text{sequency} = 2 \end{aligned}$$

$$i = \frac{N}{2} + 1, \qquad \text{sequency} = 3$$

$$\frac{N}{2} + 2 \le i \le N - 1, \qquad \text{sequency} = \begin{cases} 2\left(i - \frac{N}{2}\right) + 1, & i = \text{even} \\ 2\left(i - \frac{N}{2}\right), & i = \text{odd} \end{cases}$$

5.11 THE KL TRANSFORM

The KL transform was originally introduced as a series expansion for continuous random processes by Karhunen [27] and Loeve [28]. For random sequences Hotelling [26] first studied what was called a method of principal components, which is the discrete equivalent of the KL series expansion. Consequently, the KL transform is also called the Hotelling transform or the method of principal components.

For a real $N \times 1$ random vector $\mathbf{u}$, the basis vectors of the KL transform (see Section 2.9) are given by the orthonormalized eigenvectors of its autocorrelation matrix $\mathbf{R}$, that is,

$$\mathbf{R}\boldsymbol{\phi}_k = \lambda_k \boldsymbol{\phi}_k, \qquad 0 \le k \le N - 1 \tag{5.132}$$

The KL transform of $\mathbf{u}$ is defined as

$$\mathbf{v} = \mathbf{\Phi}^{*T}\mathbf{u} \tag{5.133}$$

and the inverse transform is

$$\mathbf{u} = \mathbf{\Phi}\mathbf{v} = \sum_{k=0}^{N-1} v(k)\boldsymbol{\phi}_k \tag{5.134}$$

where $\boldsymbol{\phi}_k$ is the kth column of $\mathbf{\Phi}$. From (2.44) we know $\mathbf{\Phi}$ reduces $\mathbf{R}$ to its diagonal form, that is,

$$\mathbf{\Phi}^{*T}\mathbf{R}\mathbf{\Phi} = \mathbf{\Lambda} = \text{Diag}\{\lambda_k\} \tag{5.135}$$

We often work with the covariance matrix rather than the autocorrelation matrix. With $\boldsymbol{\mu} \triangleq E[\mathbf{u}]$, then

$$\mathbf{R}_0 \triangleq \text{cov}[\mathbf{u}] \triangleq E[(\mathbf{u} - \boldsymbol{\mu})(\mathbf{u} - \boldsymbol{\mu})^T] = E[\mathbf{u}\mathbf{u}^T] - \boldsymbol{\mu}\boldsymbol{\mu}^T = \mathbf{R} - \boldsymbol{\mu}\boldsymbol{\mu}^T \tag{5.136}$$

If the vector $\boldsymbol{\mu}$ is known, then the eigenmatrix of $\mathbf{R}_0$ determines the KL transform of the zero mean random process $\mathbf{u} - \boldsymbol{\mu}$. In general, the KL transform of $\mathbf{u}$ and $\mathbf{u} - \boldsymbol{\mu}$ need not be identical.

Note that whereas the image transforms considered earlier were functionally independent of the data, the KL transform depends on the (second-order) statistics of the data.

Example 5.4 (KL Transform of Markov-1 Sequences)

The covariance matrix of a zero mean Markov sequence of N elements is given by (2.68). Its eigenvalues λ_k and the eigenvectors $\boldsymbol{\phi}_k$ are given by

$$\lambda_k = \frac{1 - \rho^2}{1 - 2\rho\cos\omega_k + \rho^2}$$

$$\phi_k(m) = \phi(m, k) \tag{5.137}$$
$$= \left(\frac{2}{N+\lambda_k}\right)^{1/2} \sin\left(\omega_k\left(m+1-\frac{N+1}{2}\right)+\frac{(k+1)\pi}{2}\right), \qquad 0 \le m, k \le N-1$$

where the $\{\omega_k\}$ are the positive roots of the equation

$$\tan(N\omega) = -\frac{(1-\rho^2)\sin\omega}{\cos\omega - 2\rho + \rho^2\cos\omega}, \qquad N \text{ even} \tag{5.138}$$

A similar result holds when N is odd. This is a transcendental equation that gives rise to nonharmonic sinusoids $\phi_k(m)$. Figure 5.1 shows the basis vectors of this 8×8 KL transform for $\rho = 0.95$. Note the basis vectors of the KLT and the DCT are quite similar. Because $\phi_k(m)$ are nonharmonic, a fast algorithm for this transform does not exist. Also note, the KL transform matrix is $\mathbf{\Phi}^T \overset{\Delta}{=} \{\phi(k, m)\}$.

Example 5.5

Since the unitary DFT reduces any circulant matrix to a diagonal form, it is the KL transform of all random sequences with circulant autocorrelation matrices, that is, for all periodic random sequences.

The DCT is the KL transform of a random sequence whose autocorrelation matrix $\mathbf{R}$ commutes with $\mathbf{Q}_c$ of (5.95) (that is, if $\mathbf{R}\mathbf{Q}_c = \mathbf{Q}_c\mathbf{R}$). Similarly, the DST is the KL transform of all random sequences whose autocorrelation matrices commute with $\mathbf{Q}$ of (5.102).

KL Transform of Images

If an $N \times N$ image $u(m, n)$ is represented by a random field whose autocorrelation function is given by

$$E[u(m, n)u(m', n')] = r(m, n; m', n'), \qquad 0 \le m, m', n, n' \le N-1 \tag{5.139}$$

then the basis images of the KL transform are the orthonormalized eigenfunctions $\psi_{k,l}(m, n)$ obtained by solving

$$\sum_{m'=0}^{N-1}\sum_{n'=0}^{N-1} r(m, n; m', n')\psi_{k,l}(m', n')$$
$$= \lambda_{k,l}\psi_{k,l}(m, n), \qquad 0 \le k, l \le N-1, 0 \le m, n \le N-1 \tag{5.140}$$

In matrix notation this can be written as

$$\mathcal{R}\boldsymbol{\psi}_i = \lambda_i\boldsymbol{\psi}_i, \qquad i = 0, \ldots, N^2-1 \tag{5.141}$$

where $\boldsymbol{\psi}_i$ is an $N^2 \times 1$ vector representation of $\psi_{k,l}(m, n)$ and $\mathcal{R}$ is an $N^2 \times N^2$ autocorrelation matrix of the image mapped into an $N^2 \times 1$ vector $\boldsymbol{u}$. Thus

$$\mathcal{R} = E[\boldsymbol{u}\boldsymbol{u}^T] \tag{5.142}$$

If $\mathcal{R}$ is separable, then the $N^2 \times N^2$ matrix $\mathbf{\Psi}$ whose columns are $\{\boldsymbol{\psi}_i\}$ becomes separable (see Table 2.7). For example, let

$$r(m, n; m', n') = r_1(m, m')r_2(n, n') \tag{5.143}$$

$$\psi_{k,l}(m, n) = \phi_1(m, k)\phi_2(n, l) \tag{5.144}$$

In matrix notation this means

$$\mathcal{R} = \mathbf{R}_1 \otimes \mathbf{R}_2, \qquad \mathbf{\Psi} = \mathbf{\Phi}_1 \otimes \mathbf{\Phi}_2 \tag{5.145}$$

where

$$\mathbf{\Phi}_j \mathbf{R}_j \mathbf{\Phi}_j^{*T} = \mathbf{\Lambda}_j, \qquad j = 1, 2 \tag{5.146}$$

and the KL transform of $\boldsymbol{u}$ is

$$\boldsymbol{v} = \mathbf{\Psi}^{*T}\boldsymbol{u} = [\mathbf{\Phi}_1^{*T} \otimes \mathbf{\Phi}_2^{*T}]\boldsymbol{u} \tag{5.147}$$

For row-ordered vectors this is equivalent to

$$\mathbf{V} = \mathbf{\Phi}_1^{*T}\mathbf{U}\mathbf{\Phi}_2^{*} \tag{5.148}$$

and the inverse KL transform is

$$\mathbf{U} = \mathbf{\Phi}_1 \mathbf{V} \mathbf{\Phi}_2^{T} \tag{5.149}$$

The advantage in modeling the image autocorrelation by a separable function is that instead of solving the $N^2 \times N^2$ matrix eigenvalue problem of (5.141), only two $N \times N$ matrix eigenvalue problems of (5.146) need to be solved. Since an $N \times N$ matrix eigenvalue problem requires $O(N^3)$ computations, the reduction in dimensionality achieved by the separable model is $O(N^6)/O(N^3) \simeq O(N^3)$, which is very significant. Also, the transformation calculations of (5.148) and (5.149) require $2N^3$ operations compared to N^4 operations required for $\mathbf{\Psi}^{*T}\boldsymbol{u}$.

Example 5.6

Consider the separable covariance function for a zero mean random field

$$r(m, n; m', n') = \rho^{|m-m'|}\rho^{|n-n'|} \tag{5.150}$$

This gives $\mathcal{R} = \mathbf{R} \otimes \mathbf{R}$, where $\mathbf{R}$ is given by (2.68). The eigenvectors of $\mathbf{R}$ are given by ϕ_k in Example 5.4. Hence $\mathbf{\Psi} = \mathbf{\Phi} \otimes \mathbf{\Phi}$ and the KL transform matrix is $\mathbf{\Phi}^T \otimes \mathbf{\Phi}^T$. Figure 5.2 shows the basis images of this 8×8 two-dimensional KL transform for $\rho = 0.95$.

Properties of the KL Transform

The KL transform has many desirable properties, which make it optimal in many signal processing applications. Some of these properties are discussed here. For simplicity we assume $\mathbf{u}$ has zero mean and a positive definite covariance matrix $\mathbf{R}$.

Decorrelation. The KL transform coefficients $\{v(k), k = 0, \ldots, N-1\}$ are uncorrelated and have zero mean, that is,

$$E[v(k)] = 0$$

$$E[v(k)v^*(l)] = \lambda_k \delta(k - l) \tag{5.151}$$

The proof follows directly from (5.133) and (5.135), since

$$E[\mathbf{v}\mathbf{v}^{*T}] \triangleq \mathbf{\Phi}^{*T}E[\mathbf{u}\mathbf{u}^T]\mathbf{\Phi} = \mathbf{\Phi}^{*T}\mathbf{R}\mathbf{\Phi} = \mathbf{\Lambda} = \text{diagonal} \tag{5.152}$$

which implies the latter relation in (5.151). It should be noted that $\mathbf{\Phi}$ is not a *unique matrix* with respect to this property. There could be many matrices (unitary and nonunitary) that would decorrelate the transformed sequence. For example, a lower triangular matrix $\mathbf{\Phi}$ could be found that satisfies (5.152).

Example 5.7

The covariance matrix $\mathbf{R}$ of (2.68) is diagonalized by the lower triangular matrix

$$\mathbf{L} \triangleq \begin{bmatrix} 1 & & & 0 \\ -\rho & \ddots & & \\ & \ddots & \ddots & \\ 0 & & -\rho & 1 \end{bmatrix} \Rightarrow \mathbf{L}^T\mathbf{R}\mathbf{L} = \begin{bmatrix} 1-\rho^2 & & & & 0 \\ & 1-\rho^2 & & & \\ & & \ddots & & \\ & & & 1-\rho^2 & \\ 0 & & & & 1 \end{bmatrix} \triangleq \mathbf{D} \tag{5.153}$$

Hence the transformation $\mathbf{v} = \mathbf{L}^T\mathbf{u}$, will cause the sequence $v(k)$ to be uncorrelated. Comparing with Example 5.4, we see that $\mathbf{L} \neq \mathbf{\Phi}$. Moreover, $\mathbf{L}$ is not unitary and the diagonal elements of $\mathbf{D}$ are not the eigenvalues of $\mathbf{R}$.

Basis restriction mean square error. Consider the operations in Fig. 5.16. The vector $\mathbf{u}$ is first transformed to $\mathbf{v}$. The elements of $\mathbf{w}$ are chosen to be the first m elements of $\mathbf{v}$ and zeros elsewhere. Finally, $\mathbf{w}$ is transformed to $\mathbf{z}$. $\mathbf{A}$ and $\mathbf{B}$ are $N \times N$ matrices and $\mathbf{I}_m$ is a matrix with 1s along the first m diagonal terms and zeros elsewhere. Hence

$$w(k) = \begin{cases} v(k), & 0 \le k \le m-1 \\ 0, & k \ge m \end{cases} \tag{5.154}$$

Therefore, whereas $\mathbf{u}$ and $\mathbf{v}$ are vectors in an N-dimensional vector space, $\mathbf{w}$ is a vector restricted to an $m \le N$-dimensional subspace. The average mean square error between the sequences $u(n)$ and $z(n)$ is defined as

$$J_m \triangleq \frac{1}{N} E\left(\sum_{n=0}^{N-1} |u(n) - z(n)|^2\right) = \frac{1}{N}\,\text{Tr}[E\{(\mathbf{u}-\mathbf{z})(\mathbf{u}-\mathbf{z})^{*T}\}] \tag{5.155}$$

This quantity is called the *basis restriction error*. It is desired to find the matrices $\mathbf{A}$ and $\mathbf{B}$ such that J_m is minimized for each and every value of $m \in [1, N]$. This minimum is achieved by the KL transform of $\mathbf{u}$.

Theorem 5.1. The error J_m in (5.155) is minimum when

$$\mathbf{A} = \mathbf{\Phi}^{*T}, \qquad \mathbf{B} = \mathbf{\Phi}, \qquad \mathbf{AB} = \mathbf{I} \tag{5.156}$$

where the columns of $\mathbf{\Phi}$ are arranged according to the decreasing order of the eigenvalues of $\mathbf{R}$.

Proof. From Fig. 5.16, we have

$$\mathbf{v} = \mathbf{A}\mathbf{u}, \qquad \mathbf{w} = \mathbf{I}_m\mathbf{v}, \quad \text{and} \quad \mathbf{z} = \mathbf{B}\mathbf{w} \tag{5.157}$$

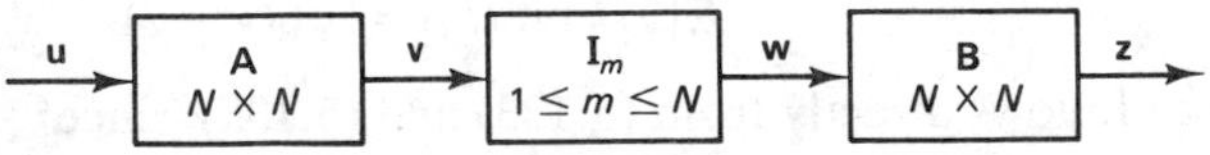

Figure 5.16 KL transform basis restriction.

Using these we can rewrite (5.155) as

$$J_m = \frac{1}{N}\,\text{Tr}[(\mathbf{I} - \mathbf{B}\mathbf{I}_m\mathbf{A})\mathbf{R}(\mathbf{I} - \mathbf{B}\mathbf{I}_m\mathbf{A})^{*T}]$$

To minimize J_m we first differentiate it with respect to the elements of $\mathbf{A}$ and set the result to zero [see Problem 2.15 for the differentiation rules]. This gives

$$\mathbf{I}_m\mathbf{B}^T(\mathbf{I} - \mathbf{B}\mathbf{I}_m\mathbf{A})^*\mathbf{R} = \mathbf{0} \tag{5.158}$$

which yields

$$J_m = \frac{1}{N}\,\text{Tr}[(\mathbf{I} - \mathbf{B}\mathbf{I}_m\mathbf{A})\mathbf{R}] \tag{5.159}$$

$$\mathbf{I}_m\mathbf{B}^{*T} = \mathbf{I}_m\mathbf{B}^{*T}\mathbf{B}I_m\mathbf{A} \tag{5.160}$$

At $m = N$, the minimum value of J_N must be zero, which requires

$$\mathbf{I} - \mathbf{B}\mathbf{A} = \mathbf{0} \quad \text{or} \quad \mathbf{B} = \mathbf{A}^{-1} \tag{5.161}$$

Using this in (5.160) and rearranging terms, we obtain

$$\mathbf{I}_m\mathbf{B}^{*T}\mathbf{B} = \mathbf{I}_m\mathbf{B}^{*T}\mathbf{B}\mathbf{I}_m, \qquad 1 \le m \le N \tag{5.162}$$

For (5.162) to be true for every m, it is necessary that $\mathbf{B}^{*T}\mathbf{B}$ be diagonal. Since $\mathbf{B} = \mathbf{A}^{-1}$, it is easy to see that (5.160) remains invariant if $\mathbf{B}$ is replaced by $\mathbf{DB}$ or $\mathbf{BD}$, where $\mathbf{D}$ is a diagonal matrix. Hence, without loss of generality we can normalize $\mathbf{B}$ so that $\mathbf{B}^{*T}\mathbf{B} = \mathbf{I}$, that is, $\mathbf{B}$ is a unitary matrix. Therefore, $\mathbf{A}$ is also unitary and $\mathbf{B} = \mathbf{A}^{*T}$. This gives

$$J_m = \frac{1}{N}\,\text{Tr}((\mathbf{I} - \mathbf{A}^{*T}\mathbf{I}_m\mathbf{A})\mathbf{R}) = \frac{1}{N}\,\text{Tr}(\mathbf{R} - \mathbf{I}_m\mathbf{A}\mathbf{R}\mathbf{A}^{*T}) \tag{5.163}$$

Since $\mathbf{R}$ is fixed, J_m is minimized if the quantity

$$\tilde{J}_m \triangleq \text{Tr}(\mathbf{I}_m\mathbf{A}\mathbf{R}\mathbf{A}^{*T}) = \sum_{k=0}^{m-1} \mathbf{a}_k^T\mathbf{R}\mathbf{a}_k^* \tag{5.164}$$

is *maximized* where $\mathbf{a}_k^T$ is the kth row of $\mathbf{A}$. Since $\mathbf{A}$ is unitary,

$$\mathbf{a}_k^T\mathbf{a}_k^* = 1 \tag{5.165}$$

To maximize $\tilde{J}_m$ subject to (5.165), we form the Lagrangian

$$\tilde{J}_m = \sum_{k=0}^{m-1} \mathbf{a}_k^T\mathbf{R}\mathbf{a}_k^* + \sum_{k=0}^{m-1} \lambda_k(1 - \mathbf{a}_k^T\mathbf{a}_k^*) \tag{5.166}$$

and differentiate it with respect to $\mathbf{a}_j$. The result gives a necessary condition

$$\mathbf{R}\mathbf{a}_j^* = \lambda_j\mathbf{a}_j^* \tag{5.167}$$

where $\mathbf{a}_j^*$ are orthonormalized eigenvectors of $\mathbf{R}$. This yields

$$\tilde{J}_m = \sum_{k=0}^{m-1} \lambda_k \tag{5.168}$$

which is maximized if $\{\mathbf{a}_j^*, 0 \le j \le m-1\}$ correspond to the largest m eigenvalues of $\mathbf{R}$. Because $\tilde{J}_m$ must be maximized for every m, it is necessary to arrange $\lambda_0 \ge \lambda_1 \ge \lambda_2 \ge \cdots \ge \lambda_{N-1}$. Then $\mathbf{a}_j^T$, the rows of $\mathbf{A}$, are the conjugate transpose of the eigenvectors of $\mathbf{R}$, that is, $\mathbf{A}$ is the KL transform of $\mathbf{u}$.

Distribution of variances. Among all the unitary transformations $\mathbf{v} = \mathbf{Au}$, the KL transform $\mathbf{\Phi}^{*T}$ packs the maximum average energy in $m \le N$ samples of $\mathbf{v}$. Define

$$\sigma_k^2 \triangleq E[|v(k)|^2], \qquad \sigma_0^2 \ge \sigma_1^2 \cdots \ge \sigma_{N-1}^2$$
$$S_m(\mathbf{A}) \triangleq \sum_{k=0}^{m-1} \sigma_k^2 \tag{5.169}$$

Then for any fixed $m \in [1, N]$

$$S_m(\mathbf{\Phi}^{*T}) \ge S_m(\mathbf{A}) \tag{5.170}$$

Proof. Note that

$$\begin{aligned} S_m(\mathbf{A}) &= \sum_{k=0}^{m-1} (\mathbf{ARA}^{*T})_{k,k} \\ &= \text{Tr}(\mathbf{I}_m \mathbf{A}^{*T}\mathbf{RA}) \\ &= \tilde{J}_m \end{aligned}$$

which, we know from the last property [see (5.164)], is maximized when $\mathbf{A}$ is the KL transform. Since $\sigma_k^2 = \lambda_k$ when $\mathbf{A} = \mathbf{\Phi}^{*T}$, from (5.168)

$$\sum_{k=0}^{m-1} \lambda_k \ge \sum_{k=0}^{m-1} \sigma_k^2, \qquad 1 \le m \le N \tag{5.171}$$

Threshold representation. The KL transform also minimizes $E[m]$, the expected number of transform coefficients required, so that their energy just exceeds a prescribed threshold (see Problem 5.26 and [33]).

A fast KL transform. In application of the KL transform to images, there are dimensionality difficulties. The KL transform depends on the statistics as well as the size of the image and, in general, the basis vectors are not known analytically. After the transform matrix has been computed, the operations for performing the transformation are quite large for images.

It has been shown that certain statistical image models yield a fast KL transform algorithm as an alternative to the conventional KL transform for images. It is based on a stochastic decomposition of an image as a sum of two random sequences. *The first random sequence is such that its KL transform is a fast transform* and the second sequence, called the *boundary response,* depends only on information at the boundary points of the image. For details see Sections 6.5 and 6.9.

The rate-distortion function. Suppose a random vector $\mathbf{u}$ is unitary transformed to $\mathbf{v}$ and transmitted over a communication channel (Fig. 5.17). Let $\mathbf{v}^\cdot$ and $\mathbf{u}^\cdot$

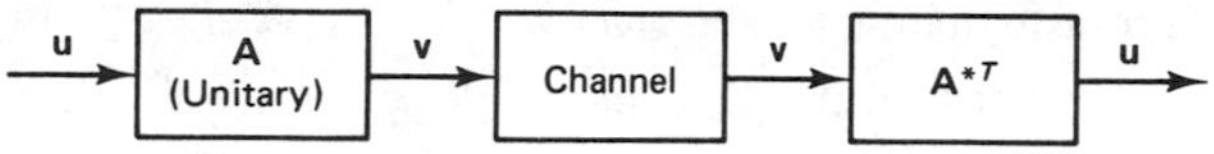

Figure 5.17 Unitary transform data transmission. Each element of **v** is coded independently.

be the reproduced values of **v** and **u**, respectively. Further, assume that **u**, **v**, $\mathbf{v}^{\cdot}$, and $\mathbf{u}^{\cdot}$ are Gaussian. The average distortion in **u** is

$$D = \frac{1}{N} E[(\mathbf{u} - \mathbf{u}^{\cdot})^{*T}(\mathbf{u} - \mathbf{u}^{\cdot})] \tag{5.172}$$

Since **A** is unitary and $\mathbf{u} = \mathbf{A}^{*T}\mathbf{v}$ and $\mathbf{u}^{\cdot} = \mathbf{A}^{*T}\mathbf{v}^{\cdot}$, we have

$$\begin{aligned} D &= \frac{1}{N} E[(\mathbf{v} - \mathbf{v}^{\cdot})^{*T}\mathbf{A}\mathbf{A}^{*T}(\mathbf{v} - \mathbf{v}^{\cdot})] \\ &= \frac{1}{N} E[(\mathbf{v} - \mathbf{v}^{\cdot})^{*T}(\mathbf{v} - \mathbf{v}^{\cdot})] = \frac{1}{N} E(\delta\mathbf{v}^{*T}\delta\mathbf{v}) \end{aligned} \tag{5.173}$$

where $\delta\mathbf{v} = \mathbf{v} - \mathbf{v}^{\cdot}$ represents the error in the reproduction of **v**. From the preceding, D is invariant under all unitary transformations. The rate-distortion function is now obtained, following Section 2.13, as

$$R = \frac{1}{N}\sum_{k=0}^{N-1} \max\left[0, \tfrac{1}{2}\log_2 \frac{\sigma_k^2}{\theta}\right] \tag{5.174}$$

$$D = \frac{1}{N}\sum_{k=0}^{N-1} \min[\theta, \sigma_k^2] \tag{5.175}$$

where

$$\sigma_k^2 = E[|v(k)|^2] = [\mathbf{A}\mathbf{R}\mathbf{A}^{*T}]_{k,k} \tag{5.176}$$

depend on the transform **A**. Therefore, the rate

$$R = R(\mathbf{A}) \tag{5.177}$$

also depends on **A**. For each fixed D, *the KL transform achieves the minimum rate among all unitary transforms,* that is,

$$R(\mathbf{\Phi}^{*T}) \le R(\mathbf{A}) \tag{5.178}$$

This property is discussed further in Chapter 11, on transform coding.

Example 5.8

Consider a 2×1 vector **u**, whose covariance matrix is

$$\mathbf{R} = \begin{bmatrix} 1 & \rho \\ \rho & 1 \end{bmatrix}, \qquad |\rho| < 1$$

The KL transform is

$$\mathbf{\Phi}^{*T} = \mathbf{\Phi} = \frac{1}{\sqrt{2}}\begin{bmatrix} 1 & 1 \\ 1 & -1 \end{bmatrix}$$

The transformation $\mathbf{v} = \mathbf{\Phi u}$ gives

$$E\{[v(0)]^2\} = \lambda_0 = 1 + \rho, \qquad E\{[v(1)]^2\} = 1 - \rho$$

$$R(\mathbf{\Phi}) = \tfrac{1}{2}\left[\max\left(0, \tfrac{1}{2}\log\frac{1+\rho}{\theta}\right) + \max\left(0, \tfrac{1}{2}\log\frac{1-\rho}{\theta}\right)\right]$$

Compare this with the case when $\mathbf{A} = \mathbf{I}$ (that is, $\mathbf{u}$ is transmitted), which gives $\sigma_0^2 = \sigma_1^2 = 1$, and

$$R(\mathbf{I}) = \tfrac{1}{4}[-2\log\theta], \qquad 0 < \theta < 1$$

Suppose we let θ be small, say $\theta < 1 - |\rho|$. Then it is easy to show that

$$R(\mathbf{\Phi}) < R(\mathbf{I})$$

This means for a fixed level of distortion, the number of bits required to transmit the KLT sequence would be less than those required for transmission of the original sequence.

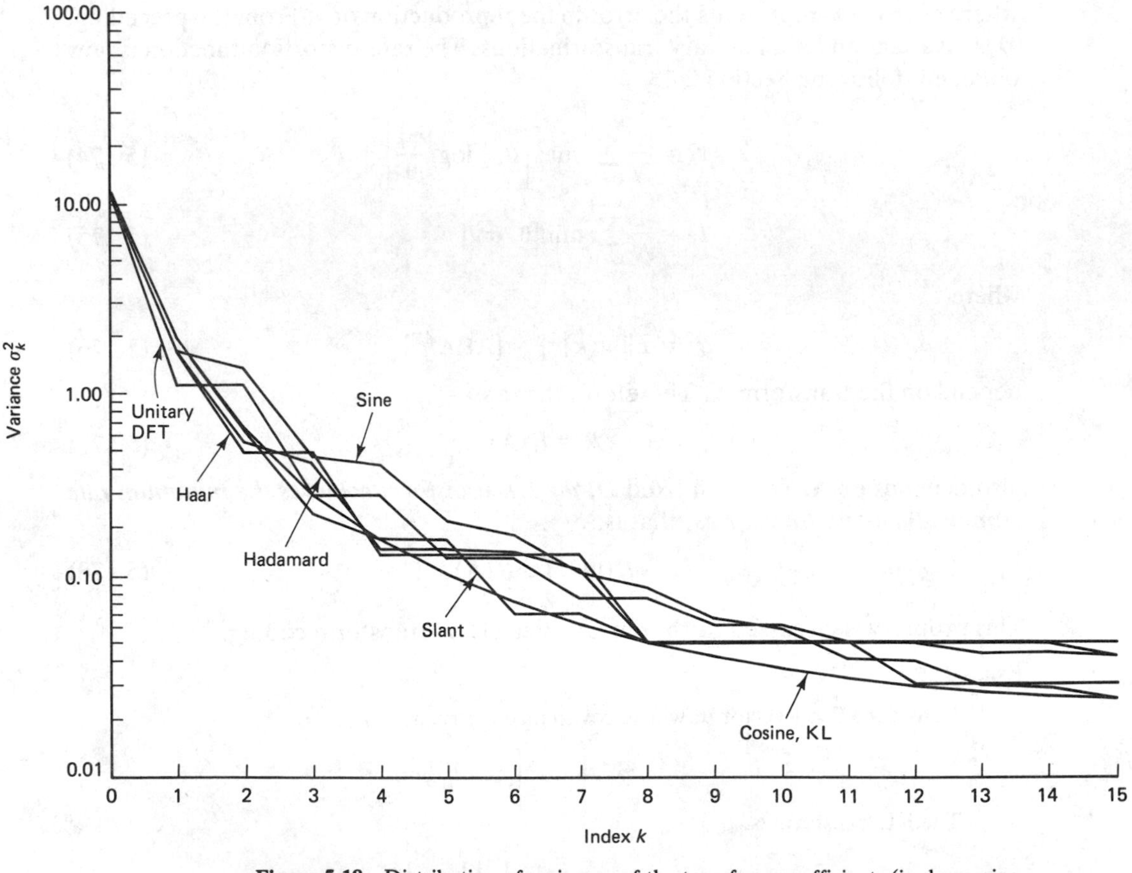

Figure 5.18 Distribution of variances of the transform coefficients (in decreasing order) of a stationary Markov sequence with $N = 16$, $\rho = 0.95$ (see Example 5.9).

TABLE 5.2 Variances σ_k^2 of Transform Coefficients of a Stationary Markov Sequence with $\rho = 0.95$ and $N = 16$. See Example 5.9.

Transform / ↓ k	KL	Cosine	Sine	Unitary DFT	Hadamard	Haar	Slant
0	12.442	12.406	11.169	12.406	12.406	12.406	12.406
1	1.946	1.943	1.688	1.100	1.644	1.644	1.904
2	0.615	0.648	1.352	0.292	0.544	0.487	0.641
3	0.292	0.295	0.421	0.139	0.431	0.487	0.233
4	0.171	0.174	0.463	0.086	0.153	0.144	0.173
5	0.114	0.114	0.181	0.062	0.152	0.144	0.172
6	0.082	0.083	0.216	0.051	0.149	0.144	0.072
7	0.063	0.063	0.098	0.045	0.121	0.144	0.072
8	0.051	0.051	0.116	0.043	0.051	0.050	0.051
9	0.043	0.043	0.060	0.045	0.051	0.050	0.051
10	0.037	0.037	0.067	0.051	0.051	0.050	0.051
11	0.033	0.033	0.040	0.062	0.051	0.050	0.051
12	0.030	0.030	0.042	0.086	0.051	0.050	0.031
13	0.028	0.028	0.031	0.139	0.051	0.050	0.031
14	0.027	0.027	0.029	0.292	0.050	0.050	0.031
15	0.026	0.026	0.026	1.100	0.043	0.050	0.031

Example 5.9 (Comparison Among Unitary Transforms for a Markov Sequence)

Consider a first-order zero mean stationary Markov sequence of length N whose covariance matrix $\mathbf{R}$ is given by (2.68) with $\rho = 0.95$. Figure 5.18 shows the distribution of variances σ_k^2 of the transform coefficients (in decreasing order) for different transforms. Table 5.2 lists σ_k^2 for the various transforms.

Define the normalized basis restriction error as

$$J_m = \frac{\sum_{k=m}^{N-1} \sigma_k^2}{\sum_{k=0}^{N-1} \sigma_k^2}, \qquad m = 0, \ldots, N-1 \tag{5.179}$$

where σ_k^2 have been arranged in decreasing order.

Figure 5.19 shows J_m versus m for the various transforms. It is seen that the cosine transform performance is indistinguishable from that of the KL transform for $\rho = 0.95$. In general it seems possible to find a fast sinusoidal transform (that is, a transform consisting of sine or cosine functions) as a good substitute for the KL transform for different values of ρ as well as for higher-order stationary random sequences (see Section 5.12).

The mean square performance of the various transforms also depends on the dimension N of the transform. Such comparisons are made in Section 5.12.

Example 5.10 (Performance of Transforms on Images)

The mean square error test of the last example can be extended to actual images. Consider an $N \times N$ image $u(m, n)$ from which its mean is subtracted out to make it zero mean. The transform coefficient variances are estimated as

$$\sigma^2(k, l) = E[|v(k, l)|^2] \cong |v(k, l)|^2$$

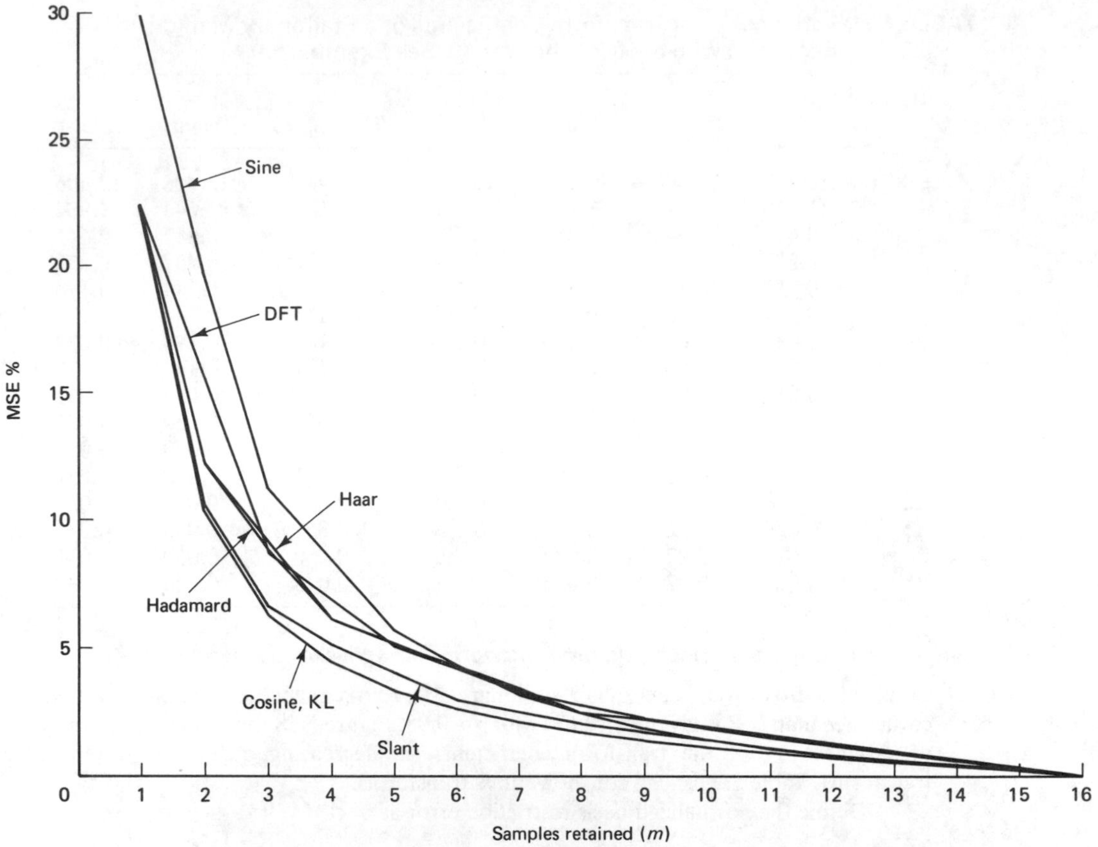

Figure 5.19 Performance of different unitary transforms with respect to basis restriction errors (J_m) versus the number of basis (m) for a stationary Markov sequence with $N = 16$, $\rho = 0.95$.

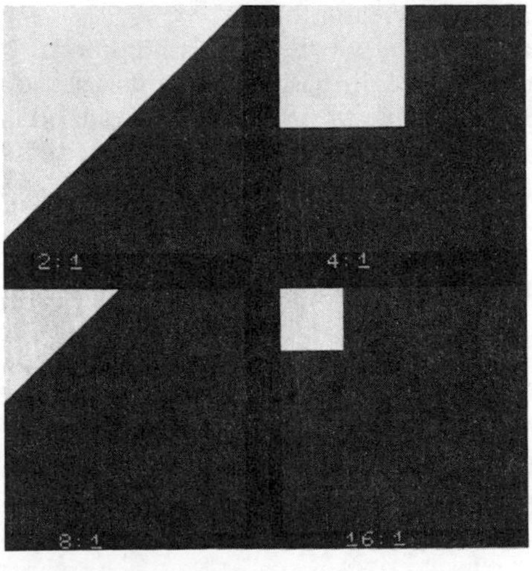

Figure 5.20 Zonal filters for 2 : 1, 4 : 1, 8 : 1, 16 : 1 sample reduction. White areas are passbands, dark areas are stopbands.

The image transform is filtered by a *zonal mask* (Fig. 5.20) such that only a fraction of the transform coefficients are retained and the remaining ones are set to zero. Define the normalized mean square error

$$J_s \triangleq \frac{\sum\sum_{k,l\,\epsilon\,\text{stopband}} |v_{k,l}|^2}{\sum\sum_{k,l=0}^{N-1} |v_{k,l}|^2} = \frac{\text{energy in stopband}}{\text{total energy}}$$

Figure 5.21 shows an original image and the image obtained after cosine transform zonal filtering to achieve various sample reduction ratios. Figure 5.22 shows the zonal filtered images for different transforms at a 4 : 1 sample reduction ratio. Figure 5.23 shows the mean square error versus sample reduction ratio for different transforms. Again we find the cosine transform to have the best performance.

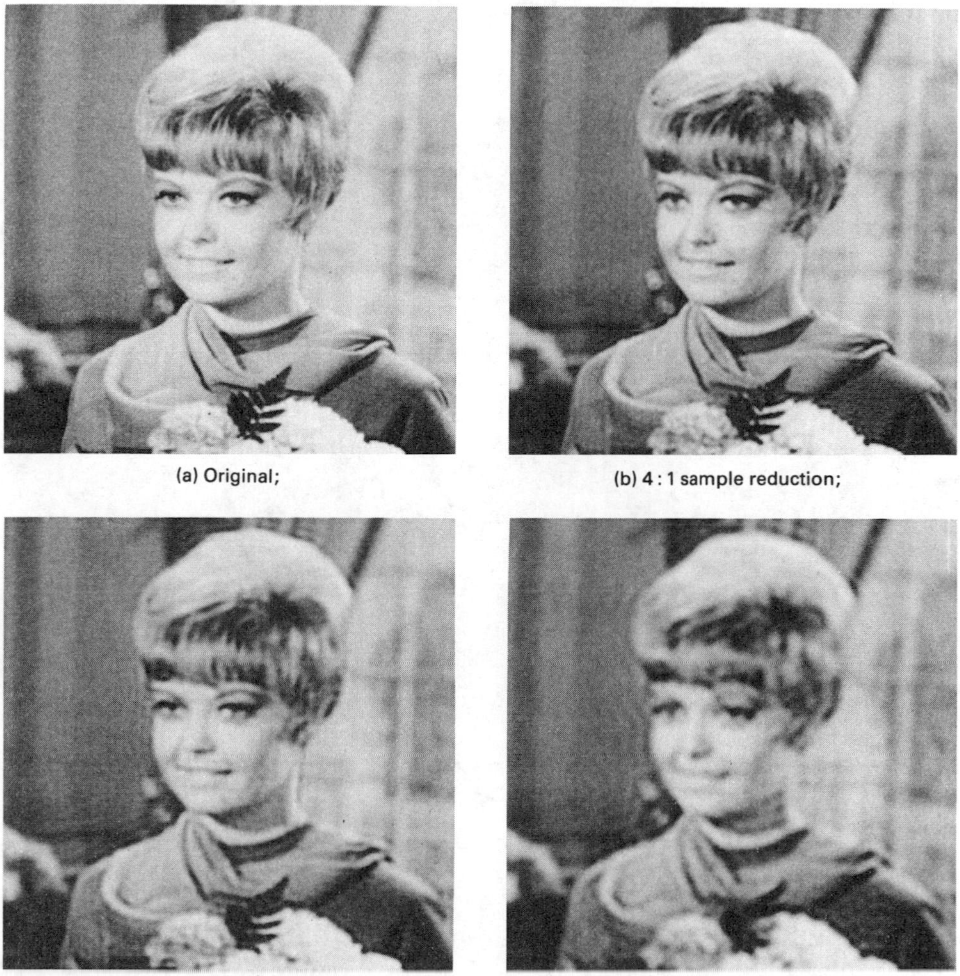

(a) Original; (b) 4 : 1 sample reduction; (c) 8 : 1 sample reduction; (d) 16 : 1 sample reduction.

Figure 5.21 Basis restriction zonal filtered images in cosine transform domain.

(a) Cosine;

(b) sine;

(c) unitary DFT;

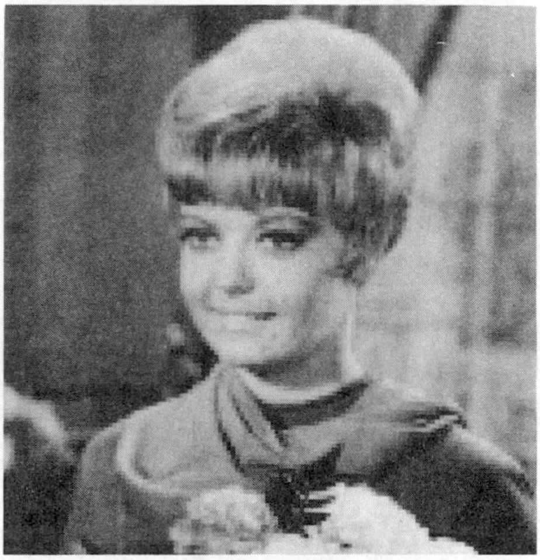

(d) Hadamard;

(e) Haar;

(f) Slant.

Figure 5.22 Basis restriction zonal filtering using different transforms with 4 : 1 sample reduction.

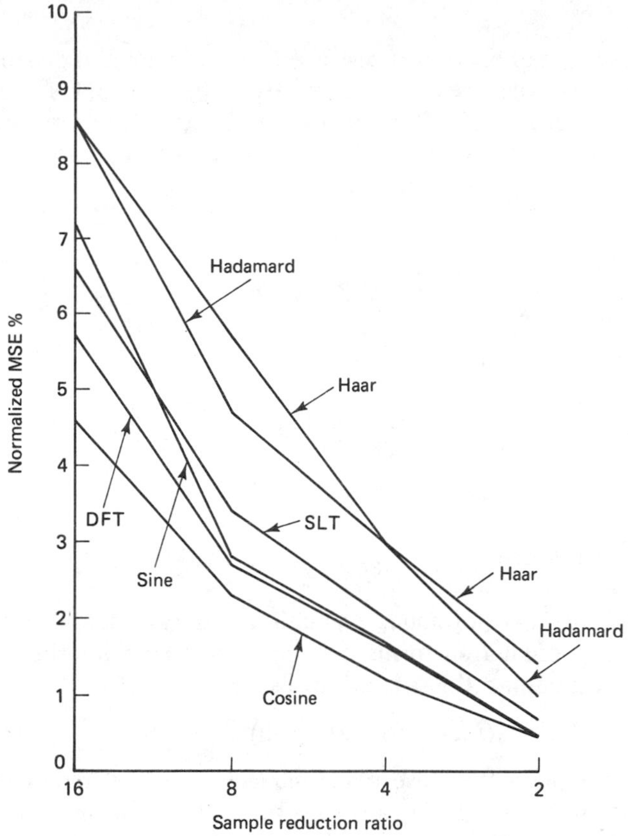

Figure 5.23 Performance comparison of different transforms with respect to basis restriction zonal filtering for 256 × 256 images.

5.12 A SINUSOIDAL FAMILY OF UNITARY TRANSFORMS

This is a class of complete orthonormal sets of eigenvectors generated by the parametric family of matrices whose structure is similar to that of $\mathbf{R}^{-1}$ [see (5.96)],

$$\mathbf{J} = \mathbf{J}(k_1, k_2, k_3) = \begin{bmatrix} 1 - k_1\alpha & -\alpha & & & \mathbf{0} & k_3\alpha \\ -\alpha & 1 & \ddots & & & \\ & \ddots & \ddots & \ddots & & \\ & & & 1 & & -\alpha \\ & & \mathbf{0} & & & \\ k_3\alpha & & & & -\alpha & 1 - k_2\alpha \end{bmatrix} \tag{5.180}$$

In fact, for $k_1 = k_2 = \rho$, $k_3 = 0$, $\beta^2 = (1 - \rho^2)/(1 + \rho^2)$, and $\alpha = \rho/(1 + \rho^2)$, we have

$$\mathbf{J}(\rho, \rho, 0) = \beta^2 \mathbf{R}^{-1} \tag{5.181}$$

Since $\mathbf{R}$ and $\beta^2\mathbf{R}^{-1}$ have an identical set of eigenvectors, the KL transform associated with $\mathbf{R}$ can be determined from the eigenvectors of $\mathbf{J}(\rho, \rho, 0)$. Similarly, it can be shown that the basis vectors of previously discussed cosine, sine, and discrete Fourier transforms are the eigenvectors of $\mathbf{J}(1, 1, 0)$, $\mathbf{J}(0, 0, 0)$, and $\mathbf{J}(1, 1, -1)$, respectively. In fact, several other fast transforms whose basis vectors are sinusoids can be generated for different combinations of k_1, k_2, and k_3. For example, for $0 \le m, k \le N - 1$, we obtain the following transforms:

1. Odd sine − 1: $k_1 = k_3 = 0, k_2 = 1$

$$\phi_m(k) = \frac{2}{\sqrt{2N+1}} \sin \frac{(k+1)(2m+1)\pi}{2N+1} \tag{5.182}$$

2. Odd cosine − 1: $k_1 = 1, k_2 = k_3 = 0$

$$\phi_m(k) = \frac{2}{\sqrt{2N+1}} \cos \frac{(2k+1)(2m+1)\pi}{2(2N+1)} \tag{5.183}$$

Other members of this family of transforms are given in [34].

Approximation to the KL Transform

The **J** matrices play a useful role in performance evaluation of the sinusoidal transforms. For example, two sinusoidal transforms can be compared with the KL transform by comparing corresponding **J**-matrix distances

$$\Delta(k_1, k_2, k_3) \triangleq \|\mathbf{J}(k_1, k_2, k_3) - \mathbf{J}(\rho, \rho, 0)\|^2 \tag{5.184}$$

This measure can also explain the close performance of the DCT and the KLT. Further, it can be shown that the DCT performs better than the sine transform for $0.5 \le \rho \le 1$ and the sine transform performs better than the cosine for other values of ρ. The **J** matrices are also useful in finding a fast sinusoidal transform approximation to the KL transform of an arbitrary random sequence whose covariance matrix is **A.** If **A** commutes with a **J** matrix, that is, $\mathbf{AJ} = \mathbf{JA}$, then they will have an identical set of eigenvectors. The best fast sinusoidal transform may be chosen as the one whose corresponding **J** matrix minimizes the *commuting distance* $\|\mathbf{AJ} - \mathbf{JA}\|^2$. Other uses of the **J** matrices are (1) finding fast algorithms for inversion of banded Toeplitz matrices, (2) efficient calculation of transform coefficient variances, which are needed in transform domain processing algorithms, and (3) establishing certain useful asymptotic properties of these transforms. For details see [34].

5.13 OUTER PRODUCT EXPANSION AND SINGULAR VALUE DECOMPOSITION

In the foregoing transform theory, we considered an $N \times M$ image **U** to be a vector in an NM-dimensional vector space. However, it is possible to represent any such image in an r-dimensional subspace where r is the rank of the matrix **U**.

Let the image be real and $M \le N$. The matrices $\mathbf{UU}^T$ and $\mathbf{U}^T\mathbf{U}$ are nonnegative, symmetric and have the identical eigenvalues, $\{\lambda_m\}$. Since $M \le N$, there are at most $r \le M$ nonzero eigenvalues. It is possible to find r orthogonal, $M \times 1$

eigenvectors $\{\boldsymbol{\phi}_m\}$ of $\mathbf{U}^T\mathbf{U}$ and r orthogonal $N \times 1$ eigenvectors $\{\boldsymbol{\psi}_m\}$ of $\mathbf{U}\mathbf{U}^T$, that is,

$$\mathbf{U}^T\mathbf{U}\boldsymbol{\phi}_m = \lambda_m \boldsymbol{\phi}_m, \qquad m = 1, \ldots, r \tag{5.185}$$

$$\mathbf{U}\mathbf{U}^T\boldsymbol{\psi}_m = \lambda_m \boldsymbol{\psi}_m, \qquad m = 1, \ldots, r \tag{5.186}$$

The matrix $\mathbf{U}$ has the representation

$$\mathbf{U} = \mathbf{\Psi}\mathbf{\Lambda}^{1/2}\mathbf{\Phi}^T \tag{5.187}$$

$$= \sum_{m=1}^{r} \sqrt{\lambda_m}\,\boldsymbol{\psi}_m \boldsymbol{\phi}_m^T \tag{5.188}$$

where $\mathbf{\Psi}$ and $\mathbf{\Phi}$ are $N \times r$ and $M \times r$ matrices whose mth columns are the vectors $\boldsymbol{\psi}_m$ and $\boldsymbol{\phi}_m$, respectively, and $\mathbf{\Lambda}^{1/2}$ is an $r \times r$ diagonal matrix, defined as

$$\mathbf{\Lambda}^{1/2} = \begin{bmatrix} \sqrt{\lambda_1} & & \mathbf{0} \\ & \ddots & \\ \mathbf{0} & & \sqrt{\lambda_r} \end{bmatrix} \tag{5.189}$$

Equation (5.188) is called the *spectral representation,* the *outer product expansion,* or the *singular value decomposition* (SVD) of $\mathbf{U}$. The nonzero eigenvalues (of $\mathbf{U}^T\mathbf{U}$), λ_m, are also called the singular values of $\mathbf{U}$. If $r \ll M$, then the image containing NM samples can be represented by $(M + N)r$ samples of the vectors $\{\lambda_m^{1/4}\boldsymbol{\psi}_m, \lambda_m^{1/4}\boldsymbol{\phi}_m; m = 1, \ldots, r\}$.

Since $\mathbf{\Psi}$ and $\mathbf{\Phi}$ have orthogonal columns, from (5.187) the *SVD transform* of the image $\mathbf{U}$ is defined as

$$\mathbf{\Lambda}^{1/2} = \mathbf{\Psi}^T\mathbf{U}\mathbf{\Phi} \tag{5.190}$$

which is a separable transform that diagonalizes the given image.The proof of (5.188) is outlined in Problem 5.31.

Properties of the SVD Transform

1. Once $\boldsymbol{\phi}_m, m = 1, \ldots, r$ are known, the eigenvectors $\boldsymbol{\psi}_m$ can be determined as

$$\boldsymbol{\psi}_m \triangleq \frac{1}{\sqrt{\lambda_m}}\mathbf{U}\boldsymbol{\phi}_m, \qquad m = 1, \ldots, r \tag{5.191}$$

   It can be shown that $\boldsymbol{\psi}_m$ are orthonormal eigenvectors of $\mathbf{U}\mathbf{U}^T$ if $\boldsymbol{\phi}_m$ are the orthonormal eigenvectors of $\mathbf{U}^T\mathbf{U}$.

2. The SVD transform as defined by (5.190) is not a unitary transform. This is because $\mathbf{\Psi}$ and $\mathbf{\Phi}$ are rectangular matrices. However, we can include in $\mathbf{\Phi}$ and $\mathbf{\Psi}$ additional orthogonal eigenvectors $\boldsymbol{\phi}_m$ and $\boldsymbol{\psi}_m$, which satisfy $\mathbf{U}\boldsymbol{\phi}_m = 0$, $m = r + 1, \ldots, M$ and $\mathbf{U}^T\boldsymbol{\psi}_m = 0, m = r + 1, \ldots, N$ such that these matrices are unitary and the *unitary SVD transform* is

$$\begin{bmatrix} \mathbf{\Lambda}^{1/2} \\ \mathbf{0} \end{bmatrix} = \mathbf{\Psi}^T\mathbf{U}\mathbf{\Phi} \tag{5.192}$$

3. The image $\mathbf{U}_k$ generated by the partial sum

$$\mathbf{U}_k \triangleq \sum_{m=1}^{k} \sqrt{\lambda_m}\, \boldsymbol{\psi}_m \boldsymbol{\phi}_m^T, \qquad k \le r \tag{5.193}$$

is the best least squares rank-k approximation of $\mathbf{U}$ if λ_m are in decreasing order of magnitude. For any $k \le r$, the least squares error

$$\epsilon_k^2 = \sum_{m=1}^{M} \sum_{n=1}^{N} |u(m,n) - u_k(m,n)|^2, \qquad k = 1,2,\ldots,r \tag{5.194}$$

reduces to

$$\epsilon_k^2 = \sum_{m=k+1}^{r} \lambda_m \tag{5.195}$$

Let $L \triangleq NM$. Note that we can always write a two-dimensional unitary transform representation as an outer product expansion in an L-dimensional space, namely,

$$\mathbf{U} = \sum_{l=1}^{L} w_l \mathbf{a}_l \mathbf{b}_l^T \tag{5.196}$$

where w_l are scalars and $\mathbf{a}_l$ and $\mathbf{b}_l$ are sequences of orthogonal basis vectors of dimensions $N \times 1$ and $M \times 1$, respectively. The least squares error between $\mathbf{U}$ and any partial sum

$$\hat{\mathbf{U}}_k \triangleq \sum_{l=1}^{k} w_l \mathbf{a}_l \mathbf{b}_l^T \tag{5.197}$$

is minimized for any $k \in [1, L]$ when the above expansion coincides with (5.193), that is, when $\hat{\mathbf{U}}_k = \mathbf{U}_k$.

This means *the energy concentrated in the transform coefficients* $w_l, l = 1, \ldots, k$ *is maximized by the SVD transform for the given image.* Recall that the KL transform, maximizes *the average energy in a given number of transform coefficients, the average being taken over the ensemble for which the autocorrelation function is defined.* Hence, on an image-to-image basis, the SVD transform will concentrate more energy in the same number of coefficients. But the SVD has to be calculated for *each* image. On the other hand the KL transform needs to be calculated only once for the whole image ensemble. Therefore, while one may be able to find a reasonable fast transform approximation of the KL transform, no such fast transform substitute for the SVD is expected to exist.

Although applicable in image restoration and image data compression problems, the usefulness of SVD in such image processing problems is severely limited because of large computational effort required for calculating the eigenvalues and eigenvectors of large image matrices. However, the SVD is a fundamental result in matrix theory that is useful in finding the generalized inverse of singular matrices and in the analysis of several image processing problems.

Example 5.11

Let

$$\mathbf{U} = \begin{bmatrix} 1 & 2 \\ 2 & 1 \\ 1 & 3 \end{bmatrix}$$

The eigenvalues of $\mathbf{U}^T\mathbf{U}$ are found to be $\lambda_1 = 18.06$, $\lambda_2 = 1.94$, which give $r = 2$, and the SVD transform of $\mathbf{U}$ is

$$\mathbf{\Lambda}^{1/2} = \begin{bmatrix} 4.25 & 0 \\ 0 & 1.39 \end{bmatrix}$$

The eigenvectors are found to be

$$\boldsymbol{\phi}_1 = \begin{bmatrix} 0.5019 \\ 0.8649 \end{bmatrix}, \qquad \boldsymbol{\phi}_2 = \begin{bmatrix} 0.8649 \\ -0.5019 \end{bmatrix}$$

(*continued on page 180*)

TABLE 5.3 Summary of Image Transforms

DFT/unitary DFT	Fast transform, most useful in digital signal processing, convolution, digital filtering, analysis of circulant and Toeplitz systems. Requires complex arithmetic. Has very good energy compaction for images.
Cosine	Fast transform, requires real operations, near optimal substitute for the KL transform of highly correlated images. Useful in designing transform coders and Wiener filters for images. Has excellent energy compaction for images.
Sine	About twice as fast as the fast cosine transform, symmetric, requires real operations; yields fast KL transform algorithm which yields recursive block processing algorithms, for coding, filtering, and so on; useful in estimating performance bounds of many image processing problems. Energy compaction for images is very good.
Hadamard	Faster than sinusoidal transforms, since no multiplications are required; useful in digital hardware implementations of image processing algorithms. Easy to simulate but difficult to analyze. Applications in image data compression, filtering, and design of codes. Has good energy compaction for images.
Haar	Very fast transform. Useful in feature extracton, image coding, and image analysis problems. Energy compaction is fair.
Slant	Fast transform. Has "image-like basis"; useful in image coding. Has very good energy compaction for images.
Karhunen-Loeve	Is optimal in many ways; has no fast algorithm; useful in performance evaluation and for finding performance bounds. Useful for small size vectors e.g., color multispectral or other feature vectors. Has the best energy compaction in the mean square sense over an ensemble.
Fast KL	Useful for designing fast, recursive-block processing techniques, including adaptive techniques. Its performance is better than independent block-by-block processing techniques.
Sinusoidal transforms	Many members have fast implementation, useful in finding practical substitutes for the KL transform, analysis of Toeplitz systems, mathematical modeling of signals. Energy compaction for the optimum-fast transform is excellent.
SVD transform	Best energy-packing efficiency for any given image. Varies drastically from image to image; has no fast algorithm or a reasonable fast transform substitute; useful in design of separable FIR filters, finding least squares and minimum norm solutions of linear equations, finding rank of large matrices, and so on. Potential image processing applications are in image restoration, power spectrum estimation and data compression.

From above $\boldsymbol{\psi}_1$ is obtained via (5.191) to yield

$$\mathbf{U}_1 = \sqrt{\lambda_1}\,\boldsymbol{\psi}_1\,\boldsymbol{\phi}_1^T = \begin{bmatrix} 1.120 & 1.94 \\ 0.935 & 1.62 \\ 1.549 & 2.70 \end{bmatrix}$$

as the best least squares rank-1 approximation of **U**. Let us compare this with the two dimensional cosine transform **U**, which is given by

$$\mathbf{V} = \mathbf{C}_3\mathbf{U}\mathbf{C}_2^T = \frac{1}{\sqrt{12}}\begin{bmatrix} \sqrt{2} & \sqrt{2} & \sqrt{2} \\ \sqrt{3} & 0 & -\sqrt{3} \\ 1 & -2 & 1 \end{bmatrix}\begin{bmatrix} 1 & 2 \\ 2 & 1 \\ 1 & 3 \end{bmatrix}\begin{bmatrix} 1 & 1 \\ 1 & -1 \end{bmatrix} = \frac{1}{\sqrt{12}}\begin{bmatrix} 10\sqrt{2} & -2\sqrt{2} \\ -\sqrt{3} & \sqrt{3} \\ -1 & -5 \end{bmatrix}$$

It is easy to see that $\sum\sum_{k,l} v^2(k, l) = \lambda_1 + \lambda_2$. The energy concentrated in the K samples of SVD, $\sum_{m=1}^{K}\lambda_m$, $K = 1, 2$, is greater than the energy concentrated in any K samples of the cosine transform coefficients (show!).

5.14 SUMMARY

In this chapter we have studied the theory of unitary transforms and their properties. Several unitary tranforms, DFT, cosine, sine, Hadamard, Haar, Slant, KL, sinusoidal family, fast KL, and SVD, were discussed. Table 5.3 summarizes the various transforms and their applications.

PROBLEMS

5.1 For given P, Q show that the error σ_e^2 of (5.8) is minimized when the series coefficients $v(k, l)$ are given by (5.3). Also show that the basis images must form a complete set for σ_e^2 to be zero for $P = Q = N$.

5.2 *(Fast transforms and Kronecker separability)* From (5.23) we see that the number of operations in implementing the matrix-vector product is reduced from $O(N^4)$ to $O(N^3)$ if $\mathcal{A}$ is a Kronecker product. Apply this idea inductively to show that if $\mathcal{A}$ is $M \times M$ and

$$\mathcal{A} = \mathbf{A}_1 \otimes \mathbf{A}_2 \otimes \ldots \otimes \mathbf{A}_m$$

where $\mathbf{A}_k$ is $n_k \times n_k$, $M = \prod_{k=1}^{m} n_k$, then the transformation of (5.23) can be implemented in $O(M \sum_{k=1}^{m} n_k)$, which equals $nM \log_n M$ if $n_k = n$. Many fast algorithms for unitary matrices can be given this interpretation which was suggested by Good [9]. Transforms possessing this property are sometimes called *Good transforms*.

5.3 For the 2×2 transform **A** and the image **U**

$$\mathbf{A} = \tfrac{1}{2}\begin{bmatrix} \sqrt{3} & 1 \\ -1 & \sqrt{3} \end{bmatrix}, \qquad \mathbf{U} = \begin{bmatrix} 2 & 3 \\ 1 & 2 \end{bmatrix}$$

calculate the transformed image **V** and the basis images.

5.4 Consider the vector $\mathbf{x}$ and an orthogonal transform $\mathbf{A}$

$$\mathbf{x} = \begin{bmatrix} x_0 \\ x_1 \end{bmatrix}, \qquad \mathbf{A} = \begin{bmatrix} \cos\theta & \sin\theta \\ -\sin\theta & \cos\theta \end{bmatrix}$$

Let $\mathbf{a}_0$ and $\mathbf{a}_1$ denote the columns of $\mathbf{A}^T$ (that is, the basis vectors of $\mathbf{A}$). The transformation $\mathbf{y} = \mathbf{A}\mathbf{x}$ can be written as $y_0 = \mathbf{a}_0^T\mathbf{x}, y_1 = \mathbf{a}_1^T\mathbf{x}$. Represent the vector $\mathbf{x}$ in Cartesian coordinates on a plane. Show that the transform $\mathbf{A}$ is a rotation of the coordinates by θ and y_0 and y_1 are the projections of $\mathbf{x}$ in the new coordinate system (see Fig. P5.4).

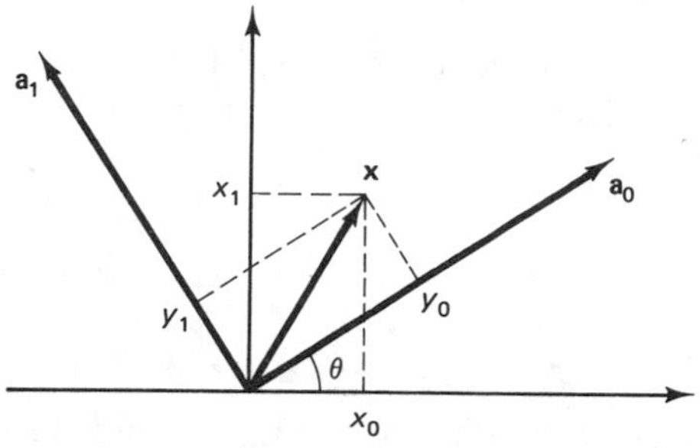

Figure P5.4

5.5 Prove that the magnitude of determinant of a unitary transform is unity. Also show that all the eigenvalues of a unitary matrix have unity magnitude.

5.6 Show that the entropy of an $N \times 1$ Gaussian random vector $\mathbf{u}$ with mean $\boldsymbol{\mu}$ and covariance $\mathbf{R}_u$ given by

$$H(\mathbf{u}) = \frac{N}{2}\log_2(2\pi e|\mathbf{R}_u|^{1/N})$$

is invariant under any unitary transformation.

5.7 Consider the zero mean random vector $\mathbf{u}$ with covariance $\mathbf{R}_u$ discussed in Example 5.2. From the class of unitary transforms

$$\mathbf{A}_\theta = \begin{bmatrix} \cos\theta & \sin\theta \\ -\sin\theta & \cos\theta \end{bmatrix}, \qquad \mathbf{v} = \mathbf{A}_\theta \mathbf{u}$$

determine the value of θ for which (a) the average energy compressed in v_0 is maximum and (b) the components of $\mathbf{v}$ are uncorrelated.

5.8 Prove the two-dimensional energy conservation relation of (5.28).

5.9 *(DFT and circulant matrices)*

a. Show that (5.60) follows directly from (5.56) if $\boldsymbol{\phi}_k$ is chosen as the kth column of the unitary DFT $\mathbf{F}$. Now write (5.58) as a circulant matrix operation $\mathbf{x}_2 = \mathbf{H}\mathbf{x}_1$. Take unitary DFT of both sides and apply (5.60) to prove the circular convolution theorem, that is, (5.59).

b. Using (5.60) show that the inverse of an $N \times N$ circulant matrix can be obtained in $O(N \log N)$ operations via the FFT by calculating the elements of its first column as

$$[\mathbf{H}^{-1}]_{n,0} = \frac{1}{N}\sum_{k=0}^{N-1} W_N^{-kn}\lambda_k^{-1} = \text{inverse DFT of } \{\lambda_k^{-1}\}$$

c. Show that the $N \times 1$ vector $\mathbf{x}_2 = \mathbf{T}\mathbf{x}_1$, where $\mathbf{T}$ is $N \times N$ Toeplitz but not circulant, can be evaluated in $O(N \log N)$ operations via the FFT.

5.10 Show that the N^2 complex elements $v(k, l)$ of the unitary DFT of a real sequence $\{u(m, n), 0 \le m, n \le N-1\}$ can be determined from the knowledge of the partial sequence

$$\left\{v(k, 0), 0 \le k \le \frac{N}{2}\right\}, \left\{v\left(k, \frac{N}{2}\right), 0 \le k \le \frac{N}{2}\right\},$$

$$\left\{v(k, l), 0 \le k \le N-1, 1 \le l \le \frac{N}{2} - 1\right\}, \qquad (N \text{ even})$$

which contains only N^2 nonzero real elements, in general.

5.11 **a.** Find the eigenvalues of the 2×2 doubly block circulant matrix

$$\mathcal{H} = \left[\begin{array}{cc|cc} 1 & 2 & 3 & 4 \\ 2 & 1 & 4 & 3 \\ \hline 3 & 4 & 1 & 2 \\ 4 & 3 & 2 & 1 \end{array}\right]$$

b. Given the arrays $x_1(m, n)$ and $x_2(m, n)$ as follows:

$x_1(m, n)$

n \ m	0	1
1	3	4
0	1	2

$x_2(m, n)$

n \ m	−1	0	1
1	0	−1	0
0	−1	4	−1
−1	0	−1	0

Write their convolution $x_3(m, n) = x_2(m, n) \circledast x_1(m, n)$ as a doubly block circulant matrix operating on a vector of size 16 and calculate the result. Verify your result by performing the convolution directly.

5.12 Show that if an image $\{u(m, n), 0 \le m, n \le N-1\}$ is multiplied by the checkerboard pattern $(-1)^{m+n}$, then its unitary DFT is centered at $(N/2, N/2)$. If the unitary DFT of $u(m, n)$ has its region of support as shown in Fig. P5.12, what would be the region of support of the unitary DFT of $(-1)^{m+n} u(m, n)$? Figure 5.6 shows the magnitude of the unitary DFTs of an image $u(m, n)$ and the image $(-1)^{m+n} u(m, n)$. This method can be used for computing the unitary DFT whose origin is at the center of the image matrix. The frequency increases as one moves away from the origin.

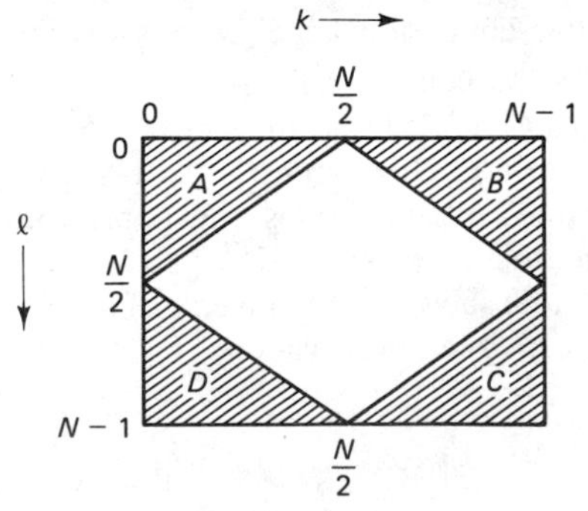

Figure P5.12

5.13 Show that the real and imaginary parts of the unitary DFT matrix are not orthogonal matrices in general.

5.14 Show that the $N \times N$ cosine transform matrix $\mathbf{C}$ is orthogonal. Verify your proof for the case $N = 4$.

5.15 Show that the $N \times N$ sine transform is orthogonal and is the eigenmatrix of $\mathbf{Q}$ given by (5.102). Verify your proof for the case $N = 3$.

5.16 Show that the cosine and sine transforms of an $N \times 1$ sequence $\{u(0), \ldots, u(N-1)\}$ can be calculated from the DFTs of the $2N \times 1$ symmetrically extended sequence $\{u(N-1), u(N-2), \ldots, u(1), u(0), u(0), u(1), \ldots, u(N-1)\}$ and of the $(2N+2) \times 1$ antisymmetrically extended sequence $\{0 - u(N-1), \ldots, -u(1), -u(0), 0, u(0), u(1), \ldots, u(N-1)\}$, respectively.

5.17 Suppose an $N \times N$ image $\mathbf{U}$ is mapped into a row-ordered $N^2 \times 1$ vector $\boldsymbol{u}$. Show that the $N^2 \times N^2$ one-dimensional Hadamard transform of $\boldsymbol{u}$ gives the $N \times N$ two-dimensional Hadamard transform of $\mathbf{U}$. Is this true for the other transforms discussed in the text? Give reasons.

5.18 Using the Kronecker product recursion (5.104), prove that a $2^n \times 2^n$ Hadamard transform is orthogonal.

5.19 Calculate and plot the energy packed in the first 1, 2, 4, 8, 16 sequency ordered samples of the Hadamard transform of a 16×1 vector whose autocorrelations are $r(k) = (0.95)^k$.

5.20 Prove that an $N \times N$ Haar transform matrix is orthogonal and can be implemented in $O(N)$ operations on an $N \times 1$ vector.

5.21 Using the recursive formula for generating the slant transforms prove that these matrices are orthogonal and fast.

5.22 If the KL transform of a zero mean $N \times 1$ vector $\mathbf{u}$ is $\boldsymbol{\Phi}$, then show that the KL transform of the sequence

$$\hat{u}(n) = u(n) + \mu, \qquad 0 \le n \le N-1$$

where μ is a constant, remains the same only if the vector $\mathbf{1} \stackrel{\Delta}{=} (1, 1, \ldots, 1)^T$ is an eigenvector of the covariance matrix of $\mathbf{u}$. Which of the fast transforms discussed in the text satisfy this property?

5.23 If $\mathbf{u}_1$ and $\mathbf{u}_2$ are random vectors whose autocorrelation matrices commute, then show that they have a common KL transform. Hence, show that the KL transforms for autocorrelation matrices $\mathbf{R}$, $\mathbf{R}^{-1}$, and $f(\mathbf{R})$, where $f(\cdot)$ is an arbitrary function, are identical. What are the corresponding eigenvalues?

5.24* The autocorrelation array of a 4×1 zero mean vector $\mathbf{u}$ is given by $\{0.95^{|m-n|}, 0 \le m, n \le 3\}$.

a. What is the KL transform of $\mathbf{u}$?

b. Compare the basis vectors of the KL transform with the basis vectors of the 4×4 unitary DFT, DCT, DST, Hadamard, Haar, and Slant transforms.

c. Compare the performance of the various transforms by plotting the basis restriction error J_m versus m.

5.25* The autocorrelation function of a zero mean random field is given by (5.150), where $\rho = 0.95$. A 16×16 segment of this random field is unitarily transformed.

a. What is the maximum energy concentrated in 16, 32, 64, and 128 transform coefficients for each of the *seven* transforms, KL, cosine, sine, unitary DFT, Hadamard, Haar, and Slant?

b. Compare the performance of these transforms for this random field by plotting the mean square error for sample reduction ratios of 2, 4, 8, and 16. (*Hint:* Use Table 5.2.)

5.26 *(Threshold representation)* Referring to Fig. 5.16, where $u(n)$ is a Gaussian random sequence, the quantity

$$C_m = \frac{1}{N}\sum_{n=0}^{N-1}[u(n) - z(n)]^2 = \frac{1}{N}\sum_{n=0}^{N-1}\left[u(n) - \sum_{j=0}^{m-1} b(n,j)v(j)\right]^2$$

is a random variable with respect to m. Let m be such that

$$C_{m-1} > \epsilon^2, \qquad C_m \le \epsilon^2 \quad \text{for any fixed} \quad \epsilon^2 > 0$$

If $\mathbf{A}$ and $\mathbf{B}$ are restricted to be unitary transforms, then show that $E[m]$ is minimized when $\mathbf{A} = \mathbf{\Phi}^{*T}$, $\mathbf{B} = \mathbf{A}^{-1}$, where $\mathbf{\Phi}^{*T}$ is the KL transform of $u(n)$. For details see [33].

5.27 *(Minimum entropy property of the KL transform)* [30] Define an entropy in the **A**-transform domain as

$$H[\mathbf{A}] = -\sum_{k=0}^{N-1} \sigma_k^2 \log_e \sigma_k^2$$

where σ_k^2 are the variances of the transformed variables $v(k)$. Show that among all the unitary transforms the KL transform minimizes this entropy, that is, $H[\mathbf{\Phi}^{*T}] \le H[\mathbf{A}]$.

5.28 **a.** Write the $N \times N$ covariance matrix $\mathbf{R}$ defined in (2.68) as

$$\beta^2 \mathbf{R}^{-1} = \mathbf{J}(k_1, k_2, k_3) - \mathbf{\Delta J}$$

where $\mathbf{\Delta J}$ is a sparse $N \times N$ matrix with nonzero terms at the four corners. Show that the above relation yields

$$\mathbf{R} = \beta^2 \mathbf{J}^{-1} + \beta^2 \mathbf{J}^{-1}\mathbf{\Delta J}\mathbf{J}^{-1} + \mathbf{J}^{-1}(\mathbf{\Delta R})\mathbf{J}^{-1}$$

where $\mathbf{\Delta R} \triangleq \mathbf{\Delta J R \Delta J}$ is also a sparse matrix, which has at most four (corner) nonzero terms. If $\mathbf{\Phi}$ diagonalizes $\mathbf{J}$, then show that the variances of the transform coefficients are given by

$$\sigma_k^2 \triangleq [\mathbf{\Phi}^{*T}\mathbf{R}\mathbf{\Phi}]_{k,k} = \frac{\beta^2}{\lambda_k} + \frac{\beta^2}{\lambda_k^2}[\mathbf{\Phi}^{*T}\mathbf{\Delta J}\mathbf{\Phi}]_{k,k} + \frac{1}{\lambda_k^2}[\mathbf{\Phi}^{*T}\mathbf{\Delta R}\mathbf{\Phi}]_{k,k} \qquad \text{P5.28-1}$$

Now verify the formulas

$$\sigma_k^2(\text{DCT}) = \frac{\beta^2}{\lambda_k} - \frac{4(1-\rho)^2\alpha^2}{N\lambda_k^2\rho}[1 - (-1)^k\rho^N]\left[\cos^2\left(\frac{k\pi}{2N}\right) - \frac{1}{2}\delta(k)\right] \qquad \text{P5.28-2}$$

where $\lambda_k = 1 - 2\alpha \cos k\pi/N$,

$$\sigma_k^2(\text{DST}) = \frac{\beta^2}{\lambda_k} + \frac{4\alpha^2}{(N+1)\lambda_k^2}[1 + (-1)^k\rho^{N+1}]\sin^2\left(\frac{(k+1)\pi}{N+1}\right), \qquad 0 \le k \le N-1 \qquad \text{P5.28-3}$$

where $\lambda_k = 1 - 2\alpha\cos(k+1)\pi/(N+1)$, and

$$\sigma_k^2(\text{DFT}) = \frac{\beta^2}{\lambda_k} - \frac{2(1-\rho^N)\alpha[\cos(2\pi k/N) - 2\alpha]}{N\lambda_k^2} \qquad \text{P5.28-4}$$

where $\lambda_k = 1 - 2\alpha \cos 2\pi k/N$, $0 \le k \le N-1$.

b. Using the formulas P5.28-2–P5.28-4 and (5.120), calculate the fraction of energy packed in $N/2$ transform coefficients arranged in decreasing order by the cosine, sine, unitary DFT, and Hadamard transforms for $N = 4$, 16, 64, 256, 1024, and 4096 for a stationary Markov sequence whose autocorrelation matrix is given by $\mathbf{R} = \{\rho^{|m-n|}\}$, $\rho = 0.95$.

5.29 **a.** For an arbitrary real stationary sequence, its autocorrelation matrix, $\mathbf{R} \triangleq \{r(m-n)\}$, is Toeplitz. Show that **A**-transform coefficient variances denoted by $\sigma_k^2(\mathbf{A})$, can be obtained in $O(N \log N)$ operations via the formulas

$$\sigma_k^2(\mathbf{F}) = \frac{1}{N}\sum_{n=-N+1}^{N-1}(N-|n|)r(n)W_N^{-nk}, \qquad \mathbf{F} = \text{unitary DFT}$$

$$\sigma_k^2(\text{DST}) = r(0) + \frac{1}{N+1}\left[2a(k) + b(k)\cot\left(\frac{\pi(k+1)}{N+1}\right)\right],$$

$$a(k) + jb(k) \triangleq \sum_{n=1}^{N-1}[r(n) + r(-n)]\exp\left[\frac{j\pi n(k+1)}{N+1}\right]$$

where $0 \le k \le N-1$. Find a similar expression for the DCT.

b. In two dimensions, for stationary random fields, (5.36) implies we have to evaluate

$$\sigma_{k,l}^2(\mathbf{A}) = \sum_m\sum_n\sum_{m'}\sum_{n'} a(k,m)a^*(k,m')r(m-m',n-n')a(l,n)a^*(l,n')$$

Show that $\sigma_{k,l}^2(\mathbf{A})$ can be evaluated in $O(N^2 \log N)$ operations, when **A** is the FFT, DST, or DCT.

5.30 Compare the maximum energy packed in k SVD transform coefficients for $k = 1, 2$, of the 2×4 image

$$\mathbf{U} = \begin{pmatrix} 1 & 2 & 5 & 6 \\ 3 & 4 & 7 & 8 \end{pmatrix}$$

with that packed by the cosine, unitary DFT, and Hadamard transforms.

5.31 *(Proof of SVD representation)* Define $\boldsymbol{\phi}_m$ such that $\mathbf{U}\boldsymbol{\phi}_m = \mathbf{0}$ for $m = r+1, \ldots, M$ so that the set $\boldsymbol{\phi}_m$, $1 \le m \le M$ is complete and orthonormal. Substituting for $\boldsymbol{\psi}_m$ from (5.191) in (5.188), obtain the following result:

$$\sum_{m=1}^{r}\sqrt{\lambda_m}\,\boldsymbol{\psi}_m\boldsymbol{\phi}_m^T = \mathbf{U}\left[\sum_{m=1}^{r}\boldsymbol{\phi}_m\boldsymbol{\phi}_m^T\right] = \mathbf{U}\left[\sum_{m=1}^{M}\boldsymbol{\phi}_m\boldsymbol{\phi}_m^T\right] = \mathbf{U}$$

BIBLIOGRAPHY

Sections 5.1, 5.2

General references on image transforms:

1. H. C. Andrews. *Computer Techniques in Image Processing*. New York: Academic Press, 1970, Chapters 5, 6.
2. H. C. Andrews. "Two Dimensional Transforms." in *Topics in Applied Physics: Picture Processing and Digital Filtering*, vol. 6, T. S. Huang (ed)., New York: Springer Verlag, 1975.
3. N. Ahmed and K. R. Rao. *Orthogonal Transforms for Digital Signal Processing*. New York: Springer Verlag, 1975.

4. W. K. Pratt. *Digital Image Processing*. New York: Wiley Interscience, 1978.
5. H. F. Harmuth. *Transmission of Information by Orthogonal Signals*. New York: Springer Verlag, 1970.
6. Proceedings Symposia on *Applications of Walsh Functions,* University of Maryland, *IEEE-EMC* (1970–73) and Cath. U. of Ameri., 1974.
7. D. F. Elliott and K. R. Rao. *Fast Transforms, Algorithms and Applications,* New York: Academic Press, 1983.

Section 5.3

For matrix theory description of unitary transforms:

8. R. Bellman. *Introduction to Matrix Analysis,* New York: McGraw-Hill, 1960.

Sections 5.4, 5.5

For DFT, FFT, and their applications:

9. I. J. Good. "The Interaction Algorithm and Practical Fourier Analysis." *J. Royal Stat. Soc.* (London) B20 (1958): 361.
10. J. W. Cooley and J. W. Tukey. "An Algorithm for the Machine Calculation of Complex Fourier Series," *Math. Comput.* 19, 90 (April 1965): 297–301.
11. *IEEE Trans. Audio and Electroacoustics.* Special Issue on the Fast Fourier Transform AU-15 (1967).
12. G. D. Bergland. "A Guided Tour of the Fast Fourier Transform." *IEEE Spectrum* 6 (July 1969): 41–52.
13. E. O. Brigham. *The Fast Fourier Transform.* Englewood Cliffs, N.J.: Prentice-Hall, 1974.
14. A. K. Jain. "Fast Inversion of Banded Toeplitz Matrices Via Circular Decomposition." *IEEE Trans. ASSP* ASSP-26, no. 2 (April 1978): 121–126.

Sections 5.6, 5.7

15. N. Ahmed, T. Natarajan, and K. R. Rao. "Discrete Cosine Transform." *IEEE Trans. on Computers* (correspondence) C-23 (January 1974): 90–93.
16. A. K. Jain. "A Fast Karhunen Loeve Transform for a Class of Random Processes." *IEEE Trans. Communications,* Vol. COM-24, pp. 1023–1029, Sept. 1976.
17. A. K. Jain. "Some New Techniques in Image Processing," *Proc. Symposium on Current Mathematical Problems in Image Science,* Monterey, California, November 10–12, 1976.
18. W. H. Chen, C. H. Smith, and S. C. Fralick. "A Fast Computational Algorithm for the Discrete Cosine Transform." *IEEE Trans. Commun.* COM-25 (September 1977): 1004–1009.
19. M. J. Narasimha and A. M. Peterson. "On the Computation of the Discrete Cosine Transform. *IEEE Trans. Commun.* COM-26, no. 6 (June 1978): 934–936.

20. P. Yip and K. R. Rao. "A Fast Computational Algorithm for the Discrete Sine Transform." *IEEE Trans. Commun.* COM-28, no. 2 (February 1980): 304–307.

Sections 5.8, 5.9, 5.10

For Walsh functions and Hadamard, Haar, and slant transforms, see [1–6] and:

21. J. L. Walsh. "A Closed Set of Orthogonal Functions." *American J. of Mathematics* 45 (1923): 5–24.
22. R. E. A. C. Paley. "A Remarkable Series of Orthogonal Functions." *Proc. London Math. Soc.* 34 (1932): 241–279.
23. H. Kitajima. "Energy Packing Efficiency of the Hadamard Transform." *IEEE Trans. Comm.* (correspondence) COM-24 (November 1976): 1256–1258.
24. J. E. Shore. "On the Applications of Haar Functions," *IEEE Trans. Communications* COM-21 (March 1973): 209–216.
25. W. K. Pratt, W. H. Chen, and L. R. Welch. "Slant Transform Image Coding." *IEEE Trans. Comm.* COM-22 (August 1974): 1075–1093. Also see W. H. Chen, "Slant Transform Image Coding." Ph.D. Thesis, University of Southern California, Los Angeles, California, 1973.

Section 5.11

For theory of KL transform and its historic development:

26. H. Hotelling. "Analysis of a Complex of Statistical Variables into Principle Components." *J. Educ. Psychology* 24 (1933): 417–441 and 498–520.
27. H. Karhunen. "Uber Lineare Methoden in der Wahrscheinlich-Keitsrechnung." *Ann. Acad. Science Fenn,* Ser. A.I. 37, Helsmki, 1947. (also see translation by I. Selin in the Rand Corp., Doc. T-131, August 11, 1960).
28. M. Loeve. "Fonctions Aleatoires de Seconde Ordre," in P. Levy, *Processus Stochastiques et Mouvement Brownien.* Paris, France: Hermann, 1948.
29. J. L. Brown, Jr. "Mean Square Truncation Error in Series Expansion of Random Functions," *J. SIAM* 8 (March 1960): 28–32.
30. S. Watanabe. "Karhunen-Loeve Expansion and Factor Analysis, Theoretical Remarks and Applications." *Trans. Fourth Prague Conf. Inform. Theory,* Statist. Decision Functions, and Random Processes, Prague, 1965, pp. 635–660.
31. H. P. Kramer and M. V. Mathews. "A Linear Coding for Transmitting a Set of Correlated Signals." *IRE Trans. Inform. Theory* IT-2 (September 1956): 41–46.
32. W. D. Ray and R. M. Driver. "Further Decomposition of the Karhunen-Loeve Series Representation of a Stationary Random Process." *IEEE Trans. Info. Theory* IT-11 (November 1970): 663–668.

For minimum mean square variance distribution and entropy properties we follow, primarily, [30] and [31]. Some other properties of the KL transform are discussed in:

33. V. R. Algazi and D. J. Sakrison. "On the Optimality of Karhunen-Loeve Expansion." (Correspondence) *IEEE Trans. Information Theory* (March 1969): 319–321.

Section 5.12

The sinusoidal family of transforms was introduced in [17] and:

34. A. K. Jain. "A Sinusoidal Family of Unitary Transforms." *IEEE Trans. Pattern Anal. Mach. Intelligence* PAMI-1, no. 6 (October 1979): 356–365.

Section 5.13

The theory and applications of outerproduct expansion (SVD) can be found in many references, such as [2, 4], and:

35. G. E. Forsythe, P. Henrici. "The Cyclic Jacobi Method for Computing the Principal Values of a Complex Matrix." *Trans. Amer. Math. Soc.* 94 (1960): 1–23.
36. G. H. Golub and C. Reinsch. "Singular Value Decomposition and Least Squares Solutions." *Numer. Math.* 14 (1970): 403–420.
37. S. Treitel and J. L. Shanks. "The Design of Multistage Separable Planar Filters." *IEEE Trans. Geoscience Elec.* Ge-9 (January 1971): 10–27.

6

Image Representation by Stochastic Models

6.1 INTRODUCTION

In stochastic representations an image is considered to be a sample function of an array of random variables called a *random field* (see Section 2.10). This characterization of an ensemble of images is useful in developing image processing techniques that are valid for an entire class and not just for an individual image.

Covariance Models

In many applications such as image restoration and data compression it is often sufficient to characterize an ensemble of images by its mean and covariance functions. Often one starts with a stationary random field representation where the mean is held constant and the covariance function is represented by the separable or the nonseparable exponential models defined in Section 2.10. The separable covariance model of (2.84) is very convenient for analysis of image processing algorithms, and it also yields computationally attractive algorithms (for example, algorithms that can be implemented line by line and then column by column). On the other hand, the nonseparable covariance function of (2.85) is a better model [21] but is not as convenient for analysis.

Covariance models have been found useful in transform image coding, where the covariance function is used to determine the variances of the transform coefficients. Autocorrelation models with spatially varying mean and variance but spatially invariant correlation have been found useful in adaptive block-by-block processing techniques in image coding and restoration problems.

Linear System Models

An alternative to representing random fields by mean and covariance functions is to characterize them as the outputs of linear systems whose inputs are random fields

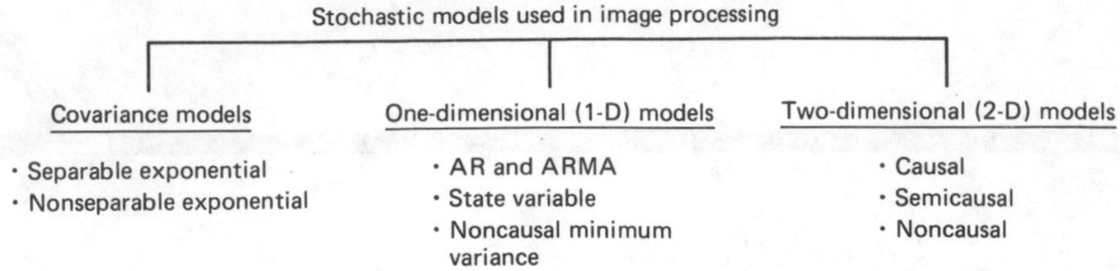

Figure 6.1 Stochastic models used in image processing.

with known or desired statistical properties (for example, white noise inputs). Such linear systems are represented by difference equations and are often useful in developing computationally efficient image processing algorithms. Also, adaptive algorithms based on updating the difference equation parameters are easier to implement than those based on updating covariance functions. The problem of finding a linear stochastic difference equation model that realizes the covariances of an ensemble is known as the *spectral factorization problem.*

Figure 6.1 summarizes the stochastic models that have been used in image processing [1]. Applications of stochastic image models are in image data compression, image restoration, texture synthesis and analysis, two-dimensional power spectrum estimation, edge extraction from noisy images, image reconstruction from noisy projections, and in several other situations.

6.2 ONE-DIMENSIONAL (1-D) CAUSAL MODELS

A simple way of characterizing an image is to consider it a 1-D signal that appears at the output of a raster scanner, that is, a sequence of rows or columns. If the interrow or inter-column dependencies are ignored then 1-D linear systems are useful for modeling such signals.

Let $u(n)$ be a real, stationary random sequence with zero mean and covariance $r(n)$. If $u(n)$ is considered as the output of a *stable,* linear shift invariant system $H(z)$ whose input is a stationary zero mean random sequence $\varepsilon(n)$, then its SDF is given by

$$S(z) = H(z)S_\varepsilon(z)H(z^{-1}), \qquad z = e^{j\omega}, -\pi < \omega \leq \pi \tag{6.1}$$

where $S_\varepsilon(z)$ is the SDF of $\varepsilon(n)$. If $H(z)$ must also be causal while remaining stable, then it must have a one-sided Laurent series

$$H(z) = \sum_{n=0}^{\infty} h(n)z^{-n} \tag{6.2}$$

and all its poles must lie inside the unit circle [2].

Autoregressive (AR) Models

A zero mean random sequence $u(n)$ is called an autoregressive (AR) process of order p (Figure 6.2) when it can be generated as the output of the system

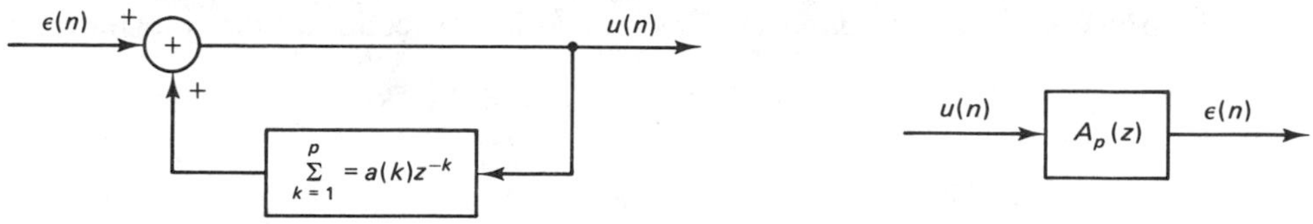

(a) AR sequence realization.

(b) Prediction-error filter.

Figure 6.2 pth-order AR model.

$$u(n) = \sum_{k=1}^{p} a(k)u(n-k) + \varepsilon(n), \qquad \forall n \tag{6.3a}$$

$$E[\varepsilon(n)] = 0, \qquad E\{[\varepsilon(n)]^2\} = \beta^2, \qquad E[\varepsilon(n)u(m)] = 0, \qquad m < n \tag{6.3b}$$

where $\varepsilon(n)$ is a stationary zero mean input sequence that is independent of past outputs. This system uses the most recent p outputs and the current input to generate recursively the next output. Autoregressive models are of special significance in signal and image processing because they possess several important properties, discussed next.

Properties of AR Models

The quantity

$$\bar{u}(n) \triangleq \sum_{k=1}^{p} a(k)u(n-k) \tag{6.4}$$

is the best linear mean square predictor of $u(n)$ based on all its past but depends only on the previous p samples. For Gaussian sequences this means a pth-order AR sequence is a Markov-p process [see eq. (2.66b)]. Thus (6.3a) can be written as

$$u(n) = \bar{u}(n) + \varepsilon(n) \tag{6.5}$$

which says the sample at n is the sum of its *minimum variance, causal, prediction estimate* plus the *prediction error* $\varepsilon(n)$, which is also called the *innovations sequence.* Because of this property an AR model is sometimes called a causal *minimum variance representation* (MVR). The causal filter defined by

$$A_p(z) \triangleq 1 - \sum_{n=1}^{p} a(n)z^{-n} \tag{6.6}$$

is called the *prediction error filter.* This filter generates the prediction error sequence $\varepsilon(n)$ from the sequence $u(n)$.

The prediction error sequence is white, that is,

$$E[\varepsilon(n)\varepsilon(m)] = \beta^2\delta(n-m) \tag{6.7}$$

For this reason, $A_p(z)$ is also called the whitening filter for $u(n)$. The proof is considered in Problem 6.1.

Except for possible zeros at $z = 0$, the transfer function and the SDF of an AR process are *all-pole models*. This follows by inspection of the transfer function

$$H(z) = \frac{1}{A_p(z)} \tag{6.8}$$

The SDF of an AR model is given by

$$S(z) = \frac{\beta^2}{A_p(z)A_p(z^{-1})}, \qquad z = e^{j\omega}, \qquad -\pi < \omega \le \pi \tag{6.9}$$

Because $r_\varepsilon(n) = \beta^2 \delta(n)$ gives $S_\varepsilon(z) = \beta^2$, this formula follows directly by applying (6.1).

For sequences with mean μ, the AR model can be modified as

$$\left.\begin{aligned} x(n) &= \sum_{k=1}^{p} a(k)x(n-k) + \varepsilon(n) \\ u(n) &= x(n) + \mu \end{aligned}\right\} \tag{6.10a}$$

where the properties of $\varepsilon(n)$ are same as before. This representation can also be written as

$$u(n) = \sum_{k=1}^{p} a(k)u(n-k) + \varepsilon(n) + \mu\left[1 - \sum_{k=1}^{p} a(k)\right] \tag{6.10b}$$

which is equivalent to assuming (6.3a) with $E[\varepsilon(n)] = \mu[1 - \sum_k a(k)]$, $\text{cov}[\varepsilon(n)\varepsilon(m)] = \beta^2 \delta(n-m)$.

Identification of AR models. Multiplying both sides of (6.3a) by $\varepsilon(m)$, taking expectations and using (6.3b) and (6.7) we get

$$E[u(n)\varepsilon(m)] = E[\varepsilon(n)\varepsilon(m)] = \beta^2 \delta(n-m), \qquad m \ge n \tag{6.11}$$

Now, multiplying both sides of (6.3a) by $u(0)$ and taking expectations, we find the AR model satisfies the relation

$$r(n) - \sum_{k=1}^{p} a(k)r(n-k) = \beta^2 \delta(n), \qquad \forall n \ge 0 \tag{6.12}$$

where $r(n) \triangleq E[u(n)u(0)]$ is the covariance function of $u(n)$. This result is important for identification of the AR model parameters $a(k)$, β^2 from a given set of covariances $\{r(n), -p \le n \le p\}$. In fact, a pth-order AR model can be uniquely determined by solving (6.12) for $n = 0, \ldots, p$. In matrix notation, this is equivalent to solving the following *normal equations*:

$$\mathbf{Ra} = \mathbf{r} \tag{6.13a}$$

$$r(0) - \mathbf{a}^T\mathbf{r} = \beta^2 \tag{6.13b}$$

where $\mathbf{R}$ is the $p \times p$ Toeplitz matrix

$$\mathbf{R} \triangleq \begin{bmatrix} r(0) & r(1) \ldots & r(p-1) \\ r(1) & \ddots & \vdots \\ \vdots & \ddots & r(1) \\ r(p-1) \ldots & r(1) & r(0) \end{bmatrix} \tag{6.13c}$$

and $\mathbf{a} \triangleq [a(1)a(2)\dots a(p)]^T$, $\mathbf{r} \triangleq [r(1)r(2)\dots r(p)]^T$. If $\mathbf{R}$ is positive definite, then the AR model is guaranteed to be stable, that is, the solution $\{a(k), 1 \le k \le p\}$ is such that the roots of $A_p(z)$ lie inside the unit circle. This procedure allows us to fit a stable AR model to any sequence $u(n)$ whose $p+1$ covariances $r(0)$, $r(1)$, $r(2), \dots, r(p)$ are known.

Example 6.1.

The covariance function of a raster scan line of an image can be obtained by considering the covariance between two pixels on the same row. Both the 2-D models of (2.84) and (2.85) reduce to a 1-D model of the form $r(n) = \sigma^2 \rho^{|n|}$. To fit an AR model of order 2, for instance, we solve

$$\sigma^2 \begin{bmatrix} 1 & \rho \\ \rho & 1 \end{bmatrix} \begin{bmatrix} a(1) \\ a(2) \end{bmatrix} = \sigma^2 \begin{bmatrix} \rho \\ \rho^2 \end{bmatrix}$$

which gives $a(1) = \rho$, $a(2) = 0$, and $\beta^2 = \sigma^2(1-\rho^2)$. The corresponding representation for a scan line of the image, having pixel mean of μ, is a first-order AR model

$$x(n) = \rho x(n-1) + \varepsilon(n), \qquad r_\varepsilon(n) = \sigma^2(1-\rho^2)\delta(n)$$

$$u(n) = x(n) + \mu \tag{6.14}$$

with $A(z) = 1 - \rho z^{-1}$, $S_\varepsilon = \sigma^2(1-\rho^2)$, and $S(z) = \sigma^2(1-\rho^2)/[(1 - \rho z^{-1})(1-\rho z)]$.

Maximum entropy extension. Suppose we are given a positive definite sequence $r(n)$ for $|n| \le p$ (that is, $\mathbf{R}$ is a positive definite matrix). Then it is possible to extrapolate $r(n)$ for $|n| > p$ by first fitting an AR model via (6.13a) and (6.13b) and then running the recursions of (6.12) for $n > p$, that is, by solving

$$\left.\begin{aligned} r(p+n) &= \sum_{k=1}^{p} a(k) r(p+n-k), \qquad && \forall n \ge 1 \\ r(-n) &= r(n), && \forall n \end{aligned}\right\} \tag{6.15}$$

This extension has the property that among all possible positive definite extensions of $\{r(n)\}$, for $|n| > p$, it maximizes the entropy

$$H \triangleq \frac{1}{2\pi} \int_{-\pi}^{\pi} \log S(\omega)\, d\omega \tag{6.16}$$

where $S(\omega)$ is the Fourier transform of $\{r(n), \forall n\}$. The AR model SDF† $S(\omega)$, which can be evaluated from the knowledge of $a(n)$ via (6.9), is also called the *maximum entropy spectrum* of $\{r(n), |n| \le p\}$. This result gives a method of estimating the power spectrum of a partially observed signal. One would start with an estimate of the $p+1$ covariances, $\{r(n), 0 \le n \le p\}$, calculate the AR model parameters β^2, $a(k)$, $k = 1, \dots, p$, and finally evaluate (6.9). This algorithm is also useful in certain image restoration problems [see Section 8.14 and Problem 8.26].

Applications of AR Models in Image Processing

As seen from Example 6.1, AR models are useful in image processing for representing image scan lines. The prediction property of the AR models has been

† Recall from Section 2.11 our notation $S(\omega) \triangleq S(z)$, $z = e^{j\omega}$.

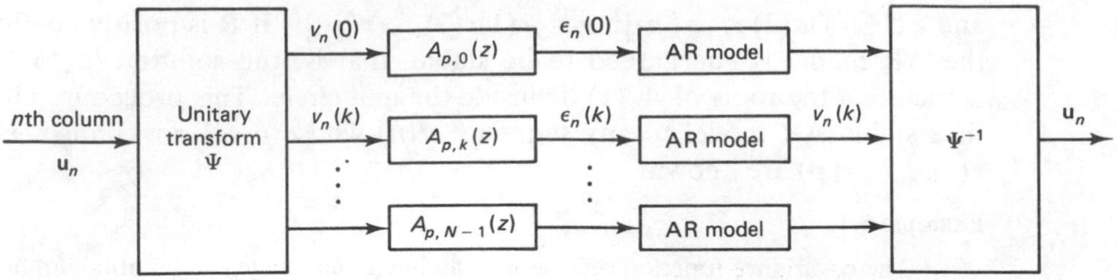

Figure 6.3 Semirecursive representation of images.

exploited in data compression of images and other signals. For example, a digitized AR sequence $u(n)$ represented by B bits/sample is completely equivalent to the digital sequence $\tilde{\varepsilon}(n) \overset{\Delta}{=} u(n) - \bar{u}^{\cdot}(n)$, where $\bar{u}^{\cdot}(n)$ is the quantized value of $\bar{u}(n)$. The quantity $\tilde{\varepsilon}(n)$ represents the unpredictable component of $u(n)$, and its entropy is generally much less than that of $u(n)$. Therefore, it can be encoded by many fewer bits per sample than B. AR models have also been found very useful in representation and *linear predictive coding* (LPC) of speech signals [7].

Another useful application of AR models is in semirecursive representation of images. Each image column $\mathbf{u}_n, n = 0, 1, 2, \ldots$, is first transformed by a unitary matrix, and each row of the resulting image is represented by an independent AR model. Thus if $\mathbf{v}_n \overset{\Delta}{=} \boldsymbol{\Psi}\mathbf{u}_n$, where $\boldsymbol{\Psi}$ is a unitary transform, then the sequence $\{v_n(k), n = 0, 1, 2 \ldots\}$ is represented by an AR model for each k (Figure 6.3), as

$$\left.\begin{aligned} v_n(k) &= \sum_{i=1}^{p} a_i(k) v_{n-i}(k) + \varepsilon_n(k), \qquad \forall n, k = 0, 1, \ldots, N-1 \\ E[\varepsilon_n(k)] &= 0, \qquad E[\varepsilon_n(k)\varepsilon_{n'}(k')] = \beta^2(k)\delta(n-n')\delta(k-k') \end{aligned}\right\} \tag{6.17}$$

The optimal choice of $\boldsymbol{\Psi}$ is the KL transform of the ensemble of all the image columns so that the elements of $\mathbf{v}_n$ are uncorrelated. In practice, the value of $p = 1$ or 2 is sufficient, and fast transforms such as the cosine or the sine transform are good substitutes for the KL transform. In Section 6.9 we will see that certain, so-called semicausal models also yield this type of representation. Such models are useful in filtering and data compression of images [see Sections 8.12 and 11.6].

Moving Average (MA) Representations

A random sequence $u(n)$ is called a moving average (MA) process of order q when it can be written as a weighted running average of uncorrelated random variables

$$u(n) = \sum_{k=0}^{q} b(k)\varepsilon(n-k) \tag{6.18}$$

where $\varepsilon(n)$ is a zero mean white noise process of variance β^2 (Fig. 6.4). The SDF of this MA is given by

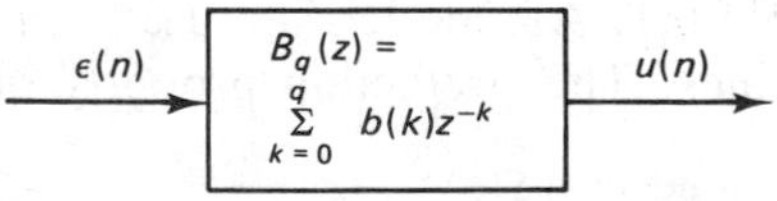

Figure 6.4 qth-order MA model.

$$S(z) = \beta^2 B_q(z) B_q(z^{-1}) \tag{6.19a}$$

$$B_q(z) = \sum_{k=0}^{q} b(k) z^{-k} \tag{6.19b}$$

From the preceding relations it is easy to deduce that the covariance sequence of a qth-order MA is zero outside the interval $[-q, q]$. In general, any covariance sequence that is zero outside the interval $[-q, q]$ can be generated by a qth-order MA filter $B_q(z)$. Note that $B_q(z)$ is an FIR filter, which means MA representations are *all-zero models*.

Example 6.2

Consider the first-order MA process

$$u(n) = \varepsilon(n) - \alpha\varepsilon(n-1), \qquad E[\varepsilon(n)\varepsilon(m)] = \beta^2 \delta(m-n)$$

Then $B_1(z) = 1 - \alpha z^{-1}$, $S(z) = \beta^2[1 + \alpha^2 - \alpha(z + z^{-1})]$. This shows the covariance sequence of $u(n)$ is $r(0) = \beta^2(1 + \alpha^2)$, $r(\pm 1) = -\alpha\beta^2$, $r(n) = 0$, $|n| > 1$.

Autoregressive Moving Average (ARMA) Representations

An AR model whose input is an MA sequence yields a representation of the type (Fig. 6.5)

$$\sum_{k=0}^{p} a(k) u(n-k) = \sum_{l=0}^{q} b(l) \varepsilon(n-l) \tag{6.20}$$

where $\varepsilon(n)$ is a zero mean white sequence of variance β^2. This is called an ARMA representation of order (p, q). Its transfer function and the SDF are given by

$$H(z) = \frac{B_q(z)}{A_p(z)} \tag{6.21}$$

$$S(z) = \frac{\beta^2 B_q(z) B_q(z^{-1})}{A_p(z) A_p(z^{-1})} \tag{6.22}$$

For $q = 0$, it is a pth-order AR and for $p = 0$, it is a qth-order MA.

State Variable Models

A state variable representation of a random vector sequence $\mathbf{y}_n$ is of the form [see Fig. 8.22(a)]

$$\mathbf{x}_{n+1} = \mathbf{A}_n \mathbf{x}_n + \mathbf{B}_n \boldsymbol{\varepsilon}_n$$

$$\mathbf{y}_n = \mathbf{C}_n \mathbf{x}_n + \boldsymbol{\eta}_n, \qquad \forall n \tag{6.23}$$

Here $\mathbf{x}_n$ is an $m \times 1$ vector whose elements $x_n(i)$, $i = 1, \ldots, m$ are called the states of the process $\mathbf{y}_n$, which may represent the image pixel at time or location n; $\boldsymbol{\varepsilon}_n$ is a

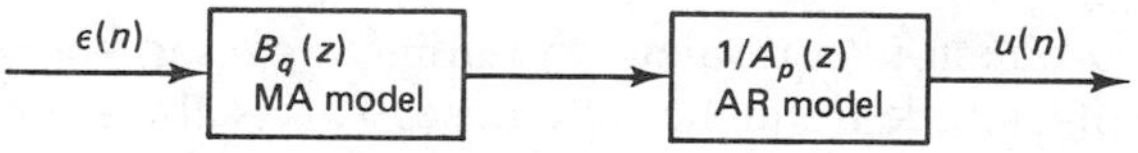

Figure 6.5 (p, q)-order ARMA model.

$p \times 1$ vector sequence of independent random variables and $\boldsymbol{\eta}_n$ is the additive white noise. The matrices $\mathbf{A}_n$, $\mathbf{B}_n$, and $\mathbf{C}_n$ are of appropriate dimensions and $\boldsymbol{\eta}_n$, $\boldsymbol{\varepsilon}_n$ satisfy

$$\left.\begin{aligned} E[\boldsymbol{\varepsilon}_n] = \mathbf{0}, \qquad & E[\boldsymbol{\eta}_n \boldsymbol{\eta}_{n'}^T] = \mathbf{Q}_n \delta(n-n'), \qquad & E[\boldsymbol{\eta}_n] = \mathbf{0} \\ E[\boldsymbol{\eta}_n \boldsymbol{\varepsilon}_{n'}^T] = \mathbf{0}, \qquad & E[\boldsymbol{\varepsilon}_n \boldsymbol{\varepsilon}_{n'}^T] = \mathbf{P}_n \delta(n-n'), \qquad & \forall n, n' \end{aligned}\right\} \tag{6.24}$$

The state variable model just defined is also a vector Markov-1 process. The ARMA models discussed earlier are special cases of state variable models. The application of state variable models in image restoration is discussed in Sections 8.10 and 8.11.

Image Scanning Models

The output $s(k)$ of a raster scanned image becomes a nonstationary sequence even when the image is represented by a stationary random field. This is because equal intervals between scanner outputs do not correspond to equal distances in their spatial locations. [Also see Problem 6.3.] Thus the covariance function

$$r_s(k, l) \triangleq E\{[s(k) - \mu][s(k-l) - \mu]\} \tag{6.25}$$

depends on k, the position of the raster, as well as the displacement variable l. Such a covariance function can only yield a time-varying realization, which would increase the complexity of associated processing algorithms. A practical alternative is to replace $r_s(k, l)$ by its average over the scanner locations [9], that is, by

$$\hat{r}_s(l) \triangleq \frac{1}{N} \sum_{k=0}^{N-1} r_s(k, l) \tag{6.26}$$

Given $\hat{r}_s(l)$, we can find a suitable order AR realization using (6.12) or the results of the following two sections. Another alternative is to determine a so-called cyclostationary state variable model, which requires a periodic initialization of the states for the scanning process. A vector scanning model, which is *Markov* and *time invariant,* can be obtained from the cyclostationary model [10]. State variable scanning models have also been generalized to two dimensions [11, 12]. The causal models considered in Section 6.6 are examples of these.

6.3 ONE-DIMENSIONAL SPECTRAL FACTORIZATION

Spectral factorization refers to the determination of a white noise driven linear system such that the power spectrum density of its output matches a given SDF. Basically, we have to find a causal and stable filter $H(z)$ whose white noise input has the spectral density K, a constant, such that

$$S(z) = KH(z)H(z^{-1}), \qquad z = e^{j\omega} \tag{6.27}$$

where $S(z)$ is the given SDF. Since, for $z = e^{j\omega}$,

$$S(e^{j\omega}) \triangleq S(\omega) = K|H(\omega)|^2 \tag{6.28}$$

the spectral factorization problem is equivalent to finding a causal, stable, linear filter that realizes a given magnitude frequency response. This is also equivalent to specifying the phase of $H(\omega)$, because its magnitude can be calculated within a constant from $S(\omega)$.

Rational SDFs

$S(z)$ is called a proper rational function if it is a ratio of polynomials that can be factored, as in (6.27), such that

$$H(z) = \frac{B_q(z)}{A_p(z)} \tag{6.29}$$

where $A_p(z)$ and $B_q(z)$ are polynomials of the form

$$A_p(z) = 1 - \sum_{k=1}^{p} a(k)z^{-k}, \qquad B_q(z) \equiv \sum_{k=0}^{q} b(k)z^{-k} \tag{6.30}$$

For such SDFs it is always possible to find a causal and stable filter $H(z)$. The method is based on a fundamental result in algebra, which states that any polynomial $P_n(x)$ of degree n has exactly n roots, so that it can be reduced to a product of first-order polynomials, that is,

$$P_n(x) = \prod_{i=1}^{n} (\alpha_i x - \beta_i) \tag{6.31}$$

For a proper rational $S(z)$, which is strictly positive and bounded for $z = e^{j\omega}$, there will be no roots (poles or zeros) on the unit circle. Since $S(z) = S(z^{-1})$, for every root inside the unit circle there is a root outside the unit circle. Hence, if $H(z)$ is *chosen* so that it is causal and all the roots of $A_p(z)$ lie inside the unit circle, then (6.27) will be satisfied, and we will have a causal and stable realization. Moreover, if $B_q(z)$ is chosen to be causal and such that its roots also lie inside the unit circle, then the inverse filter $1/H(z)$, which is $A_p(z)/B_q(z)$, is also causal and stable. *A filter that is causal and stable and has a causal, stable inverse is called a minimum-phase filter.*

Example 6.3

Let

$$S(z) = \frac{4.25 - (z + z^{-1})}{2.5 - (z + z^{-1})} \tag{6.32}$$

The roots of the numerator are $z_1 = 0.25$ and $z_2 = 4$ and those of the denominator are $z_3 = 0.5$ and $z_4 = 2$. Note the roots occur in reciprocal pairs. Now we can write

$$S(z) = \frac{K(1 - 0.25z^{-1})(1 - 0.25z)}{(1 - 0.5z^{-1})(1 - 0.5z)}$$

Comparing this with (6.32), we obtain $K = 2$. Hence a filter with $H(z) = (1 - 0.25z^{-1})/(1 - 0.5z^{-1})$ whose input is zero mean white noise with variance of 2, will be a minimum phase realization of $S(z)$. The representation of this system will be

$$u(n) = 0.5u(n-1) + \varepsilon(n) - 0.25\varepsilon(n-1)$$

$$E[\varepsilon(n)] = 0, \qquad E[\varepsilon(n)\varepsilon(m)] = 2\delta(n-m) \tag{6.33}$$

This is an ARMA model of order (1, 1).

Remarks

It is not possible to find finite-order ARMA realizations when $S(z)$ is not rational. The spectral factors are irrational, that is, they have infinite Laurent series. In practice, suitable approximations are made to obtain finite-order models. There is a subtle difference between the terms *realization* and *modeling* that should be pointed out here. Realization refers to an exact matching of the SDF or the covariances of the model output to the given quantities. Modeling refers to an approximation of the realization such that the match is close or as close as we wish.

One method of finding minimum phase realizations when $S(z)$ is not rational is by the Wiener-Doob spectral decomposition technique [5, 6]. This is discussed for 2-D case in Section 6.8. The method can be easily adapted to 1-D signals (see Problem 6.6).

6.4 AR MODELS, SPECTRAL FACTORIZATION, AND LEVINSON ALGORITHM

The theory of AR models offers an attractive method of approximating a given SDF arbitrarily closely by a finite order AR spectrum. Specifically, (6.13a) and (6.13b) can be solved for a sufficiently large p such that the SDF $S_p(z) \triangleq \beta_p^2/[A_p(z)A_p(z^{-1})]$ is as close to the given $S(z)$, $z = \exp(j\omega)$, as we wish under some mild restrictions on $S(z)$ [1]. This gives the spectral factorization of $S(z)$ as $p \to \infty$. If $S(z)$ happens to have the rational form of (6.9), then $a(n)$ will turn out to be zero for all $n > p$. An efficient method of solving (6.13a) and (6.13b) is given by the following algorithm.

The Levinson-Durbin Algorithm

This algorithm recursively solves the sequence of problems $\mathbf{R}_n \mathbf{a}_n = \mathbf{r}_n$, $r(0) - \mathbf{a}_n^T \mathbf{r}_n = \beta_n^2$, $n = 1, 2, \ldots,$ where n denotes the size of the Toeplitz matrix $\mathbf{R}_n \triangleq \{r(i-j), 1 < i, j < n\}$. The $\mathbf{r}_n$ and $\mathbf{a}_n$ are $n \times 1$ vectors of elements $r(k)$ and $a_n(k)$, $k = 1, \ldots, n$. For any Hermitan Toeplitz matrix $\mathbf{R}_n$, the solution at step n is given by the recursions

$$a_n(k) = \begin{cases} a_{n-1}(k) - \rho_n a_{n-1}^*(n-k);\ a_n(0) = 1, & 1 \le k \le n-1, \quad n \ge 2 \\ \rho_n, & k = n,\ n \ge 1 \end{cases}$$

$$\beta_n^2 = \beta_{n-1}^2(1 - \rho_n \rho_n^*), \qquad \beta_0^2 = r(0) \tag{6.34}$$

$$\rho_{n+1} = \frac{1}{\beta_n^2}\left[r(n+1) - \sum_{k=1}^{n} a_n(k) r(n+1-k)\right], \qquad \rho_1 = \frac{r(1)}{r(0)}$$

where $n = 1, 2, \ldots, p$. The AR model coefficients are given by $a(k) = a_p(k)$, $\beta^2 = \beta_p^2$. One advantage of this algorithm is that (6.13a) can now be solved in $O(p^2)$ multiplication and addition operations, compared to $O(p^3)$ operations required by conventional matrix inversion methods such as Gaussian elimination. Besides giving a fast algorithm for solving Toeplitz equations, the Levinson-Durbin recursions reveal many important properties of AR models (see Problem 6.8).

The quantities $\{\rho_n, 1 \le n \le p\}$ are called the *partial correlations,* and their negatives, $-\rho_n$, are called the *reflection coefficients* of the pth-order AR model. The quantity ρ_n represents correlation between $u(m)$ and $u(m+n)$ if $u(m+1), \ldots, u(m+n-1)$ are held fixed. It can be shown that the AR model is stable, that is, the roots of $A_p(z)$ are inside the unit circle, if $|\rho_n| < 1$. This condition is satisfied when $\mathbf{R}$ is positive definite.

The Levinson recursions give a useful algorithm for determining a finite-order stable AR model whose spectrum fits a given SDF arbitrarily closely (Fig. 6.6). Given $r(0)$ and any one of the sequences $\{r(n)\}$, $\{a(n)\}$, $\{\rho_n\}$, the remaining sequences can be determined uniquely from these recursions. This property is useful in developing the stability tests for 2-D systems. The Levinson algorithm is also useful in modeling 2-D random fields, where the SDFs rarely have rational factors.

Example 6.4

The result of AR modeling problem discussed in Example 6.1 can also be obtained by applying the Levinson recursions for $p = 2$. Thus $r(0) = \sigma^2$, $r(1) = \sigma^2\rho$, $r(2) = \sigma^2\rho^2$ and we get

$$\rho_1 = \frac{\sigma^2\rho}{\sigma^2} = \rho, \qquad \beta_1^2 = \sigma^2(1-\rho^2), \qquad a_1(1) = \rho_1 = \rho$$

$$\rho_2 = \frac{1}{\beta_1^2}[\sigma^2\rho^2 - \rho\sigma^2\rho] = 0$$

This gives $\beta^2 = \beta_2^2 = \beta_1^2$, $a(1) = a_2(1) = a_1(1) = \rho$, and $a(2) = a_2(2) = 0$, which leads to (6.14). Since $|\rho_1|$ and $|\rho_2|$ are less than 1, this AR model is stable.

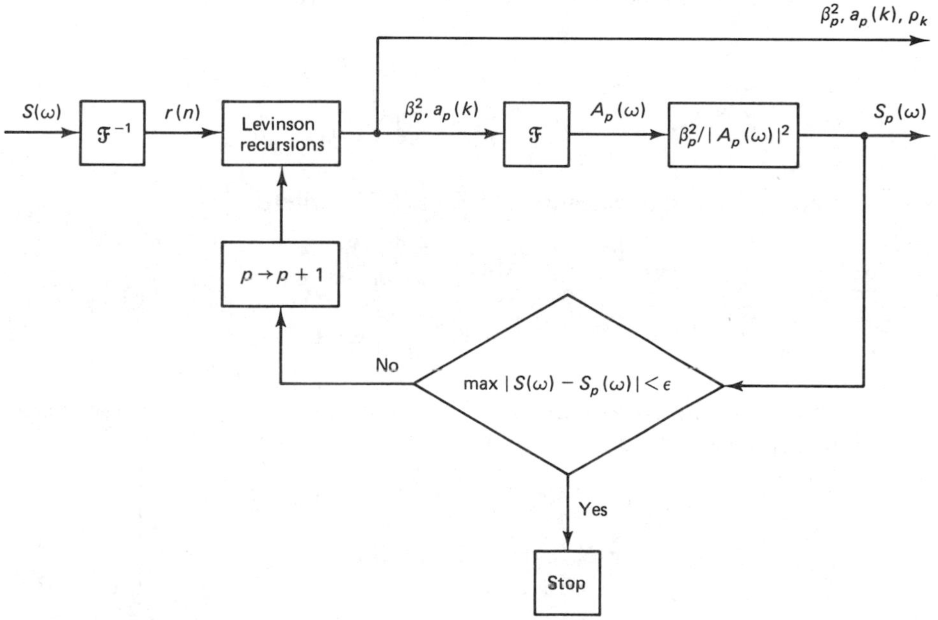

Figure 6.6 AR modeling via Levinson recursions. $A_p(\omega) \triangleq 1 - \sum_{k=1}^{p} a_p(k)\,e^{jk\omega}$

6.5 NONCAUSAL REPRESENTATIONS

Causal prediction, as in the case of AR models, is motivated by the fact that a scanning process orders the image pixels in a time sequence. As such, images represent spatial information and have no causality associated in those coordinates. It is, therefore, natural to think of prediction processes that do not impose the causality constraints, that is, noncausal predictors. A linear noncausal predictor $\bar{u}(n)$ depends on the past as well as the future values of $u(n)$. It is of the form

$$\bar{u}(n) = \sum_{k \neq n} \alpha(k) u(n-k) \tag{6.35}$$

where $\alpha(k)$ are determined to minimize the variance of the prediction error $u(n) - \bar{u}(n)$. The noncausal MVR of a random sequence $u(n)$ is then defined as

$$u(n) = \bar{u}(n) + \nu(n) = \sum_{\substack{k=-\infty \\ k \neq 0}}^{\infty} \alpha(k) u(n-k) + \nu(n) \tag{6.36}$$

where $\nu(n)$ is the noncausal prediction error sequence. Figure 6.7a shows the noncausal MVR system representation. The sequence $u(n)$ is the output of a noncausal system whose input is the noncausal prediction error sequence $\nu(n)$. The transfer function of this system is $1/A(z)$, where $A(z) \triangleq 1 - \Sigma_{n \neq 0} \alpha(n) z^{-n}$ is called the *noncausal MVR prediction error filter*. The filter coefficients $\alpha(n)$ can be determined according to the following theorem.

Theorem 6.1 (Noncausal MVR theorem). Let $u(n)$ be a zero mean, stationary random sequence whose SDF is $S(z)$. If $1/S(e^{j\omega})$ has the Fourier series

$$\frac{1}{S(e^{j\omega})} \triangleq \sum_{n=-\infty}^{\infty} r^+(n) e^{-j\omega n}$$

$$r^+(n) = \frac{1}{2\pi} \int_{-\pi}^{\pi} S^{-1}(e^{j\omega}) e^{j\omega n} \, d\omega \tag{6.37}$$

then $u(n)$ has the noncausal MVR of (6.36), where

$$\alpha(n) = \frac{-r^+(n)}{r^+(0)}$$

$$\beta^2 \triangleq E\{[\nu(n)]^2\} = \frac{1}{r^+(0)} \tag{6.38}$$

Moreover, the covariances $r_\nu(n)$ and the SDF $S_\nu(z)$ of the noncausal prediction error $\nu(n)$ are given by

$$\left.\begin{aligned} r_\nu(n) &= -\beta^2 \alpha(n), \qquad \alpha(0) \triangleq -1 \\ S_\nu(z) &= \beta^2 A(z) \triangleq \beta^2 \Bigg[1 - \sum_{\substack{n=-\infty \\ n \neq 0}}^{\infty} \alpha(n) z^{-n}\Bigg] \end{aligned}\right\} \tag{6.39}$$

The proof is developed in Problem 6.10.

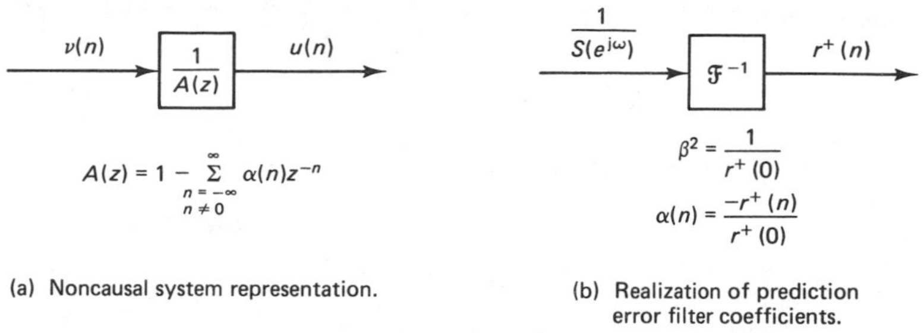
v(n)
1/A(z)
u(n)
A(z) = 1 − Σ α(n)z^{−n}
n = −∞
n ≠ 0
∞
(a) Noncausal system representation.
1/S(e^{jω})
𝔉^{−1}
r⁺(n)
β² = 1/r⁺(0)
α(n) = −r⁺(n)/r⁺(0)
(b) Realization of prediction error filter coefficients.

Figure 6.7 Noncausal MVRs.

Remarks

1. The noncausal prediction error sequence is not white. This follows from (6.39). Also, since $r_\nu(-n) = r_\nu(n)$, the filter coefficient sequence $\alpha(n)$ is even, that is, $\alpha(-n) = \alpha(n)$, which implies $A(z^{-1}) = A(z)$.
2. For the linear noncausal system of Fig. 6.7a, the output SDF is given by $S(z) = S_\nu(z)/[A(z)A(z^{-1})]$. Using (6.39) and the fact that $A(z) = A(z^{-1})$, this becomes

$$S(z) = \frac{\beta^2 A(z)}{[A(z)A(z^{-1})]} = \frac{\beta^2}{A(z)} \tag{6.40}$$

3. Figure 6.7b shows the algorithm for realization of noncausal MVR filter coefficients. Eq. (6.40) and Fig. 6.7b show spectral factorization of $S(z)$ is *not required* for realizing noncausal MVRs. To obtain a finite-order model, the Fourier series coefficients $r^+(n)$ should be truncated to a sufficient number of terms, say, p. Alternatively, we can find the optimum pth-order minimum variance noncausal prediction error filter. For sufficiently large p, these methods yield finite-order, stable, noncausal MVR-models while matching the given covariances to a desired accuracy.

Noncausal MVRs of Autoregressive Sequences

The foregoing results can be used to determine the noncausal MVRs of autoregressive sequences. For example, a zero mean random sequence with covariance $\rho^{|n|}$ having the AR representation of (6.14) has the SDF (see Example 6.1)

$$\left.\begin{aligned} S(z) &= \frac{\beta^2}{[1 - \alpha(z + z^{-1})]} \\ \beta^2 &= \frac{1-\rho^2}{1+\rho^2}, \qquad \alpha = \frac{\rho}{1+\rho^2} \end{aligned}\right\} \tag{6.41}$$

This gives the Fourier series

$$\frac{1}{S(e^{j\omega})} = \frac{1}{\beta^2} - \frac{\alpha}{\beta^2}(e^{j\omega} + e^{-j\omega})$$

which means

$$\alpha(0) = -1, \qquad \alpha(1) = \alpha(-1) = \alpha, \qquad r^+(0) = \frac{1}{\beta^2}$$

The resulting noncausal MVR is

$$\left.\begin{aligned} u(n) &= \alpha[u(n-1) + u(n+1)] + \nu(n) \\ A(z) &= 1 - \alpha(z + z^{-1}) \\ r_\nu(n) &= \beta^2\{1 - \alpha[\delta(n-1) + \delta(n+1)]\} \\ S_\nu(z) &= \beta^2[1 - \alpha(z + z^{-1})] \end{aligned}\right\} \tag{6.42}$$

where $\nu(n)$ is a first-order MA sequence. On the other hand, the AR representation of (6.14) is a causal MVR whose input $\varepsilon(n)$ is a white noise sequence. The generalization of (6.42) yields the noncausal MVR of a pth-order AR sequence as

$$\begin{aligned} u(n) &= \sum_{k=1}^{p} \alpha(k)[u(n-k) + u(n+k)] + \nu(n) \\ A(z) &= 1 - \sum_{k=1}^{p} \alpha(k)(z^{-k} + z^{k}) \\ S_\nu(z) &= \beta_\nu^2 A(z) \end{aligned} \tag{6.43}$$

where $\nu(n)$ is a pth-order MA with zero mean and variance β_ν^2 (see Problem 6.11).

A Fast KL Transform [13]

The foregoing noncausal MVRs give the following interesting result.

Theorem 6.2. The KL transform of a stationary first-order AR sequence $\{u(n), 1 \le n \le N\}$ whose boundary variables $u(0)$ and $u(N+1)$ are known is the sine transform, which is a fast transform.

Proof. Writing the noncausal representation of (6.42) in matrix notation, where $\mathbf{u}$ and $\boldsymbol{\nu}$ are $N \times 1$ vectors consisting of elements $\{u(n), 1 \le n \le N\}$ and $\{\nu(n), 1 \le n \le N\}$, respectively, we get

$$\mathbf{Qu} = \boldsymbol{\nu} + \mathbf{b} \tag{6.44}$$

where $\mathbf{Q}$ is defined in (5.102) and $\mathbf{b}$ contains only the two boundary values, $u(0)$ and $u(N+1)$. Specifically,

$$b(1) = \alpha u(0), \qquad b(N) = \alpha u(N+1), \qquad b(n) = 0, \qquad 2 \le n \le N-1 \tag{6.45}$$

The covariance matrix of $\boldsymbol{\nu}$ is obtained from (6.42), as

$$\mathbf{R}_\nu \triangleq E[\boldsymbol{\nu}\boldsymbol{\nu}^T] = \{r_\nu(m-n)\} = \beta^2 \mathbf{Q} \tag{6.46}$$

The orthogonality condition for minimum variance requires that $\nu(n)$ must be orthogonal to $u(k)$, for all $k \ne n$ and $E[\nu(n)u(n)] = E[\nu^2(n)] = \beta^2$. This gives

$$E[\boldsymbol{\nu}\mathbf{b}^T] = \mathbf{0}, \qquad E[\mathbf{u}\boldsymbol{\nu}^T] = \beta^2 \mathbf{I} \tag{6.47}$$

Multiplying both sides of (6.44) by $\mathbf{Q}^{-1}$ and defining

$$\mathbf{u}^0 \triangleq \mathbf{Q}^{-1}\boldsymbol{\nu}, \qquad \mathbf{u}^b \triangleq \mathbf{Q}^{-1}\mathbf{b} \tag{6.48}$$

we obtain an *orthogonal decomposition* of $\mathbf{u}$ as

$$\mathbf{u} = \mathbf{u}^0 + \mathbf{u}^b \tag{6.49}$$

Note that $\mathbf{u}^b$ is orthogonal to $\mathbf{u}^0$ due to the orthogonality of $\boldsymbol{\nu}$ and $\mathbf{b}$ [see (6.47)] and is completely determined by the boundary variables $u(0)$ and $u(N+1)$. Figure 6.8 shows how this decomposition can be realized. The covariance matrix of $\mathbf{u}^0$ is given by

$$\mathbf{R}_0 \triangleq \mathbf{E}[\mathbf{u}^0\mathbf{u}^{0T}] = \mathbf{Q}^{-1}E[\boldsymbol{\nu}\boldsymbol{\nu}^T]\mathbf{Q}^{-1} = \beta^2\mathbf{Q}^{-1} \tag{6.50}$$

Because β^2 is a scalar, the eigenvectors of $\mathbf{R}_0$ are the same as those of $\mathbf{Q}$, which we know to be the column vectors of the sine transform (see Section 5.11). Hence, the KL transform of $\mathbf{u}^0$ is a fast transform. Moreover, since $\mathbf{u}^b$ and $\mathbf{u}^0$ are orthogonal, the conditional mean of $\mathbf{u}$ given $u(0), u(N+1)$ is simply $\mathbf{u}^b$, that is,

$$\begin{aligned}\boldsymbol{\mu}_b \triangleq E[\mathbf{u}|u(0), u(N+1)] &= E[\mathbf{u}^0 + \mathbf{u}^b|u(0), u(N+1)] \\ &= E[\mathbf{u}^0] + E[\mathbf{u}^b|u(0), u(N+1)] = \mathbf{u}^b\end{aligned} \tag{6.51}$$

Therefore,

$$\text{Cov}[\mathbf{u}|u(0), u(N+1)] = E[(\mathbf{u}-\boldsymbol{\mu}_b)(\mathbf{u}-\boldsymbol{\mu}_b)^T] = E[\mathbf{u}^0\mathbf{u}^{0T}] = \mathbf{R}_0 \tag{6.52}$$

Hence the KL transform of $\mathbf{u}$ conditioned on $u(0)$ and $u(N+1)$ is the eigenmatrix of $\mathbf{R}_0$, that is, the sine transform.

In Chapter 11 we use this result for developing a (recursive) block-by-block transform coding algorithm, which is more efficient than the conventional block-by-block KL transform coding method. Orthogonal decompositions such as (6.49) can be obtained for higher order AR sequences also (see Problem 6.12). In general, the KL transform of $\mathbf{u}^0$ is determined by the eigenvectors of a banded Toeplitz matrix whose eigenvectors can be approximated by an appropriate transform from the *sinusoidal family* of orthogonal transforms discussed in Chapter 5.

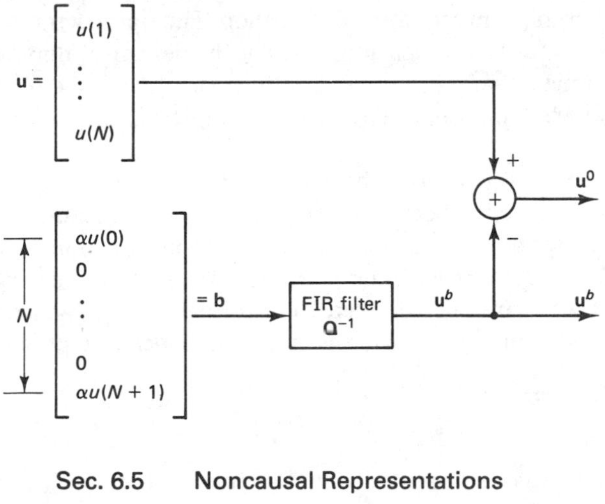

Figure 6.8 Noncausal orthogonal decomposition for first-order AR sequences.

Optimum Interpolation of Images

The noncausal MVRs are also useful in finding the optimum interpolators for random sequences. For example, suppose a line of an image, represented by a first-order AR model, is subsampled so that N samples are missing between given samples. Then the best mean square estimate of a missing sample $u(n)$ is $\hat{u}(n) \triangleq E[u(n)|u(0), u(N+1)]$, which is precisely $u^b(n)$, that is,

$$\hat{\mathbf{u}} = \mathbf{Q}^{-1}\mathbf{b} \Rightarrow \hat{u}(n) = \alpha[\mathbf{Q}^{-1}]_{n,1}u(0) + \alpha[\mathbf{Q}^{-1}]_{n,N}u(N+1) \tag{6.53}$$

When the interpixel correlation $\rho \to 1$, it can be shown that $\hat{u}(n)$ is the straight-line interpolator between $u(0)$ and $u(N+1)$, that is,

$$\hat{u}(n) = u(0) + \frac{n}{N+1}[u(N+1) - u(0)] \tag{6.54}$$

For values of ρ near 1, the interpolation formula of (6.53) becomes a cubic polynomial in $n/(N+1)$ (Problem 6.13).

6.6 LINEAR PREDICTION IN TWO DIMENSIONS

The notion of causality does not extend naturally to two or higher dimensions. Line-by-line processing techniques that utilize the simple 1-D algorithms do not exploit the 2-D structure and the interline dependence. Since causality has no intrinsic importance in two dimensions, it is natural to consider other data structures to characterize 2-D models. There are three canonical forms, namely, causal, semicausal, and noncausal, that we shall consider here in the framework of linear prediction.

These three types of stochastic models have application in many image processing problems. For example, causal models yield recursive algorithms in data compression of images by the *differential pulse code modulation* (DPCM) technique and in recursive filtering of images.

Semicausal models are causal in one dimension and noncausal in the other and lead themselves naturally to *hybrid algorithms,* which are recursive in one dimension and unitary transform based (nonrecursive) in the other. The unitary transform decorrelates the data in the noncausal dimension, setting up the causal dimension for processing by 1-D techniques. Such techniques combine the advantages of high performance of transform-based methods and ease of implementation of 1-D algorithms.

Noncausal models give rise to transform-based algorithms. For example, the notion of fast KL transform discussed in Section 6.5 arises from the noncausal MVR of Markov sequences. Many spatial image processing operators are noncausal—that is, finite impulse response deblurring filters and gradient masks. The coefficients of such filters or *masks,* generally derived by intuitive reasoning, can be obtained more accurately and quite rigorously by invoking noncausal prediction concepts discussed here.

The linear prediction models considered here can also be used effectively for 2-D spectral factorization and spectral estimation. Details are considered in Section 6.7.

Let $u(m, n)$ be a stationary random field with zero mean and covariance $r(k, l)$. Let $\bar{u}(m, n)$ denote a linear prediction estimate of $u(m, n)$, defined as

$$\bar{u}(m, n) = \sum_{(k, l)} \sum_{\in \hat{S}_x} a(k, l) u(m - k, n - l) \tag{6.55}$$

where $a(k, l)$ are called the predictor coefficients and $\hat{S}_x$, a subset of the 2-D lattice, is called the prediction region.

The samples included in $\hat{S}_x$ depend on the type of prediction considered, namely, causal ($x = 1$), semicausal ($x = 2$), or noncausal ($x = 3$). With a hypothetical scanning mechanism that *scans sequentially from top to bottom and left to right,* the three prediction regions are defined as follows.

Causal Prediction

A *causal predictor* is a function of only the elements that arrive before it. Thus the causal prediction region is (Fig. 6.9a)

$$\hat{S}_1 = \{l \geq 1, \forall k\} \cup \{l = 0, k \geq 1\} \tag{6.56}$$

This definition of causality includes the special case of single-quadrant causal predictors.

$$\bar{u}(m, n) = \sum_{k=0}^{\infty} \sum_{\substack{l=0 \\ (k,l) \neq (0,0)}}^{\infty} a(k, l) u(m - k, n - l) \tag{6.57}$$

This is called a *strongly causal* predictor. In signal processing literature the term *causal* is sometimes used for strongly causal models only, and (6.56) is also called the nonsymmetric half-plane (NSHP) model [22].

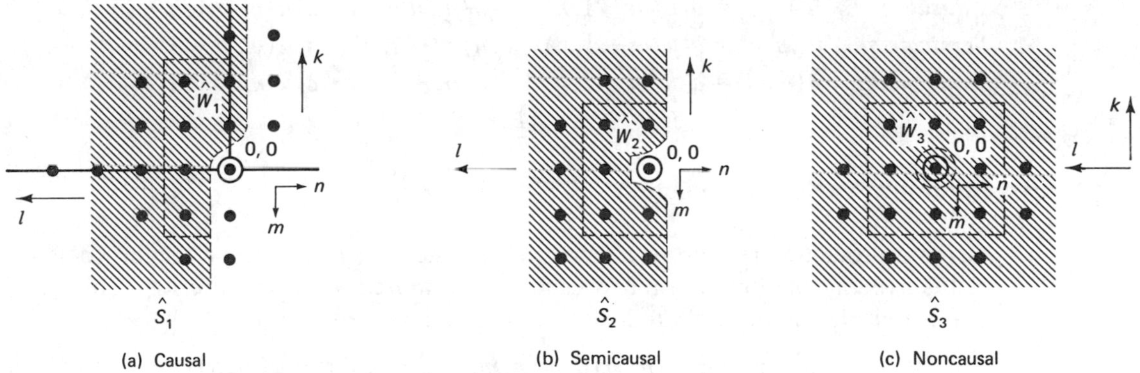

Figure 6.9 Three canonical prediction regions $\hat{S}_x$ and the corresponding finite prediction windows $\hat{W}_x$, $x = 1, 2, 3$.

Semicausal Prediction

A *semicausal predictor* is causal in one of the coordinates and noncausal in the other. For example, the *semicausal prediction* region that is causal in n and noncausal in m is (Fig. 6.9b)

$$\hat{S}_2 = \{l \ge 1, \forall k\} \cup \{l = 0, k \ne 0\} \tag{6.58}$$

Noncausal Prediction

A *noncausal predictor* $\bar{u}(m, n)$ is a function of possibly all the variables in the random field except $u(m, n)$ itself. The noncausal prediction region is (Fig. 6.9c)

$$\hat{S}_3 = \{\forall (k, l) \ne (0, 0)\} \tag{6.59}$$

In practice, only a finite neighborhood, called a prediction window, $\hat{W}_x \subset \hat{S}_x$, can be used in the prediction process, so that

$$\bar{u}(m, n) = \sum\sum_{(k,l) \in \hat{W}_x} a(k, l)u(m - k, n - l), \qquad (m, n) \in \hat{S}_x, \qquad x = 1, 2, 3 \tag{6.60}$$

Some commonly used $\hat{W}_x$ are (Fig. 6.9)

$$\begin{aligned} &\text{Causal:} && \hat{W}_1 \triangleq \{-p \le k \le p, 1 \le l \le q\} \cup \{1 \le k \le p, l = 0\} \\ &\text{Semicausal:} && \hat{W}_2 \triangleq \{-p \le k \le p, 0 \le l \le q, (k, l) \ne (0, 0)\} \\ &\text{Noncausal:} && \hat{W}_3 \triangleq \{-p \le k \le p, -q \le l \le q, (k, l) \ne (0, 0)\} \end{aligned} \tag{6.61a}$$

We also define

$$W_x \triangleq \hat{W}_x \cup (0, 0), \qquad x = 1, 2, 3 \tag{6.61b}$$

Example 6.5

The following are examples of causal, semicausal, and noncausal predictors.

$$\begin{aligned} &\textit{Causal:} && \bar{u}(m, n) = a_1 u(m - 1, n) + a_2 u(m, n - 1) + a_3 u(m - 1, n - 1) \\ &\textit{Semicausal:} && \bar{u}(m, n) = a_1 u(m - 1, n) + a_2 u(m + 1, n) + a_3 u(m, n - 1) \\ &\textit{Noncausal:} && \bar{u}(m, n) = a_1 u(m - 1, n) + a_2 u(m + 1, n) + a_3 u(m, n - 1) \\ & && \qquad + a_4 u(m, n + 1) \end{aligned}$$

Minimum Variance Prediction

Given a prediction region for forming the estimate $\bar{u}(m, n)$, the prediction coefficients $a(m, n)$ can be determined using the *minimum variance* criterion. This requires that the variance of prediction error be minimized, that is,

$$\beta^2 \triangleq \min E[\varepsilon^2(m, n)], \qquad \varepsilon(m, n) \triangleq u(m, n) - \bar{u}(m, n) \tag{6.62}$$

The orthogonality condition associated with this minimum variance prediction is

$$E[\varepsilon(m, n)u(m - k, n - l)] = 0, \qquad (k, l) \in \hat{S}_x, \forall (m, n) \tag{6.63}$$

Using the definitions of $\varepsilon(m, n)$ and $\bar{u}(m, n)$, this yields

$$E\left\{\left[u(m,n) - \sum\sum_{(i,j)\in \hat{W}_x} a(i,j)u(m-i, n-j)\right]u(m-k, n-l)\right\}$$
$$= \beta^2\delta(k,l), \qquad (k,l) \in S_x, \qquad \forall(m,n) \tag{6.64}$$

$$S_x \triangleq \hat{S}_x \cup (0,0), \qquad x = 1,2,3 \tag{6.65}$$

from which we get

$$r(k,l) - \sum\sum_{(i,j)\in \hat{W}_x} a(i,j)r(k-i, l-j) = \beta^2\delta(k,l), \quad (k,l)\in S_x, x = 1,2,3 \tag{6.66}$$

The solution of the above simultaneous equations gives the predictor coefficients $a(i,j)$ and the prediction error variance β^2. Using the symmetry property $r(k,l) = r(-k,-l)$, it can be deduced from (6.66) that

$$\left.\begin{aligned} a(-i, 0) &= a(i, 0) && \text{for semicausal predictors} \\ a(-i, -j) &= a(i,j) && \text{for noncausal predictors} \end{aligned}\right\} \tag{6.67}$$

Stochastic Representation of Random Fields

In general, the random field $u(m, n)$ can be characterized as

$$u(m,n) = \bar{u}(m,n) + \varepsilon(m,n) \tag{6.68}$$

where $\bar{u}(m, n)$ is an arbitrary prediction of $u(m, n)$ and $\varepsilon(m, n)$ is another random field such that (6.68) realizes the covariance properties of $u(m, n)$. There are three types of representations that are of interest in image processing:

1. Minimum variance representations (MVR)
2. White noise driven representations (WNDR)
3. Autoregressive moving average (ARMA) representations

For minimum variance representations, $\bar{u}(m, n)$ is chosen to be a minimum variance predictor and $\varepsilon(m, n)$ is the prediction error. For white noise–driven representations $\varepsilon(m, n)$ is chosen to be a white noise field. In ARMA representations, $\varepsilon(m, n)$ is a two-dimensional moving average, that is, a random field with a truncated covariance function:

$$E[\varepsilon(i,j)\varepsilon(m,n)] = 0, \qquad \forall |i-m| > K, |j-n| > L \tag{6.69}$$

for some fixed integers $K > 0$, $L > 0$.

Example 6.6

The covariance function

$$r(k,l) = \delta(k,l) - \alpha[\delta(k-1,l) + \delta(k+1,l) + \delta(k,l-1) + \delta(k,l+1)]$$

has a finite region of support and represents a moving average random field. Its spectral density function is a (finite-order) two-dimensional polynomial given by

$$S(z_1, z_2) = 1 - \alpha(z_1 + z_1^{-1} + z_2 + z_2^{-1}), \qquad |\alpha| < \tfrac{1}{4}$$

Finite-Order MVRs

The finite order predictors of (6.60) yield MVRs of the form

$$u(m, n) = \sum\sum_{(k, l) \in \hat{W}_x} a(k, l)u(m-k, n-l) + \varepsilon(m, n) \tag{6.70}$$

A random field characterized by this MVR must satisfy (6.63) and (6.64). Multiplying both sides of (6.70) by $\varepsilon(m, n)$, taking expectations and using (6.63), we obtain

$$E[u(m, n)\varepsilon(m, n)] = \beta^2 \tag{6.71}$$

Using the preceding results we find

$$\begin{aligned} r_\varepsilon(k, l) &\triangleq E[\varepsilon(m, n)\varepsilon(m-k, n-l)] \\ &= E\{\varepsilon(m, n)[u(m-k, n-l) - \bar{u}(m-k, n-l)]\} \\ &= \beta^2\left[\delta(k, l) - \sum\sum_{(i, j) \in \hat{W}_x} a(i, j)\delta(k+i, l+j)\right] \end{aligned} \tag{6.72}$$

With $a(0, 0) \triangleq -1$ and using (6.67), the covariance function of the prediction error $\varepsilon(m, n)$ is obtained as

$$r_\varepsilon(k, l) = \begin{cases} \beta^2\delta(k, l), & \forall (k, l) \text{ for causal MVRs} \\ -\beta^2\delta(l)\sum_{i=-p}^{p} a(i, 0)\delta(k+i), & \forall (k, l) \text{ for semicausal MVRs} \\ -\beta^2\sum_{i=-p}^{p}\sum_{j=-q}^{q} a(i, j)\delta(k-i)\delta(l-j), & \forall (k, l) \text{ for noncausal MVRs} \end{cases} \tag{6.73}$$

The filter represented by the two-dimensional polynomial

$$A(z_1, z_2) \triangleq 1 - \sum\sum_{(m, n) \in \hat{W}_x} a(m, n)z_1^{-m} z_2^{-n} \tag{6.74}$$

is called the *prediction error filter*. The prediction error $\varepsilon(m, n)$ is obtained by passing $u(m, n)$ through this filter. The transfer function of (6.70) is given by

$$H(z_1, z_2) = \frac{1}{A(z_1, z_2)} = \left[1 - \sum\sum_{(m, n) \in \hat{W}_x} a(m, n)z_1^{-m} z_2^{-n}\right]^{-1} \tag{6.75}$$

Taking the 2-D Z-transform of (6.73) and using (6.74), we obtain

$$S_\varepsilon(z_1, z_2) = \begin{cases} \beta^2, & \text{causal MVRs} \\ -\beta^2\sum_{m=-p}^{p} a(m, 0)z_1^{-m} = \beta^2 A(z_1, \infty), & \text{semicausal MVRs} \\ \beta^2\left[1 - \sum\sum_{(m, n) \in W_3} a(m, n)z_1^{-m}z_2^{-n}\right] = \beta^2 A(z_1, z_2) & \text{noncausal MVRs} \end{cases} \tag{6.76}$$

Using (6.75), (6.76) and the symmetry condition $A(z_1, z_2) = A(z_1^{-1}, z_2^{-1})$ for noncausal MVRs, we obtain the following expressions for the SDFs:

$$S_u(z_1, z_2) = \frac{S_\varepsilon(z_1, z_2)}{A(z_1, z_2)A(z_1^{-1}, z_2^{-1})} = \begin{cases} \dfrac{\beta^2}{A(z_1, z_2)A(z_1^{-1}, z_2^{-1})}, & \text{causal MVRs} \\ \dfrac{\beta^2 A(z_1, \infty)}{[A(z_1, z_2)A(z_1^{-1}, z_2^{-1})]}, & \text{semicausal MVRs} \\ \dfrac{\beta^2 A(z_1, z_2)}{A(z_1 z_2)A(z_1^{-1}, z_2^{-1})} = \dfrac{\beta^2}{A(z_1, z_2)}, & \text{noncausal MVRs} \end{cases} \tag{6.77}$$

Thus the SDFs of all MVRs are determined completely by their prediction error filters $A(z_1, z_2)$ and the prediction error variances β^2. From (6.73), we note the causal MVRs are also white noise–driven models, just like the 1-D AR models. The semicausal MVRs are driven by random fields, which are white in the causal dimension and moving averages in the noncausal dimension. The noncausal MVRs are driven by 2-D moving average fields.

Remarks

1. *Definition:* A two-dimensional sequence $x(m, n)$ is called *causal, semicausal,* or *noncausal* if its region of support is contained in S_1, S_2, or S_3, respectively. Based on this definition, we call a filter causal, semicausal, or noncausal if its impulse response is causal, semicausal, or noncausal, respectively.
2. If the prediction error filters $A(z_1, z_2)$ are causal, semicausal, or noncausal, then their inverses $1/A(z_1, z_2)$ are likewise, respectively.
3. The causal models are recursive, that is, the output sample $u(m, n)$ can be uniquely computed recursively from the past outputs and inputs—from $\{u(m, n), \varepsilon(m, n) \text{ for } (m, n) \in \hat{S}_1\}$. Therefore, causal MVRs are difference equations that can be solved as initial value problems.

   The semicausal models are semirecursive, that is, they are recursive only in one dimension. The full vector $\mathbf{u}_n = \{u(m, n), \forall m\}$ can be calculated from the past output vectors $\{\mathbf{u}_j, j < n\}$ and all the past and present input vectors $\{\boldsymbol{\varepsilon}_j, j \leq n\}$. Therefore, semicausal MVRs are difference equations that have to be solved as initial value problems in one of the dimensions and as boundary value problems in the other dimension.

   The noncausal models are nonrecursive because the output $u(m, n)$ depends on all the past and future inputs. Noncausal MVRs are boundary value difference equations.
4. The causal, semicausal, and noncausal models are related to the hyperbolic, parabolic, and elliptic classes, respectively, of partial differential equations [21].
5. Every finite-order causal MVR also has a finite-order noncausal minimum variance representation, although the converse is not true. This is because the SDF of a causal MVR can always be expressed in the form of the SDF of a noncausal MVR.

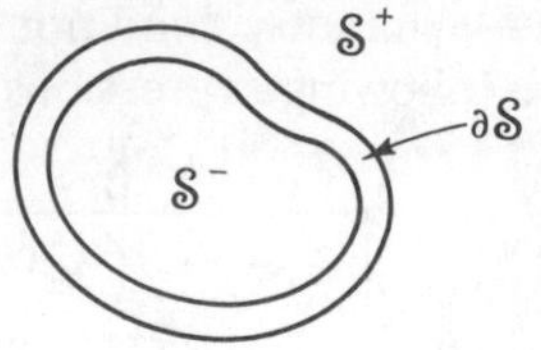

Figure 6.10 Partition for Markovianness of random fields.

6. (*Markov random fields*). A two-dimensional random field is called *Markov* if at every pixel location we can find a partition $\mathcal{S}^+$ (future), $\partial\mathcal{S}$ (present) and $\mathcal{S}^-$ (past) of the two-dimensional lattice $\{m, n\}$ (Fig. 6.10) providing support to the sets of random variables U^+, ∂U, and U^-, respectively, such that

$$P[U^+|U^-, \partial U] = P[U^+|\partial U]$$

This means, given the present (∂U), the future (U^+) is independent of the past (U^-). It can be shown that every Gaussian noncausal MVR is a Markov random field [19]. If the noncausal prediction window is $[-p, p] \times [-q, q]$, then the random field is called Markov $[p \times q]$. Using property 5, it can be shown that every causal MVR is also a Markov random field.

Example 6.7

For the models of Example 6.5, their prediction error filters are given by

$$A(z_1, z_2) = 1 - a_1 z_1^{-1} - a_2 z_2^{-1} - a_3 z_1^{-1} z_2^{-1} \quad \text{(causal model)}$$

$$A(z_1, z_2) = 1 - a_1 z_1^{-1} - a_2 z_1 - a_3 z_2^{-1} \quad \text{(semicausal model)}$$

$$A(z_1, z_2) = 1 - a_1 z_1^{-1} - a_2 z_1 - a_3 z_2^{-1} - a_4 z_2 \quad \text{(noncausal model)}$$

Let the prediction error variance be β^2. If these are to be MVR filters, then the following must be true.

Causal model. $S_\varepsilon(z_1, z_2) = \beta^2$, which means $\varepsilon(m, n)$ is a white noise field. This gives

$$S_u(z_1, z_2) = \frac{\beta^2}{(1 - a_1 z_1^{-1} - a_2 z_2^{-1} - a_3 z_1^{-1} z_2^{-1})(1 - a_1 z_1 - a_2 z_2 - a_3 z_1 z_2)}$$

Semicausal model. $S_\varepsilon(z_1, z_2) = \beta^2[1 - a_1 z_1^{-1} - a_2 z_1] = \beta^2 A(z_1, \infty)$. Because the SDF $S_\varepsilon(z_1, z_2)$ must equal $S_\varepsilon(z_1^{-1}, z_2^{-1})$, we must have $a_1 = a_2$ and

$$S_u(z_1, z_2) = \frac{\beta^2(1 - a_1 z_1^{-1} - a_1 z_1)}{(1 - a_1 z_1^{-1} - a_1 z_1 - a_3 z_2^{-1})(1 - a_1 z_1 - a_1 z_1^{-1} - a_3 z_2)}.$$

Clearly $\varepsilon(m, n)$ is a moving average in the m (noncausal) dimension and is white in the n (causal) dimension.

Noncausal model.

$$S_\varepsilon(z_1, z_2) = \beta^2[1 - a_1 z_1^{-1} - a_2 z_1 - a_3 z_2^{-1} - a_4 z_2] = \beta^2 A(z_1, z_2)$$

Once again, for S_ε to be an SDF, we must have $a_1 = a_2$, $a_3 = a_4$, and

$$S_u(z_1, z_2) = \frac{\beta^2 A(z_1, z_2)}{A(z_1, z_2) A(z_1^{-1}, z_2^{-1})} = \frac{\beta^2}{1 - a_1 z_1^{-1} - a_1 z_1 - a_3 z_2^{-1} - a_3 z_2}$$

Now $\varepsilon(m, n)$ is a two-dimensional moving average field. However, it is a special moving average whose SDF is proportional to the frequency response of the prediction error filter. This allows cancellation of the $A(z_1, z_2)$ term in the numerator and the denominator of the preceding expression for $S_u(z_1, z_2)$.

Example 6.8

Consider the linear system

$$A(z_1, z_2)\, U(z_1, z_2) = F(z_1, z_2)$$

where $A(z_1, z_2)$ is a causal, semicausal, or noncausal prediction filter discussed in Example 6.7 and $F(z_1, z_2)$ represents the Z-transform of an arbitrary input $f(m, n)$. For the causal model, we can write the difference equation

$$u(m, n) - a_1 u(m-1, n) - a_2 u(m, n-1) - a_3 u(m-1, n-1) = f(m, n)$$

This equation can be solved recursively for all $m \geq 0$, $n \geq 0$, for example, if the initial values $u(m, 0)$, $u(0, n)$ and the input $f(m, n)$ are given (see Fig. 6.11a). To obtain the

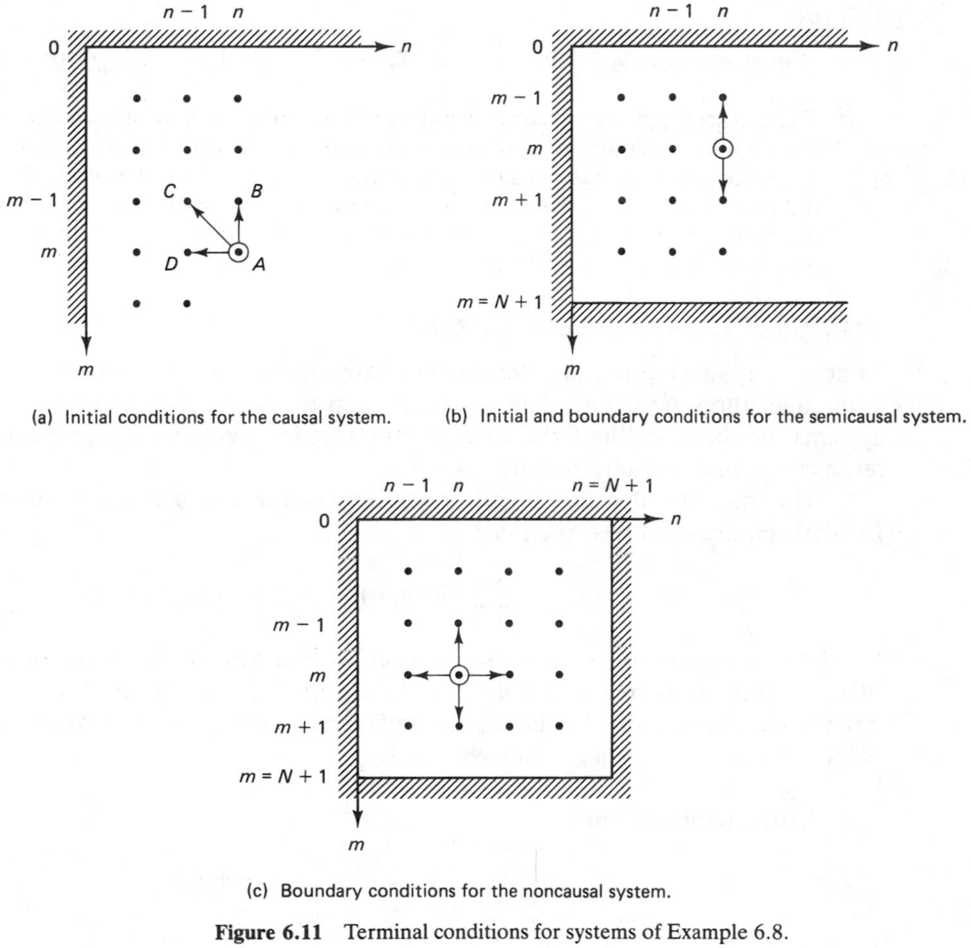

n − 1
n
0
n
m − 1
m
C
B
D
A
m
(a) Initial conditions for the causal system.
n − 1
n
0
n
m − 1
m
m + 1
m = N + 1
m
(b) Initial and boundary conditions for the semicausal system.
n − 1
n
n = N + 1
0
n
m − 1
m
m + 1
m = N + 1
m
(c) Boundary conditions for the noncausal system.

Figure 6.11 Terminal conditions for systems of Example 6.8.

solution at location A, we need only the previously determined solutions at locations B, C, and D and the input at A. Such problems are called initial-value problems.

The semicausal system satisfies the difference equation

$$u(m,n) - a_1 u(m-1,n) - a_2 u(m+1,n) - a_3 u(m,n-1) = f(m,n)$$

Now the solution at (m, n) (see Fig.6.11b) needs the solution at a future location $(m+1,n)$. However, a unique solution for a column vector $\mathbf{u}_n \triangleq [u(1,n), \ldots, u(N,n)]^T$ can be obtained recursively for all $n \geq 0$ if the initial values $u(m, 0)$ and the boundary values $u(0,n)$ and $u(N+1,n)$ are known and if the coefficients a_1, a_2, a_3 satisfy certain stability conditions discussed shortly. Such problems are called initial-boundary-value problems and can be solved semirecursively, that is, recursively in one dimension and nonrecursively in the other.

For the noncausal system, we have

$$u(m,n) - a_1(m-1,n) - a_2 u(m+1,n) - a_3(m,n-1) - a_4 u(m,n+1) = f(m,n)$$

which becomes a boundary-value problem because the solution at (m, n) requires the solutions at $(m+1,n)$ and $(m, n+1)$ (see Fig. 6.11c).

Note that it is possible to reindex, for example, the semicausal system equation as

$$u(m,n) - \frac{1}{a_2} u(m-1,n) + \frac{a_1}{a_2} u(m-2,n) + \frac{a_3}{a_2} u(m-1,n-1) = \frac{-1}{a_2} f(m-1,n)$$

This can seemingly be solved recursively for all $m \geq 0, n \geq 0$ as an initial-value problem if the initial conditions $u(m, 0), u(0,n), u(-1,n)$ are known. Similarly, we could reindex the noncausal system equation and write it as an initial-value problem. However, this procedure is nonproductive because semicausal and noncausal systems can become unstable if treated as causal systems and may not yield unique solutions as initial value problems. The stability of these models is considered next.

Stability of Two-Dimensional Systems

In designing image processing techniques, care must be taken to assure the underlying algorithms are stable. The algorithms can be viewed as 2-D systems. These systems should be stable. Otherwise, small errors in calculations (input) can cause large errors in the result (output).

The stability of a linear shift invariant system whose impulse response is $h(m,n)$ requires that (see Section 2.5)

$$\sum_m \sum_n |h(m,n)| < \infty \tag{6.78}$$

We define a system to be causal, semicausal, or noncausal if the region of support of its impulse response is contained in S_1, S_2, or S_3, respectively. Based on this definition, the stability conditions for different models whose transfer function $H(z_1, z_2) = 1/A(z_1, z_2)$, are as follows.

Noncausal systems

$$A(z_1, z_2) = 1 - \sum_{m=-p}^{p} \sum_{\substack{n=-q \\ (m,n) \neq (0,0)}}^{q} a(m,n) z_1^{-m} z_2^{-n} \tag{6.79}$$

These are stable as nonrecursive filters if and only if

$$A(z_1, z_2) \neq 0, \qquad |z_1| = 1, \qquad |z_2| = 1$$

Semicausal systems

$$A(z_1, z_2) = 1 - \sum_{\substack{m=-p \\ m \neq 0}}^{p} a(m, 0) z_1^{-m} - \sum_{m=-p}^{p} \sum_{n=1}^{q} a(m, n) z_1^{-m} z_2^{-n}$$

These are semirecursively stable if and only if

$$A(z_1, z_2) \neq 0, \qquad |z_1| = 1, \qquad |z_2| \geq 1 \tag{6.80}$$

Causal systems

$$A(z_1, z_2) = 1 - \sum_{m=1}^{p} a(m, 0) z_1^{-m} - \sum_{m=-p}^{p} \sum_{n=1}^{q} a(m, n) z_1^{-m} z_2^{-n}$$

These are recursively stable if and only if

$$\begin{aligned} A(z_1, z_2) &\neq 0, \qquad |z_1| \geq 1, \qquad z_2 = \infty, \\ A(z_1, z_2) &\neq 0, \qquad |z_1| = 1, \qquad |z_2| \geq 1 \end{aligned} \tag{6.81}$$

These conditions assure H to have a uniformly convergent series in the appropriate regions in the z_1, z_2 hyperplane so that (6.78) is satisfied subject to the causality conditions. Proofs are considered in Problem 6.17.

The preceding stability conditions for the causal and semicausal systems require that for each $\omega_1 \in (-\pi, \pi)$, the one-dimensional prediction error filter $A(e^{j\omega_1}, z_2)$ be stable and causal, that is, all its roots should lie within the unit circle $|z_2| = 1$. This condition can be tested by finding, for each ω_1, the partial correlations $\rho(\omega_1)$ associated with $A(e^{j\omega_1}, z_2)$ via the Levinson recursions and verifying that $|\rho(\omega_1)| < 1$ for every ω_1. For semicausal systems it is also required that the one-dimensional polynomial $A(z_1, \infty)$ be stable as a noncausal prediction error filter, that is, it should have no zeros on the unit circle. For causal systems, $A(z_1, \infty)$ should be stable as a causal prediction error filter, that is, its zeros should be inside the unit circle.

6.7 TWO-DIMENSIONAL SPECTRAL FACTORIZATION AND SPECTRAL ESTIMATION VIA PREDICTION MODELS

Now we consider the problem of realizing the foregoing three types of representations given the spectral density function or, equivalently, the covariance function of the image. Let $H(z_1, z_2)$ represent the transfer function of a two-dimensional stable linear system. The SDF of the output $u(m, n)$ when forced by a stationary random field $\varepsilon(m, n)$ is given by

$$S_u(z_1, z_2) = H(z_1, z_2) H(z_1^{-1}, z_2^{-1}) S_\varepsilon(z_1, z_2) \tag{6.82}$$

The problem of finding a stable system $H(z_1, z_2)$ given $S_u(z_1, z_2)$ and $S_\varepsilon(z_1, z_2)$ is called the *two-dimensional spectral factorization problem.* In general, it is not possible to reduce a two-dimensional polynomial as a product of lower order factors. Thus, unlike in 1-D, it may not be possible to find a suitable finite-order linear system realization of a 2-D rational SDF. To obtain finite-order models our only recourse is to relax the requirement of an exact match of the model SDF with the given SDF. As in the one-dimensional case [see (6.28)] the problem of designing a stable filter $H(z_1, z_2)$ whose magnitude of the frequency response is given is essentially the same as spectral factorization to obtain white noise–driven representations.

Example 6.9

Consider the SDF

$$S(z_1, z_2) = 1 - \alpha(z_1 + z_1^{-1} + z_2 + z_2^{-1}), \qquad |\alpha| < \tfrac{1}{4} \tag{6.83}$$

Defining $A(z_2) = 1 - \alpha(z_2 + z_2^{-1})$, $S(z_1, z_2)$ can be factored as

$$\left.\begin{aligned} &S(z_1, z_2) = A(z_2) - \alpha(z_1 + z_1^{-1}) = H(z_1, z_2)H(z_1^{-1}, z_2^{-1}) \\ &\text{where} \\ &H(z_1, z_2) \overset{\Delta}{=} (1 - p(z_2)z_1^{-1})\sqrt{\frac{\alpha}{p(z_2)}}, \qquad p(z_2) \overset{\Delta}{=} \frac{A(z_2) + \sqrt{A^2(z_2) - 4\alpha^2}}{2\alpha} \end{aligned}\right\} \tag{6.84}$$

Note that H is not rational. In fact, rational factors satisfying (6.82) are not possible. Therefore, a finite-order linear shift invariant realization of $S(z_1, z_2)$ is not possible.

Separable Models

If the given SDF is separable, that is, $S(z_1, z_2) = S_1(z_1)S_2(z_2)$ or, equivalently, $r(k, l) = r_1(k)r_2(l)$ and $S_1(z_1)$ and $S_2(z_2)$ have the one-dimensional realizations $[H_1(z_1), S_{\varepsilon_1}(z_1)]$ and $[H_2(z_2), S_{\varepsilon_2}(z_2)]$ [see (6.1)], then $S(z_1, z_2)$ has the realization

$$\begin{aligned} H(z_1, z_2) &= H_1(z_1)H_2(z_2) \\ S_\varepsilon(z_1, z_2) &= S_{\varepsilon_1}(z_1)S_{\varepsilon_2}(z_2) \end{aligned} \tag{6.85}$$

Example 6.10 (Realizations of the separable covariance function)

The separable covariance function of (2.84) can be factored as $r_1(k) = \sigma^2\rho_1^{|k|}$, $r_2(l) = \rho_2^{|l|}$. Now the covariance function $r(k) \overset{\Delta}{=} \sigma^2\rho^{|k|}$ has (1) a causal first-order AR realization (see Example 6.1) $A(z) \overset{\Delta}{=} 1 - \rho z^{-1}$, $S_\varepsilon(z) = \sigma^2(1 - \rho^2)$ and (2) a noncausal MVR (see Section 6.5) with $A(z) \overset{\Delta}{=} 1 - \alpha(z + z^{-1})$, $S_\nu(z) = \sigma^2\beta^2 A(z)$, $\beta^2 \overset{\Delta}{=} (1 - \rho^2)/(1 + \rho^2)$, $\alpha \overset{\Delta}{=} \rho/(1 + \rho^2)$. Applying these results to $r_1(k)$ and $r_2(l)$, we can obtain the following three different realizations, where α_i, β_i, $i = 1, 2$, are defined by replacing ρ by ρ_i, $i = 1, 2$ in the previous definitions of α and β.

Causal MVR (C1 model). Both $r_1(k)$ and $r_2(l)$ have causal realizations. This gives

$$\begin{aligned} A(z_1, z_2) &= (1 - \rho_1 z_1^{-1})(1 - \rho_2 z_2^{-1}), \\ S_\varepsilon(z_1, z_2) &= \sigma^2(1 - \rho_1^2)(1 - \rho_2^2) \\ u(m, n) &= \rho_1 u(m-1, n) + \rho_2 u(m, n-1) \\ &\quad - \rho_1\rho_2 u(m-1, n-1) + \varepsilon(m, n) \end{aligned} \tag{6.86}$$

$$r_\varepsilon(k, l) = \sigma^2(1-\rho_1^2)(1-\rho_2^2)\delta(k, l)$$

Semicausal MVR (SC1 model). Here $r_1(k)$ has noncausal realization, and $r_2(l)$ has causal realization. This yields the semicausal prediction-error filter

$$\begin{aligned} A(z_1, z_2) &= [1-\alpha_1(z_1+z_1^{-1})](1-\rho_2 z_2^{-1}), \\ S_\varepsilon(z_1, z_2) &= \sigma^2\beta_1^2(1-\rho_2^2)[1-\alpha_1(z_1+z_1^{-1})] \\ u(m, n) &= \alpha_1[u(m-1, n)+u(m+1, n)] \\ &\quad + \rho_2 u(m, n-1) - \rho_2\alpha_1[u(m-1, n-1) \\ &\quad + u(m+1, n-1)] + \varepsilon(m, n) \end{aligned} \tag{6.87}$$

Noncausal MVR (NC1 model). Now both $r_1(k)$ and $r_2(l)$ have noncausal realizations, giving

$$\begin{aligned} A(z_1, z_2) &= [1-\alpha_1(z_1+z_1^{-1})][1-\alpha_2(z_2+z_2^{-1})], \\ S_\varepsilon(z_1, z_2) &= \sigma^2\beta_1^2\beta_2^2 A(z_1, z_2) \\ u(m, n) &= \alpha_1[u(m-1, n)+u(m+1, n)] \\ &\quad + \alpha_2[u(m, n-1)+u(m, n+1)] \\ &\quad - \alpha_1\alpha_2[u(m-1, n-1) \\ &\quad + u(m+1, n-1)+u(m-1, n+1) \\ &\quad + u(m+1, n+1)] + \varepsilon(m, n) \end{aligned} \tag{6.88}$$

This example shows that all the three canonical forms of minimum variance representations are realizable for separable rational SDFs. For nonseparable and/or irrational SDFs, only approximate realizations can be achieved, as explained next.

Realization of Noncausal MVRs

For nonseparable SDFs, the problem of determining the prediction-error filters becomes more difficult. In general, this requires solution of (6.66) with $p \to \infty$, $q \to \infty$. In this limiting casc for noncausal MVRs, we find [see (6.77)]

$$S_u(z_1, z_2) = \frac{\beta^2}{A(z_1, z_2)} = \frac{\beta^2}{1 - \sum\sum_{(m,n)\in\hat{S}_3} a(m, n) z_1^{-m} z_2^{-n}} \tag{6.89}$$

For the given SDF $S(z_1, z_2)$, suppose $1/S(z_1, z_2)$ has the Fourier series

$$\frac{1}{S(z_1, z_2)} = \sum_{m=-\infty}^{\infty}\sum_{n=-\infty}^{\infty} r^+(m, n) z_1^{-m} z_2^{-n}, \qquad z_1 = e^{j\omega_1}, \qquad z_2 = e^{j\omega_2} \tag{6.90}$$

then the quantities desired in (6.89) to match S and S_u are obtained as

$$\beta^2 = \frac{1}{r^+(0, 0)}, \qquad a(m, n) = \frac{-r^+(m, n)}{r^+(0, 0)} \tag{6.91}$$

This is the two-dimensional version of the one-dimensional noncausal MVRs considered in Section 6.5.

In general, this representation will be of infinite order unless the Fourier series of $1/S(z_1, z_2)$ is finite. An approximate finite-order realization can be obtained by truncating the Fourier series to a finite number of terms such that the truncated series remains positive for every (ω_1, ω_2). By keeping the number of terms sufficiently large, the SDF and the covariances of the finite-order noncausal MVR can be matched arbitrarily closely to the respective given functions [16].

Example 6.11

Consider the SDF

$$S(z_1, z_2) = \frac{1}{[1 - \alpha(z_1 + z_1^{-1} + z_2 + z_2^{-1})]}, \qquad 0 < \alpha < \frac{1}{4} \tag{6.92}$$

Clearly, S^{-1} has a finite Fourier series that gives $\beta^2 = 1, a(m, n) = \alpha$ for $(m, n) = (\pm 1, 0), (0, \pm 1)$ and $a(m, n) = 0$, otherwise. Hence the noncausal MVR is

$$u(m, n) = \alpha[u(m + 1, n) + u(m - 1, n) + u(m, n + 1) + u(m, n - 1)] + \varepsilon(m, n)$$

$$r_\varepsilon(k, l) = \begin{cases} 1, & (k, l) = (0, 0) \\ -\alpha, & (k, l) = (\pm 1, 0), (0, \pm 1) \\ 0, & \text{otherwise} \end{cases}$$

Realization of Causal and Semicausal MVRs

Inverse Fourier transforming $S(e^{j\omega_1}, e^{j\omega_2})$ with respect to ω_1 gives a covariance sequence $r_l(e^{j\omega_2})$, l = integers, which is parametric in ω_2. Hence for each ω_2, we can find an AR model realization of order q, for instance, via the Levinson recursion (Section 6.4), which will match $r_l(e^{j\omega_2})$ for $-q \leq l \leq q$.

Let $\beta^2(e^{j\omega_2})$, $a_n(e^{j\omega_2})$, $1 \leq n \leq q$ be the AR model parameters where the prediction error $\beta^2(e^{j\omega_2}) > 0$ for every ω_2. Now $\beta^2(e^{j\omega_2})$ is a one-dimensional SDF and can be factored by one-dimensional spectral factorization techniques. It has been shown that the causal and semicausal MVRs can be realized when $\beta^2(e^{j\omega_2})$ is factored by causal (AR) and noncausal MVRs, respectively. To obtain finite-order models $a_n(e^{j\omega_2})$ and $\beta^2(e^{j\omega_2})$ are replaced by suitable rational approximations of order p, for instance. Stability of the models is ascertained by requiring the reflection coefficients associated with the rational approximations of $a_n(e^{j\omega_2})$, $n = 1, 2, \dots, q$ to be less than unity in magnitude. The SDFs realized by these finite-order MVRs can be made arbitrarily close to the given SDF by increasing p and q sufficiently. Details may be found in [16].

Realization via Orthogonality Condition

The foregoing techniques require working with infinite sets of equations, which have to be truncated in practical situations. A more practical alternative is to start with the covariance sequence $r(k, l)$ and solve the subset of equations (6.66) for $(k, l) \in W_x \subset S_x$, that is,

$$r(k, l) = \sum_{(m, n) \in \hat{W}_x} \sum a_{p,q}(m, n) r(k - m, l - n) + \beta^2_{p,q} \delta(k, l); \qquad (k, l) \in W_x \tag{6.93}$$

where the dependence of the model parameters on the window size is explicitly shown. These equations are such that the solution for prediction coefficients on $\hat{W}_x$ requires covariances $r(k, l)$ from a larger window. Consequently, unlike the 1-D AR models, the covariances generated by the model need not match the given covariances used originally to solve for the model coefficients, and *there can be many different sets of covariances which yield the same MVR predictors.* Also, stability of the model is not guaranteed for a chosen order.

In spite of the said shortcomings, the advantages of the foregoing method are (1) only a finite number of linear equations need to be solved, and (2) by solving these equations for increasing orders (p, q), it is possible to obtain eventually a finite-order stable model whose covariances match the given $r(k, l)$ to a desired accuracy. Moreover, there is a 2-D Toeplitz structure in (6.93) that can be exploited to compute recursively $a_{p,q}(m, n)$ from $a_{p-1,q}(m, n)$ or $a_{p,q-1}(m, n)$, and so on. This yields an efficient computational procedure, which is similar to the Levinson-Durbin algorithm discussed in Section 6.4.

If the given covariances do indeed come from a finite-order MVR, then the solution of (6.93) would automatically yield that finite-order MVR. Finally, given the solution of (6.93), we can determine the model SDF via (6.77). This feature gives an attractive algorithm for estimating the SDF of a 2-D random field from a limited number of covariances.

Example 6.12

Consider the isotropic covariance function $r(k, l) = 0.9^{\sqrt{k^2 + l^2}}$. Figure 6.12 shows the impulse responses $\{-a(m, n)\}$ of the prediction error filters for causal, semicausal, and

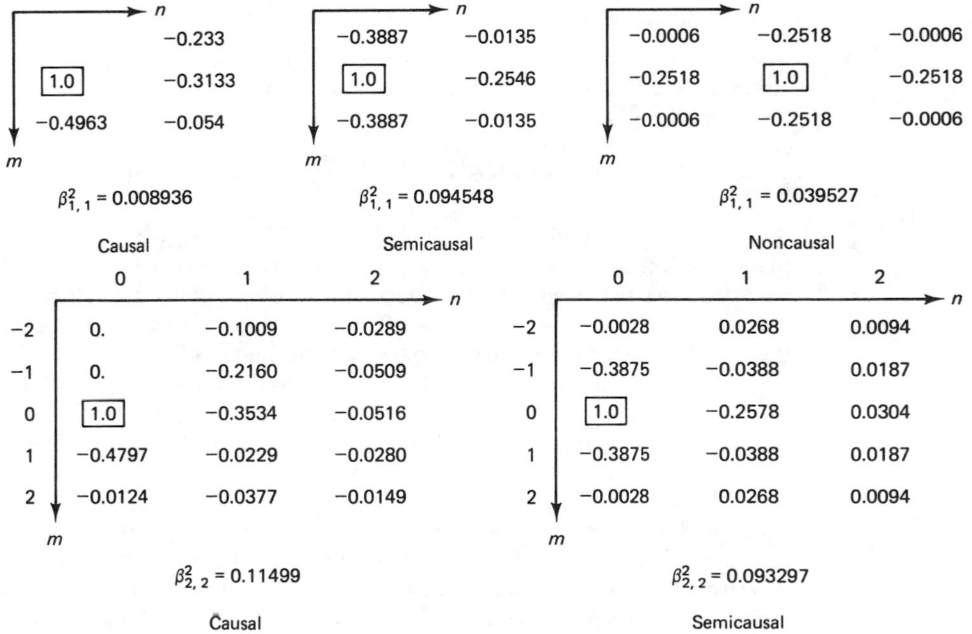

Figure 6.12 Prediction error filter impulse responses of different MVRs. The origin (0, 0) is at the location of the boxed elements.

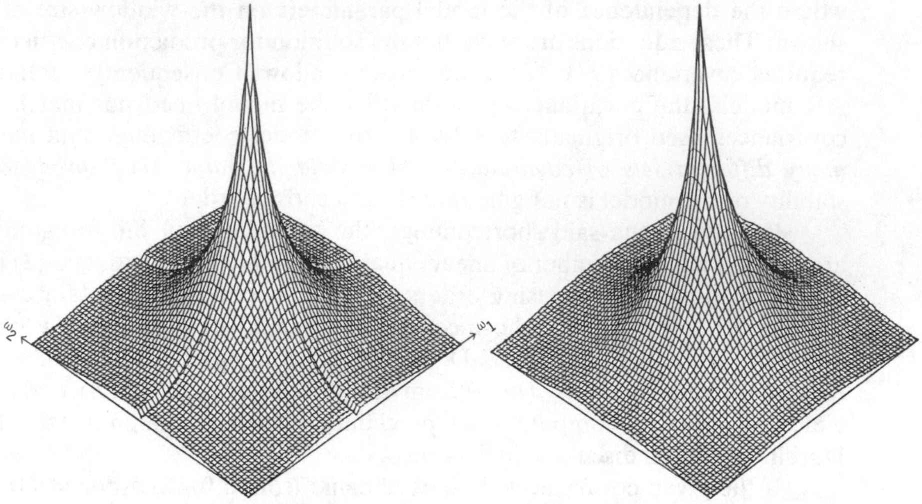

(a) Given spectrum (b) Semicausal model spectrum

Figure 6.13 Spectral match obtained by semicausal MVR of order $p = q = 2$.

noncausal MVRs of different orders. The difference between the given covariances and those generated by the models has been found to decrease when the model order is increased [16]. Figure 6.13 shows the spectral density match obtained by the semicausal MVR of order (2, 2) is quite good.

Example 6.13 (Two-Dimensional Spectral Estimation)

The spectral estimation problem is to find the SDF estimate of a random field given either a finite number of covariances or a finite number of samples of the random field. As an example, suppose the 2-D covariance sequence

$$r(k, l) = \cos \pi\left(\frac{k}{8} + \frac{l}{4}\right) + \frac{\cos 3\pi(k + l)}{16} + 0.05\delta(k, l)$$

is available on a small grid $\{-16 \le k, l \le 16\}$. These covariances correspond to two plane waves of small frequency separation in 10-dB noise. The two frequencies are not resolved when the given data is padded with zeros on a 256×256 grid and the DFT is taken (Fig. 6.14). The low resolution afforded by Fourier methods can be attributed to the fact that the data are assumed to be zero outside their known extent.

The causal and semicausal models improve the resolution of the SDF estimate beyond the Fourier limit by implicitly providing an extrapolation of the covariance data outside its known extent. The method is to fit a suitable order model that is expected to characterize the underlying random field accurately. To this end, we fit $(p, q) = (2, 2)$ order models by solving (6.66) for $(k, l) \in W'_x$ where $W'_x \subset S_x, x = 1, 2$ are subsets of causal and semicausal prediction regions corresponding to $(p, q) = (6, 12)$ and defined in a similar manner as W_x in (6.61b). Since $W_x \subset W'_x$, the resulting system of equations is overdetermined and was solved by least squares techniques (see Section 8.9). Note that by solving (6.66) for $(k, l) \in W'_x$, we are enforcing the orthogonality condition of (6.63) over W'_x. Ideally, we should let $W'_x = S_x$. Practically, a reasonably large region suffices. Once the model parameters are obtained, the SDF is calculated using (6.77)

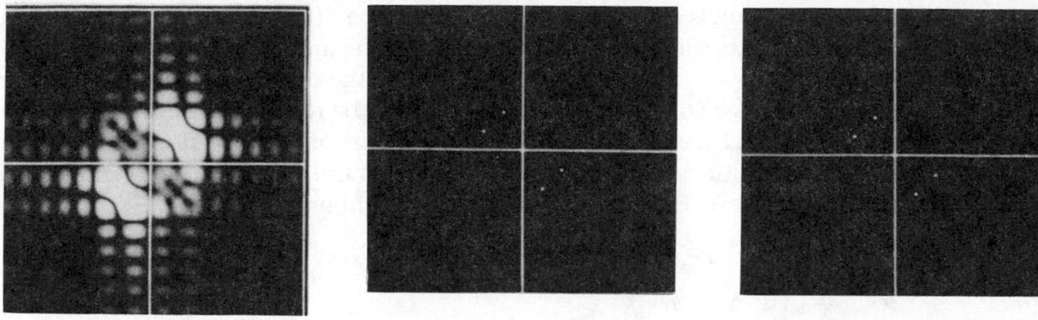

(a) DFT spectrum; (b) causal MVR spectrum; (c) semicausal MVR spectrum.

Figure 6.14 Two-dimensional spectral estimation.

with $z_1 = \exp(j\omega_1)$, $z_2 = \exp(j\omega_2)$. Results given in Fig. 6.14 show both the causal and the semicausal MVRs resolve the two frequencies quite well. This approach has also been employed successfully for spectral estimation when observations of the random field rather than its covariances are available. Details of algorithms for identifying model parameters from such data are given in [17].

6.8 SPECTRAL FACTORIZATION VIA THE WIENER-DOOB HOMOMORPHIC TRANSFORMATION

Another approach to spectral factorization for causal and semicausal models is through the Wiener-Doob homomorphic transformation method. The principle behind this method is to map the poles and zeros of the SDF into singularities of its logarithm. Assume that the SDF S is positive and continuous and the Fourier series of $\log S$ is absolutely convergent. Then

$$\log S(z_1, z_2) = \sum_{m=-\infty}^{\infty} \sum_{n=-\infty}^{\infty} c(m, n) z_1^{-m} z_2^{-n}, \qquad z_1 = e^{j\omega_1}, z_2 = e^{j\omega_2} \tag{6.94}$$

where the *cepstrum* $c(m, n)$ is absolutely summable

$$\sum_m \sum_n |c(m, n)| < \infty \tag{6.95}$$

Suppose $\log S$ is decomposed as a sum of three components

$$\log S \triangleq \hat{S}_\varepsilon + C^+ + C^- \tag{6.96}$$

Then

$$S = \frac{S_\varepsilon}{A^+(z_1, z_2) A^-(z_1, z_2)} \triangleq S_\varepsilon H^+ H^- \tag{6.97}$$

is a product of three factors, where

$$S_\varepsilon \triangleq \exp(\hat{S}_\varepsilon), \quad A^+ \triangleq \exp(-C^+), \quad A^- \triangleq \exp(-C^-), \quad H^+ \triangleq \frac{1}{A^+}, \quad H^- \triangleq \frac{1}{A^-} \tag{6.98}$$

If the decomposition is such that $A^+(z_1, z_2)$ equals $A^-(z_1^{-1}, z_2^{-1})$ and $S_\varepsilon(z_1, z_2)$ is an SDF, then there exists a stable, two-dimensional linear system with transfer function $H^+(z_1, z_2) = 1/A^+(z_1, z_2)$ such that the SDF of the output is S if the SDF of the input is S_ε. The causal or the semicausal models can be realized by decomposing the cepstrum so that C^+ has causal or semicausal regions of support, respectively. The specific decompositions for causal MVRs, semicausal WNDRs, and semicausal MVRs are given next. Figure 6.15a shows the algorithms.

Causal MVRs [22]

Partition the cepstrum as

$$c(m, n) = c^+(m, n) + c^-(m, n) + \hat{\gamma}_\varepsilon(m, n)$$

where $c^+(m, n) = 0, (m, n) \notin \hat{S}_1; c^-(m, n) = 0, (m, n) \in S_1; c^+(0, 0) = c^-(0, 0) = 0$, $\gamma_\varepsilon(m, n) = c(0, 0)\delta(m, n)$. $\hat{S}_1$ and S_1 are defined in (6.56) and (6.65), respectively. Hence we define

$$C^+ = C^+(z_1, z_2) \overset{\Delta}{=} \sum_{m=1}^{\infty} c(m, 0) z_1^{-m} + \sum_{m=-\infty}^{\infty} \sum_{n=1}^{\infty} c(m, n) z_1^{-m} z_2^{-n} \tag{6.99a}$$

$$C^- = C^-(z_1, z_2) \overset{\Delta}{=} \sum_{m=-\infty}^{-1} c(m, 0) z_1^{-m} + \sum_{m=-\infty}^{\infty} \sum_{n=-\infty}^{-1} c(m, n) z_1^{-m} z_2^{-n} \tag{6.99b}$$

$$\hat{S}_\varepsilon = \hat{S}_\varepsilon(z_1, z_2) \overset{\Delta}{=} c(0, 0) \tag{6.99c}$$

Using (6.95) it can be shown that $C^+(z_1, z_2)$ is analytic in the region $\{|z_1| = 1, |z_2| \geq 1\} \cup \{|z_1| \geq 1, z_2 = \infty\}$. Since e^x is a monotonic function, A^+ and H^+ are also analytic in the same region and, therefore, have no singularities in that region. The region of support of their impulse responses $a^+(m, n)$ and $h^+(m, n)$ will be the same as that of $c^+(m, n)$, that is, S_1. Hence A^+ and H^+ will be causal filters.

Semicausal WNDRs

Here we let $c^+(m, n) = 0, (m, n) \notin S_2$ and define

$$C^+(z_1, z_2) = \tfrac{1}{2} \sum_{m=-\infty}^{\infty} c(m, 0) z_1^{-m} + \sum_{m=-\infty}^{\infty} \sum_{n=1}^{\infty} c(m, n) z_1^{-m} z_2^{-n} \tag{6.100a}$$

$$C^-(z_1, z_2) = \tfrac{1}{2} \sum_{m=-\infty}^{\infty} c(m, 0) z_1^{-m} + \sum_{m=-\infty}^{\infty} \sum_{n=-\infty}^{-1} c(m, n) z_1^{-m} z_2^{-n} \tag{6.100b}$$

$$\hat{S}_\varepsilon(z_1, z_2) = 0 \tag{6.100c}$$

Now C^+ and hence A^+ and H^+ will be analytic in the region $\{|z_1| = 1, |z_2| \geq 1\}$ and the impulse responses $a^+(m, n)$ and $h^+(m, n)$ will be zero for $\{n < 0, \forall m\}$, that is, these filters are semicausal. Also $S_\varepsilon = 1$, which means the model is white noise driven.

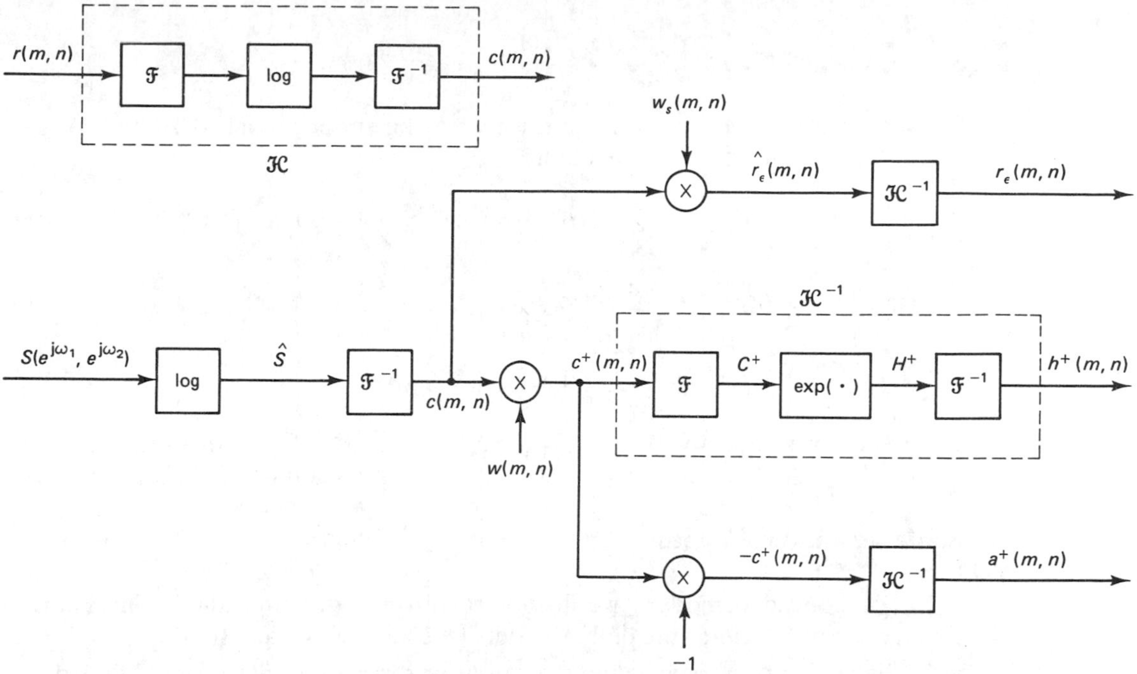

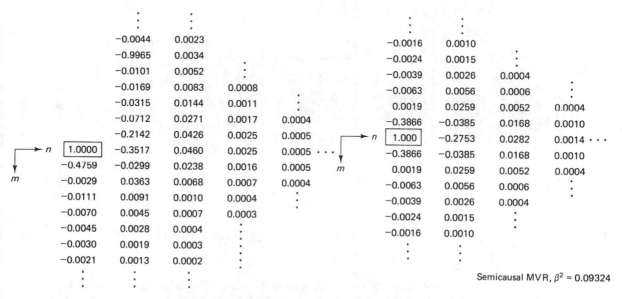

Causal MVR

$$w(m, n) = \begin{cases} 1, & (m, n) \in \hat{S}_1 \\ 0, & \text{otherwise} \end{cases}$$

$$w_s(m, n) = \delta(m)\delta(n)$$

Semicausal WNDR

$$w(m, n) = \begin{cases} 1 - \tfrac{1}{2}\delta(n), & (m, n) \in S_2 \\ 0, & \text{otherwise} \end{cases}$$

$$w_s(m, n) = 0$$

Semicausal MVR

$$w(m, n) = \begin{cases} 1, & (m, n) \in S_2 \\ 0, & \text{otherwise} \end{cases}$$

$$w_s(m, n) = -\delta(n)$$

(a) Wiener-Doob homomorphic transform method. $\mathcal{H}$ is the two dimensional homomorphic transform;

		⋮	⋮			
		−0.0044	0.0023			
		−0.9965	0.0034	⋮		
		−0.0101	0.0052			
		−0.0169	0.0083	0.0008		
		−0.0315	0.0144	0.0011	⋮	
		−0.0712	0.0271	0.0017	0.0004	
		−0.2142	0.0426	0.0025	0.0005	
$\rightarrow n$	1.0000	−0.3517	0.0460	0.0025	0.0005	…
$\downarrow m$	−0.4759	−0.0299	0.0238	0.0016	0.0005	
	−0.0029	0.0363	0.0068	0.0007	0.0004	
	−0.0111	0.0091	0.0010	0.0004	⋮	
	−0.0070	0.0045	0.0007	0.0003		
	−0.0045	0.0028	0.0004	⋮		
	−0.0030	0.0019	0.0003			
	−0.0021	0.0013	0.0002			
	⋮	⋮	⋮			

Causal MVR, $\beta^2 = 0.11438$

	⋮	⋮			
	−0.0016	0.0010			
	−0.0024	0.0015	⋮		
	−0.0039	0.0026	0.0004		
	−0.0063	0.0056	0.0006	⋮	
	0.0019	0.0259	0.0052	0.0004	
	−0.3866	−0.0385	0.0168	0.0010	
$\rightarrow n$	1.000	−0.2753	0.0282	0.0014	…
$\downarrow m$	−0.3866	−0.0385	0.0168	0.0010	
	0.0019	0.0259	0.0052	0.0004	
	−0.0063	0.0056	0.0006	⋮	
	−0.0039	0.0026	0.0004		
	−0.0024	0.0015	⋮		
	−0.0016	0.0010			
	⋮	⋮			

Semicausal MVR, $\beta^2 = 0.09324$

(b) Infinite order prediction error filters obtained by Wiener-Doob factorization for isotropic covariance model with $\rho = 0.9$, $\sigma = 1$.

Figure 6.15 Two-dimensional spectral factorization

Semicausal MVRs

For semicausal MVRs, it is necessary that S_ε be proportional to $A^+(z_1, \infty)$ [see (6.76)]. Using this condition we obtain

$$C^+(z_1, z_2) = \sum_{m=-\infty}^{\infty} c(m, 0) z_1^{-m} + \sum_{m=-\infty}^{\infty} \sum_{n=1}^{\infty} c(m, n) z_1^{-m} z_2^{-n} \tag{6.101a}$$

$$C^-(z_1, z_2) = \sum_{m=-\infty}^{\infty} c(m, 0) z_1^{-m} + \sum_{m=-\infty}^{\infty} \sum_{n=-\infty}^{-1} c(m, n) z_1^{-m} z_2^{-n} \tag{6.101b}$$

$$\hat{S}_\varepsilon(z_1, z_2) = -\sum_{m=-\infty}^{\infty} c(m, 0) z_1^{-m} \tag{6.101c}$$

The region of analyticity of C^+ and A^+ is $\{|z_1| = 1, |z_2| \geq 1\}$, and that of $\hat{S}_\varepsilon$ and S_ε is $\{|z_1| = 1, \forall z_2\}$. Also, $a^+(m, n)$ and $h^+(m, n)$ will be semicausal, and S_ε is a valid SDF.

Remarks and Examples

1. For noncausal models, we do not go through this procedure. The Fourier series of S^{-1} yields the desired model [see Section 6.7 and (6.91)].
2. Since $C^+(z_1, z_2)$ is analytic, the causal or semicausal filters just obtained are stable and have stable causal or semicausal inverse filters.
3. A practical algorithm for implementing this method requires numerical approximations via the DFT in the evaluation of $\hat{S}$ and A^+ or H^+.

Example 6.14

Consider the SDF of (6.83). Assuming $0 < \alpha \ll 1$, we obtain the Fourier series $\log S \simeq \alpha(z_1 + z_1^{-1} + z_2 + z_2^{-1}) + O(\alpha^2)$. Ignoring $O(\alpha^2)$ terms and using the preceding results, we get

Causal MVR

$$C^+ = \alpha(z_1^{-1} + z_2^{-1}), \qquad \hat{S}_\varepsilon(z_1, z_2) = 0$$

$$\Rightarrow \quad A^+(z_1, z_2) = \exp(-C^+) \cong 1 - \alpha(z_1^{-1} + z_2^{-1}), \qquad S_\varepsilon(z_1, z_2) = 1$$

Semicausal MVR

$$C^+ = \alpha(z_1 + z_1^{-1}) + \alpha z_2^{-1}, \qquad \hat{S}_\varepsilon = -\alpha(z_1 + z_1^{-1})$$

$$\Rightarrow \quad A^+(z_1, z_2) \cong 1 - \alpha(z_1 + z_1^{-1}) - \alpha z_2^{-1}, \qquad S_\varepsilon(z_1, z_2) \cong 1 - \alpha(z_1 + z_1^{-1})$$

Semicausal WNDR

$$C^+ = \frac{\alpha}{2}(z_1 + z_1^{-1}) + \alpha z_2^{-1}, \qquad \hat{S}_\varepsilon = 0$$

$$\Rightarrow \quad A^+(z_1, z_2) \cong 1 - \frac{\alpha}{2}(z_1 + z_1^{-1}) - \alpha z_2^{-1}, \qquad S_\varepsilon(z_1, z_2) = 1$$

Example 6.15

Figure 6.15b shows the results of applying the Wiener-Doob factorization algorithm to the isotropic covariance function of Example 6.12. These models theoretically achieve perfect covariance match. Comparing these with Fig. 6.12 we see that the causal and semicausal MVRs of orders (2, 2) are quite close to their infinite-order counterparts in Figure 6.15b.

6.9 IMAGE DECOMPOSITION, FAST KL TRANSFORMS, AND STOCHASTIC DECOUPLING

In Section 6.5 we saw that a one-dimensional noncausal MVR yields an orthogonal decomposition, which leads to the notion of a fast KL transform. These ideas can be generalized to two dimensions by perturbing the boundary conditions of the stochastic image model such that the KL transform of the resulting random field becomes a fast transform. This is useful in developing efficient image processing algorithms. For certain semicausal models this technique yields uncorrelated sets of one-dimensional AR models, which can be processed independently by one-dimensional algorithms.

Periodic Random Fields

A convenient representation for images is obtained when the image model is *forced* to lie on a doubly periodic grid, that is, a torroid (like a doughnut). In that case, the sequences $u(m, n)$ and $\varepsilon(m, n)$ are doubly periodic, that is,

$$\begin{aligned} \varepsilon(m, n) &= \varepsilon(m + M, n + N) \\ u(m, n) &= u(m + M, n + N) \qquad \forall m, n \end{aligned} \tag{6.102}$$

where (M, N) are the periods of (m, n) dimensions. For stationary random fields the covariance matrices of $u(m, n)$ and $\varepsilon(m, n)$ will become doubly block-circulant, and their KL transform will be the two-dimensional unitary DFT, which is a fast transform. The periodic grid assumption is recommended only when the grid size is very large compared to the model order. Periodic models are useful for obtaining asymptotic performance bounds but are quite restrictive in many image processing applications where small (typically 16×16) blocks of data are processed at a time. Properties of periodic random field models are discussed in Problem 6.21.

Example 6.16

Suppose the causal MVR of Example 6.7, written as

$$u(m, n) = \rho_1 (m - 1, n) + \rho_2 u(m, n - 1) + \rho_3 u(m - 1, n - 1) + \varepsilon(m, n)$$

$$E[\varepsilon(m, n)] = 0, \qquad r_\varepsilon(m, n) = \beta^2 \delta(m, n)$$

is defined on an $N \times N$ periodic grid. Denoting $v(k, l)$ and $e(k, l)$ as the two-dimensional unitary DFTs of $u(m, n)$ and $\varepsilon(m, n)$, respectively, we can transform both sides of the model equation as

$$v(k, l) = (\rho_1 W_N^k + \rho_2 W_N^l + \rho_3 W_N^k W_N^l) v(k, l) + e(k, l), \qquad W_N \triangleq \exp\left(\frac{-j2\pi}{N}\right)$$

where we have used the fact that $u(m, n)$ is periodic. This can be written as

$v(k, l) = e(k, l)/A(k, l)$, where

$$A(k, l) = 1 - \rho_1 W_N^k - \rho_2 W_N^l - \rho_3 W_N^k W_N^l$$

Since $\varepsilon(m, n)$ is a stationary zero mean sequence uncorrelated over the $N \times N$ grid, $e(k, l)$ is also an uncorrelated sequence with variance β^2. This gives

$$E[v(k, l)v^*(k', l')] = \frac{E[e(k, l)e^*(k'l')]}{|A(k, l)|^2} = \frac{\beta^2}{|A(k, l)|^2}\delta(k - k', l - l')$$

that is, $v(k, l)$ is an uncorrelated sequence. *This means the unitary DFT is the KL transform of $u(m, n)$.*

Noncausal Models and Fast KL Transforms

Example 6.17

The noncausal (NC2) model defined by the equation (also see Example 6.7)

$$u(m, n) - \alpha[u(m - 1, n) + u(m + 1, n) + u(m, n - 1) + u(m, n + 1)] = \varepsilon(m, n) \tag{6.103a}$$

becomes an ARMA representation when $\varepsilon(m, n)$ is a moving average with covariance

$$r_\varepsilon(k, l) \triangleq \beta^2 \begin{cases} 1, & (k, l) = (0, 0) \\ -\alpha_1, & (k, l) = (\pm 1, 0) \quad \text{or} \quad (0, \pm 1) \\ 0, & \text{otherwise} \end{cases} \tag{6.103b}$$

For an $N \times N$ image $\mathbf{U}$, (6.103a) can be written as

$$\mathbf{QU} + \mathbf{UQ} = \boldsymbol{\varepsilon} + \mathbf{B}_1 + \mathbf{B}_2$$

$$\mathbf{B}_1 \triangleq \begin{bmatrix} \alpha\mathbf{b}_1^T \\ \cdots \\ \mathbf{0}^T \\ \cdots \\ \alpha\mathbf{b}_3^T \end{bmatrix}, \qquad B_2 \triangleq \alpha[\mathbf{b}_2 : \mathbf{0} : \mathbf{b}_4] \tag{6.104}$$

where $\mathbf{b}_1$, $\mathbf{b}_2$, $\mathbf{b}_3$, and $\mathbf{b}_4$ are $N \times 1$ vectors containing the boundary elements of the image (Fig. 6.16), $\mathbf{Q}$ is a symmetric, tridiagonal, Toeplitz matrix with values $\frac{1}{2}$ along the

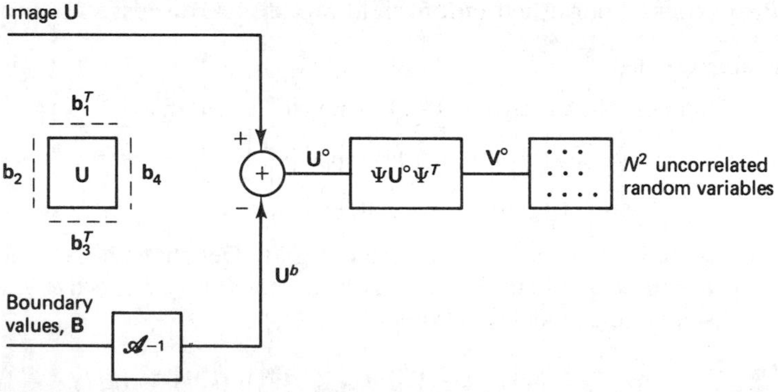

Figure 6.16 Realization of noncausal model decomposition and the concept of fast KL transform algorithms. Ψ is the fast sine transform.

main diagonal and $-\alpha$ along the two subdiagonals. This random field has the decomposition

$$\mathbf{U} = \mathbf{U}^0 + \mathbf{U}^b$$

where $\mathbf{U}^b$ is determined from the boundary values and the KL transform of $\mathbf{U}^0$ is the (fast) sine transform. Specifically,

$$\left.\begin{aligned} \mathcal{u}^b &= \mathcal{A}^{-1}(\mathcal{b}_1 + \mathcal{b}_2), \qquad \mathcal{A} \triangleq (\mathbf{I} \otimes \mathbf{Q} + \mathbf{Q} \otimes \mathbf{I}) \\ \mathcal{u}^0 &= \mathcal{u} - \mathcal{u}^b \end{aligned}\right\} \tag{6.105}$$

where $\mathcal{u}$, $\mathcal{u}^0$, $\mathcal{u}^b$, $\mathcal{b}_1$ and $\mathcal{b}_2$ are $N^2 \times 1$ vectors obtained by lexicographic ordering of the $N \times N$ matrices $\mathbf{U}$, $\mathbf{U}^0$, $\mathbf{U}^b$, $\mathbf{B}_1$, and $\mathbf{B}_2$, respectively. Figure 6.16 shows a realization of the noncausal model decomposition and the concept of the fast KL transform algorithm. The appropriate boundary variables of the random field are processed first to obtain the boundary response $u^b(m, n)$. Then the residual $u^0(m, n) \triangleq u(m, n) - u^b(m, n)$ can be processed by its KL transform, which is the (fast) sine transform. Application of this model in image data compression is discussed in Section 11.5. Several other low-order noncausal models also yield this type of decomposition [21].

Semicausal Models and Stochastic Decoupling

Certain semicausal models can be decomposed into a set of uncorrelated one-dimensional random sequences (Fig. 6.17). For example, consider the semicausal (SC2) white noise driven representation

$$u(m, n) - \alpha[u(m-1, n) + u(m+1, n)] = \gamma u(m, n-1) + \varepsilon(m, n) \tag{6.106}$$

where $\varepsilon(m, n)$ is a white noise random field with $r_\varepsilon(k, l) = \beta^2 \delta(k, l)$. For an image with N pixels per column, where $\mathbf{u}_n, \boldsymbol{\varepsilon}_n, \ldots,$ represent $N \times 1$ vectors, (6.106) can be written as

$$\mathbf{Q}\mathbf{u}_n = \gamma \mathbf{u}_{n-1} + \mathbf{b}_n + \boldsymbol{\varepsilon}_n, \qquad n \geq 1 \tag{6.107}$$

where $\mathbf{Q}$ is defined in (5.102). The $N \times 1$ vector $\mathbf{b}_n$ contains only two boundary

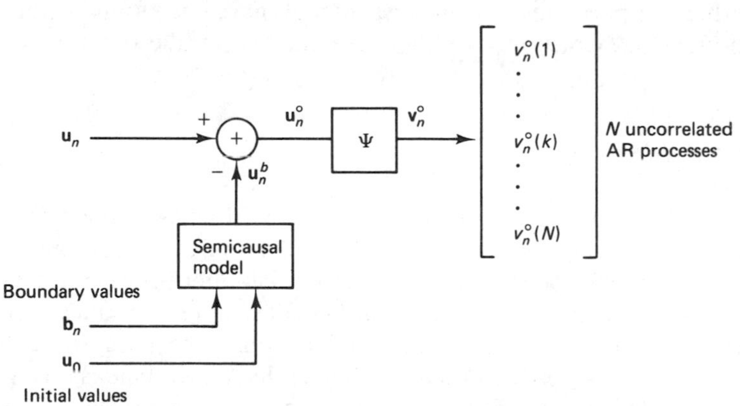

Figure 6.17 Semicausal model decomposition.

terms, $b_n(1) = \alpha u(0, n), b_n(N) = \alpha u(N+1, n), b_n(m) = 0, 2 \leq m \leq N-1$. Now $\mathbf{u}_n$ can be decomposed as

$$\mathbf{u}_n = \mathbf{u}_n^0 + \mathbf{u}_n^b \tag{6.108}$$

where

$$\mathbf{Q}\mathbf{u}_n^b = \gamma \mathbf{u}_{n-1}^b + \mathbf{b}_n, \qquad \mathbf{u}_0^b = \mathbf{u}_0 \tag{6.109}$$

$$\mathbf{Q}\mathbf{u}_n^0 = \gamma \mathbf{u}_{n-1}^0 + \boldsymbol{\varepsilon}_n, \qquad \mathbf{u}_0^0 = \mathbf{0} \tag{6.110}$$

Clearly, $\mathbf{u}_n^b$ is a vector sequence generated by the boundary values $\mathbf{b}_n$ and the initial vector $\mathbf{u}_0$. Multiplying both sides of (6.110) by $\boldsymbol{\Psi}$, the sine transform, and remembering that $\boldsymbol{\Psi}^T \boldsymbol{\Psi} = I$, $\boldsymbol{\Psi}\mathbf{Q}\boldsymbol{\Psi}^T = \boldsymbol{\Lambda} = \text{Diag}\{\lambda(k)\}$, we obtain

$$\boldsymbol{\Psi}\mathbf{Q}\boldsymbol{\Psi}^T \boldsymbol{\Psi}\mathbf{u}_n^0 = \gamma \boldsymbol{\Psi}\mathbf{u}_{n-1}^0 + \boldsymbol{\Psi}\boldsymbol{\varepsilon}_n, \qquad \boldsymbol{\Psi}\mathbf{u}_0^0 = \mathbf{0} \quad \Rightarrow \boldsymbol{\Lambda}\mathbf{v}_n^0 = \gamma \mathbf{v}_{n-1}^0 + \mathbf{e}_n \tag{6.111}$$

where $\mathbf{v}_n^0$ and $\mathbf{e}_n$ are the sine transforms of $\mathbf{u}_n^0$ and $\boldsymbol{\varepsilon}_n$, respectively. This reduces to a set of equations decoupled in k,

$$\lambda(k) v_n^0(k) = \gamma v_{n-1}^0(k) + e_n(k), \qquad v_0^0(k) = 0, \qquad n \geq 1, 1 \leq k \leq N$$

where

$$E[e_n(k) e_{n'}(k')] = \beta^2 \delta(n - n') \delta(k - k') \tag{6.112}$$

Since $\boldsymbol{\varepsilon}_n$ is a stationary white noise vector sequence, its transform coefficients $\mathbf{e}_n(k)$ are uncorrelated in the k-dimension. Therefore, $v_n^0(k)$ are also uncorrelated in the k-dimension and (6.112) is a set of uncorrelated AR sequences. The semicausal model of (6.87) also yields this type of decomposition [21]. Figure 6.17 shows the realization of semicausal model decompositions. Disregarding the boundary effects, (6.112) suggests that the rows of a column-by-column transformed image using a suitable unitary transform may be represented by AR models, as mentioned in Section 6.2. This is indeed a useful image representation, and techniques based on such models lead to what are called *hybrid algorithms* [1, 23, 24], that is, algorithms that are recursive in one dimension and unitary transform based in the other. Applications of semicausal models have been found in image coding, restoration, edge extraction, and high-resolution spectral estimation in two dimensions.

6.10 SUMMARY

In this chapter we have considered several stochastic models for images. The one-dimensional AR and state variable models are useful for line-by-line processing of images. Such models will be found useful in filtering and restoration problems. The causal, semicausal, and noncausal models were introduced as different types of realizations of 2-D spectral density functions. One major difficulty in the identification of 2-D models arises due to the fact that a two-dimensional polynomial may not be reducible to a product of a finite number of lower-order polynomials. Therefore, two-dimensional spectral factorization algorithms generally yield infinite-order models. We considered several methods of finding finite order approximate realizations that are stable. Applications of these stochastic models are in many image processing problems found in the subsequent chapters.

PROBLEMS

6.1 *(AR model properties)* To prove (6.4) show that $\bar{u}(n)$ satisfies the orthogonality condition $E[(u(n) - \bar{u}(n))u(m)] = 0$ for every $m < n$. Using this and (6.3a), prove (6.7).

6.2 An image is scanned line by line. The mean value of a pixel in a scan line is 70. The autocorrelations for $n = 0, 1, 2$ are 6500, 6468, and 6420, respectively.

a. Find the first- and second-order AR models for a scan line. Which is a better model? Why?

b. Find the first five covariances and the autocorrelations generated by the second-order AR model and verify that the given autocorrelations are matched by the model.

6.3 Show that the covariance matrix of a row-ordered vector obtained from an $N \times N$ array of a stationary random field is not fully Toeplitz. Hence, a row scanned two-dimensional stationary random field does not yield a one-dimensional stationary random sequence.

6.4 One easy method of solving (6.28) when we are given $S(\omega)$ is to let $H(\omega) = \sqrt{S(\omega)}$. For $K = 1$ and $S(\omega) = (1 + \rho^2) - 2\rho \cos \omega$, show that this algorithm will not yield a finite-order ARMA realization. Can this filter be causal or stable?

6.5 What are the necessary and sufficient conditions that an ARMA system be minimum phase? Find if the following filters are (i) causal and stable, (ii) minimum phase.

a. $H(z) = 1 - 0.8z^{-1}$

b. $H(z) = (1 - z^{-1})/[1.81 - 0.9(z + z^{-1})]$

c. $H(z) = (1 - 0.2z^{-1})/(1 - 0.9z^{-1})$

6.6 Following the Wiener-Doob decomposition method outlined in Section 6.8, show that a 1-D SDF $S(z)$ can be factored to give $H(z) \triangleq \exp[C^+(z)]$, $C^+(z) \triangleq \sum_{n=1}^{\infty} c(n)z^{-n}$, $K = \exp[c(0)]$, where $\{c(n), \forall n\}$ are the Fourier series coefficients of $\log S(\omega)$. Show that $H(z)$ is a minimum-phase filter.

6.7 Using the identity for a Hermitian Toeplitz matrix

$$\mathbf{R}_{n+1} \equiv \left[\begin{array}{c|c} \mathbf{R}_n & \hat{\mathbf{b}}_n^* \\ \hline \hat{\mathbf{b}}_n^T & r(0) \end{array}\right]$$

where $\hat{\mathbf{b}}_n \triangleq [r(n), r(n-1), \ldots, r(1)]^T$, prove the Levinson recursions.

6.8 Assume that $\mathbf{R}$ is real and positive definite. Then $a_n(k)$ and ρ_n will also be real. Using the Levinson recursions, prove the following.

a. $|\rho_n| < 1, \forall n$.

b. Given $r(0)$ and any one of the sequences $\{r(n)\}, \{a(n)\}, \{\rho_n\}$, the remaining sequences can be determined uniquely from it.

6.9 For the third-order AR model

$$u(n) = 0.1u(n-1) + 0.782u(n-2) + 0.1u(n-3) + \varepsilon(n), \qquad \beta^2 = 0.067716$$

find the partial correlations and determine if this system is stable. What are the first four covariances of this sequence? Assume $u(n)$ to be a zero mean, unity variance random sequence.

6.10 *(Proof of noncausal MVR theorem)* First show that the orthogonality condition for minimum-variance noncausal estimate $\bar{u}(n)$ yields

$$r(k) - \sum_{l \neq 0} \alpha(l) r(k-l) = \beta^2 \delta(k), \qquad \forall k$$

Solve this via the Fourier transform to arrive at (6.38) and show that $S(z) = \beta^2/A(z)$. Apply this to the relation $S(z) = S_\nu(z)/A(z)A(z^{-1})$ to obtain (6.39).

6.11 Show that the parameters of the noncausal MVR of (6.43) for pth-order AR sequences defined by (6.3a) are given by

$$\alpha(k) = \sum_{n=0}^{p-k} \frac{a(n)a(n+k)}{C^2}, \qquad \beta_\nu^2 = \frac{\beta^2}{C^2}, \qquad C^2 \triangleq \sum_{n=0}^{p} |a(n)|^2$$

where $a(0) \triangleq -1$.

6.12 Show that for a stationary pth-order AR sequence of length N, its noncausal MVR can be written as

$$\mathbf{Hu} = \boldsymbol{\nu} + \mathbf{b}, \qquad E[\boldsymbol{\nu}\boldsymbol{\nu}^T] = \beta^2 \mathbf{H}, \qquad E[\boldsymbol{\nu}\mathbf{b}^T] = \mathbf{0}$$

where $\mathbf{H}$ is an $N \times N$ symmetric banded Toeplitz matrix with $2p$ subdiagonals, and the $N \times 1$ vector $\mathbf{b}$ contains $2p$ nonzero terms involving the boundary values $\{u(1-k), u(N+k), 1 \le k \le p\}$. Show that this yields an orthogonal decomposition similar to (6.49).

6.13 For a zero mean Gaussian random sequence with covariance $\rho^{|n|}$, show that the optimum interpolator $\hat{u}(n)$ based on $u(0)$ and $u(N+1)$ is given by

$$\hat{u}(n) = [(\rho^{N+1-n} - \rho^{-N-1+n})u(0) + (\rho^n - \rho^{-n})u(N+1)]/(\rho^{N+1} - \rho^{-N-1})$$

Show that this reduces to the straight-line interpolator of (6.53) when $\rho \to 1$ and a cubic polynomial interpolation formula when $\rho = 1$.

6.14 *(Fast KL transform for a continuous random process)* Let $u(x)$ be a zero mean, stationary Gaussian random process whose covariance function is $r(\tau) = \exp(-\alpha|\tau|)$, $\alpha > 0$. Let $x \in [-L, L]$ and define $u_0 \triangleq u(-L), u_1 \triangleq u(L)$. Show that $u(x)$ has the orthogonal decomposition

$$u(x) = u^0(x) + u^b(x), \qquad -L \le x \le L$$

$$u^b(x) = E[u(x)|u_0, u_1] = a_0(x)u_0 + a_1(x)u_1$$

$$a_0(x) = \frac{\sinh \alpha(L-x)}{\sinh 2\alpha L}, \qquad a_1(x) = \frac{\sinh \alpha(L+x)}{\sinh 2\alpha L}$$

such that $u_b(x)$, which is determined by the boundary values u_0 and u_1, is orthogonal to $u^0(x)$. Moreover, the KL expansion of $u^0(x)$ is given by the *harmonic sinusoids*

$$\Psi_k(x) = \begin{cases} \sqrt{\dfrac{1}{L}} \cos \dfrac{k\pi x}{2L}, & k = 1, 3, 5 \ldots, \quad -L \le x \le L \\[2ex] \sqrt{\dfrac{1}{L}} \sin \dfrac{k\pi x}{2L}, & k = 2, 4, 6 \ldots \end{cases}$$

6.15 Determine whether the following filters are causal, semicausal or noncausal for a vertically scanned, left-to-right image.

a. $H(z_1, z_2) = z_1 + z_1^{-1} + z_1^{-1} z_2^{-1}$

b. $H(z_1, z_2) = 1 + z_1^{-1} + z_2^{-1} z_1 + z_2^{-1} z_2^{-1}$

c. $H(z_1, z_2) = \dfrac{1}{(2 - z_1^{-1} - z_2^{-1})}$

d. $H(z_1, z_2) = \dfrac{1}{4 - z_1 - z_1^{-1} - z_2 - z_2^{-1}}$

Sketch their regions of support.

6.16 Given the prediction error filters

$$A(z_1, z_2) = 1 - a_1 z_1^{-1} - a_2 z_2^{-1} - a_3 z_1^{-1} z_2^{-1} - a_4 z_1 z_2^{-1} \qquad \text{(causal)}$$

$$A(z_1, z_2) = 1 - a_1 (z_1 + z_1^{-1}) - a_2 z_2^{-1} - a_3 z_2^{-1} (z_1 + z_1^{-1}) \qquad \text{(semicausal)}$$

$$\begin{aligned} A(z_1, z_2) = 1 &- a_1 (z_1 + z_1^{-1}) - a_2 (z_2 + z_2^{-1}) \\ &- a_3 z_2^{-1} (z_1 + z_1^{-1}) - a_4 z_1^{-1} (z_2 + z_2^{-1}) \qquad \text{(noncausal)} \end{aligned}$$

a. Assuming the prediction error has zero mean and variance β^2, find the SDF of the prediction error sequences if these are to be minimum variance prediction-error filters.

b. What initial or boundary values are needed to solve the filter equations $A(z_1, z_2) U(z_1, z_2) = F(z_1, z_2)$ over an $N \times N$ grid in each case?

6.17 The stability condition (6.78) is equivalent to the requirement that $|H(z_1, z_2)| < \infty$, $|z_1| = 1, |z_2| = 1$.

a. Show that this immediately gives the general stability condition of (6.79) for any two-dimensional system, in particular for the noncausal systems.

b. For a semicausal system $H(m, n) = 0$ for $n < 0$, for every m. Show that this restriction yields the stability conditions of (6.80).

c. For a causal system we need $h(m, n) = 0$ for $n < 0, \forall m$ and $h(m, n) = 0$ for $n = 0, m < 0$. Show that this restriction yields the stability conditions of (6.81).

6.18 Assuming the prediction-error filters of Example 6.7 represent MVRs, find conditions on the predictor coefficients so that the associated MVRs are stable.

6.19 If a transfer function $H(z_1, z_2) = H_1(z_1) H_2(z_2)$, then show that the system is stable and (a) causal, (b) semicausal or (c) noncausal provided $H_1(z_1)$ and $H_2(z_2)$ are transfer functions of one-dimensional stable systems that are (a) both causal, (b) one causal and one noncausal, or (c) both noncausal, respectively.

6.20 Assuming the cepstrum $c(m, n)$ is absolutely summable, prove the stability conditions for the causal and semicausal models.

6.21 **a.** Show that the KL transform of any periodic random field that is also stationary is the two-dimensional (unitary) DFT.

b. Suppose the finite-order causal, semicausal, and noncausal MVRs given by (6.70) are defined on a periodic grid with period (M, N). Show that the SDF of these random fields is given by

$$S(\omega_1, \omega_2) = \sum_{k=0}^{M-1} \sum_{l=0}^{N-1} \hat{S}(k, l) \delta\left(\omega_1 - \frac{2\pi k}{M}\right) \delta\left(\omega_2 - \frac{2\pi l}{N}\right), \qquad -\pi \le \omega_1, \omega_2 \le \pi$$

What is $\hat{S}(k, l)$?

BIBLIOGRAPHY

Section 6.1

For a survey of mathematical models and their relevance in image processing and related bibliography:

1. A. K. Jain. "Advances in mathematical models for image processing." *Proceedings IEEE* 69, no. 5 (May 1981): 502–528.

Sections 6.2–6.4

Further discussions on spectral factorization and state variable models, ARMA models, and so on, are available in:

2. A. V. Oppenheim and R. W. Schafer. *Digital Signal Processing*. Englewood Cliffs, N.J.: Prentice-Hall, 1975.
3. A. H. Jazwinsky. *Stochastic Processes and Filtering Theory*. New York: Academic Press, 1970, pp. 70–92.
4. P. Whittle. *Prediction and Regulation by Linear Least-Squares Methods*. London: English University Press, 1954.
5. N. Wiener. *Extrapolation, Interpolation and Smoothing of Stationary Time Series*. New York: John Wiley, 1949.
6. C. L. Rino. "Factorization of Spectra by Discrete Fourier Transforms." *IEEE Transactions on Information Theory* IT-16 (July 1970): 484–485.
7. J. Makhoul. "Linear Prediction: A Tutorial Review." *Proceedings IEEE* 63 (April 1975): 561–580.
8. *IEEE Trans. Auto. Contr.* Special Issue on System Identification and Time Series Analysis. T. Kailath, D. O. Mayne and R. K. Mehra (eds.), Vol. AC-19, December 1974.
9. N. E. Nahi and T. Assefi. "Bayesian Recursive Image Estimation." *IEEE Trans. Comput.* (Short Notes) C-21 (July 1972): 734–738.
10. S. R. Powell and L. M. Silverman. "Modeling of Two Dimensional Covariance Functions with Application to Image Restoration." *IEEE Trans. Auto. Contr.* AC-19 (February 1974): 8–12.
11. R. P. Roesser. "A Discrete State Space Model for Linear Image Processing." *IEEE Trans. Auto. Contr.* AC-20 (February 1975): 1–10.
12. E. Fornasini and G. Marchesini. "State Space Realization Theory of Two-Dimensional Filters." *IEEE Trans. Auto. Contr.* AC-21 (August 1976): 484–492.

Section 6.5

Noncausal representations and fast KL transforms for discrete random processes are discussed in [1, 21] and:

13. A. K. Jain. "A Fast Karhunen Loeve Transform for a Class of Random Processes." *IEEE Trans. Comm.* COM-24 (September 1976): 1023–1029. Also see *IEEE Trans. Comput.* C-25 (November 1977): 1065–1071.

Sections 6.6 and 6.7

Here we follow [1]. The linear prediction models discussed here can also be generalized to nonstationary random fields [1]. For more on random fields:

14. P. Whittle. "On Stationary Processes in the Plane." *Biometrika* 41 (1954): 434–449.
15. T. L. Marzetta. "A Linear Prediction Approach to Two-Dimensional Spectral Factorization and Spectral Estimation." Ph.D. Thesis, Department Electrical Engineering and Computer Science, MIT, February 1978.
16. S. Ranganath and A. K. Jain. "Two-Dimensional Linear Prediction Models Part I: Spectral Factorization and Realization." *IEEE Trans. ASSP* ASSP-33, no. 1 (February 1985): 280–299. Also see S. Ranganath. "Two-Dimensional Spectral Factorization, Spectral Estimation and Applications in Image Processing." Ph.D. Dissertation, Department Electrical and Computer Engineering, UC Davis, March 1983.
17. A. K. Jain and S. Ranganath. "Two-Dimensional Linear Prediction Models and Spectral Estimation," Ch. 7 in *Advances in Computer Vision and Image Processing*. (T. S. Huang, ed.). Vol. 2, Greeenwich, Conn.: JAI Press Inc., 1986, pp. 333–372.
18. R. Chellappa. "Two-Dimensional Discrete Gaussian Markov Random Field Models for Image Processing," in *Progress in Pattern Recognition,* L. Kanal and A. Rosenfeld (eds). Vol. 2, New York, N.Y.: North Holland, 1985, pp. 79–112.
19. J. W. Woods. "Two-Dimensional Discrete Markov Fields." *IEEE Trans. Inform. Theory* IT-18 (March 1972): 232–240.

For stability of two-dimensional systems:

20. D. Goodman. "Some Stability Properties of Two-dimensional Linear Shift Invariant Filters," *IEEE Trans. Cir. Sys.* CAS-24 (April 1977): 201–208.

The relationship between the three types of prediction models and partial differential equations is discussed in:

21. A. K. Jain. "Partial Differential Equations and Finite Difference Methods in Image Processing, Part I—Image Representation." *J. Optimization Theory and Appl.* 23, no. 1 (September 1977): 65–91. Also see *IEEE Trans. Auto. Control* AC-23 (October 1978): 817–834.

Section 6.8

Here we follow [1] and have applied the method of [6] and:

22. M. P. Ekstrom and J. W. Woods. "Two-Dimensional Spectral Factorization with Application in Recursive Digital Filtering." *IEEE Trans. on Acoust. Speech and Signal Processing* ASSP-24 (April 1976): G115–128.

Section 6.9

For fast KL transform decomposition in two dimensions, stochastic decoupling, and related results see [13], [21], and:

23. S. H. Wang. "Applications of Stochastic Models for Image Data Compression." Ph.D. Dissertation, Department of Electrical Engineering, SUNY Buffalo, September 1979. Also Technical Report SIPL-79-6, Signal and Image Processing Laboratory, Department Electrical and Computer Engineering, UC Davis, September 1979.
24. A. K. Jain. "A Fast Karhunen-Loeve Transform for Recursive Filtering of Images Corrupted by White and Colored Noise." *IEEE Trans. Comput.* C-26 (June 1977): 560–571.

7

Image Enhancement

7.1 INTRODUCTION

Image enhancement refers to accentuation, or sharpening, of image features such as edges, boundaries, or contrast to make a graphic display more useful for display and analysis. The enhancement process does not increase the inherent information content in the data. But it does increase the dynamic range of the chosen features so that they can be detected easily. Image enhancement includes gray level and contrast manipulation, noise reduction, edge crispening and sharpening, filtering, interpolation and magnification, pseudocoloring, and so on. The greatest difficulty in image enhancement is quantifying the criterion for enhancement. Therefore, a large number of image enhancement techniques are empirical and require interactive procedures to obtain satisfactory results. However, image enhancement remains a very important topic because of its usefulness in virtually all image processing applications. In this chapter we consider several algorithms commonly used for enhancement of images. Figure 7.1 lists some of the common image enhancement techniques.

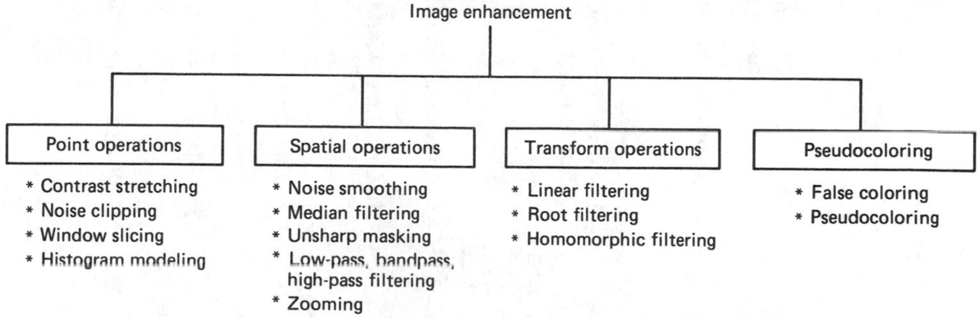

Figure 7.1 Image enhancement.

TABLE 7.1 Zero-memory Filters for Image Enhancement. Input and output gray levels are distributed between [0, L]. Typically, $L = 255$

1. Contrast stretching	$f(u) = \begin{cases} \alpha u, & 0 \le u < a \\ \beta(u-a)+v_a, & a \le u < b \\ \gamma(u-b)+v_b, & b \le u < L \end{cases}$	The slopes α, β, γ determine the relative contrast stretch. See Fig. 7.2.
2. Noise clipping and thresholding	$f(u) = \begin{cases} 0, & 0 \le u < a \\ \alpha u, & a \le u \le b \\ L, & u \ge b \end{cases}$	Useful for binary or other images that have bimodal distribution of gray levels. The a and b define the valley between the peaks of the histogram. For $a = b = t$, this is called *thresholding*.
3. Gray scale reversal	$f(u) = L - u$	Creates *digital negative* of the image.
4. Gray-level window slicing	$f(u) = \begin{cases} L, & a \le u \le b \\ 0, & \text{otherwise} \end{cases}$	Fully illuminates pixels lying in the interval $[a, b]$ and removes the background.
5. Bit extraction	$f(u) = (i_n - 2i_{n-1})L$ $i_n = \text{Int}\left[\frac{u}{2^{B-n}}\right], \; n = 1, 2, \ldots, B$	B = number of bits used to represent u as an integer. This extracts the nth most-significant bit.
6. Bit removal	$f(u) = 2u \text{ modulo } (L+1), \quad 0 \le u \le L$	Most-significant-bit removal.
	$f(u) = 2\text{Int}\left[\frac{u}{2}\right]$	Least-significant-bit removal.
7. Range compression	$v = c \log_{10}(1+u), \quad u \ge 0$ $c \triangleq \frac{L}{\log_{10}(1+L)}$	Intensity to contrast transformation.

7.2 POINT OPERATIONS

Point operations are zero memory operations where a given gray level $u \in [0, L]$ is mapped into a gray level $v \in [0, L]$ according to a transformation

$$v = f(u) \tag{7.1}$$

Table 7.1 lists several of these transformations.

Contrast Stretching

Low-contrast images occur often due to poor or nonuniform lighting conditions or due to nonlinearity or small dynamic range of the imaging sensor. Figure 7.2 shows a typical contrast stretching transformation, which can be expressed as

$$v = \begin{cases} \alpha u, & 0 \le u < a \\ \beta(u - a) + v_a, & a \le u < b \\ \gamma(u - b) + v_b, & b \le u < L \end{cases} \tag{7.2}$$

The slope of the transformation is chosen greater than unity in the region of stretch.

The parameters a and b can be obtained by examining the histogram of the image. For example, the gray scale intervals where pixels occur most frequently would be stretched most to improve the overall visibility of a scene. Figure 7.3 shows examples of contrast stretched images.

Clipping and Thresholding

A special case of contrast stretching where $\alpha = \gamma = 0$ (Fig. 7.4) is called clipping. This is useful for noise reduction when the input signal is known to lie in the range $[a, b]$.

Thresholding is a special case of clipping where $a = b \triangleq t$ and the output becomes binary (Fig. 7.5). For example, a seemingly binary image, such as a printed page, does not give binary output when scanned because of sensor noise and background illumination variations. Thresholding is used to make such an image binary. Figure 7.6 shows examples of clipping and thresholding on images.

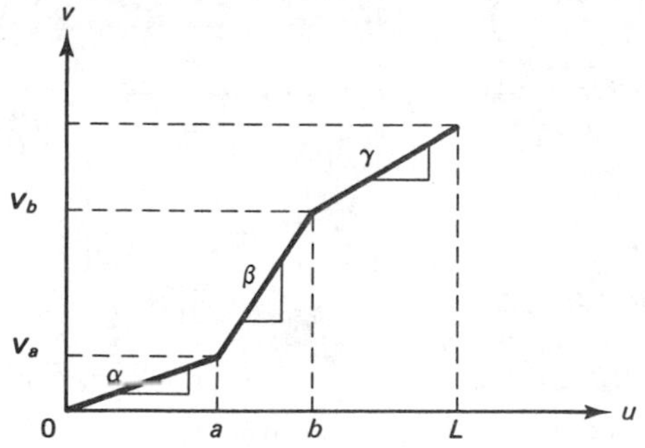

Figure 7.2 Contrast stretching transformation. For dark region stretch $\alpha > 1$, $a \simeq L/3$; midregion stretch, $\beta > 1$, $b \simeq \frac{2}{3}L$; bright region stretch $\gamma > 1$.

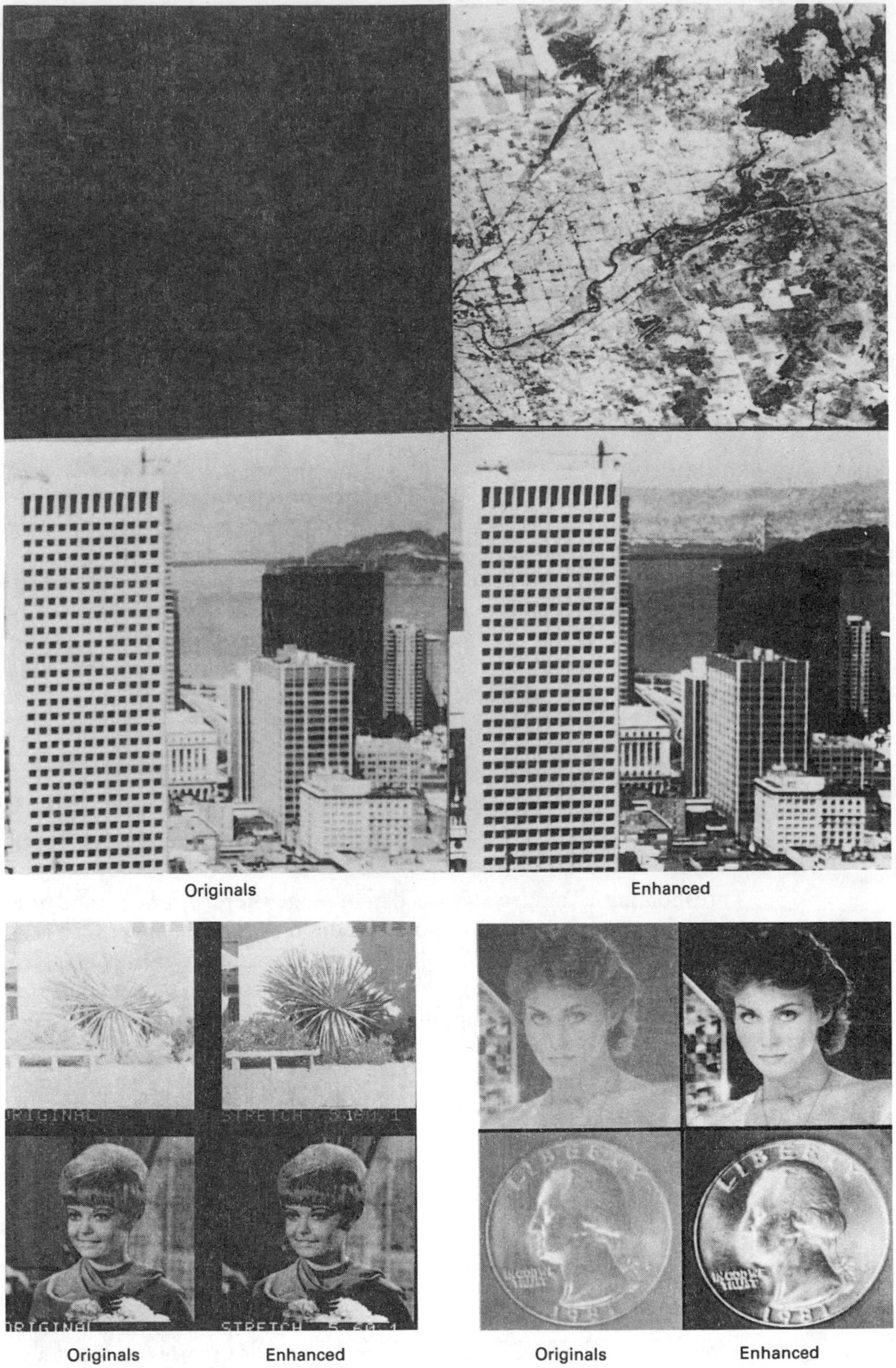

Figure 7.3 Contrast stretching.

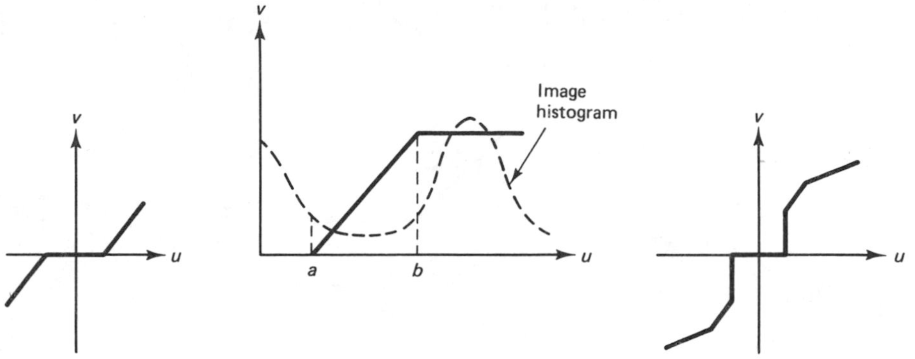

Figure 7.4 Clipping transformations.

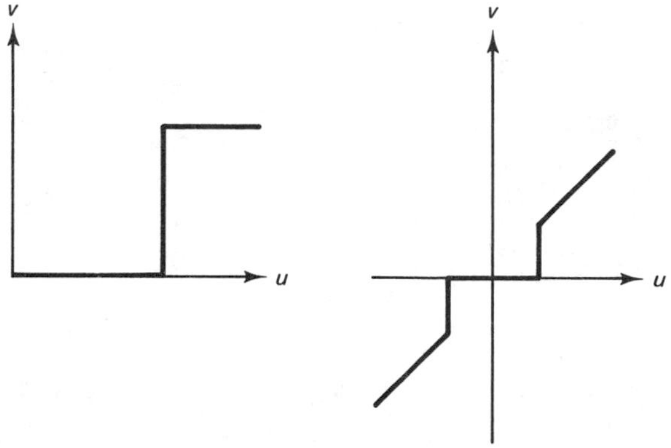

Figure 7.5 Thresholding transformations.

Figure 7.6 Clipping and thresholding.

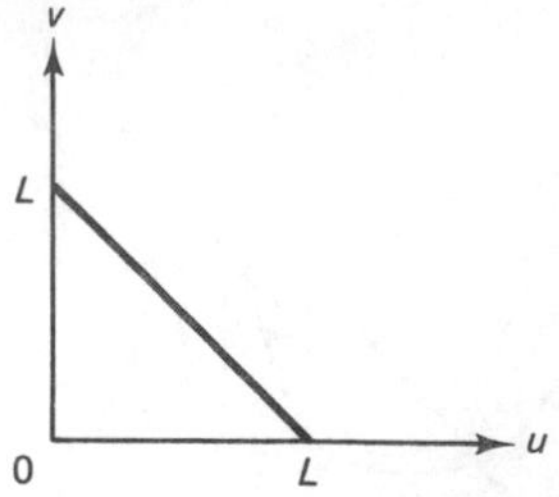

Figure 7.7 Digital negative transformation.

Digital Negative

A *negative* image can be obtained by reverse scaling of the gray levels according to the transformation (Fig. 7.7)

$$v = L - u \tag{7.3}$$

Figure 7.8 shows the digital negatives of different images. Digital negatives are useful in the display of medical images and in producing negative prints of images.

Intensity Level Slicing (Fig. 7.9)

Without background:
$$v = \begin{cases} L, & a \le u \le b \\ 0, & \text{otherwise} \end{cases} \tag{7.4}$$

With background:
$$v = \begin{cases} L, & a \le u \le b \\ u, & \text{otherwise} \end{cases} \tag{7.5}$$

These transformations permit segmentation of certain gray level regions from the rest of the image. This technique is useful when different features of an image are contained in different gray levels. Figure 7.10 shows the result of intensity window slicing for segmentation of low-temperature regions (clouds, hurricane) of two images where high-intensity gray levels are proportional to low temperatures.

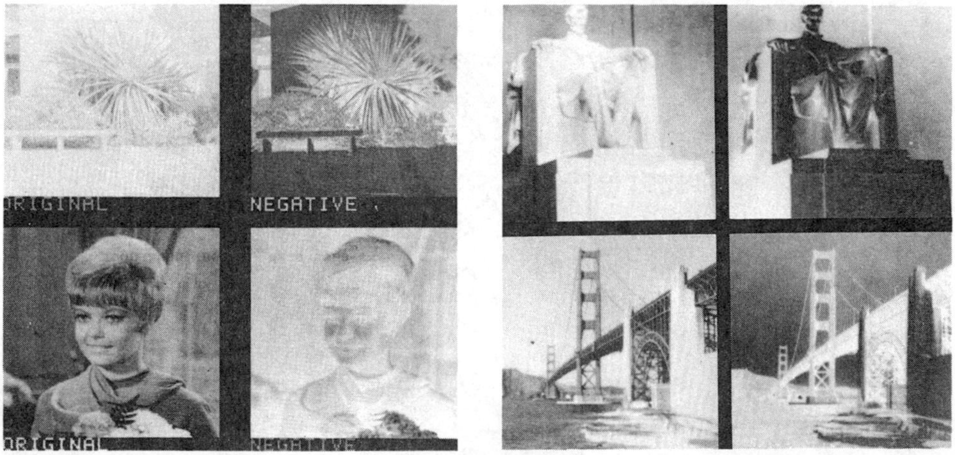

(a) (b)

Figure 7.8 Digital negatives.

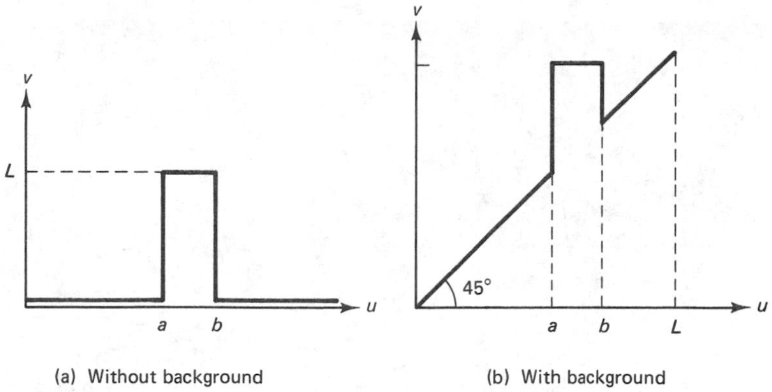

Figure 7.9 Intensity level slicing.

Bit Extraction

Suppose each image pixel is uniformly quantized to B bits. It is desired to extract the nth most-significant bit and display it. Let

$$u = k_1 2^{B-1} + k_2 2^{B-2} + \cdots + k_n 2^{B-n} + \cdots + k_{B-1} 2 + k_B \tag{7.6}$$

Then we want the output to be

$$v = \begin{cases} L, & \text{if } k_n = 1 \\ 0, & \text{otherwise} \end{cases} \tag{7.7}$$

It is easy to show that

$$k_n = i_n - 2i_{n-1} \tag{7.8}$$

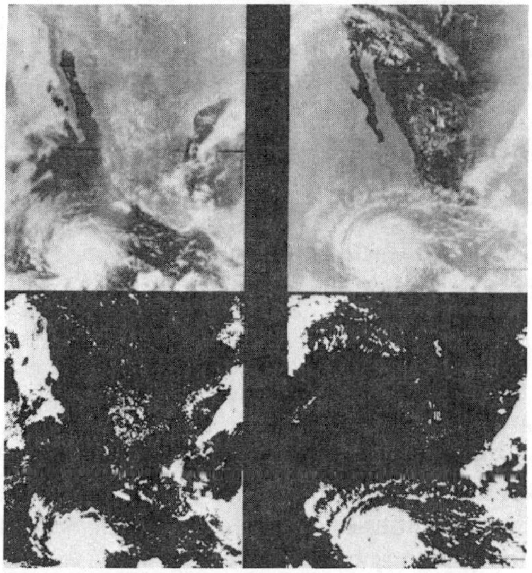

Figure 7.10 Level slicing of intensity window [175, 250]. Top row: visual and infrared (IR) images; bottom row: segmented images.

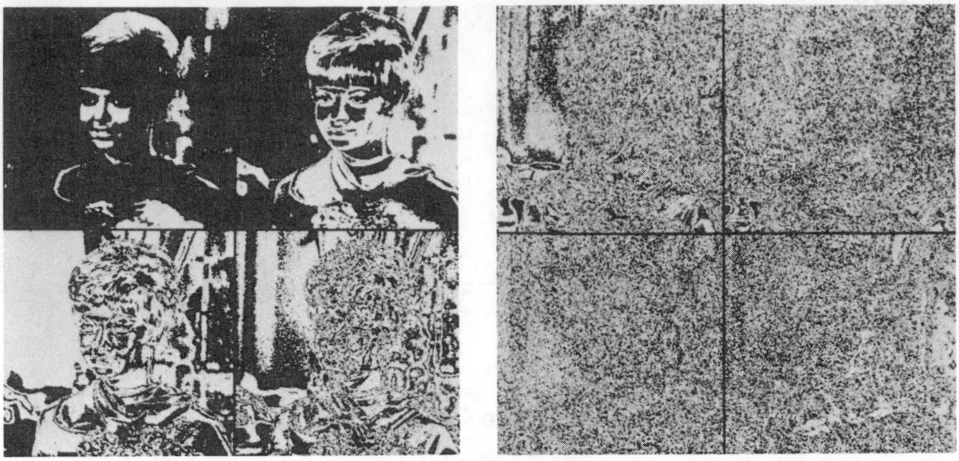

(a) First four significant bit images

0	1
2	3

;

(b) last four significant bit images

4	5
6	7

Figure 7.11 8-bit planes of a digital image.

where

$$i_n \triangleq \operatorname{Int}\left[\frac{u}{2^{B-n}}\right], \quad \operatorname{Int}[x] \triangleq \text{integer part of } x \tag{7.9}$$

Figure 7.11 shows the first 8 most-significant bit images of an 8-bit image. This transformation is useful in determining the number of visually significant bits in an image. In Fig. 7.11 only the first 6 bits are visually significant, because the remaining bits do not convey any information about the image structure.

Range Compression

Sometimes the dynamic range of the image data may be very large. For example, the dynamic range of a typical unitarily transformed image is so large that only a few pixels are visible. The dynamic range can be compressed via the logarithmic transformation

$$v = c \log_{10}(1 + |u|) \tag{7.10}$$

where c is a scaling constant. This transformation enhances the small magnitude pixels compared to those pixels with large magnitudes (Fig. 7.12).

Image Subtraction and Change Detection

In many imaging applications it is desired to compare two complicated or busy images. A simple but powerful method is to align the two images and subtract them. The difference image is then enhanced. For example, the missing components on a circuit board can be detected by subtracting its image from that of a properly assembled board. Another application is in imaging of the blood vessels and arteries

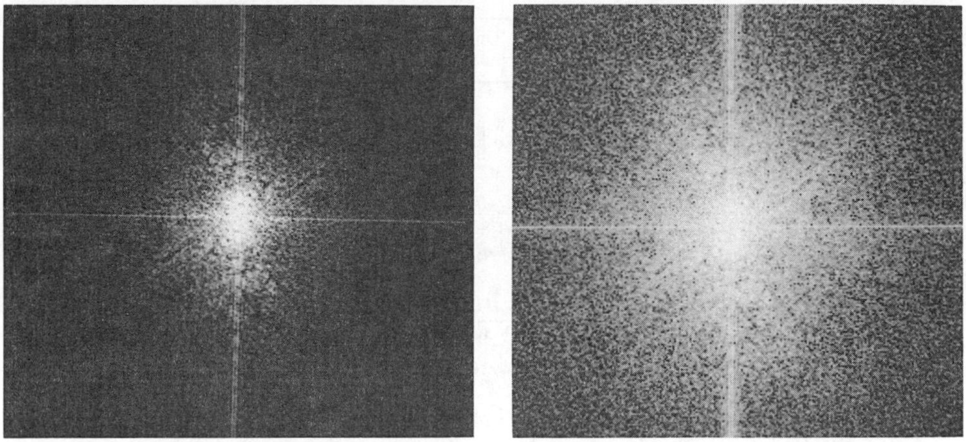

(a) Original (b) log

Figure 7.12 Range compression.

in a body. The blood stream is injected with a radio-opaque dye and X-ray images are taken before and after the injection. The difference of the two images yields a clear display of the blood-flow paths (see Fig. 9.44). Other applications of change detection are in security monitoring systems, automated inspection of printed circuits, and so on.

7.3 HISTOGRAM MODELING

The histogram of an image represents the relative frequency of occurrence of the various gray levels in the image. Histogram-modeling techniques modify an image so that its histogram has a desired shape. This is useful in stretching the low-contrast levels of images with narrow histograms. Histogram modeling has been found to be a powerful technique for image enhancement.

Histogram Equalization

In histogram equalization, the goal is to obtain a uniform histogram for the output image. Consider an image pixel value $u \geq 0$ to be a random variable with a continuous probability density function $p_u(\mathcal{u})$ and cumulative probability distribution $F_u(\mathcal{u}) \triangleq P[u \leq \mathcal{u}]$. Then the random variable

$$v \triangleq F_u(u) \triangleq \int_0^u p_u(\mathcal{u})\,d\mathcal{u} \tag{7.11}$$

will be uniformly distributed over $(0, 1)$ (Problem 7.3). To implement this transformation on digital images, suppose the input u has L gray levels $x_i, i = 0, 1, \ldots, L-1$ with probabilities $p_u(x_i)$. These probabilities can be determined from the

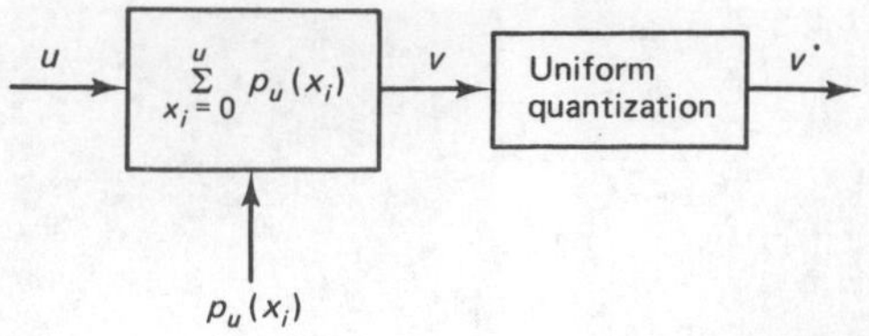

Figure 7.13 Histogram equalization transformation.

histogram of the image that gives $h(x_i)$, the number of pixels with gray level value x_i. Then

$$p_u(x_i) = \frac{h(x_i)}{\sum_{i=0}^{L-1} h(x_i)}, \qquad i = 0, 1, \ldots, L-1 \tag{7.12}$$

The output $v^\cdot$, also assumed to have L levels, is given as follows:[†]

$$v \triangleq \sum_{x_i=0}^{u} p_u(x_i) \tag{7.13a}$$

$$v^\cdot \triangleq \text{Int}\left[\frac{(v - v_{\min})}{1 - v_{\min}}(L-1) + 0.5\right] \tag{7.13b}$$

where $v_{\min}$ is the smallest positive value of v obtained from (7.13a). Now $v^\cdot$ will be uniformly distributed only approximately because v is not a uniformly distributed variable (Problem 7.3). Figure 7.13 shows the histogram-equalization algorithm for digital images. From (7.13a) note that v is a discrete variable that takes the value

$$v_k = \sum_{i=0}^{k} p_u(x_i) \tag{7.14}$$

if $u = x_k$. Equation (7.13b) simply uniformly requantizes the set $\{v_k\}$ to $\{v_k^\cdot\}$. Note that this requantization step is necessary because the probabilities $p_u(x_k)$ and $p_v(v_k)$ are the same. Figure 7.14 shows some results of histogram equalization.

Histogram Modification

A generalization of the procedure of Fig. 7.13 is given in Fig. 7.15. The input gray level u is first transformed nonlinearly by $f(u)$, and the output is uniformly quantized. In histogram equalization, the function

$$f(u) = \sum_{x_i=0}^{u} p_u(x_i) \tag{7.15}$$

typically performs a compression of the input variable. Other choices of $f(u)$ that have similar behavior are

$$f(u) = \frac{\sum_{x_i=0}^{u} p_u^{1/n}(x_i)}{\sum_{x_i=0}^{x_{L-1}} p_u^{1/n}(x_i)}, \qquad n = 2, 3, \ldots \tag{7.16}$$

[†] We assume $x_0 = 0$.

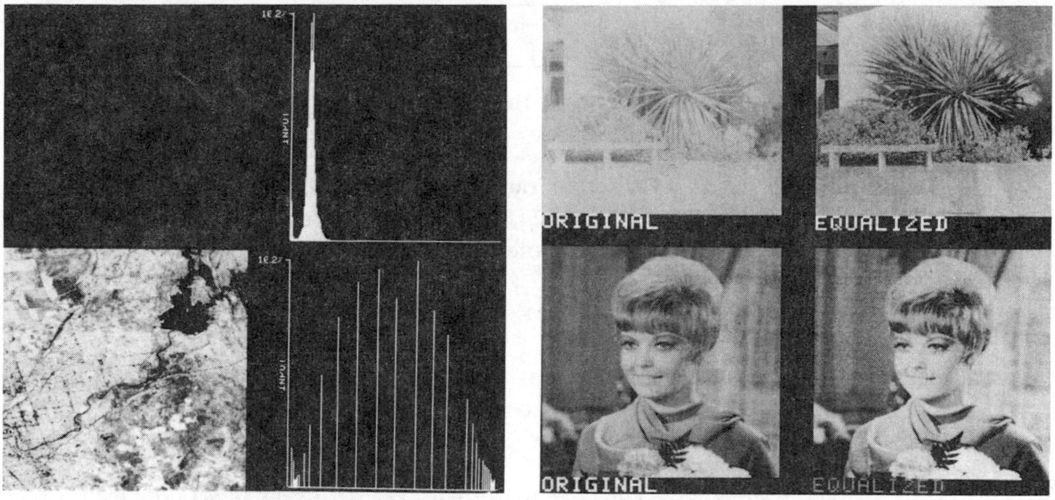

(a) Top row: input image, its histogram; bottom row: processed image, its histogram;

(b) left columns: input images; right columns: processed images.

Figure 7.14 Histogram equalization.

$$f(u) = \log(1 + u), \qquad u \geq 0 \tag{7.17}$$

$$f(u) = u^{1/n}, \qquad u \geq 0, \quad n = 2, 3, \ldots \tag{7.18}$$

These functions are similar to the companding transformations used in image quantization.

Histogram Specification

Suppose the random variable $u \geq 0$ with probability density $p_u(u)$ is to be transformed to $v \geq 0$ such that it has a specified probability density $p_v(v)$. For this to be true, we define a uniform random variable

$$w \triangleq \int_0^u p_u(x)\,dx = F_u(u) \tag{7.19}$$

that also satisfies the relation

$$w = \int_0^v p_v(y)\,dy = F_v(v) \tag{7.20}$$

Eliminating w, we obtain

$$v = F_v^{-1}(F_u(u)) \tag{7.21}$$

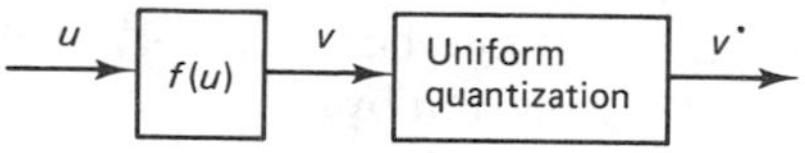

Figure 7.15 Histogram modification.

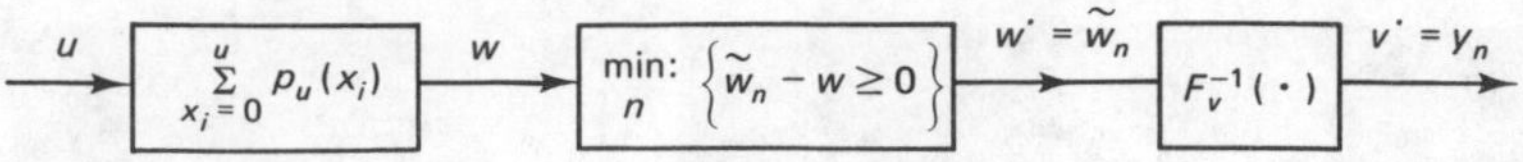

Figure 7.16 Histogram specification.

If u and v are given as discrete random variables that take values x_i and y_i, $i = 0, \ldots, L-1$, with probabilities $p_u(x_i)$ and $p_v(y_i)$, respectively, then (7.21) can be implemented approximately as follows. Define

$$w \triangleq \sum_{x_i=0}^{u} p_u(x_i), \qquad \tilde{w}_k \triangleq \sum_{i=0}^{k} p_v(y_i), \qquad k = 0, \ldots, L-1 \tag{7.22}$$

Let w^* denote the value $\tilde{w}_n$ such that $\tilde{w}_n - w \geq 0$ for the smallest value of n. Then $v^* = y_n$ is the output corresponding to u. Figure 7.16 shows this algorithm.

Example 7.1

Given $x_i = y_i = 0, 1, 2, 3$, $p_u(x_i) = 0.25$, $i = 0, \ldots, 3$, $p_v(y_0) = 0$, $p_v(y_1) = p_v(y_2) = 0.5$, $p_v(y_3) = 0$. Find the transformation between u and v. The accompanying table shows how this mapping is developed.

u	$p_u(x_i)$	w	$\tilde{w}_k$	w^*	n	v^*	
0	0.25	0.25	0.00	0.50	1	1	
1	0.25	0.50	0.50	0.50	1	1	
2	0.25	0.75	1.00	1.00	2	2	$p_{v^*}(x_i) = ?$
3	0.25	1.00	1.00	1.00	2	2	

7.4 SPATIAL OPERATIONS

Many image enhancement techniques are based on spatial operations performed on local neighborhoods of input pixels. Often, the image is convolved with a finite impulse response filter called *spatial mask*.

Spatial Averaging and Spatial Low-pass Filtering

Here each pixel is replaced by a weighted average of its neighborhood pixels, that is,

$$v(m, n) = \sum\sum_{(k, l) \in W} a(k, l) y(m-k, n-l) \tag{7.23}$$

where $y(m, n)$ and $v(m, n)$ are the input and output images, respectively, W is a suitably chosen window, and $a(k, l)$ are the filter weights. A common class of spatial averaging filters has all equal weights, giving

$$v(m, n) = \frac{1}{N_w} \sum\sum_{(k, l) \in W} y(m-k, n-l) \tag{7.24}$$

where $a(k, l) = 1/N_w$ and N_w is the number of pixels in the window W. Another spatial averaging filter used often is given by

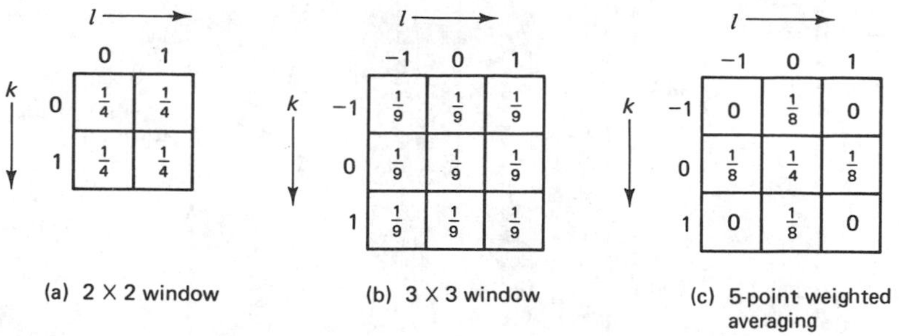

Figure 7.17 Spatial averaging masks $a(k, l)$.

$$v(m, n) = \tfrac{1}{2}[y(m, n) + \tfrac{1}{4}\{y(m-1, n) + y(m+1, n) + y(m, n-1) + y(m, n+1)\}] \quad (7.25)$$

that is, each pixel is replaced by its average with the average of its nearest four pixels. Figure 7.17 shows some spatial averaging masks.

Spatial averaging is used for noise smoothing, low-pass filtering, and subsampling of images. Suppose the observed image is given as

$$y(m, n) = u(m, n) + \eta(m, n) \quad (7.26)$$

where $\eta(m, n)$ is white noise with zero mean and variance σ_η^2. Then the spatial average of (7.24) yields

$$v(m, n) = \frac{1}{N_w} \sum\sum_{(k, l) \in W} u(m-k, n-l) + \bar{\eta}(m, n) \quad (7.27)$$

where $\bar{\eta}(m, n)$ is the spatial average of $\eta(m, n)$. It is a simple matter to show that $\bar{\eta}(m, n)$ has zero mean and variance $\bar{\sigma}_\eta^2 = \sigma_\eta^2 / N_w$, that is, the noise power is reduced by a factor equal to the number of pixels in the window W. If the noiseless image $u(m, n)$ is constant over the window W, then spatial averaging results in an improvement in the output signal-to-noise ratio by a factor of N_w. In practice the size of the window W is limited due to the fact that $u(m, n)$ is not really constant, so that spatial averaging introduces a distortion in the form of blurring. Figure 7.18 shows examples of spatial averaging of an image containing Gaussian noise.

Directional Smoothing

To protect the edges from blurring while smoothing, a directional averaging filter can be useful. Spatial averages $v(m, n : \theta)$ are calculated in several directions (see Fig. 7.19) as

$$v(m, n : \theta) = \frac{1}{N_\theta} \sum\sum_{(k, l) \in W_\theta} y(m-k, n-l) \quad (7.28)$$

and a direction θ^* is found such that $|y(m, n) - v(m, n : \theta^*)|$ is minimum. Then

$$v(m, n) = v(m, n : \theta^*) \quad (7.29)$$

gives the desired result. Figure 7.20 shows an example of this method.

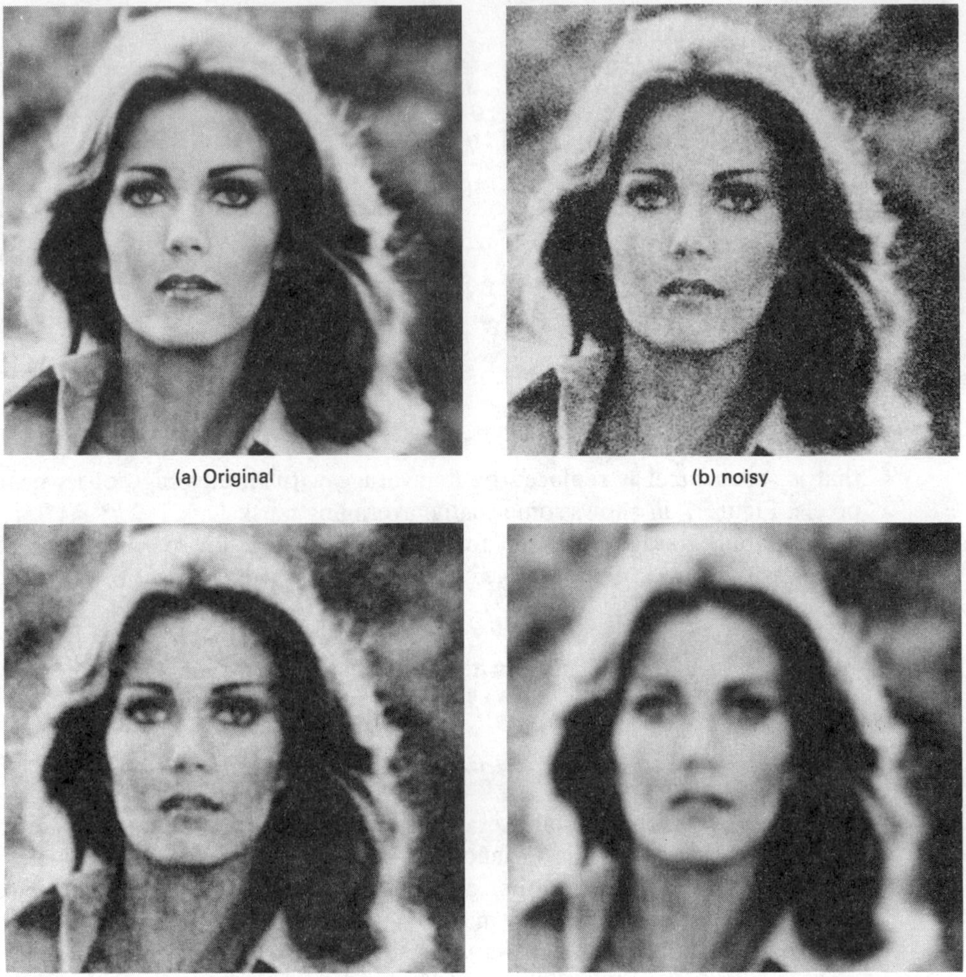

(a) Original (b) noisy

(c) 3 × 3 filter (d) 7 × 7 filter

Figure 7.18 Spatial averaging filters for smoothing images containing Gaussian noise.

Median Filtering

Here the input pixel is replaced by the median of the pixels contained in a window around the pixel, that is,

$$v(m, n) = \text{median}\{y(m-k, n-l), (k, l) \in W\} \tag{7.30}$$

where W is a suitably chosen window. The algorithm for median filtering requires arranging the pixel values in the window in increasing or decreasing order and picking the middle value. Generally the window size is chosen so that N_w is odd. If N_w is even, then the median is taken as the average of the two values in the middle.

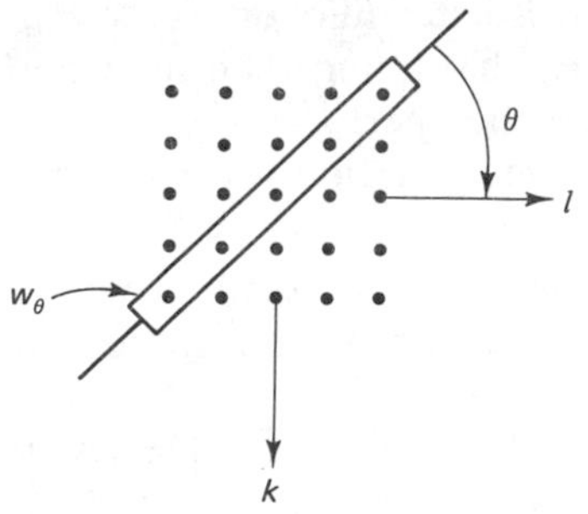

Figure 7.19 Directional smoothing filter.

Typical windows are 3×3, 5×5, 7×7, or the five-point window considered for spatial averaging in Fig. 7.17c.

Example 7.2

Let $\{y(m)\} = \{2, 3, 8, 4, 2\}$ and $W = [-1, 0, 1]$. The median filter output is given by

$$v(0) \overset{\Delta}{=} 2 \quad \text{(boundary value)}, \qquad v(1) = \text{median}\,\{2, 3, 8\} = 3$$

$$v(2) = \text{median}\,\{3, 8, 4\} = 4, \qquad v(3) = \text{median}\,\{8, 4, 2\} = 4$$

$$v(4) = 2 \quad \text{(boundary value)}$$

Hence $\{v(m)\} = \{2, 3, 4, 4, 2\}$. If W contains an even number of pixels—for example, $W = [-1, 0, 1, 2]$—then $v(0) = 2$, $v(1) = 3$, $v(2) = \text{median}\,\{2, 3, 8, 4\} = (3 + 4)/2 = 3.5$, and so on, gives $\{v(m)\} = \{2, 3, 3.5, 3.5, 2\}$.

The median filter has the following properties:

1. It is a nonlinear filter. Thus for two sequences $x(m)$ and $y(m)$

$$\text{median}\{x(m) + y(m)\} \neq \text{median}\{x(m)\} + \text{median}\{y(m)\}$$

(a) 5×5 spatial smoothing

(b) directional smoothing

Figure 7.20

2. It is useful for removing isolated lines or pixels while preserving spatial resolutions. Figure 7.21 shows that the median filter performs very well on images containing binary noise but performs poorly when the noise is Gaussian. Figure 7.22 compares median filtering with spatial averaging for images containing binary noise.
3. Its performance is poor when the number of noise pixels in the window is greater than or half the number of pixels in the window.

Since the median is the $(N_w + 1)/2$ largest value ($N_w = $ odd), its search requires $(N_w - 1) + (N_w - 2) + \cdots + (N_w - 1)/2 = 3(N_w^2 - 1)/8$ comparisons. This number equals 30 for 3×3 windows and 224 for 5×5 windows. Using a more efficient

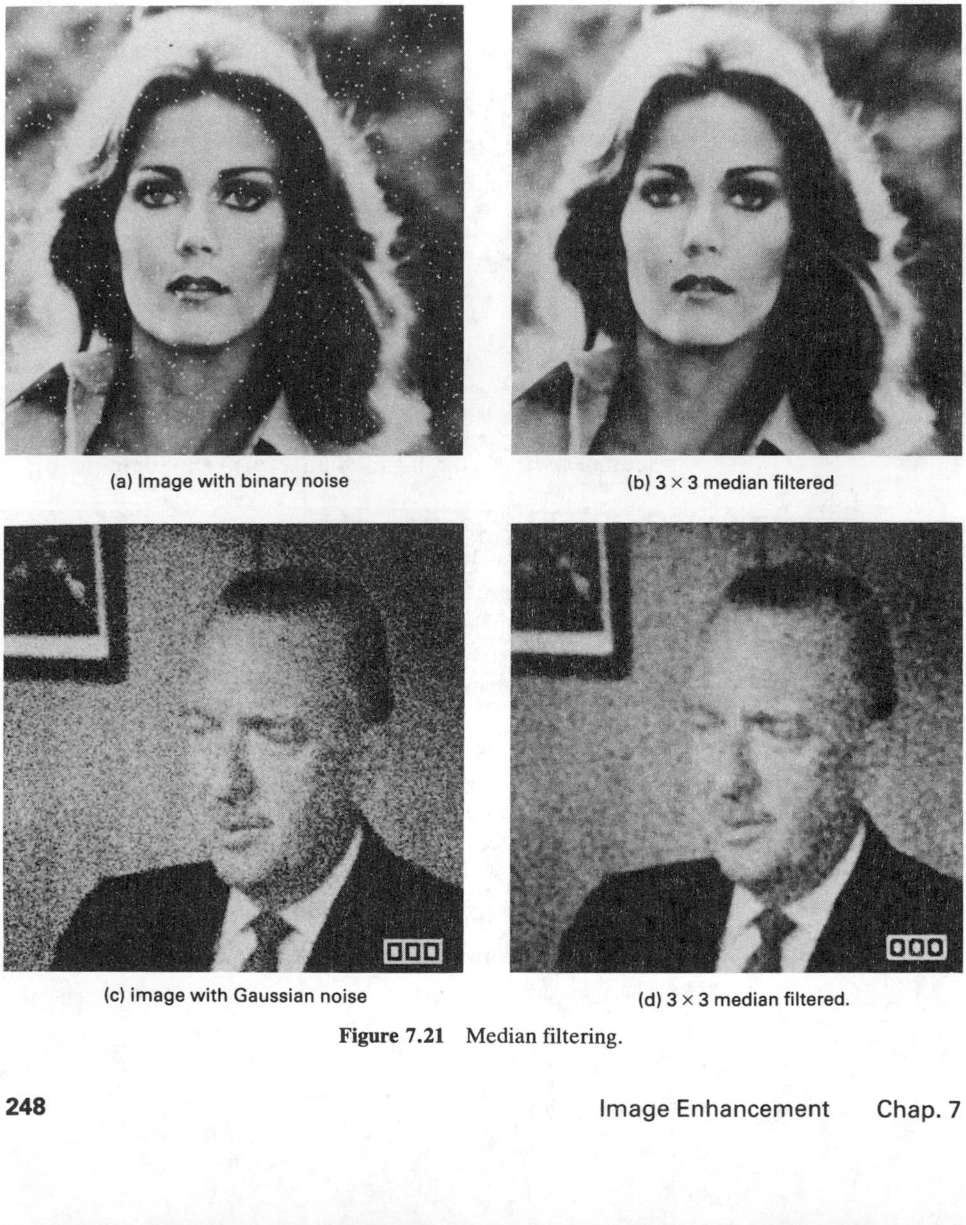

Figure 7.21 Median filtering.

Figure 7.22 Spatial averaging versus median filtering.

a	*b*
c	*d*

(a) Original
(b) with binary noise
(c) five nearest neighbors spatial average
(d) 3×3 median filtered

search technique, the number of comparisons can be reduced to approximately $\frac{1}{2} N_w \log_2 N_w$ [5]. For moving window medians, the operation count can be reduced further. For example, if k pixels are deleted and k new pixels are added to a window, then the new median can be found in no more than $k(N_w + 1)$ comparisons. A practical two-dimensional median filter is the *separable median filter,* which is obtained by successive one-dimensional median filtering of rows and columns.

Other Smoothing Techniques

An alternative to median filtering for removing binary or isolated noise is to find the spatial average according to (7.24) and replace the pixel at m, n by this average whenever the noise is large, that is, the quantity $|v(m, n) - y(m, n)|$ is greater than some prescribed threshold. For additive Gaussian noise, more sophisticated smoothing algorithms are possible. These algorithms utilize the statistical properties of the image and the noise fields. Adaptive algorithms that adjust the filter response according to local variations in the statistical properties of the data are also possible. In many cases the noise is multiplicative. Noise-smoothing algorithms for such images can also be designed. These and other algorithms are considered in Chapter 8.

Unsharp Masking and Crispening

The unsharp masking technique is used commonly in the printing industry for crispening the edges. A signal proportional to the unsharp, or low-pass filtered, version of the image is subtracted from the image. This is equivalent to adding the gradient, or a high-pass signal, to the image (see Fig. 7.23). In general the unsharp masking operation can be represented by

$$v(m, n) = u(m, n) + \lambda g(m, n) \tag{7.31}$$

where $\lambda > 0$ and $g(m, n)$ is a suitably defined gradient at (m, n). A commonly used gradient function is the discrete Laplacian

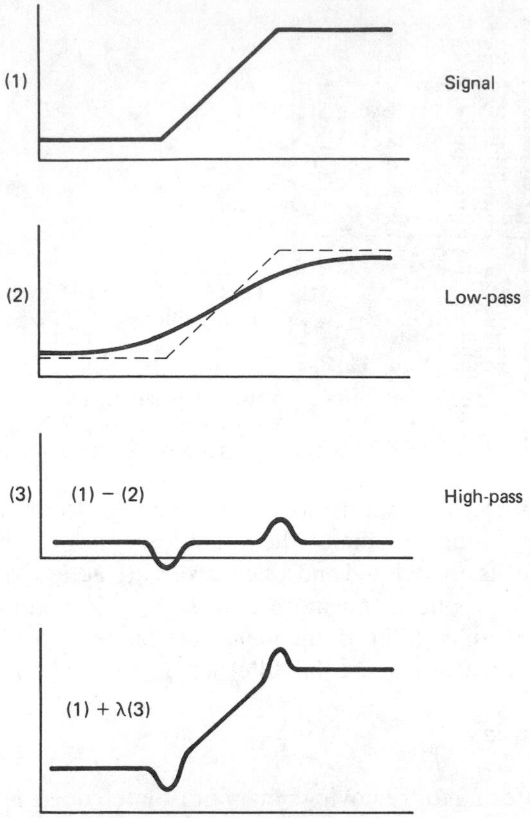

Figure 7.23 Unsharp masking operations.

$$g(m, n) \triangleq u(m, n) - \tfrac{1}{4}[u(m-1, n) + u(m, n-1) + u(m+1, n) + u(m, n-1)] \tag{7.32}$$

Figure 7.24 shows an example of unsharp masking using the Laplacian operator.

Spatial Low-pass, High-pass, and Band-pass Filtering

Earlier it was mentioned that a spatial averaging operation is a low-pass filter (Fig. 7.25a). If $h_{LP}(m, n)$ denotes a FIR low-pass filter, then a FIR high-pass filter, $h_{HP}(m, n)$, can be defined as

$$h_{HP}(m, n) = \delta(m, n) - h_{LP}(m, n) \tag{7.33}$$

Such a filter can be implemented by simply subtracting the low-pass filter output from its input (Fig. 7.25b). Typically, the low-pass filter would perform a relatively long-term spatial average (for example, on a 5×5, 7×7, or larger window).

A spatial band-pass filter can be characterized as (Fig. 7.25c)

$$h_{BP}(m, n) = h_{L_1}(m, n) - h_{L_2}(m, n) \tag{7.34}$$

where $h_{L_1}(m, n)$ and $h_{L_2}(m, n)$ denote the FIRs of low-pass filters. Typically, h_{L_1} and h_{L_2} would represent short-term and long-term averages, respectively.

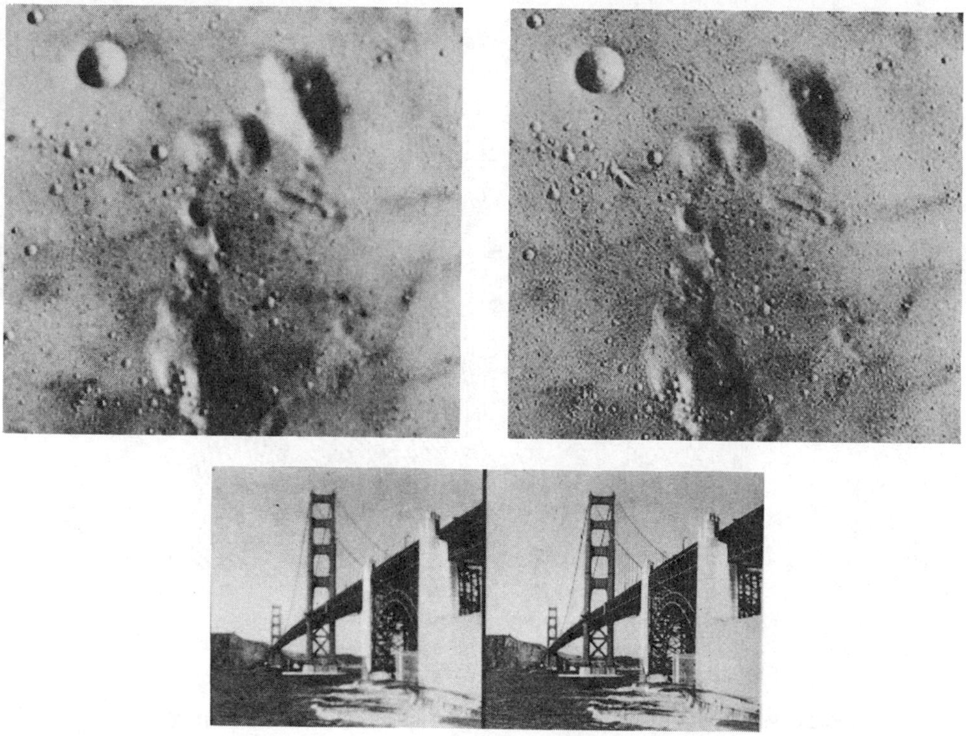

Figure 7.24 Unsharp masking. Original (left), enhanced (right).

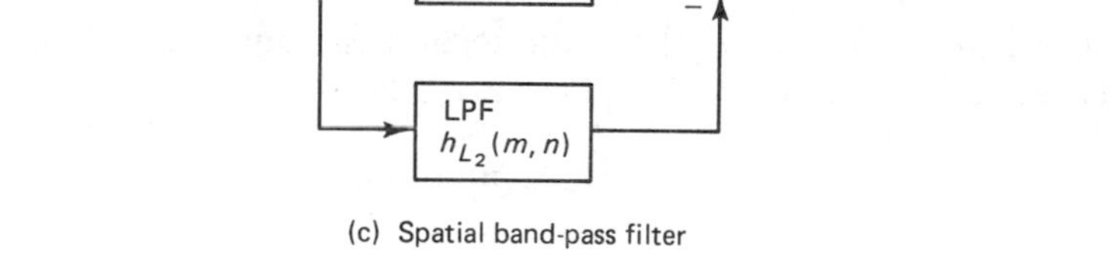

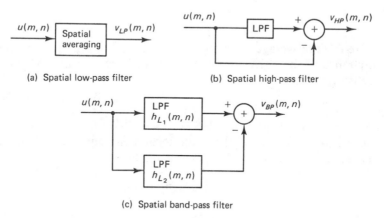

Figure 7.25 Spatial filters.

Low-pass filters are useful for noise smoothing and interpolation. High-pass filters are useful in extracting edges and in sharpening images. Band-pass filters are useful in the enhancement of edges and other high-pass image characteristics in the presence of noise. Figure 7.26 shows examples of high-pass, low-pass and band-pass filters.

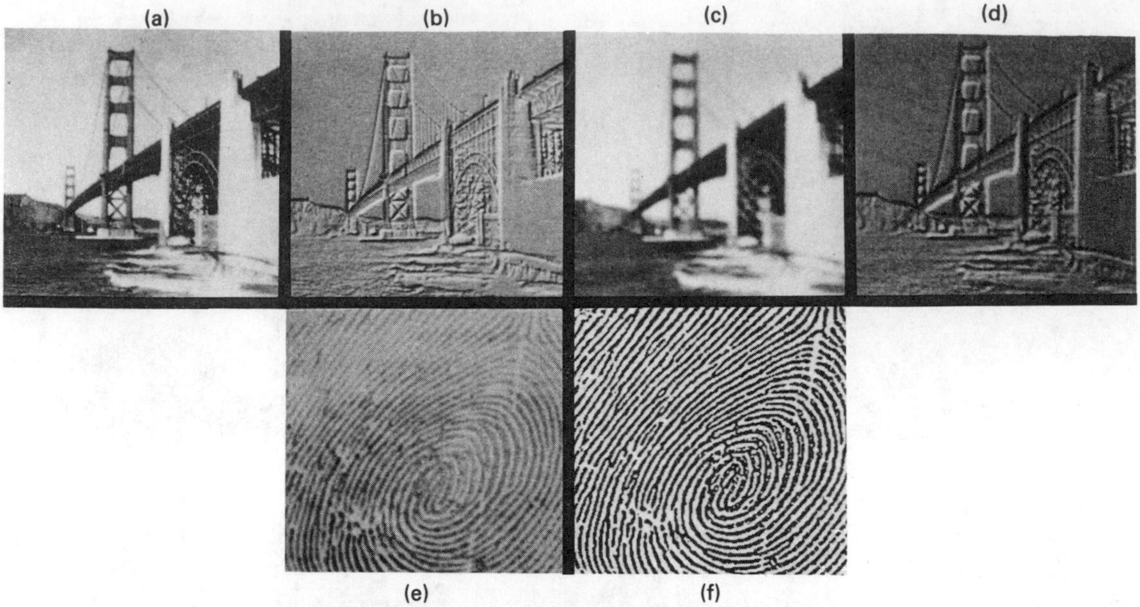

Figure 7.26 Spatial filtering examples.
Top row: original, high-pass, low-pass and band-pass filtered images.
Bottom row: original and high-pass filtered images.

Inverse Contrast Ratio Mapping and Statistical Scaling

The ability of our visual system to detect an object in a uniform background depends on its size (resolution) and the contrast ratio, γ, which is defined as

$$\gamma = \frac{\sigma}{\mu} \tag{7.35}$$

where μ is the average luminance of the object and σ is the standard deviation of the luminance of the object plus its surround. Now consider the *inverse contrast ratio* transformation

$$v(m, n) = \frac{\mu(m, n)}{\sigma(m, n)} \tag{7.36}$$

where $\mu(m, n)$ and $\sigma(m, n)$ are the local mean and standard deviation of $u(m, n)$ measured over a window W and are given by

$$\mu(m, n) = \frac{1}{N_w} \sum\sum_{(k, l) \in W} u(m-k, n-l) \tag{7.37a}$$

$$\sigma(m, n) = \left\{ \frac{1}{N_w} \sum\sum_{(k, l) \in W} [u(m-k, n-l) - \mu(m, n)]^2 \right\}^{1/2} \tag{7.37b}$$

This transformation generates an image where the weak (that is, low-contrast) edges are enhanced. A special case of this is the transformation

$$v(m, n) = \frac{u(m, n)}{\sigma(m, n)} \tag{7.38}$$

Figure 7.27 Inverse contrast ratio mapping of images. Faint edges in the originals have been enhanced. For example, note the bricks on the patio and suspension cables on the bridge.

which scales each pixel by its standard deviation to generate an image whose pixels have unity variance. This mapping is also called *statistical scaling* [13]. Figure 7.27 shows examples of inverse contrast ratio mapping.

Magnification and Interpolation (Zooming)

Often it is desired to zoom on a given region of an image. This requires taking an image and displaying it as a larger image.

Replication. Replication is a zero-order hold where each pixel along a scan line is repeated once and then each scan line is repeated. This is equivalent to taking an $M \times N$ image and interlacing it by rows and columns of zeros to obtain a $2M \times 2N$ matrix and convolving the result with an array **H**, defined as

$$\mathbf{H} = \begin{bmatrix} 1 & 1 \\ 1 & 1 \end{bmatrix} \tag{7.39}$$

This gives

$$v(m, n) = u(k, l), \qquad k \triangleq \text{Int}\left[\frac{m}{2}\right], \qquad l = \text{Int}\left[\frac{n}{2}\right], \qquad m, n = 0, 1, 2, \ldots \tag{7.40}$$

Figure 7.28 shows examples of interpolation by replication.

Linear Interpolation. Linear interpolation is a first order hold where a straight line is first fitted in between pixels along a row. Then pixels along each column are interpolated along a straight line. For example, for a 2×2 magnification, linear interpolation along rows gives

$$\left.\begin{aligned} v_1(m, 2n) &= u(m, n), \\ &\qquad 0 \le m \le M-1, 0 \le n \le N-1 \\ v_1(m, 2n+1) &= \tfrac{1}{2}[u(m, n) + u(m, n+1)], \\ &\qquad 0 \le m \le M-1, 0 \le n \le N-1 \end{aligned}\right\} \tag{7.41}$$

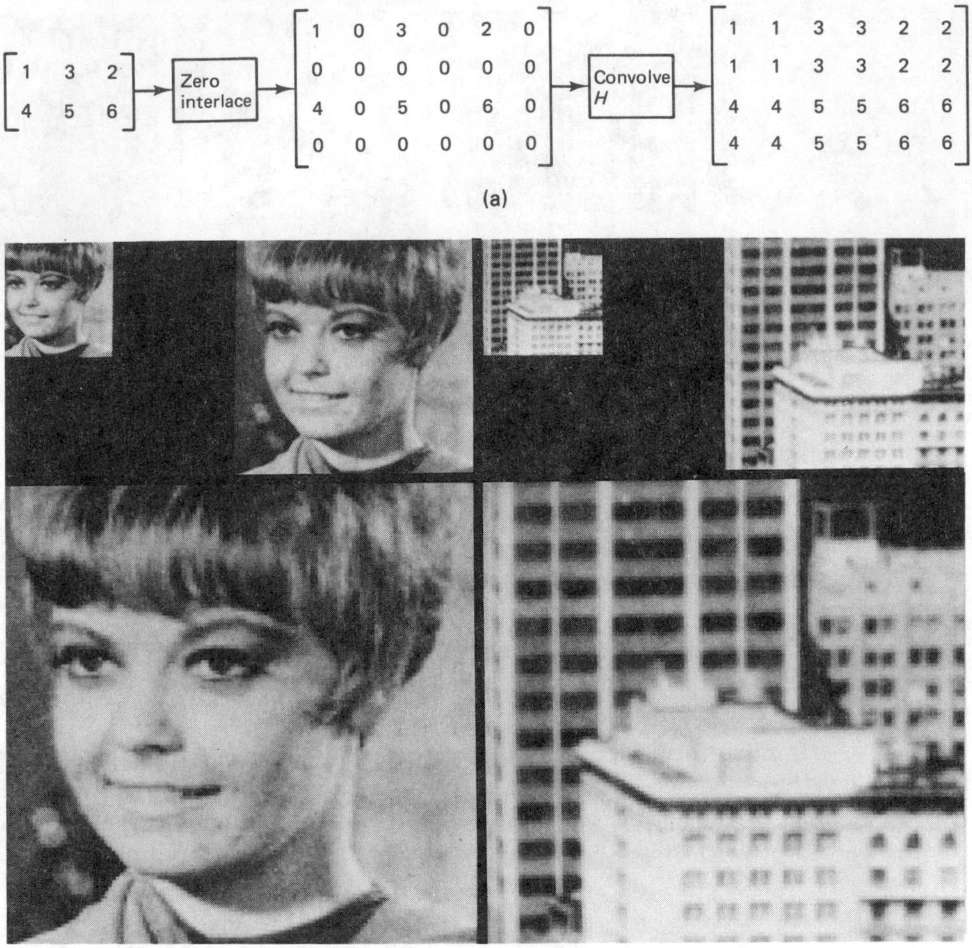

Figure 7.28 Zooming by replication from 128 × 128 to 256 × 256 and 512 × 512 images.

Linear interpolation of the preceding along columns gives the first result as

$$\left.\begin{aligned} v(2m, n) &= v_1(m, n) \\ v(2m+1, n) &= \tfrac{1}{2}[v_1(m, n) + v_1(m+1, n)], \\ &\qquad 0 \le m \le M-1, 0 \le N \le 2N-1 \end{aligned}\right\} \tag{7.42}$$

Here it is assumed that the input image is zero outside $[0, M-1] \times [0, N-1]$. The above result can also be obtained by convolving the $2M \times 2N$ zero interlaced image with the array

$$\mathbf{H} = \begin{bmatrix} \frac{1}{4} & \frac{1}{2} & \frac{1}{4} \\ \frac{1}{2} & \boxed{1} & \frac{1}{2} \\ \frac{1}{4} & \frac{1}{2} & \frac{1}{4} \end{bmatrix} \tag{7.43}$$

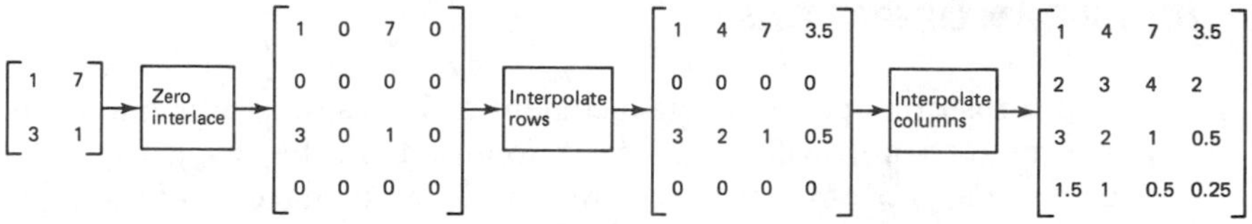

1 7
3 1
Zero interlace
1 0 7 0
0 0 0 0
3 0 1 0
0 0 0 0
Interpolate rows
1 4 7 3.5
0 0 0 0
3 2 1 0.5
0 0 0 0
Interpolate columns
1 4 7 3.5
2 3 4 2
3 2 1 0.5
1.5 1 0.5 0.25

(a)

(b)

Figure 7.29 Zooming by linear interpolation from 128×128 to 256×256 and 512×512 images.

whose origin ($m = 0, n = 0$) is at the center of the array, that is, the boxed element. Figure 7.29 contains examples of linear interpolation. In most of the image processing applications, linear interpolation performs quite satisfactorily. Higher-order (say, p) interpolation is possible by padding each row and each column of the input image by p rows and p columns of zeros, respectively, and convolving it p times with $\mathbf{H}$ (Fig. 7.30). For example $p = 3$ yields a cubic spline interpolation in between the pixels.

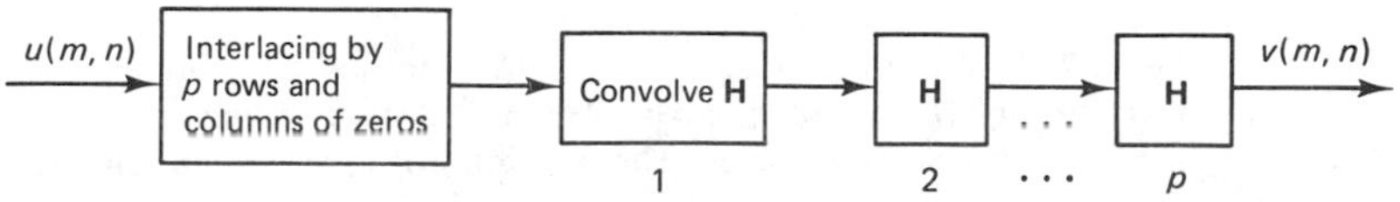

u(m, n)
Interlacing by p rows and columns of zeros
Convolve H
H
H
v(m, n)
1
2
p

Figure 7.30 pth order interpolation.

7.5 TRANSFORM OPERATIONS

In the transform operation enhancement techniques, zero-memory operations are performed on a transformed image followed by the inverse transformation, as shown in Fig. 7.31. We start with the transformed image $\mathbf{V} = \{v(k, l)\}$ as

$$\mathbf{V} = \mathbf{AUA}^T \tag{7.44}$$

where $\mathbf{U} = \{u(m, n)\}$ is the input image. Then the inverse transform of

$$v^{\cdot}(k, l) = f(v(k, l)) \tag{7.45}$$

gives the enhanced image as

$$\mathbf{U}^{\cdot} = \mathbf{A}^{-1}\mathbf{V}^{\cdot}[\mathbf{A}^T]^{-1} \tag{7.46}$$

Generalized Linear Filtering

In generalized linear filtering, the zero-memory transform domain operation is a pixel-by-pixel multiplication

$$v^{\cdot}(k, l) = g(k, l)v(k, l) \tag{7.47}$$

where $g(k, l)$ is called a *zonal mask*. Figure 7.32 shows zonal masks for low-pass, band-pass and high-pass filters for the DFT and other orthogonal transforms.

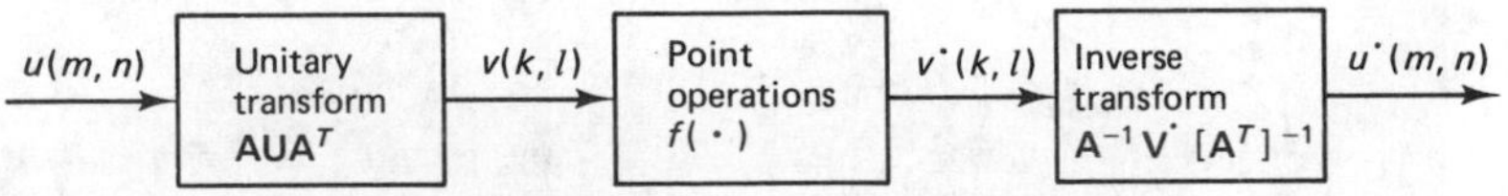

Figure 7.31 Image enhancement by transform filtering.

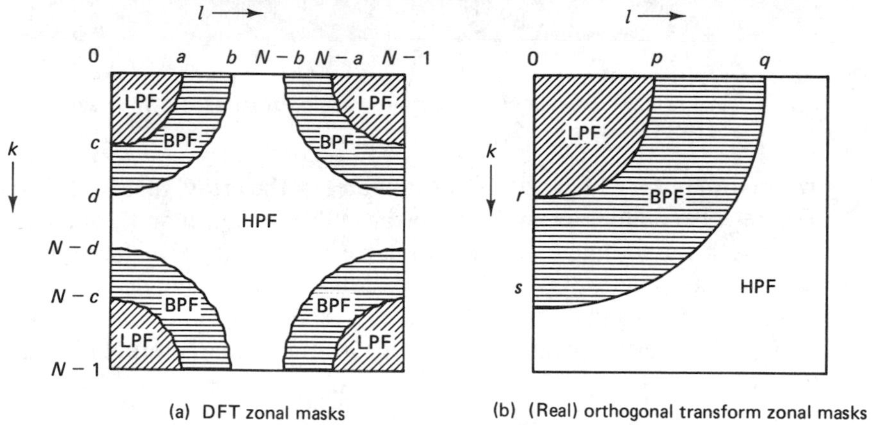

Figure 7.32 Examples of zonal masks $g(k, l)$ for low-pass filtering (LPF), band-pass filtering (BPF), and high-pass (HPF) filtering in (complex) DFT and (real) orthogonal transform domains. The function $g(k, l)$ is zero outside the region of support shown for the particular filter.

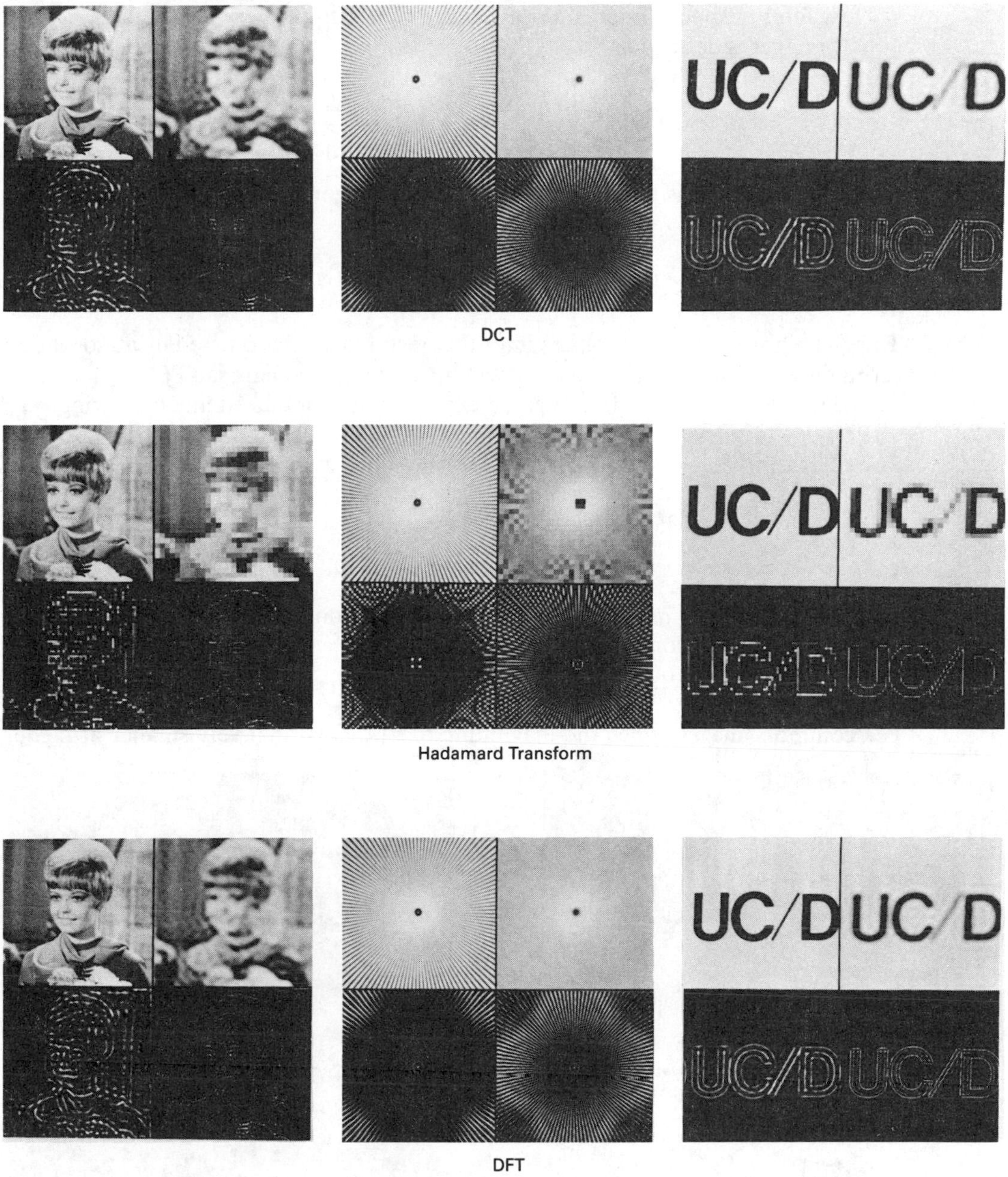

Figure 7.33 Generalized linear filtering using DCT, and Hadamard transform and DFT. In each case *a b* / *c d* (a) original; (b) low-pass filtered; (c) band-pass filtered; (d) high-pass filtered.

A filter of special interest is the *inverse Gaussian* filter, whose zonal mask for $N \times N$ images is defined as

$$g(k, l) = \begin{cases} \exp\left\{\dfrac{(k^2 + l^2)}{2\sigma^2}\right\}, & 0 \le k,\ l \le \dfrac{N}{2} \\ g(N-k,\ N-l), & \text{otherwise} \end{cases} \tag{7.48}$$

when **A** in (7.44) is the DFT. For other orthogonal transforms discussed in Chapter 5,

$$g(k, l) = \exp\frac{(k^2 + l^2)}{2\sigma^2}, \qquad 0 \le k, l \le N-1 \tag{7.49}$$

This is a high-frequency emphasis filter that restores images blurred by atmospheric turbulence or other phenomena that can be modeled by Gaussian PSFs.

Figures 7.33 and 7.34 show some examples of generalized linear filtering.

Root Filtering

The transform coefficients $v(k, l)$ can be written as

$$v(k, l) = |v(k, l)|e^{j\theta(k, l)} \tag{7.50}$$

In root filtering, the α-root of the magnitude component of $v(k, l)$ is taken, while retaining the phase component, to yield

$$v^{\cdot}(k, l) = |v(k, l)|^{\alpha} e^{j\theta(k, l)}, \qquad 0 \le \alpha \le 1 \tag{7.51}$$

For common images, since the magnitude of $v(k, l)$ is relatively smaller at higher

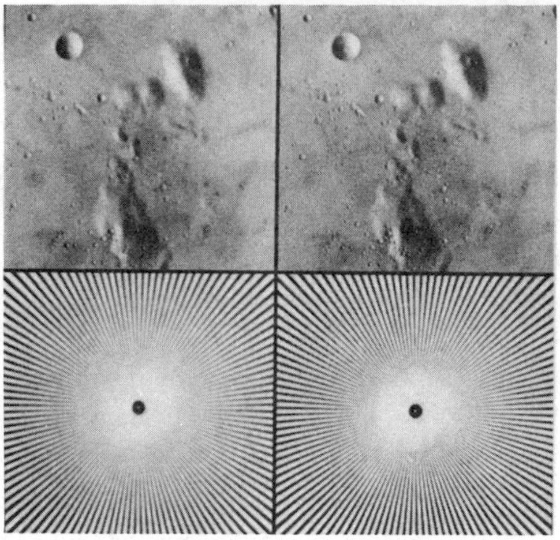

Originals Enhanced

Inverse Gaussian filtering

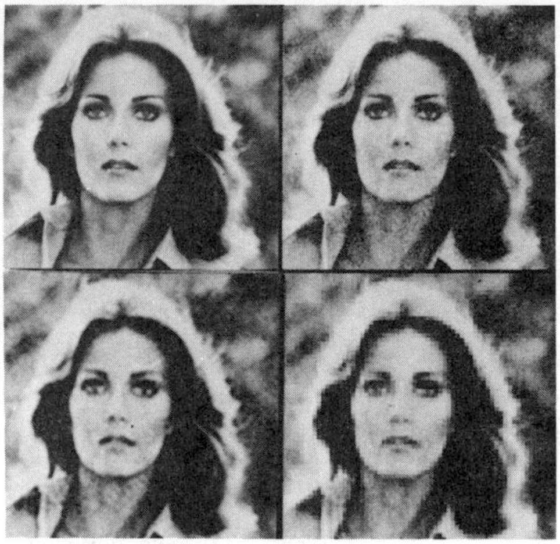

Lowpass filtering for noise smoothing

a	b
c	d

(a) Original
(b) noisy
(c) DCT filter
(d) Hadamard transform filter

Figure 7.34 Examples of transform based linear filtering.

(a) Original; (b) $\alpha = 0$ (phase only); (c) $\alpha = 0.5$; (d) $\alpha = 0.7$.

Figure 7.35 Root filtering

spatial frequencies, the effect of α-rooting is to enhance higher spatial frequencies (low amplitudes) relative to lower spatial frequencies (high amplitudes). Figure 7.35 shows the effect of these filters. (Also see Fig. 8.19.)

Generalized Cepstrum and Homomorphic Filtering

If the magnitude term in (7.51) is replaced by the logarithm of $|v(k, l)|$ and we define

$$s(k, l) \triangleq [\log |v(k, l)|] e^{j\theta(k, l)}, \qquad |v(k, l)| > 0 \tag{7.52}$$

then the inverse transform of $s(k, l)$, denoted by $c(m, n)$, is called the *generalized cepstrum* of the image (Fig. 7.36). In practice a positive constant is added to $|v(k, l)|$ to prevent the logarithm from going to negative infinity. The image $c(m, n)$ is also

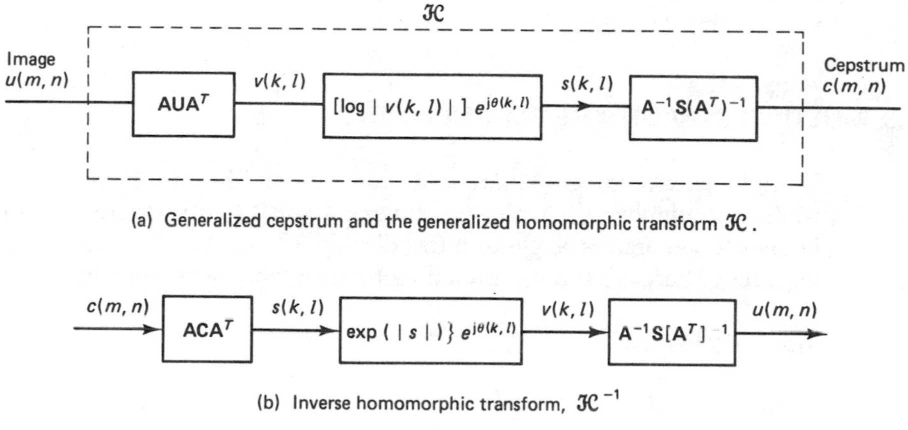

(a) Generalized cepstrum and the generalized homomorphic transform $\mathcal{H}$.

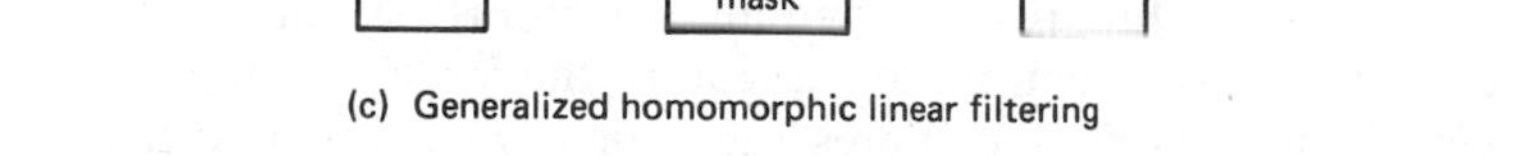

(b) Inverse homomorphic transform, $\mathcal{H}^{-1}$

(c) Generalized homomorphic linear filtering

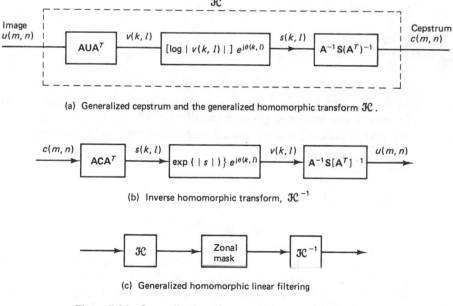

Figure 7.36 Generalized cepstrum and homomorphic filtering.

cepstrum of the building image

generalized cepstra
(a) original
(b) DFT
(c) DCT
(d) Hadamard transform

a	*b*
c	*d*

Figure 7.37 Generalized cepstra:

called the generalized homomorphic transform, $\mathcal{H}$, of the image $u(m, n)$. The generalized homomorphic linear filter performs zero-memory operations on the $\mathcal{H}$-transform of the image followed by inverse $\mathcal{H}$-transform, as shown in Fig. 7.36. Examples of cepstra are given in Fig. 7.37. The homomorphic transformation reduces the dynamic range of the image in the transform domain and increases it in the cepstral domain.

7.6 MULTISPECTRAL IMAGE ENHANCEMENT

In multispectral imaging there is a sequence of I images $U_i(m, n), i = 1, 2, \ldots, I$, where the number I is typically between 2 and 12. It is desired to combine these images to generate a single or a few display images that are representative of their features. There are three common methods of enhancing such images.

Intensity Ratios

Define the ratios

$$R_{i,j}(m, n) \triangleq \frac{u_i(m, n)}{u_j(m, n)}, \qquad i \neq j \tag{7.53}$$

where $u_i(m, n)$ represents the intensity and is assumed to be positive. This method gives $I^2 - I$ combinations for the ratios, the few most suitable of which are chosen by visual inspection. Sometimes the ratios are defined with respect to the average image $(1/I) \sum_{i=1}^{I} u_i(m, n)$ to reduce the number of combinations.

Log-Ratios

Taking the logarithm on both sides of (7.53), we get

$$L_{i,j} \triangleq \log R_{i,j} = \log u_i(m, n) - \log u_j(m, n) \tag{7.54}$$

The log-ratio $L_{i,j}$ gives a better display when the dynamic range of $R_{i,j}$ is very large, which can occur if the spectral features at a spatial location are quite different.

Principal Components

For each (m, n) define the $I \times 1$ vector

$$\mathbf{u}(m, n) = \begin{bmatrix} u_1(m, n) \\ u_2(m, n) \\ \vdots \\ u_I(m, n) \end{bmatrix} \tag{7.55}$$

The $I \times I$ KL transform of $\mathbf{u}(m, n)$, denoted by $\boldsymbol{\Phi}$, is determined from the autocorrelation matrix of the ensemble of vectors $\{\mathbf{u}_i(m, n), i = 1, \ldots, I\}$. The rows of $\boldsymbol{\Phi}$, which are eigenvectors of the autocorrelation matrix, are arranged in decreasing order of their associated eigenvalues. Then for any $I_0 \le I$, the images $v_i(m, n)$, $i = 1, \ldots, I_0$ obtained from the KL transformed vector

$$\mathbf{v}(m, n) = \boldsymbol{\Phi}\mathbf{u}(m, n) \tag{7.56}$$

are the first I_0 principal components of the multispectral images.

Figure 7.38 contains examples of multispectral image enhancement.

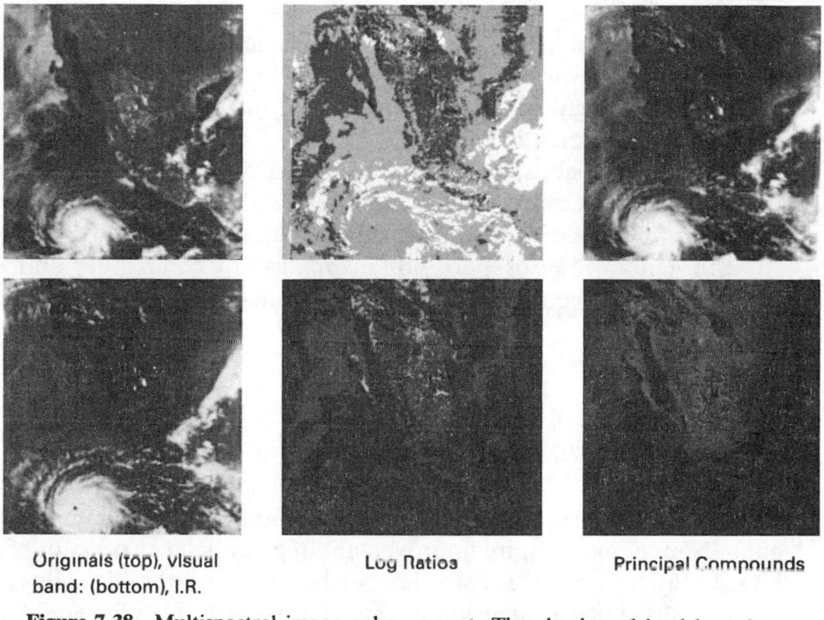

Originals (top), visual band: (bottom), I.R. Log Ratios Principal Compounds

Figure 7.38 Multispectral image enhancement. The clouds and land have been separated.

7.7 FALSE COLOR AND PSEUDOCOLOR

Since we can distinguish many more colors than gray levels, the perceptual dynamic range of a display can be effectively increased by coding complex information in color. False color implies mapping a color image into another color image to provide a more striking color contrast (which may not be natural) to attract the attention of the viewer.

Pseudocolor refers to mapping a set of images $u_i(m, n), i = 1, \dots, I$ into a color image. Usually the mapping is determined such that different features of the data set can be distinguished by different colors. Thus a large data set can be presented comprehensively to the viewer.

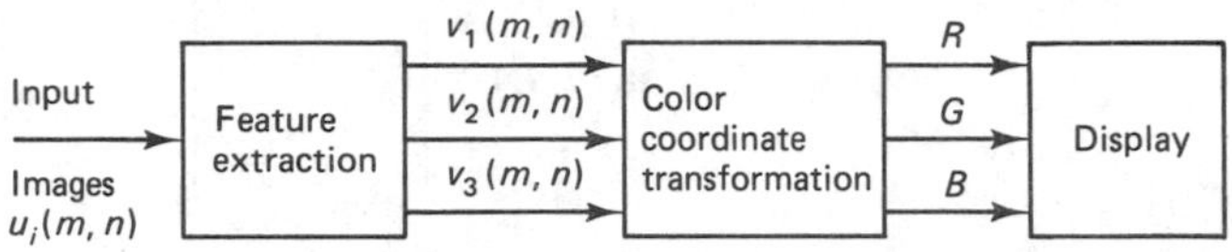

Figure 7.39 Pseudocolor image enhancement.

Figure 7.39 shows the general procedure for determining pseudocolor mappings. The given input images are mapped into three feature images, which are then mapped into the three color primaries. Suppose it is desired to pseudocolor a monochrome image. Then we have to map the gray levels onto a suitably chosen curve in the color space. One way is to keep the saturation (S^*) constant and map the gray level values into brightness (W^*) and the local spatial averages of gray levels into hue (θ^*).

Other methods are possible, including a pseudorandom mapping of gray levels into R, G, B coordinates, as is done in some image display systems. The comet Halley image shown on the cover page of this text is an example of pseudocolor image enhancement. Different densities of the comet surface (which is ice) were mapped into different colors.

For image data sets where the number of images is greater than or equal to three, the data set can be reduced to three ratios, three log-ratios, or three principal components, which are then mapped into suitable colors.

In general, the pseudocolor mappings are nonunique, and extensive interactive trials may be required to determine an acceptable mapping for displaying a given set of data.

7.8 COLOR IMAGE ENHANCEMENT

In addition to the requirements of monochrome image enhancement, color image enhancement may require improvement of color balance or color contrast in a color image. Enhancement of color images becomes a more difficult task not only because of the added dimension of the data but also due to the added complexity of color perception. A practical approach to developing color image enhancement

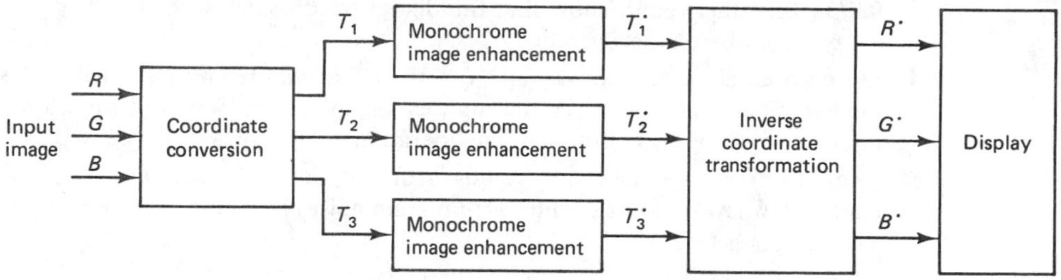

Figure 7.40 Color image enhancement.

algorithms is shown in Fig. 7.40. The input color coordinates of each pixel are independently transformed into another set of color coordinates, where the image in each coordinate is enhanced by its own (monochrome) image enhancement algorithm, which could be chosen suitably from the foregoing set of algorithms. The enhanced image coordinates $T_1^\cdot, T_2^\cdot, T_3^\cdot$ are inverse transformed to $R^\cdot, G^\cdot, B^\cdot$ for display. Since each image plane $T_k(m, n), k = 1, 2, 3$, is enhanced independently, care has to be taken so that the enhanced coordinates $T_k^\cdot$ are within the color gamut of the R-G-B system. The choice of color coordinate system $T_k, k = 1, 2, 3$, in which enhancement algorithms are implemented may be problem-dependent.

7.9 SUMMARY

In this chapter we have presented several image enhancement techniques accompanied by examples. Image enhancement techniques can be improved if the enhancement criterion can be stated precisely. Often such criteria are application-dependent, and the final enhancement algorithm can only be obtained by trial and error. Modern digital image display systems offer a variety of control function switches, which allow the user to enhance interactively an image for display.

PROBLEMS

7.1 (Enhancement of a low-contrast image) Take a 25¢ coin; scan and digitize it to obtain a 512×512 image.

- **a.** Enhance it by a suitable contrast stretching transformation and compare it with histogram equalization.
- **b.** Perform unsharp masking and spatial high-pass operations and contrast stretch the results. Compare their performance as edge enhancement operators.

7.2 Using (7.6) and (7.9), prove the formula (7.8) for extracting the nth bit of a pixel.

7.3 **a.** Show that the random variable v defined via (7.11) satisfies the condition $\text{Prob}[v \le \sigma] = \text{Prob}[u \le F^{-1}(\sigma)] = F(F^{-1}(\sigma)) = \sigma$, where $0 < \sigma < 1$. This means v is uniformly distributed over (0, 1).

b. On the other hand, show that the digital transformation of (7.12) gives $p_v(v_k) = p_u(x_k)$, where v_k is given by (7.14).

7.4 Develop an algorithm for (a) an $M \times M$ median filter, and (b) an $M \times 1$ separable median filter that minimizes the number of operations required for filtering $N \times N$ images, where $N \gg M$. Compare the operation counts for $M = 3, 5, 7$.

7.5* (Adaptive unsharp masking) A powerful method of sharpening images in the presence of low levels of noise (such as film grain noise) is via the following algorithm [15]. The high-pass filter

$$\frac{1}{16}\begin{bmatrix} -1 & -2 & -1 \\ -2 & 12 & -2 \\ -1 & -2 & -1 \end{bmatrix}$$

which can be used for unsharp masking, can be written as $\frac{1}{16}$ times the sum of the following eight directional masks (H_θ):

0 0 0	-1 0 0	0-2 0	0 0-1	0 0 0	0 0 0	0 0 0	0 0 0
-2 2 0	0 1 0	0 2 0	0 1 0	0 2-2	0 1 0	0 2 0	0 1 0
0 0 0	0 0 0	0 0 0	0 0 0	0 0 0	0 0-1	0-2 0	-1 0 0

The input image is filtered by each of these masks and the outputs that exceed a threshold are summed and mixed with the input image (see Fig. P7.5). Perform this algorithm on a scanned photograph and compare with nonadaptive unsharp masking.

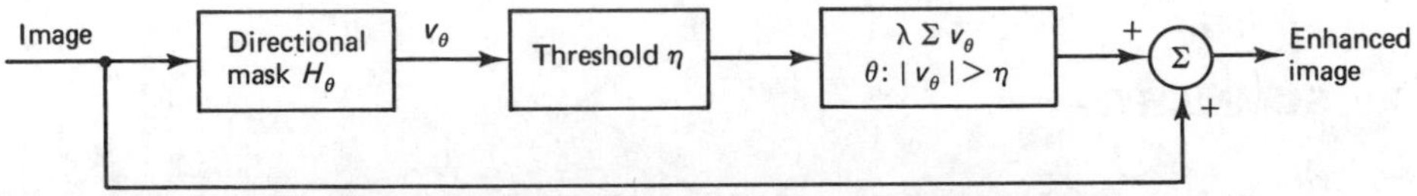

Figure P7.5 Adaptive unsharp masking.

7.6* (Filtering using phase) One of the limitations of the noise-smoothing linear filters is that their frequency response has zero phase. This means the phase distortions due to noise remain unaffected by these algorithms. To see the effect of phase, enhance a noisy image (with, for instance, SNR = 3 dB) by spatial averaging and transform processing such that the phase of the enhanced image is the same as that of the original image (see Fig. P7.6). Display $\hat{u}(m, n)$, $\tilde{u}(m, n)$ and $u(m, n)$ and compare results at different noise levels. If, instead of preserving $\theta(k, l)$, suppose we preserve 10% of the samples $\psi(m, n) \triangleq [\text{IDFT}\{\exp j\theta(k, l)\}]$ that have the largest magnitudes. Develop an algorithm for enhancing the noisy image now.

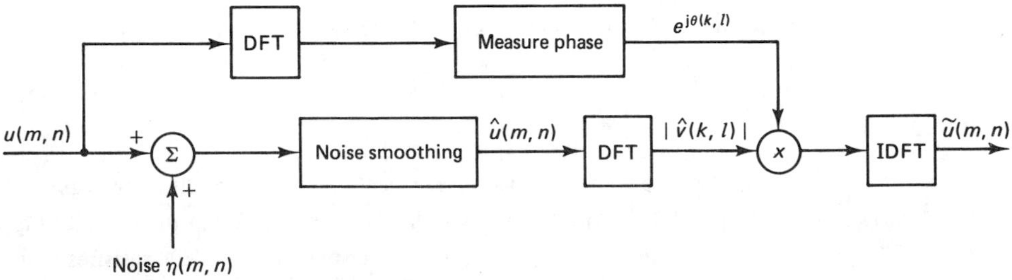

Figure P7.6 Filtering using phase.

7.7 Take a 512×512 image containing noise. Design low-pass, band-pass, and high-pass zonal masks in different transform domains such that their passbands contain equal energy. Map the three filtered images into R, G, B color components at each pixel and display the resultant pseudocolor image.

7.8 Scan and digitize a low-contrast image (such as fingerprints or a coin). Develop a technique based on contrast ratio mapping to bring out the faint edges in such images.

BIBLIOGRAPHY

Section 7.1

General references on image enhancement include the several books on digital image processing cited in Chapter 1. Other surveys are given in:

1. H. C. Andrews. *Computer Techniques in Image Processing.* New York: Academic Press, 1970.
2. H. C. Andrews, A. G. Tescher and R. P. Kruger. "Image Processing by Digital Computers." *IEEE Spectrum* 9 (1972): 20–32.
3. T. S. Huang. "Image Enhancement: A Review." *Opto-Electronics* 1 (1969): 49–59.
4. T. S. Huang, W. F. Schreiber and O. J. Tretiak. "Image Processing." *Proc. IEEE* 59 (1971): 1586–1609.
5. J. S. Lim. "Image Enhancement." In *Digital Image Processing Techniques* (M. P. Ekstrom, ed.) Chapter 1, pp. 1–51. New York: Academic Press, 1984.
6. T. S. Huang (ed.). *Two-Dimensional Digital Signal Processing* I and II. *Topics in Applied Physics,* vols. 42–43. Berlin: Springer Verlag, 1981.

Sections 7.2 and 7.3

For gray level and histogram modification techniques:

7. R. Nathan. "Picture Enhancement for the Moon, Mars, and Man" in *Pictorial Pattern Recognition* (G. C. Cheng, ed.). Washington, D.C.: Thompson, pp. 239–266, 1968.
8. F. Billingsley. "Applications of Digital Image Processing." *Appl. Opt.* 9 (February 1970): 289–299.
9. D. A. O'Handley and W. B. Green. "Recent Developments in Digital Image Processing at the Image Processing Laboratory at the Jet Propulsion Laboratory." *Proc. IEEE* 60 (1972): 821–828.
10. R. E. Woods and R. C. Gonzalez. "Real Time Digital Image Enhancement," *Proc. IEEE* 69 (1981): 643–654.

Section 7.4

Further results on median filtering and other spatial neighborhood processing techniques can be found in:

11. J. W. Tukey. *Exploratory Data Analysis.* Reading, Mass.: Addison Wesley, 1971.

12. T. S. Huang and G. Y. Tang. "A Fast Two-Dimensional Median Filtering Algorithm." *IEEE Trans. Accoust. Speech, Signal Process.* ASSP-27 (1979): 13–18.
13. R. H. Wallis. "An Approach for the Space Variant Restoration and Enhancement of Images." *Proc. Symp. Current Math. Problems in Image Sci.* (1976).
14. W. F. Schreiber. "Image Processing for Quality Improvement." *Proc. IEEE,* 66 (1978): 1640–1651.
15. P. G. Powell and B. E. Bayer. "A Method for the Digital Enhancement of Unsharp, Grainy Photographic Images." *Proc. Int. Conf. Electronic Image Proc.,* IEEE, U.K. (July 1982): 179–183.

Section 7.5

For cepstral and homomorphic filtering based approaches see [4, 5] and:

16. T. G. Stockham, Jr. "Image Processing in the Context of a Visual Model." *Proc. IEEE* 60 (1972): 828–842.

Sections 7.6–7.8

For multispectral and pseudocolor enhancement techniques:

17. L. W. Nichols and J. Lamar. "Conversion of Infrared Images to Visible in Color." *Appl. Opt.* 7 (September 1968): 1757.
18. E. R. Kreins and L. J. Allison. "Color Enhancement of Nimbus High Resolution Infrared Radiometer Data." *Appl. Opt.* 9 (March 1970): 681.
19. A. K. Jain, A. Nassir, and D. Nelson. "Multispectral Feature Display via Pseudo Coloring." Proc. 24th Annual SPSE Meeting, p. K-3.

8

Image Filtering and Restoration

8.1 INTRODUCTION

Any image acquired by optical, electro-optical or electronic means is likely to be degraded by the sensing environment. The degradations may be in the form of sensor noise, blur due to camera misfocus, relative object-camera motion, random atmospheric turbulence, and so on. Image restoration is concerned with filtering the observed image to minimize the effect of degradations (Fig. 8.1). The effectiveness of image restoration filters depends on the extent and the accuracy of the knowledge of the degradation process as well as on the filter design criterion. A frequently used criterion is the mean square error. Although, as a global measure of visual fidelity, its validity is questionable (see Chapter 3) it is a reasonable local measure and is mathematically tractable. Other criteria such as weighted mean square and maximum entropy are also used, although less frequently.

Image restoration differs from image enhancement in that the latter is concerned more with accentuation or extraction of image features rather than restoration of degradations. Image restoration problems can be quantified precisely, whereas enhancement criteria are difficult to represent mathematically. Consequently, restoration techniques often depend only on the class or ensemble

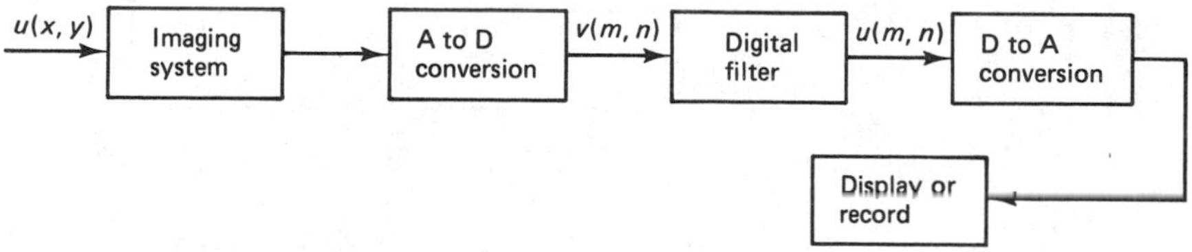

Figure 8.1 Digital image restoration system.

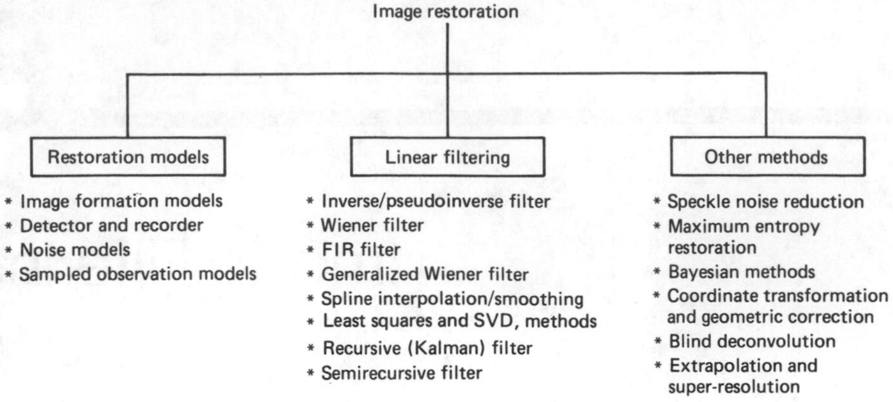

Figure 8.2 Image restoration.

properties of a data set, whereas image enhancement techniques are much more image dependent. Figure 8.2 summarizes several restoration techniques that are discussed in this chapter.

8.2 IMAGE OBSERVATION MODELS

A typical imaging system consists of an image formation system, a detector, and a recorder. For example, an electro-optical system such as the television camera contains an optical system that focuses an image on a photoelectric device, which is scanned for transmission or recording of the image. Similarly, an ordinary camera uses a lens to form an image that is detected and recorded on a photosensitive film. A general model for such systems (Fig. 8.3) can be expressed as

$$v(x, y) = g[w(x, y)] + \eta(x, y) \tag{8.1}$$

$$w(x, y) = \iint_{-\infty}^{\infty} h(x, y; x', y')u(x', y')\,dx'\,dy' \tag{8.2}$$

$$\eta(x, y) = f[g(w(x, y))]\eta_1(x, y) + \eta_2(x, y) \tag{8.3}$$

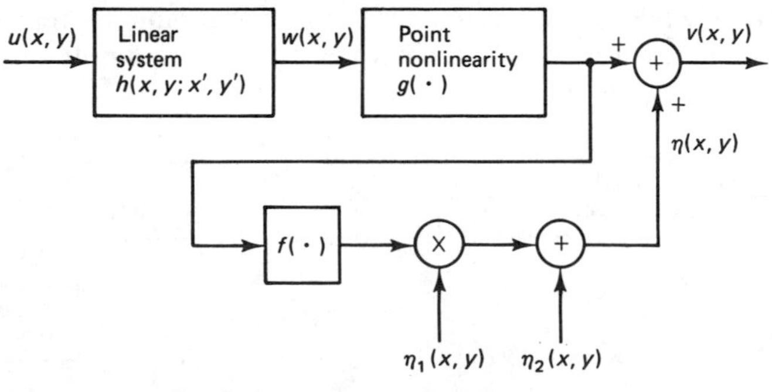

Figure 8.3 Image observation model.

where $u(x, y)$ represents the object (also called the *original image*), and $v(x, y)$ is the observed image. The image formation process can often be modeled by the linear system of (8.2), where $h(x, y; x', y')$ is its impulse response. For space invariant systems, we can write

$$h(x, y; x', y') = h(x - x', y - y'; 0, 0) \triangleq h(x - x', y - y') \tag{8.4}$$

The functions $f(\cdot)$ and $g(\cdot)$ are generally nonlinear and represent the characteristics of the detector/recording mechanisms. The term $\eta(x, y)$ represents the additive noise, which has an image-dependent random component $f[g(w)]\eta_1$ and an image-independent random component η_2.

Image Formation Models

Table 8.1 lists impulse response models for several spatially invariant systems. Diffraction-limited coherent systems have the effect of being ideal low-pass filters. For an incoherent system, this means band-limitedness and a frequency response obtained by convolving the coherent transfer function (CTF) with itself (Fig. 8.4). Degradations due to phase distortion in the CTF are called *aberrations* and manifest themselves as distortions in the pass-band of the incoherent optical transfer function (OTF). For example, a severely out-of-focus lens with rectangular aperture causes an aberration in the OTF, as shown in Fig. 8.4.

Motion blur occurs when there is relative motion between the object and the camera during exposure. Atmospheric turbulence is due to random variations in the refractive index of the medium between the object and the imaging system. Such degradations occur in the imaging of astronomical objects. Image blurring also occurs in image acquisition by scanners in which the image pixels are integrated over the scanning aperture. Examples of this can be found in image acquisition by radar, beam-forming arrays, and conventional image display systems using tele-

TABLE 8.1 Examples of Spatially Invariant Models

Type of system	Impulse response $h(x, y)$	Frequency response $H(\xi_1, \xi_2)$
Diffraction limited, coherent (with rectangular aperture)	$ab\ \text{sinc}(ax)\ \text{sinc}(by)$	$\text{rect}\left(\frac{\xi_1}{a}, \frac{\xi_2}{b}\right)$
Diffraction limited, incoherent (with rectangular aperture)	$\text{sinc}^2(ax)\ \text{sinc}^2(by)$	$\text{tri}\left(\frac{\xi_1}{a}, \frac{\xi_2}{b}\right)$
Horizontal motion	$\frac{1}{\alpha_0}\ \text{rect}\left(\frac{x}{\alpha_0} - \frac{1}{2}\right)\delta(y)$	$e^{-j\pi\xi_1\alpha_0}\ \text{sinc}(\xi_1\alpha_0)$
Atmospheric turbulence	$\exp\{-\pi\alpha^2(x^2 + y^2)\}$	$\frac{1}{\alpha^2}\exp\left[\frac{-\pi(\xi_1^2 + \xi_2^2)}{\alpha^2}\right]$
Rectangular scanning aperture	$\text{rect}\left(\frac{x}{\alpha}, \frac{y}{\beta}\right)$	$\alpha\beta\ \text{sinc}(\alpha\xi_1)\ \text{sinc}(\beta\xi_2)$
CCD interactions	$\sum\sum_{k,l=-1}^{1} \alpha_{k,l}\delta(x - k\Delta, y - l\Delta)$	$\sum\sum_{k,l=-1}^{1} \alpha_{k,l} e^{-j2\pi\Delta(\xi_1 k + \xi_2 l)}$

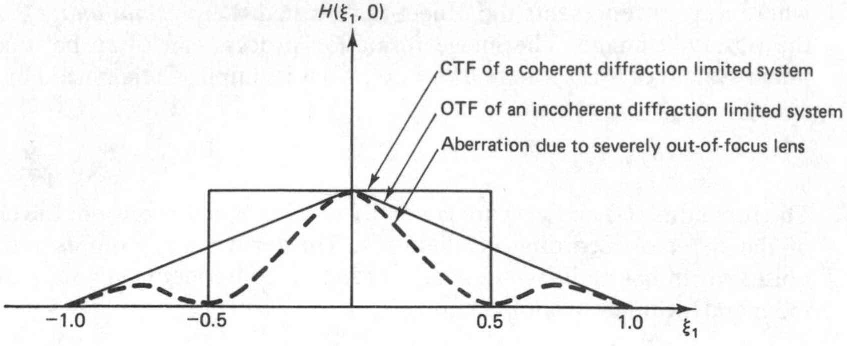

Figure 8.4 Degradations due to diffraction limitedness and lens aberration.

vision rasters. In the case of CCD arrays used for image acquisition, local interactions between adjacent array elements blur the image.

Figure 8.5 shows the PSFs of some of these degradation phenomena. For an ideal imaging system, the impulse response is the infinitesimally thin Dirac delta function having an infinite passband. Hence, the extent of blur introduced by a system can be judged by the shape and width of the PSF. Alternatively, the passband of the frequency response can be used to judge the extent of blur or the resolution of the system. Figures 2.2 and 2.5 show the PSFs and MTFs due to atmospheric turbulence and diffraction-limited systems. Figure 8.6 shows examples of blurred images.

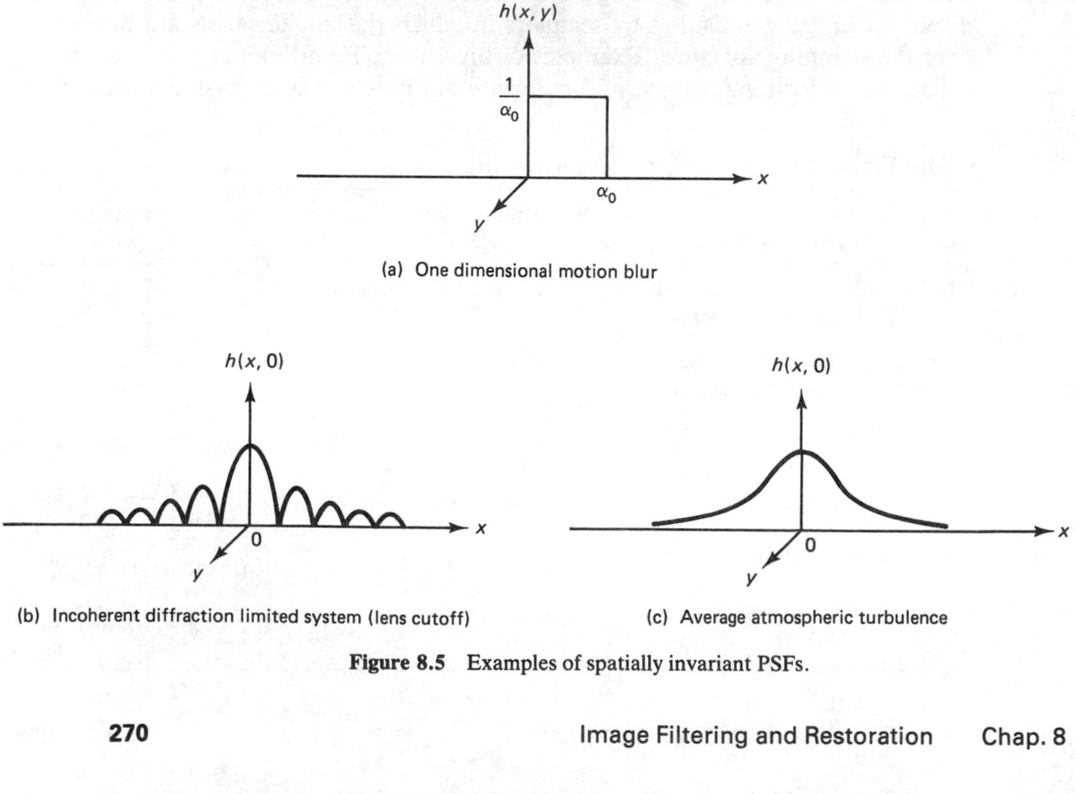

Figure 8.5 Examples of spatially invariant PSFs.

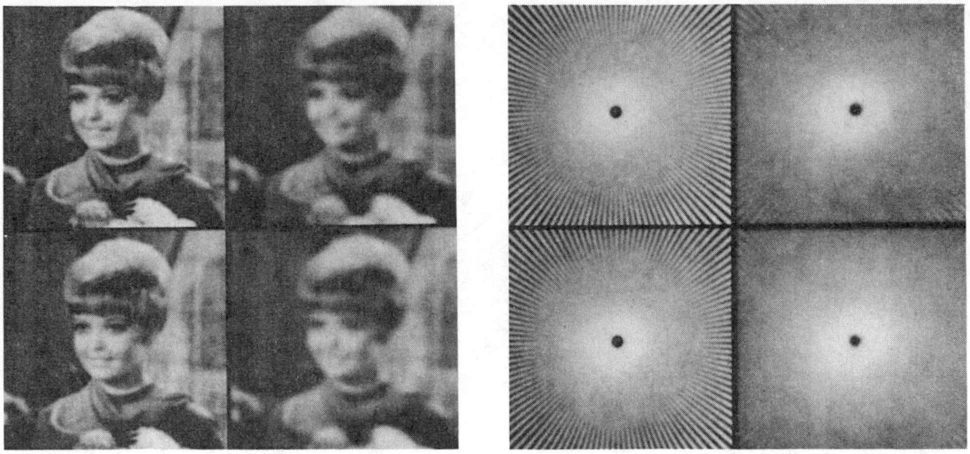

Figure 8.6 Examples of blurred images (panels: a b / c d). (a) Small exponential PSF blur; (b) large exponential PSF blur; (c) small sinc^2 PSF blur; (d) large sinc^2 PSF blur.

Example 8.1

An object $u(x, y)$ being imaged moves uniformly in the x-direction at a velocity υ. If the exposure time is T, the observed image can be written as

$$v(x, y) = \frac{1}{T}\int_0^T u(x - \upsilon t, y)\, dt = \frac{1}{\alpha_0}\int_0^{\alpha_0} u(x - \alpha, y)\, d\alpha$$

$$= \frac{1}{\alpha_0}\iint_{-\infty}^{\infty} \operatorname{rect}\left(\frac{\alpha}{\alpha_0} - \frac{1}{2}\right)\delta(\beta)u(x - \alpha, y - \beta)\, d\alpha\, d\beta$$

where $\alpha \triangleq \upsilon t$, $\alpha_0 \triangleq \upsilon T$. This shows the system is shift invariant and its PSF is given by

$$h(x, y) = \frac{1}{\alpha_0}\operatorname{rect}\left(\frac{x}{\alpha_0} - \frac{1}{2}\right)\delta(y)$$

(See Fig. 8.5a.)

Example 8.2 (A Spatially Varying Blur)

A forward looking radar (FLR) mounted on a platform at altitude h sends radio frequency (RF) pulses and scans around the vertical axis, resulting in a doughnut-shaped coverage of the ground (see Fig. 8.7a). At any scan position, the area illuminated can be considered to be bounded by an elliptical contour, which is the radar antenna half-power gain pattern. The received signal at any scan position is the sum of the returns from all points in a resolution cell, that is, the region illuminated at distance r during the pulse interval. Therefore,

$$v_p(r, \phi) = \int_{-\phi_0(r)}^{\phi_0(r)}\int_{l_1(r)}^{l_2(r)} u_p(r + s, \phi + \theta')s\, ds\, d\theta' \tag{8.5}$$

where $\phi_0(r)$ is the angular width of the elliptical contour from its major axis, $l_1(r)$ and $l_2(r)$ correspond to the inner and outer ground intercepts of radar pulse width around the point at radial distance r, and $u_p(r, \phi)$ and $v_p(r, \phi)$ are the functions $u(x, y)$ and $v(x, y)$, respectively, expressed in polar coordinates. Figure 8.7c shows the effect of scanning by a forward-looking radar. Note that the PSF associated with (8.5) is *not* shift invariant. (show!)

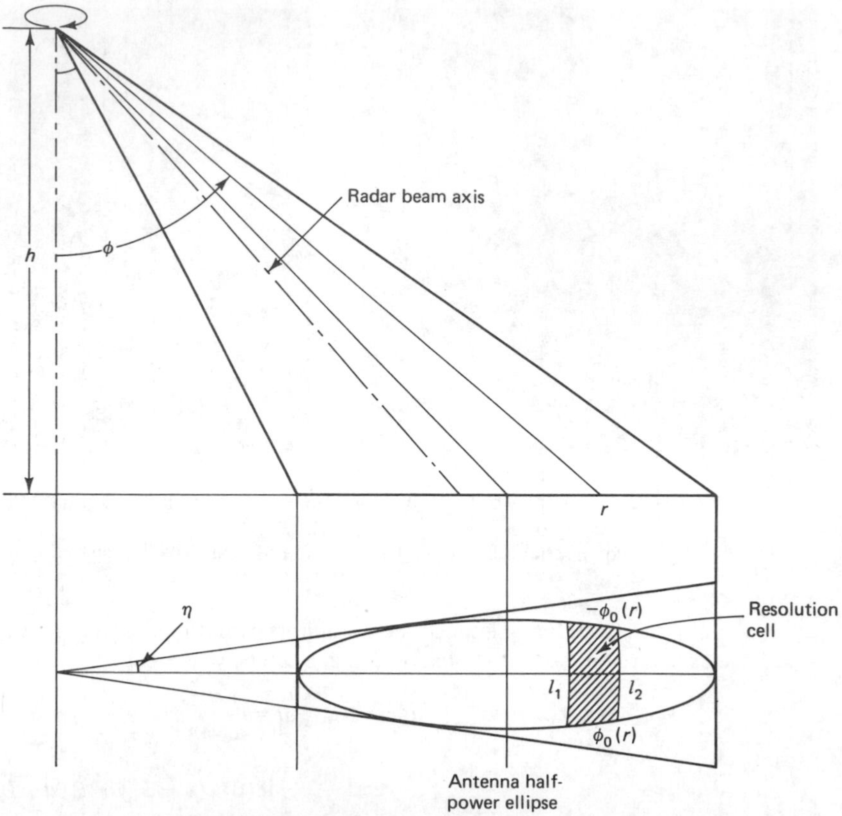

Cross section of radar beam

(a) scanning geometry;

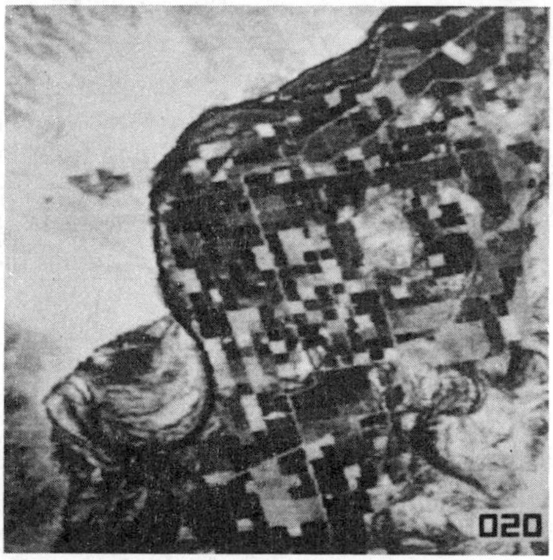

(b) object;

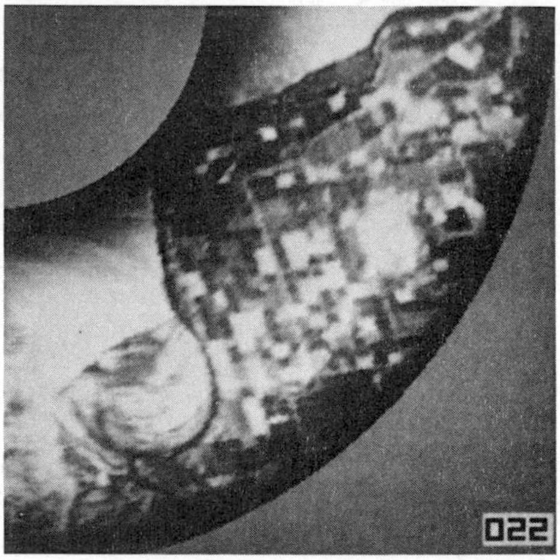

(c) FLR image (simulated).

Figure 8.7 FLR imaging.

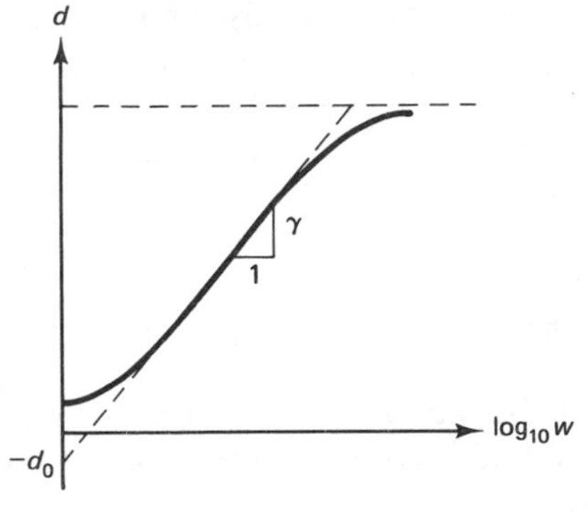

Figure 8.8 Typical response of a photographic film.

Detector and Recorder Models

The response of image detectors and recorders is generally nonlinear. For example, the response of photographic films, image scanners, and display devices can be written as

$$g = \alpha w^{\beta} \tag{8.6}$$

where α and β are device-dependent constants and w is the input variable. For photographic films, however, a more useful form of the model is (Fig. 8.8)

$$d = \gamma \log_{10} w - d_0 \tag{8.7}$$

where γ is called the *gamma of the film.* Here w represents the incident light intensity and d is called the optical density. A film is called *positive* if it has negative γ. For $\gamma = -1$, one obtains a linear model between w and the reflected or transmitted light intensity, which is proportional to $g \triangleq 10^{-d}$. For photoelectronic devices, w represents the incident light intensity, and the output g is the scanning beam current. The quantity β is generally positive and around 0.75.

Noise Models

The general noise model of (8.3) is applicable in many situations. For example, in photoelectronic systems the noise in the electron beam current is often modeled as

$$\eta(x, y) = \sqrt{g(x, y)}\, \eta_1(x, y) + \eta_2(x, y) \tag{8.8}$$

where g is obtained from (8.6) and η_1 and η_2 are zero-mean, mutually independent, Gaussian white noise fields. The signal-dependent term arises because the detection and recording processes involve random electron emission (or silver grain deposition in the case of films) having a Poisson distribution with a mean value of g. This distribution is approximated by the Gaussian distribution as a limiting case. Since the mean and variance of a Poisson distribution are equal, the signal-dependent term has a standard deviation $\sqrt{g}$ if it is assumed that η_1 has unity variance. The other term, η_2, represents wideband thermal noise, which can be modeled as Gaussian white noise.

In the case of films, there is no thermal noise and the noise model is

$$\eta(x, y) = \sqrt{g(x, y)}\, \eta_1(x, y) \tag{8.9}$$

where g now equals d, the optical density given by (8.7). A more-accurate model for film grain noise takes the form

$$\eta(x, y) = \varepsilon(g(x, y))^{\vartheta}\, \eta_1(x, y) \tag{8.10}$$

where ε is a normalization constant depending on the average film grain area and ϑ lies in the interval $\frac{1}{3}$ to $\frac{1}{2}$.

The presence of the signal-dependent term in the noise model makes restoration algorithms particularly difficult. Often, in the function $f[g(w)]$, w is replaced by its spatial average μ_w, giving

$$\eta(x, y) = f[g(\mu_w)]\eta_1(x, y) + \eta_2(x, y) \tag{8.11}$$

which makes $\eta(x, y)$ a Gaussian white noise random field. If the detector is operating in its linear region, we obtain, for photoelectronic devices, a linear observation model of the form

$$v(x, y) = w(x, y) + \sqrt{\mu_w}\, \eta_1(x, y) + \eta_2(x, y) \tag{8.12}$$

where we have set $\alpha = 1$ in (8.6) without loss of generality. For photographic films (with $\gamma = -1$), we obtain

$$v(x, y) = -\log_{10} w + a\eta_1(x, y) - d_0 \tag{8.13}$$

where a is obtained by absorbing the various quantities in the noise model of (8.10). The constant d_0 has the effect of scaling w by a constant and can be ignored, giving

$$v(x, y) = -\log_{10} w + a\eta_1(x, y) \tag{8.14}$$

where $v(x, y)$ represents the observed optical density. The light intensity associated with v is given by

$$\begin{aligned} i(x, y) &= 10^{-v(x, y)} \\ &= w(x, y)10^{-a\eta_1(x, y)} \\ &= w(x, y)n(x, y) \end{aligned} \tag{8.15}$$

where $n \triangleq 10^{-a\eta_1}$ now appears as multiplicative noise having a log-normal distribution.

A different type of noise that occurs in the coherent imaging of objects is called *speckle noise*. For low-resolution objects, it is multiplicative and occurs whenever the surface roughness of the object being imaged is of the order of the wavelength of the incident radiation. It is modeled as

$$v(x, y) = u(x, y)s(x, y) + \eta(x, y) \tag{8.16}$$

where $s(x, y)$, the speckle noise intensity, is a white noise random field with exponential distribution, that is,

$$p_s(\xi) = \begin{cases} \dfrac{1}{\sigma^2} \exp\left(\dfrac{-\xi}{\sigma^2}\right), & \xi \geq 0 \\ 0, & \text{otherwise} \end{cases} \tag{8.17}$$

Digital processing of images with speckle is discussed in section 8.13.

Sampled Image Observation Models

With uniform sampling the observation model of (8.1)–(8.3) can be reduced to a discrete approximation of the form

$$v(m, n) = g[w(m, n)] + \eta(m, n), \qquad \forall(m, n) \tag{8.18}$$

$$w(m, n) = \sum\sum_{k, l = -\infty}^{\infty} h(m, n; k, l)u(k, l) \tag{8.19}$$

$$\eta(m, n) = f[g(w(m, n))]\eta'(m, n) + \eta''(m, n) \tag{8.20}$$

where $\eta'(m, n)$ and $\eta''(m, n)$ are discrete white noise fields, $h(m, n; k, l)$ is the impulse response of the sampled system, and $u(m, n), v(m, n)$ are the average values of $u(x, y)$ and $v(x, y)$ over a pixel area in the sampling grid (Problem 8.3).

8.3 INVERSE AND WIENER FILTERING

Inverse Filter

Inverse filtering is the process of recovering the input of a system from its output. For example, in the absence of noise the inverse filter would be a system that recovers $u(m, n)$ from the observations $v(m, n)$ (Fig. 8.9). This means we must have

$$g^I(x) = g^{-1}(x), \quad \text{or} \quad g^I[g(x)] = x \tag{8.21}$$

$$h^I(m, n; k, l) = h^{-1}(m, n; k, l) \tag{8.22}$$

that is,

$$\sum\sum_{k', l' = -\infty}^{\infty} h^I(m, n; k', l')h(k', l'; k, l) = \delta(m - k, n - l) \tag{8.23}$$

Inverse filters are useful for precorrecting an input signal in anticipation of the degradations caused by the system, such as correcting the nonlinearity of a display. Design of physically realizable inverse filters is difficult because they are often

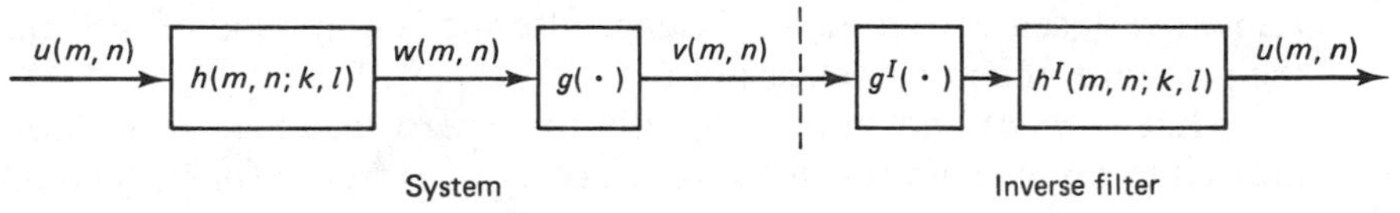

Figure 8.9 Inverse filter.

unstable. For example, for spatially invariant systems (8.23) can be written as

$$\sum_{k', l' = -\infty}^{\infty}\sum h^I(m - k', n - l')h(k', l') = \delta(m, n), \qquad \forall (m, n) \tag{8.24}$$

Fourier transforming both sides, we obtain $H^I(\omega_1, \omega_2)H(\omega_1, \omega_2) = 1$, which gives

$$H^I(\omega_1, \omega_2) = \frac{1}{H(\omega_1, \omega_2)} \tag{8.25}$$

that is, the inverse filter frequency response is the reciprocal of the frequency response of the given system. However, $H^I(\omega_1, \omega_2)$ will not exist if $H(\omega_1, \omega_2)$ has any zeros.

Pseudoinverse Filter

The pseudoinverse filter is a stabilized version of the inverse filter. For a linear shift invariant system with frequency response $H(\omega_1, \omega_2)$, the pseudoinverse filter is defined as

$$H^-(\omega_1, \omega_2) = \begin{cases} \dfrac{1}{H(\omega_1, \omega_2)}, & H \neq 0 \\ 0, & H = 0 \end{cases} \tag{8.26}$$

Here, $H^-(\omega_1, \omega_2)$ is also called the *generalized inverse* of $H(\omega_1, \omega_2)$, in analogy with the definition of the generalized inverse of matrices. In practice, $H^-(\omega_1, \omega_2)$ is set to zero whenever $|H|$ is less than a suitably chosen positive quantity ϵ.

Example 8.3

Figure 8.10 shows a blurred image simulated digitally as the output of a noiseless linear system. Therefore, $W(\omega_1, \omega_2) = H(\omega_1, \omega_2)U(\omega_1, \omega_2)$. The inverse filtered image is obtained as $\hat{U}(\omega_1 \omega_2) \triangleq W(\omega_1, \omega_2)/H(\omega_1, \omega_2)$. In the presence of additive noise, the inverse filter output can be written as

$$\hat{U} = \frac{W}{H} + \frac{N}{H} = U + \frac{N}{H} \tag{8.27}$$

where $N(\omega_1, \omega_2)$ is the noise term. Even if N is small, N/H can assume large values resulting in amplification of noise in the filtered image. This is shown in Fig. 8.10c, where the small amount of noise introduced by computer round-off errors has been amplified by the inverse filter. Pseudoinverse filtering reduces this effect (Fig. 8.10d).

The Wiener Filter

The main limitation of inverse and pseudoinverse filtering is that these filters remain very sensitive to noise. Wiener filtering is a method of restoring images in the presence of blur as well as noise.

Let $u(m, n)$ and $v(m, n)$ be arbitrary, zero mean, random sequences. It is desired to obtain an estimate, $\hat{u}(m, n)$, of $u(m, n)$ from $v(m, n)$ such that the mean square error

$$\sigma_e^2 = E\{[u(m, n) - \hat{u}(m, n)]^2\} \tag{8.28}$$

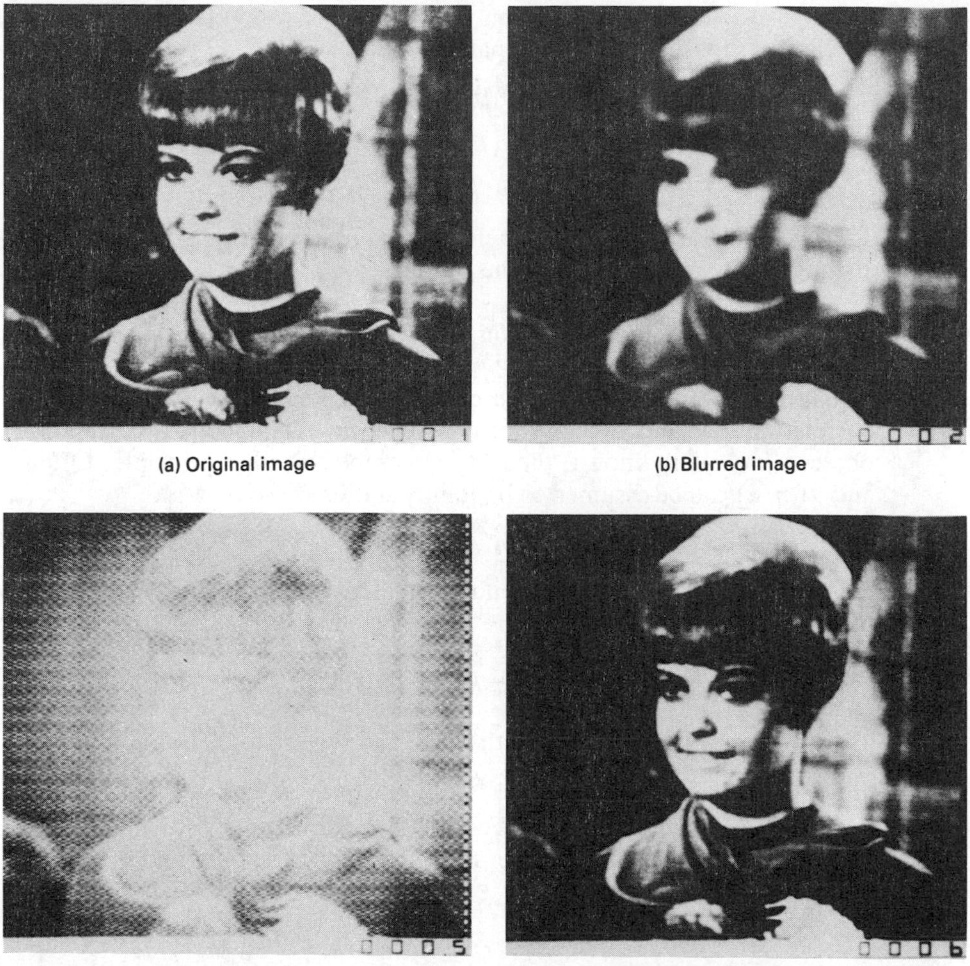

(a) Original image

(b) Blurred image

(c) Inverse filtered

(d) Pseudo-inverse filtered

Figure 8.10 Inverse and pseudo-inverse filtered images.

is minimized. The best estimate $\hat{u}(m, n)$ is known to be the conditional mean of $u(m, n)$ given $\{v(m, n)$, for every $(m, n)\}$, that is,

$$\hat{u}(m, n) = E[u(m, n)|v(k, l), \quad \forall(k, l)] \tag{8.29}$$

Equation (8.29), simple as it looks, is quite difficult to solve in general. This is because it is nonlinear, and the conditional probability density $p_{u|v}$, required for solving (8.29) is difficult to calculate. Therefore, one generally settles for the *best linear estimate* of the form

$$\hat{u}(m, n) = \sum_{k,l=-\infty}^{\infty}\sum g(m, n; k, l)v(k, l) \tag{8.30}$$

where the filter impulse response $g(m, n; k, l)$ is determined such that the mean square error of (8.28) is minimized. It is well known that if $u(m, n)$ and $v(m, n)$ are jointly Gaussian sequences, then the solution of (8.29) is linear. Minimization of (8.28) requires that the orthogonality condition

$$E[\{u(m,n) - \hat{u}(m,n)\}v(m',n')] = 0, \qquad \forall (m,n),(m',n') \tag{8.31}$$

be satisfied. Using the definition of cross-correlation

$$r_{ab}(m,n;k,l) \triangleq E[a(m,n)b(k,l)] \tag{8.32}$$

for two arbitrary random sequences $a(m, n)$ and $b(k, l)$, and given (8.30), the orthogonality condition yields the equation

$$\sum_{k,l=-\infty}^{\infty}\sum g(m,n;k,l)r_{vv}(k,l;m',n') = r_{uv}(m,n;m',n') \tag{8.33}$$

Equations (8.30) and (8.33) are called the Wiener filter equations. If the autocorrelation function of the observed image $v(m, n)$ and its cross-correlation with the object $u(m, n)$ are known, then (8.33) can be solved in principle. Often, $u(m, n)$ and $v(m, n)$ can be assumed to be jointly stationary so that

$$r_{ab}(m,n;m',n') = r_{ab}(m-m', n-n') \tag{8.34}$$

for $(a, b) = (u, u)$, (u, v), (v, v), and so on. This simplifies $g(m, n; k, l)$ to a spatially invariant filter, denoted by $g(m-k, n-l)$, and (8.33) reduces to

$$\sum_{k,l=-\infty}^{\infty}\sum g(m-k,n-l)r_{vv}(k,l) = r_{uv}(m,n) \tag{8.35}$$

Taking Fourier transforms of both sides, and solving for $G(\omega_1, \omega_2)$, we get

$$G(\omega_1,\omega_2) = S_{uv}(\omega_1,\omega_2)S_{vv}^{-1}(\omega_1,\omega_2) \tag{8.36}$$

where G, S_{uv}, and S_{vv} are the Fourier transforms of g, r_{uv}, and r_{vv} respectively. Equation (8.36) gives the Wiener filter frequency response and the filter equations become

$$\hat{u}(m,n) = \sum_{k,l=-\infty}^{\infty}\sum g(m-k,n-l)v(k,l) \tag{8.37}$$

$$\hat{U}(\omega_1,\omega_2) = G(\omega_1,\omega_2)V(\omega_1,\omega_2) \tag{8.38}$$

where U and V are the Fourier transforms of u and v respectively. Suppose the $v(m, n)$ is modeled by a linear observation system with additive noise, that is,

$$v(m,n) = \sum_{k,l=-\infty}^{\infty}\sum h(m-k,n-l)u(k,l) + \eta(m,n) \tag{8.39}$$

where $\eta(m, n)$ is a stationary noise sequence uncorrelated with $u(m, n)$ and which has the power spectral density $S_{\eta\eta}$. Then

$$\begin{aligned} S_{vv}(\omega_1,\omega_2) &= |H(\omega_1,\omega_2)|^2 S_{uu}(\omega_1,\omega_2) + S_{\eta\eta}(\omega_1,\omega_2) \\ S_{uv}(\omega_1,\omega_2) &= H^*(\omega_1,\omega_2)S_{uu}(\omega_1,\omega_2) \end{aligned} \tag{8.40}$$

This gives

$$G(\omega_1,\omega_2) = \frac{H^*(\omega_1,\omega_2)S_{uu}(\omega_1,\omega_2)}{|H(\omega_1,\omega_2)|^2 S_{uu}(\omega_1,\omega_2) + S_{\eta\eta}(\omega_1,\omega_2)} \tag{8.41}$$

which is also called the Fourier-Wiener filter. This filter is completely determined by the power spectra of the object and the noise and the frequency response of the imaging system. The mean square error can also be written as

$$\sigma_e^2 = \frac{1}{4\pi^2} \iint_{-\pi}^{\pi} S_e(\omega_1, \omega_2)\, d\omega_1\, d\omega_2 \tag{8.42a}$$

where S_e, the power spectrum density of the error, is

$$S_e(\omega_1, \omega_2) \triangleq |1 - GH|^2 S_{uu} + |G|^2 S_{\eta\eta} \tag{8.42b}$$

With G given by (8.41), this gives

$$S_e = \frac{S_{uu} S_{\eta\eta}}{|H|^2 S_{uu} + S_{\eta\eta}} \tag{8.42c}$$

Remarks

In general, the Wiener filter is not separable even if the PSF and the various covariance functions are. This means that two-dimensional Wiener filtering is not equivalent to row-by-row followed by column-by-column one-dimensional Wiener filtering.

Neither of the error fields, $e(m, n) = u(m, n) - \hat{u}(m, n)$, and $\varepsilon(m, n) \triangleq v(m, n) - h(m, n) \circledast \hat{u}(m, n)$, is white even if the observation noise $\eta(m, n)$ is white.

Wiener Filter for Nonzero Mean Images. The foregoing development requires that the observation model of (8.39) be zero mean. There is no loss of generality in this because if $u(m, n)$ and $\eta(m, n)$ are not zero mean, then from (8.39)

$$\mu_v(m, n) = h(m, n) \circledast \mu_u(m, n) + \mu_\eta(m, n), \qquad \mu_x \triangleq E[x] \tag{8.43}$$

which gives the zero mean model

$$\tilde{v}(m, n) = h(m, n) \circledast \tilde{u}(m, n) + \tilde{\eta}(m, n) \tag{8.44}$$

where $\tilde{x} \triangleq x - \mu_x$. In practice, μ_v can be estimated as the sample mean of the observed image and the Wiener filter may be implemented on $\tilde{v}(m, n)$. On the other hand, if μ_u and μ_η are known, then the Wiener filter becomes

$$\left.\begin{aligned} \hat{U} &= G(V - M_v) + M_u \\ &= GV + \frac{S_{\eta\eta}}{|H^2|S_{uu} + S_{\eta\eta}} M_u - GM_\eta \end{aligned}\right\} \tag{8.45}$$

where $M(\omega_1, \omega_2) \triangleq \mathcal{F}\{\mu(m, n)\}$ is the Fourier transform of the mean. Note that the above filter allows spatially varying means for $u(m, n)$ and $\eta(m, n)$. Only the covariance functions are required to be spatially invariant. The spatially varying mean may be estimated by local averaging of the image. In practice, this spatially varying filter can be quite effective. If μ_u and μ_η are constants, then M_u and M_η are Dirac delta functions at $\xi_1 = \xi_2 = 0$. This means that a constant is added to the processed image which does not affect its dynamic range and hence its display.

Phase of the Wiener Filter. Equation (8.41) can be written as

$$\left.\begin{aligned} G &= |G| e^{j\theta_G} \\ |G| &= \frac{|H| S_{uu}}{|H|^2 S_{uu} + S_{\eta\eta}} \\ \theta_G &= \theta_G(\omega_1, \omega_2) = \theta_{H^*} = -\theta_H = \theta_{H^{-1}} \end{aligned}\right\} \tag{8.46}$$

that is, *the phase of the Wiener filter is equal to the phase of the inverse filter* (in the frequency domain). Therefore, the Wiener filter or, equivalently, the mean square criterion of (8.28), does not compensate for phase distortions due to noise in the observations.

Wiener Smoothing Filter. In the absence of any blur, $H = 1$ and the Wiener filter becomes

$$G|_{H=1} = \frac{S_{uu}}{S_{uu} + S_{\eta\eta}} = \frac{S_{nr}}{S_{nr} + 1} \tag{8.47}$$

where $S_{nr} \triangleq S_{uu}/S_{\eta\eta}$ defines the signal-to-noise ratio at the frequencies (ω_1, ω_2). This is also called the *(Wiener) smoothing filter*. It is a zero-phase filter that depends only

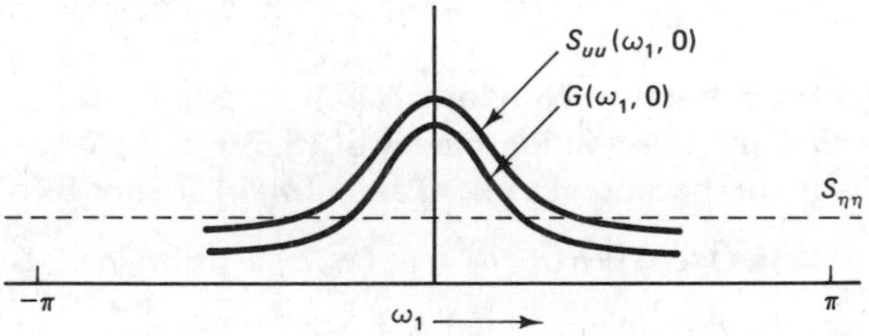

(a) Noise smoothing ($H = 1$)

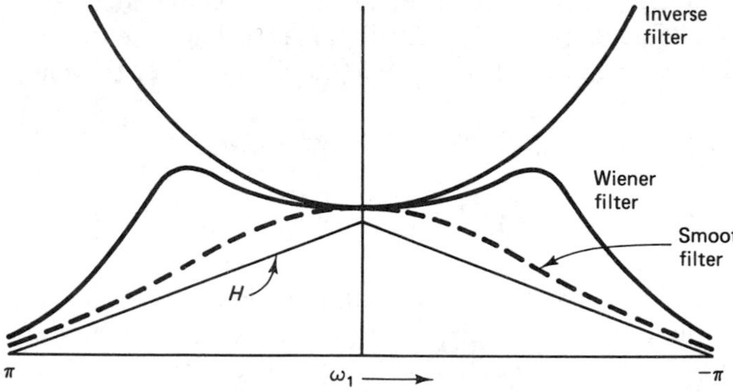

(b) Deblurring

Figure 8.11 Wiener filter characteristics.

on the signal-to-noise ratio S_{nr}. For frequencies where $S_{nr} \gg 1$, G becomes nearly equal to unity which means that all these frequency components are in the passband. When $S_{nr} \ll 1$, $G = S_{nr}$; that is, all frequency components where $S_{nr} \ll 1$, are attenuated in proportion to their signal-to-noise ratio. For images, S_{nr} is usually high at lower spatial frequencies. Therefore, the noise smoothing filter is a low-pass filter (see Fig. 8.11a).

Relation with Inverse Filtering. In the absence of noise, we set $S_{\eta\eta} = 0$ and the Wiener filter reduces to

$$G|_{S_{\eta\eta}=0} = \frac{H^* S_{uu}}{|H|^2 S_{uu}} = \frac{1}{H} \tag{8.48}$$

which is the inverse filter. On the other hand, taking the limit $S_{\eta\eta} \to 0$, we obtain

$$\lim_{S_{\eta\eta}\to 0} G = \begin{cases} \dfrac{1}{H}, & H \neq 0 \\ 0, & H = 0 \end{cases} = H^- \tag{8.49}$$

which is the pseudoinverse filter. Since the blurring process is usually a low-pass filter, the Wiener filter acts as a high-pass filter at low levels of noise.

Interpretation of Wiener Filter Frequency Response. When both noise and blur are present, the Wiener filter achieves a compromise between the low-pass noise smoothing filter and the high-pass inverse filter resulting in a band-pass filter (see Fig. 8.11b). Figure 8.12 shows Wiener filtering results for noisy blurred images. Observe that the deblurring effect of the Wiener filter diminishes rapidly as the noise level increases.

Wiener Filter for Diffraction Limited Systems. The Wiener filter for the continuous observation model, analogous to (8.39)

$$v(x, y) = \iint_{-\infty}^{\infty} h(x - x', y - y')u(x', y')dx'dy' + \eta(x, y) \tag{8.50}$$

is given by

$$G(\xi_1, \xi_2) = \frac{S_{uu}(\xi_1, \xi_2)H^*(\xi_1, \xi_2)}{|H(\xi_1, \xi_2)|^2 S_{uu}(\xi_1, \xi_2) + S_{\eta\eta}(\xi_1, \xi_2)} \tag{8.51}$$

For a diffraction limited system, $H(\xi_1, \xi_2)$ will be zero outside a region, say $\mathcal{R}$, in the frequency plane. From (8.51), G will also be zero outside $\mathcal{R}$. Thus, *the Wiener filter cannot resolve beyond the diffraction limit.*

Wiener Filter Digital Implementation. The impulse response of the Wiener filter is given by

$$g(m, n) = \frac{1}{4\pi^2} \iint_{-\pi}^{\pi} G(\omega_1, \omega_2) \exp\{j(m\omega_1 + n\omega_2)\}\, d\omega_1\, d\omega_2 \tag{8.52}$$

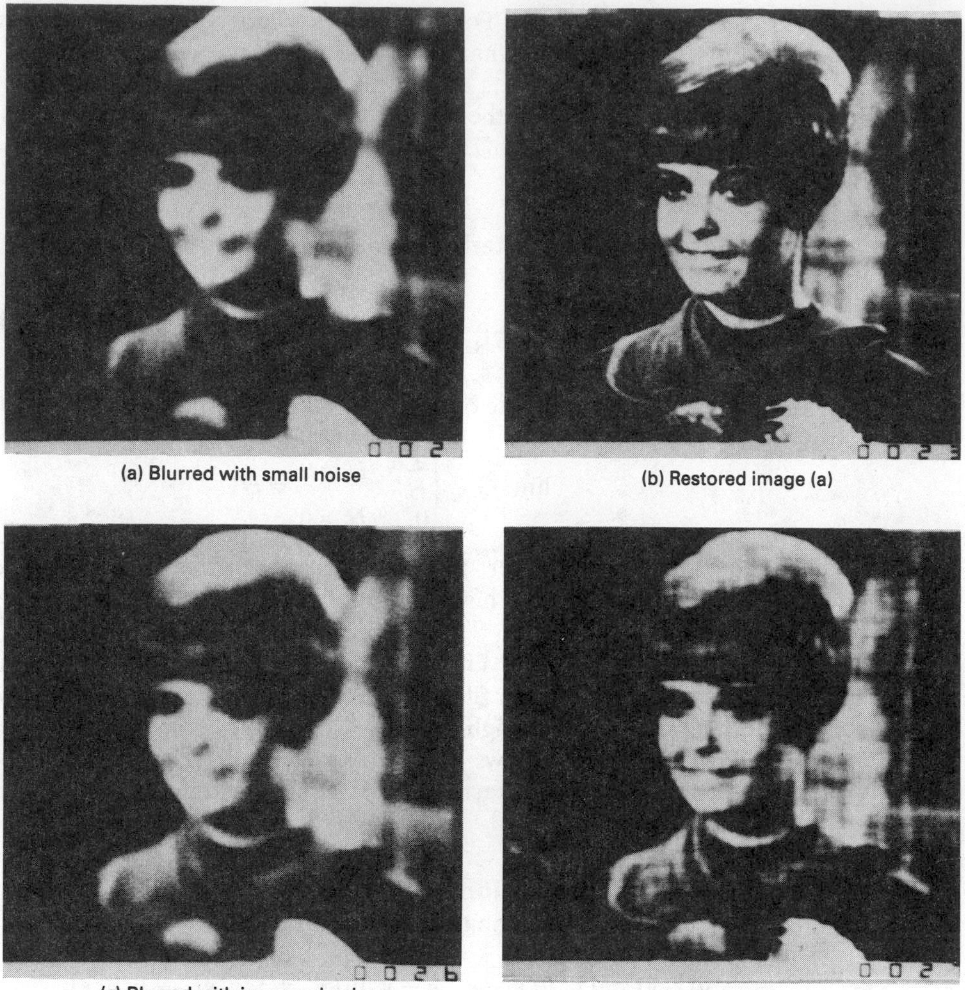

(a) Blurred with small noise

(b) Restored image (a)

(c) Blurred with increased noise

(d) Restored image (c)

Figure 8.12 Wiener filtering of noisy blurred images.

This integral can be computed approximately on a uniform $N \times N$ grid as

$$\tilde{g}(m, n) = \frac{1}{N^2} \sum_{(k, l) = -N/2}^{N/2 - 1} \sum \tilde{G}(k, l) W^{-(mk + nl)}, \qquad -\frac{N}{2} \leq m, n \leq \frac{N}{2} - 1 \tag{8.53}$$

where $\tilde{G}(k, l) \triangleq G(2\pi k/N, 2\pi l/N)$, $W \triangleq \exp(-j2\pi/N)$. The preceding series can be calculated via the two-dimensional FFT from which we can approximate $g(m, n)$ by $\tilde{g}(m, n)$ over the $N \times N$ grid defined above. Outside this grid $g(m, n)$ is assumed to be zero. Sometimes the region of support of $g(m, n)$ is much smaller than the $N \times N$ grid. Then the convolution of $g(m, n)$ with $v(m, n)$ could be implemented directly in the spatial domain. Modern image processing systems provide such a facility in hardware allowing high-speed implementation. Figure 8.13a shows this algorithm.

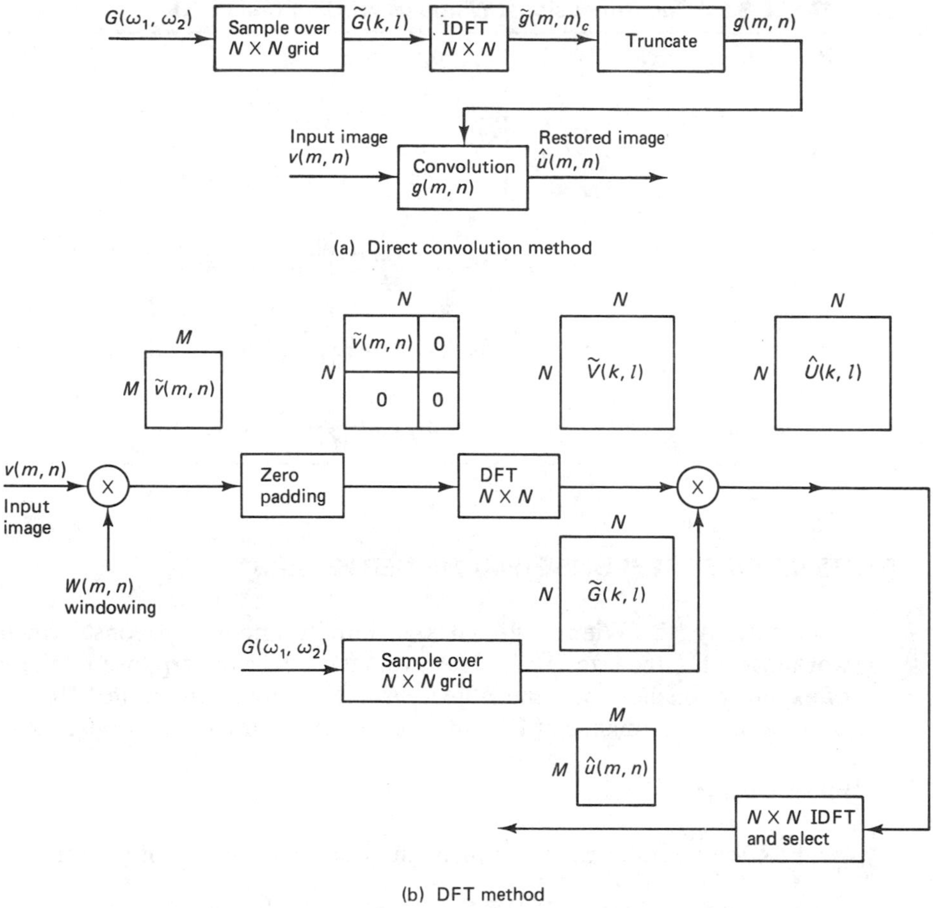

Figure 8.13 Digital implementations of the Wiener filter.

Alternatively, the Wiener filter can be implemented in the DFT domain as shown in Fig. 8.13b. Now the frequency response is sampled as $\tilde{G}(k, l)$ and is multiplied by the DFT of the zero-padded, windowed observations. An appropriate region of the inverse DFT then gives the restored image. This is simply an implementation of the convolution of $g(m, n)$ with windowed $v(m, n)$ via the DFT. The windowing operation is useful when an infinite sequence is replaced by a finite sequence in convolution of two sequences. Two-dimensional windows are often taken to be separable product of one-dimensional windows as

$$w(m, n) = w_1(m)w_2(n) \tag{8.54}$$

where $w_1(m)$ and $w_2(n)$ are one-dimensional windows (see Table 8.2). For images of size 256×256 or larger, the rectangular window has been found to be appropriate. For smaller size images, other windows are more useful. If the image size is large and the filter $g(m, n)$ can be approximated by a small size array, then their convolution can be realized by filtering the image in small, overlapping blocks and summing the outputs of the block filters [9].

TABLE 8.2 One-Dimensional Windows $w(n)$ for $0 \le n \le M-1$

Rectangular	$w(n) = 1, \quad 0 \le n \le M-1$
Bartlett (Triangular)	$w(n) = \begin{cases} \dfrac{2n}{M-1}, & 0 \le n \le \dfrac{M-1}{2} \\ 2 - \dfrac{2n}{M-1}, & \dfrac{M-1}{2} \le n \le M-1 \end{cases}$
Hanning	$w(n) = \frac{1}{2}\left[1 - \cos\left(\frac{2\pi n}{M-1}\right)\right], 0 \le n \le M-1$
Hamming	$w(n) = 0.54 - 0.46 \cos\left(\frac{2\pi n}{M-1}\right), 0 \le n \le M-1$
Blackman	$w(n) = 0.42 - 0.5 \cos\left(\frac{2\pi n}{M-1}\right) + 0.08 \cos\left(\frac{4\pi n}{M-1}\right), 0 \le n \le M-1$

8.4 FINITE IMPULSE RESPONSE (FIR) WIENER FILTERS

Theoretically, the Wiener filter has an infinite impulse response which requires working with large size DFTs. However, the effective response of these filters is often much smaller than the object size. Therefore, optimum FIR filters could achieve the performance of IIR filters but with lower computational complexity.

Filter Design

An FIR Wiener filter can be implemented as a convolution of the form

$$\hat{u}(m, n) = \sum\sum_{(i,j) \in W} g(i, j) v(m - i, n - j) \tag{8.55}$$

$$W = \{-M \le i, j \le M\} \tag{8.56}$$

where $g(i, j)$ are the optimal filter weights that minimize the mean square error $E[\{u(m, n) - \hat{u}(m, n)\}^2]$. The associated orthogonality condition

$$E[\{u(m, n) - \hat{u}(m, n)\} v(m - k, n - l)] = 0, \qquad \forall (k, l) \in W \tag{8.57}$$

reduces to a set of $(2M + 1)^2$ simultaneous equations

$$r_{uv}(k, l) - \sum\sum_{(i,j) \in W} g(i, j) r_{vv}(k - i, l - j) = 0, \qquad (k, l) \in W \tag{8.58}$$

Using (8.39) and assuming $\eta(m, n)$ to be zero mean white noise of variance σ_η^2, it is easy to deduce that

$$r_{vv}(k, l) = r_{uu}(k, l) \circledast a(k, l) + \sigma_\eta^2 \delta(k, l) \tag{8.59}$$

$$a(k, l) \triangleq h(k, l) \star h(k, l) = \sum\sum_{i,j=-\infty}^{\infty} h(i, j) h(i + k, j + l) \tag{8.60}$$

$$r_{uv}(k, l) = h(k, l) \star r_{uu}(k, l) = \sum\sum_{i,j=-\infty}^{\infty} h(i, j) r_{uu}(i + k, j + l) \tag{8.61}$$

Defining the correlation function

$$r_0(k, l) \triangleq \frac{r_{uu}(k, l)}{r_{uu}(0, 0)} = \frac{r_{uu}(k, l)}{\sigma^2} \tag{8.62}$$

where σ^2 is the variance of the image, and using the fact that $r_0(k, l) = r_0(-k, -l)$, (8.58) reduces to

$$\left[\frac{\sigma_\eta^2}{\sigma^2}\delta(k, l) + r_0(k, l) \circledast a(k, l)\right] \circledast g(k, l) \\ = h(k, l) \circledast r_0(k, l), \qquad (k, l) \in W \tag{8.63}$$

In matrix form this becomes a block Toeplitz system

$$\left[\frac{\sigma_\eta^2}{\sigma^2}\mathbf{I} + \mathcal{R}\right] \boldsymbol{g} = \boldsymbol{r}_{uv} \tag{8.64}$$

where $\mathcal{R}$ is a $(2M + 1) \times (2M + 1)$ block Toeplitz matrix of basic dimension $(2M + 1) \times (2M + 1)$ and $\boldsymbol{r}_{uv}$ and $\boldsymbol{g}$ are $(2M + 1)^2 \times 1$ vectors containing the knowns $h(k, l) \circledast r_0(k, l)$ and the unknowns $g(k, l)$, respectively.

Remarks

The number of unknowns in (8.58) can be reduced if the PSF $h(m, n)$ and the image covariance $r_{uu}(m, n)$ have some symmetry. For example, in the often-encountered case where $h(m, n) = h(|m|, |n|)$ and $r_{uu}(m, n) = r_{uu}(|m|, |n|)$, we will have $g(m, n) = g(|m|, |n|)$ and (8.63) can be reduced to $(M + 1)^2$ simultaneous equations.

The quantities $a(k, l)$ and $r_{uv}(k, l)$ can also be determined as the inverse Fourier transforms of $|H(\omega_1, \omega_2)|^2$ and $S_{uu}(\omega_1, \omega_2)H^*(\omega_1, \omega_2)$, respectively.

When there is no blur (that is, $H = 1$), $h(k, l) = \delta(k, l)$, $r_{uv}(k, l) = r_{uu}(k, l)$ and the resulting filter is the optimum FIR smoothing filter.

The size of the FIR Wiener filter grows with the amount of blur and the additive noise. Experimental results have shown that FIR Wiener filters of sizes up to 15×15 can be quite satisfactory in a variety of cases with different levels of blur and noise. In the special case of no blur, (8.63) becomes

$$\left[\frac{\sigma_\eta^2}{\sigma^2}\delta(k, l) + r_0(k, l)\right] \circledast g(k, l) = r_0(k, l), \quad (k, l) \in W \tag{8.65}$$

From this, it is seen that as the $\text{SNR} = \sigma^2/\sigma_\eta^2 \to \infty$, the filter response $g(k, l) \to \delta(k, l)$, that is, the FIR filter support goes down to one pixel. On the other hand, if $\text{SNR} \simeq 0$, $g(k, l) \simeq (\sigma^2/\sigma_\eta^2) r_0(k, l)$. This means the filter support would effectively be the same as the region outside of which the image correlations are negligible. For images with $r_0(k, l) = 0.95^{\sqrt{k^2 + l^2}}$, this region happens to be of size 32×32.

In general, as $\sigma_\eta^2 \to 0$, $g(m, n)$ becomes the optimum FIR inverse filter. As $M \to \infty$, $g(m, n)$ would converge to the pseudoinverse filter.

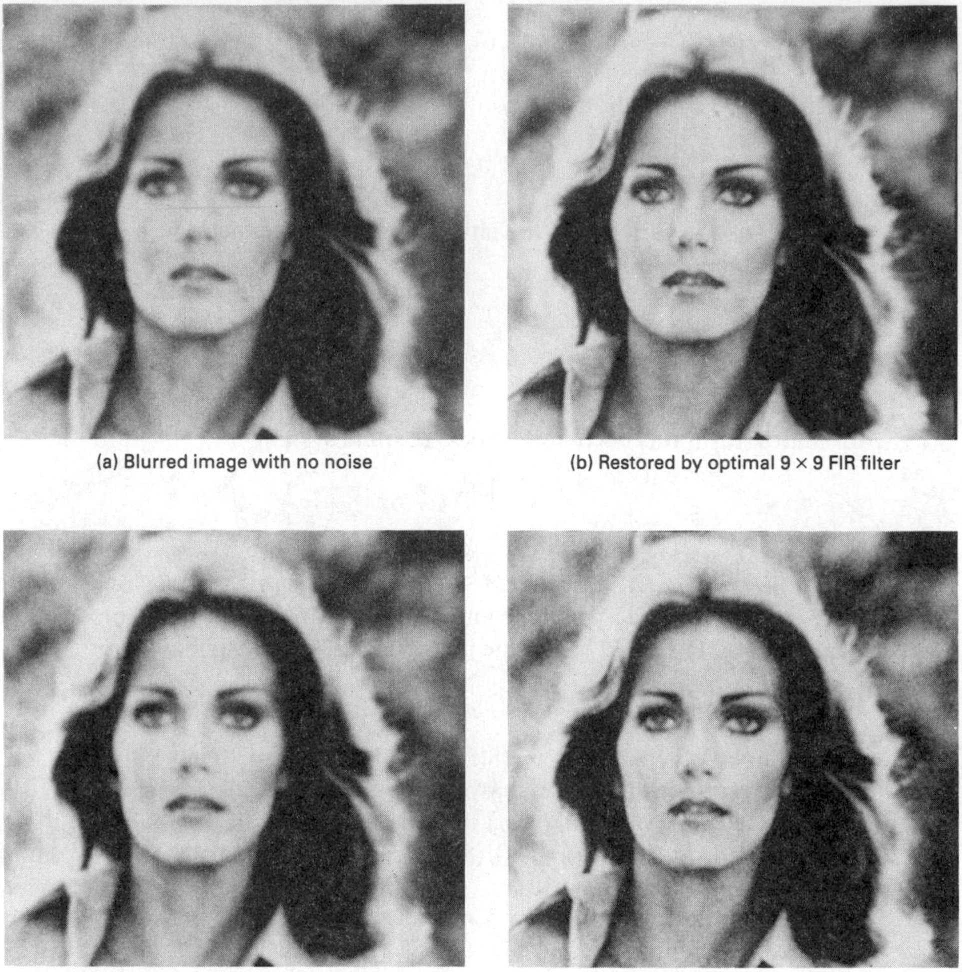

(a) Blurred image with no noise

(b) Restored by optimal 9 × 9 FIR filter

(c) Blurred image with small additive noise

(d) Restored by 9 × 9 optimal FIR filter

Figure 8.14 FIR Wiener filtering of blurred images.

Example 8.4

Figure 8.14 shows examples of FIR Wiener filtering of images blurred by the Gaussian PSF

$$h(m, n) = \exp\{-\alpha(m^2 + n^2)\}, \qquad \alpha > 0 \tag{8.66}$$

and modeled by the covariance function $r(m, n) = \sigma^2 0.95^{\sqrt{m^2+n^2}}$. If $\alpha \simeq 0$, and $h(m, n)$ is restricted to a finite region, say $-p \le m, n \le p$, then the $h(m, n)$ defined above can be used to model aberrations in a lens with finite aperture. Figs. 8.15 and 8.16 show examples of FIR Wiener filtering of blurred images that were intentionally misfocused by the digitizing camera.

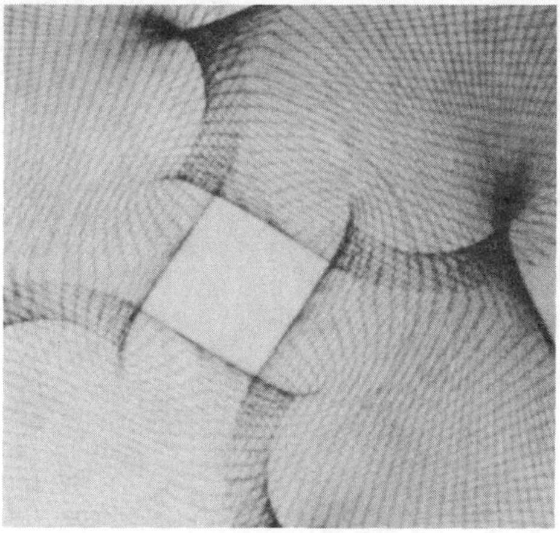

(a) Digitized image

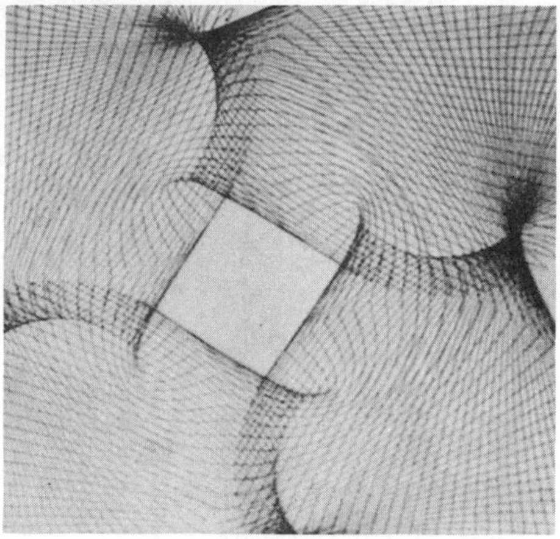

(b) FIR filtered

(c) Digitized image

(d) FIR filtered

Figure 8.15 FIR Wiener filtering of improperly digitized images.

Spatially Varying FIR Filters

The local structure and the ease of implementation of the FIR filters makes them attractive candidates for the restoration of nonstationary images blurred by spatially varying PSFs. A simple but effective nonstationary model for images is one that has spatially varying mean and variance functions and shift-invariant correlations, that is,

$$\left.\begin{aligned} &E[u(m, n)] = \mu(m, n) \\ &E[\{u(m, n) - \mu(m, n)\}\{u(m-k, n-l) - \mu(m-k, n-l)\}] \\ &\qquad = \sigma^2(m, n) r_0(k, l) \end{aligned}\right\} \tag{8.67}$$

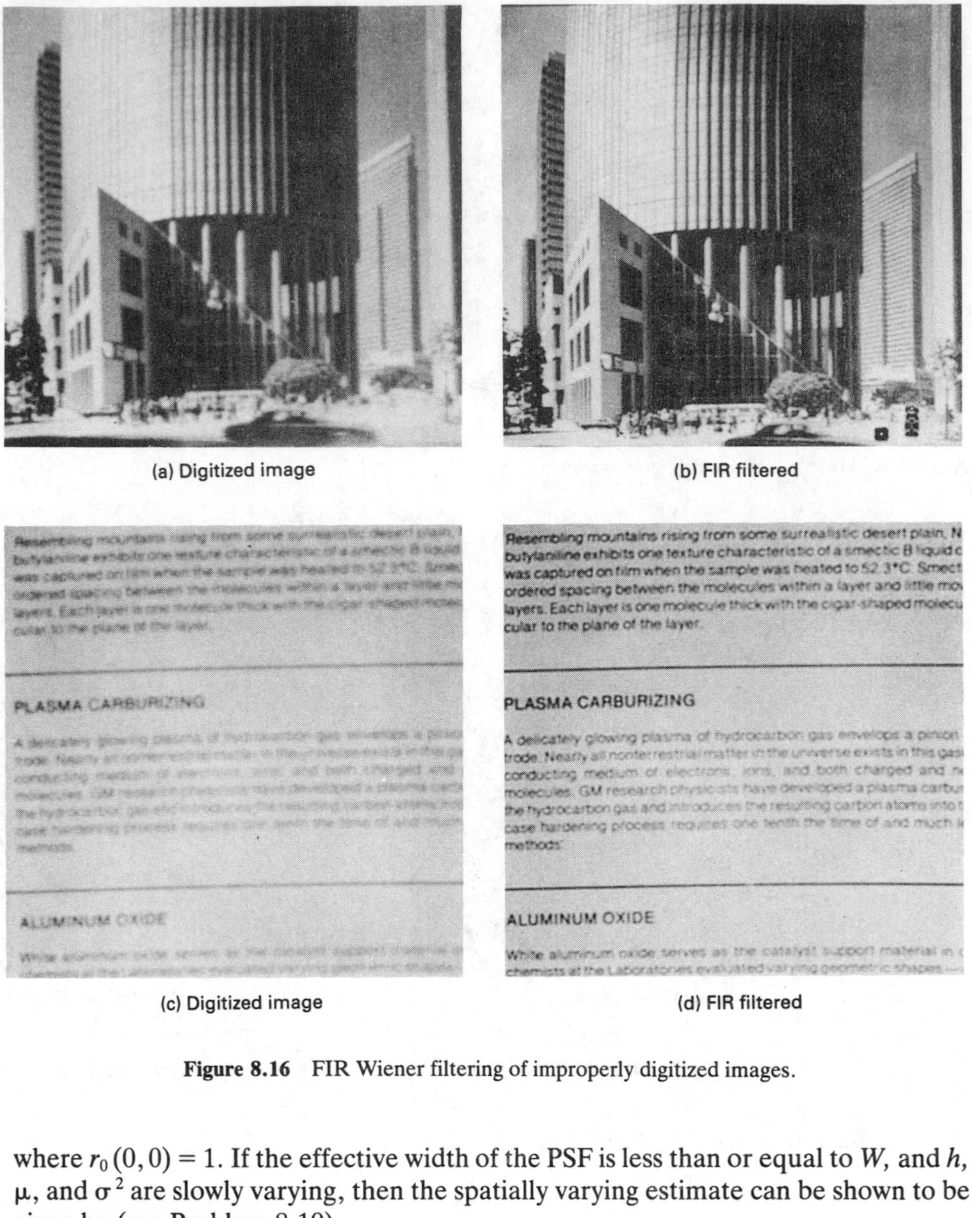

(a) Digitized image (b) FIR filtered

(c) Digitized image (d) FIR filtered

Figure 8.16 FIR Wiener filtering of improperly digitized images.

where $r_0(0, 0) = 1$. If the effective width of the PSF is less than or equal to W, and h, μ, and σ^2 are slowly varying, then the spatially varying estimate can be shown to be given by (see Problem 8.10)

$$\hat{u}(m, n) = \sum\sum_{(i,j) \in W} \tilde{g}_{m,n}(i, j) v(m - i, n - j) \tag{8.68}$$

$$\tilde{g}_{m,n}(i, j) \triangleq g_{m,n}(i, j) + \frac{1}{(2M + 1)^2}\left[1 - \sum\sum_{(k,l) \in W} g_{m,n}(k, l)\right] \tag{8.69}$$

where $g_{m,n}(i, j)$ is the solution of (8.64) with $\sigma^2 = \sigma^2(m, n)$. The quantity $\sigma^2(m, n)$ can be estimated from the observation model. The second term in (8.69) adds a constant with respect to (i, j) to the FIR filter which is equivalent to adding to the estimate a quantity proportional to the local average of the observations.

Example 8.5

Consider the case of noisy images with no blur. Figures 8.17 and 8.18 show noisy images with SNR = 10 dB and their Wiener filter estimates. A spatially varying FIR Wiener filter called *Optmask* was designed using $\sigma^2(m, n) = \sigma_v^2(m, n) - \sigma^2$, where $\sigma_v^2(m, n)$ was estimated as the local variance of the pixels in the neighborhood of (m, n). The size of the Optmask window W has changed from region to region such that the output SNR was nearly constant [8]. This criterion gives a uniform visual quality to the filtered image, resulting in a more favorable appearance. The images obtained by

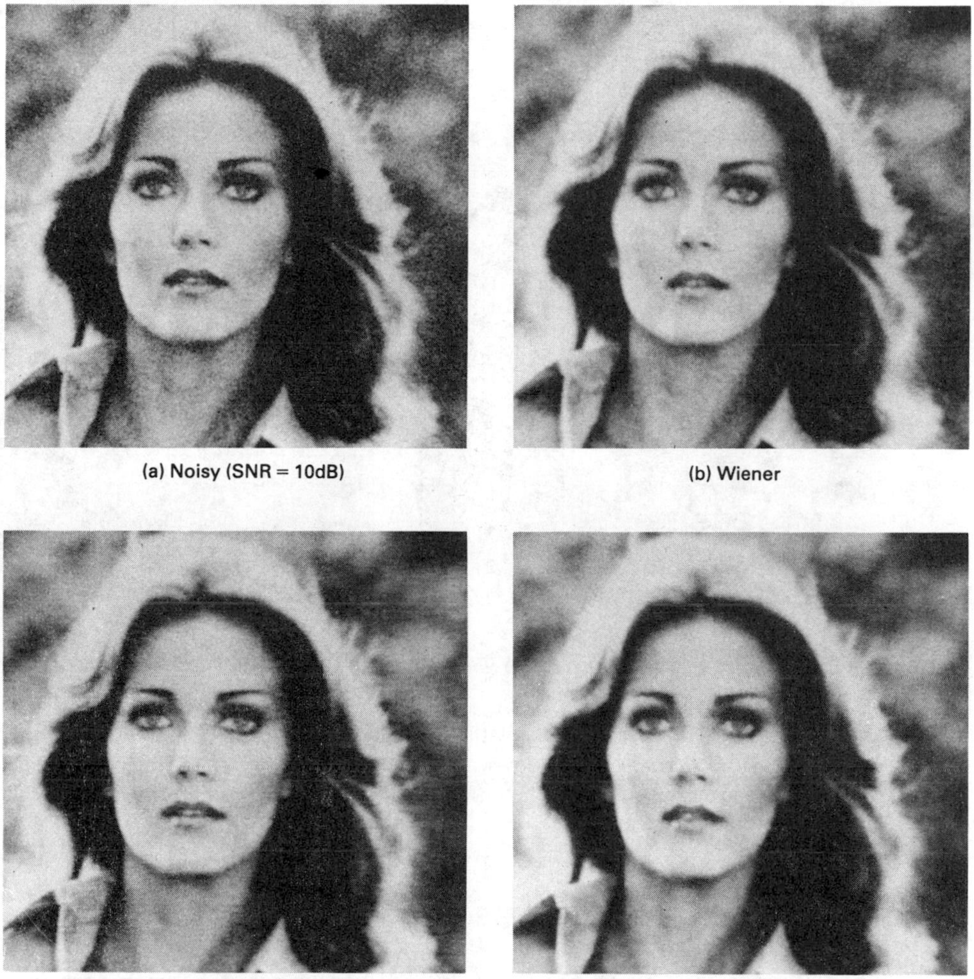

(a) Noisy (SNR = 10dB) (b) Wiener (c) Cosar (d) Optmask

Figure 8.17 Noisy image (SNR = 10 dB) and restored images.

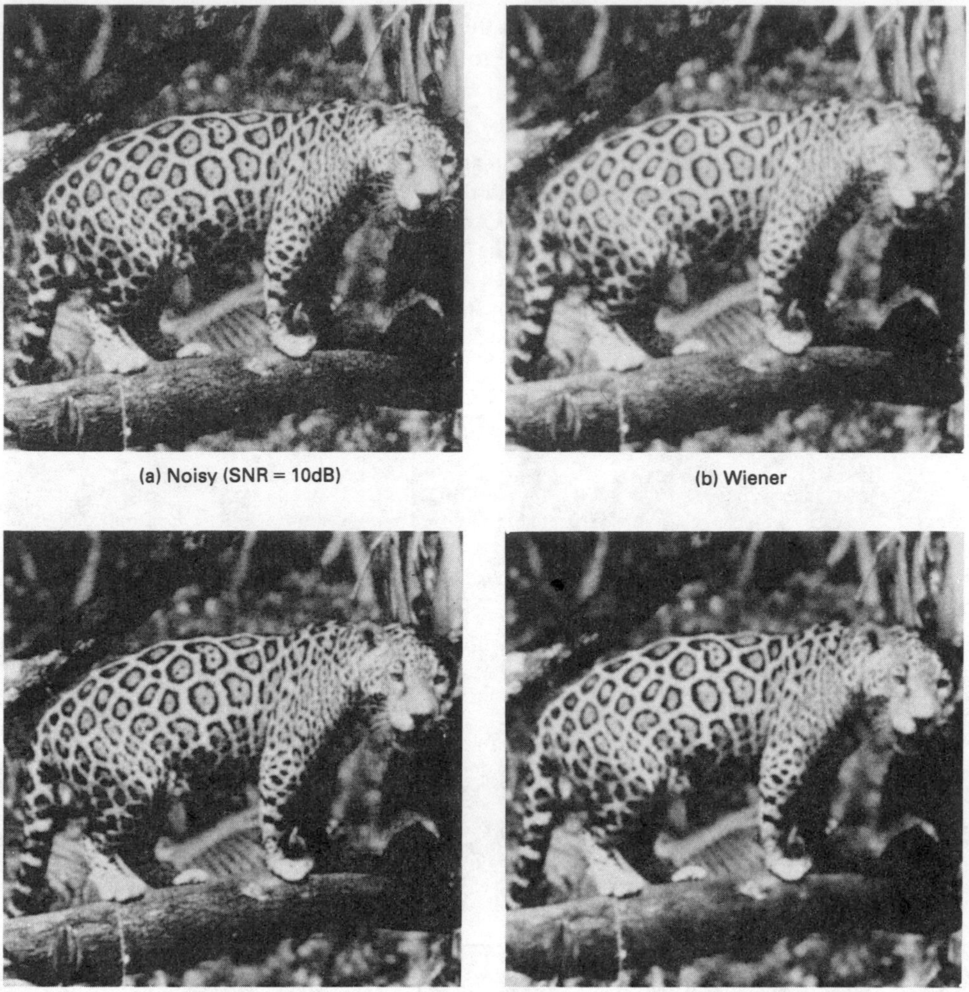

(a) Noisy (SNR = 10dB) (b) Wiener

(c) Cosar (d) Optmask

Figure 8.18 Noisy image (SNR = 10 dB) and restored images.

the *Cosar* filter are based on a spatially varying AR model, which is discussed in Section 8.12. Comparison of the spatially adaptive FIR filter output with the conventional Wiener filter shows a significant improvement. The comparison appears much more dramatic on a good quality CRT display unit compared to the pictures printed in this text.

8.5 OTHER FOURIER DOMAIN FILTERS

The Wiener filter of (8.41) motivates the use of other Fourier domain filters as discussed below.

Geometric Mean Filter

This filter is the geometric mean of the pseudoinverse and the Wiener filters, that is,

$$G_s = (H^-)^s \left(\frac{S_{uu} H^*}{S_{uu} |H|^2 + S_{\eta\eta}} \right)^{1-s}, \qquad 0 \le s \le 1 \tag{8.70}$$

For $s = \frac{1}{2}$, this becomes

$$G_{1/2} = \left(\frac{S_{uu}}{|H|^2 S_{uu} + S_{\eta\eta}} \right)^{1/2} |HH^-|^{1/2} \exp\{-j\theta_H\} \tag{8.71}$$

where $\theta_H(\omega_1, \omega_2)$ is the phase of $H(\omega_1, \omega_2)$. Unlike the Wiener filter, the power spectral density of the output of this filter is identical to that of the object when $H \neq 0$. As $S_{\eta\eta} \to 0$, this filter also becomes the pseudoinverse filter (see Problem 8.11).

Nonlinear Filters

These filters are of the form

$$\hat{U}(\omega_1, \omega_2) = G(V, \omega_1, \omega_2) \tag{8.72}$$

where $G(x, \omega_1, \omega_2)$ is a nonlinear function of x. Digital implementation of such nonlinear filters is quite easy. A simple nonlinear filter, called the *root-filter,* is defined as

$$\hat{U}(\omega_1, \omega_2) \triangleq |V|^\alpha \exp\{j\theta_v\} \tag{8.73}$$

Depending on the value of α this could be a low-pass, a high-pass, or a band-pass filter. For values of $\alpha \ll 1$, it acts as a high-pass filter (Fig. 8.19) for typical images

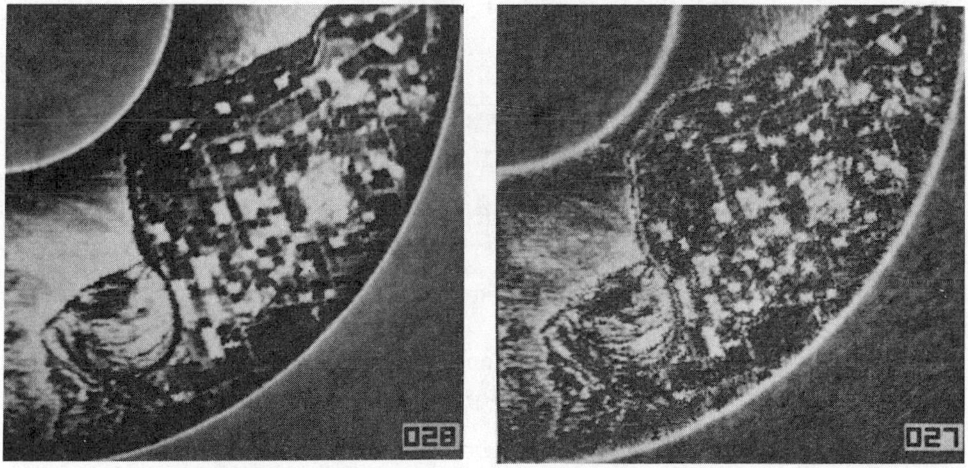

(a) $\alpha = 0.8$ (b) $\alpha = 0.6$

Figure 8.19 Root filtering of FLR image of Fig. 8.7c.

(having small energy at high spatial frequencies) because the samples of low amplitudes of $V(\omega_1, \omega_2)$ are enhanced relative to the high-amplitude samples. For $\alpha \gg 1$, the large-magnitude samples are amplified relative to the low-magnitude ones, giving a low-pass-filter effect for typical images. Another useful nonlinear filter is the complex logarithm of the observations

$$\hat{U}(\omega_1, \omega_2) = \begin{cases} \{\log|V|\}\exp\{j\theta_v\}, & |V| \ge \varepsilon \\ 0, & \text{otherwise} \end{cases} \tag{8.74}$$

where ε is some preselected small positive number. The sequence $\hat{u}(m, n)$ is also called the *cepstrum* of $v(m, n)$. (Also see Section 7.5.)

8.6 FILTERING USING IMAGE TRANSFORMS

Wiener Filtering

Consider the zero mean image observation model

$$v(m, n) = \sum_i \sum_j h(m, n; i, j)u(i, j) + \eta(m, n), \qquad 0 \le m \le N_1 - 1, 0 \le n \le N_2 - 1 \tag{8.75}$$

Assuming the object $u(m, n)$ is of size $M_1 \times M_2$ the preceding equation can be written as (see Chapter 2)

$$\boldsymbol{v} = \mathcal{H}\boldsymbol{u} + \boldsymbol{n} \tag{8.76}$$

where $\boldsymbol{v}$ and $\boldsymbol{n}$ are $N_1 N_2 \times 1$, $\boldsymbol{u}$ is $M_1 M_2 \times 1$ and $\mathcal{H}$ is a block matrix. The best linear estimate

$$\hat{\boldsymbol{u}} = \mathcal{G}\boldsymbol{v} \tag{8.77}$$

that minimizes the *average mean square error*

$$\sigma_e^2 = \frac{1}{M_1 M_2} E[(\boldsymbol{u} - \hat{\boldsymbol{u}})^T(\boldsymbol{u} - \hat{\boldsymbol{u}})] \tag{8.78}$$

is obtained by the orthogonality relation

$$E[(\boldsymbol{u} - \hat{\boldsymbol{u}})\boldsymbol{v}^T] = \mathbf{0} \tag{8.79}$$

This gives the Wiener filter as an $M_1 M_2 \times N_1 N_2$ matrix

$$\mathcal{G} = E[\boldsymbol{u}\boldsymbol{v}^T]\{E[\boldsymbol{v}\boldsymbol{v}^T]\}^{-1} = \mathcal{R}\mathcal{H}^T[\mathcal{H}\mathcal{R}\mathcal{H}^T + \mathcal{R}_n]^{-1} \tag{8.80}$$

where $\mathcal{R}$ and $\mathcal{R}_n$ are the covariance matrices of $\boldsymbol{u}$ and $\boldsymbol{n}$, respectively, which are assumed to be uncorrelated, that is,

$$\mathcal{R} = E[\boldsymbol{u}\boldsymbol{u}^T], \qquad \mathcal{R}_n = E[\boldsymbol{n}\boldsymbol{n}^T], \qquad E[\boldsymbol{u}\boldsymbol{n}^T] = \mathbf{0} \tag{8.81}$$

The resulting error is found to be

$$\sigma_e^2 = \frac{1}{M_1 M_2} \operatorname{Tr}[(\mathbf{I} - \mathcal{G}\mathcal{H})\mathcal{R}] \tag{8.82}$$

Assuming $\mathcal{R}$ and $\mathcal{R}_n$ to be positive definite, (8.80) can be written via the ABCD lemma for inversion of a matrix $\mathbf{A} - \mathbf{BCD}$ (see Table 2.6) as

$$\mathcal{G} = [\mathcal{H}^T \mathcal{R}_n^{-1} \mathcal{H} + \mathcal{R}^{-1}]^{-1} \mathcal{H}^T \mathcal{R}_n^{-1} \tag{8.83}$$

For the stationary observation model where the PSF is spatially invariant, the object is a stationary random field, and the noise is white with zero mean and variance σ_n^2, the Wiener filter becomes

$$\mathcal{G} = \mathcal{R}\mathcal{H}^T[\mathcal{H}\mathcal{R}\mathcal{H}^T + \sigma_n^2 \mathbf{I}]^{-1} = [\mathcal{H}^T\mathcal{H} + \sigma_n^2 \mathcal{R}^{-1}]^{-1}\mathcal{H}^T \tag{8.84}$$

where $\mathcal{R}$ and $\mathcal{H}$ are doubly block Toeplitz matrices.

In the case of no blur, $\mathcal{H} = \mathbf{I}$, and $\mathcal{G}$ becomes the optimum noise smoothing filter

$$\mathcal{G} = \mathcal{R}[\mathcal{R} + \mathcal{R}_n]^{-1} = [\mathcal{R}_n^{-1} + \mathcal{R}^{-1}]^{-1}\mathcal{R}_n^{-1} \tag{8.85}$$

Remarks

If the noise power goes to zero, then the Wiener filter becomes the pseudoinverse of $\mathcal{H}$ (see Section 8.9), that is,

$$\mathcal{G}^- \triangleq \lim_{\sigma_n^2 \to 0} \mathcal{G} = \begin{cases} [\mathcal{H}^T\mathcal{H}]^{-1}\mathcal{H}^T, & \text{if } N_1N_2 \ge M_1M_2, \ \text{rank}[\mathcal{H}^T\mathcal{H}] = M_1M_2 \\ \mathcal{R}\mathcal{H}^T[\mathcal{H}\mathcal{R}\mathcal{H}^T]^{-1}, & \text{if } N_1N_2 \le M_1M_2, \ \text{rank}[\mathcal{H}\mathcal{R}\mathcal{H}^T] = N_1N_2 \end{cases} \tag{8.86}$$

The Wiener filter is not separable (that is, $\mathcal{G} \neq \mathbf{G}_1 \otimes \mathbf{G}_2$) even if the PSF and the object and noise covariance functions are separable. However, the pseudo-inverse filter $\mathcal{G}^-$ is separable if $\mathcal{R}$ and $\mathcal{H}$ are.

The Wiener filter for nonzero mean random variables $\boldsymbol{u}$ and $\boldsymbol{n}$ becomes

$$\hat{\boldsymbol{u}} = \boldsymbol{\mu}_u + \mathcal{G}(\boldsymbol{v} - \boldsymbol{\mu}_v) = \mathcal{G}\boldsymbol{v} + [\mathcal{H}^T\mathcal{R}_n^{-1}\mathcal{H} + \mathcal{R}^{-1}]^{-1}\mathcal{R}^{-1}\boldsymbol{\mu}_u - \mathcal{G}\boldsymbol{\mu}_n \tag{8.87}$$

This form of the filter is useful only when the mean of the object and/or the noise are spatially varying.

Generalized Wiener Filtering

The size of the filter matrix $\mathcal{G}$ becomes quite large even when the arrays $u(m, n)$ and $v(m, n)$ are small. For example, for 16×16 arrays, $\mathcal{G}$ is 256×256. Moreover, to calculate $\mathcal{G}$, a matrix of size $M_1 M_2 \times M_1 M_2$ or $N_1 N_2 \times N_1 N_2$ has to be inverted. Generalized Wiener filtering gives an efficient method of approximately implementing the Wiener filter using fast unitary transforms. For instance, using (8.84) in (8.77) and defining

$$\mathcal{P} \triangleq [\mathcal{H}^T\mathcal{H} + \sigma_n^2\mathcal{R}^{-1}]^{-1} \tag{8.88}$$

we can write

$$\left.\begin{aligned} \hat{\boldsymbol{u}} &= \mathcal{A}^{*T}[\mathcal{A}\mathcal{P}\mathcal{A}^{*T}]\mathcal{A}[\mathcal{H}^T\boldsymbol{v}] \\ &\triangleq \mathcal{A}^{*T}\tilde{\mathcal{P}}\boldsymbol{w} \end{aligned}\right\} \tag{8.89}$$

where $\mathcal{A}$ is a unitary matrix, $\tilde{\mathcal{P}} \triangleq \mathcal{A}\mathcal{P}\mathcal{A}^{*T}$, $\boldsymbol{w} \triangleq \mathcal{A}[\mathcal{H}^T \boldsymbol{v}]$. This equation suggests the implementation of Fig. 8.20 and is called generalized Wiener filtering. For stationary observation models, $\mathcal{H}^T \boldsymbol{v}$ is a convolution operation and $\tilde{\mathcal{P}}$ turns out to be nearly diagonal for many unitary transforms. Therefore, approximating

$$\hat{\boldsymbol{w}} \triangleq \tilde{\mathcal{P}}\boldsymbol{w} \simeq [\text{Diag}\, \tilde{\mathcal{P}}]\boldsymbol{w} \tag{8.90}$$

and mapping $\boldsymbol{w}$ and $\hat{\boldsymbol{w}}$ back into $N_1 \times N_2$ arrays, we get

$$\hat{w}(k, l) \simeq \tilde{p}(k, l) w(k, l) \tag{8.91}$$

where $\tilde{p}(k, l)$ come from the diagonal elements of $\tilde{\mathcal{P}}$. Figure 8.20 shows the implementation of generalized Wiener filtering when $\tilde{\mathcal{P}}$ is approximated by its diagonal. Now, if $\mathcal{A}$ is a fast transform requiring $O(N_1 N_2 \log N_1 N_2)$ operations for transforming an $N_1 \times N_2$ array, the total number of operations are reduced from $O(M_1^3 M_2^3)$ to $O(M_1 M_2 \log M_1 M_2)$. For 16×16 arrays, this means a reduction from $O(2^{24})$ to $O(2^{11})$. Among the various fast transforms, the cosine and sine transforms have been found useful for generalized Wiener filtering of images [12, 13].

The generalized Wiener filter implementation of Fig. 8.20 is exact when $\mathcal{A}$ diagonalizes $\mathcal{P}$. However, this implementation is impractical because this diagonalizing transform is generally not a fast transform. For a fast transform $\mathcal{A}$, the diagonal elements of $\mathcal{P}$ may be approximated as follows

$$\begin{aligned} \text{Diag}\{\tilde{\mathcal{P}}\} &= \text{Diag}\{\mathcal{A}[\mathcal{H}^T \mathcal{H} + \sigma_n^2 \mathcal{R}^{-1}]^{-1} \mathcal{A}^{*T}\} \\ &\simeq [\text{Diag}\{\mathcal{A}\mathcal{H}^T \mathcal{H}\mathcal{A}^{*T}\} + \sigma_n^2 (\text{Diag}\{\mathcal{A}\mathcal{R}\mathcal{A}^{*T}\})^{-1}]^{-1} \end{aligned}$$

where we have approximated $\text{Diag}\{\mathcal{A}\mathcal{R}^{-1}\mathcal{A}^{*T}\}$ by $[\text{Diag}\{\mathcal{A}\mathcal{R}\mathcal{A}^{*T}\}]^{-1}$. This gives

$$\tilde{p}(k, l) \simeq \frac{1}{(\tilde{h}_2(k, l) + \sigma_n^2 / \tilde{\gamma}(k, l))} \tag{8.92}$$

where $\tilde{h}_2(k, l)$ and $\tilde{\gamma}(k, l)$ are the elements on the diagonal of $\mathcal{A}\mathcal{H}^T \mathcal{H}\mathcal{A}^{*T}$ and $\mathcal{A}\mathcal{R}\mathcal{A}^{*T}$, respectively. Note that $\tilde{\gamma}(k, l)$ are the mean square values of the transform coefficients, that is, if $\tilde{\boldsymbol{u}} \triangleq \mathcal{A}\boldsymbol{u}$, then $\tilde{\gamma}(k, l) = E[|\tilde{u}(k, l)|^2]$.

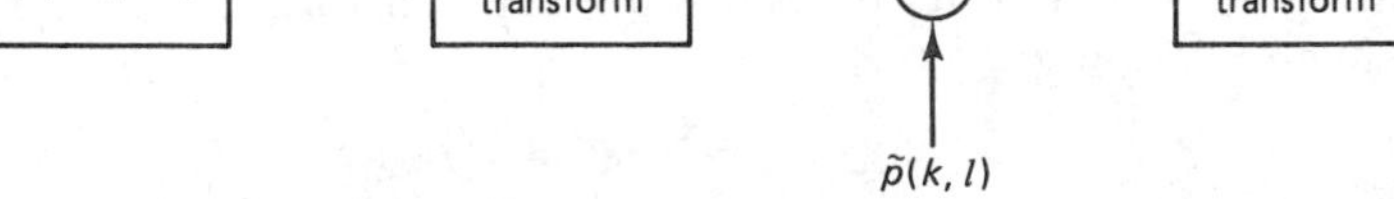

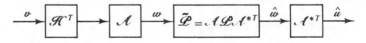

(a) Wiener filter in $\mathcal{A}$-transform domain

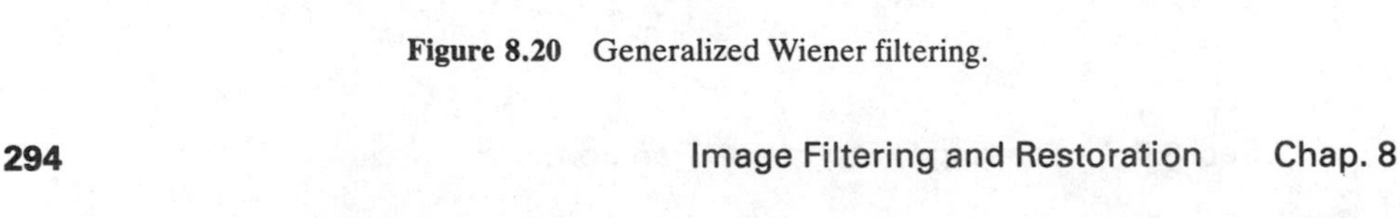

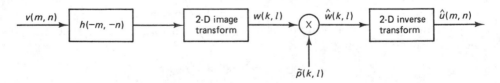

(b) Implementation by approximation of $\tilde{\mathcal{P}}$ by its diagonal

Figure 8.20 Generalized Wiener filtering.

Filtering by Fast Decompositions [13]

For narrow PSFs and stationary observation models, it is often possible to decompose the filter gain matrix $\mathcal{G}$ as

$$\mathcal{G} = \mathcal{G}_0 + \mathcal{G}_b, \qquad \mathcal{G}_0 \triangleq \mathcal{P}_0 \mathcal{H}^T, \qquad \mathcal{G}_b \triangleq \mathcal{P}_b \mathcal{H}^T \tag{8.93}$$

where $\mathcal{P}_0$ can be diagonalized by a fast transform and $\mathcal{P}_b$ is sparse and of low rank. Hence, the filter equation can be written as

$$\hat{u} = \mathcal{G} v = \mathcal{G}_0 v + \mathcal{G}_b v \triangleq \hat{u}^0 + \hat{u}^b \tag{8.94}$$

When $\tilde{\mathcal{P}}_0 \triangleq \mathcal{A}\mathcal{P}_0\mathcal{A}^{*T}$ is diagonal for some fast unitary transform $\mathcal{A}$, the component $\hat{u}^0$ can be written as

$$\hat{u}^0 = \mathcal{P}_0 \mathcal{H}^T v = \mathcal{A}^{*T}[\mathcal{A}\mathcal{P}_0\mathcal{A}^{*T}]\mathcal{A}[\mathcal{H}^T v] = \mathcal{A}^{*T} \tilde{\mathcal{P}}_0 w \tag{8.95}$$

and can be implemented as an exact generalized Wiener filter. The other component, $\hat{u}^b$, generally requires performing operations on only a few boundary values of $\mathcal{H}^T v$. The number of operations required for calculating $\hat{u}^b$ depends on the width of the PSF and the image correlation model. Usually these require fewer operations than in the direct computation of $\hat{u}$ from $\mathcal{G}$. Therefore, the overall filter can be implemented quite efficiently. If the image size is large compared to the width of the PSF, the boundary response can be ignored. Then, $\mathcal{G}$ is approximately $\mathcal{G}_0$, which can be implemented via its diagonalizing fast transform [13].

8.7 SMOOTHING SPLINES AND INTERPOLATION [15–17]

Smoothing splines are curves used to estimate a continuous function from its sample values available on a grid. In image processing, spline functions are useful for magnification and noise smoothing. Typically, pixels in each horizontal scan line are first fit by smoothing splines and then finely sampled to achieve a desired magnification in the horizontal direction. The same procedure is then repeated along the vertical direction. Thus the image is smoothed and interpolated by a separable function.

Let $\{y_i, 0 \le i \le N\}$ be a given set of $(N+1)$ observations of a continuous function $f(x)$ sampled uniformly (this is only a convenient assumption) such that

$$x_i = x_0 + ih, \qquad h > 0$$

and

$$y_i = f(x_i) + n(x_i) \tag{8.96}$$

where $n(x)$ represents the errors (or noise) in the observation process. Smoothing splines fit a smooth function $g(x)$ through the available set of observations such that its "roughness," measured by the energy in the second derivatives (that is, $\int [d^2 g(x)/dx^2]^2 \, dx$), over $[x_0, x_N]$ is minimized. Simultaneously, the least squares error at the observation points is restricted, that is, for $g_i \triangleq g(x_i)$,

$$F \triangleq \sum_{i=0}^{N} \left(\frac{g_i - y_i}{\sigma_n}\right)^2 \le S \tag{8.97}$$

For $S = 0$, this means an absolute fit through the observation points is required. Typically, σ_n^2 is the mean square value of the noise and S is chosen to lie in the range $(N+1) \pm \sqrt{2(N+1)}$, which is also called the confidence interval of S. The minimization problem has two solutions:

1. When S is sufficiently large, the constraint of (8.97) is satisfied by a straight-line fit

$$\left.\begin{aligned} g(x) &= a + bx, \qquad x_0 \le x \le x_N \\ b &= \frac{(\mu_{xy} - \mu_x \mu_y)}{(\mu_{xx} - \mu_x^2)}, \qquad a = \mu_y - b\mu_x \end{aligned}\right\} \tag{8.98}$$

where μ denotes sample average, for instance, $\mu_x \stackrel{\Delta}{=} \left(\sum_{i=0}^{N} x_i\right)/(N+1)$, and so on.

2. The constraint in (8.97) is tight, so only the equality constraint can be satisfied. The solution becomes a set of piecewise continuous third-order polynomials called *cubic splines*, given by

$$g(x) = a_i + b_i(x - x_i) + c_i(x - x_i)^2 + d_i(x - x_i)^3, \qquad x_i \le x < x_{i+1} \tag{8.99}$$

The coefficients of these spline functions are obtained by solving

$$\left.\begin{aligned} &[\mathbf{P} + \lambda\mathbf{Q}]\mathbf{c} = \lambda\mathbf{v}, \qquad \mathbf{v} \stackrel{\Delta}{=} \mathbf{L}^T\mathbf{y}, \qquad c_0 = c_N = 0 \\ &\mathbf{a} = \mathbf{y} - \frac{\sigma_n^2}{\lambda}\mathbf{L}\mathbf{c} \\ &d_i = \frac{c_{i+1} - c_i}{3h}, \qquad 1 \le i \le N-1, d_0 = d_N = 0 \\ &b_i = \frac{a_{i+1} - a_i}{h} - hc_i - h^2 d_i, \qquad b_N = 0 \qquad 0 \le i \le N-1 \end{aligned}\right\} \tag{8.100}$$

where $\mathbf{a}$ and $\mathbf{y}$ are $(N+1) \times 1$ vectors of the elements $\{a_i, 0 \le i \le N\}, \{y_i, 0 \le i \le N\}$, respectively, and $\mathbf{c}$ is the $(N-1) \times 1$ vector of elements $[c_i, 1 \le i \le N-1]$. The matrices $\mathbf{Q}$ and $\mathbf{L}$ are, respectively, $(N-1) \times (N-1)$ tridiagonal Toeplitz and $(N+1) \times (N-1)$ lower triangular Toeplitz, namely,

$$\mathbf{Q} \stackrel{\Delta}{=} \frac{h}{3}\begin{bmatrix} 4 & 1 & & \mathbf{0} \\ 1 & \ddots & \ddots & \\ & \ddots & \ddots & 1 \\ \mathbf{0} & & 1 & 4 \end{bmatrix}, \qquad \mathbf{L} \stackrel{\Delta}{=} \frac{1}{h}\begin{bmatrix} 1 & & \mathbf{0} \\ -2 & \ddots & \\ 1 & \ddots & 1 \\ & \ddots & -2 \\ \mathbf{0} & & 1 \end{bmatrix} \tag{8.101}$$

and $\mathbf{P} \stackrel{\Delta}{=} \sigma_n^2 \mathbf{L}^T\mathbf{L}$. The parameter λ is such that the equality constraint

$$F(\lambda) \stackrel{\Delta}{=} \frac{\|\mathbf{a} - \mathbf{y}\|^2}{\sigma_n^2} = \mathbf{v}^T\mathbf{A}\mathbf{P}\mathbf{A}\mathbf{v} = S \tag{8.102}$$

where $\mathbf{A} \triangleq [\mathbf{P} + \lambda \mathbf{Q}]^{-1}$, is satisfied. The nonlinear equation $F(\lambda) = S$ is solved iteratively—for instance, by Newton's method—as

$$\lambda_{k+1} = \lambda_k - \frac{F(\lambda_k) - S}{F'(\lambda_k)}, \qquad F'(\lambda) \triangleq \frac{dF(\lambda)}{d\lambda} = 2\mathbf{v}^T \mathbf{AQAPAv} \tag{8.103}$$

Remarks

The solution of (8.99) gives $g_i \triangleq g(x_i) = a_i$, that is, $\mathbf{a}$ is the best least squares estimate of $\mathbf{y}$. It can be shown that $\mathbf{a}$ can also be considered as the Wiener filter estimate of $\mathbf{y}$ based on appropriate autocorrelation models (see Problem 8.15).

The special case $S = 0$ gives what are called the *interpolating splines,* where the splines must pass through the given data points.

Example 8.6

Suppose the given data is 1, 3, 4, 2, 1 and let $h = 1$ and $\sigma_n = 1$. Then $N = 4, x_0 = 0$, $x_4 = 4$. For the case of straight-line fit, we get $\mu_x = 2$, $\mu_y = 2.2$, $\mu_{xy} = 4.2$, and $\mu_{xx} = 6$, which gives $b = -0.1$, $a = 2.4$, and $g(x) = 2.4 - 0.1x$. The least square error $\sum_i (y_i - g_i)^2 = 6.7$. The confidence interval for S is $[1.84, 8.16]$. However, if S is chosen to be less than 6.7, say $S = 5$, then we have to go to the cubic splines. Now

$$\mathbf{P} \triangleq \mathbf{L}^T\mathbf{L} = \begin{bmatrix} 6 & -4 & 1 \\ -4 & 6 & -4 \\ 1 & -4 & 6 \end{bmatrix}, \qquad \mathbf{v} \triangleq \mathbf{L}^T\mathbf{y} = \begin{bmatrix} -1 \\ -3 \\ 1 \end{bmatrix}$$

Solution of $F(\lambda) - S = 0$ gives $\lambda \simeq 0.0274$ for $S = 5$. From (8.100) we get

$$\mathbf{a} = \begin{bmatrix} 2.199 \\ 2.397 \\ 2.415 \\ 2.180 \\ 1.808 \end{bmatrix}, \quad \mathbf{b} = \begin{bmatrix} 0.198 \\ 0.057 \\ -0.194 \\ -0.357 \\ 0.000 \end{bmatrix}, \quad \mathbf{c} = \begin{bmatrix} 0.000 \\ -0.033 \\ -0.049 \\ -0.022 \\ 0.000 \end{bmatrix}, \quad \mathbf{d} = \begin{bmatrix} 0.000 \\ -0.005 \\ 0.009 \\ 0.007 \\ 0.000 \end{bmatrix}$$

The least squares error checks out to be $4.998 \simeq 5$.

8.8 LEAST SQUARES FILTERS [18, 19]

In the previous section we saw that the smoothing splines solve a least squares minimization problem. Many problems in linear estimation theory can also be reduced to least squares minimization problems. In this section we consider such problems in the context of image restoration.

Constrained Least Squares Restoration

Consider the spatially invariant image observation model of (8.39). The constrained least squares restoration filter output $\hat{u}(m, n)$, which is an estimate of $u(m, n)$, minimizes a quantity

$$J \triangleq \|q(m, n) \circledast \hat{u}(m, n)\|^2 \tag{8.104}$$

subject to the constraint

$$\|v(m,n) - h(m,n) \circledast \hat{u}(m,n)\|^2 \le \varepsilon^2 \tag{8.105}$$

where $\varepsilon^2 \ge 0$,

$$\|a(m,n)\|^2 \triangleq \sum_m \sum_n |a(m,n)|^2 = \frac{1}{4\pi^2} \iint_{-\pi}^{\pi} |A(\omega_1,\omega_2)|^2 \, d\omega_1 \, d\omega_2 \triangleq \|A(\omega_1,\omega_2)\|^2 \tag{8.106}$$

and $q(m,n)$ is an operator that measures the "roughness" of $\hat{u}(m,n)$. For example, if $q(m,n)$ is the impulse response of a high-pass filter, then minimization of J implies smoothing of high frequencies or rough edges. Using the Parseval relation this implies minimization of

$$J = \|Q(\omega_1,\omega_2)\hat{U}\,\omega_1,\omega_2)\|^2 \tag{8.107}$$

subject to

$$\|V(\omega_1,\omega_2) - H(\omega_1,\omega_2)\hat{U}(\omega_1,\omega_2)\|^2 \le \varepsilon^2 \tag{8.108}$$

The solution obtained via the Lagrange multiplier method gives

$$\hat{U}(\omega_1,\omega_2) = G_{ls}(\omega_1,\omega_2)V(\omega_1,\omega_2) \tag{8.109}$$

$$G_{ls} \triangleq \frac{H^*(\omega_1,\omega_2)}{|H(\omega_1,\omega_2)|^2 + \gamma|Q(\omega_1,\omega_2)|^2} \tag{8.110}$$

The Lagrange multiplier γ is determined such that $\hat{U}$ satisfies the equality in (8.108) subject to (8.109) and (8.110). This yields a nonlinear equation for γ:

$$f(\gamma) \triangleq \frac{\gamma^2}{4\pi^2} \iint_{-\pi}^{\pi} \frac{|Q|^4|V|^2}{(|H|^2 + \gamma|Q|^2)^2} \, d\omega_1 \, d\omega_2 - \varepsilon^2 = 0 \tag{8.111}$$

In least squares filtering, $q(m,n)$ is commonly chosen as a finite difference approximation of the Laplacian operator $\partial^2/\partial x^2 + \partial^2/\partial y^2$. For example, on a square grid with spacing $\Delta x = \Delta y = 1$, and $\alpha = \frac{1}{4}$ below, one obtains

$$\begin{aligned} q(m,n) \triangleq -\delta(m,n) + \alpha[\delta(m-1,n) + \delta(m+1,n) \\ + \delta(m,n-1) + \delta(m,n+1)] \end{aligned} \tag{8.112}$$

which gives $Q(\omega_1,\omega_2) = -1 + 2\alpha \cos\omega_1 + 2\alpha\cos\omega_2$.

Remarks

For the vector image model problem,

$$\boldsymbol{v} = \mathcal{H}\boldsymbol{u} + \boldsymbol{n} \tag{8.113}$$

$$\min_{\hat{\boldsymbol{u}}} \|\mathcal{L}\hat{\boldsymbol{u}}\|^2, \qquad \|\boldsymbol{v} - \mathcal{H}\hat{\boldsymbol{u}}\|^2 \le \varepsilon^2 \tag{8.114}$$

where $\mathcal{L}$ is a given matrix, the least squares filter is obtained by solving

$$\left.\begin{aligned} &\hat{\boldsymbol{u}} = (\gamma\mathcal{L}^T\mathcal{L} + \mathcal{H}^T\mathcal{H})^{-1}\mathcal{H}^T\boldsymbol{v} \\ &\gamma^2\|[\mathcal{H}[\mathcal{L}^T\mathcal{L}]^{-1}\mathcal{H}^T + \gamma\mathbf{I}]^{-1}\boldsymbol{v}\|^2 - \varepsilon^2 = 0 \end{aligned}\right\} \tag{8.115}$$

In the absence of blur, one obtains the least squares smoothing filter

$$\hat{u} = (\gamma \mathcal{L}^T \mathcal{L} + \mathbf{I})^{-1} v \tag{8.116}$$

This becomes the two-dimensional version of the smoothing spline solution if $\mathcal{L}$ is obtained from the discrete Laplacian operator of (8.112). A recursive solution of this problem is considered in [32].

Comparison of (8.110) and (8.116) with (8.41) and (8.84), respectively, shows that the least squares filter is in the class of Wiener filters. For example, if we specify $S_{\eta\eta} \overset{\Delta}{=} \gamma$ and $S_{uu} = 1/|Q|^2$, then the two filters are identical. In fact, specifying $g(m, n)$ is equivalent to modeling the object by a random field whose power spectral density function is $1/|Q|^2$ (see Problem 8.17).

8.9 GENERALIZED INVERSE, SVD, AND ITERATIVE METHODS

The foregoing least squares and mean square restoration filters can also be realized by direct minimization of their quadratic cost functionals. Such direct minimization techniques are most useful when little is known about the statistical properties of the observed image data and when the PSF is spatially variant. Consider the image observation model

$$\mathbf{v} \simeq \mathbf{H}\mathbf{u} \tag{8.117}$$

where $\mathbf{v}$ and $\mathbf{u}$ are vectors of appropriate dimensions and $\mathbf{H}$ is a rectangular (say $M \times N$) PSF matrix. The unconstrained least squares estimate $\hat{\mathbf{u}}$ minimizes the norm

$$J = \|\mathbf{v} - \mathbf{H}\hat{\mathbf{u}}\|^2 \tag{8.118}$$

The Pseudoinverse

A solution vector $\hat{\mathbf{u}}$ that minimizes (8.118) must satisfy

$$\mathbf{H}^T\mathbf{H}\hat{\mathbf{u}} = \mathbf{H}^T\mathbf{v} \tag{8.119}$$

If $\mathbf{H}^T\mathbf{H}$ is nonsingular, this gives

$$\hat{\mathbf{u}} = \mathbf{H}^-\mathbf{v}, \qquad \mathbf{H}^- \overset{\Delta}{=} (\mathbf{H}^T\mathbf{H})^{-1}\mathbf{H}^T \tag{8.120}$$

as the unique least squares solution. $\mathbf{H}^-$ is called the *pseudoinverse* of $\mathbf{H}$. This pseudoinverse has the interesting property

$$\mathbf{H}^-\mathbf{H} = \mathbf{I} \tag{8.121}$$

Note, however, that $\mathbf{H}\mathbf{H}^- \neq \mathbf{I}$. If $\mathbf{H}$ is $M \times N$, then $\mathbf{H}^-$ is $N \times M$. A necessary condition for $\mathbf{H}^T\mathbf{H}$ to be nonsingular is that $M \geq N$ and the rank of $\mathbf{H}$ should be N. If $M < N$ and the rank of $\mathbf{H}$ is M, then the pseudoinverse of $\mathbf{H}$ is defined as a matrix H^-, which satisfies

$$\mathbf{H}\mathbf{H}^- = \mathbf{I} \tag{8.122}$$

In this case, H^- is not unique. One solution is

$$\mathbf{H}^- = \mathbf{H}^T(\mathbf{H}\mathbf{H}^T)^{-1}. \tag{8.123}$$

In general, whenever the rank of $\mathbf{H}$ is $r < N$, (8.119) does not have a unique solution. Additional constraints on $\hat{\mathbf{u}}$ are then necessary to make it unique.

Minimum Norm Least Squares (MNLS) Solution and the Generalized Inverse

A vector $\mathbf{u}^+$ that has the minimum norm $\|\hat{\mathbf{u}}\|^2$ among all the solutions of (8.119) is called the *MNLS* (minimum norm least squares) solution. Thus

$$\mathbf{u}^+ = \min_{\hat{\mathbf{u}}}\{\|\hat{\mathbf{u}}\|^2;\ \mathbf{H}^T\mathbf{H}\hat{\mathbf{u}} = \mathbf{H}^T\mathbf{v}\} \tag{8.124}$$

Clearly, if rank $[\mathbf{H}^T\mathbf{H}] = N$, then $\mathbf{u}^+ = \hat{\mathbf{u}}$ is the least squares solution. Using the singular value expansion of $\mathbf{H}$, it can be shown that the transformation between $\mathbf{v}$ and $\mathbf{u}^+$ is linear and unique and is given by

$$\mathbf{u}^+ = \mathbf{H}^+\mathbf{v} \tag{8.125}$$

The matrix $\mathbf{H}^+$ is called the *generalized inverse* of $\mathbf{H}$. If the $M \times N$ matrix $\mathbf{H}$ has the SVD expansion [see (5.188)]

$$\mathbf{H} = \sum_{m=1}^{r} \lambda_m^{1/2}\boldsymbol{\psi}_m\boldsymbol{\phi}_m^T \tag{8.126a}$$

then, $\mathbf{H}^+$ is an $N \times M$ matrix with an SVD expansion

$$\mathbf{H}^+ = \sum_{m=1}^{r} \lambda_m^{-1/2}\boldsymbol{\phi}_m\boldsymbol{\psi}_m^T \tag{8.126b}$$

where $\boldsymbol{\phi}_m$ and $\boldsymbol{\psi}_m$ are, respectively, the eigenvectors of $\mathbf{H}^T\mathbf{H}$ and $\mathbf{H}\mathbf{H}^T$ corresponding to the singular values $\{\lambda_m, 1 \le m \le r\}$.

Using (8.126b), it can be shown $\mathbf{H}^+$ satisfies the following relations:

1. $\mathbf{H}^+ = (\mathbf{H}^T\mathbf{H})^{-1}\mathbf{H}^T$, if $r = N$
2. $\mathbf{H}^+ = \mathbf{H}^T(\mathbf{H}\mathbf{H}^T)^{-1}$, if $r = M$
3. $\mathbf{H}\mathbf{H}^+ = (\mathbf{H}\mathbf{H}^+)^T$
4. $\mathbf{H}^+\mathbf{H} = (\mathbf{H}^+\mathbf{H})^T$
5. $\mathbf{H}\mathbf{H}^+\mathbf{H} = \mathbf{H}$
6. $\mathbf{H}^+\mathbf{H}\mathbf{H}^T = \mathbf{H}^T$

The first two relations show that $\mathbf{H}^+$ is a pseudoinverse if $r = N$ or M. The MNLS solution is

$$\mathbf{u}^+ = \sum_{m=1}^{r} \lambda_m^{-1/2}\boldsymbol{\phi}_m\boldsymbol{\psi}_m^T\mathbf{v} \tag{8.127}$$

This method is quite general and is applicable to arbitrary PSFs. The major difficulty is computational because it requires calculation of $\boldsymbol{\psi}_k$ and $\boldsymbol{\phi}_k$ for large matrices. For example, for $M = N = 256$, $\mathbf{H}$ is $65{,}536 \times 65{,}536$. For images with separable PSF, that is, $\mathbf{V} \simeq \mathbf{H}_1 \mathbf{U} \mathbf{H}_2$, the generalized inverse is also separable, giving $\mathbf{U}^+ = \mathbf{H}_1^+ \mathbf{V} \mathbf{H}_2^+$.

One-Step Gradient Methods

When the ultimate aim is to obtain the restored solution $\mathbf{u}^+$, rather than the explicit pseudoinverse $\mathbf{H}^+$, then iterative gradient methods are useful. One-step gradient algorithms are of the form

$$\mathbf{u}_{n+1} = \mathbf{u}_n - \sigma_n \mathbf{g}_n, \qquad \mathbf{u}_0 = 0 \tag{8.128}$$

$$\mathbf{g}_n \triangleq -\mathbf{H}^T(\mathbf{v} - \mathbf{H}\mathbf{u}_n) = \mathbf{g}_{n-1} - \sigma_{n-1}\mathbf{H}^T\mathbf{H}\mathbf{g}_{n-1} \tag{8.129}$$

where $\mathbf{u}_n$ and $\mathbf{g}_n$ are, respectively, the trial solution and the gradient of J at iteration step n and σ_n is a scalar quantity. For the interval $0 \le \sigma_n < 2/\lambda_{\max}(\mathbf{H}^T\mathbf{H})$, $\mathbf{u}_n$ converges to the MNLS solution $\mathbf{u}^+$ as $n \to \infty$. If σ_n is chosen to be a constant, then its optimum value for fastest convergence is given by [22]

$$\sigma_{\text{opt}} = \frac{2}{[\lambda_{\max}(\mathbf{H}^T\mathbf{H}) + \lambda_{\min}(\mathbf{H}^T\mathbf{H})]} \tag{8.130}$$

For a highly ill conditioned matrix $\mathbf{H}^T\mathbf{H}$, *the condition number*, $\lambda_{\max}/\lambda_{\min}$ is large. Then σ_{opt} is close to its upper bound, $2/\lambda_{\max}$ and the error at iteration n, for $\sigma_n = \sigma$, obeys

$$\mathbf{e}_n \triangleq \mathbf{u}^+ - \mathbf{u}_n = (\mathbf{I} - \sigma\mathbf{H}^T\mathbf{H})\mathbf{e}_{n-1} \tag{8.131}$$

This implies that $\|\mathbf{e}_n\|$ is proportional to $\|\mathbf{e}_{n-1}\|$, that is, the convergence is linear and can be very slow, because σ is bounded from above. To improve the speed of convergence, σ is optimized at each iteration, which yields the *steepest descent* algorithm, with

$$\sigma_n = \frac{\mathbf{g}_n^T\mathbf{g}_n}{\mathbf{g}_n^T\mathbf{A}\mathbf{g}_n}, \qquad \mathbf{A} \triangleq \mathbf{H}^T\mathbf{H} \tag{8.132}$$

However, even this may not significantly help in speeding up the convergence when the condition number of $\mathbf{A}$ is high.

Van Cittert Filter [4]

From (8.128) and (8.129), the solution at iteration $n = i$, with $\sigma_n = \sigma$ can be written as

$$\mathbf{u}_{i+1} = \mathbf{G}_i\mathbf{v}, \qquad 0 < \sigma < \frac{2}{\lambda_{\max}}$$

where $\mathbf{G}_i$ is a power series:

$$\mathbf{G}_i = \sigma \sum_{k=0}^{i} (\mathbf{I} - \sigma\mathbf{H}^T\mathbf{H})^k\mathbf{H}^T \tag{8.133}$$

This is called the *Van Cittert filter*. Physically it represents passing the modified observation vector $\sigma \mathbf{H}^T \mathbf{v}$ through i stages of identical filters and summing their outputs (Fig. 8.21). For a spatially invariant PSF, this requires first passing the observed image $v(m, n)$ through a filter whose frequency response is $\sigma H^*(\omega_1, \omega_2)$ and then through i stages of identical filters, each having the frequency response $1 - \sigma|H(\omega_1, \omega_2)|^2$.

One advantage of this method is that the region of support of each filter stage is only twice (in each dimension) that of the PSF. Therefore, if the PSF is not too broad, each filter stage can be conveniently implemented by an FIR filter. This can be attractive for real-time pipeline implementations.

The Conjugate Gradient Method [22–24]

The conjugate gradient method is based on finding conjugate directions, which are vectors $\mathbf{d}_i \neq \mathbf{0}$, such that

$$\mathbf{d}_i^T \mathbf{A} \mathbf{d}_j = 0, \qquad i \neq j,\ 0 \leq i, j \leq N - 1 \tag{8.134}$$

When $\mathbf{A}$ is positive definite, such a set of vectors exists and forms a basis in the N-dimensional vector space. In terms of this basis, the solution can be written as

$$\mathbf{u}^+ = \sum_{i=0}^{N-1} \alpha_i \mathbf{d}_i \tag{8.135}$$

The scalars α_i and vectors $\mathbf{d}_i$ can be calculated conveniently via the recursions

$$\left.\begin{aligned}
&\mathbf{u}_{n+1} = \mathbf{u}_n + \alpha_n \mathbf{d}_n, \qquad \alpha_n \triangleq -\frac{\mathbf{g}_n^T \mathbf{d}_n}{\mathbf{d}_n^T \mathbf{A} \mathbf{d}_n}, \qquad \mathbf{u}_0 \triangleq \mathbf{0} \\
&\mathbf{d}_n = -\mathbf{g}_n + \beta_{n-1} \mathbf{d}_{n-1}, \qquad \beta_{n-1} \triangleq \frac{\mathbf{g}_n^T \mathbf{A} \mathbf{d}_{n-1}}{\mathbf{d}_{n-1}^T \mathbf{A} \mathbf{d}_{n-1}}, \qquad \mathbf{d}_0 \triangleq -\mathbf{g}_0 \\
&\mathbf{g}_n = -\mathbf{H}^T \mathbf{v} + \mathbf{A} \mathbf{u}_n = \mathbf{g}_{n-1} + \alpha_{n-1} \mathbf{A} \mathbf{d}_{n-1}, \qquad \mathbf{g}_0 \triangleq -\mathbf{H}^T \mathbf{v}
\end{aligned}\right\} \tag{8.136}$$

In this algorithm, the direction vectors $\mathbf{d}_n$ are conjugate and α_n minimizes J, not only along the nth direction but also over the subspace generated by $\mathbf{d}_0, \mathbf{d}_1, \ldots, \mathbf{d}_{n-1}$. Hence, the minimum is achieved in at most N steps, and the method is better than the one-step gradient method at each iteration. For image restoration problems where N is a large number, one need not run the algorithm to completion since large reductions in error are achieved in the first few steps. When rank $\mathbf{H}$ is $r < N$, then it is sufficient to run the algorithm up to $n = r - 1$ and $\mathbf{u}_r \to \mathbf{u}^+$.

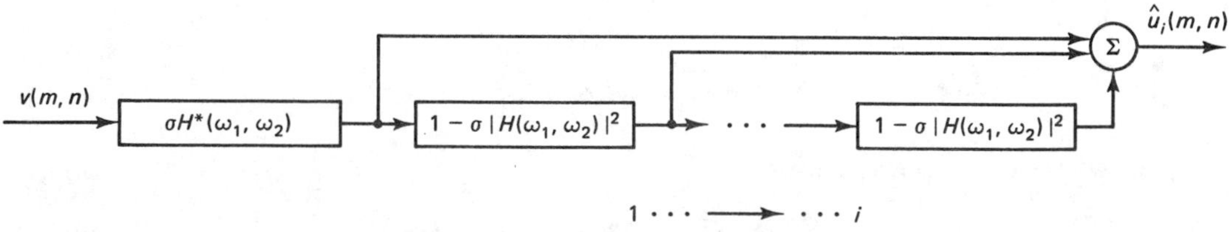

Figure 8.21 Van Cittert filter.

The major computation at each step requires the single matrix-vector product $\mathbf{Ad}_n$. All other vector operations require only N multiplications. If the PSF is narrow compared to the image size, we can make use of the sparseness of $\mathbf{A}$ and all the operations can be carried out quickly and efficiently.

Example 8.7

Consider the solution of

$$\begin{bmatrix} 1 & 2 \\ 2 & 1 \\ 1 & 3 \end{bmatrix} \begin{bmatrix} u_1 \\ u_2 \end{bmatrix} = \begin{bmatrix} 1 \\ 2 \\ 3 \end{bmatrix}$$

This gives

$$\mathbf{A} = \mathbf{H}^T \mathbf{H} = \begin{bmatrix} 6 & 7 \\ 7 & 14 \end{bmatrix}$$

whose eigenvalues are $\lambda_1 = 10 + \sqrt{65}$ and $\lambda_2 = 10 - \sqrt{65}$. Since $\mathbf{A}$ is nonsingular,

$$\mathbf{H}^+ = \mathbf{A}^{-1}\mathbf{H}^T = \tfrac{1}{35}\begin{bmatrix} 14 & -7 \\ -7 & 6 \end{bmatrix}\begin{bmatrix} 1 & 2 & 1 \\ 2 & 1 & 3 \end{bmatrix} = \tfrac{1}{35}\begin{bmatrix} 0 & 21 & -7 \\ 5 & -8 & 11 \end{bmatrix}$$

This gives

$$\mathbf{u}^+ = \mathbf{H}^+\mathbf{v} = \tfrac{1}{35}\begin{bmatrix} 21 \\ 22 \end{bmatrix} = \begin{bmatrix} 0.600 \\ 0.628 \end{bmatrix}$$

and

$$\|\mathbf{v} - \mathbf{H}\mathbf{u}^+\|^2 = 1.02857$$

For the one-step gradient algorithm, we get

$$\sigma = \sigma_{\text{opt}} = \tfrac{2}{20} = 0.1$$

and

$$\mathbf{g}_0 = \begin{bmatrix} -8 \\ -13 \end{bmatrix}, \qquad \mathbf{u}_1 = \begin{bmatrix} 0.8 \\ 1.3 \end{bmatrix}, \qquad J_1 = 9.46$$

After 12 iterations, we get

$$g_{11} = \begin{bmatrix} 0.685 \\ 1.253 \end{bmatrix}, \qquad \mathbf{u}_{12} = \begin{bmatrix} 0.555 \\ 0.581 \end{bmatrix}, \qquad J_{12} = 1.102$$

The steepest descent algorithm gives

$$\mathbf{u}_3 = \begin{bmatrix} 0.5992 \\ 0.6290 \end{bmatrix}, \qquad J_3 = 1.02857.$$

The conjugate gradient algorithm converges in two steps, as expected, giving $\mathbf{u}_2 = \mathbf{u}^+$. Both conjugate gradient and steepest descent are much faster than the one-step gradient algorithm.

Separable Point Spread Functions

If the two-dimensional point spread function is separable, then the generalized inverse is also separable. Writing $\mathbf{V} \simeq \mathbf{H}_1 \mathbf{U} \mathbf{H}_2^T$, we can rederive the various iterative algorithms. For example, the matrix form of the conjugate gradient algorithm becomes

$$\left.\begin{aligned}
&\mathbf{U}_{n+1}=\mathbf{U}_n+\alpha_n\mathbf{D}_n, &&\alpha_n=\frac{-\langle\mathbf{G}_n,\mathbf{D}_n\rangle}{\langle\mathbf{D}_n,\mathbf{A}_1\mathbf{D}_n\mathbf{A}_2\rangle}, \qquad \mathbf{U}_0=\mathbf{0}\\
&\mathbf{D}_n=-\mathbf{G}_n+\beta_{n-1}\mathbf{D}_{n-1}, &&\beta_{n-1}=\frac{\langle\mathbf{G}_n,\mathbf{A}_1\mathbf{D}_{n-1}\mathbf{A}_2\rangle}{\langle\mathbf{D}_{n-1},\mathbf{A}_1\mathbf{D}_{n-1}\mathbf{A}_2\rangle}\\
&\mathbf{G}_n=\mathbf{G}_{n-1}+\alpha_{n-1}\mathbf{A}_1\mathbf{D}_{n-1}\mathbf{A}_2, &&\mathbf{D}_0=\mathbf{G}_0=\mathbf{H}_1^T\mathbf{V}\mathbf{H}_2
\end{aligned}\right\} \tag{8.137}$$

where $\mathbf{A}_1 \triangleq \mathbf{H}_1^T\mathbf{H}_1$, $\mathbf{A}_2=\mathbf{H}_2^T\mathbf{H}_2$ and $\langle\mathbf{X},\mathbf{Y}\rangle \triangleq \sum_m\sum_n x(m,n)y(m,n)$. This algorithm has been found useful for restoration of images blurred by spatially variant PSFs [24].

8.10 RECURSIVE FILTERING FOR STATE VARIABLE SYSTEMS

Recursive filters realize an infinite impulse response with finite memory and are particularly useful for spatially varying restoration problems. In this section we consider the Kalman filtering technique, which is of fundamental importance in linear estimation theory.

Kalman Filtering [25, 26]

Consider a state variable system

$$\left.\begin{aligned}
\mathbf{x}_{n+1}&=\mathbf{A}_n\mathbf{x}_n+\mathbf{B}_n\boldsymbol{\varepsilon}_n, \qquad n=0,1,\ldots\\
\mathbf{z}_n&=\mathbf{C}_n\mathbf{x}_n\\
E[\mathbf{x}_0\mathbf{x}_0^T]&=\mathbf{R}^0, \qquad E[\boldsymbol{\varepsilon}_n\boldsymbol{\varepsilon}_{n'}]=\mathbf{P}_n\delta(n-n')
\end{aligned}\right\} \tag{8.138}$$

where $\mathbf{x}_n, \boldsymbol{\varepsilon}_n, \mathbf{z}_n$ are $m\times 1$, $p\times 1$, and $q\times 1$ vectors, respectively, all being Gaussian, zero mean, random sequences. $\mathbf{R}^0$ is the covariance matrix of the initial state vector $\mathbf{x}_0$, which is assumed to be uncorrelated with $\boldsymbol{\varepsilon}_0$. Suppose $\mathbf{z}_n$ is observed in the presence of additive white Gaussian noise as

$$\left.\begin{aligned}
\mathbf{y}_n&=\mathbf{z}_n+\boldsymbol{\eta}_n=\mathbf{C}_n\mathbf{x}_n+\boldsymbol{\eta}_n, \qquad n=0,1,\ldots\\
E[\boldsymbol{\eta}_n\boldsymbol{\eta}_{n'}]&=\mathbf{Q}_n\delta(n-n')
\end{aligned}\right\} \tag{8.139}$$

where $\boldsymbol{\eta}_n$ and $\boldsymbol{\varepsilon}_n$ are uncorrelated sequences. We define $\mathbf{s}_n$, $\mathbf{g}_n$, and $\hat{\mathbf{x}}_n$ as the best mean square estimates of the state variable $\mathbf{x}_n$ when the observations are available up to $n-1$, n, and $N>n$, respectively. Earlier, we have seen that mean square estimation of a sequence $\mathbf{x}(n)$ from observations $\mathbf{y}(n), 0\le n\le N$, gives us the Wiener filter. In this filter the entire estimated sequence is calculated from all the observations simultaneously [see (8.77)]. Kalman filtering theory gives recursive realizations of these estimates as the observation data arrives sequentially. This reduces the computational complexity of the Wiener filter calculations, especially when the state variable model is time varying. However, Kalman filtering requires a stochastic model of the quantity to be estimated (that is, object) in the form of

(8.138), whereas Wiener filtering is totally based on the knowledge of the autocorrelation of the observations and their cross-correlation with the object [see (8.80)].

One-Step Predictor. The one-step predictor is defined as $\mathbf{s}_n \triangleq E[\mathbf{x}_n|\mathbf{y}_{n'}, 0 \le n' \le n-1]$, which is the best mean square prediction of $\mathbf{x}_n$ based on observations up to $n-1$. It is given by the recursions

$$\left.\begin{aligned} &\mathbf{s}_{n+1} = \mathbf{A}_n \mathbf{s}_n + \mathbf{G}_n \mathbf{q}_n^{-1} \boldsymbol{\nu}_n, \qquad \mathbf{s}_0 = \mathbf{0} \\ &\boldsymbol{\nu}_n = \mathbf{y}_n - \mathbf{C}_n \mathbf{s}_n \\ &\mathbf{G}_n \triangleq \mathbf{A}_n \mathbf{R}_n \mathbf{C}_n^T, \qquad \mathbf{q}_n \triangleq \mathbf{C}_n \mathbf{R}_n \mathbf{C}_n^T + \mathbf{Q}_n \end{aligned}\right\} \tag{8.140}$$

$$\mathbf{R}_{n+1} = \mathbf{A}_n \mathbf{R}_n \mathbf{A}_n^T + \mathbf{B}_n \mathbf{P}_n \mathbf{B}_n^T - \mathbf{G}_n \mathbf{q}_n^{-1} \mathbf{G}_n^T, \qquad \mathbf{R}_0 = \mathbf{R}^0 \tag{8.141}$$

The sequence $\boldsymbol{\nu}_n$ is called the *innovations process*. It represents the new information obtained when the observation sample $\mathbf{y}_n$ arrives. It can be shown [25, 26] that $\boldsymbol{\nu}_n$ is a zero mean white noise sequence with covariances $\mathbf{q}_n$, that is,

$$E[\boldsymbol{\nu}_n] = 0, \qquad E[\boldsymbol{\nu}_n \boldsymbol{\nu}_{n'}^T] = \mathbf{q}_n \delta(n - n') \tag{8.142}$$

The nonlinear equation in (8.141) is called the *Riccati equation,* where $\mathbf{R}_n$ represents the covariance of the prediction error $\mathbf{e}_n \triangleq \mathbf{x}_n - \mathbf{s}_n$, which is also a white sequence:

$$E[\mathbf{e}_n] = \mathbf{0}, \qquad E[\mathbf{e}_n \mathbf{e}_{n'}^T] = \mathbf{R}_n \delta(n - n') \tag{8.143}$$

The $m \times q$ matrix $\mathbf{G}_n$ is called the Kalman gain. Note that $\mathbf{R}_n$, $\mathbf{G}_n$, and $\mathbf{q}_n$ depend only on the state variable model parameters and not on the data. Examining (8.140) and Fig. 8.22 we see that the model structure of $\mathbf{s}_n$ is recursive and is similar to the state variable model. The predictor equation can also be written as

$$\left.\begin{aligned} \mathbf{s}_{n+1} &= \mathbf{A}_n \mathbf{s}_n + \mathbf{G}_n \mathbf{q}_n^{-1} \boldsymbol{\nu}_n \\ \mathbf{y}_n &= \mathbf{C}_n \mathbf{s}_n + \boldsymbol{\nu}_n \end{aligned}\right\} \tag{8.144}$$

which is another state variable realization of the output sequence $\mathbf{y}_n$ from a white sequence $\boldsymbol{\nu}_n$ of covariances $\mathbf{q}_n$. This is called the *innovations representation* and is useful because it is causally invertible, that is, given the system output $\mathbf{y}_n$, one can find the input sequence $\boldsymbol{\nu}_n$ according to (8.140). Such representations have application in designing predictive coders for noisy data (see [6] of Chapter 11).

Online filter. The online filter is the best estimate of the current state based on all the observations received currently, that is,

$$\begin{aligned} \mathbf{g}_n &\triangleq E[\mathbf{x}_n|\mathbf{y}_{n'}, 0 \le n' \le n] \\ \mathbf{g}_n &= \mathbf{R}_n \mathbf{C}_n^T \mathbf{q}_n^{-1} \boldsymbol{\nu}_n + \mathbf{s}_n \end{aligned} \tag{8.145}$$

This estimate simply updates the predicted value $\mathbf{s}_n$ by the new information $\boldsymbol{\nu}_n$ with a gain factor $\mathbf{R}_n \mathbf{C}_n^T \mathbf{q}_n^{-1}$ to obtain the most current estimate. From (8.140), we now see that

$$\mathbf{s}_{n+1} = \mathbf{A}_n \mathbf{g}_n \tag{8.146}$$

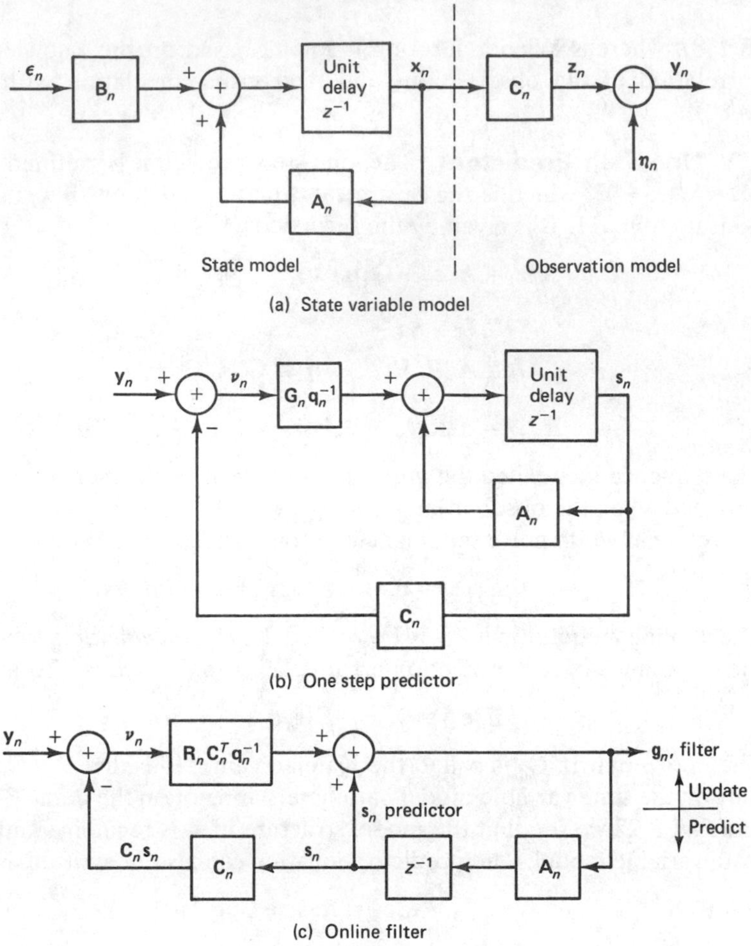

Figure 8.22 Kalman filtering.

Therefore, Kalman filtering can be thought of as having two steps (Fig. 8.22c). The first step is to update the previous prediction by the innovations. The next step is to predict from the latest update.

Fixed-interval smoother (Wiener filter). Suppose the observations are available over the full interval $[0, N]$. Then, by definition,

$$\hat{\mathbf{x}}_n = E[\mathbf{x}_n | \mathbf{y}_{n'}, 0 \le n' \le N]$$

is the Wiener filter for $\mathbf{x}_n$. With $\mathbf{x}_n$ given by (8.138), $\hat{\mathbf{x}}_n$ can be realized via the so-called *recursive backwards filter*

$$\hat{\mathbf{x}}_n = \mathbf{R}_n \boldsymbol{\lambda}_n + \mathbf{s}_n \tag{8.147}$$

$$\boldsymbol{\lambda}_n = \mathbf{A}_n^T \boldsymbol{\lambda}_{n+1} - \mathbf{C}_n^T \mathbf{q}_n^{-1} \mathbf{G}_n^T \boldsymbol{\lambda}_{n+1} + \mathbf{C}_n^T \mathbf{q}_n^{-1} \boldsymbol{\nu}_n, \qquad \boldsymbol{\lambda}_{N+1} = \mathbf{0} \tag{8.148}$$

To obtain $\hat{\mathbf{x}}_n$, the one-step predictor equations (8.140) are first solved recursively in the forward direction for $n = 0, 1, \ldots, N$ and the results $\mathbf{R}_n, \mathbf{s}_n$ (and possibly

$\mathbf{q}_n^{-1}, \mathbf{G}_n, \boldsymbol{\nu}_n$) are stored. Then the preceding two equations are solved recursively backwards from $n = N$ to $n = 0$.

Remarks

The number of computations and storage requirements for the various filters are roughly $O(m^3N^2) + O(m^2N^2) + O(q^3N)$. If $\mathbf{C}_n \mathbf{R}_n$, $\mathbf{G}_n$, and $\mathbf{q}_n$ have been precomputed and stored, then the number of online computations is $O(m^2N^2)$. It is readily seen that the Riccati equation has the largest computational complexity. Moreover, it is true that the Riccati equation and the filter gain calculations have numerical deficiencies that can cause large computational errors. Several algorithms that improve the computational accuracy of the filter gains are available and may be found in [26–28]. If we are primarily interested in the smooth estimate $\hat{\mathbf{x}}_n$, which is usually the case in image processing, then for shift invariant (or piecewise shift invariant) systems we can avoid the Riccati equation all together by going to certain FFT-based algorithms [29].

Example 8.8 (Recursive restoration of blurred images)

A blurred image observed with additive white noise is given by

$$y(n) = \sum_{k=-l_1}^{l_2} h(n, k)u(n-k) + \eta(n) \tag{8.149}$$

where $y(n), n = 1, 2, \ldots$ represents one scan line. The PSF $h(n, k)$ is assumed to be spatially varying here in order to demonstrate the power of recursive methods. Assume that each scan line is represented by a qth-order AR model

$$u(n) = \sum_{k=1}^{q} a(k)u(n-k) + \varepsilon(n), \quad E[\varepsilon(n)\varepsilon(n')] = \beta^2\delta(n-n') \tag{8.150}$$

Without loss of generality let $l_1 + l_2 \geq q$ and define a state variable $\mathbf{x}_n = [u(n+l_1) \ldots u(n+1), u(n) \ldots u(n-l_2)]^T$, which yields an $(l_1 + l_2 + 1)$-order state variable system as in (8.138) and (8.139), where

$$\mathbf{A} \triangleq \begin{bmatrix} a(1)\,a(2)\ldots a(q) & \vdots & 0\ldots 0 \\ 1 \quad 0 \ldots\ldots\ldots & \vdots & \mathbf{0} \quad \vdots \\ 0 \quad 1 \quad \mathbf{0} \ldots\ldots & \vdots & \vdots \\ \quad \ddots & & 0 \\ \mathbf{0} \quad\quad \ddots & & \vdots \\ \quad\quad\quad\quad \ddots \; 1 & & 0 \end{bmatrix}, \qquad \mathbf{B} \triangleq \begin{bmatrix} 1 \\ 0 \\ \vdots \\ 0 \end{bmatrix}, \qquad \mathbf{C}_n \triangleq [h(-l_1, n) \ldots h(l_2, n)]$$

and $\mathbf{e}_n \triangleq \varepsilon(n + l_1 + 1)$. This formulation is now in the framework of Kalman filtering, and the various recursive estimates can be obtained readily. Extension of these ideas to two-dimensional blurs is also possible. For example, it has been shown that an image blurred by a Gaussian PSFs representing atmospheric turbulence can be restored by a Kalman filter associated with a diffusion equation [66].

8.11 CAUSAL MODELS AND RECURSIVE FILTERING [32–39]

In this section we consider the recursive filtering of images having causal representations. Although the method presented next is valid for all causal MVRs, we restrict ourselves to the special case

$$u(m, n) = \rho_1 u(m-1, n) + \rho_2 u(m, n-1) - \rho_3 u(m-1, n-1) + \rho_4 u(m+1, n-1) + \varepsilon(m, n) \tag{8.151}$$

$$E[\varepsilon(m, n)] = 0, \qquad E[\varepsilon(m, n)\varepsilon(m-k, n-l)] = \beta^2 \delta(k)\delta(l) \tag{8.152}$$

We recall that for $\rho_4 = 0$, $\rho_3 = \rho_1 \rho_2$, and $\beta^2 = \sigma^2 (1 - \rho_1^2)(1 - \rho_2^2)$, this model is a realization of the separable covariance function of (2.84). The causality of (8.151) is with respect to column by column scanning of the image.

A Vector Recursive Filter

Let us denote $\mathbf{u}_n$ and $\boldsymbol{\varepsilon}_n$ as $N \times 1$ columns of their respective arrays. Then (8.151) can be written as

$$\mathbf{L}_1 \mathbf{u}_n = \mathbf{L}_2 \mathbf{u}_{n-1} + \boldsymbol{\varepsilon}_n, \qquad \operatorname{cov}[\boldsymbol{\varepsilon}_n] \triangleq \mathbf{P} = \beta^2 \mathbf{I}, E[\mathbf{u}_0 \mathbf{u}_0^T] \triangleq \mathbf{R}^0 \tag{8.153}$$

where, for convenience, the image $u(m, n)$ is assumed to be zero outside the length of the column $\mathbf{u}_n$, and $\mathbf{L}_1$ and $\mathbf{L}_2$ are Toeplitz matrices given by

$$\mathbf{L}_1 = \begin{bmatrix} 1 & & \mathbf{0} \\ -\rho_1 & \ddots & \\ \mathbf{0} & -\rho_1 & 1 \end{bmatrix}, \qquad \mathbf{L}_2 = \begin{bmatrix} \rho_2 & \rho_4 & \mathbf{0} \\ -\rho_3 & \ddots & \rho_4 \\ \mathbf{0} & -\rho_3 & \rho_2 \end{bmatrix} \tag{8.154}$$

Now (8.153) can be written as a vector AR process:

$$\mathbf{u}_n = \mathbf{L}\mathbf{u}_{n-1} + \mathbf{L}_1^{-1} \boldsymbol{\varepsilon}_n \tag{8.155}$$

where $\mathbf{L} = \mathbf{L}_1^{-1} \mathbf{L}_2$. Let the observations be given as

$$y(m, n) = u(m, n) + \eta(m, n) \tag{8.156}$$

where $\eta(m, n)$ is a stationary white noise field with zero mean and variance σ_η^2. In vector form, this becomes

$$\mathbf{y}_n = \mathbf{u}_n + \boldsymbol{\eta}_n \tag{8.157}$$

Equations (8.155) and (8.157) are now in the proper state variable form to yield the Kalman filter equations

$$\left.\begin{aligned} \mathbf{s}_{n+1} &= \mathbf{L}\mathbf{g}_n, \qquad \mathbf{s}_0 = \mathbf{0} \\ \mathbf{g}_n &= \mathbf{R}_n (\mathbf{R}_n + \sigma_\eta^2 \mathbf{I})^{-1} (\mathbf{y}_n - \mathbf{s}_n) + \mathbf{s}_n \\ \mathbf{R}_{n+1} &= \mathbf{L}\tilde{\mathbf{R}}_n \mathbf{L}^T + \mathbf{L}_1^{-1} \mathbf{P} (\mathbf{L}_1^T)^{-1}, \qquad \mathbf{R}_0 = \mathbf{R}^0 \\ \tilde{\mathbf{R}}_n &= [\mathbf{I} - \mathbf{R}_n (\mathbf{R}_n + \sigma_\eta^2 \mathbf{I})^{-1}] \mathbf{R}_n = \sigma_\eta^2 (\mathbf{R}_n + \sigma_\eta^2 \mathbf{I})^{-1} \mathbf{R}_n \end{aligned}\right\} \tag{8.158}$$

Using the fact that $\mathbf{L} = \mathbf{L}_1^{-1} \mathbf{L}_2$, this gives the following recursions.

Observation update:

$$g(m, n) = s(m, n) + \sum_{i=1}^{N} k_n(m, i)[y(i, n) - s(i, n)] \tag{8.159}$$

Prediction: $\quad \mathbf{L}_1 \mathbf{s}_{n+1} = \mathbf{L}_2 \mathbf{g}_n$

$$\Rightarrow \quad s(m, n+1) = \rho_1 s(m-1, n+1) + \rho_2 g(m, n) - \rho_3 g(m-1, n) + \rho_4 g(m+1, n) \tag{8.160}$$

where $k_n(m, i)$ are the elements of $\mathbf{K}_n \triangleq \mathbf{R}_n(\mathbf{R}_n + \sigma_\eta^2 \mathbf{I})^{-1}$. The Riccati equation can be implemented as $\mathbf{L}_1 \mathbf{R}_{n+1} \mathbf{L}_1^T = \hat{\mathbf{R}}_n$, where

$$\hat{\mathbf{R}}_n \triangleq \mathbf{L}_2 \tilde{\mathbf{R}}_n \mathbf{L}_2^T + \mathbf{P}, \qquad \tilde{\mathbf{R}}_n = [\mathbf{I} - \mathbf{K}_n]\mathbf{R}_n \tag{8.161}$$

This gives the following:

Forward recursion:

$$\mathbf{L}_1 \mathbf{R}_{n+1} = \mathbf{Q}_n \Rightarrow r_{n+1}(i, j) = \rho_1 r_{n+1}(i-1, j) + q_n(i, j) \tag{8.162}$$

Backward recursion:

$$\mathbf{Q}_n \mathbf{L}_1^T = \hat{\mathbf{R}}_n \Rightarrow q_n(i, j) = \rho_1 q_n(i, j+1) + \hat{r}_n(i, j) \tag{8.163}$$

From Kalman filtering theory we know that $\mathbf{R}_n$ is the covariance matrix of $\mathbf{e}_n \triangleq \mathbf{u}_n - \mathbf{s}_n$. Now we can write

$$\begin{aligned} \boldsymbol{\nu}_n &\triangleq \mathbf{y}_n - \mathbf{s}_n = \mathbf{u}_n - \mathbf{s}_n + \boldsymbol{\eta}_n = \mathbf{e}_n + \boldsymbol{\eta}_n \\ \text{cov}[\mathbf{e}_n] &= \mathbf{R}_n, \qquad \text{cov}[\boldsymbol{\eta}_n] = \sigma_\eta^2 \mathbf{I} \end{aligned} \tag{8.164}$$

This means that $\mathbf{K}_n$, defined before, is the one-dimensional Wiener filter for the noisy vector $\boldsymbol{\nu}_n$ and that the summation term in (8.159) represents the output of this Wiener filter, that is, $\mathbf{K}_n \boldsymbol{\nu}_n = \hat{\mathbf{e}}_n$, where $\hat{\mathbf{e}}_n$ is the best mean square estimate of $\mathbf{e}_n$ given $\boldsymbol{\nu}_n$. This gives

$$g(m, n) = s(m, n) + \hat{e}(m, n) \tag{8.165}$$

where $\hat{e}(m, n)$ is obtained by processing the elements of the $\boldsymbol{\nu}_n$. In (8.159), the estimate at the nth column is updated as $\mathbf{y}_n$ arrives. Equation (8.160) predicts recursively the next column from this update (Fig. 8.23). Examining the preceding

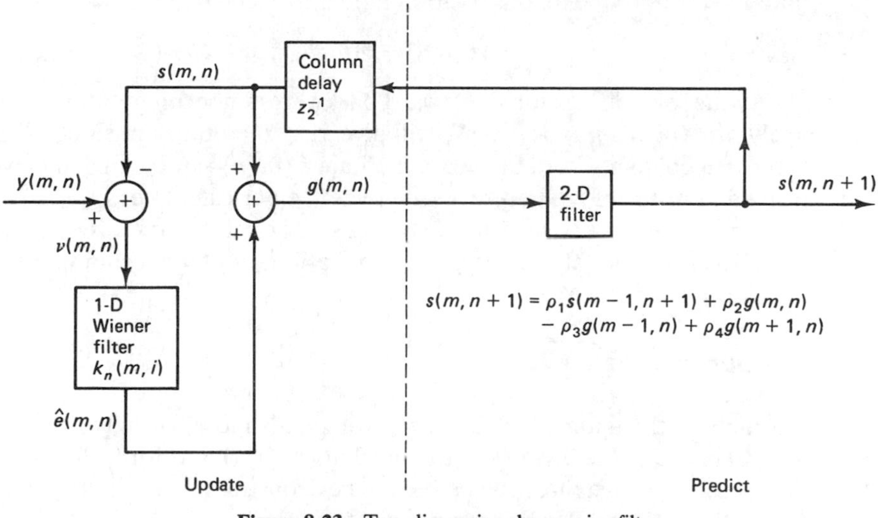

Figure 8.23 Two-dimensional recursive filter.

equations, we see that the Riccati equation is computationally the most complex, requiring $O(N^3)$ operations at each vector step, or $O(N^2)$ operations per pixel. For practical image sizes such a large number of operations is unacceptable. The following simplifications lead to more-practical algorithms.

Stationary Models

For stationary models and large image sizes, $\mathbf{R}_n$ will be nearly Toeplitz so that the matrix operations in (8.158) can be approximated by convolutions, which can be implemented via the FFT. This will reduce the number of operations to $(O(N \log N)$ per column or $O(\log N)$ per pixel.

Steady-State Filter

For smooth images, $\mathbf{R}_n$ achieves steady state quite rapidly, and therefore $\mathbf{K}_n$ in (8.159) may be replaced by its steady-state value $\mathbf{K}$. Given $\mathbf{K}$, the filter equations need $O(N)$ operations per pixel.

A Two-Stage Recursive Filter [35]

If the steady-state gain is used, then from the steady-state solution of the Riccati equation, it may be possible to find a low-order, approximate state variable model for $e(m, n)$, such as

$$\left.\begin{aligned} \mathbf{x}_{m+1} &= \mathbf{A}\mathbf{x}_m + \mathbf{B}\boldsymbol{\varepsilon}_m \\ e(m, n) &= \mathbf{C}\mathbf{x}_m \end{aligned}\right\} \tag{8.166}$$

where the dimension of the state vector $\mathbf{x}_m$ is small compared to N. This means that for each n, the covariance matrix of the sequence $e(m, n)$, $m = 1, \ldots, N$, is approximately $\mathbf{R} = \lim_{n \to \infty} \mathbf{R}_n$. The observation model for each fixed n is

$$v(m, n) = e(m, n) + \eta(m, n) \tag{8.167}$$

Then, the one-dimensional forward/backward smoothing filter operating recursively on $v(m, n)$, $m = 1, \ldots, N$, will give $\hat{\mathbf{x}}_{m,n}$, the optimum smooth estimate of $\mathbf{x}_m$ at the nth column. From this we can obtain $\hat{e}(m, n) = \mathbf{C}\hat{\mathbf{x}}_{m,n}$, which gives the vector $\hat{\mathbf{e}}_n$ needed in (8.165). Therefore, the overall filter calculates $\hat{e}(m, n)$ recursively in m, and $s(m, n)$ recursively in n (Fig. 8.23). The model of (8.166) may be obtained from $\mathbf{R}$ via AR modeling, spectral factorization, or other techniques that have been discussed in Chapter 6.

A Reduced Update Filter

In practice, the updated value $e(m, n)$ depends most strongly on the observations [i.e., $v(m, n)$] in the vicinity of the pixel at (m, n). Therefore, the dimensionality of the vector recursive filter can be reduced by constraining $\hat{e}(m, n)$ to be the output of a one-dimensional FIR Wiener filter of the form

$$\hat{e}(m, n) = \sum_{k=-p}^{p} a(k)v(m-k, n) \tag{8.168}$$

In steady state, the coefficients $a(k)$ are obtained by solving

$$\sum_{k=-p}^{p} a(k)[r(m-k) + \sigma_\eta^2 \delta(k)] = r(m), \qquad -p \le m \le p \tag{8.169}$$

where $r(m-k)$ are the elements of $\mathbf{R}$, the Toeplitz covariance matrix of $\mathbf{e}_n$ used previously. Substituting (8.165) in (8.160), the reduced update recursive filter becomes

$$\begin{aligned} s(m, n+1) = \rho_1 s(m-1, n+1) - \rho_3 s(m-1, n) + \rho_2 s(m, n) \\ + \rho_4 s(m+1, n) - \rho_3 \hat{e}(m-1, n) + \rho_2 \hat{e}(m, n) + \rho_4 \hat{e}(m+1, n) \end{aligned} \tag{8.170}$$

where $\hat{e}(m, n)$ is given by (8.168). A variant of this method has been considered in [36].

Remarks

The recursive filters just considered are useful only when a causal stochastic model such as (8.151) is available. If we start with a given covariance model, then the FIR Wiener Filter discussed earlier is more practical.

8.12 SEMICAUSAL MODELS AND SEMIRECURSIVE FILTERING

It was shown in Section 6.9 that certain random fields represented by semicausal models can be decomposed into a set of uncorrelated, one-dimensional random sequences. Such models yield semirecursive filtering algorithms, where each image column is first unitarily transformed and each transform coefficient is then passed through a recursive filter (Fig. 8.24). The overall filter is a combination of fast transform and recursive algorithms. We start by writing the observed image as

$$v(m, n) = \sum_{k=-p_1}^{p_2} \sum_{l=-q_1}^{q_2} h(k, l)u(m-k, n-l) + \eta(m, n), \qquad 1 \le m \le N,\ n = 0, 1, 2, \ldots \tag{8.171}$$

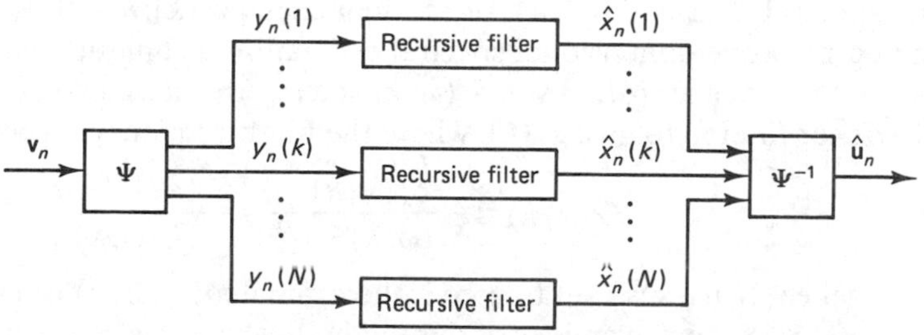

Figure 8.24 Semirecursive filtering.

where $\eta(m, n)$ is stationary white noise. In vector notation this becomes

$$\mathbf{v}_n = \sum_{l=-q_1}^{q_2} \mathbf{H}_l \mathbf{u}_{n-l} + \boldsymbol{\eta}_n + \sum_{l=-q_1}^{q_2} \tilde{\mathbf{b}}_{n-l} \tag{8.172}$$

where $\mathbf{v}_n$, $\mathbf{u}_n$, $\tilde{\mathbf{b}}_n$, and $\boldsymbol{\eta}_n$ are $N \times 1$ vectors and $\tilde{\mathbf{b}}_n$ depends only on $u(-p_2+1, n) \ldots u(0, n), u(N+1, n), u(N+p_1, n)$, which are boundary elements of the nth column of $u(m, n)$. The $\mathbf{H}_l$ are banded Toeplitz matrices.

Filter Formulation

Let $\boldsymbol{\Psi}$ be a fast unitary transform such that $\boldsymbol{\Psi}\mathbf{H}_n\boldsymbol{\Psi}^{*T}$, for every n, is nearly diagonal. From Chapter 5 we know that many sinusoidal transforms tend to diagonalize Toeplitz matrices. Therefore, defining

$$\mathbf{y}_n \triangleq \boldsymbol{\Psi}\mathbf{v}_n, \qquad \mathbf{x}_n \triangleq \boldsymbol{\Psi}\mathbf{u}_n, \qquad \tilde{\mathbf{c}}_n \triangleq \boldsymbol{\Psi}\tilde{\mathbf{b}}_n, \qquad \boldsymbol{\nu}_n \triangleq \boldsymbol{\Psi}\boldsymbol{\eta}_n$$
$$\boldsymbol{\Psi}\mathbf{H}_n\boldsymbol{\Psi}^{*T} \simeq \text{Diag}[\boldsymbol{\Psi}\mathbf{H}_n\boldsymbol{\Psi}^{*T}] \triangleq \boldsymbol{\Gamma}_n \triangleq \text{Diag}[\gamma_n(k)] \tag{8.173}$$

and multiplying both sides of (8.172) by $\boldsymbol{\Psi}$, we can reduce it to a set of scalar equations, decoupled in k, as

$$y_n(k) = \sum_{l=-q_1}^{q_2} \gamma_l(k) x_{n-l}(k) + \nu_n(k) + \sum_{l=-q_1}^{q_2} \tilde{c}_{n-l}(k), \qquad k = 1, \ldots, N \tag{8.174}$$

In most situations the image background is known or can be estimated quite accurately. Hence $\tilde{c}_n(k)$ can be assumed to be known and can be absorbed in $y_n(k)$ to give the observation system for each row of the transformed vectors as

$$y_n(k) = \sum_{l=-q_1}^{q_2} \gamma_l(k) x_{n-l}(k) + \nu_n(k), \qquad k = 1, \ldots, N \tag{8.175}$$

Now for each row, $x_n(k)$, is represented by an AR model

$$x_n(k) = \sum_{l=1}^{p} a_l(k) x_{n-l}(k) + \varepsilon_n(k), \qquad k = 1, \ldots, N \tag{8.176}$$

which together with (8.175) can be set up in the framework of Kalman filtering, as shown in Example 8.8. Alternatively, each line $[y_n(k), n = 0, 1 \ldots]$ can be processed by its one-dimensional Wiener filter. This method has been found useful in adaptive filtering of noisy images (Fig. 8.25) using the cosine transform. The entire image is divided into small blocks of size $N \times N$ (typically $N = 16$ or 32). For each k, the spectral density $S_y(\omega, k)$, of the sequence $[y_n(k), n = 0, 1, \ldots, N-1]$, is estimated by a one-dimensional spectral estimation technique, which assumes $\{y_n\}$ to be an AR sequence [8]. Given $S_y(\omega, k)$ and σ_η^2, the noise power, the sequence $y_n(k)$ is Wiener filtered to give $\hat{x}_n(k)$, where the filter frequency response is given by

$$G(\omega, k) \triangleq \frac{S_x(\omega, k)}{S_x(\omega, k) + \sigma_\eta^2} = \frac{S_y(\omega, k) - \sigma_\eta^2}{S_y(\omega, k)} \tag{8.177}$$

In practice, $S_x(\omega, k)$ is set to zero if the estimated $S_y(\omega, k)$ is less than σ_η^2. Figures 8.17 and 8.18 show examples of this method, where it is called the COSAR (cosine-AR) algorithm.

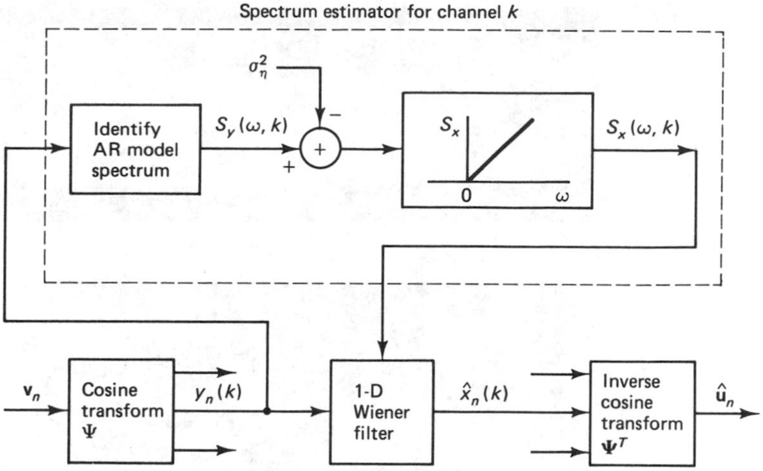

Figure 8.25 COSAR algorithm for adaptive filtering using semicausal models.

8.13 DIGITAL PROCESSING OF SPECKLE IMAGES

When monochromatic radiation is scattered from a surface whose roughness is of the order of a wavelength, interference of the waves produces a noise called *speckle.* Such noise is observed in images produced by coherent radiation from the microwave to visible regions of the spectrum. The presence of speckle noise in an imaging system reduces its resolution, particularly for low-contrast images. Therefore, suppression of speckle noise is an important consideration in design of coherent imaging systems. The problem of speckle reduction is quite different from additive noise smoothing because speckle noise is not additive. Figure 8.26b shows a speckled test pattern image (Fig. 8.26a).

Speckle Representation

In free space, speckle can be considered as an infinite sum of independent, identical phasors with random amplitude and phase [41, 42]. This yields a representation of its complex amplitude as

$$a(x, y) = a_R(x, y) + ja_I(x, y) \tag{8.178}$$

where a_R and a_I are zero mean, independent Gaussian random variables (for each x, y) with variance σ_a^2. The intensity field is simply

$$s = s(x, y) = |a(x, y)|^2 = a_R^2 + a_I^2 \tag{8.179}$$

which has the exponential distribution of (8.17) with variance $\sigma^2 \triangleq 2\sigma_a^2$ and mean $\mu_s = E[s] = \sigma^2$. A white noise field with these statistics is called the *fully developed speckle.*

For any speckle, the contrast ratio is defined as

$$\gamma = \frac{\text{standard deviation of } s}{\text{mean value of } s} \tag{8.180}$$

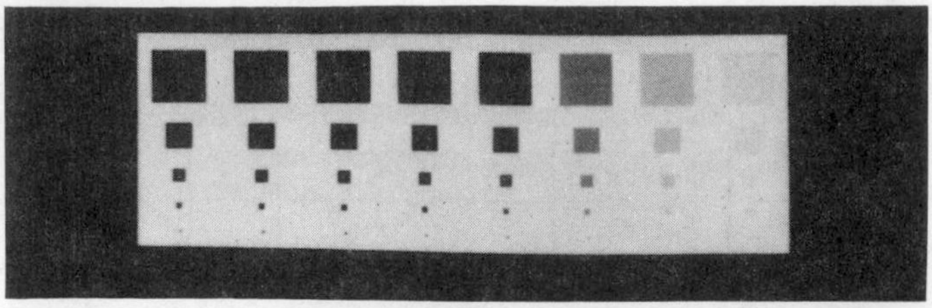

(a)

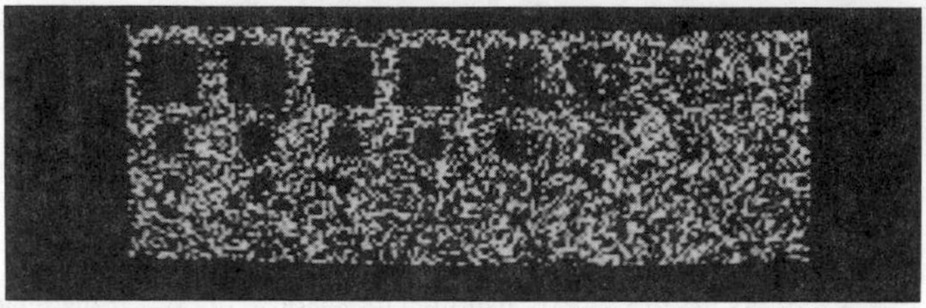

(b)

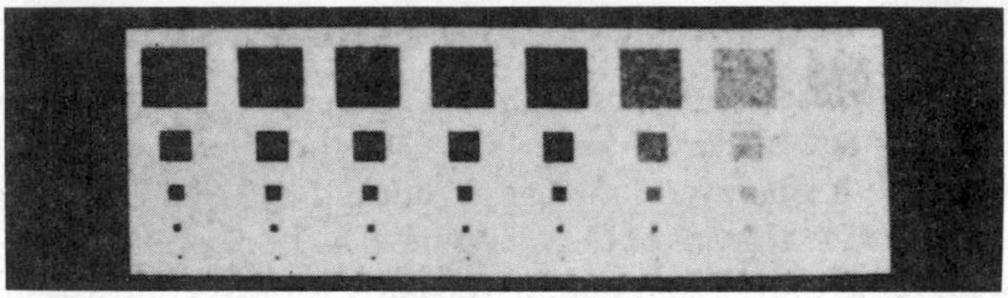

(c)

Figure 8.26 Speckle images.

For fully developed speckle, $\gamma = 1$.

When an object with complex amplitude distribution $g(x, y)$ is imaged by a coherent linear system with impulse response $K(x, y; x', y')$, the observed image intensity can be written as

$$v(x, y) = \left| \iint_{-\infty}^{\infty} K(x, y; x', y') g(x', y') e^{j\phi(x', y')} \, dx' \, dy' \right|^2 + \eta(x, y) \tag{8.181}$$

where $\eta(x, y)$ is the additive detector noise and $\phi(x, y)$ represents the phase distortion due to scattering. If the impulse response decays rapidly outside a region $R_{\text{cell}}(x, y)$, called the resolution cell, and $g(x, y)$ is nearly constant in this region, then [44]

$$\left. \begin{aligned} v(x, y) &\simeq |g(x, y)|^2 |a(x, y)|^2 + \eta(x, y) \\ &= u(x, y) s(x, y) + \eta(x, y) \end{aligned} \right\} \tag{8.182}$$

where

$$u(x, y) \triangleq |g(x, y)|^2, \qquad a(x, y) \triangleq \iint_{R_{\text{cell}}} K(x, y; x', y') e^{j\phi(x', y')} \, dx' \, dy' \tag{8.183}$$

The $u(x, y)$ represents the object intensity distribution (reflectance or transmittance) and $s(x, y)$ is the speckle intensity distribution. The random field $a(x, y)$ is Gaussian, whose autocorrelation function has support on a region twice the size of R_{cell}. Equation (8.182) shows that speckle appears as a multiplicative noise in the coherent imaging of low-resolution objects. Note that there will be no speckle in an ideal imaging system. A uniformly sampled speckle field with pixel spacing equal to or greater than the width of its correlation function will be uncorrelated.

Speckle Reduction [46–47]: *N*-Look Method

A simple method of speckle reduction is to take several statistically independent intensity images of the object and average them (Fig. 8.26c). Assuming the detector noise to be low and writing the lth image as

$$v_l(x, y) = u(x, y)s_l(x, y), \qquad l = 1, \ldots, N \tag{8.184}$$

then the temporal average of N looks is simply

$$\hat{v}_N(x, y) \triangleq \frac{1}{N}\sum_{l=1}^{N} v_l(x, y) = u(x, y)\hat{s}_N(x, y) \tag{8.185}$$

where $\hat{s}_N(x, y)$ is the N-look average of the speckle fields. This is also the maximum likelihood estimate of $[v_l(x, y), l = 1, \ldots, N]$, which yields

$$E[\hat{v}_N] = \frac{\mu_s u}{N}, \quad \text{var}[\hat{v}_N] = \frac{u^2 \mu_s^2}{N} \tag{8.186}$$

This gives the contrast ratio $\gamma = 1/N$ for $\hat{v}_N$. Therefore, the contrast improves by a factor of $\sqrt{N}$ for N-look averaging.

Spatial Averaging of Speckle

If the available number of looks, N, is small, then it is desirable to perform some kind of spatial filtering to reduce speckle. A standard technique used in synthetic aperture radar systems (where speckle noise occurs) is to average the intensity values of several adjacent pixels. The improvement in contrast ratio for spatial averaging is consistent with the N-look method except that there is an accompanying loss of resolution.

Homomorphic Filtering

The multiplicative nature of speckle suggests performing a logarithmic transformation on (8.185), giving

$$\log \hat{v}_N(x, y) = \log u(x, y) + \log \hat{s}_N(x, y) \tag{8.187}$$

Defining $w_N \triangleq \log \hat{v}_N$, $z \triangleq \log u$, and $\eta_N \triangleq \log \hat{s}_N$, we get the additive noise observation model

$$w_N(x, y) = z(x, y) + \eta_N(x, y) \tag{8.188}$$

where $\eta_N(x, y)$ is stationary white noise.

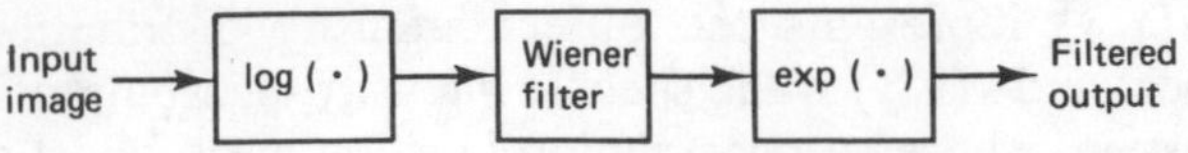

(a) The algorithm

(b) Filtering Example

Figure 8.27 Homomorphic filtering of speckle.

For $N \geq 2$, η_N can be modeled reasonably well by a Gaussian random field [45], whose spectral density function is given by

$$S_\eta(\xi_1, \xi_2) = \sigma_n^2 = \begin{cases} \dfrac{\pi^2}{6}, & N = 1 \\ \dfrac{1}{N}, & N \geq 2 \end{cases} \tag{8.189}$$

Now $z(x, y)$ can be easily estimated from $w_N(x, y)$ using Wiener filtering techniques. This gives the overall filter algorithm of Fig. 8.27, which is also called the *homomorphic filter*. Experimental studies have shown that the homomorphic Wiener filter performs quite well compared to linear filtering or other homomorphic linear filters [46]. Figure 8.27 shows the performance of an adaptive FIR Wiener filter used in the homomorphic mode.

8.14 MAXIMUM ENTROPY RESTORATION

The inputs, outputs, and the PSFs of incoherent imaging systems (the usual case) are nonnegative. The least squares or mean square criteria based restoration algorithms do not yield images with nonnegative pixel values. A restoration method based on the *maximum entropy* criterion gives nonnegative solutions. Since entropy is a measure of uncertainty, the general argument behind this criterion is that it assumes the least about the solution and gives it the maximum freedom within the limits imposed by constraints.

Distribution-Entropy Restoration

For an image observed as

$$\boldsymbol{v} \simeq \mathcal{H}\boldsymbol{u} \tag{8.190}$$

where $\mathcal{H}$ is the PSF matrix, and $\boldsymbol{u}$ and $\boldsymbol{v}$ are the object and observation arrays mapped into vectors, a maximum entropy restoration problem is to maximize

$$\mathcal{E}(\boldsymbol{u}) \triangleq -\sum_n u(n) \log u(n) \tag{8.191}$$

subject to the constraint

$$\tfrac{1}{2}\|\boldsymbol{v} - \mathcal{H}\boldsymbol{u}\|^2 = \sigma_e^2 \tag{8.192}$$

where $\sigma_e^2 > 0$ is a specified quantity. Because $u(n)$ is nonnegative and can be normalized to give $\sum_n u(n) = 1$, it can be treated as a probability distribution whose entropy is $\mathcal{E}(\boldsymbol{u})$. Using the usual Lagrangian method of optimization, the solution $\hat{\boldsymbol{u}}$ is given by the implicit equation

$$\hat{\boldsymbol{u}} = \exp\{-\mathbf{1} - \lambda\mathcal{H}^T(\boldsymbol{v} - \mathcal{H}\hat{\boldsymbol{u}})\} \tag{8.193}$$

where $\exp\{\mathbf{x}\}$ denotes a vector of elements $\exp[x(k)], k = 0, 1, \ldots, \mathbf{1}$ is a vector of all 1s and λ is a scalar Lagrange multiplier such that $\hat{\boldsymbol{u}}$ satisfies the constraint of (8.192). Interestingly, a Taylor series expansion of the exponent, truncated to the first two terms, yields the constrained least squares solution

$$\hat{\boldsymbol{u}} = (\mathcal{H}^T\mathcal{H} + \lambda\mathbf{I})^{-1}\mathcal{H}^T\boldsymbol{v} \tag{8.194}$$

Note that the solution of (8.193) is guaranteed to be nonnegative. Experimental results show that this method gives sharper restorations than the least squares filters when the image contains a small number of point objects (such as in astronomy images) [48].

A stronger restoration result is obtained by maximizing the entropy defined by (8.191) subject to the constraints

$$\left.\begin{aligned} &u(n) \geq 0, \qquad n = 0, \ldots, N-1 \\ &\mathcal{H}\boldsymbol{u} = \boldsymbol{v} \quad \text{or} \quad \sum_{j=0}^{N-1} h(m, j)u(j) = v(m), \qquad m = 0, \ldots, M-1 \end{aligned}\right\} \tag{8.195}$$

Now the solution is given by

$$\hat{u}(n) = \frac{1}{e}\exp\left[\sum_{l=0}^{M-1} h(l, n)\lambda(l)\right], \qquad n = 0, \ldots, N-1 \tag{8.196}$$

where $\lambda(l)$ are Lagrange multipliers (also called dual variables) that maximize the functional

$$J(\lambda) \triangleq \sum_{n=0}^{N-1} \hat{u}(n) - \sum_{l=0}^{M-1} \lambda(l)v(l) \tag{8.197}$$

The above problem is now unconstrained in $\lambda(n)$ and can be solved by invoking several different algorithms from optimization theory. One example is a coordinate ascent algorithm, where constraints are enforced one by one in cyclic iterations, giving [49]

$$\left.\begin{aligned} x_j(n) &= x_{j-1}(n) + h(m, n) \log \frac{v(m)}{\left[\sum_{k=0}^{N-1} h(m, k) x_{j-1}(k)\right]} \\ \hat{u}_j(n) &= \exp\{x_j(n)\} \end{aligned}\right\} \quad (8.198)$$

where $m = j$ modulo M, $k = 1, \ldots, N$ and $j = 0, 1, \ldots$. At the jth iteration, $x_j(n)$ is updated for all n and a fixed m. After $m = M$, the iterations continue cyclically, updating the constraints. Convergence to the true solution is often slow but is assured as $j \to \infty$, for $0 \le h(m, n) \le 1$. Since the PSF is nonnegative, this condition is easily satisfied by scaling the observations appropriately.

Log-Entropy Restoration

There is another maximum entropy restoration problem, which maximizes

$$\mathscr{E} = \sum_{n=0}^{N-1} \log u(n) \quad (8.199)$$

subject to the constraints of (8.195). The solution now is obtained by solving the nonlinear equations

$$\hat{u}(n) = \frac{1}{\sum_{l=0}^{M-1} h(l, n)\lambda(l)} \quad (8.200)$$

where $\lambda(l)$ maximizes

$$J = -\sum_{n=0}^{N-1} \log\left[\sum_{m=0}^{M-1} h(m, n)\lambda(m)\right] + \sum_{m=0}^{M-1} v(m)\lambda(m) \quad (8.201)$$

Once again an iterative gradient or any other suitable method may be chosen to maximize (8.201). A coordinate ascent method similar to (8.198) yields the iterative solution [50]

$$\hat{u}_{j+1}(n) = \frac{\hat{u}_j(n)}{1 + \alpha_j h(m, n)\hat{u}_j(n)}, \quad m = j \quad \text{modulo } M, \quad n = 0, 1, \ldots, N-1 \quad (8.202)$$

where α_j is determined such that the denominator term is positive and the constraint

$$\sum_{n=0}^{N-1} h(m, n)\hat{u}_{j+1}(n) = v(m) \quad (8.203)$$

is satisfied at each iteration. This means we must solve for the positive root of the nonlinear equation

$$f(\alpha_j) \triangleq \sum_{n=0}^{N-1} \frac{h(m,n)\hat{u}_j(n)}{[1+\alpha_j h(m,n)\hat{u}_j(n)]} - v(m) = 0 \tag{8.204}$$

As before, the convergence, although slow, is assured as $j \to \infty$. For $h(m,n) > 0$, which is true for PSFs, this algorithm guarantees a positive estimate at any iteration step. The speed of convergence can be improved by going to the gradient algorithm [51]

$$\lambda_{j+1}(m) = \lambda_j(m) + \alpha_j g_j(m), \qquad j = 0, 1, \ldots \tag{8.205}$$

$$g_j(m) = v(m) - \sum_{n=0}^{N-1} h(m,n)\hat{u}_j(n) \tag{8.206}$$

$$\hat{u}_j(n) = \frac{1}{\left[\sum_{m=0}^{M-1} h(m,n)\lambda_j(m)\right]} \tag{8.207}$$

where $\lambda_0(m)$ are chosen so that $\hat{u}_0(n)$ is positive and α_j is a positive root of the equation

$$f(\alpha_j) \triangleq \sum_{k=0}^{N-1} G_j(k)[\Lambda_j(k) + \alpha_j G_j(k)]^{-1} = 0 \tag{8.208}$$

where

$$G_j(k) \triangleq \sum_{m=0}^{M-1} h(m,k)g_j(m), \qquad \Lambda_j(k) = \sum_{m=0}^{M-1} h(m,k)\lambda_j(m) \tag{8.209}$$

The search for α_j can be restricted to the interval $[0, \max_k\{G_j(k)/\Lambda_j(k)\}]$.

This maximum entropy problem appears often in the theory of spectral estimation (see Problem 8.26b). The foregoing algorithms are valid in multidimensions if $u(n)$ and $v(m)$ are sequences obtained by suitable ordering of elements of the multidimensional arrays $u(i, j, \ldots)$ and $v(i, j, \ldots)$, respectively.

8.15 BAYESIAN METHODS

In many imaging situations—for instance, image recording by film—the observation model is nonlinear of the form

$$\mathbf{v} = f(\mathcal{H}\mathbf{u}) + \boldsymbol{\eta} \tag{8.210}$$

where $f(x)$ is a nonlinear function of x. The a posteriori conditional density given by Bayes' rule

$$p(\mathbf{u}|\mathbf{v}) = \frac{p(\mathbf{v}|\mathbf{u})p(\mathbf{u})}{p(\mathbf{v})} \tag{8.211}$$

is useful in finding different types of estimates of the random vector $\mathbf{u}$ from the observation vector $\mathbf{v}$. The minimum mean square estimate (MMSE) of $\mathbf{u}$ is the mean of this density. The *maximum a posteriori* (MAP) and the *maximum likelihood* (ML) estimates are the modes of $p(\mathbf{u}|\mathbf{v})$ and $p(\mathbf{v}|\mathbf{u})$, respectively. When the observation model is nonlinear, it is difficult to obtain the marginal density $p(\boldsymbol{v})$ even when $\boldsymbol{u}$ and $\boldsymbol{\eta}$ are Gaussian. (In the linear case $p(\boldsymbol{u}|\boldsymbol{v})$ is easily obtained since it is Gaussian if $\boldsymbol{u}$ and $\boldsymbol{\eta}$ are). However, the MAP and ML estimates do not require $p(\boldsymbol{v})$ and are therefore easier to obtain.

Under the assumption of Gaussian statistics for $\boldsymbol{u}$ and $\boldsymbol{\eta}$, with covariances $\mathcal{R}_u$ and $\mathcal{R}_n$, respectively, the ML and MAP estimates can be shown to be the solution of the following equations:

$$\text{ML estimate, } \hat{\boldsymbol{u}}_{ML}: \quad \mathcal{H}^T \mathcal{D} \mathcal{R}_n^{-1}[\boldsymbol{v} - f(\mathcal{H}\hat{\boldsymbol{u}}_{ML})] = \mathbf{0} \tag{8.212}$$

where

$$\mathcal{D} \triangleq \text{Diag}\left\{\frac{\partial f(x)}{\partial x}\bigg|_{x=\hat{\boldsymbol{w}}_i}\right\} \tag{8.213}$$

and $\hat{\boldsymbol{w}}_i$ are the elements of the vector $\hat{\boldsymbol{w}} \triangleq \mathcal{H}\hat{\boldsymbol{u}}_{ML}$.

$$\text{MAP estimate, } \hat{\boldsymbol{u}}_{MAP}: \quad \hat{\boldsymbol{u}}_{MAP} = \boldsymbol{\mu}_u + \mathcal{R}_u \mathcal{H}^T \mathcal{D} \mathcal{R}_n^{-1}[\boldsymbol{v} - f(\mathcal{H}\hat{\boldsymbol{u}}_{MAP})] \tag{8.214}$$

where $\boldsymbol{\mu}_u$ is the mean of $\boldsymbol{u}$ and $\mathcal{D}$ is defined in (8.213) but now $\hat{\boldsymbol{w}} \triangleq \mathcal{H}\hat{\boldsymbol{u}}_{MAP}$.

Since these equations are nonlinear, an alternative is to maximize the appropriate log densities. For example, a gradient algorithm for $\hat{\boldsymbol{u}}_{MAP}$ is

$$\hat{\boldsymbol{u}}_{j+1} = \hat{\boldsymbol{u}}_j - \alpha_j\{\mathcal{H}^T \mathcal{D}_j \mathcal{R}_n^{-1}[\boldsymbol{v} - f(\mathcal{H}\hat{\boldsymbol{u}}_j)] - \mathcal{R}_u^{-1}[\hat{\boldsymbol{u}}_j - \boldsymbol{\mu}_u]\} \tag{8.215}$$

where $\alpha_j > 0$, and $\mathcal{D}_j$ is evaluated at $\hat{\boldsymbol{w}}_j \triangleq \mathcal{H}\hat{\boldsymbol{u}}_j$.

Remarks

If the function $f(x)$ is linear, say $f(x) = x$, and $\mathcal{R}_n = \sigma_n^2 \mathbf{I}$, then $\hat{\boldsymbol{u}}_{ML}$ reduces to the least squares solution

$$\mathcal{H}^T \mathcal{H}\hat{\boldsymbol{u}}_{ML} = \mathcal{H}^T \boldsymbol{v} \tag{8.216}$$

and the MAP estimate reduces to the Wiener filter output for zero mean noise [see (8.87)],

$$\hat{\boldsymbol{u}}_{MAP} = \boldsymbol{\mu}_u + \mathcal{G}(\boldsymbol{v} - \boldsymbol{\mu}_v) \tag{8.217}$$

where $\mathcal{G} = (\mathcal{R}_u^{-1} + \mathcal{H}^T \mathcal{R}_n^{-1} \mathcal{H})^{-1} \mathcal{H}^T \mathcal{R}_n^{-1}$.

In practice, $\boldsymbol{\mu}_v$ may be estimated as a local average of $\boldsymbol{v}$ and $\boldsymbol{\mu}_u \simeq \mathcal{H}^+ f^{-1}(\boldsymbol{\mu}_v)$, where $\mathcal{H}^+$ is the generalized inverse of $\mathcal{H}$.

8.16 COORDINATE TRANSFORMATION AND GEOMETRIC CORRECTION

In many situations a geometric transformation of the image coordinates is required. An example is in the remote sensing of images via satellites, where the earth's rotation relative to the scanning geometry of the sensor generates an image on a distorted raster [55]. The problem then is to estimate a function $f(x', y')$ given at discrete locations of (x, y), where $x' = h_1(x, y), y' = h_2(x, y)$ describe the geometric transformation between the two coordinate systems. Common examples of geometric transformations are translation, scaling, rotation, skew, and reflection, all of which can be represented by the *affine transformation*

$$\begin{bmatrix} x' \\ y' \end{bmatrix} = \begin{bmatrix} a & b \\ c & d \end{bmatrix} \begin{bmatrix} x \\ y \end{bmatrix} + \begin{bmatrix} \alpha \\ \beta \end{bmatrix} \tag{8.218}$$

In principle, the image function in (x', y') coordinates can be obtained from its values on the (x_i, y_i) grid by an appropriate interpolation method followed by resampling on the desired grid. Some commonly used algorithms for interpolation at a point Q (Fig. 8.28) from samples at P_1, P_2, P_3, and P_4 are as follows.

1. Nearest neighbor:

$$F(Q) = F(P_k), \qquad k : \min_i \{d_i\} = d_k \tag{8.219}$$

that is, P_k is the nearest neighbor of Q.

2. Linear interpolation:

$$F(Q) = \frac{\sum_{k=1}^{4} F(P_k)/d_k}{\sum_{k=1}^{4} \frac{1}{d_k}} \tag{8.220}$$

3. Bilinear interpolation:

$$F(Q) = \frac{F(Q_1)/d_5' + F(Q_2)/d_6'}{(1/d_5') + (1/d_6')} = \frac{F(Q_1)d_6' + F(Q_2)d_5'}{d_5' + d_6'} \tag{8.221a}$$

where

$$F(Q_1) = \frac{F(P_1)d_4' + F(P_4)d_1'}{d_1' + d_4'}, \qquad F(Q_2) = \frac{F(P_2)d_3' + F(P_3)d_2'}{d_2' + d_3'} \tag{8.221b}$$

These methods are local and require minimal computation. However, these methods would be inappropriate if there was significant noise in the data. Smoothing splines or global interpolation methods, which use all the available data, would then be more suitable.

For many imaging systems the PSF is spatially varying in Cartesian coordinates but becomes spatially invariant in a different coordinate system, for example, in systems with spherical aberrations, coma, astigmatism, and the like [56, 57]. These and certain other distortions (such as that due to rotational motion) may be

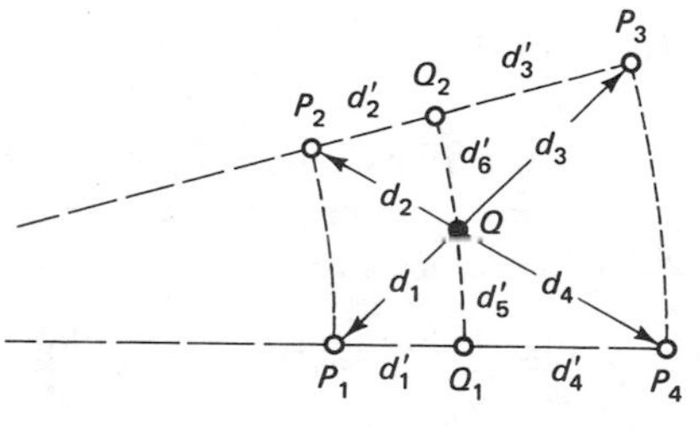

Figure 8.28 Interpolation at Q.

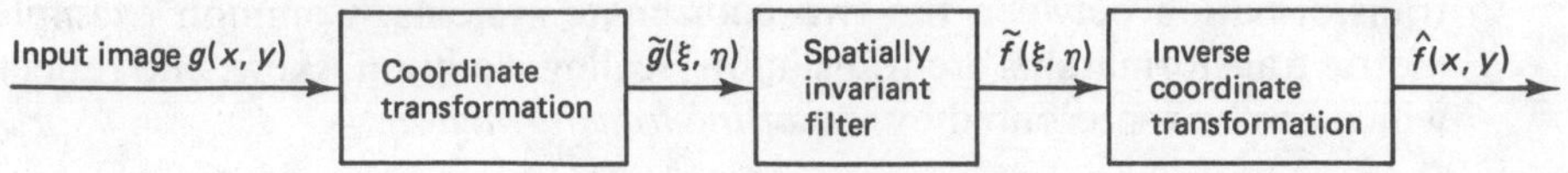

Figure 8.29 Spatially variant filtering by coordinate transformation.

corrected by the coordinate transformation method shown in Fig. 8.29. The input image is transformed from (x, y) to (ξ, η), coordinates where it is possible to filter by a spatially invariant system. The filter output is then inverse transformed to obtain the estimate in the original coordinates.

For example, the image of an object $f(r, \theta)$ obtained by an axially symmetric imaging system with coma aberration is

$$g(r, \theta) = \int_0^\infty \int_0^{2\pi} a(r_0) h\left(\frac{r}{r_0^n}, \theta - \theta_0\right) f(r_0, \theta_0) r_0 \, dr_0 \, d\theta_0 \tag{8.222}$$

where $n \geq 1$ and (r, θ) are the polar coordinates. The PSF is spatially varying in (x, y). In (r, θ) it is shift invariant in θ but spatially variant in r. Under the logarithmic transformations

$$\xi = \ln r, \qquad \xi_0 = n \ln r_0 \tag{8.223}$$

the ratio r/r_0^n becomes a function of the displacement $\xi - \xi_0$ and (8.222) can be written as a convolution integral:

$$\tilde{g}(\xi, \theta) \triangleq g(e^\xi, \theta) = \int_{-\infty}^\infty \int_0^{2\pi} \tilde{h}(\xi - \xi_0, \theta - \theta_0) \tilde{f}(\xi_0, \theta_0) \, d\xi_0 \, d\theta_0 \tag{8.224}$$

where

$$\tilde{h}(\xi - \xi_0, \theta - \theta_0) \triangleq h(e^{\xi - \xi_0}, \theta - \theta_0), \qquad \tilde{f}(\xi_0, \theta_0) \triangleq f(e^{\xi_0/n}, \theta_0) a(e^{\xi_0/n}) \frac{e^{2\xi_0/n}}{n}$$

Spatially invariant filters can now be designed to restore $\tilde{f}(\xi_0, \theta_0)$ from $\tilde{g}(\xi, \theta)$. Generalizations of this idea to other types of blurred images may be found in [56, 57].

The comet Halley shown on the front cover page of this text was reconstructed from data gathered by NASA's *Pioneer Venus Orbiter* in 1986. The observed data was severely distorted with several samples missing due to the activity of solar flares. The restored image was obtained by proper coordinate transformation, bilinear interpolation, and pseudocoloring.

8.17 BLIND DECONVOLUTION [58, 59]

Image restoration when the PSF is unknown is a difficult nonlinear restoration problem. For spatially invariant imaging systems, the power spectral density of the observed image obeys (8.40), which gives

$$\log|H|^2 = \log(S_{vv} - S_{\eta\eta}) - \log S_{uu} \tag{8.225}$$

If the additive noise is small, we can estimate

$$\log|H| \simeq \frac{1}{2M}\sum_{k=1}^{M}(\log|V_k|^2 - \log|U_k|^2) = \frac{1}{M}\sum_{k=1}^{M}[\log|V_k| - \log|U_k|] \qquad (8.226)$$

where V_k and U_k, $k = 1, \ldots, M$ are obtained by dividing images $v(m, n)$ and $u(m, n)$ into M blocks and then Fourier transforming them. Therefore, identification of H requires power spectrum estimation of the object and the observations. Restoration methods that are based on unknown H are called *blind deconvolution* methods. Note that this method gives only the magnitude of H. In many imaging situations the phase of H is zero or unimportant, such as when H represents average atmospheric turbulence, camera misfocus (or lens aberration), or uniform motion (linear phase or delay). In such cases it is sufficient to estimate the MTF, which can then be used in the Wiener filter equation. Techniques that also identify the phase are possible in special situations, but, in general, phase estimation is a difficult task.

8.18 EXTRAPOLATION OF BANDLIMITED SIGNALS

Extrapolation means extending a signal outside a known interval. Extrapolation in the spatial coordinates could improve the spectral resolution of an image, whereas frequency domain extrapolation could improve the spatial resolution. Such problems arise in power spectrum estimation, resolution of closely spaced objects in radio-astronomy, radar target detection and geophysical exploration, and the like.

Analytic Continuation

A bandlimited signal $f(x)$ can be determined completely from the knowledge of it over an arbitrary finite interval $[-\alpha, \alpha]$. This follows from the fact that a bandlimited function is an analytic function because its Taylor series

$$f(x + \Delta) = f(x) + \sum_{n=1}^{\infty}\frac{\Delta^n}{n!}\frac{d^n f(x)}{dx^n} \qquad (8.227)$$

is convergent for all x and Δ. By letting $x \in [-\alpha, \alpha]$ and $x + \Delta > \alpha$, (8.227) can be used to extrapolate $f(x)$ anywhere outside the interval $[-\alpha, \alpha]$.

Super-resolution

The foregoing ideas can also be applied to a space-limited function (i.e., $f(x) = 0$ for $|x| > \alpha$) whose Fourier transform is given over a finite frequency band. This means, theoretically, that a finite object imaged by a diffraction limited system can be perfectly resolved by extrapolation in the Fourier domain. Extrapolation of the spectrum of an object beyond the diffraction limit of the imaging system is called *super-resolution.*

Extrapolation via Prolate Spheroidal Wave Functions (PSWFs) [60]

The high-order derivatives in (8.227) are extremely sensitive to noise and truncation errors. This makes the analytic continuation method impractical for signal extrapolation. An alternative is to evaluate $f(x)$ by the series expansion

$$f(x) = \sum_{n=0}^{\infty} a_n \phi_n(x), \qquad \forall x \tag{8.228}$$

$$a_n = \frac{1}{\lambda_n} \int_{-\alpha}^{\alpha} f(x)\phi_n(x)\,dx \tag{8.229}$$

where $\phi_n(x)$ are called the *prolate spheroidal wave functions* (PSWFs). These functions are bandlimited, *orthonormal* over $[-\alpha, \alpha]$, and complete in the class of bandlimited functions. Moreover, in the interval $-\alpha \le x \le \alpha$, $\phi_n(x)$ are complete and *orthogonal*, with $\langle \phi_n, \phi_m \rangle = \lambda_n \delta(n-m)$, where $\lambda_n > 0$ is the norm $\|\phi_n\|^2$. Using this property in (8.228), a_n can be obtained from the knowledge of $f(x)$ over $[-\alpha, \alpha]$ via (8.229). Given a_n in (8.228), $f(x)$ can be extrapolated for all values of x.

In practice, we would truncate the above series to a finite but sufficient number of terms. In the presence of noise, the extrapolation error increases rapidly with the number of terms in the series (Problem 8.28). Also, the numerical computation of the PSWFs themselves is a difficult task, which is marred by its own truncation and round-off errors. Because of these difficulties, the preceding extrapolation algorithm also is quite impractical. However, the PSWFs remain fundamentally important for analysis of bandlimited signals.

Extrapolation by Error Energy Reduction [61, 62]

An interesting and more practical extrapolation algorithm is based on a principle of successive energy reduction (Fig. 8.30). First the given function $g(x) \triangleq g_0(x) = f(x)$, $x \in [-\alpha, \alpha]$, is low-pass filtered by truncating its Fourier transform to zero outside the interval $(-\xi_0, \xi_0)$. This reduces the error energy in $f_1(x)$ because the signal is known to be bandlimited. To prove this, we use the Parseval formula to obtain

$$\int_{-\infty}^{\infty} |f(x) - g(x)|^2\,dx = \int_{-\infty}^{\infty} |F(\xi) - G_0(\xi)|^2\,d\xi = \int_{-\xi_0}^{\xi_0} |F(\xi) - F_1(\xi)|^2\,d\xi$$
$$+ \int_{|\xi| > \xi_0} |G_0(\xi)|^2\,d\xi > \int_{-\xi_0}^{\xi_0} |F(\xi) - F_1(\xi)|^2\,d\xi = \int_{-\infty}^{\infty} |f(x) - f_1(x)|^2\,dx \tag{8.230}$$

Now $f_1(x)$ is bandlimited but does not match the observations over $[-\alpha, \alpha]$. The error energy is reduced once again if $f_1(x)$ is substituted by $f(x)$ over $-\alpha \le x \le \alpha$. Letting $\mathcal{S}$ denote this space-limiting operation, we obtain

$$g_1(x) \triangleq f_1(x) - \mathcal{S} f_1(x) + g_0(x) \tag{8.231}$$

and

$$\int_{-\infty}^{\infty} |f(x) - f_1(x)|^2\,dx = \int_{-\alpha}^{\alpha} |f(x) - f_1(x)|^2\,dx + \int_{|x| > \alpha} |f(x) - g_1(x)|^2\,dx \tag{8.232}$$

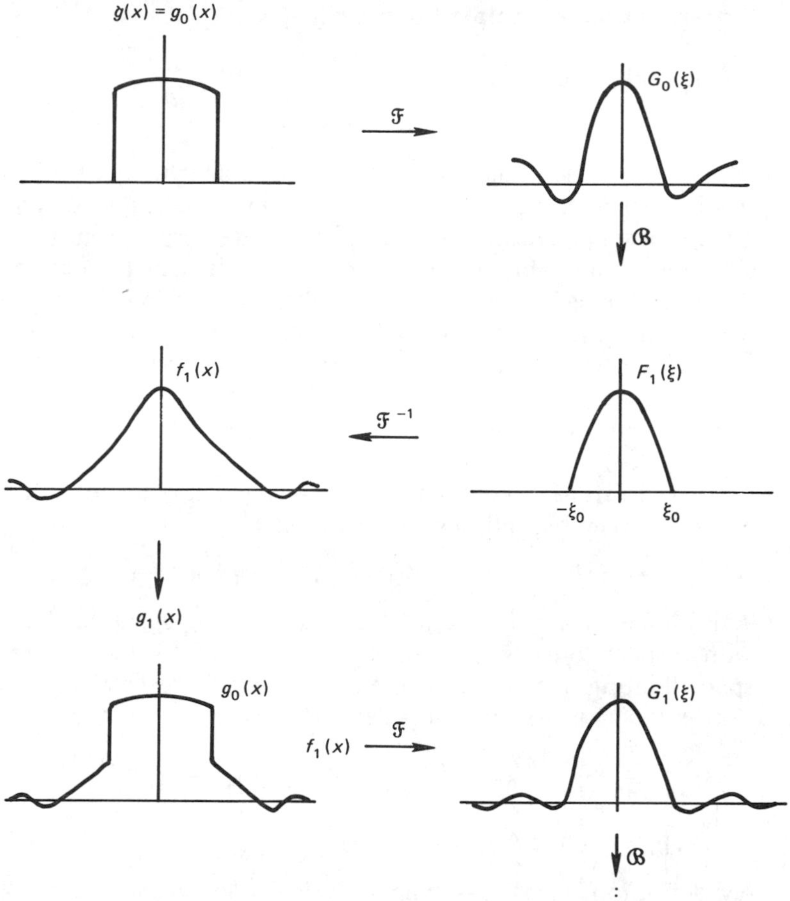

Figure 8.30 Extrapolation by successive energy reduction.

$$> \int_{|x|>\alpha} |f(x) - g_1(x)|^2 \, dx = \int_{-\infty}^{\infty} |f(x) - g_1(x)|^2 \, dx$$

Now $g_1(x)$, not being bandlimited anymore, is low-pass filtered, and the preceding procedure is repeated. This gives the iterative algorithm

$$\left.\begin{aligned} f_n(x) &= \mathscr{B} g_{n-1}(x), \qquad g_0(x) = g(x) \triangleq \mathscr{S} f(x) \\ g_n(x) &= g_0(x) + (\mathscr{I} - \mathscr{S}) f_n(x), \qquad n = 1, 2 \ldots \end{aligned}\right\} \tag{8.233}$$

where $\mathscr{I}$ is the identity operator and $\mathscr{B}$ is the bandlimiting operator. In the limit as $n \to \infty$, both $f_n(x)$ and $g_n(x)$ converge to $f(x)$ in the mean square sense [62]. It can be shown that this algorithm is a special case of a gradient algorithm associated with a least squares minimization problem [65]. This algorithm is also called the *method of alternating projections* because the iterates are projected alternately in the space of bandlimited and space-limited functions. Such algorithms are useful for solving image restoration problems that include a certain class of constraints [53, 63, 64].

Extrapolation of Sampled Signals [65]

For a small perturbation,

$$\tilde{f}(x) = f(x) + \varepsilon\eta(x), \quad \varepsilon \neq 0 \tag{8.234}$$

where $\eta(x)$ is not bandlimited, the desired analyticity of $\tilde{f}(x)$ is lost. Then it is possible to find a large number of functions that approximate $f(x)$ very closely on the observation interval $[-\alpha, \alpha]$ but differ greatly outside this interval. This situation is inevitable when one tries to implement the extrapolation algorithms digitally. Typically, the observed bandlimited function is oversampled, so it can be estimated quite accurately by interpolating the *finite* number of samples over $[-\alpha, \alpha]$. However, the interpolated signal cannot be bandlimited. Recognizing this difficulty, we consider extrapolation of sampled bandlimited signals. This approach leads to more-practical extrapolation algorithms.

Definitions. A sequence $y(n)$ is called bandlimited if its Fourier transform $Y(\omega)$, $-\pi \leq \omega < \pi$, satisfies the condition

$$Y(\omega) = 0, \qquad \omega_1 < |\omega| \leq \pi \tag{8.235}$$

This implies that $y(n)$ comes from a bandlimited signal that has been oversampled with respect to its Nyquist rate. Analogous to $\mathscr{B}$ and $\mathscr{S}$, the bandlimiting and space-limiting operators, denoted by **L** and **S**, respectively, are now $\infty \times \infty$ and $(2M+1) \times \infty$ matrix operators defined as

$$[\mathbf{Ly}]_m = \sum_{n=-\infty}^{\infty} \frac{\sin(m-n)\omega_1}{\pi(m-n)} y(n) \Rightarrow \mathscr{F}\{[\mathbf{Ly}]_m\} = \begin{cases} Y(\omega), & |\omega| < \omega_1 \\ 0, & \omega_1 < |\omega| \leq \pi \end{cases} \tag{8.236}$$

$$[\mathbf{Sy}]_m = y(m), \qquad -M \leq m \leq M \tag{8.237}$$

By definition, then, **L** is symmetric and *idempotent*, that is, $\mathbf{L}^T = \mathbf{L}$ and $\mathbf{L}^2 = \mathbf{L}$ (repeated ideal low-pass filtering produces the same result).

The Extrapolation Problem. Let $y(m)$ be a bandlimited sequence. We are given a set of space-limited noise-free observations

$$z(m) = y(m), \qquad -M \leq m \leq M \tag{8.238}$$

Given $z(m)$, extrapolate $y(m)$ outside the interval $[-M, M]$.

Minimum Norm Least Squares (MNLS) Extrapolation

Let **z** denote the $(2M+1) \times 1$ vector of observations and let **y** denote the infinite vector of $\{y(n), \forall n\}$. then $\mathbf{z} = \mathbf{Sy}$. Since $y(n)$ is a bandlimited sequence, $\mathbf{Ly} = \mathbf{y}$, and we can write

$$\mathbf{z} = \mathbf{SLy} = \mathbf{Ay}, \qquad \mathbf{A} \stackrel{\Delta}{=} \mathbf{SL} \tag{8.239}$$

This can be viewed as an underdetermined image restoration problem, where **A** represents a $(2M+1) \times \infty$ PSF matrix. A unique solution that is bandlimited and reproduces the (noiseless) signal over the interval $[-M, M]$ is the MNLS solution. It is given explicitly as

$$\begin{aligned}\mathbf{y}^+ &\triangleq \mathbf{A}^T[\mathbf{A}\mathbf{A}^T]^{-1}\mathbf{z} = \mathbf{L}^T\mathbf{S}^T[\mathbf{S}\mathbf{L}\mathbf{L}^T\mathbf{S}^T]^{-1}\mathbf{z} \\ &= \mathbf{L}\mathbf{S}^T[\mathbf{S}\mathbf{L}\mathbf{S}^T]^{-1}\mathbf{z} \triangleq \mathbf{L}\mathbf{S}^T\hat{\mathbf{L}}^{-1}\mathbf{z}\end{aligned} \tag{8.240}$$

where $\hat{\mathbf{L}}$ is a $(2M+1) \times (2M+1)$, positive definite, Toeplitz matrix with elements $\{\sin \omega_1 (m-n)/\pi(m-n), -M \le m, n \le M\}$. The matrix

$$\mathbf{A}^+ \triangleq \mathbf{A}^T[\mathbf{A}\mathbf{A}^T]^{-1} = \mathbf{L}\mathbf{S}^T\hat{\mathbf{L}}^{-1} \tag{8.241}$$

is the pseudoinverse of $\mathbf{A}$ and is called the *pseudoinverse extrapolation filter*. The extrapolation algorithm requires first obtaining $\mathbf{x} \triangleq \hat{\mathbf{L}}^{-1}\mathbf{z}$ and then low-pass filtering the sequence $\{x(m), -M \le m \le M\}$ to obtain the extrapolation as

$$y^+(m) = \sum_{j=-M}^{M} \frac{\sin \omega_1 (m-j)}{\pi(m-j)} x(j), \qquad |m| > M \tag{8.242}$$

This means the MNLS extrapolator is a time-varying FIR filter $[\hat{\mathbf{L}}^{-1}]_{m,m'}$ followed by a zero padder ($\mathbf{S}^T$) and an ideal low-pass filter ($\mathbf{L}$) (Fig. 8.31).

Iterative Algorithms

Although $\hat{\mathbf{L}}$ is positive definite, it becomes increasingly ill-conditioned as M increases. In such instances, iterative algorithms that give a stabilized inverse of $\hat{\mathbf{L}}$ are useful [65]. An example is the conjugate gradient algorithm obtained by substituting $\mathbf{A} = \hat{\mathbf{L}}$ and $\mathbf{g}_0 = -\mathbf{z}$ into (8.136). At $n = 2M + 1$, let $\hat{\mathbf{z}} \triangleq \mathbf{u}_n$. Then $\hat{\mathbf{y}}_n \triangleq \mathbf{L}\mathbf{S}^T\hat{\mathbf{z}}$ converges to $\mathbf{y}^+$. Whenever $\hat{\mathbf{L}}$ is ill-conditioned, the algorithm is terminated when β_n becomes small for $n < 2M + 1$. Compared to the energy reduction algorithm, the iterations here are performed on finite-size vectors, and only a finite number of iterations are required for convergence.

Discrete Prolate Spheroidal Sequences (DPSS)

Similar to the PSWF expansion in the continuous case, it is possible to obtain the MNLS extrapolation via the expansion

$$y^+(m) = \sum_{k=1}^{2M+1} a_k \phi_k(m), \qquad \forall m \tag{8.243}$$

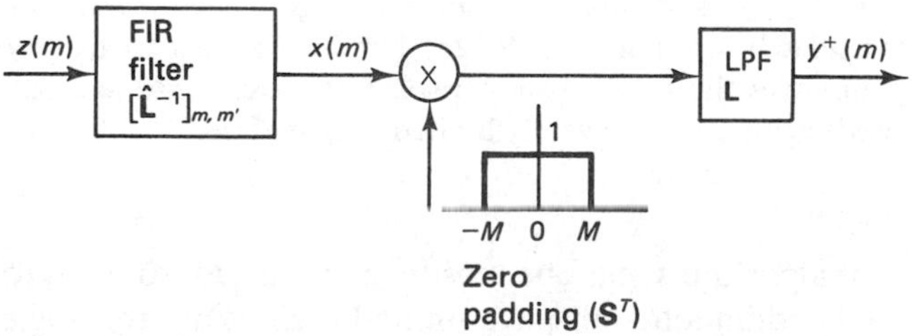

Figure 8.31 MNLS extrapolation.

where the sequence of vectors $\boldsymbol{\phi}_k \triangleq \{\phi_k(m), \forall m\}$ is called *discrete prolate spheroidal sequence* (DPSS). It is obtained from the eigenvalues λ_k and *eigenvectors* $\boldsymbol{\psi}_k$ of $\hat{\mathbf{L}}$ as

$$\boldsymbol{\phi}_k = \frac{1}{\sqrt{\lambda_k}} \mathbf{L}\mathbf{S}^T \boldsymbol{\psi}_k, \qquad k = 1, \ldots, 2M+1 \tag{8.244}$$

Like the PSWFs, the DPSS $\phi_k(m)$ are bandlimited (that is, $\mathbf{L}\boldsymbol{\phi}_k = \boldsymbol{\phi}_k$), complete, and orthogonal in the interval $-M \le m \le M$. They are complete and orthonormal in the infinite interval. Using these properties, we can obtain $\mathbf{S}\boldsymbol{\phi}_k = \sqrt{\lambda_k}\boldsymbol{\psi}_k$ and simplify (8.243) to give the algorithm

$$\left.\begin{aligned} \mathbf{z}^+ &\triangleq \sum_{k=1}^{K} \frac{\alpha_k}{\lambda_k} \boldsymbol{\phi}_k, \qquad \alpha_k = \boldsymbol{\phi}_k^T \mathbf{z} \\ \mathbf{y}^+ &\triangleq \mathbf{L}\mathbf{S}^T \mathbf{z}^+ \end{aligned}\right\} \tag{8.245}$$

In practice, the series summation is carried up to some $K \le 2M+1$, where the neglected terms correspond to the smallest values of λ_k.

Mean Square Extrapolation

In the presence of additive independent noise, the observation equation becomes

$$\mathbf{z} = \mathbf{A}\mathbf{y} + \boldsymbol{\eta} = \mathbf{S}\mathbf{L}\mathbf{y} + \boldsymbol{\eta} \tag{8.246}$$

The best linear mean square extrapolator is then given by the Wiener filter

$$\hat{\mathbf{y}} = \mathbf{R}_y \mathbf{S}^T [\mathbf{S}\mathbf{R}_y \mathbf{S}^T + \mathbf{R}_\eta]^{-1} \mathbf{z} \tag{8.247}$$

where $\mathbf{R}_y$ and $\mathbf{R}_\eta$ are the autocorrelation matrices of $\mathbf{y}$ and $\boldsymbol{\eta}$, respectively. If the autocorrelation of $\mathbf{y}$ is unknown, it is convenient to assume $\mathbf{R}_y = \sigma^2 \mathbf{L}$ (that is, power spectrum of $\mathbf{y}$ is bandlimited and constant). Then, assuming the noise to be white with $\mathbf{R}_\eta = \sigma_\eta^2 \mathbf{I}$, we obtain

$$\hat{\mathbf{y}} = \mathbf{L}\mathbf{S}^T \left[\mathbf{S}\mathbf{L}\mathbf{S}^T + \frac{\sigma_\eta^2}{\sigma^2}\mathbf{I}\right]^{-1} \mathbf{z} = \mathbf{L}\mathbf{S}^T \left[\hat{\mathbf{L}} + \frac{\sigma_\eta^2}{\sigma^2}\mathbf{I}\right]^{-1} \mathbf{z} \tag{8.248}$$

If $\sigma_\eta^2 \to 0$, then $\hat{\mathbf{y}} \to \mathbf{y}^+$, the MNLS extrapolation. A recursive Kalman filter implementation of (8.248) is also possible [65].

Example 8.9

Figure 8.32a shows the signal $y(m) = \sin(0.0792\pi m) + \sin(0.068\pi m)$, which is given for $-8 \le m \le 8$ (Fig. 8.32b) and is assumed to have a bandwidth of less than $\omega_1 = 0.1\pi$. Figures 8.32c and 8.32d show the extrapolations obtained via the iterative energy reduction and conjugate gradient algorithms. As expected, the latter algorithm has superior convergence. When the observations contain noise (13 dB below the signal power), these algorithms tend to be unstable (Fig. 8.32e), but the mean square extrapolation filter (Fig. 8.32f) improves the result. Comparison of Figures 8.32d and 8.32f shows that the extrapolated region can be severely limited due to noise.

Generalization to Two Dimensions

The foregoing extrapolation algorithms can be easily generalized to two (or higher) dimensions when the bandlimited and space-limited regions are rectangles

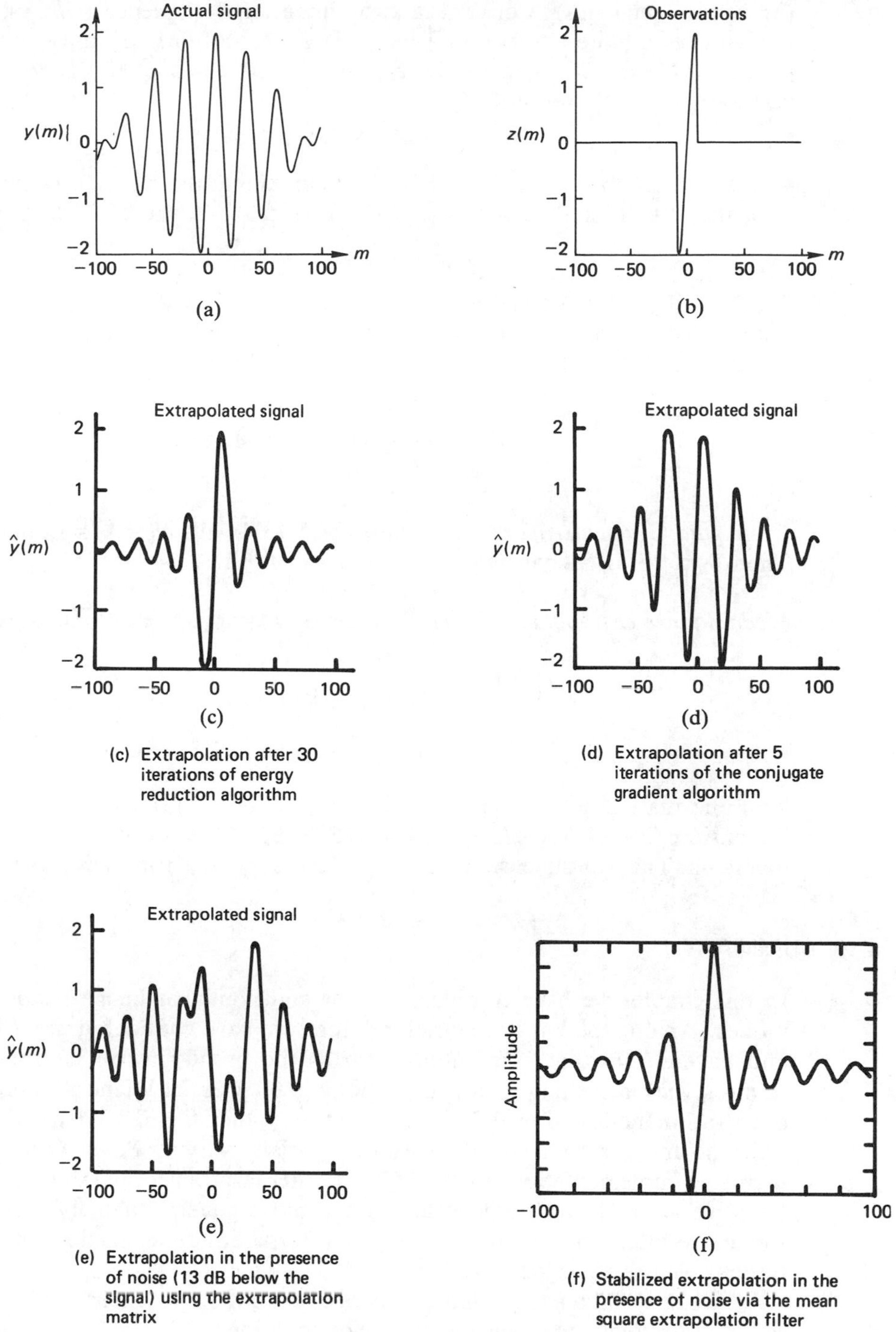

(a)

(b)

(c)

(d)

(c) Extrapolation after 30 iterations of energy reduction algorithm

(d) Extrapolation after 5 iterations of the conjugate gradient algorithm

(e)

(f)

(e) Extrapolation in the presence of noise (13 dB below the signal) using the extrapolation matrix

(f) Stabilized extrapolation in the presence of noise via the mean square extrapolation filter

Figure 8.32 Comparison of extrapolation algorithms.

(or hyper-rectangles). Consider a two-dimensional sequence $y(m, n)$, which is known over a finite observation window $[-M, M] \times [-M, M]$ and bandlimited in $[-\omega_1, \omega_1] \times [-\omega_1, \omega_1]$. Let $z(m, n) = y(m, n), -M \leq m, n \leq M$. Then, using the operators **L** and **S**, we can write

$$\mathbf{Z} = \mathbf{SLYLS}^T \tag{8.249}$$

where **Z** is a $(2M + 1) \times (2M + 1)$ matrix containing the $z(m, n)$. Defining $\boldsymbol{z}$ and $\boldsymbol{y}$ as the row-ordered mappings of **Z** and **Y**, $\mathcal{S} \triangleq \mathbf{S} \otimes \mathbf{S}$, and $\mathcal{L} \triangleq \mathbf{L} \otimes \mathbf{L}$, we get

$$\boldsymbol{z} = \mathcal{S}\mathcal{L}\boldsymbol{y} \triangleq \mathcal{A}\boldsymbol{y} \qquad (\mathcal{A} \triangleq \mathcal{S}\mathcal{L}) \tag{8.250}$$

Similar to **L**, the two-dimensional low-pass filter matrix $\mathcal{L}$ is symmetric and idempotent. All the foregoing one-dimensional algorithms can be recast in terms of $\mathcal{S}$ and $\mathcal{L}$, from which the following two-dimensional versions follow.

MNLS extrapolation:

$$\left.\begin{aligned} \mathbf{Y}^+ &= \mathbf{LS}^T[\hat{\mathbf{L}}^{-1}\mathbf{Z}\hat{\mathbf{L}}^{-1}]\mathbf{SL} \\ &= \mathbf{A}^+\mathbf{Z}\mathbf{E}^{+T} \end{aligned}\right\} \tag{8.251}$$

Conjugate Gradient Algorithm. Same as (8.137) with $\mathbf{A}_1 = \mathbf{A}_2 \triangleq \hat{\mathbf{L}}$ and $\mathbf{G}_0 \triangleq -\mathbf{Z}$. Then $\hat{\mathbf{Y}}_n \triangleq \mathbf{LS}^T\mathbf{U}_n\mathbf{SL}$ converges to $\mathbf{Y}^+$ at $n = 2M + 1$.

Mean Square Extrapolation Filter. Assume $\mathcal{R}_y = \sigma^2\mathcal{L}, \mathcal{R}_\eta = \sigma_\eta^2(\mathbf{I} \otimes \mathbf{I})$.

$$\left.\begin{aligned} \hat{\boldsymbol{z}} &\triangleq \left[(\hat{\mathbf{L}} \otimes \hat{\mathbf{L}}) + \frac{\sigma_\eta^2}{\sigma^2}(\mathbf{I} \otimes \mathbf{I})\right]^{-1} \boldsymbol{y} \\ \hat{\boldsymbol{y}} &= [(\mathbf{LS}^T) \otimes (\mathbf{LS}^T)]\hat{\boldsymbol{z}} \quad \Rightarrow \hat{\mathbf{Y}} = \mathbf{LS}^T\hat{\mathbf{Z}}\mathbf{SL} \end{aligned}\right\} \tag{8.252}$$

Now the matrix to be inverted is $(2M + 1) \times (2M + 1)$ block Toeplitz with basic dimension $(2M + 1) \times (2M + 1)$. The DPSS $\boldsymbol{\phi}_k$ of (8.244) can be useful for this inversion. The two-dimensional DPSS are given by the Kronecker product $\boldsymbol{\phi}_k \otimes \boldsymbol{\phi}_l$.

8.19 SUMMARY

In this chapter we have considered linear and nonlinear image restoration techniques. Among the linear restoration filters, we have considered the Wiener filter and have shown that other filters such as the pseudoinverse, constrained least squares, and smoothing splines also belong to the class of Wiener filters. For linear observation models, if the PSF is not very broad, then the FIR Wiener filter is quite efficient and can be adapted to handle spatially varying PSFs. Otherwise, nonrecursive implementations via the FFT (or other fast transforms) should be suitable. Iterative methods are most useful for the more general spatially varying filters. Recursive filters and semirecursive filters offer alternate realizations of the Wiener filter as well as other local estimators.

We saw that a large number of image restoration problems can be reduced to solving (approximately) a linear system of equations

$$\mathbf{Hu} \simeq \mathbf{v}$$

for the unknowns **u** from the observations **v**. Some special types of restoration problems include extrapolation of bandlimited signals and image reconstruction from projections. The latter class of problems will be studied in chapter 10.

Among the nonlinear techniques, we have considered homomorphic filtering for speckle reduction, maximum entropy restoration, ML and MAP estimation for nonlinear imaging models, and a blind deconvolution method for restoration with unknown blur.

PROBLEMS

8.1 *(Coherent image formation)* According to Fresnel theory of diffraction, the complex amplitude field of an object $u(x', y')$, illuminated by a uniform monochromatic source of light at wavelength λ at a distance z, is given by

$$v(x, y) = c_1 \iint_{-\infty}^{\infty} u(x', y') \exp\left\{\frac{jk}{2z}[(x - x')^2 + (y - y')^2]\right\} dx'\, dy'$$

where c_1 is a complex quantity with $|c_1| = 1$, $k = 2\pi/\lambda$. Show that if z is much greater than the size of the object, then $v(x, y)$ is a coherent image whose intensity is proportional to the magnitude squared of the Fourier transform of the object. This is also called the Fraunhofer diffraction pattern of the object.

8.2 *(Optical filtering)* An object $u(x, y)$ illuminated by coherent light of wavelength λ and imaged by a lens can be modeled by the spatially invariant system shown in Fig. P8.2, where

$$h_k(x, y) = \frac{1}{j\lambda d_k} \exp\left\{j\frac{\pi}{\lambda d_k}(x^2 + y^2)\right\}, k = 0, 1, 2$$

$$l(x, y) = h_2^*(x, y)p(x, y),$$

$p(x, y)$ is a square aperture of width a_1, and d_2 is the focal length of the lens. Find the incoherent impulse response of this system when $1/d_0 + 1/d_1 = 1/d_2$ and show how this method may be used to filter an image optically. (*Hint:* Recall that the incoherent impulse response is the squared magnitude of the coherent impulse response.)

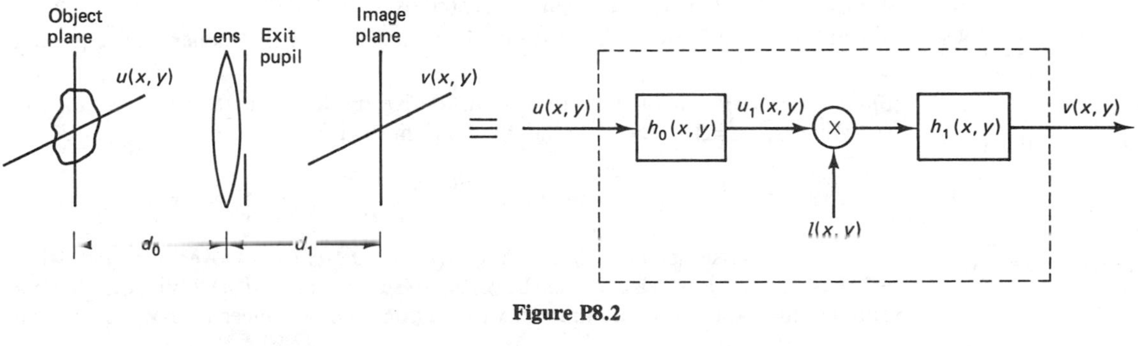

Figure P8.2

8.3 In the observation model of Equations (8.1)–(8.3), suppose the image $v(x, y)$ is sampled by a uniform square grid with spacing Δ, such that

$$v(m, n) \triangleq \frac{1}{\Delta^2} \iint_0^\Delta v(m\Delta + x, n\Delta + y)\, dx\, dy$$

If $u(m, n)$, $w(m, n)$, $\eta'(m, n)$ and $\eta''(m, n)$ are defined similarly and $u(x, y)$, $v(x, y)$, and $w(x, y)$ are approximated as piecewise constant functions over this grid, show that (8.1)–(8.3) can be approximated by (8.18)–(8.20) where

$$h(m, n; k, l) = \frac{1}{\Delta^4} \iiiint_0^\Delta h(m\Delta + x, n\Delta + y; k\Delta + x', l\Delta + y')\, dx\, dy\, dx'\, dy'$$

and $\eta'(m, n)$ and $\eta''(m, n)$ are white noise discrete random fields.

8.4 Show that the inverse filter of a spatially invariant system will also be spatially invariant.

8.5 A digital image stored in a computer is to be recorded on a photographic film using a flying spot scanner. The effect of spot size of the scanner and the film nonlinearity is equivalent to first passing the image through the model of Fig. P8.5 and then recording it on a perfect recording system. Find the inverse filter that should be inserted before recording the image to compensate for (a) scanner spot size, assuming $\gamma = -1$, and (b) film nonlinearity, ignoring the effect of spot size.

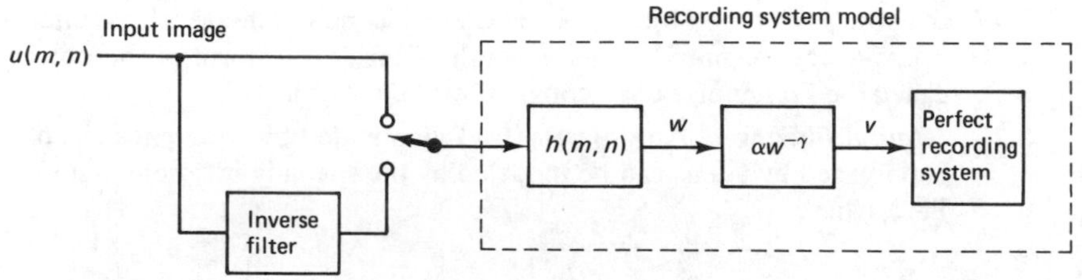

$$h(m, n) = \delta(m, n) + \tfrac{1}{4}\,[\delta(m-1, n) + \delta(m+1, n) + \delta(m, n-1) + \delta(m, n+1)]$$

Figure P8.5

8.6 Prove the first two statements made under Remarks in Section 8.3.

8.7 A motion blurred image (see Example 8.1) is observed in the presence of additive white noise. What is the Wiener filter equation if the covariance function of $u(x, y)$ is $r(x, y) = \sigma^2 \exp\{-0.05|x| - .05|y|\}$? Assume the mean of the object is known. Give an algorithm for digital implementation of this filter.

8.8 Show exactly how (8.53) may be implemented using the FFT.

8.9 Starting from (8.57) show all the steps which lead to the FIR Wiener filter equation (8.64).

8.10 *(Spatially varying FIR filters)* In order to derive the formulas (8.68), (8.69) for the spatially varying FIR filter, note that the random field

$$\tilde{u}(m, n) \triangleq \frac{u(m, n) - \mu(m, n)}{\sigma(m, n)}$$

is stationary. Assume that $\mu(m, n)$ and $\sigma(m, n)$ are constant over a region $W_0 = \{-2M \le i, j \le 2M\}$, which contains the region of support of $h(m, n)$. Using this, write a spatially invariant observation model with stationary covariances over W_0. Using the results of Section 8.4, assuming $\Sigma\Sigma h(m, n) = 1$ and using the fact $\hat{u}(m, n)$ should be unbiased, prove (8.68) and (8.69) under the assumption

$$\mu(m, n) = \mu_{\hat{u}}(m, n) = \mu_v(m, n) \simeq \frac{1}{(2M+1)^2} \sum\sum_{(i,j) \in W} v(i-m, j-n)$$

8.11 Show that the Wiener filter does not restore the power spectral density of the object, whereas the geometric mean filter does when $s = \frac{1}{2}$. (*Hint:* $S_{\hat{u}\hat{u}} = |G|^2 S_{vv}$ for any filter G). Compare the mean square errors of the two filters.

8.12* Take a low-contrast image and root filter it to enhance its high spatial frequencies.

8.13 **a.** If $\mathcal{G}$ is an arbitrary filter in (8.77) show that the average mean square error is given by

$$\sigma_e^2 = \frac{1}{M_1 M_2} \operatorname{Tr}[\mathcal{R} + \mathcal{G}(\mathcal{H}\mathcal{R}\mathcal{H}^T + \mathcal{R}_n)\mathcal{G}^T - 2\mathcal{G}\mathcal{H}\mathcal{R}]$$

b. If $\mathcal{G}$ is the Wiener filter, then show the minimum value of σ_e^2 is given by (8.82).

8.14 (*Sine/cosine transform-based Wiener filtering*) Consider the white noise–driven model for $N \times N$ images

$$u(m, n) = \alpha[u(m-1, n) + u(m+1, n) + u(m, n-1) + u(m, n+1)] \\ + \varepsilon(m, n), \qquad |\alpha| < \tfrac{1}{4}, 0 \le m, n \le N-1$$

$$S_\varepsilon(z_1, z_2) = \beta^2$$

where $u(-1, n) = u(0, n)$, $u(N, n) = u(N-1, n)$, $u(m, -1) = u(m, 0)$, $u(m, N) = u(m, N-1)$.

a. Show that the cosine transform is the KL transform of $u(m, n)$, which yields the generalized Wiener filter gain for the noise smoothing problem as

$$\tilde{p}(k, l) = \frac{\beta^2}{\beta^2 + \sigma_\eta^2[1 - 2\alpha(\cos k\pi/N + \cos l\pi/N)]^2}, \qquad 0 \le k, l \le N-1$$

b. If $u(-1, n) = u(N, n) = u(m, -1) - u(m, N) = 0$, show that the sine transform is the KL transform and find the generalized filter gain.

8.15 **a.** Show that the spline coefficient vector $\mathbf{a}$ can be obtained directly:

$$\mathbf{a} = \left(\mathbf{I} + \frac{\sigma_n^2}{\lambda}\mathbf{L}\mathbf{Q}^{-1}\mathbf{L}^T\right)^{1} \mathbf{y}$$

which is a Wiener filter if the noise in (8.96) is assumed to be white with zero mean and variance σ_n^2 and if the autocorrelation matrix of $\mathbf{y}$ is $\lambda[\mathbf{L}\mathbf{Q}^{-1}\mathbf{L}^T]^{-1}$.

b. The interpolating splines can be obtained by setting $S = 0$ or, equivalently, letting $\lambda \to \infty$ in (8.100). For the data of Example 8.6 find these splines.

8.16 Prove that (8.110) and (8.111) give the solution of the constrained least squares restoration problem stated in the text.

8.17 Suppose the object $u(m, n)$ is modeled as the output of a linear system driven by a zero mean unit variance white noise random field $\varepsilon(m, n)$, namely,

$$q(m, n) * u(m, n) = \varepsilon(m, n)$$

If $u(m, n)$ is observed via (8.39) with $S_{\eta\eta} = \gamma$, show that its Wiener filter is identical to the least squares filter of (8.110). Write down the filter equation when $q(m, n)$ is given by (8.112) and show that the object model is a white noise–driven noncausal model.

8.18 Show that (8.115) is the solution of the least squares problem defined in (8.114).

8.19 If the sequences $u(m, n), h(m, n), q(m, n)$ are periodic over an $N \times N$ grid with DFTs $U(k, l), H(k, l), Q(k, l), \ldots$, then show the impulse response of the constrained least squares filter is given by the inverse DFT of

$$G_{ls}(k, l) \triangleq \frac{H^*(k, l)}{|H(k, l)|^2 + \gamma |Q(k, l)|^2}, \qquad 0 \le k, l \le N - 1$$

where γ is obtained by solving

$$\frac{\gamma^2}{N^2} \sum_k \sum_l \frac{|Q(k, l)|^4 |V(k, l)|^2}{[|H(k, l)|^2 + \gamma |Q(k, l)|^2]^2} = \varepsilon^2$$

8.20 Show that $\mathscr{G}^-$ defined in (8.86) is the pseudoinverse of $\mathscr{H}$ and is nonunique when $N_1 N_2 \le M_1 M_2$.

8.21 Show that for separable PSFs, for which the observation equation can be written as $\boldsymbol{v} = (\mathbf{H}_1 \otimes \mathbf{H}_2)\boldsymbol{u}$, (8.136) yields the two-dimensional conjugate gradient algorithm of (8.137).

8.22 Write the Kalman filtering equations for a first-order AR sequence $u(n)$, which is observed as

$$y(n) = u(n) + \tfrac{1}{2}[u(n - 1) + u(n + 1)] + \eta(n)$$

8.23 *(K-step interpolator)* In many noise smoothing applications it is of interest to obtain the estimate which lags the observations by K steps, that is, $\hat{\mathbf{x}}_{n,K} \triangleq E[\mathbf{x}_n | \mathbf{y}_{n'}, 0 \le n' \le n + K]$. Give a recursive algorithm for obtaining this estimate. In image processing applications, often the *one-step interpolator* performs quite close to the optimum smoother. Show that it is given by

$$\hat{\mathbf{x}}_{n,1} = \mathbf{R}_n \mathbf{A}_n^T \mathbf{C}_{n+1}^T \mathbf{q}_{n+1}^{-1} \boldsymbol{\nu}_{n+1} + \mathbf{R}_n \mathbf{C}_n^T \mathbf{q}_n^{-1} [\boldsymbol{\nu}_n - \mathbf{G}_n^T \mathbf{C}_{n+1}^T \mathbf{q}_{n+1}^{-1} \boldsymbol{\nu}_{n+1}] + \mathbf{s}_n$$

8.24 In the semicausal model of (6.106) assume $u(0, N), u(N + 1, n)$ are known (that is, image background is given) and the observation model of (8.171) has no blur, that is, $h(k, l) = \delta(k, l)$. Give the complete semirecursive filtering algorithm and identify the fast transform used.

8.25 Show that there is no speckle in images obtained by an ideal imaging system. Show that, for a practical imaging system, the *speckle size* measured by its correlation distance can be used to estimate the resolution (that is, R_{cell}) of the imaging system.

8.26 **a.** Show that (8.193) is the solution of $\min_{\boldsymbol{u}}[\mathscr{E}(\boldsymbol{u}) + \lambda\{\tfrac{1}{2}\|\boldsymbol{v} - \mathscr{H}\boldsymbol{u}\|^2 - \sigma_e^2\}]$.

b. *(Maximum entropy spectrum estimation)* A special case of log-entropy restoration is the problem of maximizing

$$\mathscr{E} \triangleq \frac{1}{2\pi} \int_{-\pi}^{\pi} \log S(\omega)\, d\omega$$

where $S(\omega)$ is observed as

$$r(n) = \frac{1}{2\pi} \int_{-\pi}^{\pi} S(\omega) e^{jn\omega}\, d\omega \qquad n = 0, \pm 1, \ldots, \pm p$$

The maximization is performed with regard to the missing observations, that is, $\{r(n), |n| > p\}$. This problem is equivalent to extrapolating the partial sequence of autocorrelations $\{r(n), |n| \le p\}$ out to infinity such that the entropy $\mathscr{E}$ is maximized and $S(\omega) = \sum_{n=-\infty}^{\infty} r(n) e^{-j\omega n}$ is the SDF associated with $r(n)$. Show that the maximum entropy solution requires that $S(z) \stackrel{\Delta}{=} S(\omega) = 1/[\sum_{n=-p}^{p} \lambda(n) z^{-n}]$, $z = e^{j\omega}$, that is, the Fourier series of $1/S(\omega)$ must be truncated to $\pm p$ terms. Noting that $S(z)$ can be factored $\beta^2/A_p(z)A_p(z^{-1})$, $A_p(z) \stackrel{\Delta}{=} 1 - \sum_{k=1}^{p} a(k) z^{-k}$ show that β^2 and $a(k), k = 1, \ldots, p$ are obtained by solving the AR model equations (6.13a) and (6.13b). What is the resulting spectral density function? For an alternate direct algorithm see [68].

8.27 Using the Gaussian assumptions for u and η prove the formulas for the ML and MAP estimates given by (8.212) and (8.214).

8.28 Suppose the bandlimited signal $f(x)$ is observed as $\tilde{f}(x) = f(x) + \eta(x), |x| < \alpha$, where $\eta(x)$ is white noise with $E[\eta(x)\eta(x')] = \sigma_\eta^2 \delta(x - x')$. If $\tilde{f}(x)$ is extrapolated by the truncated PSWF series $\hat{f}(x) = \sum_{n=0}^{N-1} \tilde{a}_n \phi_n(x)$, for every x, then show that the minimum integral mean square error is $\int_{-\infty}^{\infty} E[|f(x) - \hat{f}(x)|^2]\, dx = \sum_{n=N}^{\infty} |a_n|^2 + \sigma_\eta^2 \sum_{n=0}^{N-1} (1/\lambda_n)$ where $\lambda_0 \geq \lambda_1 \geq \lambda_2 \geq \cdots \geq \lambda_n \geq \lambda_{n+1} \ldots$ and a_n is given by (8.229). This means the error due to noise increases with the number of terms in the PSWF expansion.

8.29 **a.** If the bandlimited function $f(x)$ is sampled at Nyquist rate $(1/\Delta)$ to yield $y(m) = f(m\Delta), -M \leq m \leq M$, what is the MNLS extrapolation of $y(m)$?

b. Show that the MNLS extrapolation of a bandlimited sequence is bandlimited and consistent with the given observations.

BIBLIOGRAPHY

Section 8.1, 8.2

For general surveys of image restoration and modeling of imaging systems:

1. T. S. Huang, W. F. Schreiber and O. J. Tretiak. "Image Processing." *Proc. IEEE* 59, no. 11 (November 1971): 1586–1609.
2. M. M. Sondhi. "Image Restoration: The Removal of Spatially Invariant Degradations." *Proc. IEEE* 60 (July 1972): 842–853.
3. T. S. Huang (ed.). "Picture Processing and Digital Filtering." In *Topics in Applied Physics,* Vol. 6. Berlin: Springer-Verlag, 1975.
4. H. C. Andrews and B. R. Hunt. *Digital Image Restoration,* Englewood Cliffs, N.J.: Prentice-Hall, 1977.
5. B. R. Frieden. "Image Enhancement and Restoration" in [3]. References on nonlinear filtering are also available here.
6. J. W. Goodman. *Introduction to Fourier Optics,* Ch. 6, 7. San Francisco: McGraw-Hill, 1968.

Section 8.3

Discussions on inverse and Wiener filtering of images are available in several places; see, for instance, [1, 2, 5] and:

7. C. W. Helstrom. "Image Restoration by the Method of Least Squares." *J. Opt. Soc. Am.* 57 (March 1967): 297–303.

Section 8.4

For FIR Wiener Filtering Theory and more examples:

8. A. K. Jain and S. Ranganath. "Applications of Two Dimensional Spectral Estimation in Image Restoration." *Proc. ICASSP—1981* (May 1981): 1113–1116. Also see *Proc. ICASSP—1982* (May 1982): 1520–1523.

For block by block filtering techniques:

9. J. S. Lim. "Image Restoration by Short Space Spectral Subtraction." *IEEE Trans. Acous. Speech, Sig., Proc.* ASSP-28, no. 2 (April 1980): 191–197.
10. H. J. Trussell and B. R. Hunt. "Sectioned Methods for Image Restoration." *IEEE Trans. Acous. Speech, Sig. Proc.* ASSP-26 (April 1978): 157–164.
11. R. Chellapa and R. L. Kashyap. "Digital Image Restoration Using Spatial Interaction Models." *IEEE Trans. Acous. Speech Sig. Proc.* ASSP-30 (June 1982): 461–472.

Section 8.6

For generalized Wiener filtering and filtering by fast decompositions:

12. W. K. Pratt. "Generalized Wiener Filter Computation Techniques." *Trans. IEEE Comput.* C-21 (July 1972): 636–641.
13. A. K. Jain. "An Operator Factorization Method for Restoration of Blurred Images." *IEEE Trans. Computers* C-26 (November 1977): 1061–1071. Also see vol. C-26 (June 1977): 560–571.
14. A. K. Jain. "Fast Inversion of Banded Toeplitz Matrices via Circular Decompositions." *IEEE Trans. Acous. Speech Sig. Proc.* ASSP-26, no. 2 (April 1978): 121–126.

Section 8.7

15. T. N. E. Greville (ed.) *Theory and Applications of Spline Functions.* New York: Academic Press, 1969.
16. M. J. Peyrovian. "Image Restoration by Spline Functions." USCIPI Report No. 680, University of Southern California, Los Angeles, August 1976. Also see *Applied Optics* 16 (December 1977): 3147–3153.
17. H. S. Hou. "Least Squares Image Restoration Using Spline Interpolation." Ph.D. Dissertation, IPI Report No. 650, University of Southern California, Los Angeles, March 1976. Also see *IEEE Trans. Computers* C-26, no. 9 (September 1977): 856–873.

Section 8.8

18. S. Twomey. "On the Numerical Solution of Fredholm Integral Equations of First Kind by the Inversion of Linear System Produced by Quadrature." *J. Assoc. Comput. Mach.* 10 (January 1963): 97–101.
19. B. R. Hunt. "The Application of Constrained Least Squares Estimation to Image Restoration by Digital Computer." *IEEE Trans. Comput.* C-22 (September 1973): 805–812.

Section 8.9

The generalized inverse of a matrix is sometimes also called the Moore-Penrose inverse. For greater details and examples:

20. C. R. Rao and S. K. Mitra. *Generalized Inverse of Matrices and its Applications.* New York: John Wiley and Sons, 1971.
21. A. Albert. *Regression and Moore-Penrose Pseudoinverse.* New York: Academic Press, 1972.

For numerical properties of the gradient algorithms and other iterative methods and their applications to space-variant image restoration problems:

22. D. G. Luenberger. *Introduction to Linear and Nonlinear Programming.* Reading, Mass.: Addison-Wesley, 1973.
23. T. S. Huang, D. A. Barker and S. P. Berger. "Iterative Image Restoration." *Applied Optics* 14, no. 5 (May 1975): 1165–1168.
24. E. S. Angel and A. K. Jain. "Restoration of Images Degraded by Spatially Varying Point Spread Functions by a Conjugate Gradient Method." *Applied Optics* 17 (July 1978): 2186–2190.

Section 8.10

For Kalman's original work and its various extensions in recursive filtering theory:

25. R. E. Kalman. "A New Approach to Linear Filtering and Prediction Problems." *Trans. ASME.* Ser. D., J. Basic Engineering, 82 (1960): 35–45.
26. B. D. O. Anderson and J. H. Moore. *Optimal Filtering.* Englewood Cliffs, N.J.: Prentice-Hall, 1979.
27. G. J. Bierman. "A Comparison of Discrete Linear Filtering Algorithms." *IEEE Trans. Aerosp. Electron. Syst.* AES-9 (January 1973): 28–37.
28. M. Morf and T. Kailath. "Square Root Algorithms for Least-Squares Estimation." *IEEE Trans. Aut. Contr.* AC-20 (August 1975): 487–497.

For FFT based algorithms for linear estimation and Riccati equations:

29. A. K. Jain and J. Jasiulek. "A Class of FFT Based Algorithms for Linear Estimation and Boundary Value Problems." *IEEE Trans. Acous. Speech Sig. Proc.* ASSP 31, no. 6 (December 1983): 1435–1446.

For state variable formulation for image estimation and smoothing and its extensions to restoration of motion degraded images:

30. N. E. Nahi. "Role of Recursive Estimation in Statistical Image Enhancement." *Proc. IEEE* 60 (July 1972): 872–877.
31. A. O. Aboutalib and L. M. Silverman. "Restoration of Motion Degraded Images." *IEEE Trans. Cir. Sys.* CAS-22 (March 1975): 278–286.

Section 8.11

Recursive algorithms for least squares filtering and linear estimation of images have been considered in:

32. A. K. Jain and E. Angel. "Image Restoration, Modeling and Reduction of Dimensionality." *IEEE Trans. Computers* C-23 (May 1974): 470–476. Also see *IEEE Trans. Aut. Contr.* AC-18 (February 1973): 59–62.
33. A. Habibi. "Two-Dimensional Bayesian Estimate of Images." *Proc. IEEE* 60 (July 1972): 878–883. Also see M. Strintzis, "Comments on Two-Dimensional Bayesian Estimate of Images." *Proc. IEEE* 64 (August 1976): 1255–1257.
34. A. K. Jain and J. R. Jain. "Partial Differential Equations and Finite Difference Methods in Image Processing, Part II: Image Restoration." *IEEE Trans. Aut. Control* AC-23 (October 1978): 817–834.
35. F. C. Schoute, M. F. Terhorst and J. C. Willems. "Hierarchic Recursive Image Enhancement." *IEEE Trans. Circuits and Systems* CAS-24 (February 1977): 67–78.
36. J. W. Woods and C. H. Radewan. "Kalman Filtering in Two Dimensions." *IEEE Trans. Inform. Theory* IT-23 (July 1977): 473–482.
37. S. A. Rajala and R. J. P. De Figueiredo. "Adaptive Nonlinear Image Restoration by a Modified Kalman Filtering Approach." *IEEE Trans. Acoust. Speech Sig. Proc.* ASSP-29 (October 1981): 1033–1042.
38. S. S. Dikshit. "A Recursive Kalman Window Approach to Image Restoration." *IEEE Trans. Acoust. Speech Sig. Proc.* ASSP-30, no. 2 (April 1982): 125–129.

The performance of recursive filters can be improved by adapting the image model to spatial variations; for example:

39. N. E. Nahi and A. Habibi. "Decision Directed Recursive Image Enhancement." *IEEE Trans. Cir. Sys.* CAS-22 (March 1975): 286–293.

Section 8.12

Semirecursive filtering algorithms for images were introduced in:

40. A. K. Jain. "A Semicausal Model for Recursive Filtering of Two-Dimensional Images." *IEEE Trans. Computers* C-26 (April 1977): 345–350.

For generalization of this approach see [8, 34].

Section 8.13

For fundamentals of speckle theory and some recent advances:

41. J. C. Dainty (ed.). *Laser Speckle.* New York: Springer Verlag, 1975.
42. J. W. Goodman. "Statistical Properties of Laser Speckle Patterns." In *Laser Speckle* [41].
43. *Speckle in Optics,* Special Issue, *J. Opt. Soc. Am.* 66 (November 1976).

For other results on speckle theory we follow:

44. M. Tur, K. C. Chin and J. W. Goodman. "When Is Speckle Noise Multiplicative." (letter) *Applied Optics,* 21 (April 1982): 1157–1159.

45. H. H. Arsenault and G. April. "Properties of Speckle Integrated with a Finite Aperture and Logarithmically Transformed." In [43], pp. 1160–1163.

For digital processing of speckle we follow:

46. A. K. Jain and C. R. Christensen. "Digital Processing of Images in Speckle Noise." *Proc. SPIE, Applications of Speckle Phenomena* 243 (July 1980): 46–50.

47. J. S. Lim and H. Nawab. "Techniques for Speckle Noise Removal." *Proc. SPIE* 243 (July 1980): 35–44.

For other homomorphic filtering applications and related bibliography see [58].

Section 8.14

For maximum entropy restoration algorithms applicable to images, see [5] and:

48. B. R. Frieden. "Restoring with Maximum Likelihood and Maximum Entropy." *J. Opt. Soc. Amer.* 62 (1972): 511–518.

49. A. Lent. "A Convergent Algorithm for Maximum Entropy Image Restoration with a Medical X-ray Application." In *Image Analysis and Evaluation,* SPSE Conf. Proc. (R. Shaw, ed.), Toronto, Canada, July 1976, pp. 221–267.

The iterative algorithms presented are based on unconstrained optimization of the dual functions and follow from:

50. A. K. Jain and J. Ranganath. "Two-Dimensional Spectral Estimation." *Proc. RADC Workshop on Spectral Estimation* (1978): 151–157.

51. S. W. Lang. "Spectral Estimation for Sensor Arrays." Ph.D. Thesis, M.I.T., August 1981. Also see *IEEE Trans. Acous. Speech Sig. Proc.* ASSP-30, no. 6 (December 1982): 880–887.

Section 8.15

For application of Bayesian methods for realizing MAP and ML estimators for nonlinear image restoration problems, see [4] and:

52. B. R. Hunt. "Bayesian Methods in Nonlinear Digital Image Restoration." *IEEE Trans. Computers* C-26, no. 3, pp. 219–229.

53. H. J. Trussell. "A Relationship between Image Restoration by the Maximum *A Posteriori* Method and a Maximum Entropy Method." *IEEE Trans. Acous. Speech Sig. Proc.* ASSP-28, no. 1 (February 1980): 114–117. Also see vol. ASSP-31, no. 1 (February 1983): 129–136.

54. J. B. Morton and H. C. Andrews. "*A Posteriori* Method of Image Restoration." *J. Opt. Soc. Amer.* 69, no. 2 (February 1979): 280–290.

Section 8.16

55. R. Bernstein (ed.). *Digital Image Processing for Remote Sensing*. New York: IEEE Press, 1978.

Coordinate transformation methods that convert certain space-varying PSFs into a space-invariant PSF are developed in:

56. G. M. Robbins. "Image Restoration for a Class of Linear Spatially Variant Degradations." *Pattern Recognition* 2 (1970): 91–103. Also see *Proc. IEEE* 60 (July 1972): 862–872.
57. A. A. Sawchuk. "Space-Variant Image Restoration by Coordinate Transformations." *J. Opt. Soc. Am.* 64 (February 1974): 138–144. Also see *J. Opt. Soc. Am.* 63 (1973): 1052–1062.

Section 8.17

For blind deconvolution methods and their application in image restoration:

58. E. R. Cole. "The Removal of Unknown Image Blurs by Hormomorphic Filtering." Ph.D. Dissertation, Department of Electrical Engineering, University of Utah, Salt Lake City, 1973.
59. K. T. Knox. "Image Retrieval from Astronomical Speckle Patterns." *J. Opt. Soc. Am.* 66 (November 1976): 1236–1239.

Section 8.18

Here we follow:

60. D. Slepian and H. O. Pollak. "Prolate Spheroidal Wave Functions, Fourier Analysis and Uncertainty-I." *BSTJ* 40 (Janaury 1961): 43–62.

For energy reduction algorithm and its generalization to method of alternating projections:

61. R. W. Gerchberg. "Super-resolution through Error Energy Reduction." *Opt. Acta* 14, no. 9 (September 1979): 709–720.
62. A. Papoulis. "A New Algorithm in Spectral Analysis and Bandlimited Extrapolation." *IEEE Trans. Cir. Sys.* CAS-22, no. 9 (September 1975).
63. D. C. Youla. "Generalized Image Restoration by the Method of Alternating Orthogonal Projections." *IEEE Trans. Cir. Sys.* CAS-25, no. 9 (September 1978): 694–702. Also see Youla and Webb. *IEEE Trans. Medical Imaging* MI-1, no. 2 (October 1982): 81–94.
64. M. I. Sezan and H. Stark. "Image Restoration by the Method of Convex Projections, Par 2—Applications and Numerical Results." *IEEE Trans. Medical Imaging* MI-1, no. 2 (October 1982).
65. A. K. Jain and S. Ranganath. "Extrapolation Algorithms for Discrete Signals with Applications in Spectral Estimation." *IEEE Trans. Acous. Speech. Sig. Proc.* ASSP-29

(August 1981): 830–845. Other references on bandlimited signal extrapolation can be found here.

Other References

66. E. S. Angel and A. K. Jain. "Frame to Frame Restoration of Diffusion Images." *IEEE Trans. Auto. Control* AC-23 (October 1978): 850–855.
67. B. L. McGlaimmery. "Restoration of Turbulence Degraded Images." *J. Opt. Soc. Am.* 57 (March 1967): 293–297.
68. J. S. Lim and N. A. Malik. "A New Algorithm for Two-Dimensional Maximum Entropy Power Spectrum Estimation." *IEEE Trans. Acoust. Speech, Signal Process.* ASSP-29, 1981, 401–413.

9

Image Analysis and Computer Vision

9.1 INTRODUCTION

The ultimate aim in a large number of image processing applications (Table 9.1) is to extract important *features* from image data, from which a description, interpretation, or understanding of the scene can be provided by the machine (Fig. 9.1). For example, a vision system may distinguish parts on an assembly line and list their features, such as size and number of holes. More sophisticated vision systems are

TABLE 9.1 Computer Vision Applications

	Applications	Problems
1	Mail sorting, label reading, supermarket-product billing, bank-check processing, text reading	Character recognition
2	Tumor detection, measurement of size and shape of internal organs, chromosome analysis, blood cell count	Medical image analysis
3	Parts identification on assembly lines, defect and fault inspection	Industrial automation
4	Recognition and interpretation of objects in a scene, motion control and execution through visual feedback	Robotics
5	Map making from photographs, synthesis of weather maps	Cartography
6	Finger-print matching and analysis of automated security systems	Forensics
7	Target detection and identification, guidance of helicopters and aircraft in landing, guidance of remotely piloted vehicles (RPV), missiles and satellites from visual cues	Radar imaging
8	Multispectral image analysis, weather prediction, classification and monitoring of urban, agricultural, and marine environments from satellite images	Remote sensing

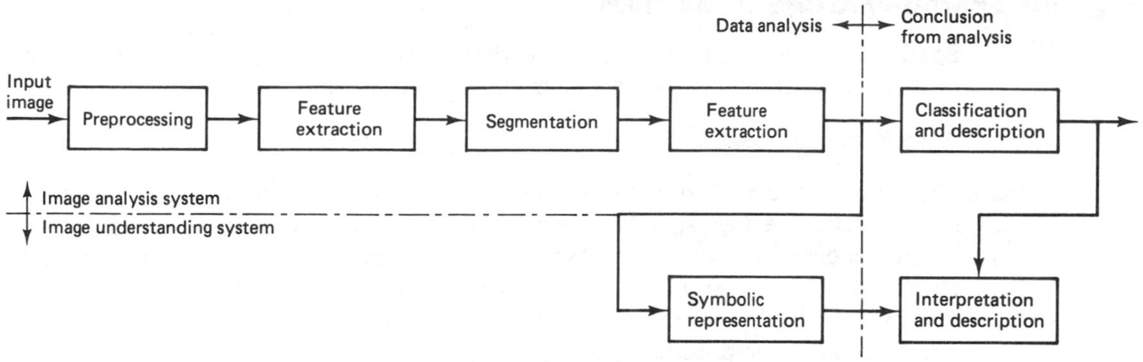

Figure 9.1 A computer vision system

able to interpret the results of analyses and describe the various objects and their relationships in the scene. In this sense image analysis is quite different from other image processing operations, such as restoration, enhancement, and coding, where the output is another image. Image analysis basically involves the study of *feature extraction, segmentation,* and *classification* techniques (Fig. 9.2).

In computer vision systems such as the one shown in Fig. 9.1, the input image is first preprocessed, which may involve restoration, enhancement, or just proper representation of the data. Then certain features are extracted for *segmentation* of the image into its components—for example, separation of different objects by extracting their boundaries. The segmented image is fed into a classifier or an image understanding system. Image classification maps different regions or segments into one of several objects, each identified by a label. For example, in sorting nuts and bolts, all objects identified as square shapes with a hole may be classified as nuts and those with elongated shapes, as bolts. Image understanding systems determine the relationships between different objects in a scene in order to provide its description. For example, an image understanding system should be able to send the report:

The field of view contains a dirt road surrounded by grass.

Such a system should be able to classify different textures such as sand, grass, or corn using prior knowledge and then be able to use predefined rules to generate a description.

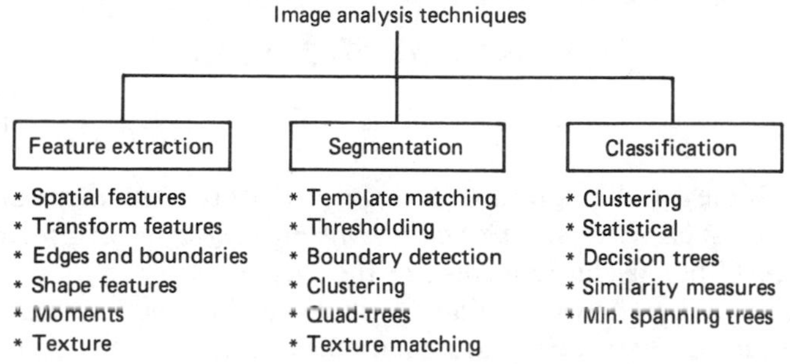

Figure 9.2

9.2 SPATIAL FEATURE EXTRACTION

Spatial features of an object may be characterized by its gray levels, their joint probability distributions, spatial distribution, and the like.

Amplitude Features

The simplest and perhaps the most useful features of an object are the amplitudes of its physical properties, such as reflectivity, transmissivity, tristimulus values (color), or multispectral response. For example, in medical X-ray images, the gray-level amplitude represents the absorption characteristics of the body masses and enables discrimination of bones from tissue or healthy tissue from diseased tissue. In infrared (IR) images amplitude represents temperature, which facilitates the segmentation of clouds from terrain (see Fig 7.10). In radar images, amplitude represents the *radar cross section,* which determines the size of the object being imaged. Amplitude features can be extracted easily by intensity window slicing or by the more general point transformations discussed in Chapter 7.

Histogram Features

Histogram features are based on the histogram of a region of the image. Let u be a random variable representing a gray level in a given region of the image. Define

$$p_u(x) \triangleq \text{Prob}[u = x] \simeq \frac{\text{number of pixels with gray level } x}{\text{total number of pixels in the region}}, \quad x = 0, \ldots, L-1 \tag{9.1}$$

Common features of $p_u(x)$ are its moments, entropy, and so on, which are defined next.

$$\text{Moments:} \quad m_i = E[u^i] = \sum_{x=0}^{L-1} x^i p_u(x), \qquad i = 1, 2, \ldots \tag{9.2}$$

$$\text{Absolute moments:} \quad \hat{m}_i = E[|u|^i] = \sum_{x=0}^{L-1} |x|^i p_u(x) \tag{9.3}$$

$$\text{Central moments:} \quad \mu_i = E\{[u - E(u)]^i\} = \sum_{x=0}^{L-1} (x - m_1)^i p_u(x) \tag{9.4}$$

$$\text{Absolute central moments:} \quad \hat{\mu}_i = E[|u - E(u)|^i] = \sum_{x=0}^{L-1} |x - m_1|^i p_u(x) \tag{9.5}$$

$$\text{Entropy:} \quad H = E[-\log_2 p_u] = -\sum_{x=0}^{L-1} p_u(x) \log_2 p_u(x) \quad \text{bits} \tag{9.6}$$

Some of the common histogram features are *dispersion* $= \hat{\mu}_1$, *mean* $= m_1$, *variance* $= \mu_2$, *mean square value or average energy* $= m_2$, *skewness* $= \mu_3$, *kurtosis* $= \mu_4 - 3$. Other useful features are the *median* and the *mode.* A narrow histogram indicates a low contrast region. Variance can be used to measure local activity in the amplitudes. Histogram features are also useful for shape analysis of objects from their projections (see Section 9.7).

Often these features are measured over a small moving window W. Some of the histogram features can be measured without explicitly determining the histogram; for example,

$$m_i(k, l) = \frac{1}{N_w} \sum\sum_{(m, n) \in W} [u(m - k, n - l)]^i \tag{9.7}$$

$$\mu_i(k, l) = \frac{1}{N_w} \sum\sum_{(m, n) \in W} [u(m - k, n - l) - m_1(k, l)]^i \tag{9.8}$$

where $i = 1, 2, \ldots$ and N_w is the number pixels in the window W. Figure 9.3 shows the spatial distribution of different histogram features measured over a 3×3 moving window. The standard deviation emphasizes the strong edges in the image and dispersion feature extracts the fine edge structure. The mean, median, and mode extract low spatial-frequency features.

Second-order joint probabilities have also been found useful in applications such as feature extraction of textures (see Section 9.11). A second-order joint probability is defined as

$$\begin{aligned} p_{\mathbf{u}}(x_1, x_2) &\overset{\Delta}{=} p_{u_1, u_2}(x_1, x_2) \overset{\Delta}{=} \text{Prob}[u_1 = x_1, u_2 = x_2], \qquad x_1, x_2 = 0, \ldots, L - 1 \\ &\simeq \frac{\text{number of pairs of pixels } u_1 = x_1, u_2 = x_2}{\text{total number of such pairs of pixels in the region}} \end{aligned} \tag{9.9}$$

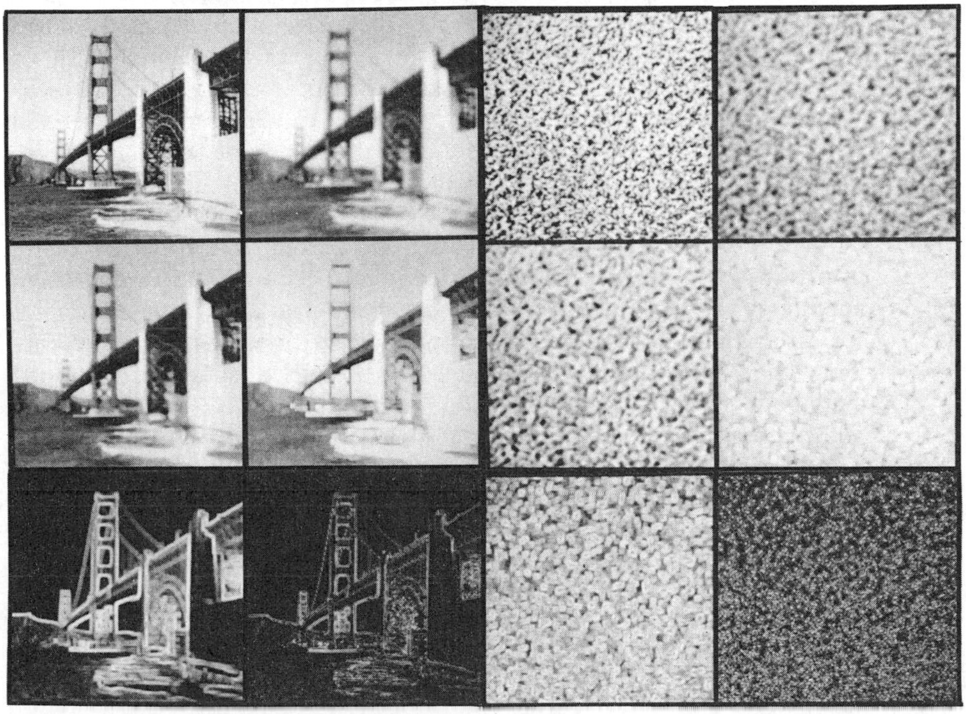

Figure 9.3 Spatial distribution of histogram features measured over a 3×3 moving window. In each case, top to bottom and left to right, original image, mean, median, mode, standard deviation, and dispersion.

where u_1 and u_2 are the two pixels in the image region specified by some relation. For example, u_2 could be specified as a pixel at distance r and angle θ from u_1. The $L \times L$ array $\{p_{\mathbf{u}}(x_1, x_2)\}$ is also called the *concurrence matrix*.

Example 9.1

For a 2-bit, 4×4 image region given as

$$\mathbf{U} = \begin{bmatrix} 0 & 1 & 0 & 2 \\ 3 & 2 & 1 & 1 \\ 2 & 1 & 0 & 1 \\ 3 & 1 & 2 & 0 \end{bmatrix}$$

we have $p_u(0) = \frac{1}{4}$, $p_u(1) = \frac{3}{8}$, $p_u(2) = \frac{1}{4}$, and $p_u(3) = \frac{1}{8}$. The second-order histogram for $u_1 = u(m, n)$, $u_2 = u(m+1, n+1)$ is the 4×4 concurrence matrix

$$\begin{matrix} & \rightarrow x_2 \\ \downarrow x_1 & \frac{1}{9}\begin{bmatrix} 1 & 1 & 1 & 0 \\ 0 & 2 & 1 & 0 \\ 1 & 1 & 0 & 0 \\ 0 & 1 & 0 & 0 \end{bmatrix} \end{matrix}$$

9.3 TRANSFORM FEATURES

Image transforms provide the frequency domain information in the data. Transform features are extracted by zonal-filtering the image in the selected transform space (Fig. 9.4). The zonal filter, also called the feature mask, is simply a slit or an aperture. Figure 9.5 shows different masks and Fourier transform features of different shapes. Generally, the high-frequency features can be used for edge and boundary detection, and angular slits can be used for detection of orientation. For example, an image containing several parallel lines with orientation θ will exhibit strong energy along a line at angle $\pi/2 + \theta$ passing through the origin of its two-dimensional Fourier transform. This follows from the properties of the Fourier transform (also see the projection theorem, Section 10.4). A combination of an angular slit with a bandlimited low-pass, band-pass or high-pass filter can be used for discriminating periodic or quasiperiodic textures. Other transforms, such as Haar and Hadamard, are also potentially useful for feature extraction. However, systematic studies remain to be done to determine their applications. Chapters 5 and 7 contain examples of image transforms and their processed outputs.

Transform-feature extraction techniques are also important when the source data originates in the transform coordinates. For example, in optical and optical-digital (hybrid) image analysis applications, the data can be acquired directly in the Fourier domain for real-time feature extraction in the focal plane.

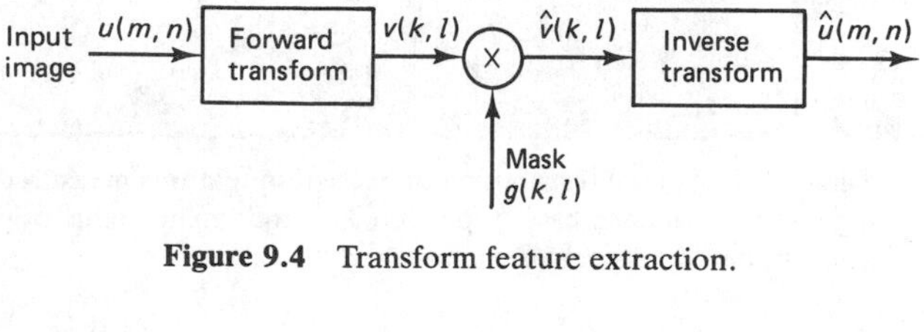

Figure 9.4 Transform feature extraction.

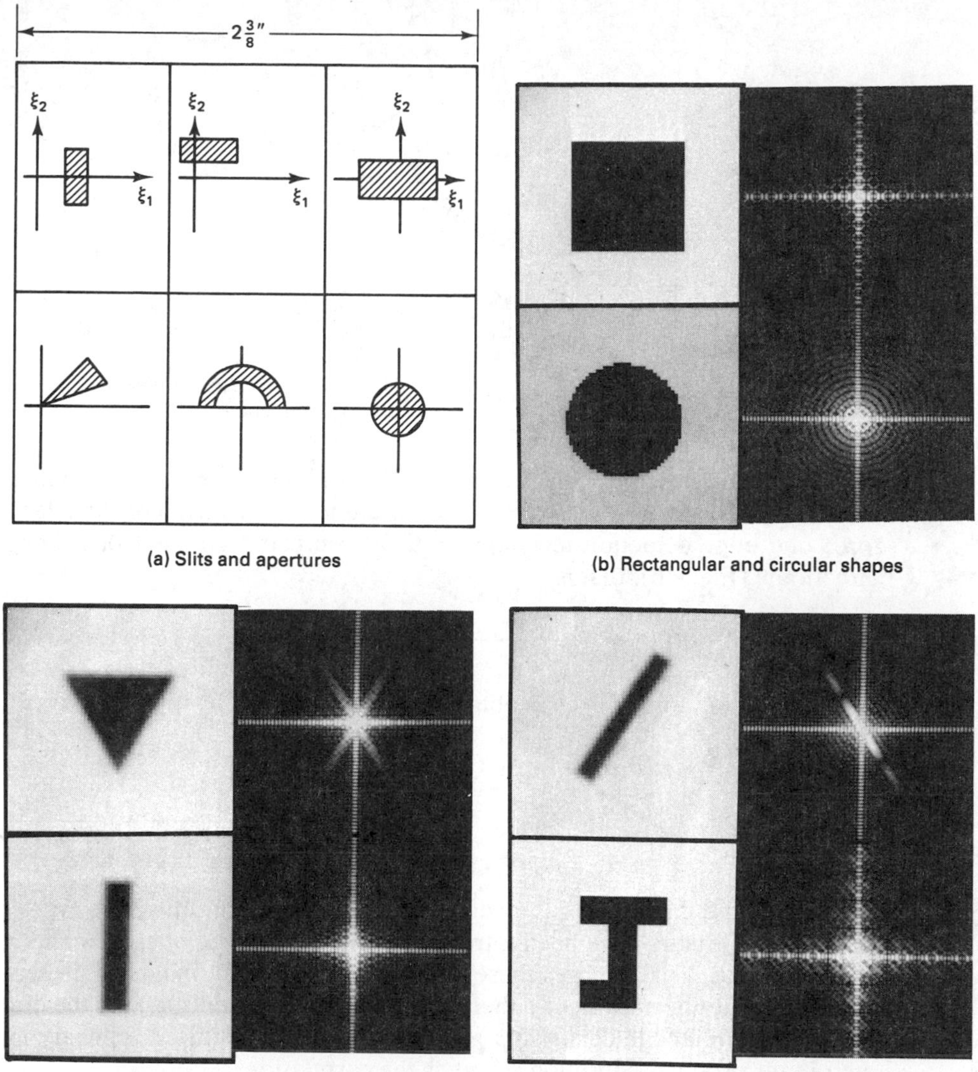

(a) Slits and apertures

(b) Rectangular and circular shapes

(c) Triangular and vertical shapes

(d) 45° orientation and the letter *J*

Figure 9.5 Fourier domain features.

9.4 EDGE DETECTION

A problem of fundamental importance in image analysis is edge detection. Edges characterize object boundaries and are therefore useful for segmentation, registration, and identification of objects in scenes. Edge points can be thought of as pixel locations of abrupt gray-level change. For example, it is reasonable to define edge points in binary images as *black pixels with at least one white nearest neighbor*, that is, pixel locations (m, n) such that $u(m, n) = 0$ and $g(m, n) = 1$, where

$$g(m, n) \triangleq [u(m, n) \oplus u(m \pm 1, n)].\text{OR}.[u(m, n) \oplus u(m, n \pm 1)] \tag{9.10}$$

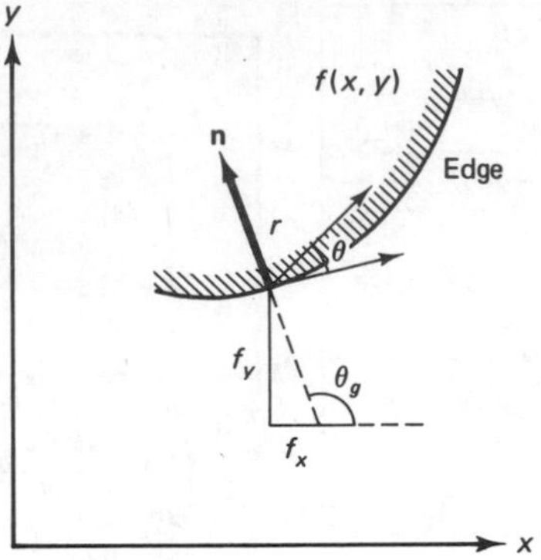

Figure 9.6 Gradient of $f(x, y)$ along r direction.

where $\oplus$ denotes the logical exclusive-OR operation. For a continuous image $f(x, y)$ its derivative assumes a local maximum in the direction of the edge. Therefore, one edge detection technique is to measure the gradient of f along r in a direction θ (Fig. 9.6), that is,

$$\frac{\partial f}{\partial r} = \frac{\partial f}{\partial x}\frac{\partial x}{\partial r} + \frac{\partial f}{\partial y}\frac{\partial y}{\partial r} = f_x \cos\theta + f_y \sin\theta \tag{9.11}$$

The maximum value of $\partial f/\partial r$ is obtained when $(\partial/\partial\theta)(\partial f/\partial r) = 0$. This gives

$$-f_x \sin\theta_g + f_y \cos\theta_g = 0 \quad\Rightarrow\quad \theta_g = \tan^{-1}\left(\frac{f_y}{f_x}\right) \tag{9.12a}$$

$$\left(\frac{\partial f}{\partial r}\right)_{\max} = \sqrt{f_x^2 + f_y^2} \tag{9.12b}$$

where θ_g is the direction of the edge. Based on these concepts, two types of edge detection operators have been introduced [6–11], *gradient operators* and *compass operators*. For digital images these operators, also called *masks*, represent finite-difference approximations of either the orthogonal gradients f_x, f_y or the directional gradient $\partial f/\partial r$. Let $\mathbf{H}$ denote a $p \times p$ mask and define, for an arbitrary image $\mathbf{U}$, their inner product at location (m, n) as the correlation

$$\langle \mathbf{U}, \mathbf{H} \rangle_{m,n} \triangleq \sum_i \sum_j h(i, j) u(i + m, j + n) = u(m, n) \circledast h(-m, -n) \tag{9.13}$$

Gradient Operators

These are represented by a pair of masks $\mathbf{H}_1, \mathbf{H}_2$, which measure the gradient of the image $u(m, n)$ in two orthogonal directions (Fig. 9.7). Defining the bidirectional gradients $g_1(m, n) \triangleq \langle \mathbf{U}, \mathbf{H}_1 \rangle_{m,n}$, $g_2(m, n) \triangleq \langle \mathbf{U}, \mathbf{H}_2 \rangle_{m,n}$ the gradient vector magnitude and direction are given by

$$g(m, n) = \sqrt{g_1^2(m, n) + g_2^2(m, n)} \tag{9.14}$$

$$\theta_g(m, n) = \tan^{-1}\frac{g_2(m, n)}{g_1(m, n)} \tag{9.15}$$

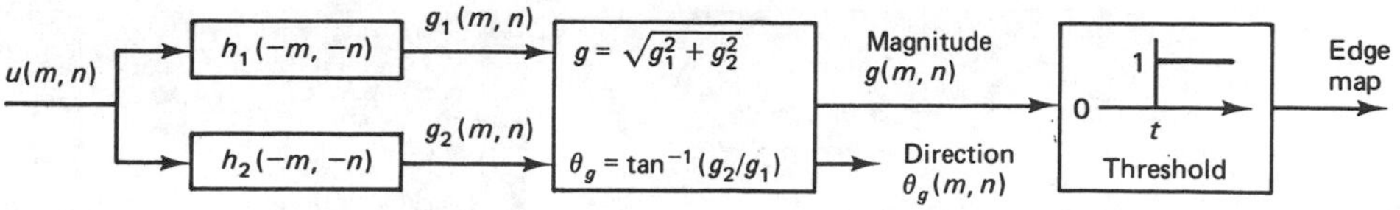

Figure 9.7 Edge detection via gradient operators.

Often the magnitude gradient is calculated as

$$g(m, n) \triangleq |g_1(m, n)| + |g_2(m, n)| \tag{9.16}$$

rather than as in (9.14). This calculation is easier to perform and is preferred especially when implemented in digital hardware.

Table 9.2 lists some of the common gradient operators. The Prewitt, Sobel, and isotropic operators compute horizontal and vertical differences of local sums. This reduces the effect of noise in the data. Note these operators have the desirable property of yielding zeros for uniform regions.

The pixel location (m, n) is declared an edge location if $g(m, n)$ exceeds some threshold t. The locations of edge points constitute an *edge map* $\varepsilon(m, n)$, which is defined as

$$\varepsilon(m, n) = \begin{cases} 1, & (m, n) \in I_g \\ 0, & \text{otherwise} \end{cases} \tag{9.17}$$

where

$$I_g \triangleq \{(m, n); g(m, n) > t\} \tag{9.18}$$

The edge map gives the necessary data for tracing the object boundaries in an image. Typically, t may be selected using the cumulative histogram of $g(m, n)$ so

TABLE 9.2 Some Common Gradient Operators. Boxed element indicates the location of the origin

	H_1	H_2
Roberts [9]	$\begin{bmatrix} \boxed{0} & 1 \\ -1 & 0 \end{bmatrix}$	$\begin{bmatrix} \boxed{1} & 0 \\ 0 & -1 \end{bmatrix}$
Smoothed (Prewitt [6])	$\begin{bmatrix} -1 & 0 & 1 \\ -1 & \boxed{0} & 1 \\ -1 & 0 & 1 \end{bmatrix}$	$\begin{bmatrix} -1 & -1 & -1 \\ 0 & \boxed{0} & 0 \\ 1 & 1 & 1 \end{bmatrix}$
Sobel [7]	$\begin{bmatrix} -1 & 0 & 1 \\ -2 & \boxed{0} & 2 \\ -1 & 0 & 1 \end{bmatrix}$	$\begin{bmatrix} -1 & -2 & -1 \\ 0 & \boxed{0} & 0 \\ 1 & 2 & 1 \end{bmatrix}$
Isotropic	$\begin{bmatrix} -1 & 0 & 1 \\ -\sqrt{2} & \boxed{0} & \sqrt{2} \\ -1 & 0 & 1 \end{bmatrix}$	$\begin{bmatrix} -1 & -\sqrt{2} & -1 \\ 0 & \boxed{0} & 0 \\ 1 & \sqrt{2} & 1 \end{bmatrix}$

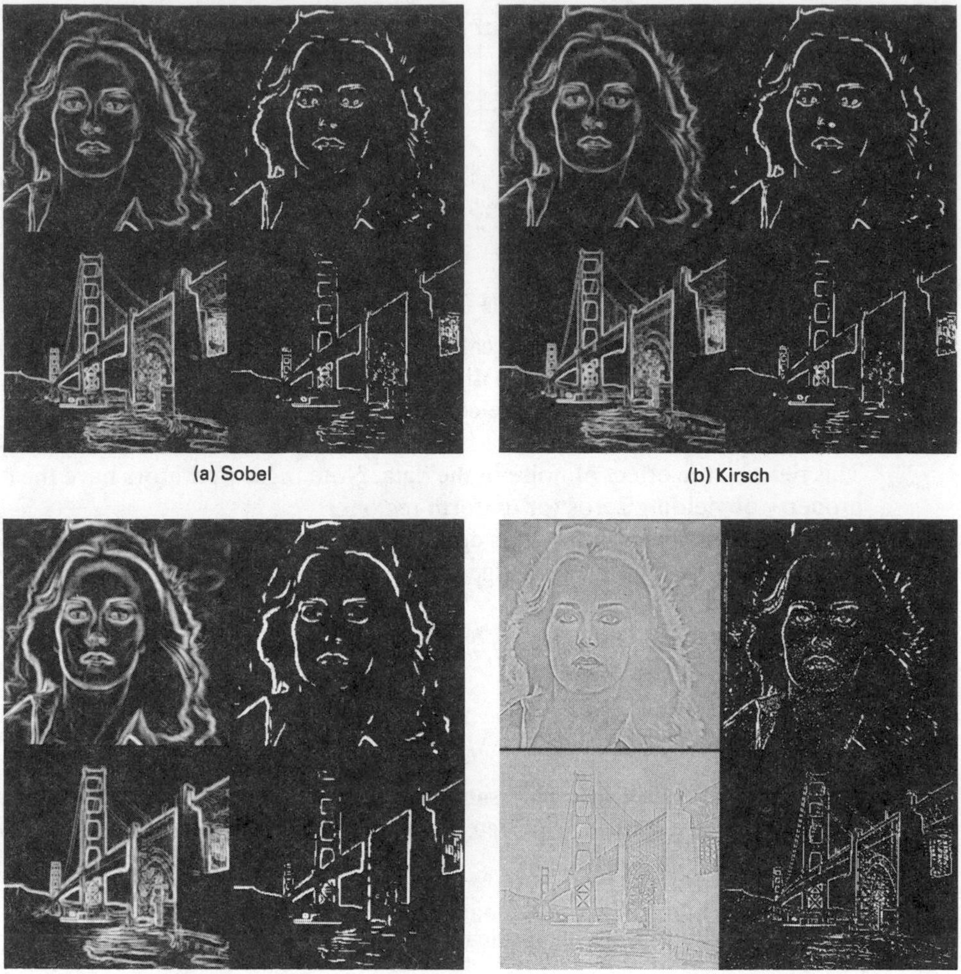

Figure 9.8 Edge detection examples. In each case, gradient images (left), edge maps (right).

that 5 to 10% of pixels with largest gradients are declared as edges. Figure 9.8a shows the gradients and edge maps using the Sobel operator on two different images.

Compass Operators

Compass operators measure gradients in a selected number of directions (Fig. 9.9). Table 9.3 shows four different compass gradients for north-going edges. An anticlockwise circular shift of the eight boundary elements of these masks gives a 45° rotation of the gradient direction. For example, the eight compass gradients corresponding to the third operator of Table 9.3 are

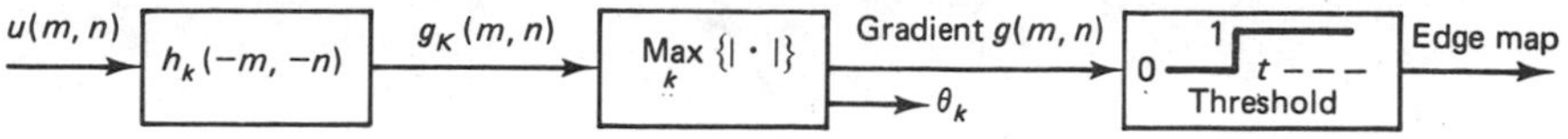

Figure 9.9 Edge detection via compass operators.

$$\begin{array}{rrrl} 1 & 1 & 1 & \uparrow \\ 0 & 0 & 0 & \text{(N)} \\ -1 & -1 & -1 & \end{array} \quad \begin{array}{rrrl} 1 & 1 & 0 & \nwarrow \\ 1 & 0 & -1 & \text{(NW)} \\ 0 & -1 & -1 & \end{array} \quad \begin{array}{rrrl} 1 & 0 & -1 & \leftarrow \\ 1 & 0 & -1 & \text{(W)} \\ 1 & 0 & -1 & \end{array} \quad \begin{array}{rrrl} 0 & -1 & -1 & \swarrow \\ 1 & 0 & -1 & \text{(SW)} \\ 1 & 1 & 0 & \end{array}$$

$$\begin{array}{rrrl} -1 & -1 & -1 & \downarrow \\ 0 & 0 & 0 & \text{(S)} \\ 1 & 1 & 1 & \end{array} \quad \begin{array}{rrrl} -1 & -1 & 0 & \searrow \\ -1 & 0 & 1 & \text{(SE)} \\ 0 & 1 & 1 & \end{array} \quad \begin{array}{rrrl} -1 & 0 & 1 & \rightarrow \\ -1 & 0 & 1 & \text{(E)} \\ -1 & 0 & 1 & \end{array} \quad \begin{array}{rrrl} 0 & 1 & 1 & \nearrow \\ -1 & 0 & 1 & \text{(NE)} \\ -1 & -1 & 0 & \end{array}$$

Let $g_k(m, n)$ denote the compass gradient in the direction $\theta_k = \pi/2 + k\pi/4$, $k = 0, \ldots, 7$. The gradient at location (m, n) is defined as

$$g(m, n) \triangleq \max_k \{|g_k(m, n)|\} \tag{9.19}$$

which can be thresholded to obtain the edge map as before. Figure 9.8b shows the results for the Kirsch operator. Note that only four of the preceding eight compass gradients are linearly independent. Therefore, it is possible to define four 3×3 arrays that are mutually orthogonal and span the space of these compass gradients. These arrays are called *orthogonal gradients* and can be used in place of the compass gradients [12]. Compass gradients with higher angular resolution can be designed by increasing the size of the mask.

TABLE 9.3 Compass Gradients (North). Each Clockwise Circular Shift of Elements about the Center Rotates the Gradient Direction by 45°

1) $\begin{bmatrix} 1 & 1 & 1 \\ 1 & \boxed{-2} & 1 \\ -1 & -1 & -1 \end{bmatrix}$	3) $\begin{bmatrix} 1 & 1 & 1 \\ 0 & \boxed{0} & 0 \\ -1 & -1 & -1 \end{bmatrix}$
2) $\begin{bmatrix} 5 & 5 & 5 \\ -3 & \boxed{0} & -3 \\ -3 & -3 & -3 \end{bmatrix}$ (Kirsch)	4) $\begin{bmatrix} 1 & 2 & 1 \\ 0 & \boxed{0} & 0 \\ -1 & -2 & -1 \end{bmatrix}$

Laplace Operators and Zero Crossings

The foregoing methods of estimating the gradients work best when the gray-level transition is quite abrupt, like a step function. As the transition region gets wider (Fig. 9.10), it is more advantageous to apply the second-order derivatives. One frequently encountered operator is the Laplacian operator, defined as

$$\nabla^2 f = \frac{\partial^2 f}{\partial x^2} + \frac{\partial^2 f}{\partial y^2} \tag{9.20}$$

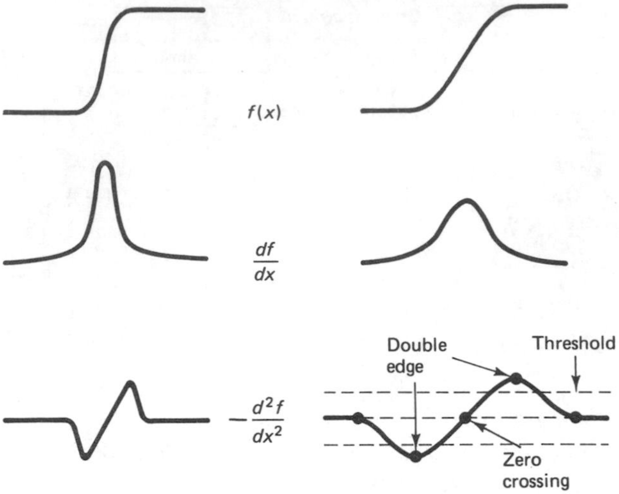

(a) First and second derivatives for edge detection

(b) An image and its zero-crossings.

Figure 9.10 Edge detection via zero-crossings.

Table 9.4 gives three different discrete approximations of this operator. Figure 9.8d shows the edge extraction ability of the Laplace mask (2). Because of the second-order derivatives, this gradient operator is more sensitive to noise than those previously defined. Also, the thresholded magnitude of $\nabla^2 f$ produces double edges. For these reasons, together with its inability to detect the edge direction, the Laplacian as such is not a good edge detection operator. A better utilization of the Laplacian is to use its zero-crossings to detect the edge locations (Fig. 9.10). A generalized Laplacian operator, which approximates the Laplacian of Gaussian functions, is a powerful zero-crossing detector [13]. It is defined as

$$h(m, n) \triangleq c\left[1 - \frac{(m^2 + n^2)}{\sigma^2}\right] \exp\left(-\frac{m^2 + n^2}{2\sigma^2}\right) \tag{9.21}$$

where σ controls the width of the Gaussian kernel and c normalizes the sum of the elements of a given size mask to unity. Zero-crossings of a given image convolved with $h(m, n)$ give its edge locations. On a two-dimensional grid, a zero-crossing is said to occur wherever there is a zero-crossing in at least one direction.

TABLE 9.4 Discrete Laplace Operators

1) $\begin{bmatrix} 0 & -1 & 0 \\ -1 & \boxed{4} & -1 \\ 0 & -1 & 0 \end{bmatrix}$	2) $\begin{bmatrix} -1 & -1 & -1 \\ -1 & \boxed{8} & -1 \\ -1 & -1 & -1 \end{bmatrix}$	3) $\begin{bmatrix} 1 & -2 & 1 \\ -2 & \boxed{4} & -2 \\ 1 & -2 & 1 \end{bmatrix}$

The $h(m, n)$ is the sampled impulse response of an analog band-pass filter whose frequency response is proportional to $(\xi_1^2 + \xi_2^2)\exp[-2\sigma^2(\xi_1^2 + \sigma_2^2)]$. Therefore, the zero-crossings detector is equivalent to a low-pass filter having a Gaussian impulse response followed by a Laplace operator. The low-pass filter serves to attenuate the noise sensitivity of the Laplacian. The parameter σ controls the amplitude response of the filter output but does not affect the location of the zero-crossings.

Directional information of the edges can be obtained by searching the zero-crossings of the second-order derivative along r for each direction θ. From (9.11), we obtain

$$\frac{\partial^2 f}{\partial r^2} = \frac{\partial f_x}{\partial r}\cos\theta + \frac{\partial f_y}{\partial r}\sin\theta = \frac{\partial^2 f}{\partial x^2}\cos^2\theta + 2\frac{\partial^2 f}{\partial x\,\partial y}\sin\theta\cos\theta + \frac{\partial^2 f}{\partial y^2}\sin^2\theta \qquad (9.22)$$

Zero-crossings are searched as θ is varied [14].

Stochastic Gradients [16]

The foregoing gradient masks perform poorly in the presence of noise. Averaging, low-pass filtering, or least squares edge fitting [15] techniques can yield some reduction of the detrimental effects of noise. A better alternative is to design edge extraction masks, which take into account the presence of noise in a controlled manner. Consider an edge model whose transition region is 1 pixel wide (Fig. 9.11).

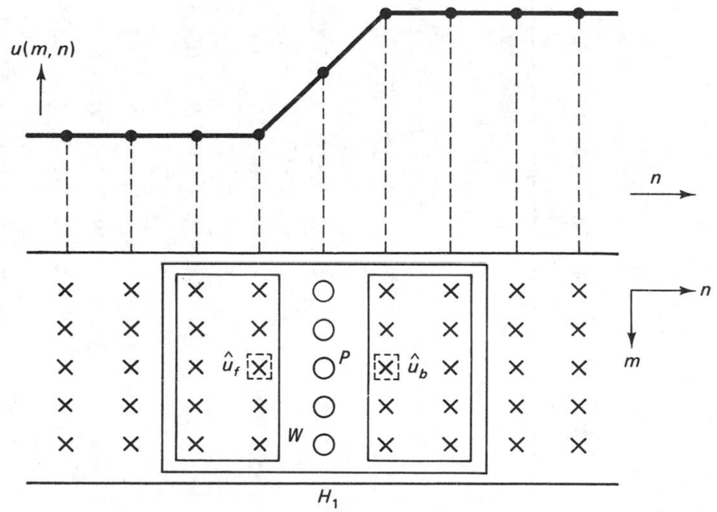

Figure 9.11 Edge model with transition region one pixel wide.

TABLE 9.5 Stochastic Gradients H_1 for Edge Extraction of Noisy Images with $r_0(k, l) = .99^{\sqrt{k^2+l^2}}$, $H_2 \triangleq H_1^T$, SNR $= \sigma^2/\sigma_\eta^2$.

	SNR = 1							SNR = 9						
3 × 3			0.97	0	−0.97					0.776	0	−0.776		
			1.00	$\boxed{0}$	−1.00					1.00	$\boxed{0}$	−1.00		
			0.97	0	−0.97					0.776	0	−0.776		
		0.802	0.836	0	−0.836	−0.802			0.267	0.364	0	−0.364	−0.267	
		0.845	0.897	0	−0.897	−0.845			0.373	0.562	0	−0.562	−0.373	
5 × 5		0.870	1.00	$\boxed{0}$	−1.00	−0.870			0.463	1.00	$\boxed{0}$	−1.00	−0.463	
		0.845	0.897	0	−0.897	−0.845			0.373	0.562	0	−0.562	−0.373	
		0.802	0.836	0	−0.836	−0.802			0.267	0.364	0	−0.364	−0.267	
	0.641	0.672	0.719	0	−0.719	−0.672	−0.641	0.073	0.240	0.283	0	−0.283	−0.140	−0.073
	0.656	0.719	0.781	0	−0.781	−0.719	−0.656	0.104	0.213	0.348	0	−0.348	−0.213	−0.104
	0.688	0.781	0.875	0	−0.875	−0.781	−0.688	0.165	0.354	0.579	0	−0.579	−0.354	−0.165
7 × 7	0.703	0.813	1.00	$\boxed{0}$	−1.00	−0.813	−0.703	0.195	0.463	1.00	$\boxed{0}$	−1.00	−0.463	−0.195
	0.688	0.781	0.875	0	−0.875	−0.781	−0.688	0.165	0.354	0.579	0	−0.579	−0.354	−0.165
	0.656	0.719	0.781	0	−0.781	−0.719	−0.656	0.104	0.213	0.348	0	−0.348	−0.213	−0.104
	0.641	0.672	0.719	0	−0.719	−0.672	−0.641	0.073	0.140	0.283	0	−0.238	−0.140	−0.073

To detect the presence of an edge at location P, calculate the horizontal gradient, for instance, as

$$g_1(m, n) \triangleq \hat{u}_f(m, n-1) - \hat{u}_b(m, n+1) \tag{9.23}$$

Here $\hat{u}_f(m, n)$ and $\hat{u}_b(m, n)$ are the optimum forward and backward estimates of $u(m, n)$ based on the noisy observations given over some finite regions W of the left- and right half-planes, respectively. Thus $\hat{u}_f(m, n)$ and $\hat{u}_b(m, n)$ are semicausal estimates (see Chapter 6). For observations $v(m, n)$ containing additive white noise, we can find the best linear mean square semicausal FIR estimate of the form

$$\hat{u}_f(m, n) \triangleq \sum_{(k,l) \in W}\sum a(k, l)v(m-k, n-l), \quad W = [(k, l): |k| \le p, 0 \le l \le q] \tag{9.24}$$

The filter weights $a(k, l)$ can be determined following Section 8.4 with the modification that W is a semicausal window. [See (8.65) and (8.69)]. The backward semicausal estimate employs the same filter weights, but backward. Using the definitions in (9.23), the stochastic gradient operator $\mathbf{H}_1$ is obtained as shown in Table 9.5. The operator $\mathbf{H}_2$ would be the 90° counterclockwise rotation of $\mathbf{H}_1$, which, due to its symmetry properties, would simply be $\mathbf{H}_1^T$. These masks have been normalized so that the coefficient $a(0, 0)$ in (9.24) is unity. Note that for high SNR the filter weights decay rapidly. Figure 9.8c shows the gradients and edge maps obtained by applying the 5×5 stochastic masks designed for SNR = 9 but applied to noiseless images. Figure 9.12 compares the edges detected from noisy images by the Sobel and the stochastic gradient masks.

Performance of Edge Detection Operators

Edge detection operators can be compared in a number of different ways. First, the image gradients may be compared visually, since the eye itself performs some sort of edge detection. Figure 9.13 displays different gradients for noiseless as well as noisy images. In the noiseless case all the operators are roughly equivalent. The stochastic gradient is found to be quite effective when noise is present. Quantitatively, the performance in noise of an edge detection operator may be measured as follows. Let n_0 be the number of edge pixels declared and n_1 be number of missed or new edge pixels after adding noise. If n_0 is held fixed for the noiseless as well as noisy images, then the edge detection error rate is

$$P_e = \frac{n_1}{n_0} \tag{9.25}$$

In Figure 9.12 the error rate for the Sobel operator used on noisy images with SNR $\simeq$ 10 dB is 24%, whereas it is only 2% for the stochastic operator.

Another figure of merit for the noise performance of edge detection operators is the quantity

$$P = \frac{1}{\max(N_I, N_D)} \sum_{i=1}^{N_D} \frac{1}{1 + \alpha d_i^2} \tag{9.26}$$

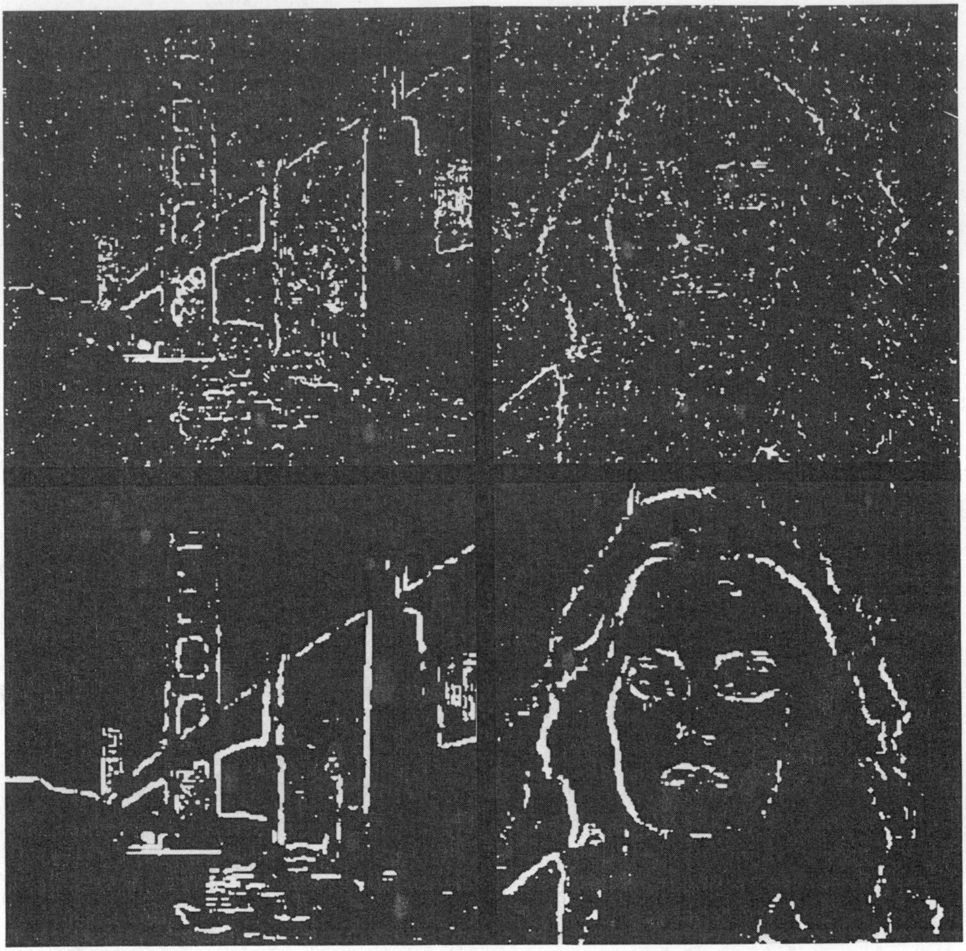

Figure 9.12 Edge detection from noisy images. Upper two, Sobel. Lower two, stochastic.

where d_i is the distance between a pixel declared as edge and the nearest ideal edge pixel, α is a calibration constant, and N_I and N_D are the number of ideal and detected edge pixels respectively. Among the gradient and compass operators of Tables 9.2 and 9.3 (not including the stochastic masks), the Sobel and Prewitt operators have been found to yield the highest performance (where performance is proportional to the value of P) [17].

Line and Spot Detection

Lines are extended edges. Table 9.6 shows compass gradients for line detection. Other forms of line detection require fitting a line (or a curve) through a set of edge points. Some of these ideas are explored in Section 9.5.

(a) Gradients for noiseless image (b) Gradients for noisy image

Figure 9.13 Comparison of edge extraction operators. In each case $\begin{array}{|c|c|}\hline 1 & 2 \\ \hline 3 & 4 \\ \hline\end{array}$ the operators are (1) smoothed gradient, (2) sobel, (3) isotropic, (4) semicausal model based 5 × 5 stochastic. Largest 5% of the gradient magnitudes were declared edges.

TABLE 9.6 Line Detection Operators

$\begin{bmatrix} -1 & -1 & -1 \\ 2 & 2 & 2 \\ -1 & -1 & -1 \end{bmatrix}$	$\begin{bmatrix} -1 & -1 & 2 \\ -1 & 2 & -1 \\ 2 & -1 & -1 \end{bmatrix}$	$\begin{bmatrix} -1 & 2 & -1 \\ -1 & 2 & -1 \\ -1 & 2 & -1 \end{bmatrix}$	$\begin{bmatrix} 2 & -1 & -1 \\ -1 & 2 & -1 \\ -1 & -1 & 2 \end{bmatrix}$
(a) E–W	(b) NE–SW	(c) N–S	(d) NW–SE

Spots are isolated edges. These are most easily detected by comparing the value of a pixel with an average or median of the neighborhood pixels.

9.5 BOUNDARY EXTRACTION

Boundaries are linked edges that characterize the shape of an object. They are useful in computation of geometry features such as size or orientation.

Connectivity

Conceptually, boundaries can be found by tracing the connected edges. On a rectangular grid a pixel is said to be *four-* or *eight-connected* when it has the same properties as one of its nearest four or eight neighbors, respectively (Fig. 9.14). There are difficulties associated with these definitions of connectivity, as shown in Fig. 9.14c. Under four-connectivity, segments 1, 2, 3, and 4 would be classified as disjoint, although they are perceived to form a connected ring. Under eight-connectivity these segments are connected, but the inside hole (for example, pixel B) is also eight-connected to the outside (for instance, pixel C). Such problems can

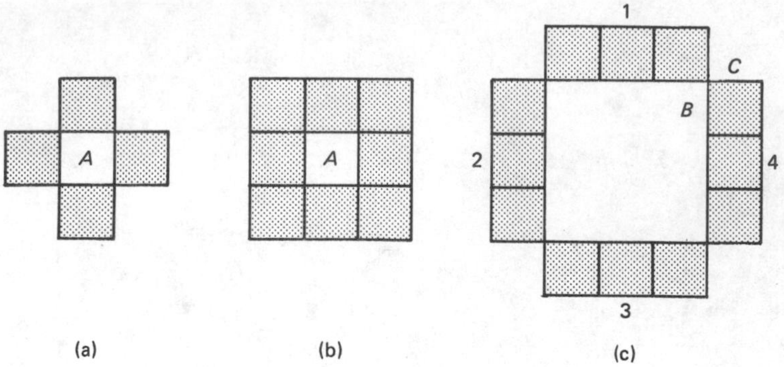

Figure 9.14 Connectivity on a rectangular grid. Pixel A and its (a) 4-connected and (b) 8-connected neighbors; (c) connectivity paradox: "Are B and C connected?"

be avoided by considering eight-connectivity for object and four-connectivity for background. An alternative is to use triangular or hexagonal grids, where three- or six-connectedness can be defined. However, there are other practical difficulties that arise in working with nonrectangular grids.

Contour Following

As the name suggests, contour-following algorithms trace boundaries by ordering successive edge points. A simple algorithm for tracing closed boundaries in binary images is shown in Fig. 9.15. This algorithm can yield a coarse contour, with some of the boundary pixels appearing twice. Refinements based on eight-connectivity tests for edge pixels can improve the contour trace [2]. Given this trace a smooth curve, such as a spline, through the nodes can be used to represent the contour. Note that this algorithm will always trace a boundary, *open or closed,* as a closed contour. This method can be extended to gray-level images by searching for edges in the 45° to 135° direction from the direction of the gradient to move from the inside to the outside of the boundary, and vice-versa [19]. A modified version of this contour-following method is called the *crack-following algorithm* [25]. In that algorithm each pixel is viewed as having a square-shaped boundary, and the object boundary is traced by following the edge-pixel boundaries.

Edge Linking and Heuristic Graph Searching [18–21]

A boundary can also be viewed as a path through a graph formed by linking the edge elements together. Linkage rules give the procedure for connecting the edge elements. Suppose a graph with node locations x_k, $k = 1, 2, \ldots$ is formed from node A to node B. Also, suppose we are given an *evaluation function* $\phi(x_k)$, which gives the value of the path from A to B constrained to go through the node x_k. In *heuristic search algorithms,* we examine the successors of the start node and select the node that maximizes $\phi(\cdot)$. The selected node now becomes the new start node and the process is repeated until we reach B. The sequence of selected nodes then consti-

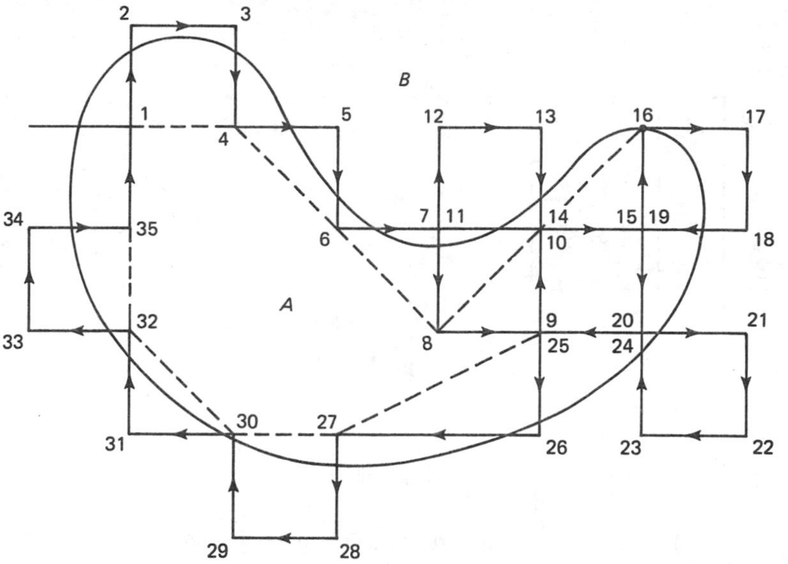

Algorithm

1. Start inside A (e.g., 1)
2. Turn left and step to next pixel if in region A, (e.g., 1 to 2), otherwise turn right and step (e.g., 2 to 3)
3. Continue until arrive at starting point 1

Figure 9.15 Contour following in a binary image.

tutes the boundary path. The speed of the algorithm depends on the chosen $\phi(\cdot)$ [20, 21]. Note that such an algorithm need not give the globally optimum path.

Example 9.2 Heuristic search algorithms [19]

Consider a 3×5 array of edges whose gradient magnitudes $|g|$ and tangential contour directions θ are shown in Fig. 9.16a. The contour directions are at 90° to the gradient directions. A pixel X is considered to be linked to Y if the latter is one of the three eight-connected neighbors (Y_1, Y_2, or Y_3 in Fig. 9.16b) *in front* of the contour direction and if $|\theta(x) - \theta(y)| < 90°$. This yields the graph of Fig. 9.16c.

As an example, suppose $\phi(x_k)$ is the sum of edge gradient magnitudes along the path from A to x_k. At A, the successor nodes are D, C, and G, with $\phi(D) = 12$, $\phi(C) = 6$, and $\phi(G) = 8$. Therefore, node D is selected, and C and G are discarded. From here on nodes E, F, and B provide the remaining path. Therefore, the boundary path is $ADEFB$. On the other hand, note that path $ACDEFB$ is the path of maximum cumulative gradient.

Dynamic Programming

Dynamic programming is a method of finding the global optimum of multistage processes. It is based on Bellman's *principal of optimality* [22], which states that *the optimum path between two given points is also optimum between any two points lying*

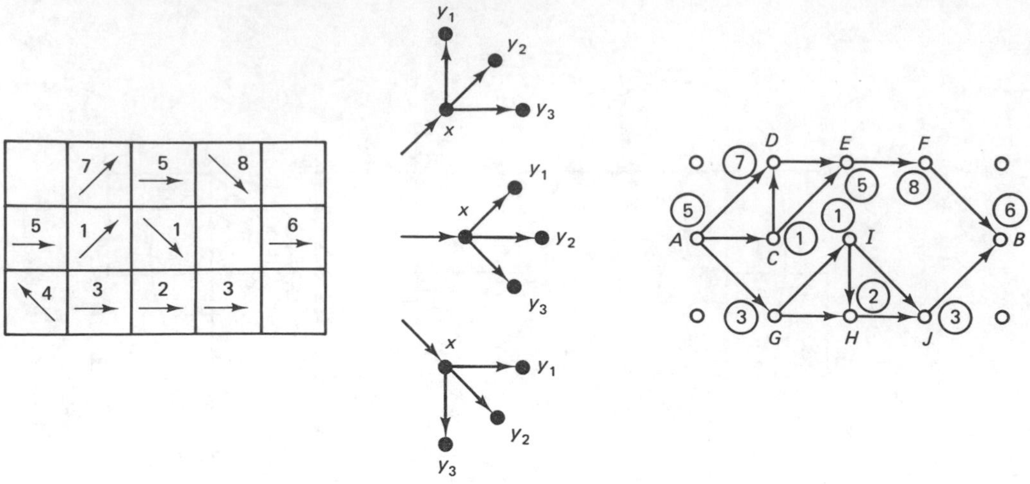

(a) Gradient magnitudes contour directions

(b) Linkage rules

(c) Graph interpretation

Figure 9.16 Heuristic graph search method for boundary extraction.

on the path. Thus if C is a point on the optimum path between A and B (Fig. 9.17), then the segment CB is the optimum path from C to B, *no matter how one arrives at* C.

To apply this idea to boundary extraction [23], suppose the edge map has been converted into a forward-connected graph of N stages and we have an evaluation function

$$S(\mathbf{x}_1, \mathbf{x}_2, \ldots, \mathbf{x}_N, N) \triangleq \sum_{k=1}^{N} |g(\mathbf{x}_k)| - \alpha \sum_{k=2}^{N} |\theta(\mathbf{x}_k) - \theta(\mathbf{x}_{k-1})| - \beta \sum_{k=2}^{N} d(\mathbf{x}_k, \mathbf{x}_{k-1}) \quad (9.27)$$

Here $\mathbf{x}_k$, $k = 1, \ldots, N$ represents the nodes (that is, the vector of edge pixel locations) in the kth stage of the graph, $d(\mathbf{x}, \mathbf{y})$ is the distance between two nodes $\mathbf{x}$ and $\mathbf{y}$; $|g(\mathbf{x}_k)|$, $\theta(\mathbf{x}_k)$ are the gradient magnitude and angles, respectively, at the node $\mathbf{x}_k$, and α and β are nonnegative parameters. The optimum boundary is given by connecting the nodes $\hat{\mathbf{x}}_k$, $k = 1, \ldots, N$, so that $S(\hat{\mathbf{x}}_1, \hat{\mathbf{x}}_2, \ldots, \hat{\mathbf{x}}_N, N)$ is maximum. Define

$$\Phi(\mathbf{x}_N, N) \triangleq \max_{\mathbf{x}_1 \ldots \mathbf{x}_{N-1}} \{S(\mathbf{x}_1, \ldots, \mathbf{x}_N, N)\} \quad (9.28)$$

Using the definition of (9.27), we can write the recursion

$$\begin{aligned} S(\mathbf{x}_1, \ldots, \mathbf{x}_N, N) &= S(\mathbf{x}_1, \ldots, \mathbf{x}_{N-1}, N-1) \\ &\quad + \{|g(\mathbf{x}_N)| - \alpha|\theta(\mathbf{x}_N) - \theta(\mathbf{x}_{N-1})| - \beta d(\mathbf{x}_N, \mathbf{x}_{N-1})\} \quad (9.29) \\ &\triangleq S(\mathbf{x}_1, \ldots, \mathbf{x}_{N-1}, N-1) + f(\mathbf{x}_{N-1}, \mathbf{x}_N) \end{aligned}$$

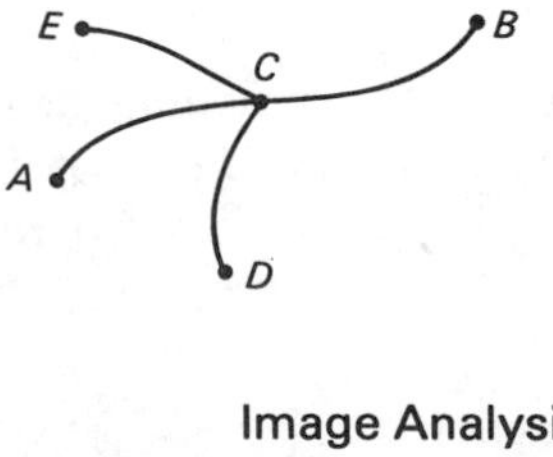

Figure 9.17 Bellman's principle of optimality. If the path AB is optimum, then so is CB no matter how you arrive at C.

where $f(\mathbf{x}_{N-1}, \mathbf{x}_N)$ represents the terms in the brackets. Letting $N = k$ in (9.28) and (9.29), it follows by induction that

$$\begin{aligned}\Phi(\mathbf{x}_k, k) &= \max_{\mathbf{x}_1 \ldots \mathbf{x}_{k-1}} \{S(\mathbf{x}_1, \ldots, \mathbf{x}_{k-1}, k-1) + f(\mathbf{x}_{k-1}, \mathbf{x}_k)\} \\ &= \max_{\mathbf{x}_{k-1}} \{\Phi(\mathbf{x}_{k-1}, k-1) + f(\mathbf{x}_{k-1}, \mathbf{x}_k)\}, \qquad k = 2, \ldots, N \\ S(\hat{\mathbf{x}}_1, \ldots, \hat{\mathbf{x}}_N, N) &= \max_{\mathbf{x}_N} \{\Phi(\mathbf{x}_N, N)\} \\ \Phi(\mathbf{x}_1, 1) &\triangleq |g(\mathbf{x}_1)|\end{aligned} \tag{9.30}$$

This procedure is remarkable in that the global optimization of $S(\mathbf{x}_1, \ldots, \mathbf{x}_N, N)$ has been reduced to N stages of two variable optimizations. In each stage, for each value of $\mathbf{x}_k$ one has to search the optimum $\Phi(\mathbf{x}_k, k)$. Therefore, if each $\mathbf{x}_k$ takes L different values, the total number of search operations is $(N-1)(L^2-1) + (L-1)$. This would be significantly smaller than the $L^N - 1$ exhaustive searches required for direct maximization of $S(\mathbf{x}_2, \mathbf{x}_2, \ldots, \mathbf{x}_N, N)$ when L and N are large.

Example 9.3

Consider the gradient image of Fig. 9.16. Applying the linkage rule of Example 9.2 and letting $\alpha = 4/\pi$, $\beta = 0$, we obtain the graph of Fig. 9.18a which shows the values of various segments connecting different nodes. Specifically, we have $N = 5$ and $\Phi(A, 1) = 5$. For $k = 2$, we get

$$\Phi(D, 2) = \max(11, 12) = 12$$

which means in arriving at D, the path ACD is chosen. Proceeding in this manner some of the candidate paths (shown by dotted lines) are eliminated. At $k = 4$, only two paths are acceptable, namely, $ACDEF$ and $AGHJ$. At $k = 5$, the path JB is eliminated, giving the optimal boundary as $ACDEFB$.

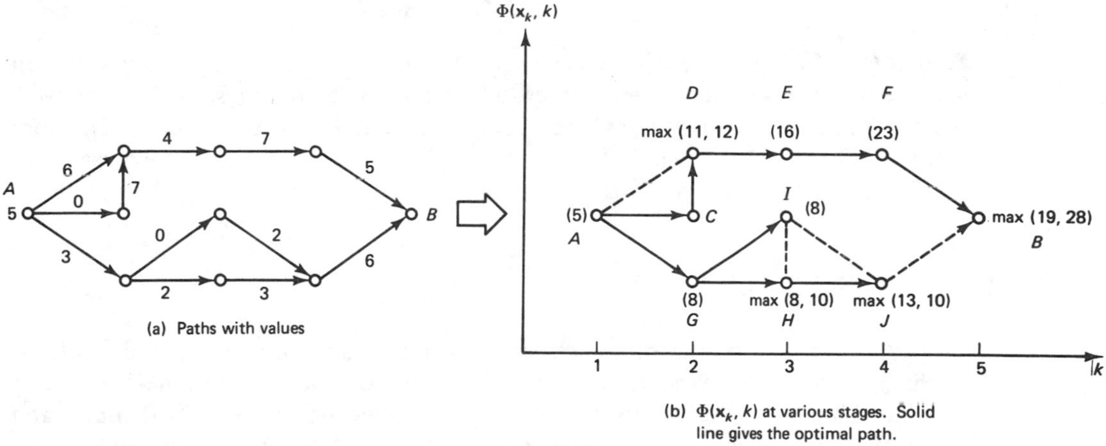

Figure 9.18 Dynamic programming for optimal boundary extraction.

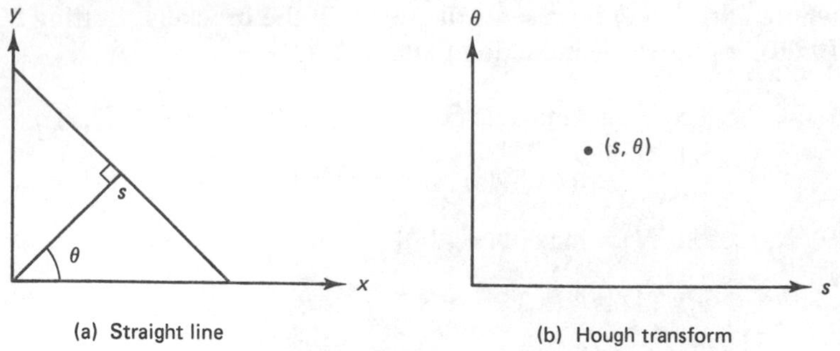

Figure 9.19 The Hough transform.

Hough Transform [1, 24]

A straight line at a distance s and orientation θ (Fig. 9.19a) can be represented as

$$s = x \cos\theta + y \sin\theta \tag{9.31}$$

The Hough transform of this line is just a point in the (s, θ) plane; that is, all the points on this line map into a single point (Fig. 9.19b). This fact can be used to detect straight lines in a given set of boundary points. Suppose we are given boundary points (x_i, y_i), $i = 1, \ldots, N$. For some chosen quantized values of parameters s and θ, map each (x_i, y_i) into the (s, θ) space and count $C(s, \theta)$, the number of edge points that map into the location (s, θ), that is, set

$$C(s_k, \theta_l) = C(s_k, \theta_l) + 1, \quad \text{if} \quad x_i \cos\theta + y_i \sin\theta = s_k \quad \text{for} \quad \theta = \theta_l \tag{9.32}$$

Then the local maxima of $C(s, \theta)$ give the different straight line segments through the edge points. This two-dimensional search can be reduced to a one-dimensional search if the gradients θ_i at each edge location are also known. Differentiating both sides of (9.31) with respect to x, we obtain

$$\frac{dy}{dx} = -\cot\theta = \tan\left(\frac{\pi}{2} + \theta\right) \tag{9.33}$$

Hence $C(s, \theta)$ need be evaluated only for $\theta = \pi/2 - \theta_i$. The Hough transform can also be generalized to detect curves other than straight lines. This, however, increases the dimension of the space of parameters that must be searched [3]. From Chapter 10, it can be concluded that the Hough transform can also be expressed as the Radon transform of a line delta function.

9.6 BOUNDARY REPRESENTATION

Proper representation of object boundaries is important for analysis and synthesis of shape. Shape analysis is often required for detection and recognition of objects in a scene. Shape synthesis is useful in computer-aided design (CAD) of parts and assemblies, image simulation applications such as video games, cartoon movies, environmental modeling of aircraft-landing testing and training, and other computer graphics problems.

Chain Codes [26]

In chain coding the direction vectors between successive boundary pixels are encoded. For example, a commonly used chain code (Fig. 9.20) employs eight directions, which can be coded by 3-bit code words. Typically, the chain code contains the start pixel address followed by a string of code words. Such codes can be generalized by increasing the number of allowed direction vectors between successive boundary pixels. A limiting case is to encode the curvature of the contour as a function of contour length t (Fig. 9.21).

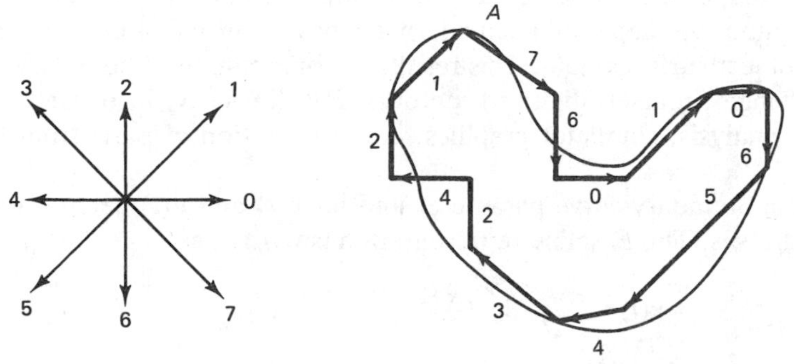

(b) Contour

Algorithm:

1. Start at any boundary pixel, *A*.
2. Find the nearest edge pixel and code its orientation. In case of a tie, choose the one with largest (or smallest) code value.
3. Continue until there are no more boundary pixels.

Boundary pixel orientations: (A), 76010655432421

Chain code: A 111 110 000 001 000 110 101 101 110 011 010 100 010 001

Figure 9.20 Chain code for boundary representation.

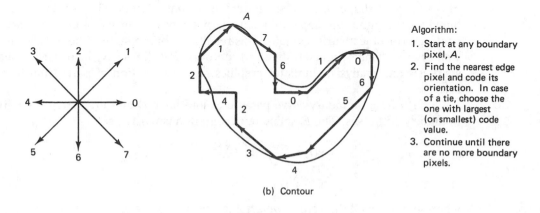

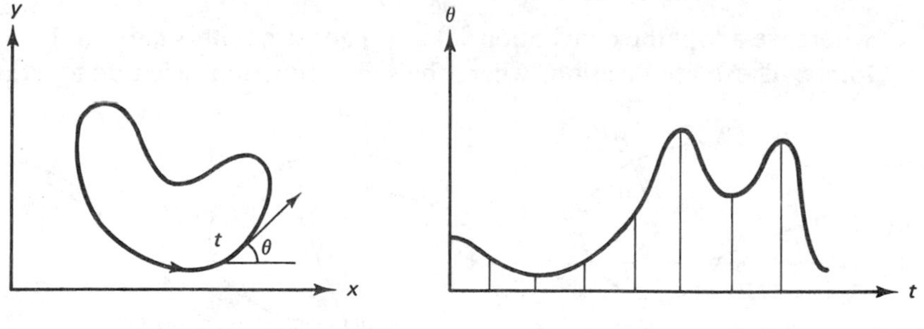

(a) Contour

(b) θ vs. t curve. Encode $\theta(t)$.

Figure 9.21 Generalized chain coding.

Fitting Line Segments [1]

Straight-line segments give simple approximation of curve boundaries. An interesting sequential algorithm for fitting a curve by line segments is as follows (Fig. 9.22).

Algorithm. Approximate the curve by the line segment joining its end points (A, B). If the distance from the farthest curve point (C) to the segment is greater than a predetermined quantity, join AC and BC. Repeat the procedure for new segments AC and BC, and continue until the desired accuracy is reached.

B-Spline Representation [27–29]

B-splines are piecewise polynomial functions that can provide local approximations of contours of shapes using a small number of parameters. This is useful because human perception of shapes is deemed to be based on curvatures of parts of contours (or object surfaces) [30]. This results in compression of boundary data as well as smoothing of coarsely digitized contours. B-splines have been used in shape synthesis and analysis, computer graphics, and recognition of parts from boundaries.

Let t be a boundary curve parameter and let $x(t)$ and $y(t)$ denote the given boundary addresses. The B-spline representation is written as

$$\mathbf{x}(t) = \sum_{i=0}^{n} \mathbf{p}_i B_{i,k}(t)$$
$$\mathbf{x}(t) \triangleq [x(t), y(t)]^T, \qquad \mathbf{p}_i \triangleq [p_{1i}, p_{2i}]^T \tag{9.34}$$

where $\mathbf{p}_i$ are called the *control points* and the $B_{i,k}(t), i = 0, 1, \ldots, n, k = 1, 2 \ldots$ are called the *normalized B-splines of order k*. In computer graphics these functions are also called *basis splines* or *blending functions* and can be generated via the recursion

$$B_{i,k}(t) \triangleq \frac{(t - t_i)B_{i,k-1}(t)}{t_{i+k-1} - t_i} + \frac{(t_{i+k} - t)B_{i+1,k-1}(t)}{(t_{i+k} - t_{i+1})}, \qquad k = 2, 3, \ldots \tag{9.35a}$$

$$B_{i,1}(t) \triangleq \begin{cases} 1, & t_i \le t < t_{i+1} \\ 0, & \text{otherwise} \end{cases} \tag{9.35b}$$

where we adopt the convention $0/0 \triangleq 0$. The parameters $t_i, i = 0, 1, \ldots$ are called the *knots*. These are locations where the spline functions are tied together. Associated

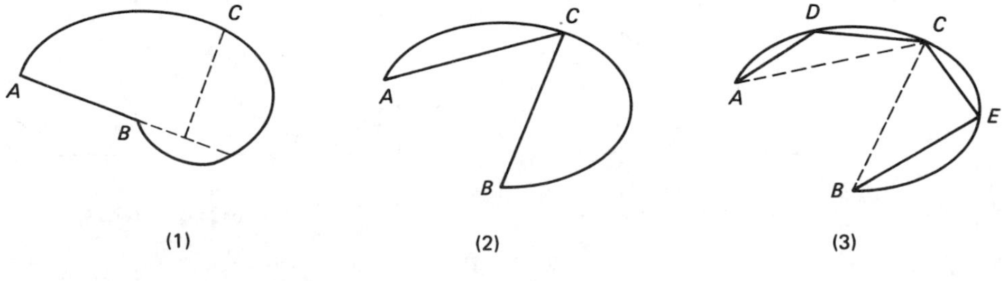

Figure 9.22 Successive approximation by line segments.

with the knots are *nodes* s_i, which are defined as the mean locations of successive $k-1$ knots, that is,

$$s_i \stackrel{\Delta}{=} \frac{1}{k-1}(t_{i+1}+t_{i+2}+\cdots+t_{i+k-1}), \qquad 0 \le i \le n, \qquad k \ge 2 \tag{9.36}$$

The variable t is also called the node parameter, of which t_i and s_i are special values. Figure 9.23 shows some of the B-spline functions. These functions are nonnegative and have finite support. In fact, for the normalized B-splines, $0 \le B_{i,k}(t) \le 1$ and the region of support of $B_{i,k}(t)$ is $[t_i, t_{i+k})$. The functions $B_{i,k}(t)$ form a basis in the space of *piecewise-polynomial functions.* These functions are called *open B-splines* or *closed (or periodic) B-splines,* depending on whether the boundary being represented is open or closed. The parameter k controls the order of continuity of the curve. For example, for $k=3$ the splines are piecewise quadratic polynomials. For $k=4$, these are cubic polynomials. In computer graphics $k=3$ or 4 is generally found to be sufficient.

When the knots are uniformly spaced, that is,

$$t_{i+1}-t_i=\Delta t, \qquad \forall i \tag{9.37a}$$

the $B_{i,k}(t)$ are called uniform splines and they become translates of $B_{0,k}(t)$, that is,

$$B_{i,k}(t)=B_{0,k}(t-i), \qquad i=k-1,k,\ldots,n-k+1 \tag{9.37b}$$

Near the boundaries $B_{i,k}(t)$ is obtained from (9.35). For uniform open B-splines with $\Delta t = 1$, the knot values can be chosen as

$$t_i=\begin{cases}0 & i<k\\ i-k+1 & k\le i\le n\\ n-k+2 & i>n\end{cases} \tag{9.38}$$

and for uniform periodic (or closed) B-splines, the knots can be chosen as

$$t_i = i\,\mathrm{mod}(n+1) \tag{9.39}$$

$$B_{i,k}(t)=B_{0,k}[(t-i)\mathrm{mod}(n+1)] \tag{9.40}$$

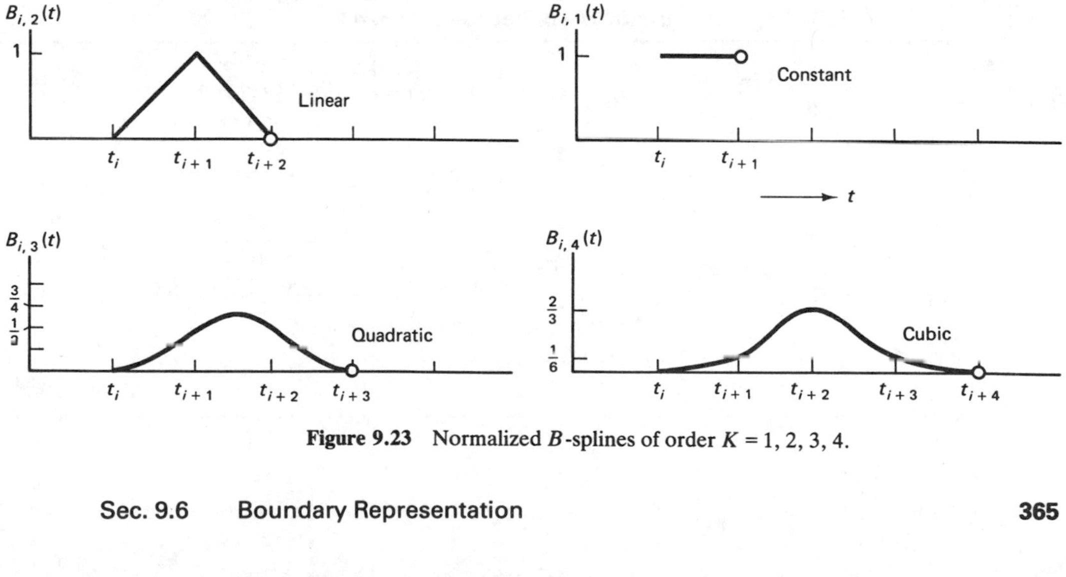

Figure 9.23 Normalized B-splines of order $K = 1, 2, 3, 4$.

For $k = 1, 2, 3, 4$ and knots given by (9.39), the analytic forms of $B_{0,k}(t)$ are provided in Table 9.7.

Control points. The control points $\mathbf{p}_i$ are not only the series coefficients in (9.34), they physically define vertices of a polygon that guides the splines to trace a smooth curve (Fig. 9.24). Once the control points are given, it is straightforward to obtain the curve trace $\mathbf{x}(t)$ via (9.34). The number of control points necessary to reproduce a given boundary accurately is usually much less than the number of points needed to trace a smooth curve. Data compression by factors of 10 to 1000 can be achieved, depending on the resolution and complexity of the shape.

A B-spline–generated boundary can be translated, scaled (zooming or shrinking), or rotated by performing corresponding control point transformations as follows:

$$\text{Translation:} \quad \tilde{\mathbf{p}}_i = \mathbf{p}_i + \mathbf{x}_0, \qquad \mathbf{x}_0 = [x_0, y_0]^T \tag{9.41}$$

$$\text{Scaling:} \quad \tilde{\mathbf{p}} = \alpha \mathbf{p}_i, \qquad \alpha = \text{scalar} \tag{9.42}$$

$$\text{Rotation:} \quad \tilde{\mathbf{p}}_i = \mathbf{R}\mathbf{p}_i, \qquad \mathbf{R} \triangleq \begin{bmatrix} \cos\theta_0 & -\sin\theta_0 \\ \sin\theta_0 & \cos\theta_0 \end{bmatrix} \tag{9.43}$$

The transformation of (9.43) gives an anticlockwise rotation by an angle θ_0. Since the object boundary can be reproduced via the control points, the latter constitute a set of *regenerative shape features*. Many useful shape parameters, such as center of mass, area, and perimeter, can be estimated easily from the control points (Problem 9.9).

Often, we are given the boundary points at discrete values of $t = s_0, s_1, \ldots, s_n$ and we must find the control points $\mathbf{p}_i$. Then

$$\mathbf{x}(s_j) = \sum_{i=0}^{n} \mathbf{p}_i B_{i,k}(s_j), \qquad j = 0, 1, \ldots, n \tag{9.44}$$

which can be written as

$$\mathbf{B}_k \mathbf{P} = \mathbf{x} \tag{9.45}$$

TABLE 9.7 Uniform Periodic B-splines for $0 \le t \le n$

$B_{0,1}(t) = \begin{cases} 1, & 0 \le t < 1 \\ 0, & \text{otherwise} \end{cases}$	$B_{0,2}(t) = \begin{cases} t, & 0 \le t < 1 \\ 2 - t, & 1 \le t < 2 \\ 0, & \text{otherwise} \end{cases}$
$B_{0,3}(t) = \begin{cases} \dfrac{t^2}{2}, & 0 \le t < 1 \\ -t^2 + 3t - 1.5, & 1 \le t < 2 \\ \dfrac{(3-t)^2}{2}, & 2 \le t < 3 \\ 0, & \text{otherwise} \end{cases}$	$B_{0,4}(t) = \begin{cases} \dfrac{t^3}{6}, & 0 \le t < 1 \\ \dfrac{-3t^3 + 12t^2 - 12t + 4}{6}, & 1 \le t < 2 \\ \dfrac{3t^3 - 24t^2 + 60t - 44}{6}, & 2 \le t < 3 \\ \dfrac{(4-t)^3}{6}, & 3 \le t < 4 \\ 0, & \text{otherwise} \end{cases}$

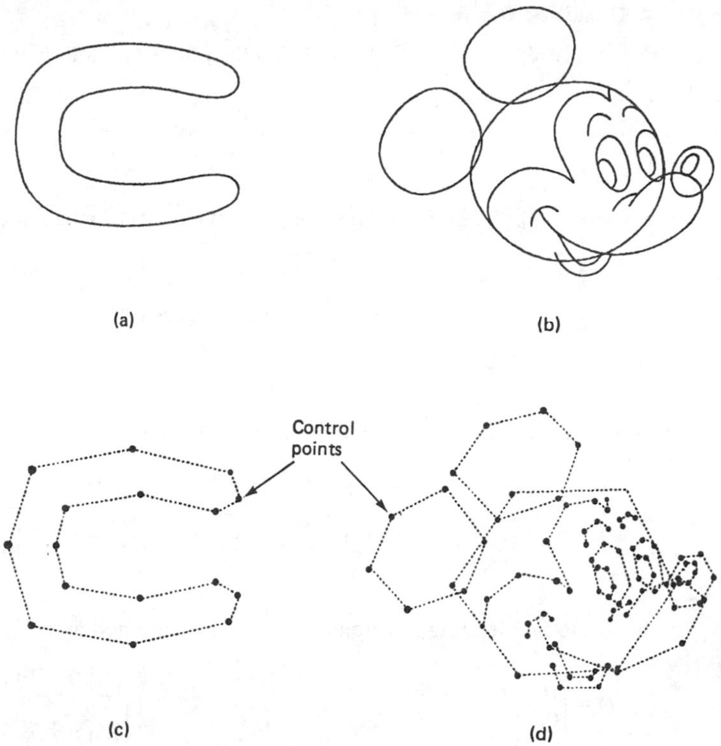

Figure 9.24 (a), (b) B-spline curves fitted through 128 and 1144 points of the original boundaries containing 1038 and 10536 points respectively to yield indistinguishable reproduction: (c), (d) corresponding 16 and 99 control points respectively. Since (a), (b) can be reproduced from (c), (d), compression ratios of greater than 100 : 1 are achieved.

where $\mathbf{B}_k$, $\mathbf{P}$, and $\mathbf{x}$ are $(n+1)\times(n+1)$, $(n+1)\times 2$, and $(n+1)\times 2$ matrices of elements $B_{i,k}(s_j)$, $\mathbf{p}_i$, $\mathbf{x}(s_j)$ respectively. When s_j are the node locations the matrix $\mathbf{B}_k$ is guaranteed to be nonsingular and the control-point array is obtained as

$$\mathbf{P} = \mathbf{B}_k^{-1}\mathbf{x} \tag{9.46}$$

For uniformly sampled closed splines, $\mathbf{B}_k$ becomes a circulant matrix, whose first row is given by

$$\mathbf{b}_0 \triangleq [b_0\, b_1 \ldots b_q, 0 \ldots 0, b_q \ldots b_1] \tag{9.47}$$

where

$$b_j \triangleq B_{0,k}(s_j), \qquad j = 0, \ldots, q, \quad \text{and} \quad q = \text{Integer}\left[\frac{(k-1)}{2}\right]$$

$$s_{j+1} - s_j = t_{j+1} - t_j = \text{constant}, \qquad \forall j$$

In the case of open B-splines, $\mathbf{B}_k$ is nearly Toeplitz when $s_j = t_j$, for every j.

Example 9.4 Quadratic B-splines, $k = 3$

(a) *Periodic case:* From (9.36) and (9.39) the nodes (sampling locations) are

$$[s_0, s_1, s_2, \ldots, s_n] = \left[\frac{3}{2}, \frac{5}{2}, \frac{7}{2}, \ldots, \frac{2n-1}{2}, \frac{n}{2}\frac{1}{2}\right]$$

Then from (9.47), the blending function $B_{0,3}(t)$ gives the circulant matrix

$$\mathbf{B}_3 = \tfrac{1}{8}\begin{bmatrix} 6 & 1 & & \mathbf{0} & 1 \\ 1 & \ddots & \ddots & & \\ & \ddots & \ddots & \ddots & \\ & & \ddots & \ddots & 1 \\ 1 & \mathbf{0} & & 1 & 6 \end{bmatrix}$$

(b) *Nonperiodic case:* From (9.36) and (9.38) the knots and nodes are obtained as

$$[t_0, t_1, \ldots, t_{n+3}] = [0, 0, 0, 1, 2, 3, \ldots, n-2, n-1, n-1, n-1]$$

$$[s_0, s_1, \ldots, s_n] = \left[0, \frac{1}{2}, \frac{3}{2}, \frac{5}{2}, \frac{7}{2}, \ldots, \frac{2n-3}{2}, n-1\right]$$

The nonperiodic blending functions for $k = 3$ are obtained as

$$B_{0,3}(t) = \begin{cases} (t-1)^2, & 0 \le t < 1 \\ 0, & 1 \le t \le n-1 \end{cases}$$

$$B_{1,3}(t) = \begin{cases} -\frac{3}{2}\left(t - \frac{2}{3}\right)^2 + \frac{2}{3}, & 0 \le t < 1 \\ \frac{1}{2}(t-2)^2, & 1 \le t < 2 \\ 0, & 2 \le t \le n-1 \end{cases}$$

$$B_{j,3}(t) = \begin{cases} 0, & 0 \le t < j-2 \\ \frac{1}{2}(t-j+2)^2, & j-2 \le t < j-1 \\ -\left(t - j + \frac{1}{2}\right)^2 + \frac{3}{4}, & j-1 \le t < j \\ \frac{1}{2}(t-j-1)^2, & j \le t < j+1 \\ 0, & j+1 \le t \le n-1, \; j = 2, 3, \ldots, n-2 \end{cases}$$

$$B_{n-1,3}(t) = \begin{cases} 0, & 0 \le t < n-3 \\ -\frac{1}{2}(t-n+3)^2, & n-3 \le t < n-2 \\ -\frac{3}{2}\left(t - n + \frac{5}{3}\right)^2 + \frac{2}{3}, & n-2 \le t \le n-1 \end{cases}$$

$$B_{n,3}(t) = \begin{cases} 0, & 0 \le t < n-2 \\ (t-n+2)^2, & n-2 \le t \le n-1 \end{cases}$$

From these we obtain

$$\mathbf{B}_3 = \frac{1}{8}\overbrace{\begin{bmatrix} 8 & & & & & \\ 2 & 5 & 1 & & & \mathbf{0} \\ & 1 & 6 & 1 & & \\ & & \ddots & \ddots & \ddots & \\ & & & 1 & 6 & 1 \\ \mathbf{0} & & & 1 & 5 & 2 \\ & & & & & 8 \end{bmatrix}}^{n+1} \; n+1$$

Figure 9.25 shows a set of uniformly sampled spline boundary points $x(s_j)$, $j = 0, \ldots, 13$. Observe that the s_j, and not $x(s_i), y(s_i)$, are required to be uniformly spaced. Periodic and open quadratic B-spline interpolations are shown in Fig. 9.25b and c. Note that in the case of open B-splines, the endpoints of the curve are also control points.

The foregoing method of extracting control points from uniformly sampled boundaries has one remaining difficulty—it requires the number of control points be equal to the number of sampled points. In practice, we often have a large number of finely sampled points on the contour and, as is evident from Fig. 9.24, the number of control points necessary to represent the contour accurately may be much smaller. Therefore, we are given $\mathbf{x}(t)$ for $t = \xi_0, \xi_1, \ldots, \xi_m$, where $m \gg n$. Then we have an overdetermined $(m + 1) \times (n + 1)$ system of equations

$$\mathbf{x}(\xi_j) = \sum_{i=0}^{n} B_{i,k}(\xi_j)\,\mathbf{p}_i, \qquad j = 0, \ldots, m \tag{9.48}$$

Least squares techniques can now be applied to estimate the control points $\mathbf{p}_i$. With proper indexing of the sampling points ξ_j and letting the ratio $(m + 1)/(n + 1)$

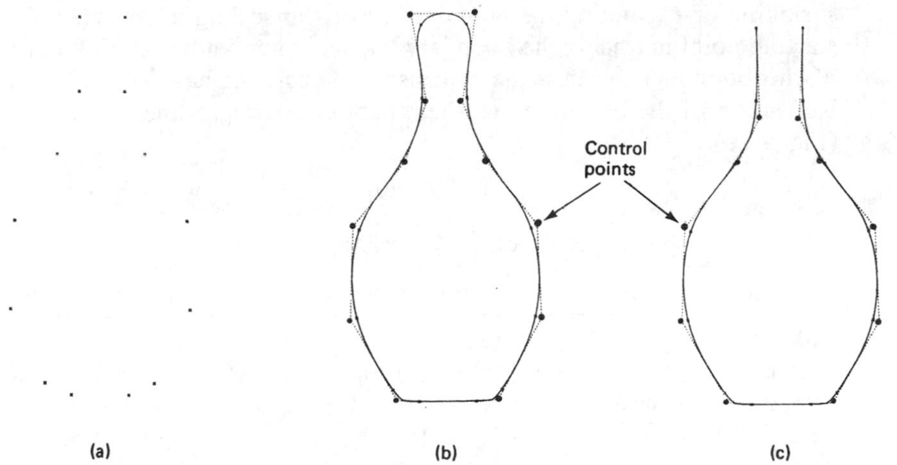

Figure 9.25 (a) Given points; (b) quadratic periodic B-spline interpolation; (c) quadratic nonperiodic B-spline interpolation.

be an integer, the least squares solution can be shown to require inversion of a circulant matrix, for which fast algorithms are available [29].

Fourier Descriptors

Once the boundary trace is known, we can consider it as a pair of waveforms $x(t), y(t)$. Hence any of the traditional one-dimensional signal representation techniques can be used. For any sampled boundary we can define

$$u(n) \triangleq x(n) + jy(n), \qquad n = 0, 1, \ldots, N-1 \tag{9.49}$$

which, for a closed boundary, would be periodic with period N. Its DFT representation is

$$\begin{aligned} u(n) &\triangleq \frac{1}{N} \sum_{k=0}^{N-1} a(k) \exp\left(\frac{j2\pi kn}{N}\right), \qquad 0 \le n \le N-1 \\ a(k) &\triangleq \sum_{n=0}^{N-1} u(n) \exp\left(\frac{-j2\pi kn}{N}\right), \qquad 0 \le k \le N-1 \end{aligned} \tag{9.50}$$

The complex coefficients $a(k)$ are called the *Fourier descriptors* (FDs) of the boundary. For a continuous boundary function, $u(t)$, defined in a similar manner to (9.49), the FDs are its (infinite) Fourier series coefficients. Fourier descriptors have been found useful in character recognition problems [32].

Effect of geometric transformations. Several geometrical transformations of a boundary or shape can be related to simple operations on the FDs (Table 9.8). If the boundary is translated by

$$u_0 \triangleq x_0 + jy_0 \tag{9.51}$$

then the new FDs remain the same except at $k = 0$. The effect of scaling, that is, shrinking or expanding the boundary results in scaling of the $a(k)$. Changing the starting point in tracing the boundary results in a modulation of the $a(k)$. Rotation of the boundary by an angle θ_0 causes a constant phase shift of θ_0 in the FDs. Reflection of the boundary (or shape) about a straight line inclined at an angle θ (Fig. 9.26),

$$Ax + By + C = 0 \tag{9.52}$$

TABLE 9.8 Properties of Fourier Descriptors

Transformation	Boundary	Fourier Descriptors
Identity	$u(n)$	$a(k)$
Translation	$\tilde{u}(n) = u(n) + u_0$	$\tilde{a}(k) = a(k) + u_0\,\delta(k)$
Scaling or Zooming	$\tilde{u}(n) = \alpha u(n)$	$\tilde{a}(k) = \alpha a(k)$
Starting Point	$\tilde{u}(n) = u(n - n_0)$	$\tilde{a}(k) = a(k)e^{-j2\pi n_0 k/N}$
Rotation	$\tilde{u}(n) = u(n)e^{j\theta_0}$	$\tilde{a}(k) = a(k)e^{j\theta_0}$
Reflection	$\tilde{u}(n) = u^*(n)e^{j2\theta} + 2\gamma$	$\tilde{a}(k) = a^*(-k)e^{j2\theta} + 2\gamma\delta(k)$

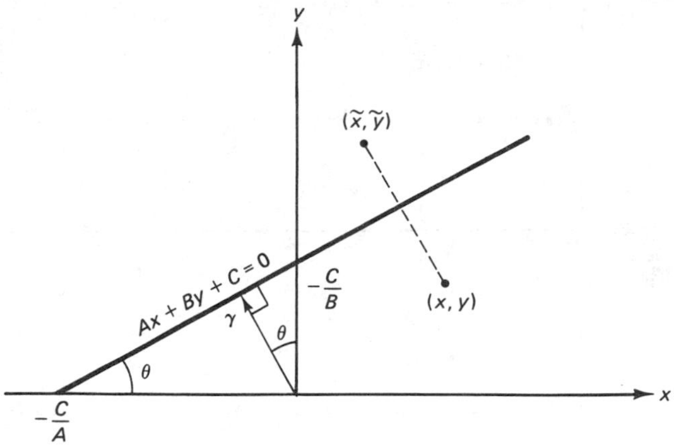

Figure 9.26 Reflection about a straight line.

gives the *new* boundary $\tilde{x}(n), \tilde{y}(n)$ as (Problem 9.11)

$$\tilde{u}(n) = u^*(n)e^{j2\theta} + 2\gamma$$

where (9.53)

$$\gamma \triangleq \frac{-(A + jB)C}{A^2 + B^2}, \qquad \exp(j2\theta) = \frac{-(A + jB)^2}{A^2 + B^2}$$

For example, if the line (9.52) is the x-axis, that is, $A = C = 0$, then $\theta = 0, \gamma = 0$ and the new FDs are the complex conjugates of the old ones.

Fourier descriptors are also *regenerative shape features.* The number of descriptors needed for reconstruction depends on the shape and the desired accuracy. Figure 9.27 shows the effect of truncation and quantization of the FDs. From Table 9.8 it can be observed that the FD magnitudes have some invariant properties. For example $|\tilde{a}(k)|, k = 1, 2, \ldots, N - 1$ are invariant to starting point, rotation, and reflection. The features $\tilde{a}(k)/|\tilde{a}(k)|$ are invariant to scaling. These properties can be used in detecting shapes regardless of their size, orientation, and so on. However, the FD magnitude or phase alone are generally inadequate for reconstruction of the original shape (Fig. 9.27).

Boundary matching. The Fourier descriptors can be used to match similar shapes even if they have different size and orientation. If $a(k)$ and $b(k)$ are the FDs of two boundaries $u(n)$ and $v(n)$, respectively, then their shapes are similar if the distance

$$d(u_0, \alpha, \theta_0, n_0) \triangleq \min_{u_0, \alpha, n_0, \theta_0} \left\{ \sum_{n=0}^{N-1} |u(n) - \alpha v(n + n_0)e^{j\theta_0} - u_0|^2 \right\} \tag{9.54}$$

is small. The parameters u_0, α, n_0, and θ_0 are chosen to minimize the effects of translation, scaling, starting points and rotation, respectively. If $u(n)$ and $v(n)$ are normalized so that $\Sigma u(n) = \Sigma v(n) = 0$, then for a given shift n_0, the above distance is

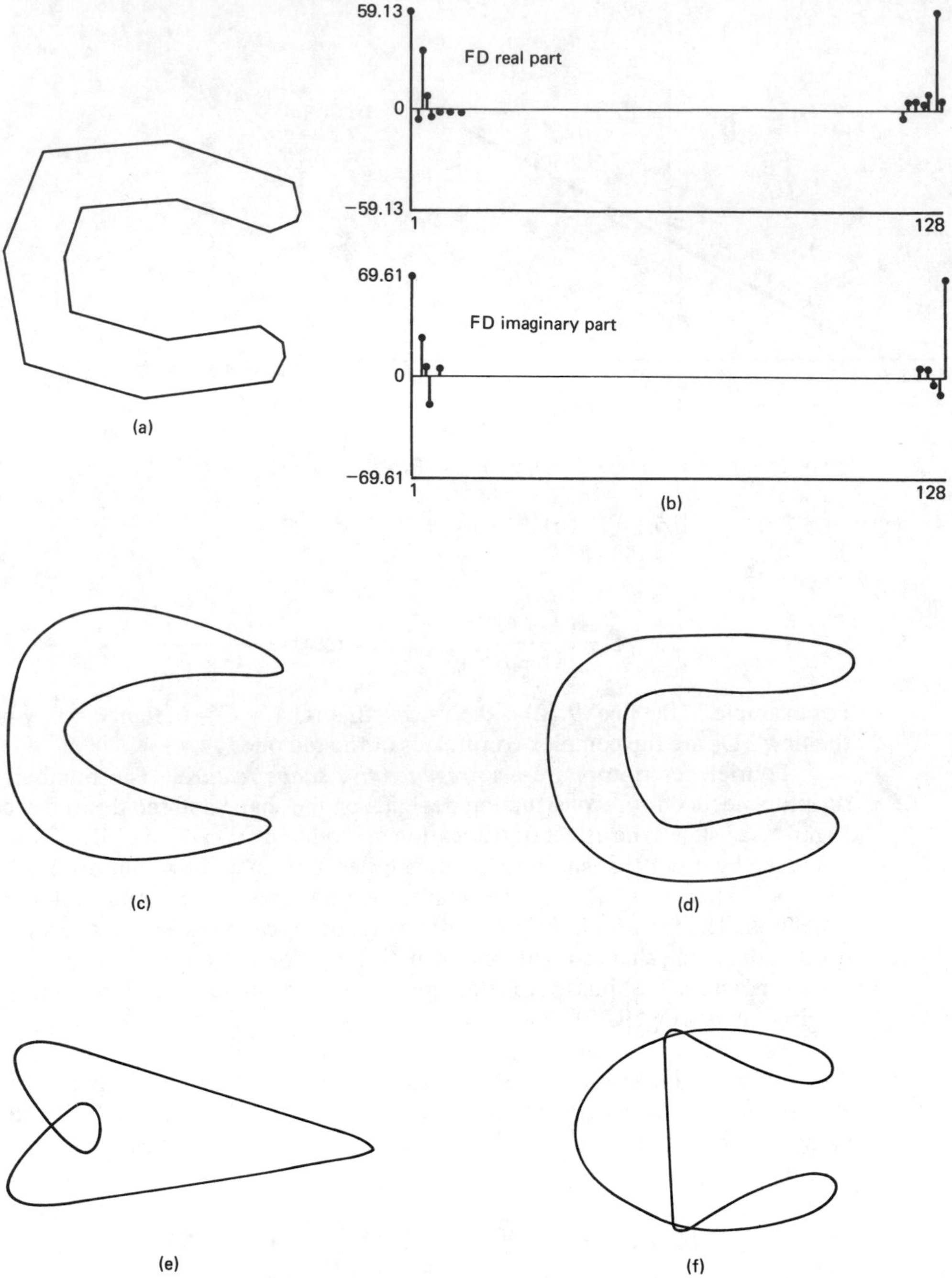

Figure 9.27 Fourier descriptors. (a) Given shape; (b) FDs, real and imaginary components; (c) shape derived from largest five FDs; (d) derived from all FDs quantized to 17 levels each; (e) amplitude reconstruction; (f) phase reconstruction.

minimum when

$$u_0 = 0$$

$$\alpha = \frac{\sum_k c(k) \cos(\psi_k + k\phi + \theta_0)}{\sum_k |b(k)|^2} \tag{9.55}$$

and

$$\tan \theta_0 = -\frac{\sum_k c(k) \sin(\psi_k + k\phi)}{\sum_k c(k) \cos(\psi_k + k\phi)}$$

where $a(k)b^*(k) = c(k)e^{j\psi_k}$, $\phi \triangleq -2\pi n_0/N$, and $c(k)$ is a real quantity. These equations give α and θ_0, from which the minimum distance d is given by

$$d = \min_\phi [d(\phi)] = \min_\phi \left\{ \sum_k |a(k) - \alpha b(k) \exp[j(k\phi + \theta_0)]|^2 \right\} \tag{9.56}$$

The distance $d(\phi)$ can be evaluated for each $\phi = \phi(n_0)$, $n_0 = 0, 1, \ldots, N-1$ and the minimum searched to obtain d. The quantity d is then a useful measure of difference between two shapes. The FDs can also be used for analysis of line patterns or open curves, skeletonization of patterns, computation of area of a surface, and so on (see Problem 9.12).

Instead of using two functions $x(t)$ and $y(t)$, it is possible to use only one function when t represents the arc length along the boundary curve. Defining the arc tangent angle (see Fig. 9.21)

$$\theta(t) = \tan^{-1}\left[\frac{dy(t)/dt}{dx(t)/dt}\right] \tag{9.57}$$

the curve can be traced if $x(0)$, $y(0)$, and $\theta(t)$ are known. Since t is the distance along the curve, it is true that

$$dt^2 = dx^2 + dy^2 \Rightarrow \left(\frac{dx}{dt}\right)^2 + \left(\frac{dy}{dt}\right)^2 = 1 \tag{9.58}$$

which gives $dx/dt = \cos\theta(t)$, $dy/dt = \sin\theta(t)$, or

$$\mathbf{x}(t) = \mathbf{x}(0) + \int_0^t \begin{bmatrix} \cos\theta(\tau) \\ \sin\theta(\tau) \end{bmatrix} d\tau \tag{9.59}$$

Sometimes, the FDs of the curvature of the boundary

$$\kappa(t) \triangleq \frac{d\theta(t)}{dt} \tag{9.60}$$

or those of the *detrended* function

$$\hat{\theta}(t) \triangleq \int_0^t \kappa(\tau)\, d\tau - \frac{2\pi t}{T} \tag{9.61}$$

are used [31]. The latter has the advantage that $\hat{\theta}(t)$ does not have the singularities at corner points that are encountered in polygon shapes. Although we have now only a real scalar set of FDs, their rate of decay is found to be much slower than those of $u(t)$.

Autoregressive Models

If we are given a class of object boundaries—for instance, screwdrivers of different sizes with arbitrary orientations—then we have an ensemble of boundaries that could be represented by a stochastic model. For instance, the boundary coordinates $x(n), y(n)$ could be represented by AR processes [33]

$$
\begin{aligned}
u_i(n) &= \sum_{k=1}^{p} a_i(k) u_i(n-k) + \varepsilon_i(n) \\
x_i(n) &= u_i(n) + \mu_i, \qquad i = 1, 2
\end{aligned}
\tag{9.62}
$$

where $x(n) \triangleq x_1(n)$ and $y(n) \triangleq x_2(n)$. Here $u_i(n)$ is a zero mean stationary random sequence, μ_i is the ensemble mean of $x_i(n)$, and $\varepsilon_i(n)$ is an uncorrelated sequence with zero mean and variance β_i^2. For simplicity we assume $\varepsilon_1(n)$ and $\varepsilon_2(n)$ to be independent, so that the coordinates $x_1(n)$ and $x_2(n)$ can be processed independently. For closed boundaries the covariances of the sequences $\{x_i(n)\}, i = 1, 2$ will be periodic. The AR model parameters $a_i(k)$, β_i^2, and μ_i can be considered as features of the given ensemble of shapes. These features can be estimated from a given boundary data set by following the procedures of Chapter 6.

Properties of AR features. Table 9.9 lists the effect of different geometric transformations on the AR model parameters. The features $a_i(k)$ are invariant under translation, scaling, and starting point. This is because the underlying correlations of the sequences $x_i(n)$, which determine the $a_i(k)$, are also invariant under these transformations. The feature β_i^2 is sensitive to scaling and μ_i is sensitive to scaling as well as translation. In the case of rotation the sum $|a_1(k)|^2 + |a_2(k)|^2$ can be shown to remain invariant.

AR models are also regenerative. Given the features $\{a_i(k), \mu_i, \beta_i^2\}$ and the residuals $\varepsilon_i(k)$, the boundary can be reconstructed. The AR model, once identified

TABLE 9.9 Properties of AR Model Parameters for Closed Boundaries

Transformation \ AR model Parameters	$\tilde{x}_i(n)$	$\tilde{\varepsilon}_i(n)$	$\tilde{a}_i(k)$	$\tilde{\beta}_i^2$	$\tilde{\mu}_i$
Identity	$x_i(n)$	$\varepsilon_i(n)$	$a_i(k)$	β_i^2	μ_i
Translation	$x_i(n) + x_{i,0}$	$\varepsilon_i(n)$	$a_i(k)$	β_i^2	$\mu_i + x_{i,0}$
Scaling/zooming	$\alpha x_i(n)$	$\alpha\varepsilon_i(n)$	$a_i(k)$	$\alpha^2\beta_i^2$	$\alpha\mu_i$
Starting point	$x_i(n + n_{0,i})$	$\varepsilon_i(n + n_{0,i})$	$a_i(k)$	β_i^2	μ_i
Rotation	$\lvert\tilde{a}_1(k)\rvert^2 + \lvert\tilde{a}_2(k)\rvert^2 = \lvert a_1(k)\rvert^2 + \lvert a_2(k)\rvert^2$				

for a class of objects, can also be used for compression of boundary data, $x_1(n), x_1(n)$ via the DPCM method (see Chapter 11).

9.7 REGION REPRESENTATION

The shape of an object may be directly represented by the region it occupies. For example, the binary array

$$u(m, n) = \begin{cases} 1, & \text{if } (m,n) \in \mathcal{R} \\ 0, & \text{otherwise} \end{cases} \tag{9.63}$$

is a simple representation of the region $\mathcal{R}$. Boundaries give an efficient representation of regions because only a subset of $u(m, n)$ is stored. Other forms of region representation are discussed next.

Run-length Codes

Any region or a binary image can be viewed as a sequence of alternating strings of 0s and 1s. Run-length codes represent these strings, or runs. For raster scanned regions, a simple run-length code consists of the start address of each string of 1s (or 0s), followed by the length of that string (Fig. 9.28). There are several forms of run-length codes that are aimed at minimizing the number of bits required to represent binary images. Details are discussed in Section 11.9. The run-length codes have the advantage that regardless of the complexity of the region, its representation is obtained in a single raster scan. The main disadvantage is that it does not give the region boundary points ordered along its contours, as in chain coding. This makes it difficult to segment different regions if several are present in an image.

Quad-trees [34]

In the quad-tree method, the given region is enclosed in a convenient rectangular area. This area is divided into four quadrants, each of which is examined if it is

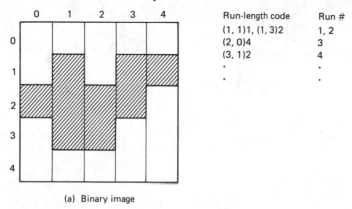

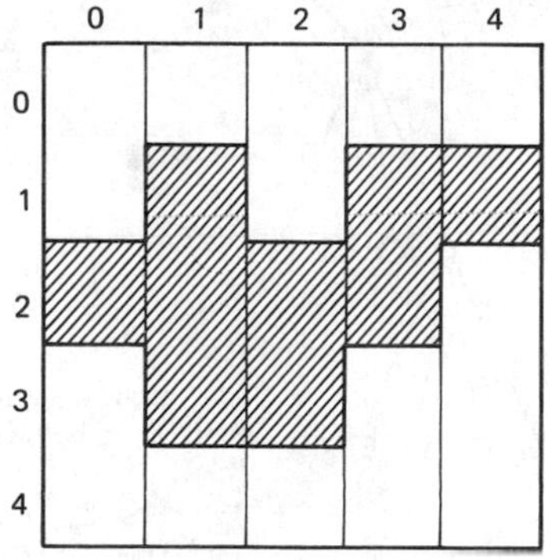

(a) Binary image

Run-length code	Run #
(1, 1)1, (1, 3)2	1, 2
(2, 0)4	3
(3, 1)2	4
.	.
.	.

Figure 9.28 Run-length coding for binary image boundary representation.

totally *black* (1s) or totally *white* (0s). The quadrant that has both black as well as white pixels is called *gray* and is further divided into four quadrants. A tree structure is generated until each subquadrant is either black only or white only. The tree can be encoded by a unique string of symbols *b* (black), *w* (white), and *g* (gray), where each *g* is necessarily followed by four symbols or groups of four symbols representing the subquadrants; see, for example, Fig. 9.29. It appears that quad-tree coding would be more efficient than run-length coding from a data-compression standpoint. However, computation of shape measurements such as perimeter and moments as well as image segmentation may be more difficult.

Projections

A two-dimensional shape or region $\mathcal{R}$ can be represented by its projections. A projection $g(s, \theta)$ is simply the sum of the run-lengths of 1s along a straight line oriented at angle θ and placed at a distance s (Fig. 9.30). In this sense a projection is simply a histogram that gives the number of pixels that project into a bin at distance s along a line of orientation θ. Features of this histogram are useful in shape analysis as well as image segmentation. For example, the first moments of $g(s, 0)$ and $g(s, \pi/2)$ give the center of mass coordinates of the region $\mathcal{R}$. Higher order moments of $g(s, \theta)$ can be used for calculating the moment invariants of shape discussed in section 9.8. Other features such as the region of support, the local maxima, and minima of $g(s, \theta)$ can be used to determine the bounding rectangles and convex hulls of shapes, which are, in turn, useful in image segmentation problems [see Ref. 11 in Chapter 10]. Projections can also serve as regenerative features of an object. The theory of reconstruction of an object from its projections is considered in detail in Chapter 10.

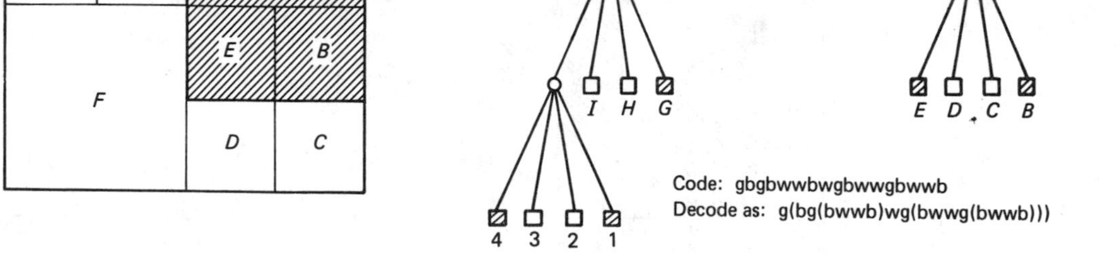

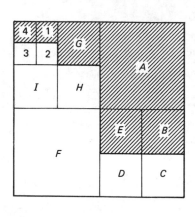

(a) Different quadrants

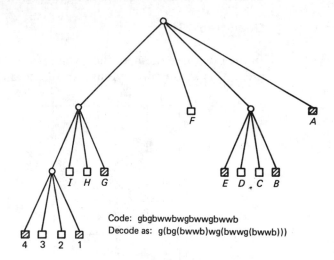

(b) Quad tree encoding

Figure 9.29 Quad-tree representation of regions.

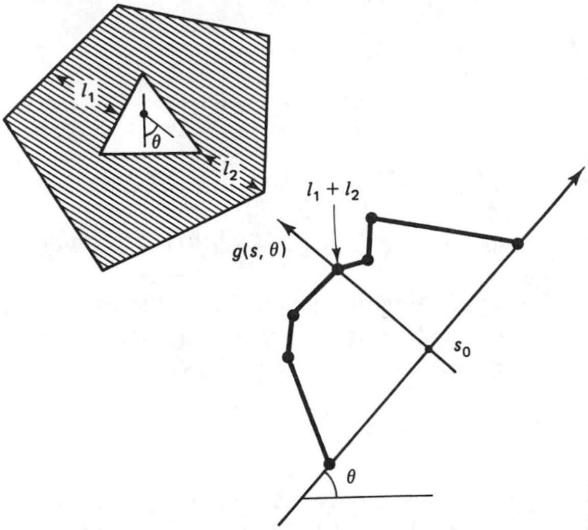

Figure 9.30 Projection $g(s, \theta)$ of a region $\mathcal{R}$. $g(s, \theta) = l_1 + l_2$.

9.8 MOMENT REPRESENTATION

The theory of moments provides an interesting and sometimes useful alternative to series expansions for representing shape of objects. Here we discuss the use of moments as features of an object $f(x, y)$.

Definitions

Let $f(x, y) \geq 0$ be a real bounded function with support on a finite region $\mathcal{R}$. We define its $(p + q)$th-order moment

$$m_{p,q} = \iint_{\mathcal{R}} f(x, y) x^p y^q \, dx \, dy, \qquad p, q = 0, 1, 2 \ldots \tag{9.64}$$

Note that setting $f(x, y) = 1$ gives the moments of the region $\mathcal{R}$ that could represent a shape. Thus the results presented here would be applicable to arbitrary objects as well as their shapes. Without loss of generality we can assume that $f(x, y)$ is nonzero only in the region $= \{x \in (-1, 1), y \in (-1, 1)\}$. Then higher-order moments will, in general, have increasingly smaller magnitudes.

The *characteristic function* of $f(x, y)$ is defined as its conjugate Fourier transform

$$F^*(\xi_1, \xi_2) \triangleq \iint_{\mathcal{R}} f(x, y) \exp\{j2\pi(x\xi_1 + y\xi_2)\} \, dx \, dy \tag{9.65}$$

The *moment-generating function* of $f(x, y)$ is defined as

$$M(\xi_1, \xi_2) \triangleq \iint_{\mathcal{R}} f(x, y) \exp(x\xi_1 + y\xi_2) \, dx \, dy \tag{9.66}$$

It gives the moments as

$$m_{p,q} = \left. \frac{\partial^{p+q} M(\xi_1, \xi_2)}{\partial \xi_1^p \, \partial \xi_2^q} \right|_{\xi_1 = \xi_2 = 0} \tag{9.67}$$

Moment Representation Theorem [35]

The infinite set of moments $\{m_{p,q}, p, q = 0, 1, \ldots\}$ uniquely determine $f(x, y)$, and vice-versa.

The proof is obtained by expanding into power series the exponential term in (9.65), interchanging the order of integration and summation, using (9.64) and taking Fourier transform of both sides. This yields the reconstruction formula

$$f(x, y) = \iint_{-\infty}^{\infty} e^{-j2\pi(x\xi_1 + y\xi_2)} \left[\sum_{p=0}^{\infty} \sum_{q=0}^{\infty} m_{p,q} \frac{(j2\pi)^{p+q}}{p!q!} \xi_1^p \xi_2^q \right] d\xi_1 \, d\xi_2 \tag{9.68}$$

Unfortunately this formula is not practical because we cannot interchange the order of integration and summation due to the fact that the Fourier transform of $(j2\pi\xi_1)^p$ is not bounded. Therefore, we cannot truncate the series in (9.68) to find an approximation of $f(x, y)$.

Moment Matching

In spite of the foregoing difficulty, if we know the moments of $f(x, y)$ up to a given order N, it is possible to find a continuous function $g(x, y)$ whose moments of order up to $p + q = N$ match those of $f(x, y)$, that is,

$$g(x, y) = \sum\sum_{0 \le i+j \le N} g_{i,j} x^i y^j \tag{9.69}$$

The coefficients $g_{i,j}$ can be found by matching the moments, that is, by setting the moments of $g(x, y)$ equal to $m_{p,q}$. A disadvantage of this approach is that the coefficients $g_{i,j}$, once determined, change if more moments are included, meaning that we must solve a coupled set of equations which grows in size with N.

Example 9.5

For $N = 3$, we obtain 10 algebraic equations ($p + q \le 3$). (Show!)

$$\begin{bmatrix} 1 & \frac{1}{3} & \frac{1}{3} \\ \frac{1}{3} & \frac{1}{5} & \frac{1}{9} \\ \frac{1}{3} & \frac{1}{9} & \frac{1}{5} \end{bmatrix} \begin{bmatrix} g_{0,0} \\ g_{2,0} \\ g_{0,2} \end{bmatrix} = \frac{1}{4} \begin{bmatrix} m_{0,0} \\ m_{2,0} \\ m_{0,2} \end{bmatrix}$$

$$\begin{bmatrix} \frac{1}{3} & \frac{1}{5} & \frac{1}{9} \\ \frac{1}{5} & \frac{1}{7} & \frac{1}{15} \\ \frac{1}{9} & \frac{1}{15} & \frac{1}{15} \end{bmatrix} \begin{bmatrix} g_{1,0} \\ g_{3,0} \\ g_{1,2} \end{bmatrix} = \frac{1}{4} \begin{bmatrix} m_{1,0} \\ m_{3,0} \\ m_{1,2} \end{bmatrix} \tag{9.70}$$

$$g_{1,1} = \tfrac{9}{4} m_{1,1}$$

where three additional equations are obtained by interchanging the indices in (9.70).

Orthogonal Moments

The moments $m_{p,q}$ are the projections of $f(x, y)$ onto monomials $\{x^p y^q\}$, which are nonorthogonal. An alternative is to use the orthogonal Legendre polynomials [36], defined as

$$\left.\begin{aligned} P_0(x) &= 1 \\ P_n(x) &= \frac{1}{n!2^n}\frac{d^n}{dx^n}(x^2-1)^n, \qquad n = 1, 2 \ldots \\ \int_{-1}^{1} P_n(x)P_m(x)\,dx &= \frac{2}{2n+1}\delta(m-n) \end{aligned}\right\} \tag{9.71}$$

Now $f(x, y)$ has the representation

$$\left.\begin{aligned} f(x, y) &= \sum_{p=0}^{\infty}\sum_{q=0}^{\infty} \lambda_{p,q} P_p(x)P_q(y) \\ \lambda_{p,q} &= \frac{(2p+1)(2q+1)}{4} \iint_{-1}^{1} f(x, y)P_p(x)P_q(y)\,dx\,dy \end{aligned}\right\} \tag{9.72}$$

where the $\lambda_{p,q}$ are called the *orthogonal moments*. Writing $P_m(x)$ as an mth-order polynomial

$$P_m(x) = \sum_{j=0}^{m} c_{m,j}x^j \tag{9.73}$$

the relationship between the orthogonal moments and the $m_{p,q}$ is obtained by substituting (9.72) in (9.73), as

$$\lambda_{p,q} = \frac{(2p+1)(2q+1)}{4}\left(\sum_{j=0}^{p}\sum_{k=0}^{q} c_{p,j}c_{q,k}m_{j,k}\right) \tag{9.74}$$

For example, this gives

$$\begin{aligned} &\lambda_{0,0} = \tfrac{1}{4}m_{0,0}, \qquad \lambda_{1,0} = \tfrac{3}{4}m_{1,0}, \qquad \lambda_{0,1} = \tfrac{3}{4}m_{0,1} \\ &\lambda_{2,0} = \tfrac{3}{8}[3m_{2,0} - m_{0,0}], \qquad \lambda_{0,2} = \tfrac{3}{8}[3m_{0,2} - m_{0,0}), \qquad \lambda_{1,1} = \tfrac{9}{4}m_{1,1}, \ldots \end{aligned} \tag{9.75}$$

The orthogonal moments depend on the usual moments, which are at most of the same order, and vice versa. Now an approximation to $f(x, y)$ *can* be obtained by truncating (9.72) at a given finite order $p + q = N$, that is,

$$f(x, y) \simeq g(x, y) = \sum_{p=0}^{N}\sum_{q=0}^{N-p} \lambda_{p,q} P_p(x)P_q(y) \tag{9.76}$$

The preceding equation is the same as (9.69) except that the different terms have been regrouped. The advantage of this representation is that the equations for $\lambda_{p,q}$ are decoupled (see Eqn. (9.74)) so that unlike $m_{p,q}$, the $\lambda_{p,q}$ are not required to be updated as the order N is increased.

Moment Invariants

These refer to certain functions of moments, which are invariant to geometric transformations such as translation, scaling, and rotation. Such features are useful in identification of objects with unique shapes regardless of their location, size, and orientation.

Translation. Under a translation of coordinates, $x' = x + \alpha, y' = y + \beta$, the central moments

$$\mu_{p,q} = \iint (x - \bar{x})^p (y - \bar{y})^q f(x, y)\, dx\, dy \tag{9.77}$$

are invariants, where $\bar{x} \overset{\Delta}{=} m_{1,0}/m_{0,0}, \bar{y} \overset{\Delta}{=} m_{0,1}/m_{0,0}$. In the sequel we will consider only the central moments.

Scaling. Under a scale change, $x' = \alpha x, y' = \alpha y$ the moments of $f(\alpha x, \alpha y)$ change to $\mu'_{p,q} = \mu_{p,q}/\alpha^{p+q+2}$. The normalized moments, defined as

$$\eta_{p,q} = \frac{\mu'_{p,q}}{(\mu'_{0,0})^{\gamma}}, \qquad \gamma = (p + q + 2)/2 \tag{9.78}$$

are then invariant to size change.

Rotation and reflection. Under a linear coordinate transformation

$$\begin{bmatrix} x' \\ y' \end{bmatrix} = \begin{bmatrix} \alpha & \beta \\ \gamma & \delta \end{bmatrix} \begin{bmatrix} x \\ y \end{bmatrix} \tag{9.79}$$

the moment-generating function will change. Via the theory of algebraic invariants [37], it is possible to find certain polynomials of $\mu_{p,q}$ that remain unchanged under the transformation of (9.79). For example, some moment invariants with respect to rotation (that is, for $\alpha = \delta = \cos\theta$, $\beta = -\gamma = \sin\theta$) and reflection ($\alpha = -\delta = \cos\theta$, $\beta = \gamma = \sin\theta$) are given as follows:

1. For first-order moments, $\mu_{0,1} = \mu_{1,0} = 0$, (always invariant).
2. For second-order moments, $(p + q = 2)$, the invariants are

$$\begin{aligned} \phi_1 &= \mu_{2,0} + \mu_{0,2} \\ \phi_2 &= (\mu_{2,0} - \mu_{0,2})^2 + 4\mu_{1,1}^2 \end{aligned} \tag{9.80}$$

3. For third-order moments $(p + q = 3)$, the invariants are

$$\begin{aligned} \phi_3 &= (\mu_{3,0} - 3\mu_{1,2})^2 + (\mu_{0,3} - 3\mu_{2,1})^2 \\ \phi_4 &= (\mu_{3,0} + \mu_{1,2})^2 + (\mu_{0,3} + \mu_{2,1})^2 \\ \phi_5 &= (\mu_{3,0} - 3\mu_{1,2})(\mu_{3,0} + \mu_{1,2})[(\mu_{3,0} + \mu_{1,2})^2 - 3(\mu_{2,1} + \mu_{0,3})^2] \\ &\quad + (\mu_{0,3} - 3\mu_{2,1})(\mu_{0,3} + \mu_{2,1})[(\mu_{0,3} + \mu_{2,1})^2 - 3(\mu_{1,2} + \mu_{3,0})^2] \end{aligned} \tag{9.81}$$

$$\phi_6 = (\mu_{2,0} - \mu_{0,2})[(\mu_{3,0} + \mu_{1,2})^2 - (\mu_{2,1} + \mu_{0,3})^2] + 4\mu_{1,1}(\mu_{3,0} + \mu_{1,2})(\mu_{0,3} + \mu_{2,1})$$

It can be shown that for Nth-order moments ($N \geq 3$), there are $(N+1)$ absolute invariant moments, which remain unchanged under both reflection and rotation [35]. A number of other moments can be found that are invariant in absolute value, in the sense that they remain unchanged under rotation but change sign under reflection. For example, for third-order moments, we have

$$\begin{aligned}\phi_7 = &(3\mu_{2,1} - \mu_{0,3})(\mu_{3,0} + \mu_{1,2})[(\mu_{3,0} + \mu_{1,2})^2 - 3(\mu_{2,1} + \mu_{0,3})^2] \\ &+ (\mu_{3,0} - 3\mu_{2,1})(\mu_{2,1} + \mu_{0,3})[(\mu_{0,3} + \mu_{2,1})^2 - 3(\mu_{3,0} + \mu_{1,2})^2]\end{aligned} \tag{9.82}$$

The relationship between invariant moments and $\mu_{p,q}$ becomes more complicated for higher-order moments. Moment invariants can be expressed more conveniently in terms of what are called *Zernike moments*. These moments are defined as the projections of $f(x, y)$ on a class of polynomials, called *Zernike polynomials* [36]. These polynomials are separable in the polar coordinates and are orthogonal over the unit circle.

Applications of Moment Invariants

Being invariant under linear coordinate transformations, the moment invariants are useful features in pattern-recognition problems. Using N moments, for instance, an image can be represented as a point in an N-dimensional vector space. This converts the pattern recognition problem into a standard decision theory problem, for which several approaches are available. For binary digital images we can set $f(x, y) = 1, (x, y) \in \mathcal{R}$. Then the moment calculation reduces to the separable computation

$$m_{p,q} = \sum_x x^p \sum_y y^q \tag{9.83}$$

These moments are useful for shape analysis. Moments can also be computed optically [38] at high speeds. Moments have been used in distinguishing between shapes of different aircraft, character recognition, and scene-matching applications [39, 40].

9.9 STRUCTURE

In many computer vision applications, the objects in a scene can be characterized satisfactorily by structures composed of line or arc patterns. Examples include handwritten or printed characters, fingerprint ridge patterns, chromosomes and biological cell structures, circuit diagrams and engineering drawings, and the like. In such situations the thickness of the pattern strokes does not contribute to the recognition process. In this section we present several transformations that are useful for analysis of structure of patterns.

Medial Axis Transform

Suppose that a fire line propagates with constant speed from the contour of a connected object towards its inside. Then all those points lying in positions where at least two wave fronts of the fire line meet during the propagation (quench points) will constitute a form of a skeleton called *the medial axis* [41] of the object.

Algorithms used to obtain the medial axis can be grouped into two main categories, depending on the kind of information preserved:

Skeleton algorithms. Here the image is described using an intrinsic coordinate system. Every point is specified by giving its distance from the nearest boundary point. The skeleton is defined as the set of points whose distance from the nearest boundary is locally maximum. Skeletons can be obtained using the following algorithm:

1. Distance transform

$$u_k(m, n) = u_0(m, n) + \min_{\Delta(m, n; i, j)} \{u_{k-1}(i, j); ((i, j):\Delta(m, n; i, j) \le 1)\}, \qquad u_0(m, n) \triangleq u(m, n) \qquad k = 1, 2, \ldots \tag{9.84}$$

where $\Delta(m, n; i, j)$ is the distance between (m, n) and (i, j). The transform is done when k equals the maximum thickness of the region.

2. The skeleton is the set of points:

$$\{(m, n) : u_k(m, n) \ge u_k(i, j), \Delta(m, n; i, j) \le 1\} \tag{9.85}$$

Figure 9.31 shows an example of the preceding algorithm when $\Delta(m, n; i, j)$ represents the Euclidean distance. It is possible to recover the original image given its skeleton and the distance of each skeleton point to its contour. It is simply obtained by taking the union of the circular neighborhoods centered on the skeleton points and having radii equal to the associated contour distance. Thus the skeleton is a regenerative representation of an object.

Thinning algorithms. Thinning algorithms transform an object to a set of simple digital arcs, which lie roughly along their medial axes. The structure ob-

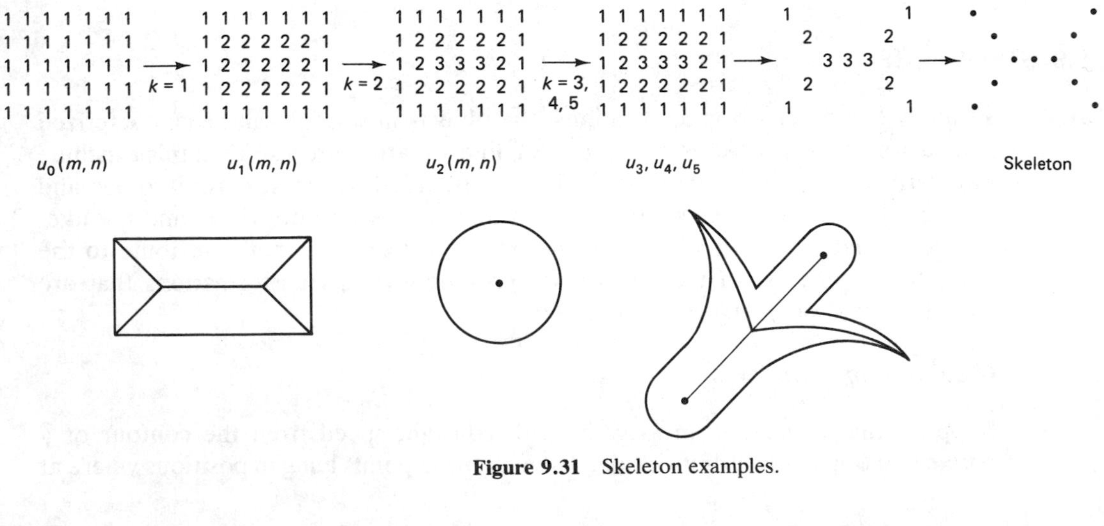

Figure 9.31 Skeleton examples.

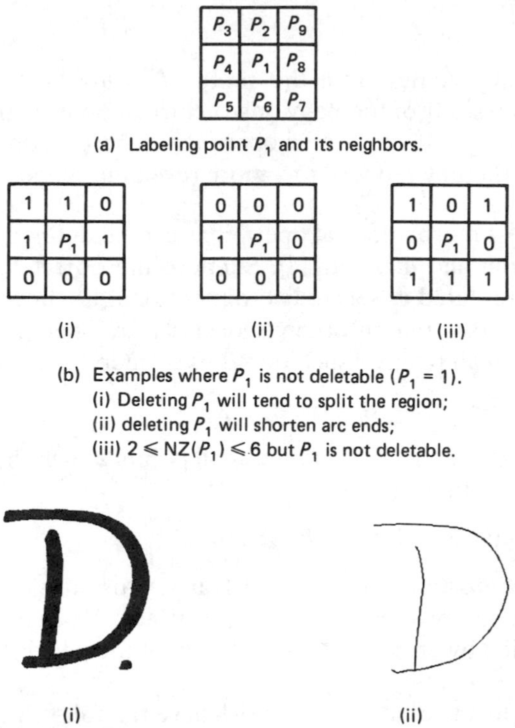

(a) Labeling point P_1 and its neighbors.

(b) Examples where P_1 is not deletable ($P_1 = 1$).
(i) Deleting P_1 will tend to split the region;
(ii) deleting P_1 will shorten arc ends;
(iii) $2 \leqslant NZ(P_1) \leqslant 6$ but P_1 is not deletable.

(c) Example of thinning.
(i) Original;
(ii) thinned.

Figure 9.32 A thinning algorithm.

tained is not influenced by small contour inflections that may be present on the initial contour. The basic approach [42] is to *delete* from the object **X** simple border points that have more than one neighbor in **X** and whose deletion does not locally disconnect **X**. Here a *connected region* is defined as one in which any two points in the region can be connected by a curve that lies entirely in the region. In this way, endpoints of thin arcs are not deleted. A simple algorithm that yields connected arcs while being insensitive to contour noise is as follows [43].

Referring to Figure 9.32a, let $Z0(P_1)$ be the number of zero to nonzero transitions in the ordered set $P_2, P_3, P_4, \ldots, P_9, P_2$. Let $NZ(P_1)$ be the number of nonzero neighbors of P_1. Then P_1 is deleted if (Fig. 9.32b)

$$\left.\begin{array}{ll} & 2 \le NZ(P_1) \le 6 \\ \text{and} & Z0(P_1) = 1 \\ \text{and} & P_2 \cdot P_4 \cdot P_8 = 0 \quad \text{or} \quad Z0(P_2) \ne 1 \\ \text{and} & P_2 \cdot P_4 \cdot P_6 = 0 \quad \text{or} \quad Z0(P_4) \ne 1 \end{array}\right\} \qquad (9.86)$$

The procedure is repeated until no further changes occur in the image. Figure 9.32c gives an example of applying this algorithm. Note that at each location such as P_1 we end up examining pixels from a 5×5 neighborhood.

Morphological Processing

The term *morphology* originally comes from the *study of forms* of plants and animals. In our context we mean study of topology or structure of objects from their images. Morphological processing refers to certain operations where an object is *hit* with a *structuring element* and thereby reduced to a more revealing shape.

Basic operations. Most morphological operations can be defined in terms of two basic operations, *erosion* and *dilation* [44]. Suppose the object **X** and the structuring element **B** are represented as sets in two-dimensional Euclidean space. Let $\mathbf{B}_x$ denote the translation of **B** so that its origin is located at x. Then the erosion of **X** by **B** is defined as the set of all points x such that $\mathbf{B}_x$ is included in **X**, that is,

$$\text{Erosion:} \quad \mathbf{X} \ominus \mathbf{B} \triangleq \{x : \mathbf{B}_x \subset \mathbf{X}\} \tag{9.87}$$

Similarly, the dilation of **X** by **B** is defined as the set of all points x such that $\mathbf{B}_x$ hits **X**, that is, they have a nonempty intersection:

$$\text{Dilation:} \quad \mathbf{X} \oplus \mathbf{B} \triangleq \{x : \mathbf{B}_x \cap \mathbf{X} \neq \phi\} \tag{9.88}$$

Figure 9.33 shows examples of erosion and dilation. Clearly, erosion is a shrinking operation, whereas dilation is an expansion operation. It is also obvious that erosion of an object is accompanied by enlargement or dilation of the background.

Properties. The erosion and dilation operations have the following properties:

1. They are translation invariant, that is, a translation of the object causes the same shift in the result.
2. They are not inverses of each other.
3. Distributivity:

$$\begin{aligned} \mathbf{X} \oplus (\mathbf{B} \cup \mathbf{B}') &= (\mathbf{X} \oplus \mathbf{B}) \cup (\mathbf{X} \oplus \mathbf{B}') \\ \mathbf{X} \ominus (\mathbf{B} \cup \mathbf{B}') &= (\mathbf{X} \ominus \mathbf{B}) \cap (\mathbf{X} \ominus \mathbf{B}') \end{aligned} \tag{9.89}$$

4. Local knowledge:

$$(\mathbf{X} \cap \mathbf{Z}) \ominus \mathbf{B} = (\mathbf{X} \ominus \mathbf{B}) \cap (\mathbf{Z} \ominus \mathbf{B}) \tag{9.90}$$

5. Iteration:

$$\left.\begin{aligned} (\mathbf{X} \ominus \mathbf{B}) \ominus \mathbf{B}' &= \mathbf{X} \ominus (\mathbf{B} \oplus \mathbf{B}') \\ (\mathbf{X} \oplus \mathbf{B}) \oplus \mathbf{B}' &= \mathbf{X} \oplus (\mathbf{B} \oplus \mathbf{B}') \end{aligned}\right\} \tag{9.91}$$

6. Increasing:

$$\text{If } \mathbf{X} \subset \mathbf{X}' \Rightarrow \left.\begin{aligned} &\mathbf{X} \ominus \mathbf{B} \subset \mathbf{X}' \ominus \mathbf{B} \quad \forall \mathbf{B} \\ &\mathbf{X} \oplus \mathbf{B} \subset \mathbf{X}' \oplus \mathbf{B} \quad \forall \mathbf{B} \end{aligned}\right\} \tag{9.92a}$$

$$\text{If } \mathbf{B} \subset \mathbf{B}' \Rightarrow \mathbf{X} \ominus \mathbf{B} \subset \mathbf{X} \ominus \mathbf{B}' \quad \forall \mathbf{X} \tag{9.92b}$$

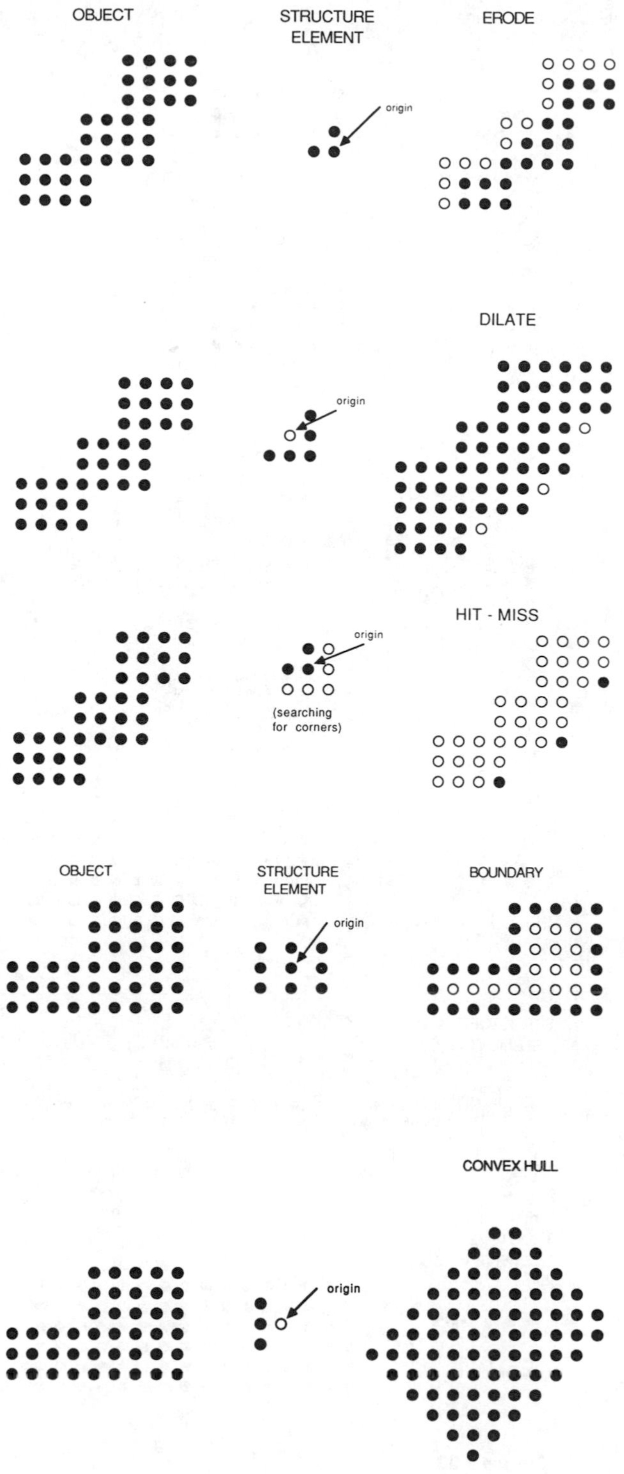

Figure 9.33 Examples of some morphological operations.

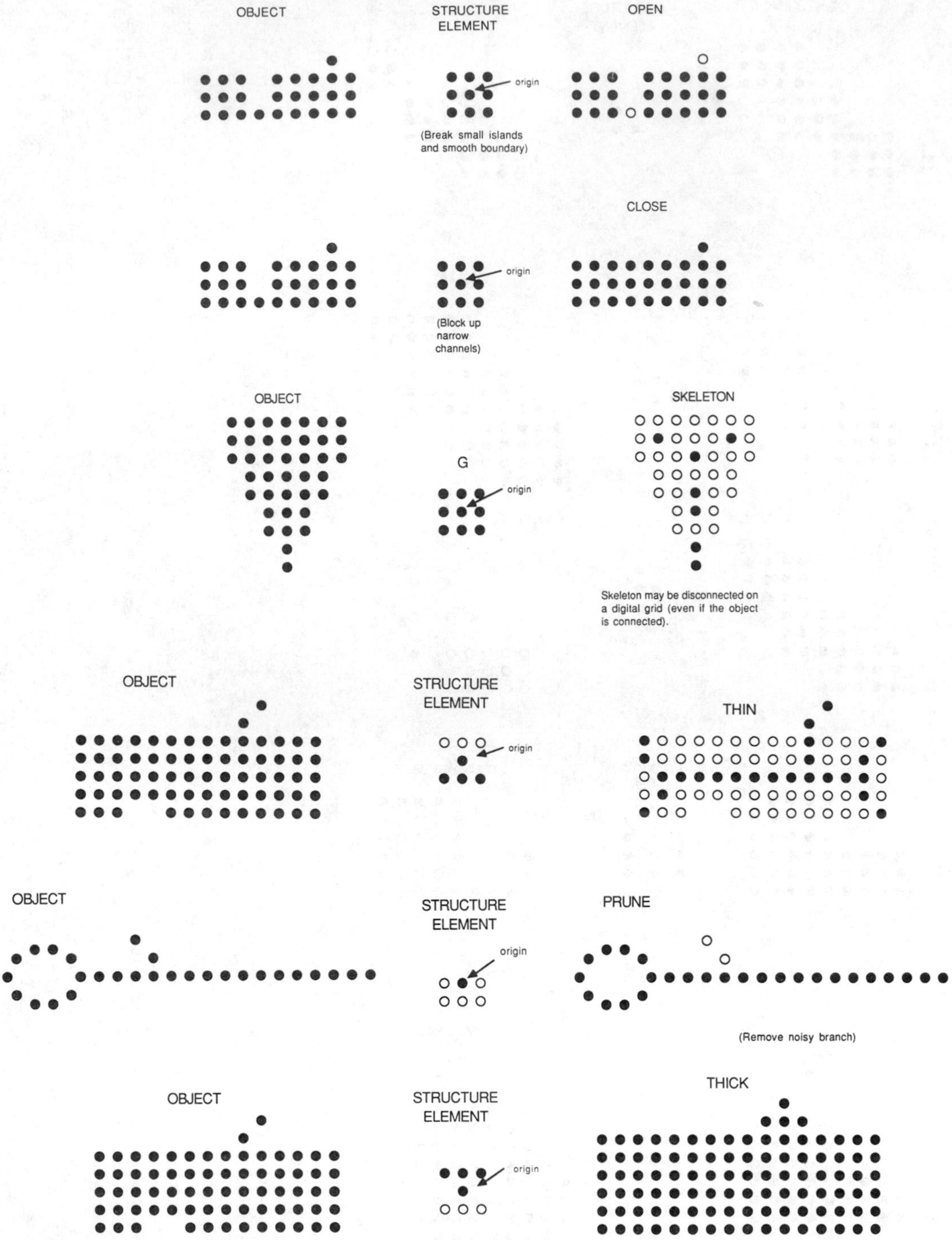

Figure 9.33 Cont'd.

7. Duality: Let $\mathbf{X}^c$ denote the complement of $\mathbf{X}$. Then

$$\mathbf{X}^c \oplus \mathbf{B} = (\mathbf{X} \ominus \mathbf{B})^c \tag{9.93}$$

This means erosion and dilation are duals with respect to the complement operation.

Morphological Transforms

The medial axes transforms and thinning operations are just two examples of morphological transforms. Table 9.10 lists several useful morphological transforms that are derived from the basic erosion and dilation operations. The *hit-miss* transform tests whether or not the structure $\mathbf{B}_{ob}$ belongs to $\mathbf{X}$ and $\mathbf{B}_{bk}$ belongs to $\mathbf{X}^c$. The *opening* of $\mathbf{X}$ with respect to $\mathbf{B}$, denoted by $\mathbf{X}_B$, defines the domain swept by all translates of $\mathbf{B}$ that are included in $\mathbf{X}$. *Closing* is the dual of opening. *Boundary* gives the boundary pixels of the object, but they are not ordered along its contour. This table also shows how the morphological operations could be used to obtain the previously defined *skeletonizing* and *thinning* transformations. *Thickening* is the dual of thinning. *Pruning* operation smooths skeletons or thinned objects by removing parasitic branches.

Figure 9.33 shows examples of morphological transforms. Figure 9.34 shows an application of morphological processing in a printed circuit board inspection application. The observed image is binarized by thresholding and is reduced to a single-pixel-wide contour image by the thinning transform. The result is pruned to obtain clean line segments, which can be used for inspection of faults such as cuts (open circuits), short circuits, and the like.

We now give the development of skeleton and thinning algorithms in the context of the basic morphological operations.

Skeletons. Let rD_x denote a disc of radius r at point x. Let $s_r(x)$ denote the set of centers of maximal discs rD_x that are contained in $\mathbf{X}$ and intersect the boundary of $\mathbf{X}$ at two or more locations. Then the skeleton $S(\mathbf{X})$ is the set of centers $s_r(x)$.

$$\begin{aligned} S(\mathbf{X}) &= \bigcup_{r>0} s_r(x) \\ &= \bigcup_{r>0} \{(\mathbf{X} \ominus rD)/(\mathbf{X} \ominus rD)_{drD}\} \end{aligned} \tag{9.94}$$

where $\cup$ and / represent the set union and set difference operations, respectively, and drD denotes opening with respect to an infinitesimal disc.

To recover the original object from its skeleton, we take the union of the circular neighborhoods centered on the skeleton points and having radii equal to the associated contour distance.

$$\mathbf{X} = \bigcup_{r>0} \{s_r(x) \oplus rD\} \tag{9.95}$$

We can find the skeleton on a digitized grid by replacing the disc rD in (9.94) by the 3×3 square grid $\mathbf{G}$ and obtain the algorithm summarized in Table 9.10. Here the operation $(\mathbf{X} \ominus n\mathbf{G})$ denotes the nth iteration $(\mathbf{X} \ominus \mathbf{G}) \ominus \mathbf{G} \ominus \cdots \ominus \mathbf{G}$ and $(\mathbf{X} \ominus n\mathbf{G})_{\mathbf{G}}$ is the opening of $(\mathbf{X} \ominus n\mathbf{G})$ with respect to $\mathbf{G}$.

TABLE 9.10 Some Useful Morphological Transforms

Operation	Definition	Properties & Usage
Hit-Miss	$\mathbf{X} \circledast \mathbf{B} = (\mathbf{X} \ominus \mathbf{B}_{ob})/(\mathbf{X} \oplus \mathbf{B}_{bk})$	Searching for a match or a specific configuration. $\mathbf{B}_{ob}$: set formed from pixels in **B** that should belong to the object. $\mathbf{B}_{bk}$: . . . background.
Open	$\mathbf{X}_B = (\mathbf{X} \ominus \mathbf{B}) \oplus \mathbf{B}$	Smooths contours, suppress small islands and sharp caps of **X**. Ideal for object size distribution study.
Close	$\mathbf{X}^B = (\mathbf{X} \oplus \mathbf{B}) \ominus \mathbf{B}$	Blocks up narrow channels and thin lakes. Ideal for the study of inter object distance.
Boundary	$\partial \mathbf{X} = \mathbf{X}/\mathbf{X} \ominus \mathbf{G}$	Gives the set of boundary points.
Convex Hull	$\mathbf{X}_1^1 = \mathbf{X}$ $\mathbf{X}_{i+1}^1 = (\mathbf{X}_i^1 \circledast \mathbf{B}^1)$ $\mathbf{X}_{CH} = \bigcup_{j=1}^{4} \mathbf{X}_\infty^j$	$\mathbf{B}^1, \mathbf{B}^2, \ldots$ are rotated versions of the structuring element **B**. **C** is an appropriate structuring element choice for **B**.
Skeleton	$S(\mathbf{X}) = \bigcup_{n=0}^{n_{\max}} s_n(x)$ $= \bigcup_{n=0}^{n_{\max}} [(\mathbf{X} \ominus n\mathbf{G})/(\mathbf{X} \ominus n\mathbf{G})_G]$ $\mathbf{X} = \bigcup_{n=0}^{n_{\max}} [s_n(x) \oplus n\mathbf{G}]$	$n_{\max}$: max size after which **X** erodes down to an empty set. The skeleton is a regenerative representation of the object.
Thin	$\mathbf{X} \bigcirc \mathbf{B} = \mathbf{X}/\mathbf{X} \circledast \mathbf{B}$ $\mathbf{X} \bigcirc \{\mathbf{B}\} = ((\ldots((\mathbf{X} \bigcirc \mathbf{B}^1) \bigcirc \mathbf{B}^2) \ldots) \bigcirc \mathbf{B}^n)$	To symmetrically thin **X** a sequence of structuring elements, $\{\mathbf{B}\} = \{\mathbf{B}^i, 1 \le i \le n\}$, is used in cascade, where $\mathbf{B}^i$ is a rotated version of $\mathbf{B}^{i-1}$. A widely used element is **L**.
Thick	$\mathbf{X} \odot \mathbf{B} = \mathbf{X} \cup \mathbf{X} \circledast \mathbf{B}$	Dual of thinning.
Prune	$\mathbf{X}_1 = \mathbf{X} \bigcirc \{\mathbf{B}\}$ $\mathbf{X}_2 = \bigcup_{j=1}^{8} (\mathbf{X}_1 \circledast \mathbf{P}^j)$ $\mathbf{X}_{pn} = \mathbf{X}_1 \cup [(\mathbf{X}_2 \oplus \{\mathbf{G}\}) \cap \mathbf{X}]$	**E** is a suitable structuring element. $\mathbf{X}_2$: end points $\mathbf{X}_{pn}$: pruned object with Parasite branches suppressed.

The symbols "/" and " ∪ " represent the set difference and the set union operations respectively. Examples of structuring elements are

$$\mathbf{G} = \begin{bmatrix} 1 & 1 & 1 \\ 1 & 1 & 1 \\ 1 & 1 & 1 \end{bmatrix}, \quad \mathbf{C} = \begin{matrix} 1 & d & d \\ 1 & 0 & d \\ 1 & d & d \end{matrix}, \quad \mathbf{L} = \begin{matrix} 0 & 0 & 0 \\ d & 1 & d \\ 1 & 1 & 1 \end{matrix}, \quad \mathbf{E} = \begin{bmatrix} d & d & d \\ 0 & 1 & 0 \\ 0 & 0 & 0 \end{bmatrix}$$

where 1, 0, and d signify the object, background and 'don't care' states, respectively.

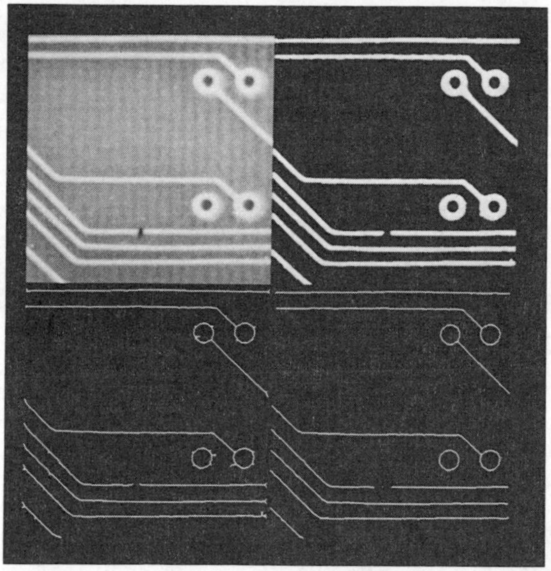

Figure 9.34 Morphological processing for printed circuit board inspection. (a) Original; (b) preprocessed (thresholded); (c) thinned; (d) pruned.

Thinning. In the context of morphological operations, thinning can be defined as

$$\mathbf{X} \bigcirc \mathbf{B} = \mathbf{X}/(\mathbf{X} \circledast \mathbf{B}) \tag{9.96}$$

where **B** is the structuring element chosen for the thinning and $\circledast$ denotes the hit-miss operation defined in Table 9.10.

To thin **X** symmetrically, a sequence of structuring elements, $\{\mathbf{B}\} \triangleq \{\mathbf{B}^i, 1 \le i \le n\}$, is used in cascade, where $\mathbf{B}^i$ is a rotated version of $\mathbf{B}^{i-1}$.

$$\mathbf{X} \bigcirc \{\mathbf{B}\} = ((\ldots((\mathbf{X} \bigcirc \mathbf{B}^1) \bigcirc \mathbf{B}^2)\ldots) \bigcirc \mathbf{B}^n) \tag{9.97}$$

A suitable structuring element for the thinning operation is the **L**-structuring element shown in Table 9.10.

The thinning process is usually followed by a *pruning* operation to trim the resulting arcs (Table 9.10). In general, the original objects are likely to have noisy boundaries, which result in unwanted parasitic branches in the thinned version. It is the job of this step to clean up these without disconnecting the arcs.

Syntactic Representation [45]

The foregoing techniques reduce an object to a set of structural elements, or *primitives.* By adding a syntax, such as connectivity rules, it is possible to obtain a *syntactic representation,* which is simply a string of symbols, each representing a primitive (Figure 9.35). The syntax allows a unique representation and interpretation of the string. The design of a syntax that transforms the symbolic and the syntactic representations back and forth is a difficult task. It requires specification of a complete and unambiguous set of rules, which have to be derived from the *understanding* of the scene under study.

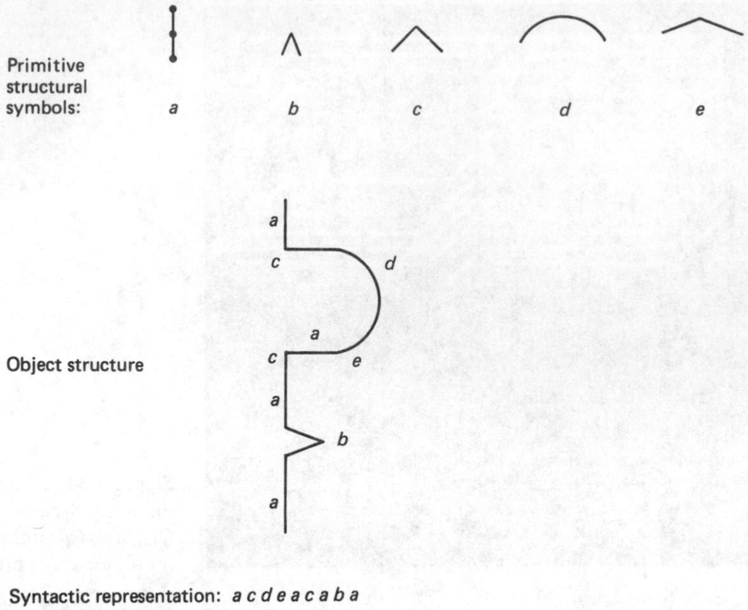

Figure 9.35 Syntactic representation.

9.10 SHAPE FEATURES

The shape of an object refers to its profile and physical structure. These characteristics can be represented by the previously discussed boundary, region, moment, and structural representations. These representations can be used for matching shapes, recognizing objects, or for making measurements of shapes. Figure 9.36 lists several useful features of shape.

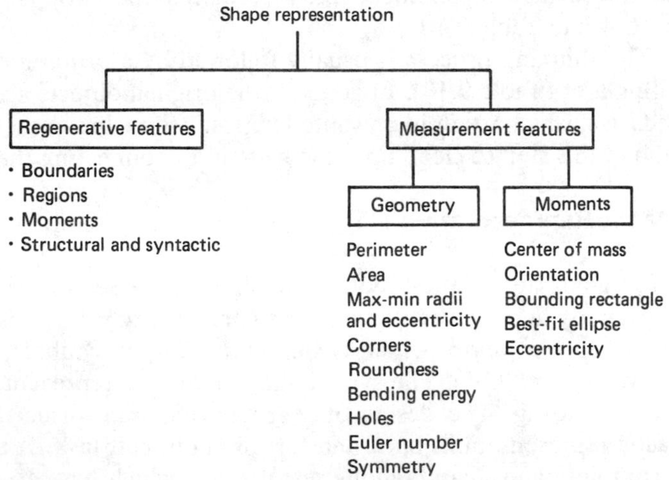

Figure 9.36

Geometry Features

In many image analysis problems the ultimate aim is to measure certain geometric attributes of the object, such as the following:

1. *Perimeter*

$$T = \int \sqrt{\dot{x}^2(t) + \dot{y}^2(t)}\, dt \qquad (9.98)$$

where t is necessarily the boundary parameter but not necessarily its length.

2. *Area*

$$A = \iint_{\mathcal{R}} dx\, dy = \int_{\partial\mathcal{R}} y(t)\frac{dx(t)}{dt}\, dt - \int_{\partial\mathcal{R}} x(t)\frac{dy(t)}{dt}\, dt \qquad (9.99)$$

where $\mathcal{R}$ and $\partial\mathcal{R}$ denote the object region and its boundary, respectively.

3. *Radii* $R_{\min}$, $R_{\max}$ are the minimum and maximum distances, respectively, to boundary from the center of mass (Fig. 9.37a). Sometimes the ratio $R_{\max}/R_{\min}$ is used as a measure of eccentricity or elongation of the object.

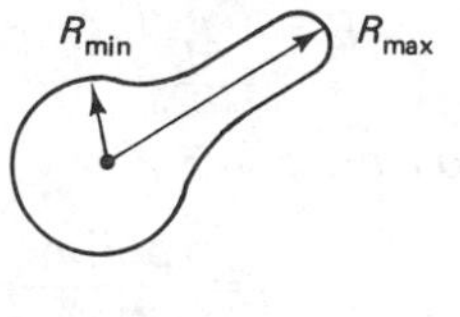

(a) Maximum and minimum radii

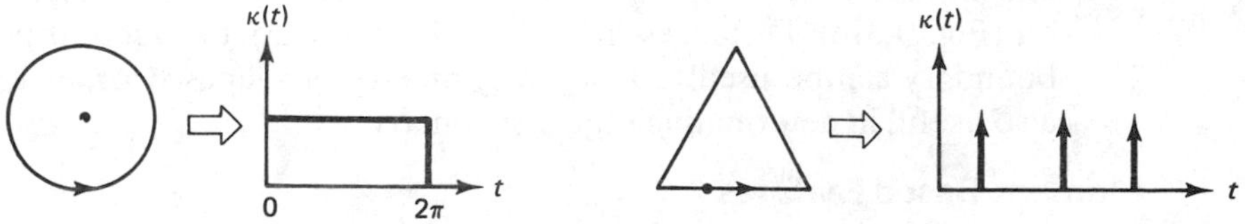

(b) Curvature functions for corner detection

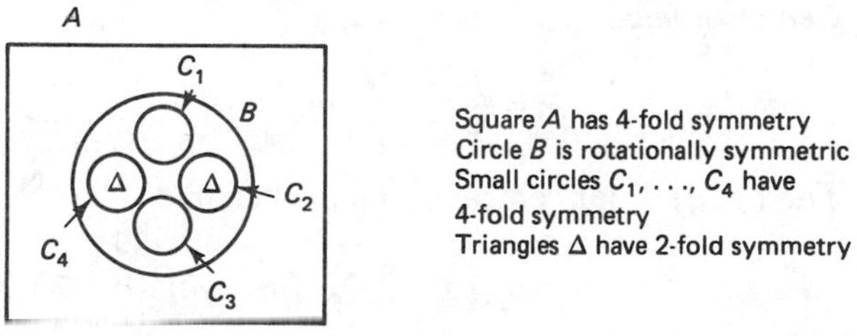

(c) Types of symmetry

Figure 9.37 Geometry features.

4. *Number of holes* n_h
5. *Euler number*

$$\mathcal{E} \triangleq \text{number of connected regions} - n_h \tag{9.100}$$

6. *Corners* These are locations on the boundary where the curvature $\kappa(t)$ becomes unbounded. When t represents distance along the boundary, then from (9.57) and (9.58), we can obtain

$$|\kappa(t)|^2 \triangleq \left(\frac{d^2 y}{dt^2}\right)^2 + \left(\frac{d^2 x}{dt^2}\right)^2 \tag{9.101}$$

In practice, a corner is declared whenever $|\kappa(t)|$ assumes a large value (Fig. 9.37b).

7. *Bending energy* This is another attribute associated with the curvature.

$$E = \frac{1}{T}\int_0^T |\kappa(t)|^2\, dt \tag{9.102}$$

In terms of $\{a(k)\}$, the FDs of $u(t)$, this is given by

$$E = \sum_{k=-\infty}^{\infty} |a(k)|^2 \left(\frac{2\pi k}{T}\right)^4 \tag{9.103}$$

8. *Roundness, or compactness*

$$\gamma = \frac{(\text{perimeter})^2}{4\pi(\text{area})} \tag{9.104}$$

For a disc, γ is minimum and equals 1.

9. *Symmetry* There are two common types of symmetry of shapes, *rotational* and *mirror*. Other forms of symmetry are twofold, fourfold, eightfold, and so on (Fig. 9.37c). Distances from the center of mass to different points on the boundary can be used to analyze symmetry of shapes. Corner locations are also useful in determining object symmetry.

Moment-Based Features

Many shape features can be conveniently represented in terms of moments. For a shape represented by a region $\mathcal{R}$ containing N pixels, we have the following:

1. *Center of mass*

$$\bar{m} = \frac{1}{N} \sum\sum_{(m,n)\in\mathcal{R}} m, \qquad \bar{n} = \frac{1}{N} \sum\sum_{(m,n)\in\mathcal{R}} n \tag{9.105}$$

The (p, q) order central moments become

$$\mu_{p,q} = \sum\sum_{(m,n)\in\mathcal{R}} (m - \bar{m})^p (n - \bar{n})^q \tag{9.106}$$

2. *Orientation* Orientation is defined as the angle of axis of the least moment of inertia. It is obtained by minimizing with respect to θ (Fig. 9.38a) the sum

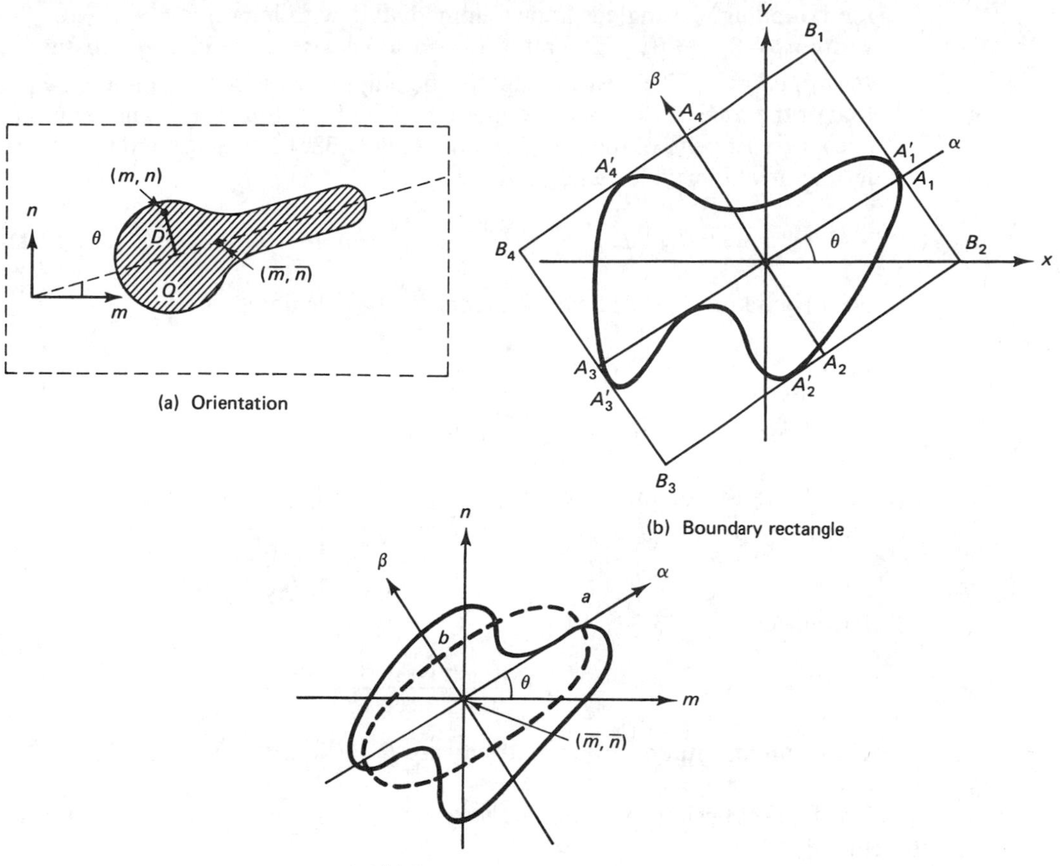

(a) Orientation

(b) Boundary rectangle

(c) Best fit ellipse

Figure 9.38 Moment-based features.

$$I(\theta) = \sum_{(m,n) \in \mathcal{R}}\sum D^2(m, n) = \sum_{(m,n) \in \mathcal{R}}\sum [(n - \bar{n}) \cos\theta - (m - \bar{m}) \sin\theta]^2 \tag{9.107}$$

The result is

$$\theta = \tfrac{1}{2} \tan^{-1}\left[\frac{2\mu_{1,1}}{\mu_{2,0} - \mu_{0,2}}\right] \tag{9.108}$$

3. *Bounding rectangle* The bounding rectangle is the smallest rectangle enclosing the object that is also aligned with its orientation (Fig. 9.38b). Once θ is known we use the transformation

$$\begin{aligned} \alpha &= x \cos\theta + y \sin\theta \\ \beta &= -x \sin\theta + y \cos\theta \end{aligned} \tag{9.109}$$

on the boundary points and search for α_{min}, α_{max}, β_{min}, and β_{max}. These give the locations of points A_3', A_1', A_2', and A_4', respectively, in Fig. 9.38b. From these the bounding rectangle is known immediately with length $l_b = \alpha_{max} - \alpha_{min}$ and width $w_b = \beta_{max} - \beta_{min}$. The ratio $(l_b w_b/\text{area})$ is also a useful shape feature.

4. *Best-fit ellipse* The best-fit ellipse is the ellipse whose second moment equals that of the object. Let a and b denote the lengths of semimajor and semiminor axes, respectively, of the best-fit ellipse (Fig. 9.38c). The least and the greatest moments of inertia for an ellipse are

$$I_{min} = \frac{\pi}{4} ab^3, \qquad I_{max} = \frac{\pi}{4} a^3 b \tag{9.110}$$

For orientation θ, the above moments can be calculated as

$$I'_{min} = \sum\sum_{(m,n)\in\mathcal{R}} [(n - \bar{n})\cos\theta - (m - \bar{m})\sin\theta]^2$$
$$I'_{max} = \sum\sum_{(m,n)\in\mathcal{R}} [(n - \bar{n})\sin\theta + (m - \bar{m})\cos\theta]^2 \tag{9.111}$$

For the best-fit ellipse we want $I_{min} = I'_{min}$, $I_{max} = I'_{max}$, which gives

$$a = \left(\frac{4}{\pi}\right)^{1/4} \left[\frac{(I'_{max})^3}{I'_{min}}\right]^{1/8}, \qquad b = \left(\frac{4}{\pi}\right)^{1/4} \left[\frac{(I'_{min})^3}{I'_{max}}\right]^{1/8} \tag{9.112}$$

5. *Eccentricity*

$$\varepsilon \triangleq \frac{(\mu_{2,0} - \mu_{0,2})^2 + 4\mu_{1,1}}{\text{area}}$$

Other representations of eccentricity are R_{max}/R_{min}, I'_{max}/I'_{min}, and a/b.

The foregoing shape features are very useful in the design of vision systems for object recognition.

9.11 TEXTURE

Texture is observed in the structural patterns of surfaces of objects such as wood, grain, sand, grass, and cloth. Figure 9.39 shows some examples of textures [46]. The term texture generally refers to repetition of basic texture elements called *texels.* A texel contains several pixels, whose placement could be periodic, quasi-periodic or random. Natural textures are generally random, whereas artificial textures are often deterministic or periodic. Texture may be coarse, fine, smooth, granulated, rippled, regular, irregular, or linear. In image analysis, texture is broadly classified into two main categories, statistical and structural [47].

Statistical Approaches

Textures that are random in nature are well suited for statistical characterization, for example, as realizations of random fields. Figure 9.40 lists several statistical measures of texture. We discuss these briefly next.

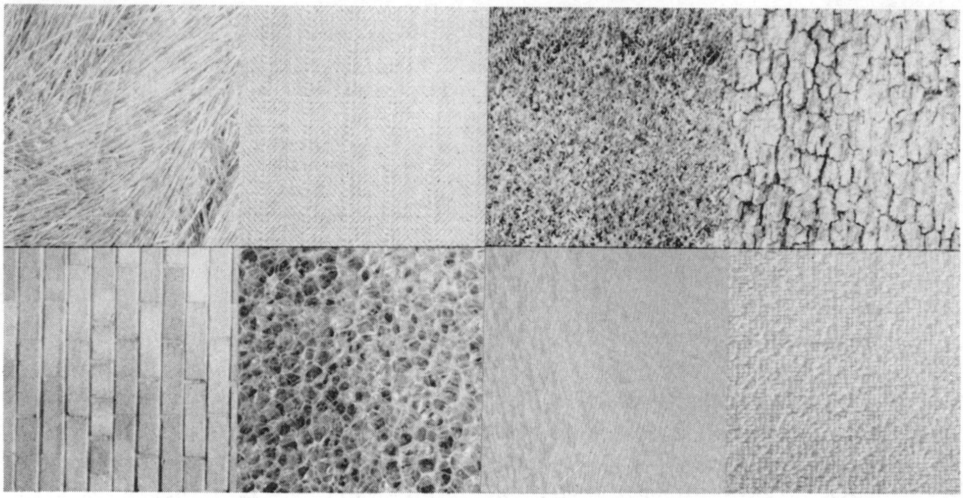

Figure 9.39 Brodatz textures.

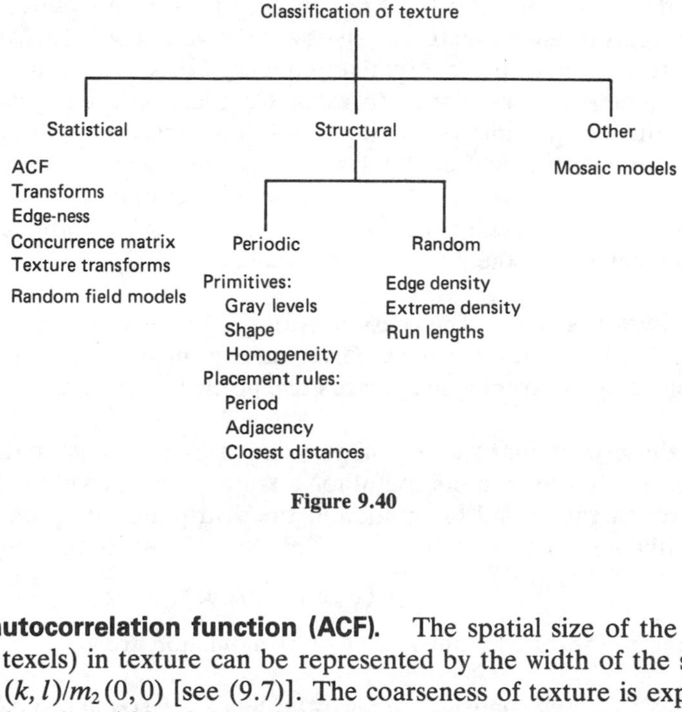

Figure 9.40

The autocorrelation function (ACF). The spatial size of the tonal primitives (i.e., texels) in texture can be represented by the width of the spatial ACF $r(k, l) = m_2(k, l)/m_2(0, 0)$ [see (9.7)]. The coarseness of texture is expected to be proportional to the width of the ACF which can be represented by distances x_0, y_0 such that $r(x_0, 0) = r(0, y_0) = \frac{1}{2}$. Other measures of spread of the ACF are obtained via the moment-generating function

$$M(k, l) \triangleq \sum_m \sum_n (m - \mu_1)^k (n - \mu_2)^l r(m, n) \tag{9.113}$$

where

$$\mu_1 \triangleq \sum_m \sum_n m r(m, n), \qquad \mu_2 \triangleq \sum_m \sum_n n r(m, n)$$

Features of special interest are the *profile spreads* $M(2,0)$ and $M(0,2)$, the *cross-relation* $M(1,1)$, and the *second-degree spread* $M(2,2)$. The calibration of the ACF spread on a fine-coarse texture scale depends on the resolution of the image. This is because a seemingly flat region (no texture) at a given resolution could appear as fine texture at higher resolution and coarse texture at lower resolution. The ACF by itself is not sufficient to distinguish among several texture fields because many different image ensembles can have the same ACF.

Image transforms. Texture features such as coarseness, fineness, or orientation can be estimated by generalized linear filtering techniques utilizing image transforms (Fig. 9.4). A two-dimensional transform $v(k, l)$ of the input image is passed through several band-pass filters or masks $g_i(k, l), i = 1, 2, 3, \ldots,$ as

$$z_i(k, l) = v(k, l) g_i(k, l) \tag{9.114}$$

Then the energy in this $z_i(k, l)$ represents a transform feature. Different types of masks appropriate for texture analysis are shown in Fig. 9.5a. With circular slits we measure energy in different spatial frequency or sequency bands. Angular slits are useful in detecting orientation features. Combinations of angular and circular slits are useful for periodic or quasi-periodic textures. Image transforms have been applied for discrimination of terrain types—for example, deserts, farms, mountains, riverbeds, urban areas, and clouds [48]. Fourier spectral analysis has been found useful in detection and classification of black lung disease by comparing the textural patterns of the diseased and normal areas [49].

Edge density. The coarseness of random texture can also be represented by the density of the edge pixels. Given an edge map [see (9.17)] the edge density is measured by the average number of edge pixels per unit area.

Histogram features. The two-dimensional histogram discussed in Section 9.2 has proven to be quite useful for texture analysis. For two pixels u_1 and u_2 at relative distance r and orientation θ, the distribution function [see (9.9)] can be explicitly written as

$$p_{\mathbf{u}}(x_1, x_2) = f(r, \theta; x_1, x_2) \tag{9.115}$$

Some useful texture features based on this function are

$$\text{Inertia:} \quad I(r, \theta) \triangleq \sum_{x_1, x_2}\sum |x_1 - x_2|^2 f(r, \theta; x_1, x_2) \tag{9.116}$$

$$\text{Mean† distribution:} \quad \mu(r; x_1, x_2) = \frac{1}{N_0} \sum_\theta f(r, \theta; x_1, x_2) \tag{9.117}$$

† The symbol N_0 represents the total number of orientations.

$$\text{Variance distribution:} \quad \sigma^2(r;x_1,x_2) = \frac{1}{N_0}\sum_{\theta}[f(r,\theta;x_1,x_2) - \mu(r;x_1,x_2)]^2 \tag{9.118}$$

$$\text{Spread distribution:} \quad \eta(r;x_1,x_2) = \max_{\theta}\{f(r,\theta;x_1,x_2)\} - \min_{\theta}\{f(r,\theta;x_1,x_2)\} \tag{9.119}$$

The inertia is useful in representing the spread of the function $f(r, \theta; x_1, x_2)$ for a given set of (r, θ) values. $I(r, \theta)$ becomes proportional to the coarseness of the texture at different distances and orientations. The mean distribution $\mu(r; x_1, x_2)$ is useful when the angular variations in textural properties are unimportant. The variance $\sigma^2(r; x_1, x_2)$ indicates the angular fluctuations of textural properties. The function $\eta(r; x_1, x_2)$ gives a measure of orientation-independent spread.

Random texture models. It has been suggested that visual perception of random texture fields may be unique only up to second-order densities [50]. It was observed that two textured fields with the same second-order probability distributions appeared to be indistinguishable. Although not always true, this conjecture has proven useful for synthesis and analysis of many types of textures. Thus two different textures can often be discriminated by comparing their second-order histograms.

A simple model for texture analysis is shown in Fig. 9.41a [51]. The texture field is first decorrelated by a filter $a(m, n)$, which can be designed from the knowledge of the ACF. Thus if $r(m, n)$ is the ACF, then

$$u(m, n) \circledast a(m, n) \triangleq \varepsilon(m, n) \tag{9.120}$$

is an uncorrelated random field. From Chapter 6 (see Section 6.6) this means that any WNDR of $u(m, n)$ would give an admissible whitening (or decorrelating) filter.

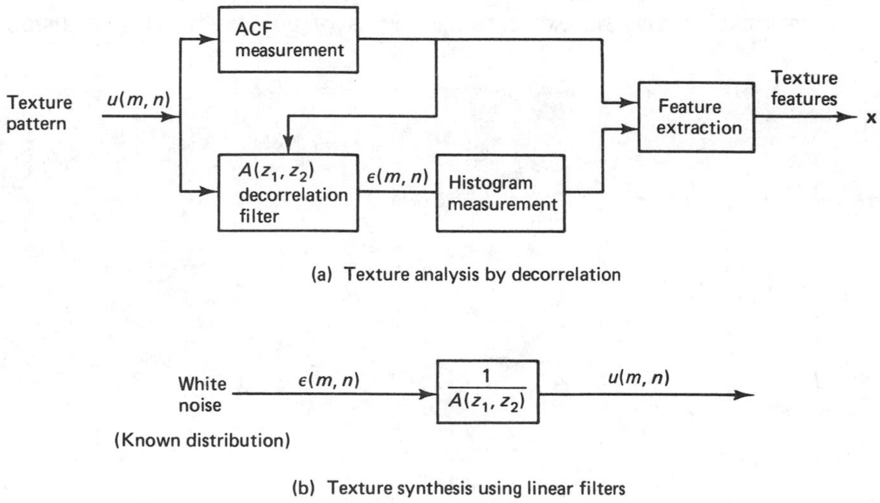

Figure 9.41 Random texture models.

Such a filter is not unique, and it could have causal, semicausal, or noncausal structure. Since the edge extraction operators have a tendency to decorrelate images, these have been used [51] as alternatives to the true whitening filters. The ACF features such as $M(0,2)$, $M(2,0)$, $M(1,1)$, and $M(2,2)$ [see (9.113)] and the features of the first-order histogram of $\varepsilon(m,n)$, such as average m_1, deviation $\sqrt{\mu_2}$, skewness μ_3, and kurtosis $\mu_4 - 3$, have been used as the elements of the texture feature vector $\mathbf{x}$ in Fig. 9.41a.

Random field representations of texture have been considered using one-dimensional time series as well as two-dimensional random field models (see [52], [53] and bibliography of Chapter 6). Following Chapter 6, such models can be identified from the given data. The model coefficients are then used as features for texture discrimination. Moreover these random field models can synthesize random texture fields when driven by the uncorrelated random field $\varepsilon(m,n)$ of known probability density (Fig. 9.41b).

Example 9.6 Texture synthesis via causal and semicausal models

Figure 9.42a shows a given 256×256 grass texture. Using estimated covariances, a $(p,q) = (3,4)$-order white Gaussian noise–driven causal model was designed and used to synthesize the texture of Fig. 9.42b. Figure 9.42c shows the texture synthesized via a $(p,q) = (3,4)$ semicausal white noise–driven model. This model was designed via the Wiener-Doob homomorphic factorization method of Section 6.8.

Structural Approaches [4, 47, 54]

Purely structural textures are deterministic texels, which repeat according to some placement rules, deterministic or random. A texel is isolated by identifying a group of pixels having certain invariant properties, which repeat in the given image. The texel may be defined by its gray level, shape, or homogeneity of some local property, such as size, orientation, or second-order histogram (concurrence matrix). The placement rules define the spatial relationships between the texels. These spatial relationships may be expressed in terms of adjacency, closest distance, period-

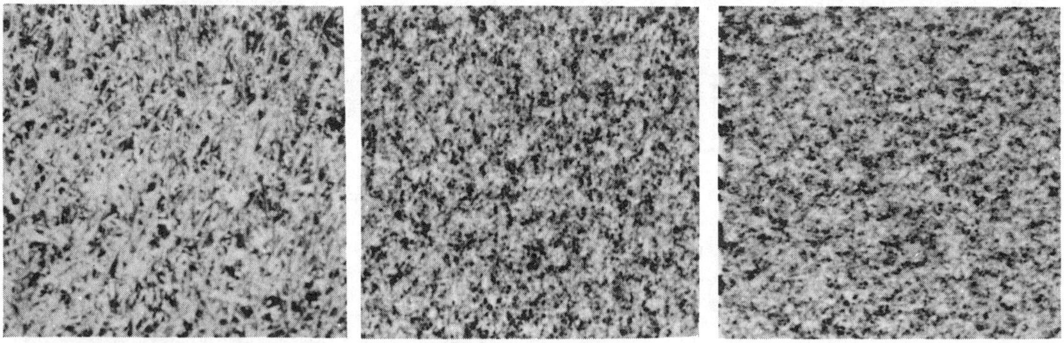

(a) Original gross texture (b) Texture synthesized by causal model (c) Texture synthesized by semicausal model

Figure 9.42 Texture synthesis using causal and semicausal models.

icities, and so on, in the case of deterministic placement rules. In such cases the texture is labeled as being *strong*.

For randomly placed texels, the associated texture is called *weak* and the placement rules may be expressed in terms of measures such as the following:

1. Edge density
2. Run lengths of maximally connected texels
3. Relative extrema density, which is the number of pixels per unit area showing gray levels that are locally maxima or minima relative to their neighbors. For example, a pixel $u(m, n)$ is a relative minimum or a relative maximum if it is, respectively, less than or greater than its nearest four neighbors. (In a region of constant gray levels, which may be a plateau or a valley, each pixel counts as an extremum.)

   This definition does not distinguish between images having a few large plateaus and those having many single extrema. An alternative is to count each plateau as one extremum. The height and the area of each extremum may also be considered as features describing the texels.

Example 9.7 Synthesis for quasiperiodic textures

The raffia texture (Fig. 9.43a) can be viewed as a quasiperiodic repetition of a deterministic pattern. The spatial covariance function of a small portion of the image was analyzed to estimate the periodicity and the randomness in the periodic rate. A 17×17 primitive was extracted from the parent texture and repeated according to the quasiperiodic placement rule to give the image of Fig. 9.43b.

Other Approaches

A method that combines the statistical and the structural approaches is based on what have been called *mosaic models* [55]. These models represent random geometrical processes. For example, regular or random tessellations of a plane into bounded convex polygons give rise to cell-structured textures. A mosaic model

(a) Original 256×256 raffia

(b) Synthesized raffia by quasi-periodic placement of a primitive

Figure 9.43 Texture synthesis by structural approach.

could define rules for partitioning a plane into different cells, where each cell contains a geometrical figure whose features (such as center or orientation) have prescribed probability distributions. For example circles of fixed radius placed according to a Poisson point process defined on a plane would constitute a mosaic texture. In general, mosaic models should provide higher resolution than the random field models. However, quantification of the underlying geometric pattern and identification of the placement process would be more difficult. For textures that exhibit regularities in primitive placements, grammatical models can be developed [3]. Such grammars give a few rules for combining certain primitive shapes or symbols to generate several complex patterns.

9.12 SCENE MATCHING AND DETECTION

A problem of much significance in image analysis is the detection of change or presence of an object in a given scene. Such problems occur in remote sensing for monitoring growth patterns of urban areas, weather prediction from satellite images, diagnosis of disease from medical images, target detection from radar images, and automation using robot vision, and the like. Change detection is also useful in alignment or spatial registration of two scenes imaged at different instants or using different sensors. For example, a large object photographed in small overlapping sections can be reconstructed by matching the overlapping parts.

Image Subtraction

Changes in a dynamic scene observed as $u_i(m, n)$, $i = 1, 2, \ldots$ are given by

$$e_i(m, n) = u_i(m, n) - u_{i-1}(m, n) \tag{9.121}$$

Although elementary, this image-subtraction technique is quite powerful in carefully controlled imaging situations. Figure 9.44 shows an example from digital radiology. The images u_1 and u_2 represent, respectively, the X-ray images before and after injection of a radio-opaque dye in a renal study. The change, not visible in u_2, can be easily detected as renal arteries after u_1 has been subtracted out. Image subtraction is also useful in motion detection based security monitoring systems, segmentation of parts from a complex assembly, and so on.

Template Matching and Area Correlation

The presence of a known object in a scene can be detected by searching for the location of match between the object template $u(m, n)$ and the scene $v(m, n)$. Template matching can be conducted by searching the displacement of $u(m, n)$, where the mismatch energy is minimum. For a displacement (p, q), we define the mismatch energy

$$\begin{aligned}\sigma_\eta^2(p, q) &\triangleq \sum_m \sum_n [v(m, n) - u(m-p, n-q)]^2 \\ &= \sum_m \sum_n |v(m, n)|^2 + \sum_m \sum_n |u(m, n)|^2 - 2\sum_m \sum_n v(m, n)u(m-p, n-q)\end{aligned} \tag{9.122}$$

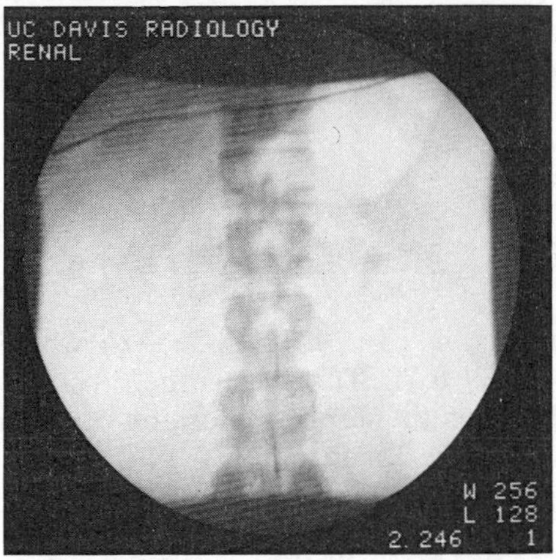

a) Precontrast

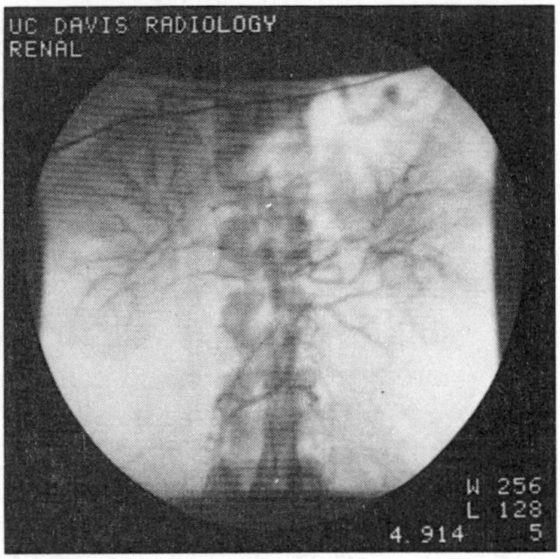

b) Postcontrast

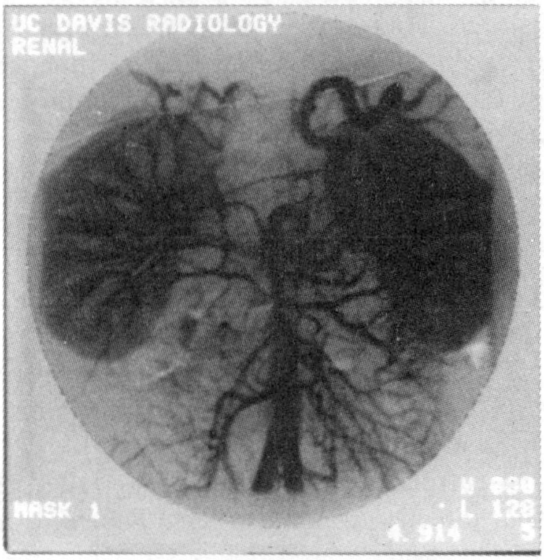

c) Difference

Figure 9.44 Change detection in digital radiography.

For $\sigma^2_\eta(p, q)$ to achieve a minimum, it is sufficient to maximize the cross-correlation

$$c_{vu}(p, q) \triangleq \sum_m \sum_n v(m, n)u(m - p, n - q), \qquad \forall(p, q) \tag{9.123}$$

From the Cauchy-Schwarz inequality, we have

$$\begin{aligned} |c_{vu}| &= \left|\sum_m \sum_n v(m, n)u(m - p, n - q)\right| \\ &\le \left[\sum_m \sum_n |v(m, n)|^2\right]^{1/2} \left[\sum_m \sum_n |u(m, n)|^2\right]^{1/2} \end{aligned} \tag{9.124}$$

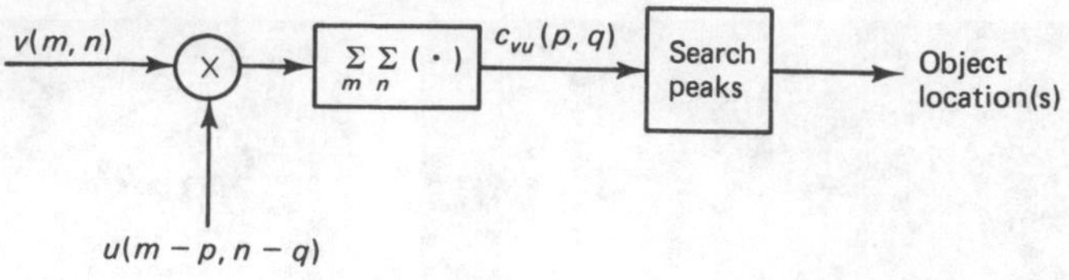

Figure 9.45 Template matching by area correlation.

where the equality occurs if and only if $v(m, n) = \alpha u(m - p, n - q)$, where α is an arbitrary constant and can be set equal to 1. This means the cross-correlation $c_{vu}(p, q)$ attains the maximum value when the displaced position of the template coincides with the observed image. Then, we obtain

$$c_{vu}(p, q) = \sum_m \sum_n |v(m, n)|^2 > 0 \tag{9.125}$$

and the desired maximum occurs when the observed image and the template are spatially registered. Therefore, a given object $u(m, n)$ can be located in the scene by searching the peaks of the cross correlation function (Fig. 9.45). Often the given template and the observed image are not only spatially translated but are also relatively scaled and rotated. For example,

$$v(m, n) = \alpha u\left(\frac{m - p'}{\gamma_1}, \frac{n - q'}{\gamma_2}; \theta\right) \tag{9.126}$$

where γ_1 and γ_2 are the scale factors, (p', q') are the displacement coordinates, and θ is the rotation angle of the observed image with respect to the template. In such cases the cross-correlation function maxima have to be searched in the parameter space $(p', q', \gamma_1, \gamma_2, \theta)$. This can become quite impractical unless reasonable estimates of γ_1, γ_2, and θ are given.

The cross-correlation $c_{vu}(p, q)$ is also called the *area correlation*. It can be evaluated either directly or as the inverse Fourier transform of

$$C_{vu}(\omega_1, \omega_2) \triangleq \mathcal{F}\{c_{vu}(p, q)\} = V(\omega_1, \omega_2)U^*(\omega_1, \omega_2) \tag{9.127}$$

The direct computation of the area correlation is useful when the template is small. Otherwise, a suitable-size FFT is employed to perform the Fourier transform calculations. Template matching is particularly efficient when the data is binary. In that case, it is sufficient to search the minima of the total binary difference

$$\gamma_{vu}(p, q) \triangleq \sum_m \sum_n [v(m, n) \oplus u(m - p, n - q)] \tag{9.128}$$

which requires only the simple logical exclusive-OR operations. The quantity $\gamma_{vu}(p, q)$ gives the number of pixels in the image that do not match with the template at location (p, q). This algorithm is useful in recognition of printed characters or objects characterized by known boundaries as in the inspection of printed circuit boards.

Matched Filtering [56–57]

Suppose a deterministic object $u(m, n)$, displaced by (m_0, n_0), is observed in the presence of a surround (for example, other objects), and observations are a colored noise field $\eta(m, n)$ with power spectral density $S_\eta(\omega_1, \omega_2)$. The observations are

$$\mathrm{v}(m, n) = u(m - m_0, n - n_0) + \eta(m, n) \tag{9.129}$$

The matched filtering problem is to find a linear filter $g(m, n)$ that maximizes the output signal-to-noise ratio (SNR)

$$\begin{aligned} \text{SNR} &\triangleq \frac{|s(0,0)|^2}{\sum_m \sum_n E[|g(m,n) \circledast \eta(m,n)|^2]}, \\ s(m,n) &\triangleq g(m,n) \circledast u(m - m_0, n - n_0) \end{aligned} \tag{9.130}$$

Here $s(m, n)$ represents the signal content in the filtered output $g(m, n) \circledast v(m, n)$. Following Problem 9.16, the matched filter frequency response is found to be

$$G(\omega_1, \omega_2) = \frac{U^*(\omega_1, \omega_2)}{S_\eta(\omega_1, \omega_2)} \exp[-j(\omega_1 m_0 + \omega_2 n_0)] \tag{9.131}$$

which gives its impulse response as

$$g(m, n) = r_\eta^-(m, n) \circledast u(-m - m_0, -n - n_0)$$

where

$$r_\eta^-(m, n) \triangleq \mathscr{F}^{-1}\left[\frac{1}{S_\eta(\omega_1, \omega_2)}\right] \tag{9.132}$$

Defining

$$\nu(m, n) \triangleq \mathrm{v}(m, n) \circledast r_\eta^-(m, n) \tag{9.133}$$

the matched filter output can be written as

$$\begin{aligned} g(m,n) \circledast \mathrm{v}(m,n) &= u(-m - m_0, -n - n_0) \circledast \nu(m,n) \\ &= \sum_i \sum_j \nu(i,j) u(i - m - m_0, j - n - n_0) \end{aligned} \tag{9.134}$$

which, according to (9.123), is $c_{\nu u}(m + m_0, n + n_0)$, the area correlation of $\nu(m, n)$ with $u(m + m_0, n + n_0)$. If (m_0, n_0) were known, then the SNR would be maximized at $(m, n) = (0, 0)$, as desired in (9.130) (show!). In practice these displacement values are unknown. Therefore, we compute the correlation $c_{\nu u}(m, n)$ and search for the location of maxima that gives (m_0, n_0). Therefore, the matched filter can be implemented as an area correlator with a preprocessing filter (Fig. 9.46a). Recall from Section 6.7 (Eq. (6.91)) that $r_\eta^-(m, n)$ would be proportional to the impulse response of the minimum variance *noncausal prediction error filter* for a random field with power spectral density $S_\eta(\omega_1, \omega_2)$. For highly correlated random fields—for instance, the usual monochrome images—$r_\eta^-(m, n)$ represents a high-pass filter. For example, if the background has object-like power spectrum [see Section 2.11]

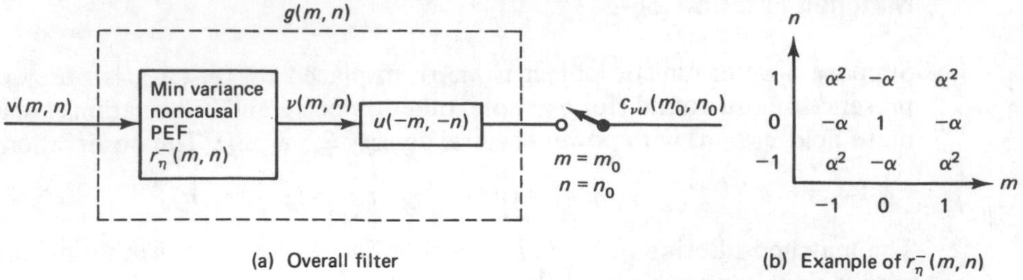

Figure 9.46 Matched filtering in the presence of colored noise. For white noise case $r_\eta^-(m,n) = \delta(m, n)$.

$$S_\eta(\omega_1, \omega_2) = \frac{1}{(1 - 2\alpha \cos \omega_1)(1 - 2\alpha \cos \omega_2)}, \qquad 0 \le \alpha < \frac{1}{2} \tag{9.135}$$

then $r_\eta^-(m, n)$ is a high-pass filter whose impulse response is shown in Fig. 9.46b (see Example 6.10). This suggests that template matching is more effective if the edges (and other high frequencies) are matched whenever a given object has to be detected in the presence of a correlated background.

If $\eta(m, n)$ is white noise, then S_η will be constant for instance, and $r^-(m, n) = \delta(m, n)$. Now the matched filter reduces to the area correlator of Fig. 9.45.

Direct Search Methods [58–59]

Direct methods of searching for an object in a scene are useful when the template size is small compared to the region of search. We discuss some efficient direct search techniques next.

Two-dimensional logarithmic search. This method reduces the search iterations to about $\log n$ for an $n \times n$ area. Consider the mean distortion function

$$D(i, j) = \frac{1}{MN} \sum_{m=1}^{M} \sum_{n=1}^{N} f(v(m, n) - u(m + i, n + j)), \qquad -p \le i, j \le p \tag{9.136}$$

where $f(x)$ is a given positive and increasing function of x, $u(m, n)$ is an $M \times N$ template and $v(m, n)$ is the observed image. The template match is restricted to a preselected $[-p, p] \times [-p, p]$ region. Some useful choices for $f(x)$ are $|x|$ and x^2. We define *direction of minimum distortion* (DMD) as the direction vector (i, j) that minimizes $D(i, j)$. Template match occurs when the DMD has been found within the search region.

Exhaustive search for DMD would require evaluation of $D(i, j)$ for $(2p + 1)^2$ directions. If the $D(i, j)$ increases monotonically as we move away from the DMD along any direction, then the search can be speeded up by successively reducing the area of search. Figure 9.47a illustrates the procedure for $p = 5$. The algorithm consists of searching five locations (marked ◇), which contain the center of the search area and the midpoints between the center and the four boundaries of the area. The locations searched at the initial step are marked 1. The optimum direction

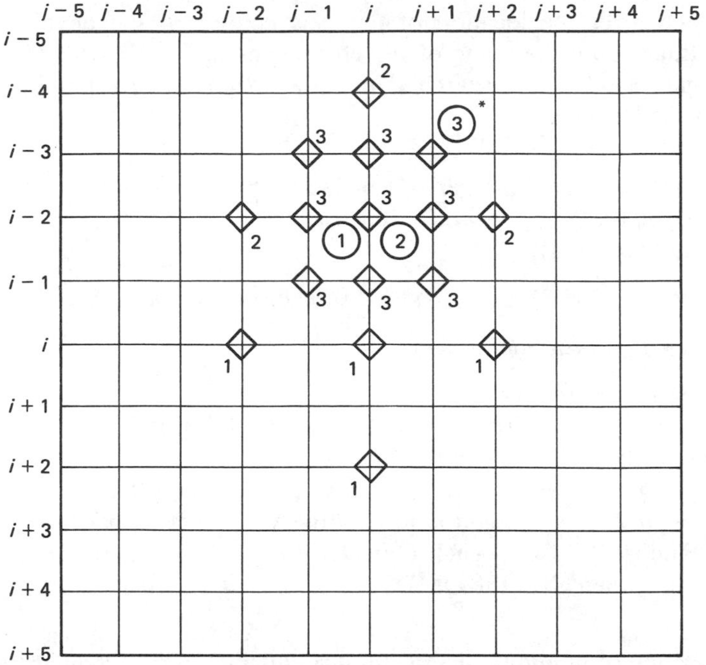

A 2-D logarithmic search procedure for the direction of minimum distortion. The figure shows the concept of the 2-D logarithmic search to find a pixel in another frame, which is registered with respect to the pixel (i, j) of a given frame, such that the mean square error over a block defined around (i, j) is minimized. The search is done step by step, with ◇ indicating the directions searched at a step number marked. The numbers circled show the optimum directions for that search step and the * shows the final optimum direction, $(i - 3, j + 1)$, in this example. This procedure requires, searching only 13 to 21 locations for the given grid, as opposed to 121 total possibilities.

(a)

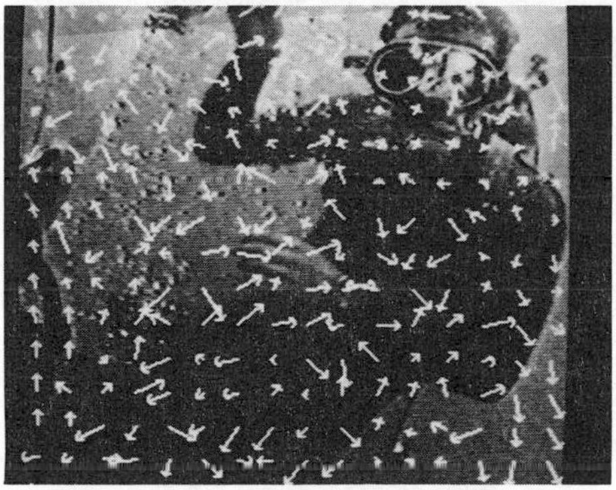

(b)

Figure 9.47 Two-dimensional logarithmic search. (a) The algorithm. (b) Example *Courtesy* Stuart Wells, Herriott-Watt Univ. U.K.

(circled numbers) gives the location of the new center for the next step. This procedure continues until the plane of search reduces to a 3×3 size. In the final step all the nine locations are searched and the location corresponding to the minimum gives the DMD.

Algorithm [58]. For any integer $m > 0$, define $\mathcal{N}(m) = \{(i, j); -m \leq i, j \leq m\}$, $\mathcal{M}(m) = \{(0,0), (m, 0), (0, m), (-m, 0), (0, -m)\}$.

Step 1 (Initialization)

$$D(i, j) = \infty, \qquad (i, j) \in \mathcal{N}(p); \; n' \triangleq \text{Integer } [\log_2 p]; \; n = \max\{2, 2^{n'-1}\}$$

$$q = l = 0 \qquad \text{(or an initial guess for DMD)}$$

Step 2 $\mathcal{M}'(n) \leftarrow \mathcal{M}(n)$

Step 3 Find $(i, j) \in \mathcal{M}'(n)$ such that $D(i + q, j + l)$ is minimum. If $i = 0$ and $j = 0$, go to Step 5; otherwise go to Step 4.

Step 4 $q \leftarrow q + i$, $l \leftarrow l + j$; $\mathcal{M}'(n) \leftarrow \mathcal{M}'(n) - (-i, -j)$; go to Step 3.

Step 5 $n \leftarrow n/2$. If $n = 1$, go to Step 6; otherwise go to Step 2.

Step 6 Find $(i, j) \in \mathcal{N}(1)$ such that $D(i + q, j + l)$ is minimum. $q \leftarrow i + q$, $l \leftarrow l + j$. (q, l) then gives the DMD.

If the direction of minimum distortion lies outside $\mathcal{N}(p)$, the algorithm converges to a point on the boundary that is closest to the DMD. This algorithm has been found useful for estimating planar-motion of objects by measuring displacements of local regions from one frame to another. Figure 9.47b shows the motion vectors detected in an underwater scene involving a diver and turbulent water flow.

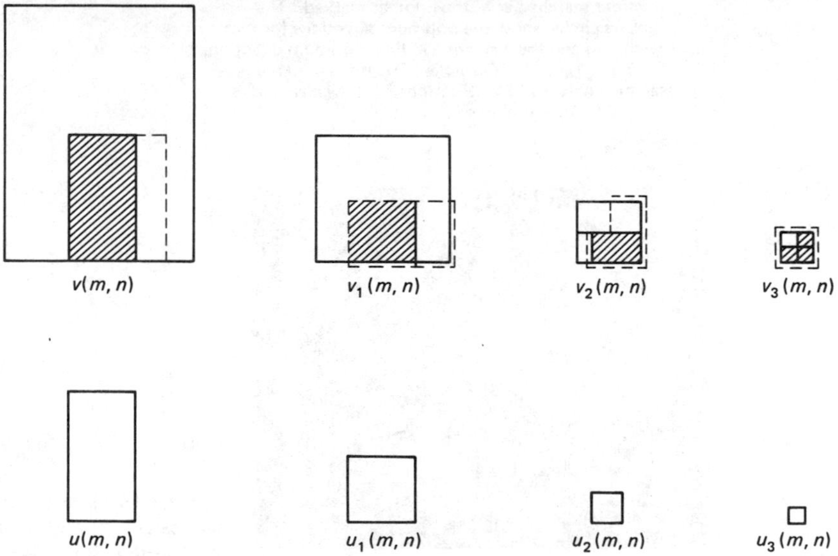

Figure 9.48 Hierarchical search. Shaded area shows the region where match occurs. Dotted lines show regions searched.

Sequential search. Another way of speeding up search is to compute the cumulative error

$$e_{p,q}(i, j) \triangleq \sum_{m=1}^{p} \sum_{n=1}^{q} |v(m, n) - u(m + i, n + j)|, \qquad p \le M, q \le N \tag{9.137}$$

and terminate the search at (i, j) if $e_{p,q}(i, j)$ exceeds some predetermined threshold. The search may then be continued only in those directions where $e_{p,q}(i, j)$ is below a threshold.

Another possibility is to search in the i direction until a minimum is found and then switch the search in the j direction. This search in *alternating conjugate directions* is continued until the location of the minimum remains unchanged.

Hierarchical search. If the observed image is very large, we may first search a low-resolution-reduced copy using a likewise reduced copy of the template. If multiple matches occur (Fig. 9.48), then the regions represented by these locations are searched using higher-resolution copies to further refine and reduce the search area. Thus the full-resolution region searched can be a small fraction of the total area. This method of *coarse-fine search* is also logarithmically efficient.

9.13 IMAGE SEGMENTATION

Image segmentation refers to the decomposition of a scene into its components. It is a key step in image analysis. For example, a document reader would first segment the various characters before proceeding to identify them. Figure 9.49 lists several image segmentation techniques which will be discussed now.

Amplitude Thresholding or Window Slicing

Amplitude thresholding is useful whenever the amplitude features (see Section 9.2) sufficiently characterize the object. The appropriate amplitude feature values are calibrated so that a given amplitude interval represents a unique object characteristic. For example, the large amplitudes in the remotely sensed IR image of Fig. 7.10b represent low temperatures or high altitudes. Thresholding the high-intensity values segments the cloud patterns (Fig. 7.10d). Thresholding techniques are also useful in segmentation of binary images such as printed documents, line drawings and graphics, multispectral and color images, X-ray images, and so on. Threshold

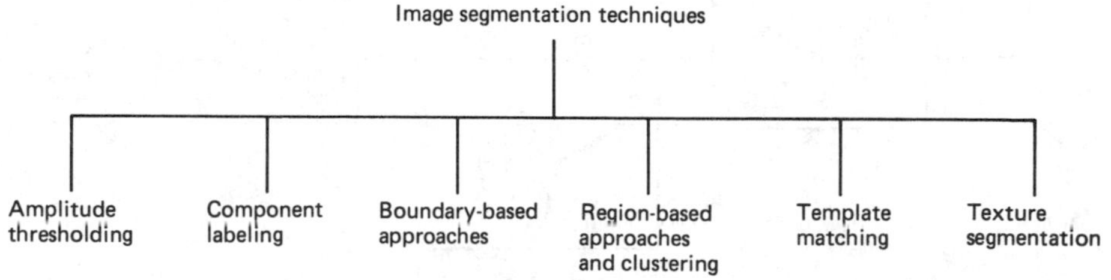

Figure 9.49

selection is an important step in this method. Some commonly used approaches are as follows:

1. The histogram of the image is examined for locating peaks and valleys. If it is multimodal, then the valleys can be used for selecting thresholds.
2. Select the threshold (t) so that a predetermined fraction (η) of the total number of samples is below t.
3. Adaptively threshold by examining local neighborhood histograms.
4. Selectively threshold by examining histograms only of those points that satisfy a chosen criterion. For example, in low-contrast images, the histogram of those pixels whose Laplacian magnitude is above a prescribed value will exhibit clearer bimodal features than that of the original image.
5. If a probabilistic model of the different segmentation classes is known, determine the threshold to minimize the probability of error or some other quantity, for instance, Bayes' risk (see Section 9.14).

Example 9.8

We want to segment the Washington Monument from the scene in Fig. 9.50. First, the low intensities are thresholded to isolate the very dark areas (trees here). Then we detect a rectangle bounding the monument by thresholding the horizontal and vertical projection signatures defined as $h(n) \triangleq \sum_m u(m, n)/\sum_m 1$, $v(m) \triangleq \sum_m u(m, n)/\sum_n 1$. Contour-following the boundary of the object inside the rectangle gives the segmented object.

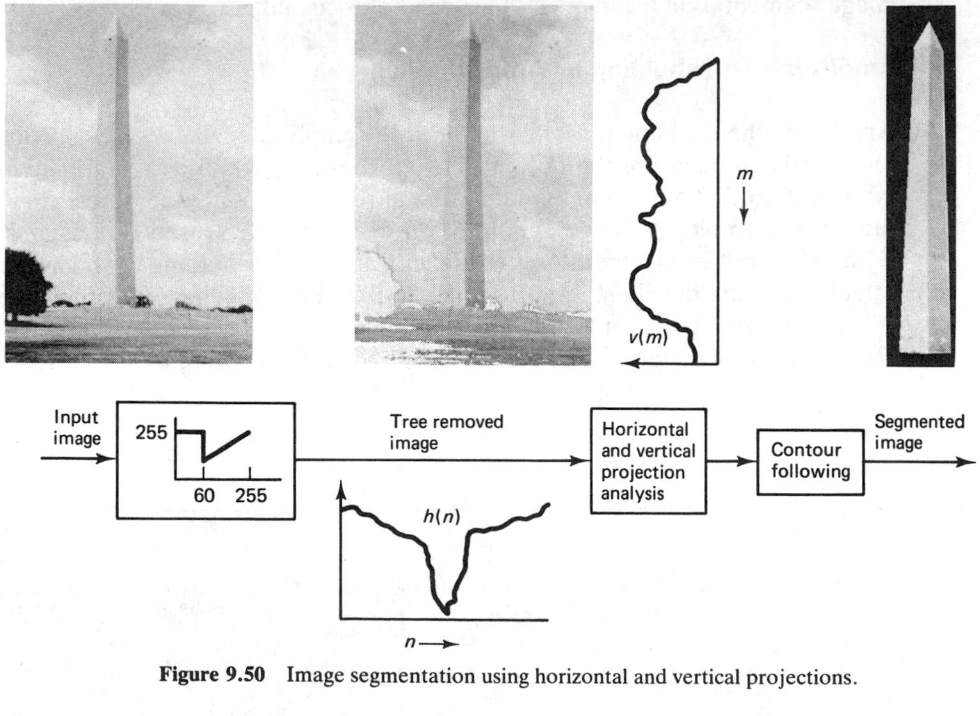

Figure 9.50 Image segmentation using horizontal and vertical projections.

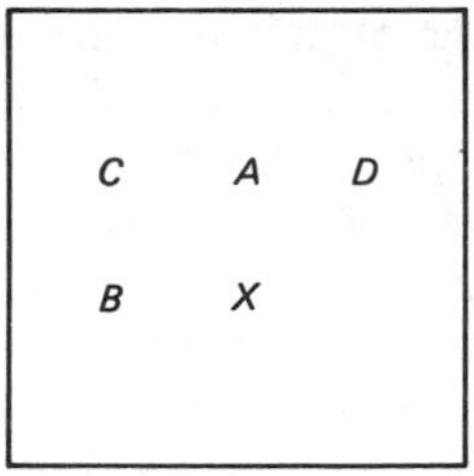

Figure 9.51 Neighborhood of pixel X in a pixel labeling algorithm.

Component Labeling

A simple and effective method of segmentation of binary images is by examining the connectivity of pixels with their neighbors and labeling the connected sets. Two practical algorithms are as follows.

Pixel labeling. Suppose a binary image is raster scanned left to right and top to bottom. The current pixel, X (Fig. 9.51), is labeled as belonging to either an object (1s) or a hole (0s) by examining its connectivity to the neighbors A, B, C, and D. For example, if $X = 1$, then it is assigned to the object(s) to which it is connected. If there are two or more qualified objects, then those objects are declared to be equivalent and are merged. A new object label is assigned when a transition from 0s to an isolated 1 is detected. Once the pixel is labeled, the features of that object are updated. At the end of the scan, features such as centroid, area, and perimeter are saved for each region of connected 1s.

Run-length connectivity analysis. An alternate method of segmenting binary images is to analyze the connectivity of run lengths from successive scan lines. To illustrate this idea, we consider Fig. 9.52, where the black or white runs are denoted by *a, b, c,* . . . . A segmentation table is created, where the run *a* of the first scan line is entered into the first column. The object of the first run *a* is named *A*. The first run of the next scan line, *b,* is of the same color as *a* and overlaps *a*. Hence

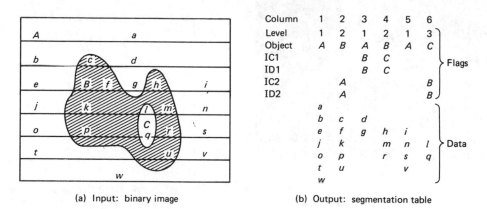

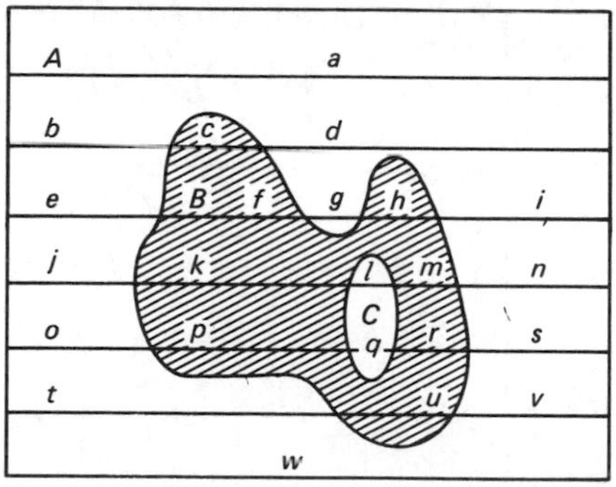

(a) Input: binary image

Column	1	2	3	4	5	6	
Level	1	2	1	2	1	3	Flags
Object	*A*	*B*	*A*	*B*	*A*	*C*	
IC1			*B*	*C*			
ID1			*B*	*C*			
IC2		*A*				*B*	
ID2		*A*				*B*	
	a						Data
	b	*c*	*d*				
	e	*f*	*g*	*h*	*i*		
	j	*k*		*m*	*n*	*l*	
	o	*p*		*r*	*s*	*q*	
	t	*u*			*v*		
	w						

(b) Output: segmentation table

Figure 9.52 Run-length connectivity algorithm.

b belongs to the object *A* and is placed underneath *a* in the first column. Since *c* is of different color, it is placed in a new column, for an object labeled *B*. The run *d* is of the same color as *a* and overlaps *a*. Since *b* and *d* both overlap *a*, *divergence* is said to have occurred, and a new column of object *A* is created, where *d* is placed. A divergence flag ID1 is set in this column to indicate that object *B* has caused this divergence. Also, the flag ID2 of *B* (column 2) is set to *A* to indicate *B* has caused divergence in *A*. Similarly, *convergence* occurs when two or more runs of 0s or 1s in a given line overlap with a run of same color in the previous line. Thus convergence occurs in run *u*, which sets the convergence flags IC1 to *C* in column 4 and IC2 to *B* in column 6. Similarly, *w* sets the convergence flag IC2 to *A* in column 2, and the column 5 is labeled as belonging to object *A*.

In this manner, all the objects with different closed boundaries are segmented in a *single pass*. The segmentation table gives the data relevant to each object. The convergence and divergence flags also give the hierarchy structure of the object. Since *B* causes divergence as well as convergence in *A* and *C* has a similar relationship with *B*, the objects *A*, *B*, and *C* are assigned levels 1, 2, and 3, respectively.

Example 9.9

A vision system based on run-length connectivity analysis is outlined in Fig. 9.53a. The input object is imaged and digitized to give a binary image. Figure 9.53b shows the run-length representation of a key and its segmentation into the outer profile and the

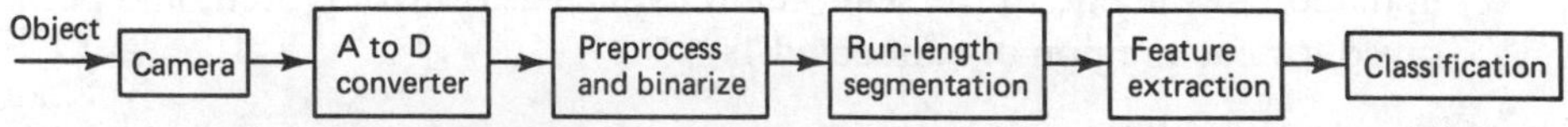

(a) Vision system

(b) Run Length Data for a Key

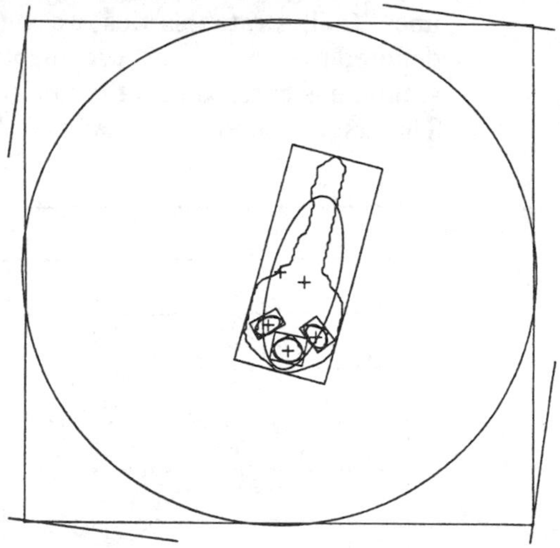

(c) Display of Best-Fit Ellipse, Bounding Rectangle and Center of Mass for the Key.

Figure 9.53 Vision system based on run-length connecitvity analysis.

three holes. For each object, features such as number of holes, area of holes, bounding rectangle, center of mass, orientation, and lengths of major and minor axes of the best-fit ellipse are calculated (Fig. 9.53c). A system trained on the basis of such features can then identify the given object from a trained vocabulary of objects [69].

Boundary-Based Approaches

Boundary extraction techniques segment objects on the basis of their profiles. Thus, contour following, connectivity, edge linking and graph searching, curve fitting, Hough transform, and other techniques of Section 9.5 are applicable to image segmentation. Difficulties with boundary-based methods occur when objects are touching or overlapping or if a break occurs in the boundary due to noise or artifacts in the image.

Example 9.10 Boundary analysis–based vision system

Figure 9.54a shows an example of an object-recognition system, which uses the boundary information for image segmentation. The edges detected from the image of the input object are linked to determine the boundary. A spline fit (Section 9.6) is performed to extract the control points (Fig. 9.54b), which are then used to determine the object location (center of mass), orientation, and other shape parameters [71].

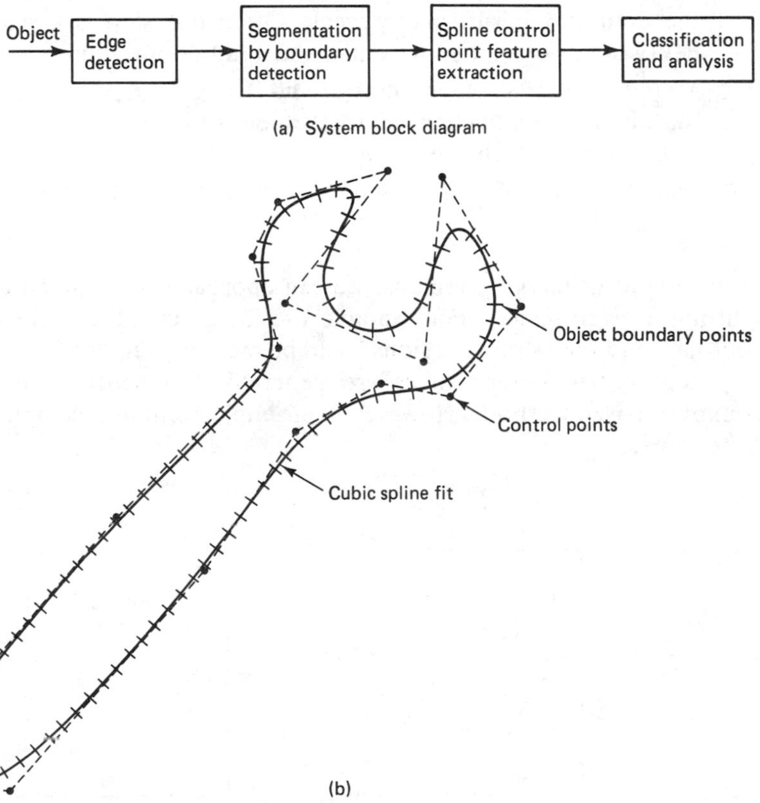

Figure 9.54 Object recognition system based on boundary analysis.

Region-Based Approaches and Clustering

The main idea in region-based segmentation techniques is to identify various regions in an image that have similar features. Clustering techniques encountered in pattern-recognition literature have similar objectives and can be applied for image segmentation. Examples of clustering are given in Section 9.14.

One class of region-based techniques involves *region growing* [72]. The image is divided into *atomic regions* of constant gray levels. Similar adjacent regions are merged sequentially until the adjacent regions become sufficiently different (Fig. 9.55). The trick lies in selecting the criterion for merging. Some merging heuristics are as follows:

1. Merge two regions $\mathcal{R}_i$ and $\mathcal{R}_j$ if $w/P_m > \theta_1$, where $P_m = \min(P_i, P_j)$, P_i and P_j are the perimeters of $\mathcal{R}_i$ and $\mathcal{R}_j$, and w is the number of weak boundary locations (pixels on either side have their magnitude difference less than some threshold σ). The parameter θ_1 controls the size of the region to be merged. For example $\theta_1 \simeq 1$ implies two regions will be merged only if one of the regions almost surrounds the other. Typically, $\theta_1 = 0.5$.
2. Merge $\mathcal{R}_i$ and $\mathcal{R}_j$ if $w/I > \theta_2$, where I is the length of the common boundary between the two regions. Typically $\theta_2 \simeq 0.75$. So the two regions are merged if the boundary is sufficiently weak. Often this step is applied after the first heuristic has been used to reduce the number of regions.
3. Merge $\mathcal{R}_i$ and $\mathcal{R}_j$ only if there are no strong edge points between them. Note that the run-length connectivity method for binary images can be interpreted as an example of this heuristic.
4. Merge $\mathcal{R}_i$ and $\mathcal{R}_j$ if their similarity distance [see Section 9.14] is less than a threshold.

Instead of merging regions, we can approach the segmentation problem by splitting a given region. For example the image could be split by the *quad-tree* approach and then similar regions could be merged (Fig. 9.56).

Region-based approaches are generally less sensitive to noise than the boundary-based methods. However, their implementation complexity can often be quite large.

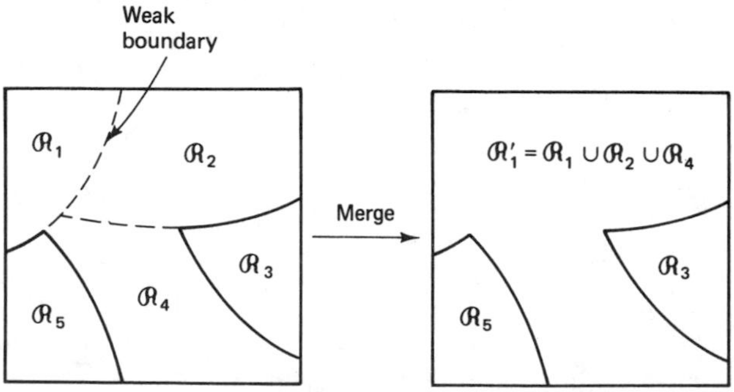

Figure 9.55 Region growing by merging.

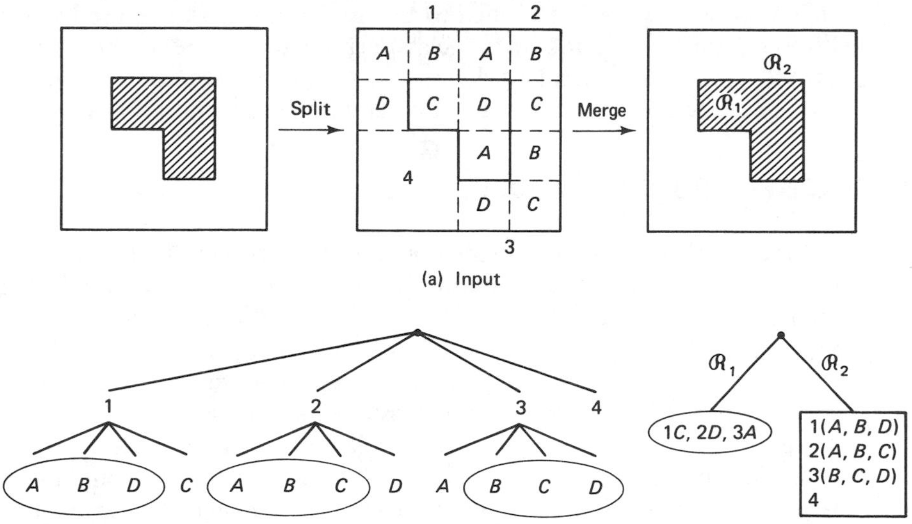

(a) Input

(b) Quad tree split

(c) Segmented regions

Figure 9.56 Region growing by split and merge techniques.

Template Matching

One direct method of segmenting an image is to match it against templates from a given list. The detected objects can then be segmented out and the remaining image can be analyzed by other techniques (Fig. 9.57). This method can be used to segment busy images, such as journal pages containing text and graphics. The text can be segmented by template-matching techniques and graphics can be analyzed by boundary following algorithms.

Texture Segmentation

Texture segmentation becomes important when objects in a scene have a textured background. Since texture often contains a high density of edges, boundary-based techniques may become ineffective unless texture is filtered out. Clustering and

(a) Template

(b) Input image

(c) Filtered image

Figure 9.57 Background segmentation (or filtering) via template matching.

region-based approaches applied to textured features can be used to segment textured regions. In general, texture classification and segmentation is quite a difficult problem. Use of a priori knowledge about the existence and kinds of textures that may be present in a scene can be of great utility in practical problems.

9.14 CLASSIFICATION TECHNIQUES

A major task after feature extraction is to classify the object into one of several categories. Figure 9.2 lists various classification techniques applicable in image analysis. Although an in-depth discussion of classification techniques can be found in the pattern-recognition literature—see, for example, [1]—we will briefly review these here to establish their relevance in image analysis.

It should be mentioned that classification and segmentation processes have closely related objectives. Classification can lead to segmentation, and vice-versa. Classification of pixels in an image is another form of component labeling that can result in segmentation of various objects in the image. For example, in remote sensing, classification of multispectral data at each pixel location results in segmentation of various regions of wheat, barley, rice, and the like. Similarly, image segmentation by template matching, as in character recognition, leads to classification or identification of each object.

There are two basic approaches to classification, *supervised* and *nonsupervised,* depending on whether or not a set of prototypes is available.

Supervised Learning

Supervised learning, also called supervised classification, can be *distribution free* or *statistical.* Distribution-free methods do not require knowledge of any a priori probability distribution functions and are based on reasoning and heuristics. Statistical techniques are based on probability distribution models, which may be parametric (such as Gaussian distributions) or nonparametric.

Distribution-free classification. Suppose there are K different objects or pattern classes $S_1, S_2, \ldots, S_k, \ldots, S_K$. Each class is characterized by M_k prototypes, which have $N \times 1$ feature vectors $\mathbf{y}_m^{(k)}, m = 1, \ldots, M_k$. Let $\mathbf{x}$ denote an $N \times 1$ feature vector obtained from the observed image. A fundamental function in pattern recognition is called the *discriminant function.* It is defined such that the kth discriminant function $g_k(\mathbf{x})$ takes the maximum value if $\mathbf{x}$ belongs to class k, that is, the decision rule is

$$g_k(\mathbf{x}) > g_i(\mathbf{x}) \qquad k \neq i \Leftrightarrow \mathbf{x} \in S_k \tag{9.138}$$

For a K class problem, we need $K - 1$ discriminant functions. These functions divide the N-dimensional feature space into K different regions with a maximum of $K(K-1)/2$ hypersurfaces. The partitions become hyperplanes if the discriminant function is *linear,* that is, if it has the form

$$g_k(\mathbf{x}) = \mathbf{a}_k^T \mathbf{x} + b_k \tag{9.139}$$

Such a function arises, for example, when $\mathbf{x}$ is classified to the class whose centroid is nearest in Euclidean distance to it (Problem 9.17). The associated classifier is called *the minimum mean (Euclidean) distance classifier.*

An alternative decision rule is to classify $\mathbf{x}$ to S_i if among a total of k nearest prototype neighbors of $\mathbf{x}$, the maximum number of neighbors belong to class S_i. This is the *k-nearest neighbor classifier,* which for $k = 1$ becomes a *minimum-distance classifier.*

When the discriminant function can classify the prototypes correctly for some linear discriminants, the classes are said to be *linearly separable.* In that case, the weights $\mathbf{a}_k$ and b_k can be determined via a *successive linear training* algorithm. Other discriminants can be *piecewise linear,* quadratic, or polynomial functions. The k-nearest neighbor classification can be shown to be equivalent to using piecewise linear discriminants.

Decision tree classification [60–61]. Another distribution-free classifier, called a *decision tree classifier,* splits the N-dimensional feature space into unique regions by a sequential method. The algorithm is such that every class need not be tested to arrive at a decision. This becomes advantageous when the number of classes is very large. Moreover, unlike many other training algorithms, this algorithm is guaranteed to converge whether or not the feature space is linearly separable.

Let $\mu_k(i)$ and $\sigma_k(i)$ denote the mean and standard deviation, respectively, measured from repeated independent observations of the kth prototype vector element $y_m^{(k)}(i), m = 1, \ldots, M_k$. Define the normalized average prototype features $z_k(i) \triangleq \mu_k(i)/\sigma_k(i)$ and an $N \times K$ matrix

$$\mathbf{Z} = \begin{bmatrix} z_1(1) & z_2(1) & \ldots z_k(1) \\ z_1(2) & z_2(2) & \ldots z_k(2) \\ \vdots & \vdots & \vdots \\ z_1(N) & z_2(N) & \ldots z_k(N) \end{bmatrix} \tag{9.140}$$

The row number of $\mathbf{Z}$ is the feature number and the column number is the object or class number. Further, let $\mathbf{Z}' \triangleq [\mathbf{Z}]$ denote the matrix obtained by arranging the elements of each row of $\mathbf{Z}$ in increasing order with the smallest element on the left and the largest on the right. Now, the algorithm is as follows.

Decision Tree Algorithm

Step 1 Convert $\mathbf{Z}$ to $\mathbf{Z}'$. Find the maximum distance between adjacent row elements in each row of $\mathbf{Z}'$. Find r, the row number with the largest maximum distance. The row r represents a feature. Set a threshold at the midpoint of the maximum distance boundaries and split row r into two parts.

Step 2 Convert $\mathbf{Z}'$ to $\tilde{\mathbf{Z}}$ such that the row r is the same in both the matrices. The elements of the other rows of $\mathbf{Z}'$ are rearranged such that each column of $\tilde{\mathbf{Z}}$ represents a prototype vector. This means, simply, that the elements of each row of $\mathbf{Z}$ are in the same order as the elements of row r. Split $\tilde{\mathbf{Z}}$ into two matrices $\mathbf{Z}_1$ and $\mathbf{Z}_2$ by splitting each row in a manner similar to row r.

Step 3 Repeat Steps 1 and 2 for the split matrices that have more than one column. Terminate the process when all the split matrices have only one column.

The preceding process produces a series of thresholds that induce questions of the form, Is feature $j >$ threshold? The questions and the two possible decisions for each question generate a series of nodes and branches of a decision tree. The terminal branches of the tree give the classification decision.

Example 9.11

The accompanying table contains the normalized average areas and perimeter lengths of five different object classes for which a vision system is to be trained.

$\rightarrow k$	1	2	3	4	5
$z(1) = \left(\frac{\mu}{\sigma}\right)$ area	6	12	20	24	27
$z(2) = \left(\frac{\mu}{\sigma}\right)$ perimeter	56	28	42	35	48

This gives

$$\eta_1 = 16$$

$$\mathbf{Z}' = \left[\begin{array}{cc|ccc} 6 & 12 & 20 & 24 & 27 \\ 28 & 35 & 42 & 48 & 56 \end{array}\right] \Rightarrow \tilde{\mathbf{Z}}_1 = \left[\begin{array}{cc|ccc} 6 & 12 & 20 & 24 & 27 \\ 56 & 28 & 42 & 35 & 48 \end{array}\right]$$

$$\qquad\qquad\qquad\qquad\qquad\qquad \mathbf{Z}_2 \qquad\qquad \mathbf{Z}_3$$

The largest adjacent difference in the first row is 8; in the second row it is 7. Hence the first row is chosen, and $z(1)$ is the feature to be thresholded. This splits $\tilde{\mathbf{Z}}_1$ into $\mathbf{Z}_2$ and $\mathbf{Z}_3$, as shown. Proceeding similarly with these matrices, we get

$$\mathbf{Z}_2' = \left[\begin{array}{c|c} 6 & 12 \\ 28 & 56 \end{array}\right] \Rightarrow \tilde{\mathbf{Z}}_2 = \left[\begin{array}{c|c} 12 & 6 \\ 28 & 56 \end{array}\right] \quad (\text{columns } 2, 1)$$

$$\eta_2 = 42$$

$$\eta_4 = 23.5$$

$$\mathbf{Z}_3' = \left[\begin{array}{c|cc} 20 & 24 & 27 \\ 35 & 42 & 48 \end{array}\right] \Rightarrow \tilde{\mathbf{Z}}_{23} = \left[\begin{array}{c|c|c} 24 & 20 & 27 \\ 35 & 42 & 48 \end{array}\right] \quad (\text{columns } 4, 3, 5)$$

$$\eta_3 = 38.5 \qquad\qquad \mathbf{Z}_4 \text{ (column 3)}$$

The thresholds partition the feature space and induce the decision tree, as shown in Fig. 9.58.

Statistical classification. In statistical classification techniques it is assumed the different object classes and the feature vector have an underlying joint probability density. Let $P(S_k)$ be the a priori probability of occurrence of class S_k and $p(\mathbf{x})$ be the probability density function of the random feature vector observed as $\mathbf{x}$.

Bayes' minimum-risk classifier. The Bayes' minimum-risk classifier minimizes the average loss or risk in assigning $\mathbf{x}$ to a wrong class. Define

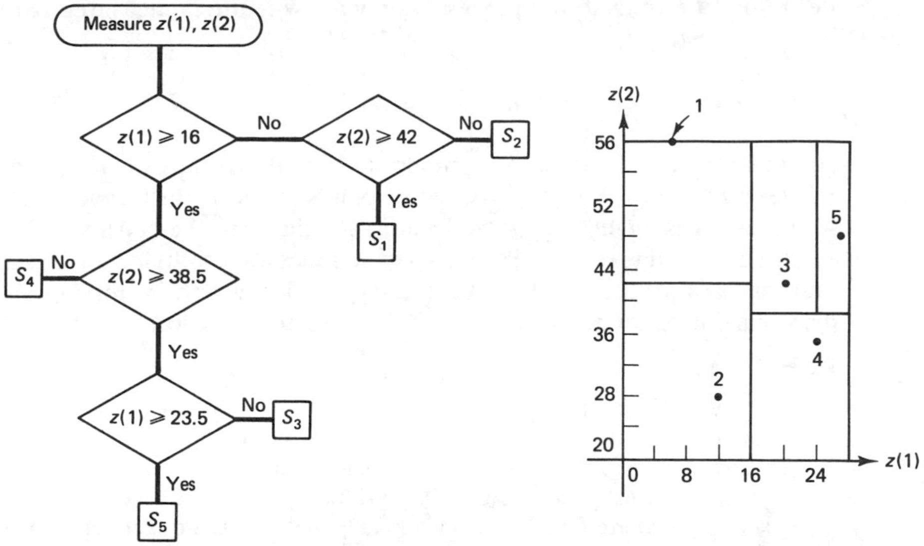

Figure 9.58 Decision tree classifier.

$$
\begin{aligned}
\text{Risk, } \mathcal{R} &\triangleq \sum_{k=1}^{K} \int_{R_k} c(\mathbf{x}|S_k)p(\mathbf{x})\,d\mathbf{x} \\
c(\mathbf{x}|S_k) &\triangleq \sum_{i=1}^{K} c_{i,k}\,p(S_i|\mathbf{x}),
\end{aligned}
\tag{9.141}
$$

where $c_{i,k}$ is the cost of assigning $\mathbf{x}$ to S_k when $\mathbf{x} \in S_i$ in fact and R_k represents the region of the feature space where $p(\mathbf{x}|S_k) > p(\mathbf{x}|S_i)$, for every $i \neq k$. The quantity $c(\mathbf{x}|S_k)$ represents the total cost of assigning $\mathbf{x}$ to S_k. It is well known the decision rule that minimizes $\mathcal{R}$ is given by

$$
\sum_{i=1}^{K} c_{i,k}P(S_i)p(\mathbf{x}|S_i) < \sum_{i=1}^{K} c_{i,j}P(S_i)p(\mathbf{x}|S_i), \qquad \forall j \neq k \quad \Rightarrow \mathbf{x} \in S_k \tag{9.142}
$$

If $c_{i,k} = 1, i \neq k$, and $c_{i,k} = 0, i = k$, then the decision rule simplifies to

$$
p(\mathbf{x}|S_k)P(S_k) > p(\mathbf{x}|S_j)P(S_j), \qquad \forall j \neq k \quad \Rightarrow \mathbf{x} \in S_k \tag{9.143}
$$

In this case the probability of error in classification is also minimized and the *minimum error classifier* discriminant becomes

$$
g_k(\mathbf{x}) = p(\mathbf{x}|S_k)P(S_k) \tag{9.144}
$$

In practice the $p(\mathbf{x}|S_k)$ are estimated from the prototype data by either *parametric or nonparametric techniques* which can yield simplified expressions for the discriminant function.

There also exist some *sequential classification* techniques such as *sequential probability ratio test* (SPRT) and *generalized* SPRT, where decisions can be made initially using fewer than N features and refined as more features are acquired sequentially [62]. The advantage lies in situations where N is large, so that it is desirable to terminate the process if only a few features measured early can yield adequate results.

Nonsupervised Learning or Clustering

In nonsupervised learning, we attempt to identify *clusters* or natural groupings in the feature space. A cluster is a set of points in the feature space for which their local density is large (relative maximum) compared to the density of feature points in the surrounding region. Clustering techniques are useful for image segmentation and for classification of raw data to establish classes and prototypes. Clustering is also a useful vector quantization technique for compression of images.

Example 9.12

The visual and IR images $u_1(m, n)$ and $u_2(m, n)$, respectively (Fig. 9.59a), are transformed pixel by pixel to give the features as $v_1(m, n) = (u_1(m, n) + u_2(m, n))/\sqrt{2}$, $v_2(m, n) = (u_1(m, n) - u_2(m, n))/\sqrt{2}$. This is simply the 2×2 Hadamard transform of the 2×1 vector $[u_1\, u_2]^T$. Figure 9.59b shows the feature images. The images $v_1(m, n)$ and $v_2(m, n)$ are found to contain mainly the clouds and land features, respectively. Thresholding these images yield the left-side images in Fig. 9.59c and d. Notice the clouds contain some land features, and vice-versa. A scatter diagram, which plots each vector $[v_1\, v_2]^T$ as a point in the v_1 versus v_2 space, is seen to have two main clusters (Fig. 9.60). Using the cluster boundaries for segmentation, we can remove the land features from clouds, and vice versa, as shown in Fig. 9.59c and d (right-side images).

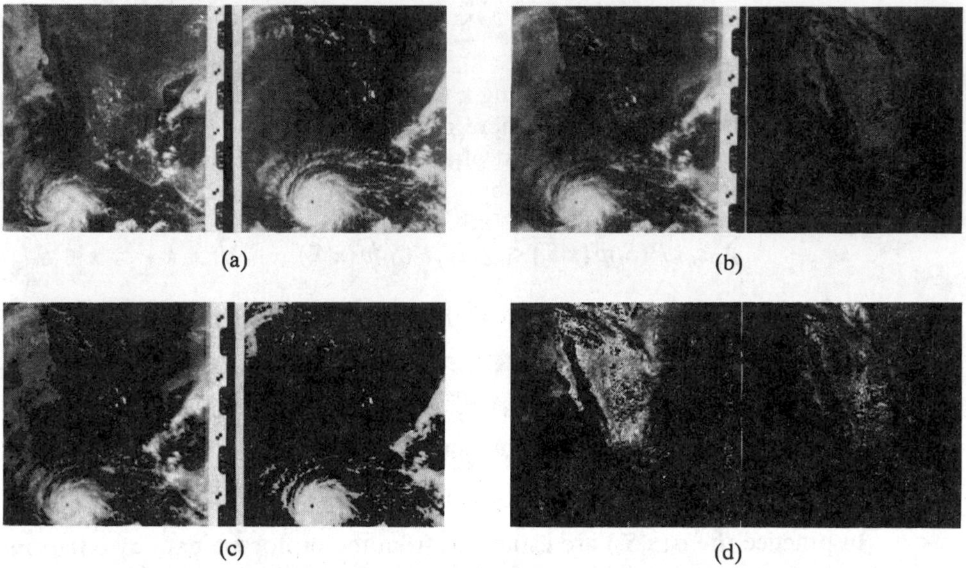

Figure 9.59 Segmentation by clustering. (a) Input images $u_1(m, n)$ and $u_2(m, n)$; (b) feature images $v_1(m, n)$ and $v_2(m, n)$; (c) segmenation of clouds by thresholding v_1 (left) and by clustering (right); (d) segmentation of land by thresholding v_2 (left) and by clustering (right).

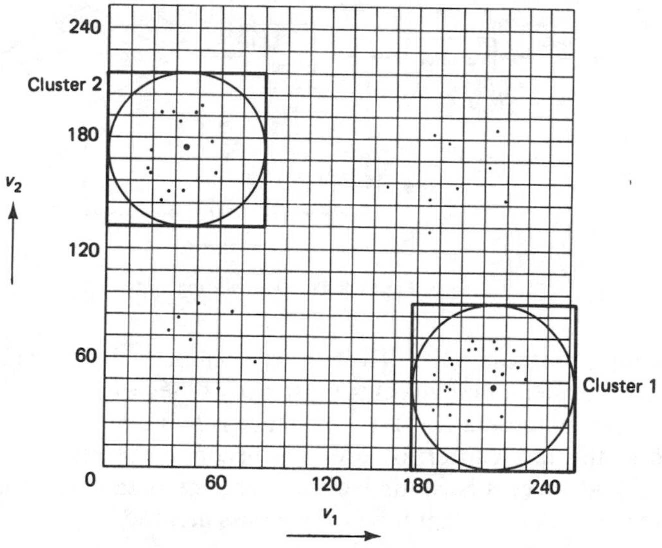

Figure 9.60 Scatter diagram in feature space.

Similarity measure approach. The success of clustering techniques rests on the partitioning of the feature space into cluster subsets. A general clustering algorithm is based on split and merge ideas (Fig. 9.61). Using a *similarity measure,* the input vectors are partitioned into subsets. Each partition is tested to check whether or not the subsets are sufficiently distinct. Subsets that are not sufficiently distinct are merged. The procedure is repeated on each of the subsets until no further subdivisions result or some other convergence criterion is satisfied. Thus, a similarity measure, a distinctiveness test, and a stopping rule are required to define a clustering algorithm. For any two feature vectors $\mathbf{x}_i$ and $\mathbf{x}_j$, some of the commonly used similarity measures are:

Dot product: $$\langle \mathbf{x}_i, \mathbf{x}_j \rangle \triangleq \mathbf{x}_i^T \mathbf{x}_j = \|\mathbf{x}_i\| \, \|\mathbf{x}_j\| \cos(\mathbf{x}_i, \mathbf{x}_j)$$

Similarity rule: $$S(\mathbf{x}_i, \mathbf{x}_j) \triangleq \frac{\langle \mathbf{x}_i, \mathbf{x}_j \rangle}{\langle \mathbf{x}_i, \mathbf{x}_i \rangle + \langle \mathbf{x}_j, \mathbf{x}_j \rangle - \langle \mathbf{x}_i, \mathbf{x}_j \rangle}$$

Weighted Euclidean distance: $$d(\mathbf{x}_i, \mathbf{x}_j) \triangleq \sum_k [x_i(k) - x_j(k)]^2 w_k$$

Normalized correlation: $$\rho(\mathbf{x}_i, \mathbf{x}_j) \triangleq \frac{\langle \mathbf{x}_i, \mathbf{x}_j \rangle}{\sqrt{\langle \mathbf{x}_i, \mathbf{x}_i \rangle \langle \mathbf{x}_j, \mathbf{x}_j \rangle}}$$

Several different algorithms exist for clustering based on similarity approach. Examples are given next.

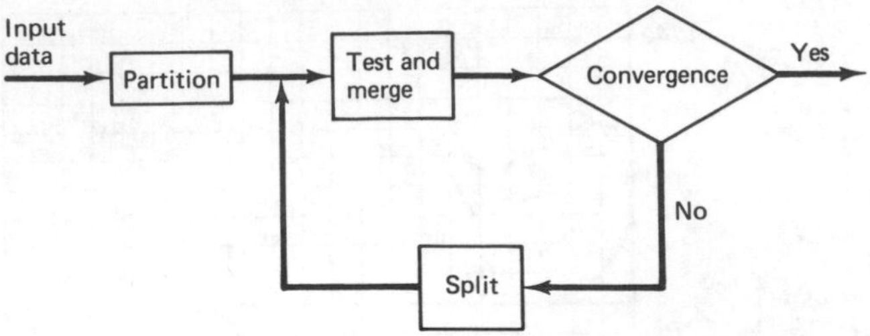

Figure 9.61 A clustering approach.

Chain method [63]. The first data sample is designated as the *representative* of the first cluster and similarity or distance of the next sample is measured from the first cluster representative. If this distance is less than a threshold, say η, then it is placed in the first cluster; otherwise it becomes the representative of the second cluster. The process is continued for each new data sample until all the data has been exhausted. Note that this is a one-pass method.

An iterative method (Isodata) [64]. Assume the number of clusters, K, is known. The partitioning of the data is done such that the average spread or variance of the partition is minimized. Let $\boldsymbol{\mu}_k(n)$ denote the kth cluster center at the nth iteration and R_k denote the region of the kth cluster at a given iteration. Initially, we assign arbitrary values to $\boldsymbol{\mu}_k(0)$. At the nth iteration take one of the data points $\mathbf{x}_i$ and assign it to the cluster whose center is closest to it, that is,

$$\mathbf{x}_i \in R_k \quad \Leftrightarrow \quad d(\mathbf{x}_i, \boldsymbol{\mu}_k(n)) = \min_{j=1,\ldots,K} [d(\mathbf{x}_i, \boldsymbol{\mu}_j(n))] \tag{9.145}$$

where $d(\mathbf{x}, \mathbf{y})$ is the distance measure used. Recompute the cluster centers by finding the point that minimizes the distance for elements within each cluster. Thus

$$\boldsymbol{\mu}_k(n+1): \quad \sum_{\mathbf{x}_i \in R_k} d(\mathbf{x}_i, \boldsymbol{\mu}_k(n+1)) = \min_{\mathbf{y}} \sum_{\mathbf{x}_i \in R_k} d(\mathbf{x}_i, \mathbf{y}), \qquad k = 1, \ldots, K \tag{9.146}$$

The procedure is repeated for each $\mathbf{x}_i$, one at a time, until the clusters and their centers remain unchanged. If $d(\mathbf{x}, \mathbf{y})$ is the Euclidean distance, then a cluster center is simply the mean location of its elements. If K is not known, we start with a large

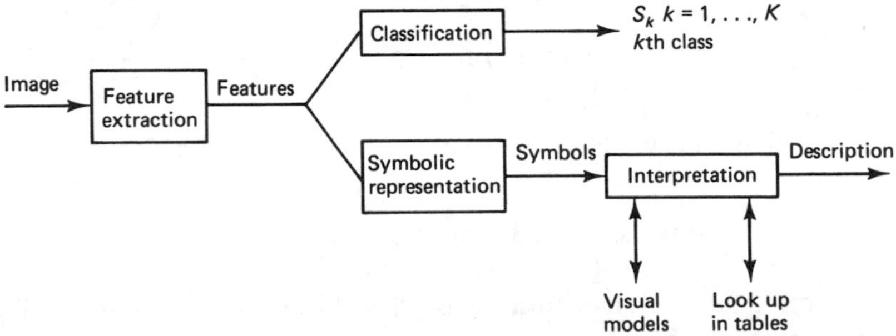

Figure 9.62 Image understanding systems.

value of K and then merge to $K - 1, K - 2, \ldots$ clusters by a suitable *cluster-distance* measure.

Other Methods

Clusters can also be viewed as being located at the nodes of the joint Nth-order histogram of the feature vector. Other clustering methods are based on statistical nonsupervised learning techniques, ranking, and intrinsic dimensionality determination, graph theory, and so on [65, 66]. Discussion of those techniques is beyond the goals of this text.

Finally it should be noted that success of clustering techniques is closely tied to feature selection. Clusters not detected in a given feature space may be easier to detect in rotated, sealed, or transformed coordinates. For images the feature vector elements could represent gray level, gradient magnitude, gradient phase, color, and/or other attributes. It may also be useful to decorrelate the elements of the feature vector.

9.15 IMAGE UNDERSTANDING

Image understanding (IU) refers to a body of *knowledge* that transforms pictorial inputs into commonly understood descriptions or symbols. Image pattern-recognition techniques we have studied classify an input into one of several categories. Interpretation to a class is provided by a priori knowledge, or *supervision.* Such pattern-recognition systems are the simplest of IU systems. In more-advanced systems (Fig. 9.62), the features are first mapped into symbols; for exam-

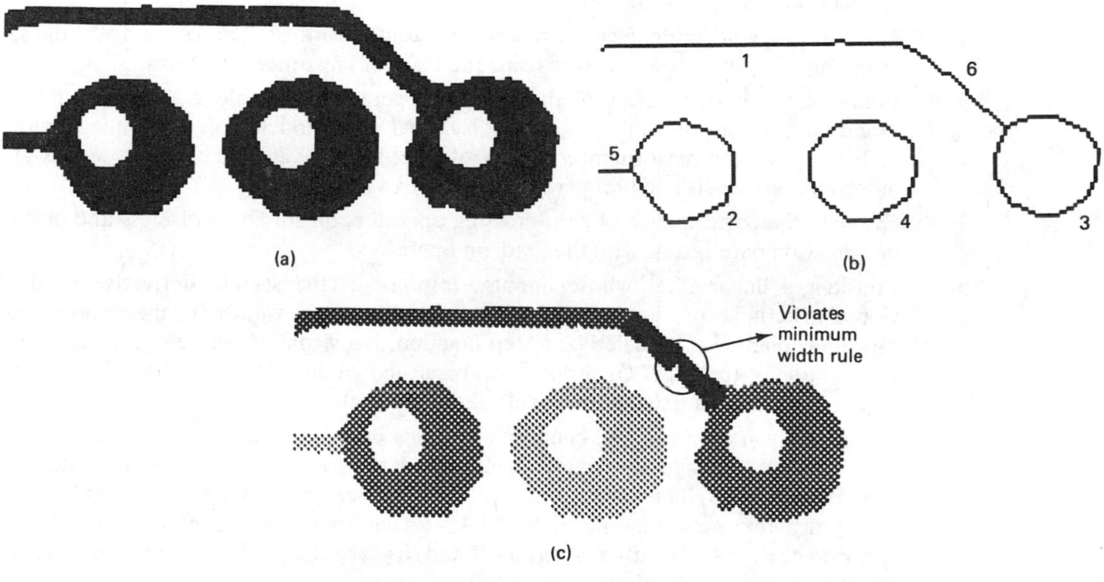

Figure 9.63 A rule-based approach for printed circuit board inspection. (a) Preprocessed image; (b) image after thinning and identifying tracks and pads; (c) segmented image (obtained by region growing). Rules can be applied to the image in (c) and violations can be detected.

ple, the shape features may be mapped into the symbols representing circles, rectangles, ellipses, and the like. Interpretation is provided to the collection of symbols to develop a description of the scene. To provide interpretation, different visual models and practical rules are adopted. For example, syntactic techniques provide grammars for strings of symbols. Other relational models provide rules for describing relations and interconnections between symbols. For example, projections at different angles of a spherical object may be symbolically represented as several circles. A relational model would provide the interpretation of a sphere or a ball. Figure 9.63 shows an example of image understanding applied to inspection of printed circuit boards [73, 74].

Much work remains to be done in formulation of problems and development of techniques for image understanding. Although the closing topic for this chapter, it offers a new beginning to a researcher interested in computer vision.

PROBLEMS

9.1 Calculate the means, autocorrelation, covariance, and inertia [see Eq. (9.116)] of the second-order histogram considered in Example 9.1.

9.2* Display the following features measured over 3×3, 5×5, 9×9, and 16×16 windows of a 512×512 image: (a) mean, (b) median, (c) dispersion, (d) standard deviation, (e) entropy, (f) skewness, and (g) kurtosis. Repeat the experiment for different images and draw conclusions about the possible use of these features in image processing applications.

9.3* From an image of your choice, extract the horizontal, vertical, 30°, 45°, and 60° edges, using the DFT and extract texture using the Haar or any other transform.

9.4* Compare the performances of the gradient operators of Table 9.2 and the 5×5 stochastic gradient of Table 9.5 on a noisy ideal edge model (Fig. 9.11) image with SNR = 9. Use the performance criteria of (9.25) and (9.26). Repeat the results at different noise levels and plot performance index versus SNR.

9.5* Evaluate the performance of zero-crossing operators on suitable noiseless and noisy images. Compare results with the gradient operators.

9.6 Consider a linear filter whose impulse response is the second derivative of the Gaussian kernel $\exp(-x^2/2\sigma^2)$. Show that, regardless of the value of σ, the response of this filter to an edge modeled by a step function, is a signal whose zero-crossing is at the location of the edge. Generalize this result in two dimensions by considering the Laplacian of the Gaussian kernel $\exp[-(x^2 + y^2)/2\sigma^2]$.

9.7 The gradient magnitude and contour directions of a 4×6 image are shown in Fig. P9.7. Using the linkage rules of Fig. 9.16b, sketch the graph interpretation and find the edge path if the evaluation function represents the sum of edge gradient magnitudes. Apply dynamic programming to Fig. P9.7 to determine the edge curve using the criterion of Eq. (9.27) with $\alpha = 4/\pi$, $\beta = 1$ and $d(\mathbf{x}, \mathbf{y})$ = Euclidean distance between $\mathbf{x}$ and $\mathbf{y}$.

9.8 **a.** Find the Hough transforms of the figures shown below in Figure P9.8.
b. *(Generalized Hough transform)* Suppose it is desired to detect a curve defined

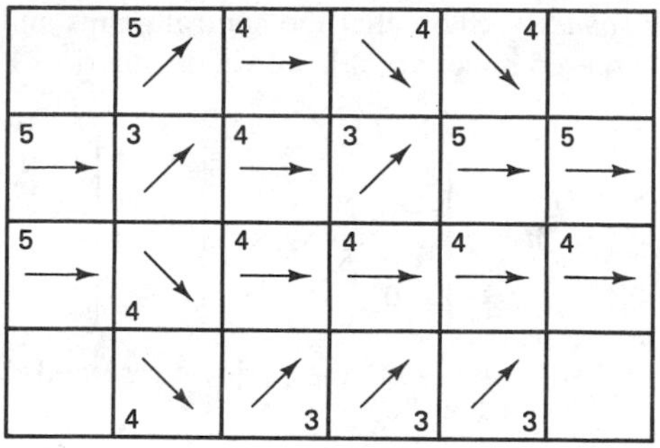

Figure P9.7

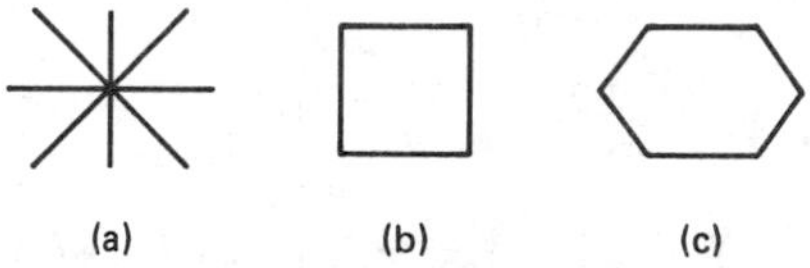

Figure P9.8

parametrically by $\phi(x, y, \mathbf{a}) = 0$, where $\mathbf{a}$ is a $p \times 1$ vector of parameters, from a set of edge point (x_i, y_i), $i = 1, \ldots, N$. Run a counter $C(\mathbf{a})$ as follows:

$$\text{Initialize:} \quad C(\mathbf{a}) = 0$$

$$\text{Do } i = 1, N: \quad C(\mathbf{a}) = C(\mathbf{a}) + 1, \qquad \text{where } \mathbf{a} \text{ is such that } \phi(x_i, y_i, \mathbf{a}) = 0$$

Then the local maxima of $C(\mathbf{a})$ gives the particular curve(s) that pass through the given edge points. If each element of $\mathbf{a}$ is quantized to L different levels, the dimension of vector $C(\mathbf{a})$ will be L^p. Write the algorithm for detecting elliptical segments described by

$$\frac{(x - x_0)^2}{a^2} + \frac{(y - y_0)^2}{b^2} = 1$$

If x_0, y_0, a, and b are represented by 8-bit words each, what is the dimension of $C(\mathbf{a})$?

c. If the gradient angles θ_i at each edge point are given, then show how the relation

$$\frac{\partial \phi}{\partial x} + \frac{\partial \phi}{\partial y} \tan \theta = 0 \qquad \text{for } (x, y, \theta) = (x_i, y_i\, \theta_i)$$

might be used to reduce the dimensionality of the search problem.

9.9 **a.** Show that the normalized uniform periodic B-splines satisfy

$$\int_0^k B_{0,k}(t)\, dt = 1 \quad \text{and} \quad \sum_{j=0}^{k-1} B_{0,k}(t + j) = 1, \qquad 0 \le t < 1$$

b. If an object of uniform density is approximated by the polygon obtained by joining the adjacent control points by straight lines, find the expressions for center of mass, perimeter, area, and moments in terms of the control points.

9.10 *(Cubic B-splines)* Show that the control points and the cubic B-splines sampled at uniformly spaced nodes are related via the matrices $\mathbf{B}_4$ as follows:

$$\mathbf{B}_4 = \frac{1}{6}\begin{bmatrix} 4 & 1 & & \mathbf{0} & 1 \\ 1 & 4 & \ddots & & \\ & \ddots & \ddots & \ddots & \\ & & \ddots & \ddots & 1 \\ 1 & \mathbf{0} & & 1 & 4 \end{bmatrix}, \qquad \frac{1}{6}\begin{bmatrix} 6 & 0 & 0 & & & \\ 1 & 4 & 1 & & \mathbf{0} & \\ & \ddots & \ddots & \ddots & & \\ & & \mathbf{0} & 1 & 4 & 1 \\ & & & 0 & 0 & 6 \end{bmatrix}$$

where the first matrix is for the periodic case and the second is for the nonperiodic case.

9.11 *(Properties of FDs)*

a. Prove the properties of the Fourier descriptors summarized in Table 9.8.

b. Using Fig. 9.26, show that the reflection of x_1, x_2 is given by

$$\tilde{x}_1 = \frac{1}{A^2 + B^2}\,[(B^2 - A^2)x_1 - 2ABx_2 - 2AC]$$

$$\tilde{x}_2 = \frac{1}{A^2 + B^2}\,[-2ABx_1 + (A^2 - B^2)x_2 - 2BC]$$

From these relations prove (9.53).

c. Show how the size, location, orientation, and symmetry of an object might be determined if its FDs and those of a prototype are given.

d. Given the FDs of $u(n)$, find the FDs of $x_1(n)$ and $x_2(n)$ and list their properties with respect to the geometrical transformations considered in the text.

9.12 *(Additional properties of FDs [32])*

a. *(FDs for a polygon curve)* For a continuous curve $u(t) \triangleq x_1(t) + jx_2(t)$ with period T, the FDs are the Fourier series coefficients $a(k) = (1/T)\int_0^T u(t) \exp(-j2\pi kt/T)\,dt$. If the object boundary is a polygon whose vertices are represented by phasors V_k, $k = 0, 1, \ldots, m-1$, show that the FDs are given by

$$a(k) = \frac{T}{(2\pi k)^2}\sum_{i=1}^{m}(b_{i-1} - b_i)\exp(-j2\pi k t_i/T)$$

where

$$b_i \triangleq \frac{V_{i+1} - V_i}{|V_{i+1} - V_i|}, \qquad t_k \triangleq \sum_{i=1}^{k}|V_i - V_{i-1}|, \qquad k > 0,\ t_0 \triangleq 0$$

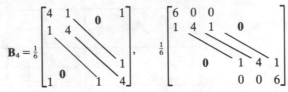

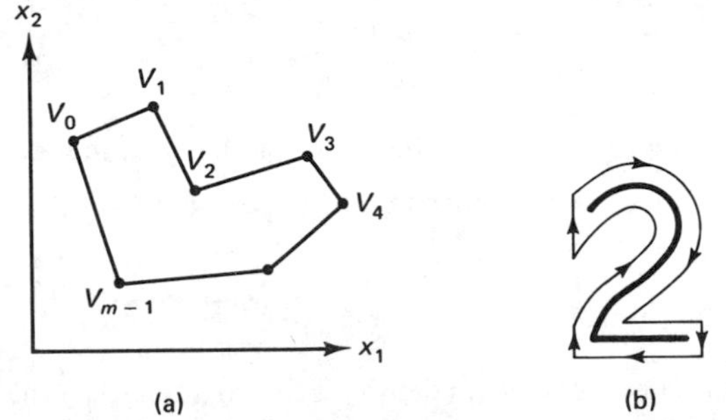

Figure P9.12

b. *(Line patterns)* If the given curve is a line pattern then a closed contour can be obtained by retracing it. Using the symmetry of the periodic curve, show that the FDs satisfy the relation

$$a(k) = a(-k)e^{-jk(2\pi/T)\beta}$$

for some β. If the trace begins at $t = 0$ at one of the endpoints of the pattern, then $\beta = 0$. Show how this property may be used to skeletonize a shape.

c. The area A enclosed by the outer boundary of a surface is given by

$$A = \tfrac{1}{2}\oint x_2\,dx_1 - \tfrac{1}{2}\oint x_1\,dx_2 = \tfrac{1}{2}\int_{t=0}^{T} x_2(t)\frac{dx_1(t)}{dt}\,dt - \tfrac{1}{2}\int_{t=0}^{T} x_1(t)\frac{dx_2}{dt}\,dt$$

In terms of FDs show that $A = -\sum_{k=\infty}^{\infty} |a(k)|^2 k\pi$. Verify this result for surface area of a line pattern.

9.13 *(Properties of AR models)*

a. Prove the translation, scaling, and rotation properties of AR model parameters listed in Table 9.9.

b. Show a closed boundary can lie reconstructed from the AR model residuals $\varepsilon_i(n)$ by inverting a circulant matrix.

c. Find the relation between AR model features and FDs of closed boundaries.

9.14* Scan and digitize the ASCII characters and find their medial axis transforms. Develop any alternative practical thinning algorithm to reduce printed characters to line shapes.

9.15 Compare the complexity of printed character recognition algorithms based on (a) template matching, (b) Fourier descriptors, and (c) moment matching.

9.16 *(Matched filtering)* Write the matched filter output SNR as

$$\text{SNR} = \left|\int_{-\infty}^{\infty}\int_{-\infty}^{\infty}[GS_\eta^{1/2}][S_\eta^{-1/2}U\exp\{-j(\omega_1 m_0 + \omega_2 n_0)\}\,d\omega_1\,d\omega_2\right|^2 \Big/ \iint_{-\infty}^{\infty}|G|^2 S_\eta\,d\omega_1\,d\omega_2$$

where G and U are Fourier transforms of $g(m, n), u(m, n)$, respectively. Apply the Schwartz inequality to show that SNR is maximized only when (9.132) is satisfied within a scaling constant that can be set to unity. What is the maximum value of SNR?

9.17 If $\boldsymbol{\mu}_k$ denotes the mean vector of class k prototypes, show that the decision rule: $\|\mathbf{x} - \boldsymbol{\mu}_k\|^2 < \|\mathbf{x} - \boldsymbol{\mu}_i\|^2$, $i \neq k \Rightarrow \mathbf{x} \in S_k$, gives a linear discriminant with $\mathbf{a}_k = 2\boldsymbol{\mu}_k$, $b_k = -\|\boldsymbol{\mu}_k\|^2$.

9.18 Find the decision tree of Example 9.11 if an object class with $z(1) = 15$, $z(2) = 30$ is added to the list of prototypes.

9.19* A printed circuit board can be modeled as a network of pathways that either merge into other paths as terminate into a node. Develop a vision system for isolating defects such as breaks (open circuits) and leaks (short circuits) in the pathways. Discuss and develop practical preprocessing, segmentation, and recognition algorithms for your system.

BIBLIOGRAPHY

Section 9.1–9.3

Some general references on feature extraction, image analysis and computer vision are:

1. R. O. Duda and P. E. Hart. *Pattern Recognition and Scene Analysis.* New York: John Wiley, 1973.
2. A. Rosenfeld and A. C. Kak. *Digital Picture Processing.* New York: Academic Press, 1976. Also see Vols. I and II, 1982.
3. D. H. Ballard and C. M. Brown. *Computer Vision.* Englewood Cliffs, N.J.: Prentice-Hall, 1982.
4. B. S. Lipkin and A. Rosenfeld (eds.). *Picture Processing and Psychopictorics.* New York: Academic Press, 1970.
5. J. K. Aggarwal, R. O. Duda and A. Rosenfeld (eds.). *Computer Methods in Image Analysis.* Los Angeles: IEEE Computer Society, 1977.

Additional literature on image analysis may be found in several texts referred to in Chapter I, in journals such as *Computer Graphics and Image Processing, Pattern Recognition, IEEE Trans. Pattern Analysis and Machine Intelligence,* and in the proceedings of conferences and workshops such as *IEEE Conferences on Pattern Recognition Image Processing, Computer Vision and Pattern Recognition, International Joint Conference Pattern Recognition,* and the like.

Section 9.4

Edge detection is a problem of fundamental importance in image analysis. Different edge detection techniques discussed here follow from:

6. J. M. S. Prewitt. "Object Enhancement and Extraction," in [4].
7. L. S. Davis. "A Survey of Edge Detection Techniques." *Computer Graphics and Image Processing,* vol. 4, pp. 248–270, 1975.
8. A. Rosenfeld and M. Thurston. "Edge and Curve Detection for Visual Scene Analysis," in [5]. Also see Vol. C-21, no. 7, (July 1972): 677–715.
9. L. G. Roberts. "Machine Perception of Three-Dimensional Solids," in [5].
10. R. Kirsch. "Computer Determination of the Constituent Structure in Biological Images." *Compt. Biomed. Res.* 4, no. 3 (1971): 315–328.
11. G. S. Robinson. "Edge Detection by Compass Gradient Masks." *Comput. Graphics Image Proc.* 6 (1977): 492–501.
12. W. Frei and C. C. Chen. "Fast Boundary Detection: A Generalization and a New Algorithm." *IEEE Trans. Computers* 26, no. 2 (October 1977): 988–998.
13. D. Marr and E. C. Hildreth. "Theory of Edge Detection." *Proc. R. Soc. Lond.* B 270 (1980): 187–217.
14. R. M. Haralick. "Zero Crossing of Second Directional Derivative Edge Detector." *Robot Vision* (A. Rosenfeld, ed.). SPIE 336, (1982): 91–96.
15. M. Heuckel. "An Operator Which Locates Edges in Digitized Pictures." *J. ACM* 18, no. 1 (January 1971): 113–125. Also see *J. ACM* 20, no. 4 (October 1973): 634–647.
16. A. K. Jain and S. Ranganath. "Image Restoration and Edge Extraction Based on 2-D Stochastic Models." *Proc. ICASSP-82,* Paris, May 1982.
17. W. K. Pratt. *Digital Image Processing.* New York: Wiley Interscience, 1978, p. 497.

Section 9.5

For various types of edge linkage rules, contour-following, boundary detection techniques, dynamic programming, and the like, we follow:

18. R. Nevatia. "Locating Object Boundaries in Textured Environments." *IEEE Trans. Comput.* C-25 (November 1976): 1170–1180.
19. A. Martelli. "Edge Detection Using Heuristic Search Methods." *Comp. Graphics Image Proc.* 1 (August 1972): 169–182. Also see Martelli in [5].
20. G. P. Ashkar and J. W. Modestino. "The Contour Extraction Problem with Biomedical Applications." *Comp. Graphics Image Proc.* 7 (1978): 331–355.
21. J. M. Lester, H. A. Williams, B. A. Weintraub, and J. F. Brenner. "Two Graph Searching Techniques for Boundary Finding in White Blood Cell Images." *Comp. Biol. Med.* 8 (1978): 293–308.
22. R. E. Bellman and S. Dreyfus. *Applied Dynamic Programming.* Princeton, N.J.: Princeton University Press, 1962.
23. U. Montanari. "On the Optimal Detection of Curves in Noisy Pictures." *Commun. ACM* 14 (May 1971): 335–345.
24. P. V. C. Hough. "Method and Means of Recognizing Complex Patterns." U.S. Patent 3,069,654, 1962.
25. R. Cederberg. "Chain-Link Coding and Segmentation for Raster Scan Devices." *Computer Graphics and Image Proc.* 10, (1979): 224–234.

Section 9.6

For chain codes, its generalizations, and run-length coding based segmentation approaches we follow:

26. H. Freeman. "Computer Processing of Line Drawing Images." *Computer Surveys* 6 (March 1974): 57–98. Also see Freeman in [5] and J. A. Saghri and H. Freeman in *IEEE Trans. PAMI* (September 1981): 533–539.

The theory of B-splines is well documented in the literature. For its applications in computer graphics:

27. W. J. Gordon and R. F. Riesenfeld. "B-spline Curves and Surfaces," in R. E. Barnhill and R. F. Riesenfeld (eds.), *Computer Aided Geometric Design,* New York: Academic Press, 1974, pp. 95–126.
28. B. A. Barsky and D. P. Greenberg. "Determining a Set of B-spline Control Vertices to Generate an Interpolating Surface." *Computer Graphics and Image Proc.* 14 (1980): 203–226.
29. D. Paglieroni and A. K. Jain. "A Control Point Theory for Boundary Representation and Matching." *Proc. ICASSP,* Vol. 4, pp. 1851–1854, Tampa, Fla. 1985.
30. D. Hoffman, "The Interpretation of Visual Illusions." *Scientific American,* Dec. 1983, pp. 151–162.

Fourier Descriptors have been applied for shape analysis of closed curves and hand-printed characters. For details see Granlund in [5] and:

31. C. T. Zahn and R. S. Roskies. "Fourier Descriptors for Plane Closed Curves." *IEEE Trans. Computers* C-21 (March 1972): 269–281.
32. E. Persoon and K. S. Fu. "Shape Discrimination Using Fourier Descriptors." *IEEE Trans. Sys. Man, Cybern.* SMC-7 (March 1977): 170–179.

For theory of AR models for boundary representation, we follow:

33. R. L. Kashyap and R. Chellappa. "Stochastic Models for Closed Boundary Analysis: Representation and Reconstruction." *IEEE Trans. Inform. Theory* IT-27 (September 1981): 627–637.

Section 9.7

Further details on quad-trees and medial axis transform:

34. H. Samet. "Region Representation: Quadtrees from Boundary Codes." *Comm. ACM* 23 (March 1980): 163–170.

Section 9.8

For basic theory of moments and its applications:

35. M. K. Hu. "Visual Pattern Recognition by Moment Invariants," in [5].
36. M. R. Teague. "Image Analysis via the General Theory of Moments." *J. of Optical Society of America* 70, no. 8 (August 1980): 920–930.
37. G. B. Gurevich. *Foundations of the Theory of Algebraic Invariants.* Groningen, The Netherlands: P. Noordhoff, 1964.
38. D. Casasent and D. Psaltis. "Hybrid Processor to Compute Invariant Moments for Pattern Recognition." *J. Optical Society of America* 5, no. 9 (September 1980): 395–397.
39. S. Dudani, K. Breeding, and R. McGhee. "Aircraft Identification by Moment Invariants." *IEEE Trans. on Computers* C-26, no. 1 (January 1977): 39–45.
40. R. Wong and E. Hall. "Scene Matching with Moment Invariants." *Computer Graphics and Image Processing* 8 (1978): 16–24.
41. H. Blum. "A Transformation for Extracting New Descriptions of Shape." Symposium on Models for the Perception of Speech and Visual Form, Cambridge: MIT Press, 1964.
42. E. R. Davies and A. P. Plummer. "Thinning Algorithms: A Critique and a New Methodology." *Pattern Recognition* 14, (1981): 53–63.
43. D. Rutovitz. "Pattern Recognition." *J. of Royal Stat. Soc.* 129, no. 66 (1966): 403–420.
44. J. Serra. *Images Analysis and Mathematical Morphology.* New York: Academic Press, 1982.
45. T. Pavlidis. "Minimum Storage Boundary Tracing Algorithm and Its Application to Automatic Inspection." *IEEE Transactions on Sys., Man. and Cybern.* 8, no. 1 (January 1978): 66–69.

Section 9.11

For surveys and further details on texture, see Hawkins in [4], Picket in [4], Haralick et al. in [5], and:

46. P. Brodatz. *Textures: A Photographic Album for Artists and Designers.* Toronto: Dover Publishing Co., 1966.
47. R. M. Haralick. "Statistical and Structural Approaches to Texture." *Proc. IEEE* 67 (May 1979): 786–809. Also see *Image Texture Analysis,* New York: Plenum, 1981.
48. G. G. Lendaris and G. L. Stanley. "Diffraction Pattern Sampling for Automatic Pattern Recognition," in [5].
49. R. P. Kruger, W. B. Thompson, and A. F. Turner. "Computer Diagnosis of Pneumoconiosis." *IEEE Trans. Sys. Man. Cybern. SMC*-4, (January 1974): 40–49.
50. B. Julesz, et al. "Inability of Humans to Discriminate Between Visual Textures that Agree in Second Order Statistics-Revisited." *Perception* 2 (1973): 391–405. Also see *IRE Trans. Inform. Theory* IT-8 (February 1962): 84–92.
51. O. D. Faugeraus and W. K. Pratt. "Decorrelation Methods of Texture Feature Extraction." *IEEE Trans. Pattern Anal. Mach. Intell.* PAMI-2 (July 1980): 323–332.
52. B. H. McCormick and S. N. Jayaramamurthy. "Time Series Model for Texture Synthesis." *Int. J. Comput. Inform. Sci* 3 (1974): 329–343. Also see vol. 4, (1975): 1–38.
53. G. R. Cross and A. K. Jain. "Markov Random Field Texture Models." *IEEE Trans. Pattern Anal. Mach. Intell.* PAMI-5, no. 1 (January 1983): 25–39.
54. T. Pavlidis. *Structural Pattern Recognition.* New York: Springer-Verlag, 1977.
55. N. Ahuja and A. Rosenfeld. "Mosaic Models for Textures." *IEEE Trans. Pattern Anal. Mach. Intell.* PAMI-3, no. 1 (January 1981): 1–11.

Section 9.12

56. G. L. Turin. "An Introduction to Matched Filtering." *IRE Trans. Inform. Theory* (June 1960): 311–329.
57. A. Vander Lugt, F. B. Rotz, and A. Kloester, Jr. "Character Reading by Optical Spatial Filtering." in J. Tippett et al. (eds), *Optical and Electro-Optical Information Processing.* Cambridge, Mass.: MIT Press, 1965, pp. 125–141. Also see pp. 5–11 in [5].
58. J. R. Jain and A. K. Jain. "Displacement Measurement and Its Application in Interframe Image Coding." *IEEE Trans. Common* COM-29 (December 1981): 1799–1808.
59. D. L. Barnea and H. F. Silverman. "A Class of Algorithms for Fast Digital Image Registration." *IEEE Trans. Computers* (February 1972): 179–186.

Section 9.13, 9.14

Details of classification and clustering techniques may be found in [1] and other texts on pattern recognition. For decision tree algorithm and other segmentation techniques:

60. C. Rosen et al. "Exploratory Research in Advanced Automation." *SRI Technical Report* First, Second and Third Reports, NSF Grant GI-38100X1, SRI Project 2591, Menlo Park, Calif.: SRI, December 1974.

61. G. J. Agin and R. O. Duda. "SRI Vision Research for Advanced Automation." *Proc. 2nd. USA Japan Computer Conf.*, Tokyo, Japan, August 1975, pp. 113–117.
62. H. C. Andrews. *Introduction to Mathematical Techniques in Pattern Recognition,* New York: John Wiley, 1972. Also see A. B. Coleman and H. C. Andrews, "Image Segmentation by Clustering." *Proc. IEEE* 67, no. 5 (May 1979): 773–785.
63. G. Nagy. "State of the Art in Pattern Recognition." *Proc. IEEE* 5, no. 5 (May 1968): 836–861.
64. G. H. Ball and D. J. Hall. "ISODATA, A Novel Method of Data Analysis and Pattern Classification." International Communication Conference, Philadelphia, June 1966.
65. C. T. Zahn. "Graph-Theoretical Methods for Detecting and Describing Gestalt Clusters." *IEEE Trans. Computers* C-20, no. 1 (January 1971): 68–86.
66. J. C. Gower and G. J. S. Ross. "Minimum Spanning Trees, and Single Linkage Cluster Analysis." *Appl. Statistics* 18, no. 1 (1969): 54–64.
67. M. R. Anderberg. *Cluster Analysis for Application.* New York: Academic Press, 1971.
68. E. B. Henrichon, Jr. and K. S. Fu. "A Nonparametric Partitioning Procedure for Pattern Classification." *IEEE Trans. Computers* C-18, no. 7 (July 1969).
69. I. Kabir. "A Computer Vision System Using Fast, One Pass Algorithms." M.S. Thesis, University of California at Davis, 1983.
70. G. Hirzinger and K. Landzattel. "A Fast Technique for Segmentation and Recognition of Binary Patterns." IEEE Conference on Pattern Recognition and Image Processing, 1981.
71. D. W. Paglieroni. "Control Point Algorithms for Contour Processing and Shape Analysis," Ph.D. Thesis, University of California, Davis, 1986.
72. C. R. Brice and C. L. Fennema. "Scene Analysis Using Regions," in [5].

Section 9.15

For further reading or image understanding research see proceedings of International Joint Conference on Artificial Intelligence, DARPA Image Understanding Workshop and the various references cited there. For pc board inspection and rule based systems:

73. A. Darwish and A. K. Jain. "A Rule Based Approach for Visual Pattern Inspection." *IEEE Trans. Pattern Anal. Mach. Intell.* PAMI-10, no. 1 (January 1988): 56–68.
74. J. R. Mandeville. "A Novel Method for Analysis of Printed Circuit Images." *IBM J. Res. Dev.* 29 (January 1985): 73–86.

10

Image Reconstruction from Projections

10.1 INTRODUCTION

An important problem in image processing is to reconstruct a cross section of an object from several images of its transaxial projections [1–11]. A projection is a *shadowgram* obtained by illuminating an object by penetrating radiation. Figure 10.1 shows a typical method of obtaining projections. Each horizontal line shown in this figure is a one-dimensional projection of a horizontal slice of the object. Each pixel on the projected image represents the total absorption of the X-ray along its path from the source to the detector. By rotating the source-detector assembly around the object, projection views for several different angles can be obtained. The goal of *image reconstruction* is to obtain an image of a cross section of the object from these projections. Imaging systems that generate such slice views are called CT (*computerized tomography*) scanners. Note that in obtaining the projections, we lose resolution along the path of the X-rays. CT restores this resolution by using information from multiple projections. Therefore, image reconstruction from projections can also be viewed as a special case of image restoration.

Transmission Tomography

For X-ray CT scanners, a simple model of the detected image is obtained as follows. Let $f(x, y)$ denote the absorption coefficient of the object at a point (x, y) in a slice at some fixed value of z (Fig. 10.1). Assuming the illumination to consist of an infinitely thin parallel beam of X-rays, the intensity of the detected beam is given by

$$I = I_0 \exp\left[-\int_L f(x, y)\, du\right] \tag{10.1}$$

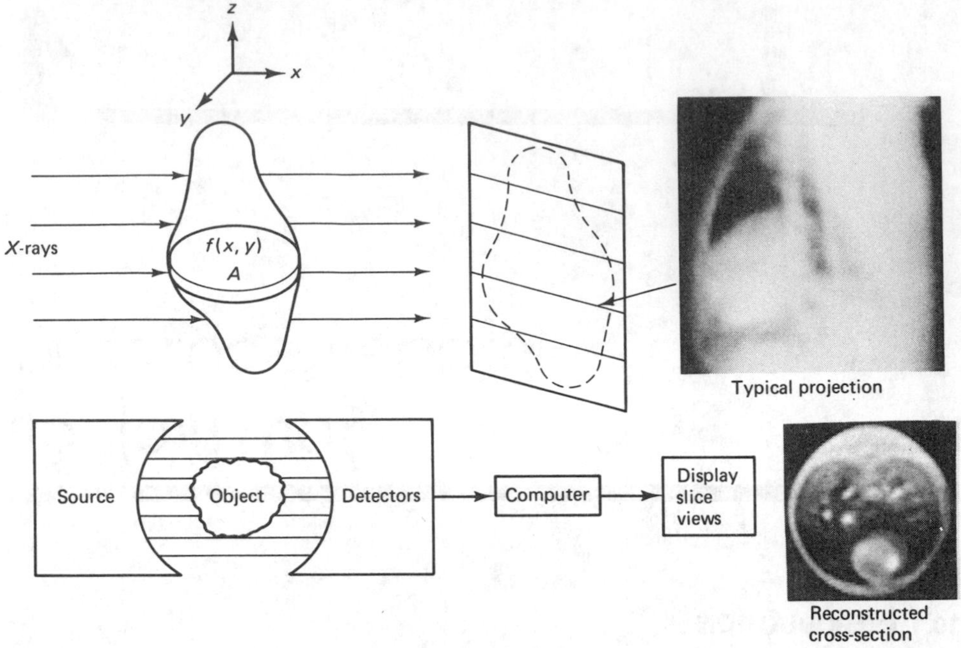

Figure 10.1 An X-ray CT scanning system.

where I_0 is the intensity of the incident beam, L is the path of the ray, and u is the distance along L (Fig. 10.2). Defining the observed signal as

$$g = \ln\left(\frac{I_0}{I}\right) \tag{10.2}$$

we obtain the linear transformation

$$g \triangleq g(s, \theta) = \int_L f(x, y)\, du, \qquad -\infty < s < \infty, 0 \le \theta < \pi \tag{10.3}$$

where (s, θ) represent the coordinates of the X-ray relative to the object. *The image reconstruction problem is to determine $f(x, y)$ from $g(s, \theta)$*. In practice we can only estimate $f(x, y)$ because only a finite number of views of $g(s, \theta)$ are available. The preceding imaging technique is called *transmission tomography* because the transmission characteristics of the object are being imaged. Figure 10.1 also shows an X-ray CT scan of a dog's thorax, that is, a cross-section slice, reconstructed from 120 such projections. X-ray CT scanners are used in *medical imaging* and *nondestructive testing* of mechanical objects.

Reflection Tomography

There are other situations where the detected image is related to the object by a transformation equivalent to (10.3). For example, in radar imaging we often obtain

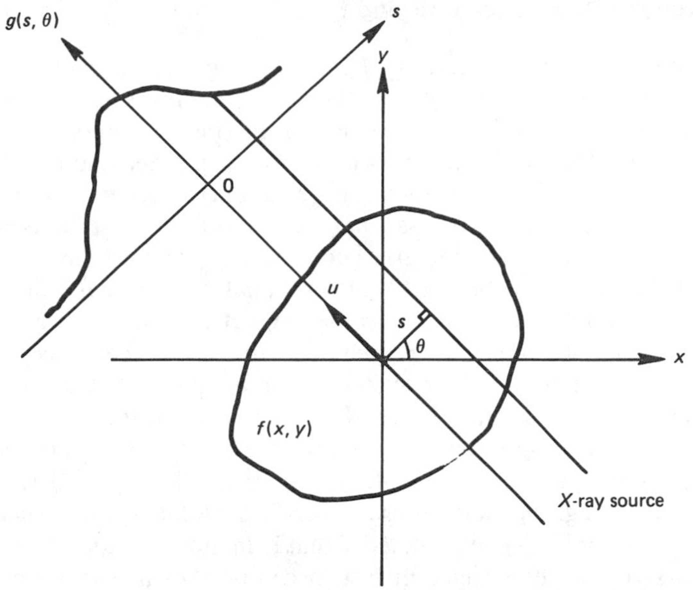

Figure 10.2 Projection imaging geometry in CT scanning.

a projection of the reflectivity of the object. This is called *reflection tomography.* For instance, in the FLR imaging geometry of Figure 8.7a, suppose the radar pulse width is infinitesimal (ideally) and the radar altitude (h) is large compared to the minor axis of the antenna half-power ellipse. Then the radar return at ground range r and scan angle ϕ can be approximated by (10.3), where $f(x, y)$ represents the ground reflectivity and L is the straight line parallel to the minor axis of the ellipse and passing through the center point of the shaded area. Other examples are found in spot mode synthetic aperture and CHIRP-doppler radar imaging [10, 36].

Emission Tomography

Another form of imaging based on the use of projections is *emission tomography,* for example, *positron emission tomography* (PET), where the emissive properties of isotopes planted within an object are imaged. Medical emission tomography exploits the fact that certain chemical compounds containing radioactive nuclei have a tendency to affix themselves to specific areas of the body, such as bone, blood, tumors, and the like. The gamma rays emitted by the decay of the isotopes are detected, from which the location of the chemical and the associated tissue within the body can be determined. In PET, the radioactive nuclei used are such that positrons (positive electrons) are emitted during decay. Near the source of emission, the positrons combine with an electron to emit two gamma rays in nearly opposite directions. Upon detection of these two rays, a measurement representing the line integral of the absorption distribution along each path is obtained.

Magnetic Resonance Imaging

Another important situation where the image reconstruction problem arises is in magnetic resonance imaging (MRI).† Being noninvasive, it is becoming increasingly attractive in medical imaging for measuring (most commonly) the density of protons (that is, hydrogen nuclei) in tissue. This imaging technique is based on the fundamental property that protons (and all other nuclei that have an odd number of protons or neutrons) possess a *magnetic moment* and *spin*. When placed in a magnetic field, the proton precesses about the magnetic field in a manner analogous to a top spinning about the earth's gravitational field. Initially the protons are aligned either parallel or antiparallel to the magnetic field. When an RF signal having an appropriate strength and frequency is applied to the object, the protons absorb energy, and more of them switch to the antiparallel state. When the applied RF signal is removed, the absorbed energy is reemitted and is detected by an RF receiver. The proton density and environment can be determined from the characteristics of this detected signal. By controlling the applied RF signal and the surrounding magnetic field, these events can be made to occur along only one line within the object. The detected signal is then a function of the line integral of the MRI signal in the object. In fact, it can be shown that the detected signal is the Fourier transform of the projection at a given angle [8, 9].

Projection-based Image Processing

In the foregoing CT problems, the projection-space coordinates (s, θ) arise naturally because of the data gathering mechanics. This coordinate system plays an important role in many other image processing applications unrelated to CT. For example, the Hough transform, useful for detection of straight-line segments of polygonal shapes (see Section 9.5), is a representation of a straight line in the projection space. Also, two-dimensional linear shift invariant filters can be realized by a set of decoupled one-dimensional filters by working in the projection space. Other applications where projections are useful are in image segmentation (see Example 9.8), geometrical analysis of objects [11] and in image processing applications requiring transformations between polar and rectangular coordinates.

We are now ready to discuss the Radon transform, which provides the mathematical framework necessary for going back and forth between the spatial coordinates (x, y) and the projection-space coordinates (s, θ).

10.2 THE RADON TRANSFORM [12, 13]

Definition

The Radon transform of a function $f(x, y)$, denoted as $g(s, \theta)$, is defined as its line integral along a line inclined at an angle θ from the y-axis and at a distance s from

† Also called nuclear magnetic resonance (NMR) imaging. To emphasize its noninvasive features, the word *nuclear* is being dropped by manufacturers of such imaging systems to avoid confusion with nuclear reactions associated with nuclear energy and radioactivity.

the origin (Fig. 10.2). Mathematically, it is written as

$$g(s, \theta) \triangleq \mathcal{R}f = \iint_{-\infty}^{\infty} f(x, y)\delta(x \cos\theta + y \sin\theta - s)\, dx\, dy, \qquad -\infty < s < \infty, 0 \le \theta < \pi \tag{10.4}$$

The symbol $\mathcal{R}$, denoting the Radon transform operator, is also called the *projection operator.* The function $g(s, \theta)$, the Radon transform of $f(x, y)$, is the one-dimensional projection of $f(x, y)$ at an angle θ. In the rotated coordinate system (s, u), where

$$\begin{aligned} s &= x \cos\theta + y \sin\theta \\ u &= -x \sin\theta + y \cos\theta \end{aligned} \quad \text{or} \quad \begin{aligned} x &= s \cos\theta - u \sin\theta \\ y &= s \sin\theta + u \cos\theta \end{aligned} \tag{10.5}$$

(10.4) can be expressed as

$$g(s, \theta) = \int_{-\infty}^{\infty} f(s \cos\theta - u \sin\theta, s \sin\theta + u \cos\theta)\, du, \qquad -\infty < s < \infty, 0 \le \theta < \pi \tag{10.6}$$

The quantity $g(s, \theta)$ is also called a *ray-sum,* since it represents the summation of $f(x, y)$ along a ray at a distance s and at an angle θ.

The Radon transform maps the spatial domain (x, y) to the domain (s, θ). Each point in the (s, θ) space corresponds to a line in the spatial domain (x, y). Note that (s, θ) are not the polar coordinates of (x, y). In fact, if (r, ϕ) are the polar coordinates of (x, y), that is,

$$x = r \cos\phi, \qquad y = r \sin\phi \tag{10.7}$$

then from Fig. 10.3a

$$s = r \cos(\theta - \phi) \tag{10.8}$$

For a fixed point (r, ϕ), this equation gives the locus of all the points in (s, θ), which is a sinusoid as shown in Fig. 10.3b. Recall from section 9.5 that the coordinate pair (s, θ) is also the Hough transform of the straight line in Fig. 10.3a.

Example 10.1

Consider a plane wave, $f(x, y) = \exp[j2\pi(4x + 3y)]$. Then its projection function is

$$\begin{aligned} g(s, \theta) &= \int_{-\infty}^{\infty} \exp[j8\pi(s \cos\theta - u \sin\theta)] \exp[j6\pi(s \sin\theta + u \cos\theta)]\, du \\ &= \exp[j2\pi s(4 \cos\theta + 3 \sin\theta)] \int_{-\infty}^{\infty} \exp[-j2\pi u(4 \sin\theta - 3 \cos\theta)]\, du \\ &= \exp[j2\pi s(4 \cos\theta + 3 \sin\theta)]\delta(4 \sin\theta - 3 \cos\theta) = (\tfrac{1}{5})e^{j10\pi s}\, \delta(\theta - \phi) \end{aligned}$$

where $\phi = \tan^{-1}(\frac{3}{4})$. Here we have used the identity

$$\delta[f(\theta)] \equiv \sum_k \left[\frac{1}{|f'(\theta_k)|}\right]\delta(\theta - \theta_k) \tag{10.9}$$

where $f'(\theta) \triangleq df(\theta)/d\theta$ and $\theta_k, k = 1, 2, \ldots,$ are the roots of $f(\theta)$.

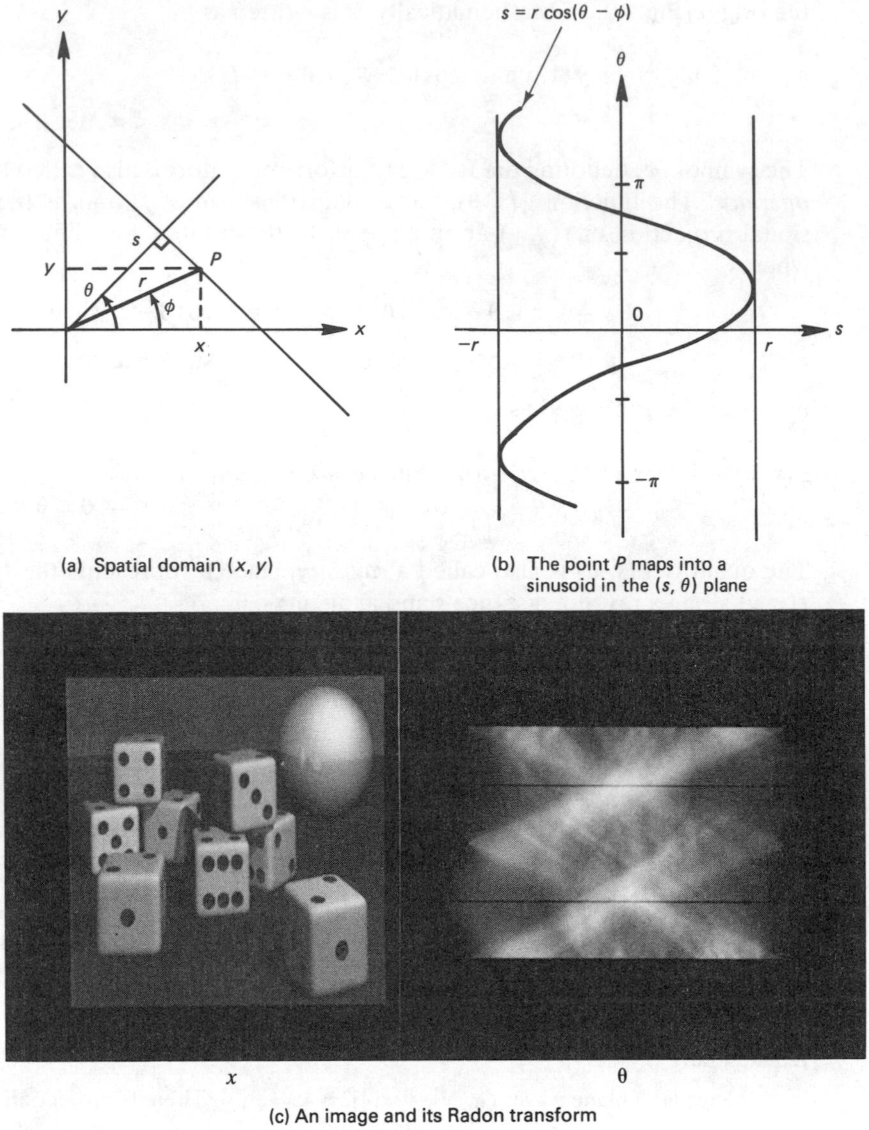

(a) Spatial domain (x, y)

(b) The point P maps into a sinusoid in the (s, θ) plane

(c) An image and its Radon transform

Figure 10.3 Spatial and Radon transform domains.

Notation

In order to avoid confusion between functions defined in different coordinates, we adopt the following notation. Let $\mathscr{U}$ be the space of functions defined on $\mathbb{R}^2$, where $\mathbb{R}$ denotes the real line. The two-dimensional Fourier transform pair for a function $f(x, y) \in \mathscr{U}$ is denoted by the relation

$$f(x, y) \overset{\mathscr{F}_2}{\longleftrightarrow} F(\xi_1, \xi_2) \tag{10.10}$$

In polar coordinates we write

$$F_p(\xi, \theta) = F(\xi \cos\theta, \xi \sin\theta) \tag{10.11}$$

The inner product in $\mathcal{U}$ is defined as

$$\langle f_1, f_2 \rangle_{\mathcal{U}} \stackrel{\Delta}{=} \iint_{-\infty}^{\infty} f_1(x, y) f_2^*(x, y)\, dx\, dy, \qquad \|f\|_{\mathcal{U}}^2 \stackrel{\Delta}{=} \langle f, f \rangle_{\mathcal{U}} \tag{10.12}$$

Let $\mathcal{V}$ be the space of functions defined on $\mathbb{R} \times [0, \pi]$. The *one-dimensional Fourier transform* of a function $g(s, \theta) \in \mathcal{V}$ is defined with respect to the variable s and is indicated as

$$g(s, \theta) \xleftrightarrow[s]{\mathcal{F}_1} G(\xi, \theta) \tag{10.13}$$

The inner product in $\mathcal{V}$ is defined as

$$\langle g_1, g_2 \rangle_{\mathcal{V}} \stackrel{\Delta}{=} \int_0^{\pi} \int_{-\infty}^{\infty} g_1(s, \theta) g_2^*(s, \theta)\, ds\, d\theta, \qquad \|g\|_{\mathcal{V}}^2 \stackrel{\Delta}{=} \langle g, g \rangle_{\mathcal{V}} \tag{10.14}$$

For simplicity we will generally consider $\mathcal{U}$ and $\mathcal{V}$ to be spaces of real functions. The notation

$$g = \mathcal{R} f \tag{10.15}$$

will be used to denote the Radon transform of $f(x, y)$, where it will be understood that $f \in \mathcal{U}, g \in \mathcal{V}$.

Properties of the Radon Transform

The Radon transform is linear and has several useful properties (Table 10.1), which can be summarized as follows. The projections $g(s, \theta)$ are space-limited in s if the object $f(x, y)$ is space-limited in (x, y), and are periodic in θ with period 2π. A translation of $f(x, y)$ results in the shift of $g(s, \theta)$ by a distance equal to the pro-

TABLE 10.1 Properties of the Radon Transform

	Function	Radon Transform
	$f(x, y) = f_p(r, \phi)$	$g(s, \theta)$
1	Linearity: $a_1 f_1(x, y) + a_2 f_2(x, y)$	$a_1 g_1(s, \theta) + a_2 g_2(s, \theta)$
2	Space limitedness: $f(x, y) = 0, \lvert x\rvert > \frac{D}{2}, \lvert y\rvert > \frac{D}{2}$	$g(s, \theta) = 0, \quad \lvert s\rvert > \frac{D\sqrt{2}}{2}$
3	Symmetry: $f(x, y)$	$g(s, \theta) = g(-s, \theta \pm \pi)$
4	Periodicity: $f(x, y)$	$g(s, \theta) = g(s, \theta + 2k\pi)$, k = integer
5	Shift: $f(x - x_0, y - y_0)$	$g(s - x_0 \cos\theta - y_0 \sin\theta, \theta)$
6	Rotation by θ_0: $f_p(r, \phi + \theta_0)$	$g(s, \theta + \theta_0)$
7	Scaling: $f(ax, ay)$	$\frac{1}{\lvert a\rvert} g(as, \theta), \quad a \neq 0$
8	Mass conservation: $M = \iint_{-\infty}^{\infty} f(x, y)\, dx\, dy$	$M = \int_{-\infty}^{\infty} g(s, \theta)\, ds, \quad \forall\theta$

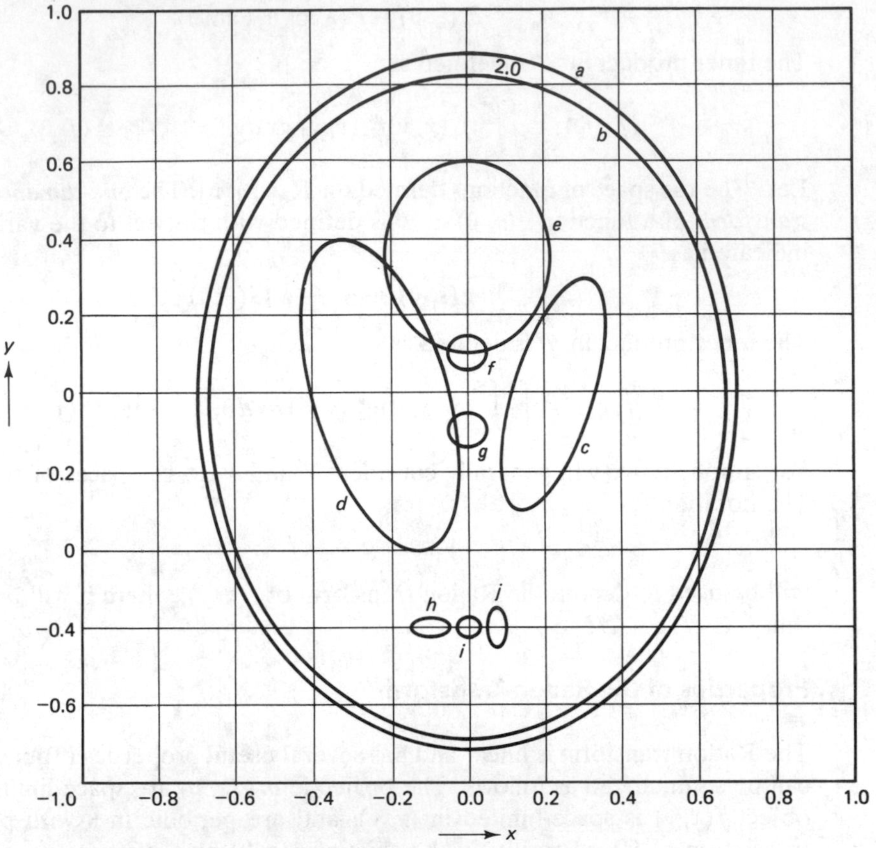

(a)

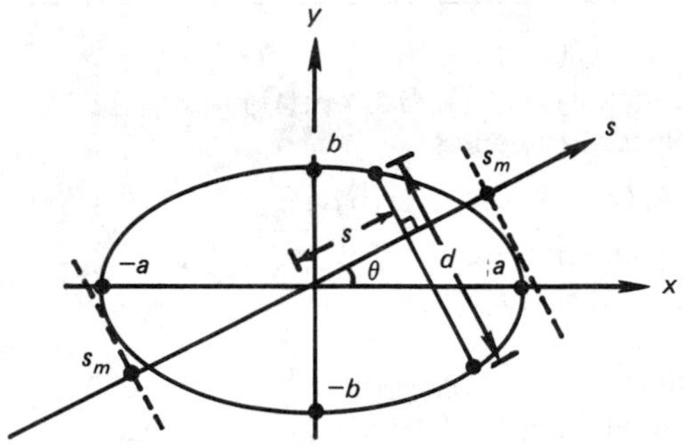

(b)

Figure 10.4 (a) Head phantom model; (b) constant-density ellipise, $f(x, y) = f_0$ for $(x^2/a^2) + (y^2/b^2) \le 1$.

TABLE 10.2 Head Phantom Components; $(\bar{x}, \bar{y})$ are the coordinates of the center of the ellipse. The densities indicated are relative to the density of water [18].

Ellipse	$\bar{x}$	$\bar{y}$	Major semiaxis	Minor semiaxis	Inclination (degrees)	Density $f_i(x, y)$
a	0.0000	0.0000	0.6900	0.9200	0.00	1.0000
b	0.0000	−0.0184	0.6624	0.8740	0.00	−0.9800
c	0.2200	0.0000	0.1100	0.3100	−18.00	−0.0200
d	−0.2200	0.0000	0.1600	0.4100	18.00	−0.0200
e	0.0000	0.3500	0.2100	0.2500	0.00	0.0100
f	0.0000	0.1000	0.0460	0.0460	0.00	0.0100
g	0.0000	−0.1000	0.0460	0.0460	0.00	0.0100
h	−0.0800	−0.6050	0.0460	0.0230	0.00	0.0100
i	0.0000	−0.6060	0.0230	0.0230	0.00	0.0100
j	0.0600	−0.6050	0.0230	0.0460	0.00	0.0100

jection of the translation vector on the line $s = x \cos\theta + y \sin\theta$. A rotation of the object by an angle θ_0 causes a translation of its Radon transform in the variable θ. A scaling of the (x, y) coordinates of $f(x, y)$ results in scaling of the s coordinate together with an amplitude scaling of $g(s, \theta)$. Finally, the total mass of a distribution $f(x, y)$ is preserved by $g(s, \theta)$ for all θ.

Example 10.2 Computer generation of projections of a phantom

In the development and evaluation of reconstruction algorithms, it is useful to simulate projection data corresponding to an idealized object. Figure 10.4a shows an object composed of ellipses, which is intended to model the human head [18, 21]. Table 10.2 gives the parameters of the component ellipses. For the ellipse shown in Fig. 10.4b, the projection at an angle θ is given by

$$g_0(s, \theta) = f_0 d = \begin{cases} \dfrac{2ab\sqrt{s_m^2 - s^2}}{s_m^2} f_0, & |s| \le s_m \\ 0, & |s| > s_m \end{cases}$$

where $s_m^2 = a^2 \cos^2\theta + b^2 \sin^2\theta$. Using the superposition, translation, and rotation properties of the Radon transform, the projection function for the object of Fig. 10.4a can be calculated (see Fig. 10.13a).

10.3 THE BACK-PROJECTION OPERATOR

Definition

Associated with the Radon transform is the *back-projection operator* $\mathcal{B}$, which is defined as

$$b(x, y) \triangleq \mathcal{B}g = \int_0^{\pi} g(x \cos\theta + y \sin\theta, \theta)\, d\theta \tag{10.16}$$

The quantity $b(x, y)$ is called the back projection of $g(s, \theta)$. In polar coordinates it can be written as

$$b(x, y) = b_p(r, \phi) = \int_0^\pi g(r \cos(\theta - \phi), \theta)\, d\theta \tag{10.17}$$

Back projection represents the accumulation of the ray-sums of all of the rays that pass through the point (x, y) or (r, ϕ). For example, if

$$g(s, \theta) = g_1(s)\delta(\theta - \theta_1) + g_2(s)\delta(\theta - \theta_2)$$

that is, if there are only two projections, then (see Fig. 10.5)

$$b_p(r, \phi) = g_1(s_1) + g_2(s_2)$$

where $s_1 = r\cos(\theta_1 - \phi)$, $s_2 = r\cos(\theta_2 - \phi)$. In general, for a fixed point (x, y) or (r, ϕ), the value of back projection $\mathcal{B}g$ is evaluated by integrating $g(s, \theta)$ over θ for all lines that pass through that point. In view of (10.8) and (10.17), the back-projection at (r, ϕ) is also the integration of $g(s, \theta)$ along the sinusoid $s = r\cos(\theta - \phi)$ in the (s, θ) plane (Fig. 10.3b).

Remarks

The back-projection operator $\mathcal{B}$ maps a function of (s, θ) coordinates into a function of spatial coordinates (x, y) or (r, ϕ).

The back-projection $b(x, y)$ at any pixel (x, y) requires projections from all directions. This is evident from (10.16).

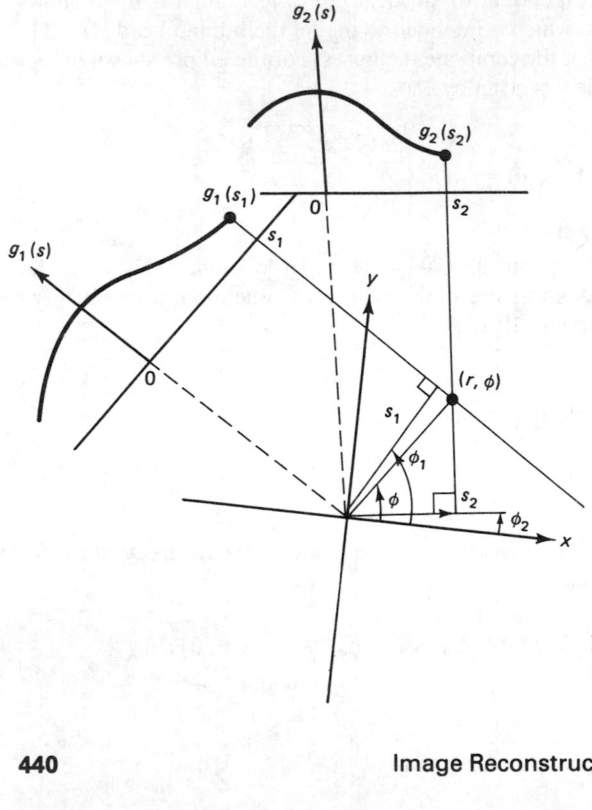

Figure 10.5 Back-projection of $g_1(s)$ and $g_2(s)$ at (r, ϕ).

It can be shown that the back-projected Radon transform

$$\tilde{f}(x, y) \overset{\Delta}{=} \mathcal{B}g = \mathcal{B}\mathcal{R}f \tag{10.18}$$

is an image of $f(x, y)$ blurred by the PSF $1/(x^2 + y^2)^{1/2}$, that is,

$$\tilde{f}(x, y) = f(x, y) \circledast (x^2 + y^2)^{-1/2} \tag{10.19}$$

where $\circledast$ denotes the two-dimensional convolution in the Cartesian coordinates. In polar coordinates

$$\tilde{f}_p(r, \phi) = f_p(r, \phi) \circledast \frac{1}{|r|} \tag{10.20}$$

where $\circledast$ now denotes the convolution expressed in polar coordinates (Problem 10.6). *Thus, the operator $\mathcal{B}$ is not the inverse of $\mathcal{R}$.* In fact, $\mathcal{B}$ is the adjoint of $\mathcal{R}$ [Problem 10.7]. Suppose the object $f(x, y)$ and its projections $g(s, \theta)$, for all θ, are discretized and mapped into vectors **f** and **g** and are related by a matrix transformation $\mathbf{g} = \mathbf{R}\mathbf{f}$. The matrix **R** then is a finite difference approximation of the operator $\mathcal{R}$. The matrix $\mathbf{R}^T$ would represent the approximation of the back-projection operator $\mathcal{B}$.

The operation $\tilde{f} = \mathcal{B}[\mathcal{R}f]$ gives the *summation algorithm* (Fig. 10.6). For a set of isolated small objects with a small number of projections, this method gives a star pattern artifact (Fig. 10.6) [15].

The object $f(x, y)$ can be restored from $\tilde{f}(x, y)$ by a two-dimensional (inverse) filter whose frequency response† is $|\xi| = \sqrt{\xi_1^2 + \xi_2^2}$, that is,

$$f(x, y) = \mathcal{F}_2^{-1}|\xi|\mathcal{F}_2[\mathcal{B}g] \tag{10.21}$$

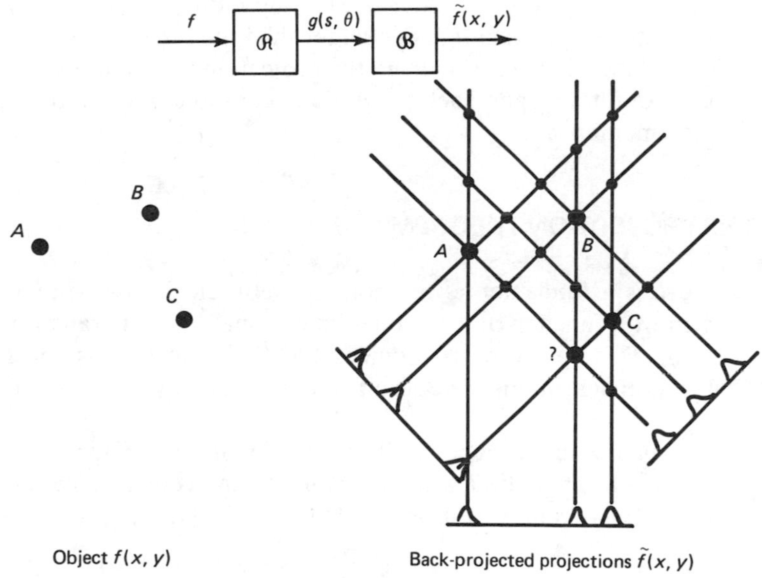

Figure 10.6 Summation algorithm for image reconstruction, $\tilde{f} \overset{\Delta}{=} \mathcal{B}g$.

† Note that the Fourier transform of $(x^2 + y^2)^{-1/2}$ is $(\xi_1^2 + \xi_2^2)^{-1/2}$.

TABLE 10.3 Filter Functions for Convolution/Filter Back-Projection Algorithms, $d \stackrel{\Delta}{=} 1/2\xi_0$

Filter	Frequency response $H(\xi)$	Impulse response $h(s)$	Discrete impulse response $\hat{h}(m) \stackrel{\Delta}{=} dh(md)$
Ram-Lak	$H_{RL}(\xi) \stackrel{\Delta}{=} \lvert\xi\rvert \operatorname{rect}(\xi d)$	$h_{RL}(s) = \xi_0^2 \cdot$ $\cdot [2 \operatorname{sinc}(2\xi_0 s) - \operatorname{sinc}^2(\xi_0 s)]$	$\hat{h}_{RL}(m) = \begin{cases} \dfrac{1}{4d}, & m = 0 \\ \dfrac{-\sin^2(\pi m/2)}{\pi^2 m^2 d}, & m \neq 0 \end{cases}$
Shepp-Logan	$\lvert\xi\rvert \operatorname{sinc}(\xi d) \operatorname{rect}(\xi d)$	$\dfrac{2(1 + \sin 2\pi\xi_0 s)}{\pi^2(d^2 - 4s^2)}$	$\dfrac{2}{\pi^2 d(1 - 4m^2)}$
Low-pass cosine	$\lvert\xi\rvert \cos(\pi\xi d) \operatorname{rect}(\xi d)$	$\frac{1}{2}\left[h_{RL}\left(s - \frac{d}{2}\right) + h_{RL}\left(s + \frac{d}{2}\right)\right]$	$\frac{1}{2}[h_{RL}(m - \frac{1}{2}) + h_{RL}(m + \frac{1}{2})]$
Generalized Hamming	$\lvert\xi\rvert [\alpha + (1 - \alpha) \cos 2\pi\xi d] \cdot$ $\operatorname{rect}(\xi d), 0 \leq \alpha \leq 1$	$\alpha h_{RL}(s) + \dfrac{1-\alpha}{2} \cdot$ $\cdot [h_{RL}(s - d) + h_{RL}(s + d)]$	$\alpha h_{RL}(m) + \left(\dfrac{1-\alpha}{2}\right) \cdot$ $\cdot [h_{RL}(m - 1) + h_{RL}(m + 1)]$
Stochastic	See eq. (10.70) and Example 10.6		

where $\mathscr{F}_2$ denotes the two-dimensional Fourier transform operator. In practice the filter $|\xi|$ is replaced by a physically realizable approximation (see Table 10.3). This method [16] is appealing because the filtering operations can be implemented approximately via the FFT. However, it has two major difficulties. First, the Fourier domain computation of $|\xi|\tilde{F}_p(\xi, \theta)$ gives $F(0, 0) = 0$, which yields the total density $\iint f(x, y)\, dx\, dy = 0$. Second, since the support of $\mathscr{B}g$ is unbounded, $\tilde{f}(x, y)$ has to be computed over a region much larger than the region of support of $f(x, y)$. A better algorithm, which follows from the projection theorem discussed next, reverses the order of filtering and back-projection operations and is more attractive for practical implementations.

10.4 THE PROJECTION THEOREM [5–7, 12, 13]

There is a fundamental relationship between the two-dimensional Fourier transform of a function and the one-dimensional Fourier transform of its Radon transform. This relationship provides the theoretical basis for several image reconstruction algorithms. The result is summarized by the following theorem.

Projection Theorem. The one-dimensional Fourier transform with respect to s of the projection $g(s, \theta)$ is equal to the central slice, at angle θ, of the two-dimensional Fourier transform of the object $f(x, y)$, that is, if

$$g(s, \theta) \xleftrightarrow[s]{\mathscr{F}_1} G(\xi, \theta)$$

then,

$$G(\xi, \theta) = F_p(\xi, \theta) \stackrel{\Delta}{=} F(\xi \cos\theta, \xi \sin\theta) \tag{10.22}$$

Figure 10.7 shows the meaning of this result. This theorem is also called the *projection-slice theorem.*

Proof. Using (10.6) in the definition of $G(\xi, \theta)$, we can write

$$G(\xi, \theta) \triangleq \int_{-\infty}^{\infty} g(s, \theta) e^{-j2\pi\xi s}\, ds$$
$$= \iint_{-\infty}^{\infty} f(s\cos\theta - u\sin\theta\ ,\ s\sin\theta + u\cos\theta) e^{-j2\pi\xi s}\, ds\, du \tag{10.23}$$

Performing the coordinate transformation from (s, u) to (x, y), [see (10.5)], this becomes

$$G(\xi, \theta) = \iint_{-\infty}^{\infty} f(x, y) \exp[-j2\pi(x\xi\cos\theta + y\xi\sin\theta)]\, dx\, dy$$
$$= F(\xi\cos\theta, \xi\sin\theta)$$

which proves (10.22).

Remarks

From the symmetry property of Table 10.1, we find that the Fourier transform slice also satisfies a similar property

$$G(-\xi, \theta + \pi) = G(\xi, \theta) \tag{10.24}$$

If $f(x, y)$ is bandlimited, then so are the projections. This follows immediately from the projection theorem.

An important consequence of the projection theorem is the following result.

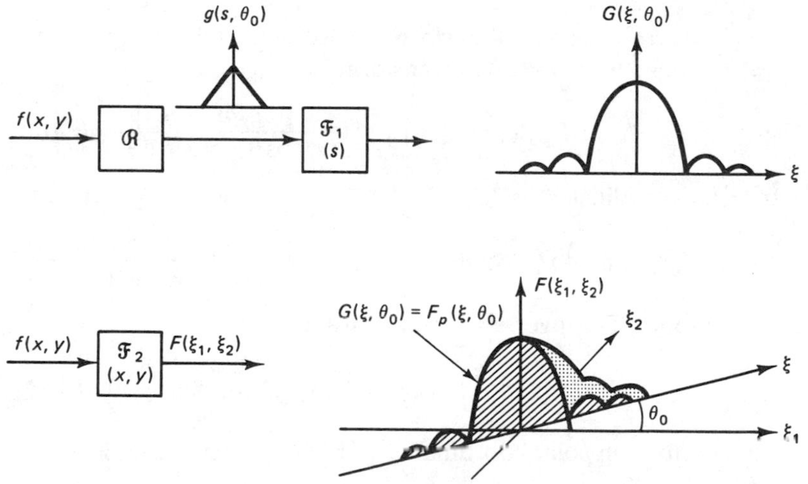

Figure 10.7 The projection theorem, $G(\xi, \theta) = F_p(\xi, \theta)$.

Convolution-Projection Theorem. The Radon transform of the two-dimensional convolution of two functions $f_1(x, y)$ and $f_2(x, y)$ is equal to the one-dimensional convolution of their Radon transforms, that is, if $g_k \triangleq \mathcal{R} f_k, k = 1, 2$, then

$$\mathcal{R}\left\{\iint_{-\infty}^{\infty} f_1(x - x', y - y')f_2(x', y')\, dx'\, dy'\right\} = \int_{-\infty}^{\infty} g_1(s - s', \theta)g_2(s', \theta)\, ds' \qquad (10.25)$$

The proof is developed in Problem 10.9. This theorem is useful in the implementation of two-dimensional linear filters by one-dimensional filters. (See Fig. 10.9 and the accompanying discussion in Section 10.5.)

Example 10.3

We will use the projection theorem to obtain the $g(s, \theta)$ of Example 10.1. The two-dimensional Fourier transform of $f(x, y)$ is $F(\xi_1, \xi_2) = \delta(\xi_1 - 4)\delta(\xi_2 - 3) = \delta(\xi \cos\theta - 4)\delta(\xi \sin\theta - 3)$. From (10.22) this gives $G(\xi, \theta) = \delta(\xi \cos\theta - 4)\delta(\xi \sin\theta - 3)$. Taking the one-dimensional inverse Fourier transform with respect to ξ and using the identity (10.9), we get the desired result

$$g(s, \theta) = \int_{-\infty}^{\infty} \delta(\xi \cos\theta - 4)\delta(\xi \sin\theta - 3)e^{j2\pi s\xi}\, d\xi$$

$$= \left(\frac{1}{|\cos\theta|}\right) \exp\left(\frac{j8\pi s}{\cos\theta}\right)\delta(4\tan\theta - 3) = (\tfrac{1}{5})e^{j10\pi s}\,\delta(\theta - \phi)$$

where $\phi \triangleq \tan^{-1}(\tfrac{3}{4})$.

10.5 THE INVERSE RADON TRANSFORM [6, 12, 13, 17]

The image reconstruction problem defined in Section 10.1 is theoretically equivalent to finding the inverse Radon transform of $g(s, \theta)$. The projection theorem is useful in obtaining this inverse. The result is summarized by the following theorem.

Inverse Radon Transform Theorem. Given $g(s, \theta) \triangleq \mathcal{R} f$, $-\infty < s < \infty$, $0 \le \theta < \pi$, its inverse Radon transform is

$$f(x, y) = \left(\frac{1}{2\pi^2}\right)\int_0^{\pi}\int_{-\infty}^{\infty} \frac{[(\partial g/\partial s)(s, \theta)]}{x\cos\theta + y\sin\theta - s}\, ds\, d\theta \qquad (10.26)$$

In polar coordinates

$$f_p(r, \phi) \triangleq f(r\cos\phi, r\sin\phi) = \left(\frac{1}{2\pi^2}\right)\int_0^{\pi}\int_{-\infty}^{\infty} \frac{[(\partial g/\partial s)(s, \theta)]}{r\cos(\theta - \phi) - s}\, ds\, d\theta \qquad (10.27)$$

Proof. The inverse Fourier transform

$$f(x, y) = \iint_{-\infty}^{\infty} F(\xi_1, \xi_2)\exp[j2\pi(\xi_1 x + \xi_2 y)\, d\xi_1\, d\xi_2$$

when written in polar coordinates in the frequency plane, gives

$$f(x, y) = \int_0^{2\pi}\int_0^{\infty} F_p(\xi, \theta)\exp[j2\pi\xi(x\cos\theta + y\sin\theta)]\xi\, d\xi\, d\theta \qquad (10.28)$$

Allowing ξ to be negative and $0 \le \theta < \pi$, we can change the limits of integration and use (10.22) to obtain (show!)

$$\begin{aligned} f(x, y) &= \int_0^{\pi} \int_{-\infty}^{\infty} |\xi| F_p(\xi, \theta) \exp[j2\pi\xi(x \cos\theta + y \sin\theta)]\, d\xi\, d\theta \\ &= \int_0^{\pi} \left\{ \int_{-\infty}^{\infty} |\xi| G(\xi, \theta) \exp[j2\pi\xi(x \cos\theta + y \sin\theta)]\, d\xi \right\} d\theta \\ &= \int_0^{\pi} \hat{g}(x \cos\theta + y \sin\theta, \theta)\, d\theta \end{aligned} \tag{10.29}$$

where

$$\hat{g}(s, \theta) \triangleq \int_{-\infty}^{\infty} |\xi| G(\xi, \theta) e^{j2\pi\xi s}\, d\xi \tag{10.30}$$

Writing $|\xi|G$ as $\xi G \operatorname{sgn}(\xi)$ and applying the convolution theorem, we obtain

$$\begin{aligned} \hat{g}(s, \theta) &= [\mathscr{F}_1^{-1}\{\xi G(\xi, \theta)\}] \circledast [\mathscr{F}_1^{-1}\{\operatorname{sgn}(\xi)\}] \\ &= \left[\left(\frac{1}{j2\pi}\right) \frac{\partial g}{\partial s}(s, \theta)\right] \circledast \left(\frac{-1}{j\pi s}\right) \\ &= \left(\frac{1}{2\pi^2}\right) \int_{-\infty}^{\infty} \left[\frac{\partial g(t, \theta)}{\partial t}\right] \frac{1}{s - t}\, dt \end{aligned} \tag{10.31}$$

where $(1/j2\pi)[\partial g(s, \theta)/\partial s]$ and $(-1/j2\pi s)$ are the Fourier inverses of $\xi G(\xi, \theta)$ and $\operatorname{sgn}(\xi)$, respectively. Combining (10.29) and (10.31), we obtain the desired result of (10.26). Equation (10.27) is arrived at by the change of coordinates $x = r \cos\phi$ and $y = r \sin\phi$.

Remarks

The inverse Radon transform is obtained in two steps (Fig. 10.8a). First, each projection $g(s, \theta)$ is filtered by a one-dimensional filter whose frequency response is $|\xi|$. The result, $\hat{g}(s, \theta)$, is then back-projected to yield $f(x, y)$. The filtering operation can be performed either in the s domain or in the ξ domain. This process yields two different methods of finding $\mathscr{R}^{-1}$, which are discussed shortly.

The integrands in (10.26), (10.27), and (10.31) have singularities. Therefore, the Cauchy principal value should be taken (via contour integration) in evaluating the integrals.

Definition. The *Hilbert transform* of a function $\phi(t)$ is defined as

$$\psi(s) \triangleq \tilde{\mathscr{H}}\phi \triangleq \phi(s) \circledast \left(\frac{1}{\pi s}\right) = \left(\frac{1}{\pi}\right) \int_{-\infty}^{\infty} \frac{\phi(t)}{s - t}\, dt \tag{10.32}$$

The symbol $\tilde{\mathscr{H}}$ represents the Hilbert transform operator. From this definition it follows that $\hat{g}(s, \theta)$ is the Hilbert transform of $(1/2\pi)\partial g(s, \theta)/\partial s$ for each θ.

Because the back-projection operation is required for finding $\mathscr{R}^{-1}$, the reconstructed image pixel at (x, y) requires projections from all directions.

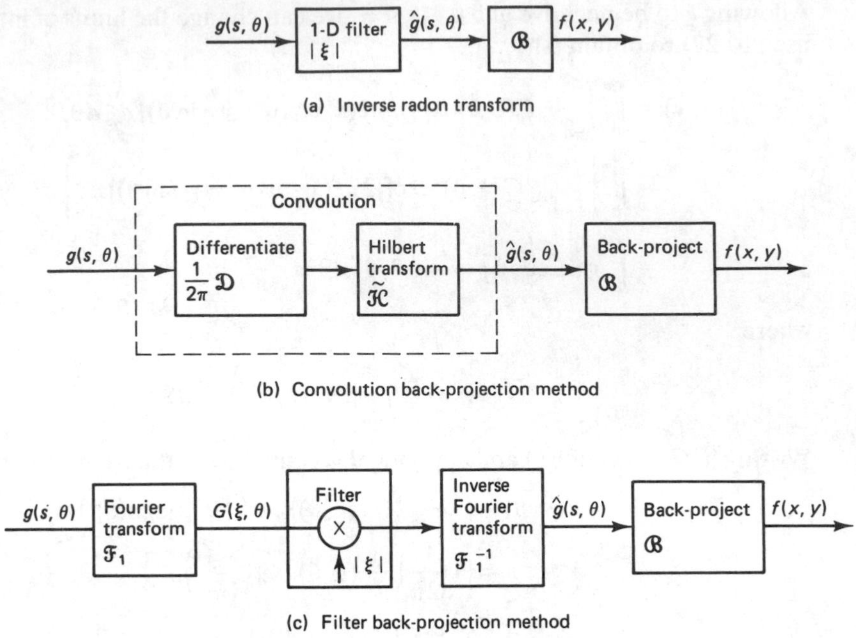

Figure 10.8 Inverse radon transform methods.

Convolution Back-Projection Method

Defining a derivative operator as

$$\mathscr{D}\phi \triangleq \frac{\partial \phi(s)}{\partial s} \tag{10.33}$$

The inverse Radon transform can be written as

$$f(x, y) = (1/2\pi)\, \mathscr{B}\, \tilde{\mathscr{H}}\, \mathscr{D} g \tag{10.34}$$

Thus the *inverse Radon transform operator* is $\mathscr{R}^{-1} = (1/2\pi)\mathscr{B}\tilde{\mathscr{H}}\mathscr{D}$. This means $\mathscr{R}^{-1}$ can also be implemented by convolving the differentiated projections with $1/2\pi s$ and back-projecting the result (Figure 10.8b).

Filter Back-Projection Method

From (10.29) and (10.30), we can also write

$$f(x, y) = \mathscr{B}\mathscr{H} g \tag{10.35}$$

where $\mathscr{H}$ is a one-dimensional filter whose frequency response is $|\xi|$, that is,

$$\begin{aligned} \hat{g} \triangleq \mathscr{H} g &\triangleq \int_{-\infty}^{\infty} |\xi| G(\xi, \theta) e^{j2\pi\xi s}\, ds \\ &= \mathscr{F}_1^{-1}\{|\xi|[\mathscr{F}_1 g]\} \end{aligned} \tag{10.36}$$

This gives

$$f(x, y) = \mathcal{B}\mathcal{F}_1^{-1}[|\xi|\mathcal{F}_1 g] \tag{10.37}$$

which can be implemented by filtering the projections in the Fourier domain and back-projecting the inverse Fourier transform of the result (Fig. 10.8c).

Example 10.4

We will find the inverse Radon transform of $g(s, \theta) = (\frac{1}{5})e^{j10\pi s}\delta(\theta - \phi)$.

Convolution back-projection method. Using $\partial g/\partial s = j2\pi e^{j10\pi s}\delta(\theta - \phi)$ in (10.26)

$$f(x, y) = \left(\frac{j2\pi}{2\pi^2}\right)\int_0^{\pi}\int_{-\infty}^{\infty} e^{j10\pi s}(x\cos\theta + y\sin\theta - s)^{-1}\delta(\theta - \phi)\,d\theta\,ds$$

$$= \left(\frac{1}{j\pi}\right)\int_{-\infty}^{\infty} e^{j10\pi s}[s - (x\cos\phi + y\sin\phi)]^{-1}\,ds$$

Since the Fourier inverse of $1/(\xi - \alpha)$ is $j\pi e^{j2\pi\alpha t}\,\text{sgn}(t)$, the preceding integral becomes $f(x, y) = \exp[j2\pi(x\cos\phi + y\sin\phi)t]\,\text{sgn}(t)|_{t=5} = \exp[j10\pi(x\cos\phi + y\sin\phi)]$.

Filter back-projection method.

$$G(\xi, \theta) = (\tfrac{1}{5})\delta(\xi - 5)\delta(\theta - \phi)$$

$$\Rightarrow \hat{g}(s, \theta) = (\tfrac{1}{5})\int_{-\infty}^{\infty} |\xi|\delta(\xi - 5)\delta(\theta - \phi)\exp(j2\pi s\xi)\,d\xi = e^{j10\pi s}\delta(\theta - \phi)$$

$$\Rightarrow f(x, y) = \int_0^{\pi}\exp[j10\pi(x\cos\theta + y\sin\theta)]\delta(\theta - \phi)\,d\theta$$

$$= \exp[j10\pi(x\cos\phi + y\sin\phi)]$$

For $\phi = \tan^1(\frac{3}{4})$, $f(x, y)$ will be the same as in Example 10.1.

Two-Dimensional Filtering via the Radon Transform

A useful application of the convolution-projection theorem is in the implementation of two-dimensional filters. Let $\tilde{A}(\xi_1, \xi_2)$ represent the frequency response of a two-dimensional filter. Referring to Fig. 10.9 and eq. (10.25), this filter can be implemented by first filtering for each θ, the one-dimensional projection $g(s, \theta)$ by

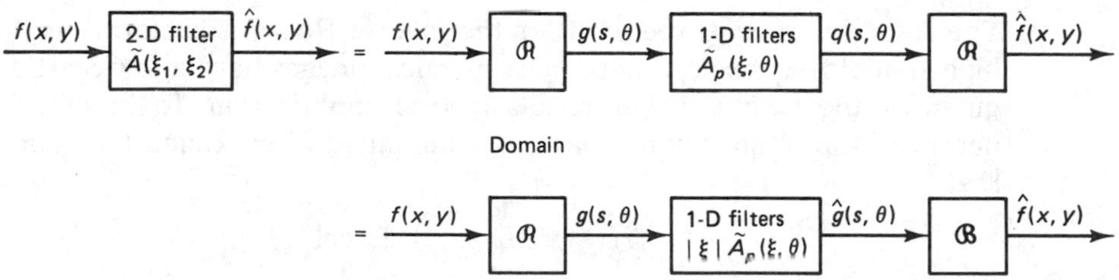

Figure 10.9 Generalized filter back projection algorithm for two-dimensional filter implementation.

a one-dimensional filter whose frequency response is $\ddot{A}_p(\xi, \theta)$ and then taking the inverse Radon transform of the result. Using the representation of $\mathcal{R}^{-1}$ in Fig. 10.8a, we obtain a *generalized filter-back-projection algorithm,* where the filter now becomes $|\xi|\tilde{A}_p(\xi, \theta)$. Hence, the two-dimensional filter $\tilde{A}(\xi_1, \xi_2)$ can be implemented as

$$\tilde{a}(x, y) \circledast f(x, y) = \mathcal{B}\mathcal{H}_\theta g \tag{10.38}$$

where $\mathcal{H}_\theta$ represents a one-dimensional filter with frequency response $A_p(\xi, \theta) \overset{\Delta}{=} |\xi|\tilde{A}_p(\xi, \theta)$.

10.6 CONVOLUTION/FILTER BACK-PROJECTION ALGORITHMS: DIGITAL IMPLEMENTATION [18–21]

The foregoing results are useful for developing practical image reconstruction algorithms. We now discuss various considerations for digital implementation of these algorithms.

Sampling Considerations

In practice, the projections are available only on a finite grid, that is, we have available

$$g_n(m) \overset{\Delta}{=} g(s_m, \theta_n) \overset{\Delta}{=} [\mathcal{R}f](s_m, \theta_n),$$
$$-\left(\frac{M}{2}\right) \le m \le \left(\frac{M}{2}\right) - 1, \qquad 0 \le n \le N - 1 \tag{10.39}$$

where, typically, $s_m = md$, $\theta_n = n\Delta$, $\Delta = \pi/N$. Thus we have N projections taken at equally spaced angles, each sampled uniformly with sampling interval d. If ξ_0 is the highest spatial frequency of interest in the given object, then d should not exceed the corresponding Nyquist interval, that is, $d \le 1/2\xi_0$. If the object is space limited, that is, $f_p(r, \phi) = 0$, $|r| > D/2$, then $D = Md$, and the number of samples should satisfy

$$M \ge 2\xi_0 D \tag{10.40}$$

Choice of Filters

The filter function $|\xi|$ required for the inverse Radon transform emphasizes the high-spatial frequencies. Since most practical images have a low SNR at high frequencies, the use of this filter results in noise amplification. To limit the unbounded nature of the frequency response, a bandlimited filter, called the Ram-Lak filter [19]

$$H(\xi) = H_{RL}(\xi) \overset{\Delta}{=} |\xi| \operatorname{rect}\left(\frac{\xi}{2\xi_0}\right) \tag{10.41}$$

has been proposed. In practice, most objects are space-limited and a bandlimiting filter with a sharp cutoff frequency ξ_0 is not very suitable, especially in the presence of noise. A small value of ξ_0 gives poor resolution and a very large value leads to noise amplification. A generalization of (10.41) is the class of filters

$$H(\xi) = |\xi| W(\xi) \tag{10.42}$$

Here $W(\xi)$ is a bandlimiting window function that is chosen to give a more-moderate high-frequency response in order to achieve a better trade-off between the filter bandwidth (that is, high-frequency response) and noise suppression. Table 10.3 lists several commonly used filters. Figure 10.10 shows the frequency and

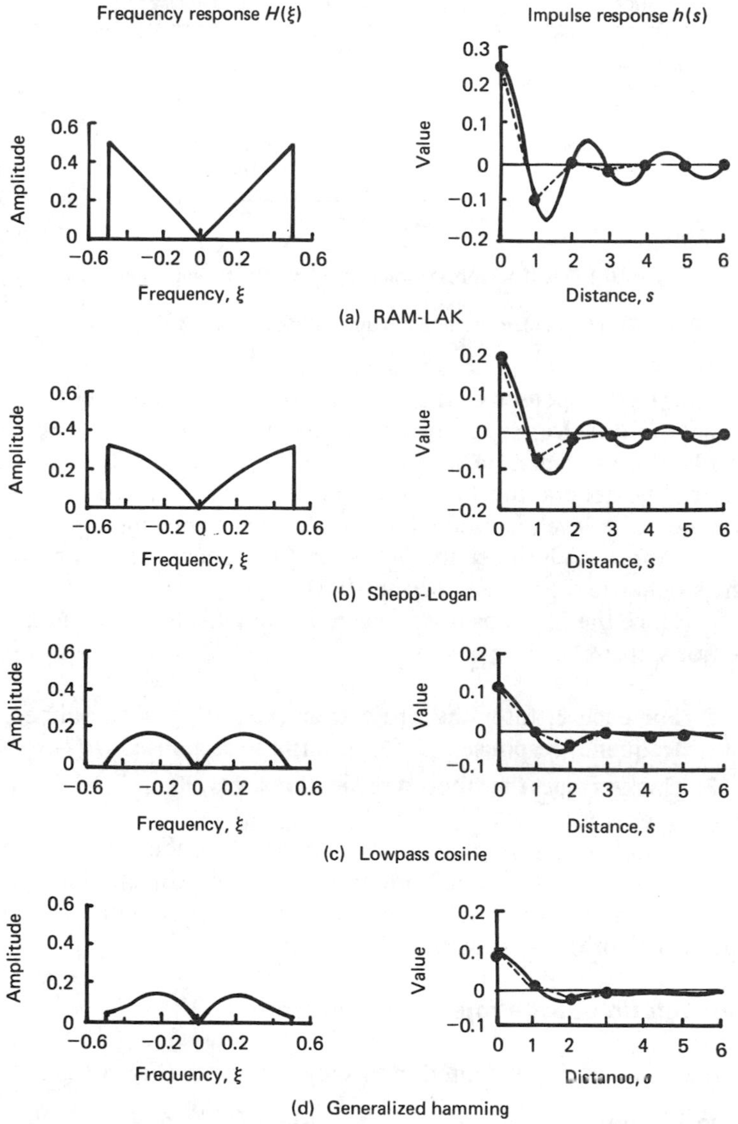

Figure 10.10 Reconstruction filters. Left column: Frequency response; right column: Impulse response; dotted lines show linearly interpolated response.

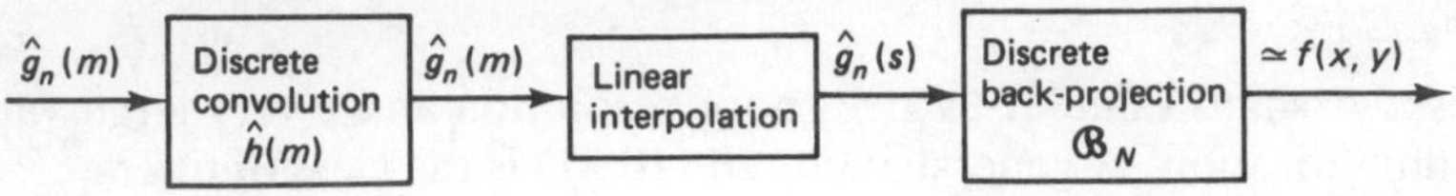

(a) Convolution back-projection algorithm: Digital implementation;

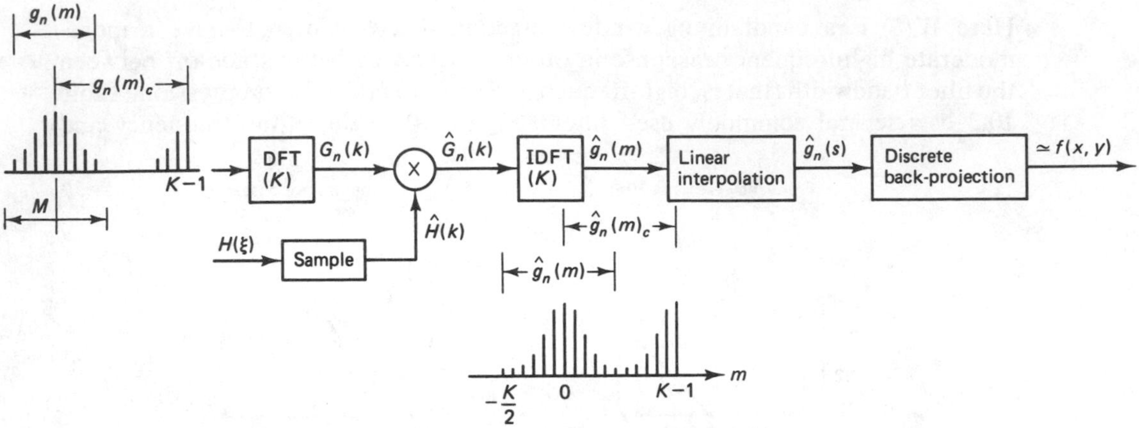

(b) Filter back-projection algorithm: Digital implementation.

Figure 10.11 Implementation of convolution/filter back projection algorithms.

the impulse responses of these filters for $d = 1$. Since these functions are real and even, the impulse responses are displayed on the positive real line only. For low levels of observation noise, the Shepp-Logan filter is preferred over the Ram-Lak filter. The generalized low-pass Hamming window, with the value of α optimized for the noise level, is used when the noise is significant. In the presence of noise a better approach is to use the optimum mean square reconstruction filter also called the stochastic filter, (see Section 10.8).

Once the filter has been selected, a practical reconstruction algorithm has two major steps:

1. For each θ, filter the projections $g(s, \theta)$ by a one-dimensional filter whose frequency response is $H(\xi)$ or impulse response is $h(s)$.
2. Back-project the filtered projections, $\hat{g}(s, \theta)$.

Depending on the implementation method of the filter, we obtain two distinct algorithms (Fig. 10.11). In both cases the back-projection integral [see eq. (10.17)] is implemented by a suitable finite-difference approximation. The steps required in the two algorithms are summarized next.

Convolution Back-Projection Algorithm

The equations implemented in this algorithm are (Fig. 10.11a)

Convolution: $$\hat{g}(s, \theta) = g(s, \theta) \circledast h(s) \tag{10.43a}$$

Back projection: $$f(x, y) = \mathcal{B}\hat{g} \tag{10.43b}$$

The filtering operation is implemented by a direct convolution in the s domain. The steps involved in the digital implementation are as follows:

1. Perform the following discrete convolution as an approximate realization of sampled values of the filtered projections, that is,

$$\hat{g}(md, n\Delta) \simeq \hat{g}_n(m) \triangleq \sum_{k=-M/2}^{M/2-1} g_n(k)\hat{h}(m-k), \qquad \frac{-M}{2} \le m \le \frac{M}{2} - 1 \tag{10.44}$$

where $\hat{h}(m) \triangleq dh(md)$ is obtained by sampling and scaling $h(s)$. Table 10.3 lists $\hat{h}(m)$ for the various filters. The preceding convolution can be implemented either directly or via the FFT as discussed in Section 5.4.

2. Linearly interpolate $\hat{g}_n(m)$ to obtain a piecewise continuous approximation of $\hat{g}(s, n\Delta)$ as

$$\hat{g}(s, n\Delta) \simeq \hat{g}_n(m) + \left(\frac{s}{d} - m\right)[\hat{g}_n(m+1) - \hat{g}_n(m)], \qquad md \le s < (m+1)d \tag{10.45}$$

3. Approximate the back-projection integral by the following operation to give

$$f(x, y) \simeq \hat{f}(x, y) \triangleq \mathscr{B}_N \hat{g} \triangleq \Delta \sum_{n=0}^{N-1} \hat{g}(x \cos n\Delta + y \sin n\Delta, n\Delta) \tag{10.46}$$

where $\mathscr{B}_N$ is called the *discrete back-projection operator*. Because of the back-projection operation, it is necessary to interpolate the filtered projections $\hat{g}_n(m)$. This is required even if the reconstructed image is evaluated on a sampled grid. For example, to evaluate

$$f(i\Delta_x, j\Delta_y) \simeq \Delta \sum_{n=0}^{N-1} \hat{g}(i\Delta_x \cos n\Delta + j\Delta_y \sin n\Delta, n\Delta) \tag{10.47}$$

on a grid with spacing (Δ_x, Δ_y), $i, j = 0, \pm 1, \pm 2, \ldots$, we still need to evaluate $\hat{g}(s, n\Delta)$ at locations in between the points md, $m = -M/2, \ldots, M/2 - 1$. Although higher-order interpolation via the Lagrange functions (see Chapter 4) is possible, the linear interpolation of (10.45) has been found to give a good trade-off between resolution and smoothing [18]. A zero-order hold is sometimes used to speed up the back-projection operation for hardware implementation.

Filter Back-Projection Algorithm

In Fig. 10.11b, the filtering operation is performed in the frequency domain according to the equation

$$\hat{g}(s, \theta) = \mathscr{F}_1^{-1}[G(\xi, \theta)H(\xi)] \tag{10.48}$$

Given $H(\xi)$, the filter frequency response, this filter is implemented approximately by using a sampled approximation of $G(\xi, \theta)$ and substituting a suitable FFT for the inverse Fourier transform. The algorithm is shown in Fig. 10.11b, which is a one-dimensional equivalent of the algorithm discussed in Section 8.3 (Fig. 8.13b). The steps of this algorithm are given next:

1. Extend the sequence $g_n(m)$, $-M/2 \le m \le (M/2) - 1$ by padding zeros and periodic repetition to obtain the sequence $g_n(m)_c$, $0 \le m \le K - 1$. Take its FFT to obtain $G_n(k)$, $0 \le k \le K - 1$. The choice of K determines the sampling resolution in the frequency domain. Typically $K \simeq 2M$ if M is large; for example, $K = 512$ if $M = 256$.
2. Sample $H(\xi)$ to obtain $\hat{H}(k) \triangleq H(k\Delta\xi)$, $\hat{H}(K-k) \triangleq \hat{H}^*(k)$, $0 \le k < K/2$, where * denotes the complex conjugate.
3. Multiply the sequences $\hat{G}_n(k)$ and $\hat{H}(k)$, $0 \le k \le K - 1$, and take the inverse FFT of the product. A periodic extension of the result gives $\hat{g}_n(m)$, $-K/2 \le m \le (K/2) - 1$. The reconstructed image is obtained via (10.45) and (10.46) as before.

Example 10.5

Figure 10.12b shows a typical projection of an object digitized on a 128×128 grid (Fig. 10.12a). Reconstructions obtained from 90 such projections, each with 256 samples per line, using the convolution back-projection algorithm with Ram-Lak and Shepp-Logan filters, are shown in Fig. 10.12c and d, respectively. Intensity plots of the object and its reconstructions along a horizontal line through its center are shown in Fig. 10.12f through h. The two reconstructions are almost identical in this (noiseless) case. The background noise that appears is due to the high-frequency response of the reconstruction filter and is typical of inverse (or pseudoinverse) filtering. The stochastic filter outputs shown in Fig. 10.12e and i show an improvement over this result. This filter is discussed in Section 10.8.

Reconstruction Using a Parallel Pipeline Processor

Recently, a powerful hardware architecture has been developed [11] that enables the high speed computation of digital approximations to the Radon transform and the back-projection operators. This allows the rapid implementation of convolution/filter back-projection algorithms as well as a large number of other image processing operations in the Radon space. Figure 10.13 shows some results of reconstruction using this processor architecture.

10.7 RADON TRANSFORM OF RANDOM FIELDS [22, 23]

So far we have considered $f(x, y)$ to be a deterministic function. In many problems, such as data compression and filtering of noise, it is useful to consider the input $f(x, y)$ to be a random field. Therefore, it becomes necessary to study the properties of the Radon transform of random fields, that is, projections of random fields.

A Unitary Transform $\tilde{\mathcal{R}}$

Radon transform theory for random fields can be understood more easily by considering the operator

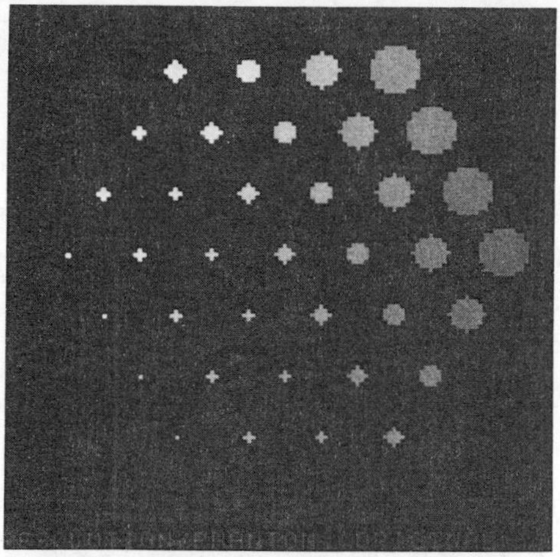

(a) Original object

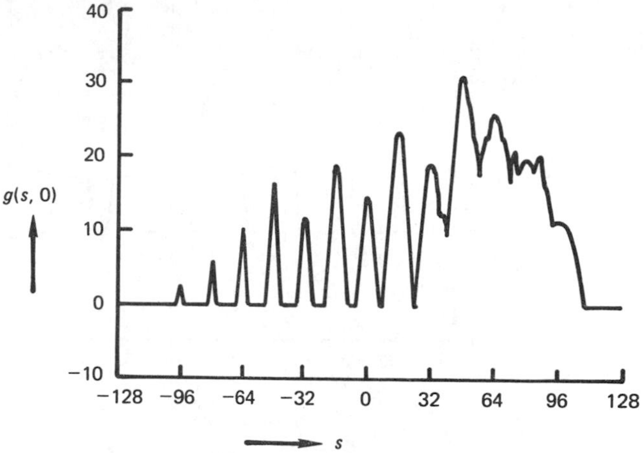

(b) A typical projection

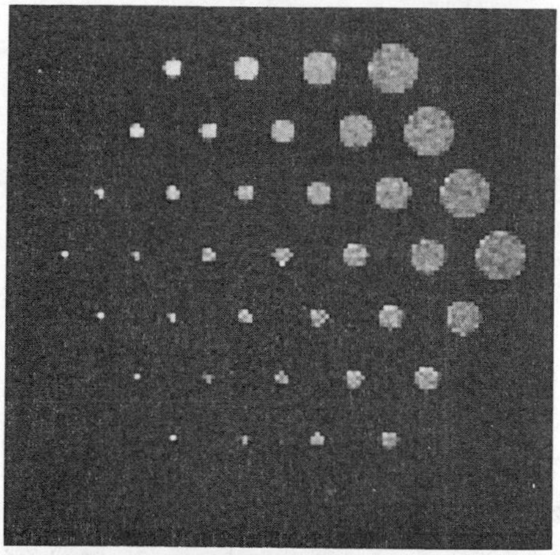

(c) Ram-Lak filter

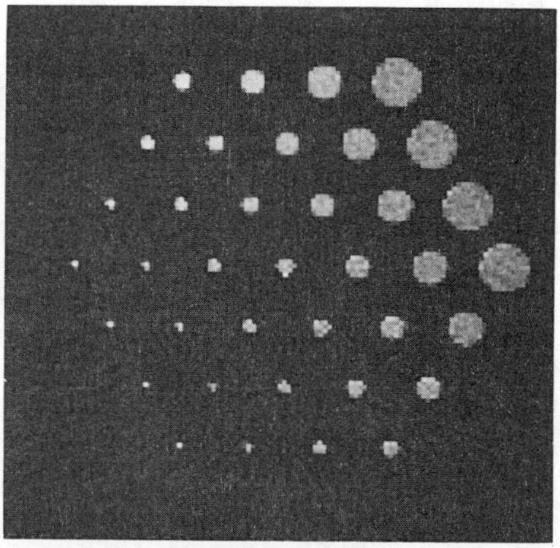

(d) Shepp-Logan filter

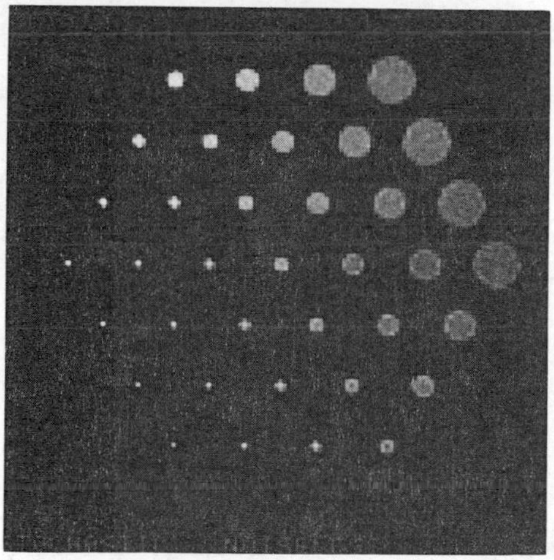

(e) Stochastic filter

Figure 10.12 Image reconstruction example.

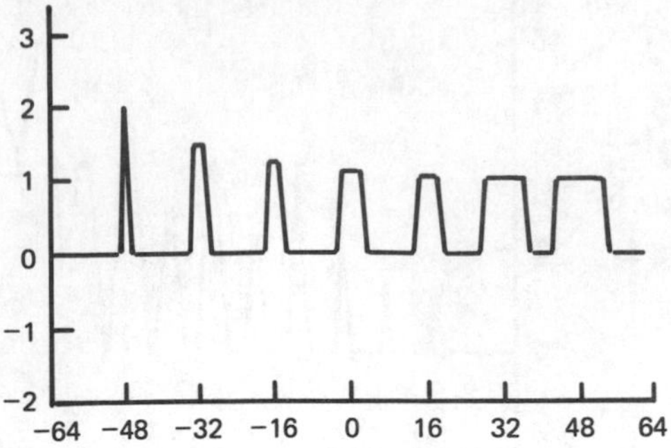

(f) Object line

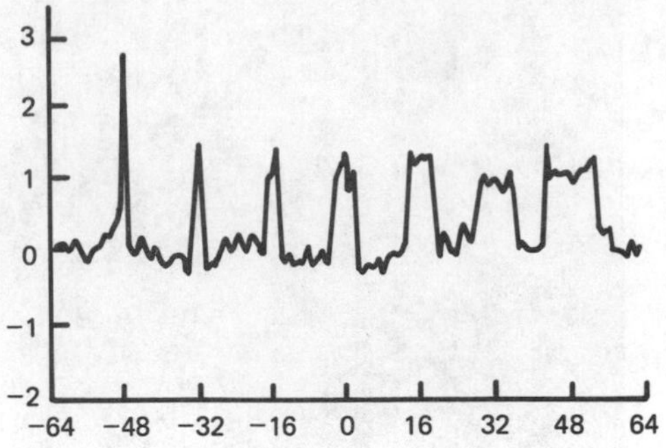

(g) Reconstruction via RAM-LAK filter

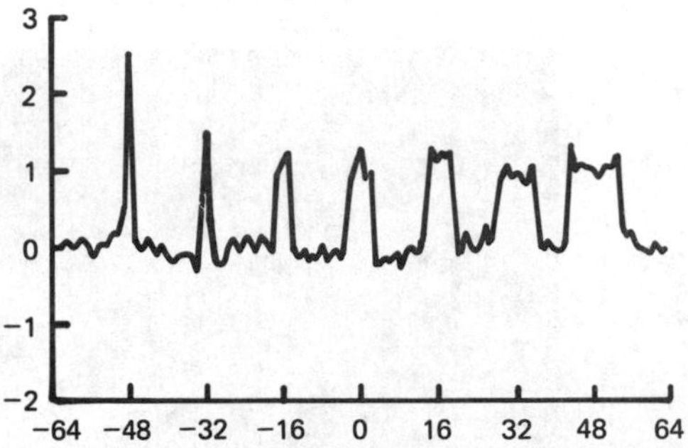

(h) Reconstruction via Shepp-Logan filter

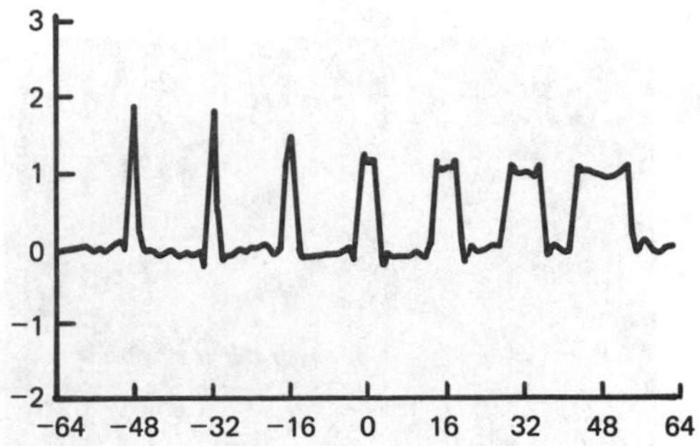

(i) Reconstruction via stochastic filter. Also see Example 10.6

Figure 10.12 Cont'd

$$\tilde{\mathcal{R}} \triangleq \mathcal{H}^{1/2} \mathcal{R} \tag{10.49}$$

where $\mathcal{H}^{1/2}$ represents a one-dimensional filter whose frequency response is $|\xi|^{1/2}$. The operation

$$\tilde{g}(s, \theta) \triangleq \tilde{\mathcal{R}} f = \mathcal{H}^{1/2} \mathcal{R} f = \mathcal{H}^{1/2} g \tag{10.50}$$

is equivalent to filtering the projections by $\mathcal{H}^{1/2}$ (Fig. 10.14). This operation can also be realized by a two-dimensional filter with frequency response $(\xi_1^2 + \xi_2^2)^{1/4}$ followed by the Radon transform.

Theorem 10.1. Let $\tilde{\mathcal{R}}^+$ denote the adjoint operation of $\tilde{\mathcal{R}}$. The operator $\tilde{\mathcal{R}}$ is unitary, that is,

$$\tilde{\mathcal{R}}^{-1} = \tilde{\mathcal{R}}^+ = \mathcal{B} \mathcal{H}^{1/2} \tag{10.51}$$

This means the inverse of $\tilde{\mathcal{R}}$ is equal to its adjoint and the $\tilde{\mathcal{R}}$ transform preserves energy, that is,

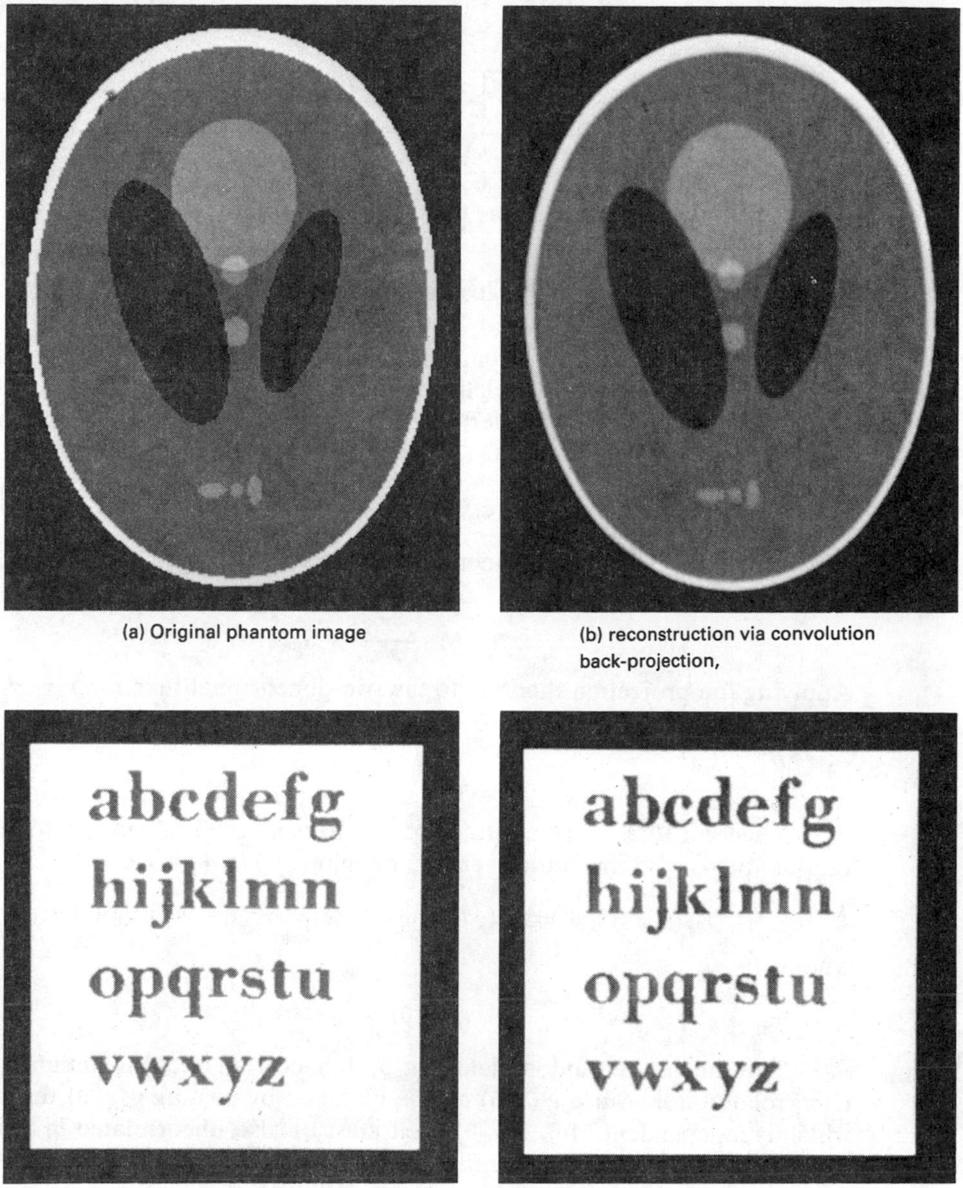

(a) Original phantom image

(b) reconstruction via convolution back-projection,

(c) original binary image

(d) reconstruction using fully constrained ART algorithm

Figure 10.13 Reconstruction examples using parallel pipeline processor.

$$\iint_{-\infty}^{\infty} |f(x, y)|^2 \, dx \, dy = \int_0^{\pi} \int_{-\infty}^{\infty} |\tilde{g}(s, \theta)|^2 \, ds \, d\theta \qquad (10.52)$$

This theorem is useful for developing the properties of the Radon transform for random fields. For proofs of this and the following theorems, see Problem 10.13.

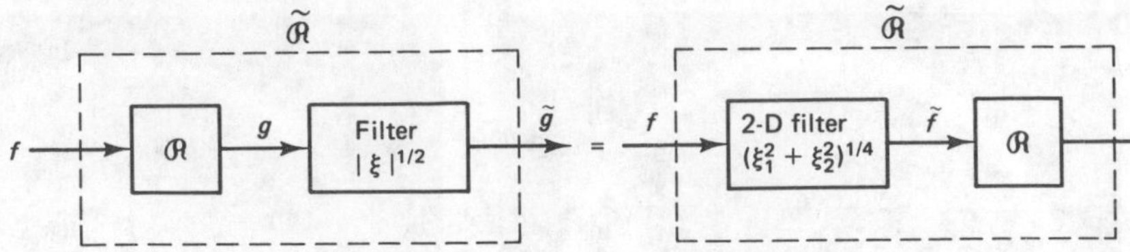

Figure 10.14 The $\tilde{\mathcal{R}}$-transform.

Radon Transform Properties for Random Fields

Definitions. Let $f(x, y)$ be a stationary random field with power spectrum density $S(\xi_1, \xi_2)$ and autocorrelation function $r(\tau_1, \tau_2)$. Then $S(\xi_1, \xi_2)$ and $r(\tau_1, \tau_2)$ form a two-dimensional Fourier transform pair. Let $S_p(\xi, \theta)$ denote the polar-coordinate representation of $S(\xi_1, \xi_2)$, that is,

$$S_p(\xi, \theta) \triangleq S(\xi \cos \theta, \xi \sin \theta) \tag{10.53}$$

Also, let $r_p(s, \theta)$ be the one-dimensional inverse Fourier transform of $S_p(\xi, \theta)$, that is,

$$r_p(s, \theta) \xleftrightarrow[s]{\mathcal{F}_1} S_p(\xi, \theta) \tag{10.54}$$

Applying the projection theorem to the two-dimensional function $r(\tau_1, \tau_2)$, we observe the relation

$$r_p(s, \theta) = \mathcal{R}r \tag{10.55}$$

Theorem 10.2. The operator $\tilde{\mathcal{R}}$ is a *whitening transform* in θ for stationary random fields, and the autocorrelation function of $\tilde{g}(s, \theta)$ is given by

$$r_{\tilde{g}\tilde{g}}(s, \theta; s', \theta') \triangleq E[\tilde{g}(s, \theta)\tilde{g}(s', \theta')] = r_{\tilde{g}}(s - s', \theta)\delta(\theta - \theta') \tag{10.56a}$$

where

$$r_{\tilde{g}}(s, \theta) = r_p(s, \theta) \tag{10.56b}$$

This means the random field $\tilde{g}(s, \theta)$ defined via (10.50) is stationary in s and uncorrelated in θ. Since $\tilde{g}(s, \theta)$ can be obtained by passing $g(s, \theta)$ through $\mathcal{H}^{1/2}$, which is independent of θ, $g(s, \theta)$ itself must be also uncorrelated in θ. Thus, *the Radon transform is also a whitening transform in θ for stationary random fields* and the autocorrelation function of $g(s, \theta)$ must be of the form

$$r_{gg}(s, \theta; s'; \theta') \triangleq E[g(s, \theta)g(s', \theta')] = r_g(s - s', \theta)\delta(\theta - \theta') \tag{10.57}$$

where $r_g(s, \theta)$ is yet to be specified. Now, for any given θ, we define the power spectrum density of $\tilde{g}(s, \theta)$ as the one-dimensional Fourier transform of its autocorrelation function with respect to s, that is,

$$S_{\tilde{g}}(\xi, \theta) \triangleq \mathcal{F}_1\{r_{\tilde{g}}(s, \theta)\} \tag{10.58}$$

From Fig. 10.14 we can write

$$S_{\tilde{g}}(\xi, \theta) = |\xi| S_g(\xi, \theta) \tag{10.59}$$

These results lead to the following useful theorem.

Projection Theorem for Random Fields

Theorem 10.3. The one-dimensional power spectrum density $S_{\tilde{g}}(\xi, \theta)$ of the $\tilde{\mathcal{R}}$ transform of a stationary random field $f(x, y)$ is the central slice at angle θ of its two-dimensional power spectrum density $S(\xi_1, \xi_2)$, that is,

$$S_{\tilde{g}}(\xi, \theta) = S_p(\xi, \theta) = S(\xi \cos\theta, \xi \sin\theta) \tag{10.60}$$

This theorem is noteworthy because it states that the central slice of the two-dimensional power spectrum density $S(\xi_1, \xi_2)$ is equal to the one-dimensional power spectrum of $\tilde{g}(s, \theta)$ and not of $g(s, \theta)$. On the other hand, the projection theorem states that the central slice of a two-dimensional *amplitude spectrum density* (that is, the Fourier transform) $F(\xi_1, \xi_2)$ is equal to the one-dimensional *amplitude spectrum density* (that is, the Fourier transform) of $g(s, \theta)$ and not of $\tilde{g}(s, \theta)$. Combining (10.59) and (10.60), we get

$$S_p(\xi, \theta) = S_{\tilde{g}}(\xi, \theta) = |\xi| S_g(\xi, \theta) \tag{10.61}$$

which gives, formally,

$$S_g(\xi, \theta) = \frac{S_p(\xi, \theta)}{|\xi|} \tag{10.62}$$

and

$$r_g(s, \theta) = \mathcal{F}_1^{-1}\left\{\frac{S_p(\xi, \theta)}{|\xi|}\right\} \tag{10.63}$$

Theorem 10.3 is useful for finding the power spectrum density of noise in the reconstructed image due to noise in the observed projections. For example, suppose $\nu(s, \theta)$ is a zero mean random field, given to be stationary in s and uncorrelated in θ, with

$$E[\nu(s, \theta)\nu(s', \theta')] = r_\nu(s - s', \theta)\delta(\theta - \theta') \tag{10.64a}$$

$$S_\nu(\xi, \theta) \xleftrightarrow[s]{\mathcal{F}_1} r_\nu(s, \theta) \tag{10.64b}$$

If $\nu(s, \theta)$ represents the additive observation noise in the projections, then the noise component in the reconstructed image will be

$$\eta(x, y) \triangleq \mathcal{B}\mathcal{H}\nu = \mathcal{B}\mathcal{H}^{1/2}\tilde{\nu} = \tilde{\mathcal{R}}^{-1}\tilde{\nu} \tag{10.65}$$

where $\tilde{\nu} \triangleq \mathcal{H}^{1/2}\nu$. Rewriting (10.65) as

$$\tilde{\nu} = \tilde{\mathcal{R}}\eta \tag{10.66}$$

and applying Theorem 10.3, we can write $S_{\eta p}(\xi, \theta)$, the power spectrum density of η, as

$$S_{\eta p}(\xi, \theta) = S_{\tilde{\nu}}(\xi, \theta) = |\xi| S_\nu(\xi, \theta) \tag{10.67}$$

This means the observation noise power spectrum density is amplified by $(\xi_1^2 + \xi_2^2)^{1/2}$ by the reconstruction process (that is, by $\mathscr{R}^{-1}$). The power spectrum S_η is bounded only if $|\xi| S_\nu(\xi, \theta)$ remains finite as $\xi \to \infty$. For example, if the random field $\nu(s, \theta)$ is bandlimited, then $\eta(x, y)$ will also be bandlimited and S_η will remain bounded.

10.8 RECONSTRUCTION FROM BLURRED NOISY PROJECTIONS [22–25]

Measurement Model

In the presence of noise, the reconstruction filters listed in Table 10.3 are not optimal in any sense. Suppose the projections are observed as

$$w(s, \theta) = \int_{-\infty}^{\infty} h_p(s - s', \theta) g(s', \theta)\, ds' + \nu(s, \theta), \qquad -\infty < s < \infty, 0 < \theta \leq \pi \tag{10.68}$$

The function $h_p(s, \theta)$ represents a shift invariant blur (with respect to s), which may occur due to the projection-gathering instrumentation, and $\nu(s, \theta)$ is additive, zero mean noise independent of $f(x, y)$ and uncorrelated in θ [see (10.64a)]. The optimum linear mean square reconstruction filter can be determined by applying the Wiener filtering ideas that were discussed in Chapter 8.

The Optimum Mean Square Filter

The optimum linear mean square estimate of $f(x, y)$, denoted by $\hat{f}(x, y)$, can be reconstructed from $w(s, \theta)$, by the filter/convolution back-projection algorithm (Problem 10.14)

$$\hat{g}(s, \theta) = \int_{-\infty}^{\infty} a_p(s - s', \theta) w(s', \theta)\, ds'$$

$$\hat{f}(x, y) = \mathscr{B}\hat{g} \tag{10.69}$$

where

$$a_p(s, \theta) \overset{\mathscr{F}_1}{\underset{s}{\longleftrightarrow}} A_p(\xi, \theta) = \frac{|\xi| H_p^*(\xi, \theta) S_p(\xi, \theta)}{[|H_p(\xi, \theta)|^2 S_p(\xi, \theta) + |\xi| S_\nu(\xi, \theta)]}$$

$$h_p(s, \theta) \overset{\mathscr{F}_1}{\underset{s}{\longleftrightarrow}} H_p(\xi, \theta) \tag{10.70}$$

Remarks

The foregoing optimum reconstruction filter can be implemented as a generalized filter/convolution back-projection algorithm using the techniques of Section 10.6. A provision has to be made for the fact that now we have a one-dimensional filter $a_p(s, \theta)$, which can change with θ.

Reconstruction from noisy projections. In the absence of blur we have $h_p(s, \theta) = \delta(s)$ and

$$w(s, \theta) = g(s, \theta) + \nu(s, \theta) \tag{10.71}$$

The reconstruction filter is then given by

$$A_p(\xi, \theta) = \frac{|\xi| S_p(\xi, \theta)}{[S_p(\xi, \theta) + |\xi| S_\nu(\xi, \theta)]} \tag{10.72}$$

Note that if there is no noise, that is, $S_\nu \to 0$, then $A_p(\xi, \theta) \to |\xi|$, which is, of course, the filter required for the inverse Radon transform.

Using (10.61) in (10.70) we can write

$$A_p(\xi, \theta) = |\xi| \tilde{A}_p(\xi, \theta) \tag{10.73}$$

where

$$\tilde{A}_p(\xi, \theta) \triangleq \frac{H_p^* S_g}{|H_p|^2 S_g + S_\nu} = \frac{H_p^* S_g}{S_w} \tag{10.74}$$

Note that $\tilde{A}_p(\xi, \theta)$ is the one-dimensional Weiner filter for $g(s, \theta)$ given $w(s, \theta)$. This means the overall optimum filter A_p is the cascade of $|\xi|$, the filter required for the inverse Radon transform, and a window function $\tilde{A}_p(\xi, \theta)$, representing the locally optimum filter for each projection. In practice, $\tilde{A}_p(\xi, \theta)$ can be estimated adaptively for each θ by estimating $S_w(\xi, \theta)$, the power spectrum density of the observed projection $w(s, \theta)$.

Example 10.6 Reconstruction from noisy projections

Suppose the covariance function of the object is modeled by the isotropic function $r(x, y) = \sigma^2 \exp(-\alpha\sqrt{x^2 + y^2})$. The corresponding power spectrum is then $S(\xi_1, \xi_2) = 2\pi\alpha\sigma^2[\alpha^2 + 4\pi^2(\xi_1^2 + \xi_2^2)]^{-3/2}$ or $S_p(\xi, \theta) = 2\pi\alpha\sigma^2[\alpha^2 + 4\pi^2\xi^2]^{-3/2}$. Assume there is no blur and let $r_\nu(s, \theta) = \sigma_\nu^2$. Then the frequency response of the optimum reconstruction filter, henceforth called *the stochastic filter,* is given by

$$A_p(\xi, \theta) = \frac{|\xi| S_p(\xi, \theta)}{S_p(\xi, \theta) + \sigma_\nu^2|\xi|} = \frac{2\pi\alpha\sigma^2|\xi|}{2\pi\alpha\sigma^2 + |\xi|\sigma_\nu^2(\alpha^2 + 4\pi^2\xi^2)^{3/2}}$$

$$= \frac{|\xi| 2\pi\alpha(\text{SNR})}{2\pi\alpha(\text{SNR}) + |\xi|(\alpha^2 + 4\pi^2\xi^2)^{3/2}}, \qquad \text{SNR} \triangleq \frac{\sigma^2}{\sigma_\nu^2}$$

This filter is independent of θ and has a frequency response much like that of a band-pass filter (Fig. 10.15a). Figure 10.15b shows the impulse response of the stochastic filter used for reconstruction from noisy projections with $\sigma_\nu^2 = 5$, $\sigma^2 = 0.0102$, and $\alpha = 0.266$. Results of reconstruction are shown in Fig. 10.15c through i. Comparisons with the Shepp-Logan filter indicate significant improvement results from the use of the stochastic filter. In terms of mean square error, the stochastic filter performs 13.5 dB better than the Shepp-Logan filter in the case of $\sigma_\nu^2 = 5$. Even in the noiseless case (Fig. 10.12) the stochastic filter designed with a high value of SNR (such as 100), provides a better reconstruction. This is because the stochastic filter tends to moderate the high-frequency components of the noise that arise from errors in computation.

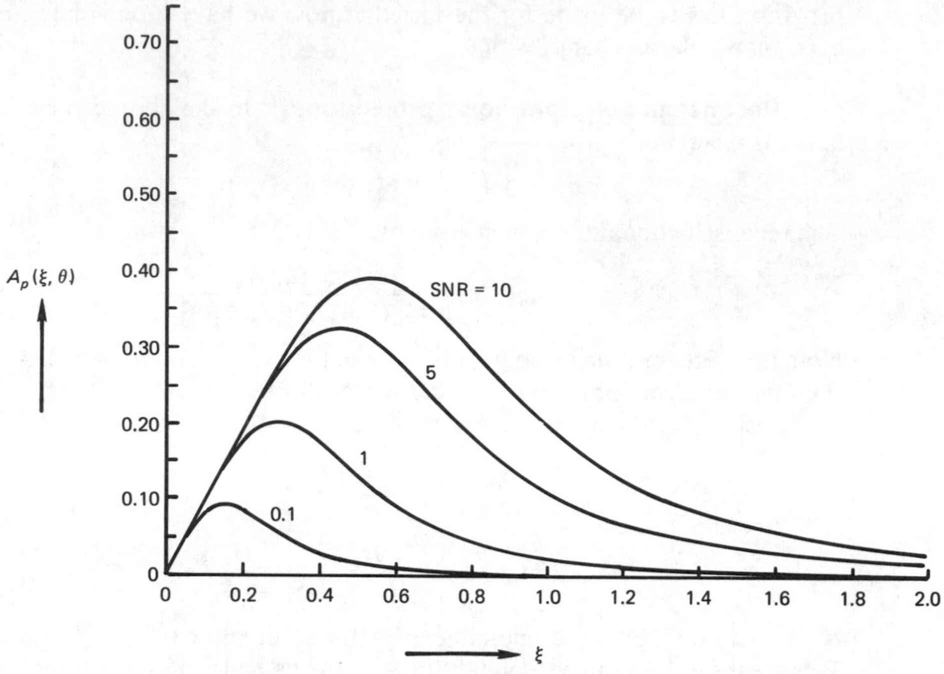

(a) A typical frequency response of a stochastic filter $A_p(\xi, \theta) = A_p(-\xi, \theta)$

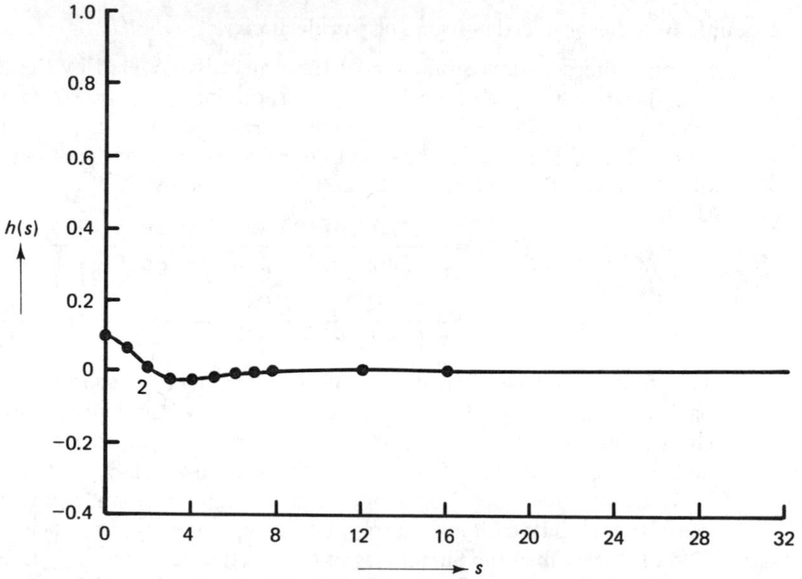

(b) Impulse response of the stochastic filter used

Figure 10.15 Reconstruction from noisy projections.

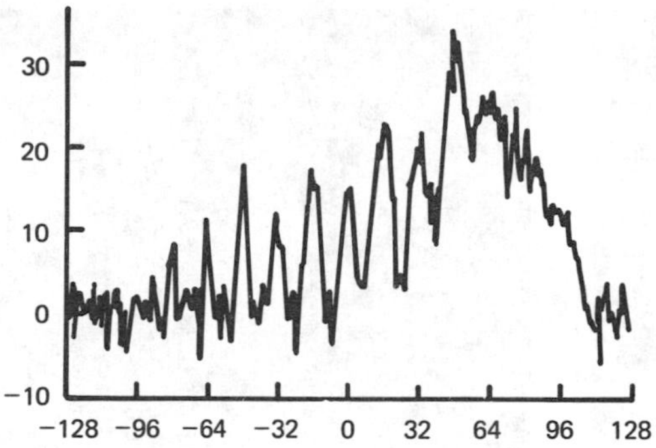

(c) Typical noisy projection, $\sigma_\nu^2 = 5$

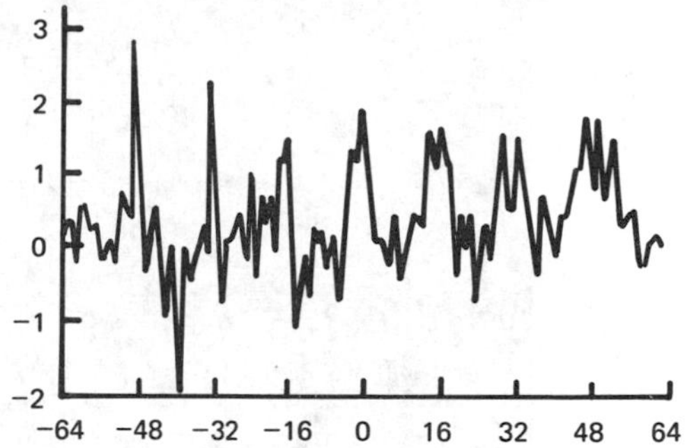

(d) Reconstruction via Shepp-Logan filter

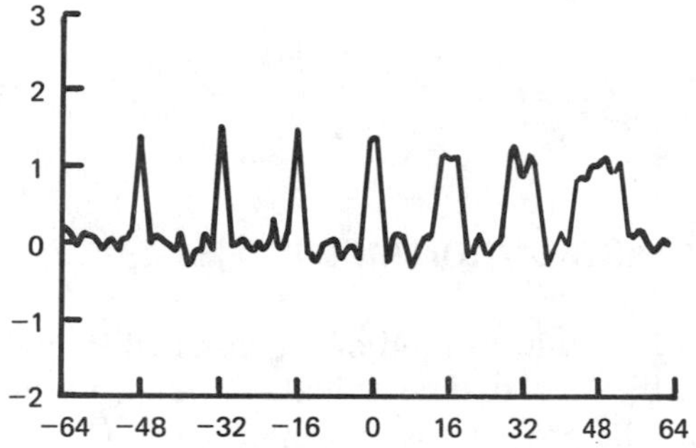

(e) Reconstruction via the stochastic filter

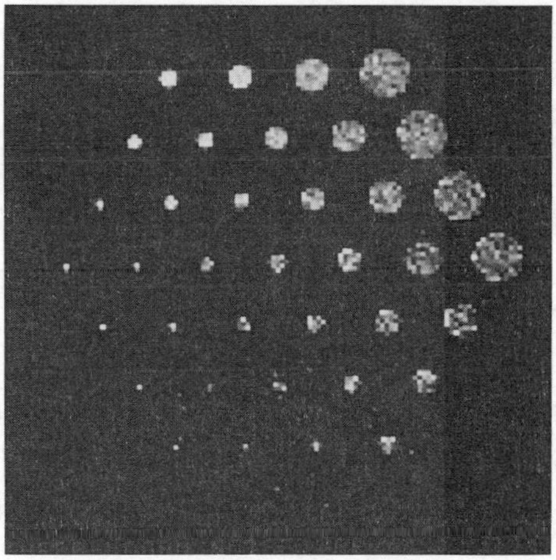

(f) Shepp-Logan filter, $\sigma_\nu^2 = 1$;

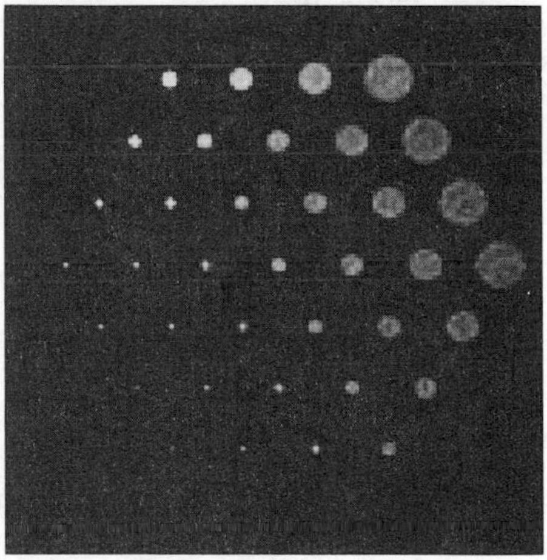

(g) Stochastic filter, $\sigma_\nu^2 = 1$;

Figure 10.15 Cont'd

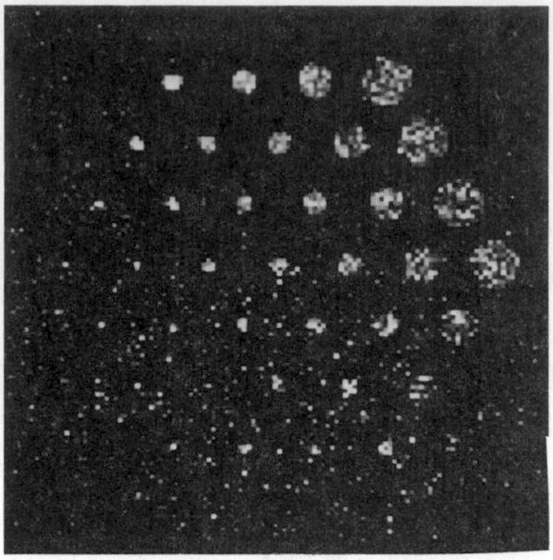

(h) Shepp-Logan filter, $\sigma_v^2 = 5$;

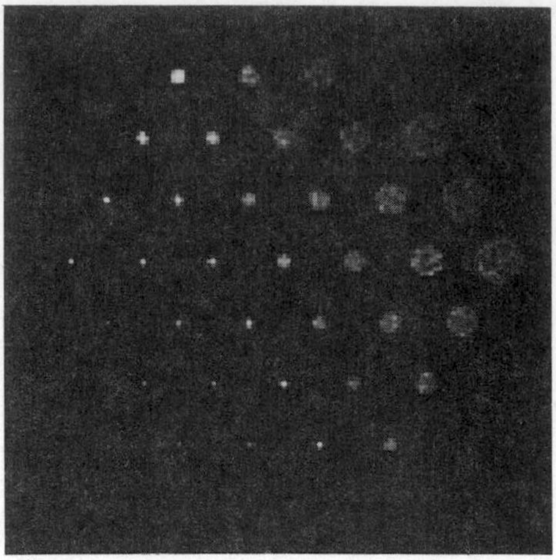

(i) Stochastic filter, $\sigma_v^2 = 5$.

Figure 10.15 Cont'd

10.9 FOURIER RECONSTRUCTION METHOD [26–29]

A conceptually simple method of reconstruction that follows from the projection theorem is to fill the two-dimensional Fourier space by the one-dimensional Fourier transforms of the projections and then take the two-dimensional inverse Fourier transform (Fig. 10.16a), that is,

$$f(x, y) = \mathscr{F}_2^{-1}[\mathscr{F}_1 g] \tag{10.75}$$

Algorithm

There are three stages of this algorithm (Fig. 10.16b). First we obtain $G_n(k) \simeq G(k\Delta\xi, n\Delta\theta), -K/2 \le k \le K/2 - 1, 0 \le n \le N - 1$, as in Fig. (10.11b). Next, the Fourier domain samples available on a polar raster are interpolated to yield estimates on a rectangular raster (see Section 8.16). In the final stage of the algorithm, the two-dimensional inverse Fourier transform is approximated by a suitable-size inverse FFT. Usually, the size of the inverse FFT is taken to be two to three times that of each dimension of the image. Further, an appropriate window is used before inverse transforming in order to minimize the effects of Fourier domain truncation and sampling.

Although there are many examples of successful implementation of this algorithm [29], it has not been as popular as the convolution back-projection algorithm. The primary reason is that the interpolation from polar to raster grid in the frequency plane is prone to aliasing effects that could yield an inferior reconstructed image.

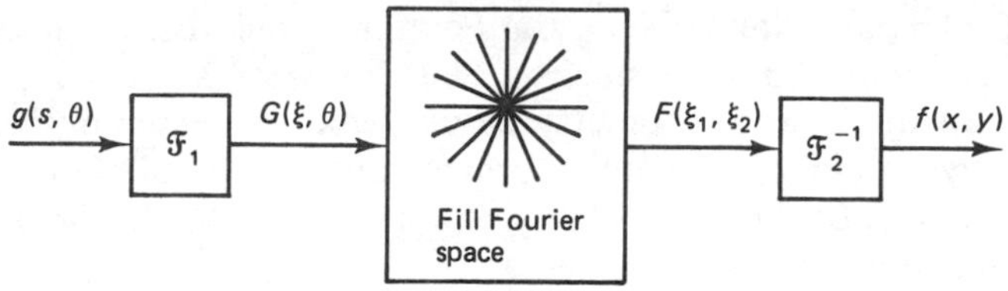

(a) The concept

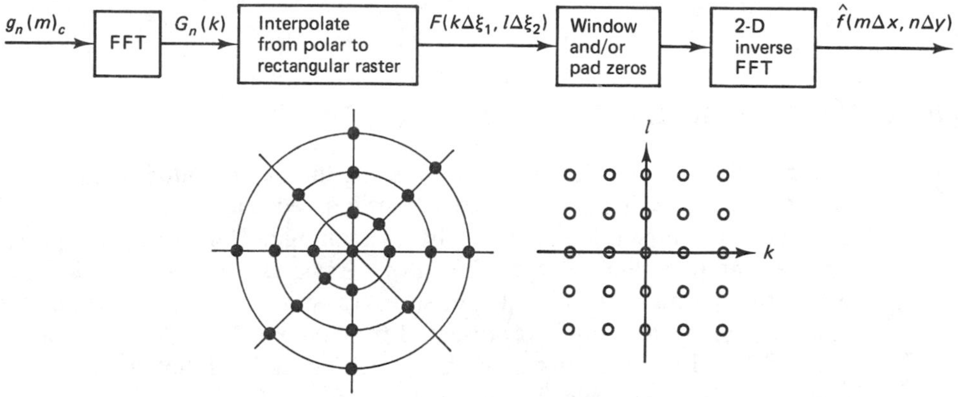

(b) A practical fourier reconstruction algorithm

Figure 10.16 Fourier reconstruction method.

Reconstruction of Magnetic Resonance Images (Fig. 10.17)

In magnetic resonance imaging there are two distinct scanning modalities, the *projection geometry* and the *Fourier geometry* [30]. In the projection geometry mode, the observed signal is $G(\xi, \theta)$, sampled at $\xi = k\Delta\xi$, $-K/2 \le k \le K/2 - 1$, $\theta = n\Delta\theta$, $0 \le n \le N - 1$, $\Delta\theta = \pi/N$. Reconstruction from such data necessitates the availability of an FFT processor, regardless of which algorithm is used. For example, the filter back-projection algorithm would require inverse Fourier transform of

(a) MRI data;

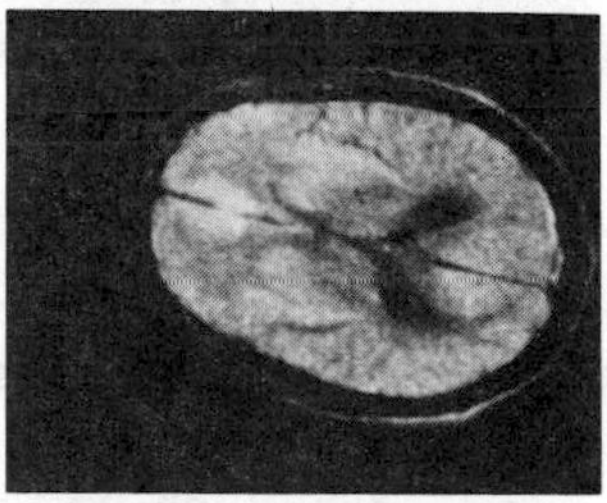

(b) Reconstructed image;

Figure 10.17 Magnetic resonance image reconstruction.

$G(\xi, \theta)H(\xi)$. Alternatively, the Fourier reconstruction algorithm just described is also suitable, especially since an FFT processor is already available.

In the Fourier geometry mode, which is becoming increasingly popular, we directly obtain samples on a rectangular raster in the Fourier domain. The reconstruction algorithm then simply requires a two-dimensional inverse FFT after windowing and zero-padding the data.

Figure 10.17a shows a 512×128 MRI image acquired in Fourier geometry mode. A 512×256 image is reconstructed (Fig. 10.17b) by a 512×256 inverse FFT of the raw data windowed by a two-dimensional Gaussian function and padded by zeros.

10.10 FAN-BEAM RECONSTRUCTION

Often the projection data is collected using fan-beams rather than parallel beams (Fig. 10.18). This is a more practical method because it allows rapid collection of projections compared to parallel beam scanning. Referring to Fig. 10.18b, the source S emits a thin divergent beam of X-rays, and a detector receives the beam after attenuation by the object. The source position is characterized by the angle β, and each projection ray is represented by the coordinates (σ, β), $-\pi/2 \le \sigma < \pi/2$, $0 \le \beta < 2\pi$. The coordinates of the (σ, β) ray are related to the parallel beam coordinates (s, θ) as (Fig. 10.18c)

$$\begin{aligned} s &= R \sin \sigma \\ \theta &= \sigma + \beta \end{aligned} \tag{10.76}$$

where R is the distance of the source from the origin of the object. For a space-limited object with maximum radius $D/2$, the angle σ lies in the interval $[-\gamma, \gamma]$, $\gamma \overset{\Delta}{=} \sin^{-1}(D/2R)$. Since a ray in the fan-beam geometry is also some ray in the parallel beam geometry, we can relate their respective projection functions $b(\sigma, \beta)$ and $g(s, \theta)$ as

$$b(\sigma, \beta) = g(s, \theta) = g(R \sin \sigma, \sigma + \beta) \tag{10.77}$$

If $b(\sigma, \beta)$ is given on a grid (σ_m, β_n), then this relation gives $g(s, \theta)$ on a grid (s_m, θ_n), $s_m \overset{\Delta}{=} R \sin \sigma_m$, $\theta_n \overset{\Delta}{=} \sigma_n + \beta_n$. This data then has to be interpolated to obtain $g(s_m, \theta_n)$ on the uniform grid $s_m = md$, $-(M/2) \le m \le (M/2) - 1$; $\theta_n = n\Delta, 0 \le n \le N - 1$. This is called *rebinning*. Alternatively, we can use

$$g(s, \theta) = b\left(\sin^{-1}\frac{s}{R}, \theta - \sin^{-1}\frac{s}{R}\right) \tag{10.78}$$

to estimate $g(s_m, \theta_n)$ on the uniform grid by interpolating $b(\sigma_m, \beta_n)$. Once $g(s, \theta)$ is available on a uniform grid, we can use the foregoing parallel beam reconstruction algorithms. Another alternative is to derive the divergent beam reconstruction algorithms directly in terms of $b(\sigma, \beta)$ by using (10.77) and (10.78) in the inverse Radon transform formulas. (See Problem 10.16.)

In practice, rebinning seems to be preferred because it is simpler and can be fed into the already developed convolution/filter back-projection algorithms (or

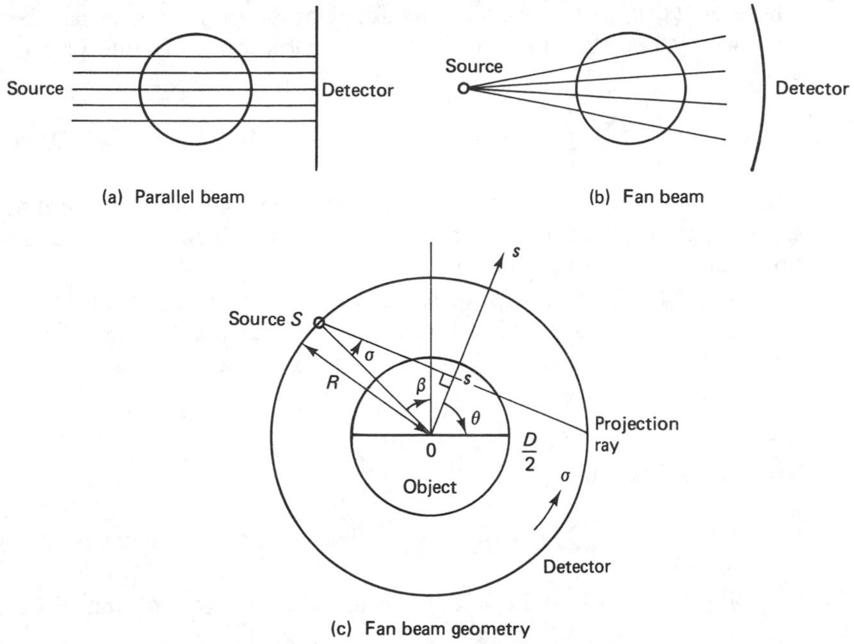

Figure 10.18 Projection data acquisition.

processors). However, there are situations where the data volume is so large that the storage requirements for rebinning assume unmanageable proportions. In such cases the direct divergent beam reconstruction algorithms would be preferable because only one projection, $b(\sigma, \beta)$, would be used at a time, in a manner characteristic of the convolution/filter back-projection algorithms.

10.11 ALGEBRAIC METHODS

All the foregoing reconstruction algorithms are based on Radon transform theory and the projection theorem. It is possible to formulate the reconstruction problem as a general image restoration problem solvable by techniques discussed in Chapter 8.

The Reconstruction Problem as a Set of Linear Equations

Suppose $f(x, y)$ is approximated by a finite series

$$f(x, y) \simeq \hat{f}(x, y) = \sum_{i=1}^{I} \sum_{j=1}^{J} a_{i,j} \phi_{i,j}(x, y) \tag{10.79}$$

where $\{\phi_{i,j}(x, y)\}$ is a set of basis functions. Then

$$g(s, \theta) \simeq \mathcal{R}\hat{f} = \sum_{i=1}^{I} \sum_{j=1}^{J} a_{i,j}[\mathcal{R}\phi_{i,j}] \triangleq \sum_{i=1}^{I} \sum_{j=1}^{J} a_{i,j} h_{i,j}(s, \theta) \tag{10.80}$$

where $h_{i,j}(s, \theta)$ is the Radon transform of $\phi_{i,j}(x, y)$, which can be computed in advance. When the observations are available on a discrete grid (s_m, θ_n), we can write

$$g(s_m, \theta_n) \simeq \sum_{i=1}^{I} \sum_{j=1}^{J} a_{i,j} h_{i,j}(s_m, \theta_n), \quad 0 \le m \le M-1, \quad 0 \le n \le N-1 \tag{10.81}$$

which can be solved for $a_{i,j}$ as a set of linear simultaneous equations via least squares, generalized inverse, or other methods. Once the $a_{i,j}$ are known, $\hat{f}(x, y)$ is obtained directly from (10.79).

A particular case of interest is when $f(x, y)$ is digitized on, for instance, an $I \times J$ grid and f is assumed to be constant in each pixel region. Then $a_{i,j}$ equals $f_{i,j}$, the sampled value of $f(x, y)$ in the (i, j)th pixel, and

$$\phi_{i,j}(x, y) = \begin{cases} 1, & \text{inside the } (i, j)\text{th pixel region} \\ 0, & \text{otherwise} \end{cases} \tag{10.82}$$

Now (10.81) becomes

$$g(s_m, \theta_n) \simeq \sum_{i=1}^{I} \sum_{j=1}^{J} f_{i,j} h_{i,j}(s_m, \theta_n), \quad 0 \le m \le M-1, \quad 0 \le n \le N-1 \tag{10.83}$$

Mapping $f_{i,j}$ into a $Q \times 1$ ($Q \triangleq IJ$) vector $\mathscr{f}$ by row (or column) ordering, we get

$$\mathscr{g} \simeq \mathscr{H}\mathscr{f} \tag{10.84}$$

where $\mathscr{g}$ and $\mathscr{H}$ are $P \times 1$, ($P \triangleq MN$) and $P \times Q$ arrays, respectively. A more-realistic observation equation is of the form

$$\mathscr{g} = \mathscr{H}\mathscr{f} + \boldsymbol{\eta} \tag{10.85}$$

where $\boldsymbol{\eta}$ represents noise. The reconstruction problem now is to estimate $\mathscr{f}$ from $\mathscr{g}$. Equations (10.84) and (10.85) are now in the framework of Wiener filtering, pseudoinverse, generalized inverse, maximum entropy, and other restoration algorithms considered in Chapter 8. The main advantage of this approach is that the algorithms needed henceforth would be independent of the scanning modality (e.g., parallel beam versus fan beam). Also, the observation model can easily incorporate a more realistic projection gathering model, which may not approximate well the Radon transform.

The main limitations of the algebraic formulation arise from the large size of the matrix $\mathscr{H}$. For example, for a 256×256 image with 100 projections each sampled to 512 points, $\mathscr{H}$ becomes a $51{,}200 \times 65{,}536$ matrix. However, $\mathscr{H}$ will be a highly sparse matrix containing only $O(I)$ or $O(J)$ nonzero entries per row. These nonzero entries correspond to the pixel locations that fall in the path of the ray (s_m, θ_n). Restoration algorithms that exploit the sparse structure of $\mathscr{H}$ are feasible.

Algebraic Reconstruction Techniques

A subset of iterative reconstruction algorithms have been historically called ART (algebraic reconstruction techniques). These algorithms iteratively solve a set of P equations

$$\langle \mathbf{h}_p, \mathcal{f} \rangle \triangleq \mathbf{h}_p^T \mathcal{f} = g_p, \qquad p = 0, \ldots, P-1 \tag{10.86}$$

where $\mathbf{h}_p^T$ is the pth row of $\mathcal{H}$ and g_p is the pth element of g. The algorithm, originally due to Kaczmarz [30], has iterations that progress cyclically as

$$\tilde{\mathcal{f}}^{(k+1)} = \mathcal{f}^{(k)} + \frac{g_{k+1} - \langle \mathbf{h}_{k+1}, \mathcal{f}^{(k)} \rangle}{\|\mathbf{h}_{k+1}\|^2} \mathbf{h}_{k+1}, \qquad k = 0, 1, \ldots \tag{10.87}$$

where $\tilde{\mathcal{f}}^{(k+1)}$ determines $\mathcal{f}^{(k+1)}$, depending on the constraints imposed on $\mathcal{f}$ (see Table 10.4), $\mathcal{f}^{(0)}$ is some initial condition, and g_k and $\mathbf{h}_k$ appear cyclically, that is,

$$\begin{aligned} \mathbf{h}_k &= \mathbf{h}_{(k \text{ modulo } P)} \\ g_k &= g_{(k \text{ modulo } P)} \end{aligned} \tag{10.88}$$

Each iteration is such that only one of the P equations (or constraints) is satisfied at a time. For example, from (10.87) we can see

$$\begin{aligned} \langle \mathbf{h}_{k+1}, \tilde{\mathcal{f}}^{(k+1)} \rangle &= \langle \mathbf{h}_{k+1}, \mathcal{f}^{(k)} \rangle + \frac{[g_{k+1} - \langle \mathbf{h}_{k+1}, \mathcal{f}^{(k)} \rangle]}{\|\mathbf{h}_{k+1}\|^2} \langle \mathbf{h}_{k+1}, \mathbf{h}_{k+1} \rangle \\ &= g_{k+1} \end{aligned} \tag{10.89}$$

that is, the $[(k+1) \text{ modulo } P]$th constraint is satisfied at $(k+1)$th iteration. This algorithm is easy to implement since it operates on *one projection sample* at a time. The operation $\langle \mathbf{h}_{k+1}, \mathcal{f}^{(k)} \rangle$ is equivalent to taking the $[(k+1) \text{ modulo } P]$th projection sample of the previous estimate. The sparse structure of $\mathbf{h}_k$ can easily be exploited to reduce the number of operations at each iteration. The speed of convergence is usually slow, and difficulties arise in deciding when to stop the iterations. Figure 10.13d shows an image reconstructed using the fully constrained ART algorithm in (10.87). The result shown was obtained after five complete passes through the image starting from an all black image.

TABLE 10.4 ART Algorithms

	Algorithm $\mathcal{f} = \{\mathcal{f}_j; 1 \le j < Q\}$	Comments
1	Unconstrained ART $\mathcal{f}_j^{(k)} = \tilde{\mathcal{f}}_j^{(k)}$	If $L_1 \triangleq \{\mathcal{f} \mid \mathcal{H}\mathcal{f} = g\}$ is nonempty, the algorithm converges to the element of L_1 with the smallest distance to $\mathcal{f}^{(0)}$.
2	Partially constrained ART: $\mathcal{f}_j^{(k)} = \begin{cases} 0, & \text{if } \tilde{\mathcal{f}}_j^{(k)} < 0 \\ \tilde{\mathcal{f}}_j^{(k)}, & \text{otherwise} \end{cases}$	If $L_2 \triangleq \{\mathcal{f} \mid \mathcal{H}\mathcal{f} = g, \mathcal{f}_j \ge 0\}$ is nonempty, the algorithm converges to an element of L_2.
3	Fully constrained ART: $\mathcal{f}_j^{(k)} = \begin{cases} 0, & \text{if } \tilde{\mathcal{f}}_j^{(k)} < 0 \\ \tilde{\mathcal{f}}_j^{(k)}, & \text{if } 0 \le \tilde{\mathcal{f}}_j^{(k)} \le 1 \\ 1, & \text{if } \tilde{\mathcal{f}}_j^{(k)} > 1 \end{cases}$	If $L_3 \triangleq \{\mathcal{f} \mid \mathcal{H}\mathcal{f} = g, 0 \le \mathcal{f}_j \le 1\}$ is nonempty, the algorithm converges to an element of L_3.

10.12 THREE-DIMENSIONAL TOMOGRAPHY

If a three-dimensional object is scanned by a parallel beam, as shown in Fig. 10.1, then the entire three-dimensional object can be reconstructed from a set of two-dimensional slices (such as the slice A), each of which can be reconstructed using the foregoing algorithms. Suppose we are given one-dimensional projections of a three-dimensional object $f(x, y, z)$. These projections are obtained by integrating $f(x, y, z)$ in a plane whose orientation is described by a unit vector $\boldsymbol{\alpha}$ (Fig. 10.19), that is,

$$g(s, \boldsymbol{\alpha}) = [\mathcal{R}f](s, \boldsymbol{\alpha}) = \iiint_{-\infty}^{\infty} f(\mathbf{x})\delta(\mathbf{x}^T \boldsymbol{\alpha} - s)\, d\mathbf{x}$$

$$= \iiint_{-\infty}^{\infty} f(x, y, z)\delta(x \sin\theta \cos\phi + y \sin\theta \sin\phi + z \cos\theta - s]\, dx\, dy\, dz \tag{10.90}$$

where $\mathbf{x} \triangleq [x, y, z]^T$, $\boldsymbol{\alpha} \triangleq [\sin\theta \cos\phi, \sin\theta \sin\phi, \cos\theta]^T$. This is also called the three-dimensional Radon transform of $f(x, y, z)$. The Radon transform theory can be readily generalized to three or higher dimensions. The following theorems provide algorithms for reconstruction of $f(x, y, z)$ from the projections $g(s, \boldsymbol{\alpha})$.

Three-Dimensional Projection Theorem. The one-dimensional Fourier transform of $g(s, \boldsymbol{\alpha})$ defined as

$$G(\xi, \boldsymbol{\alpha}) \triangleq \int_{-\infty}^{\infty} g(s, \boldsymbol{\alpha}) e^{-j2\pi\xi s}\, ds \tag{10.91}$$

is the central slice $F(\xi\boldsymbol{\alpha})$ in the direction $\boldsymbol{\alpha}$ of the three-dimensional Fourier transform of $f(\mathbf{x})$, that is,

$$G(\xi, \boldsymbol{\alpha}) = F(\xi\boldsymbol{\alpha}) \triangleq F(\xi \sin\theta \cos\phi, \xi \sin\theta \sin\phi, \xi \cos\theta) \tag{10.92}$$

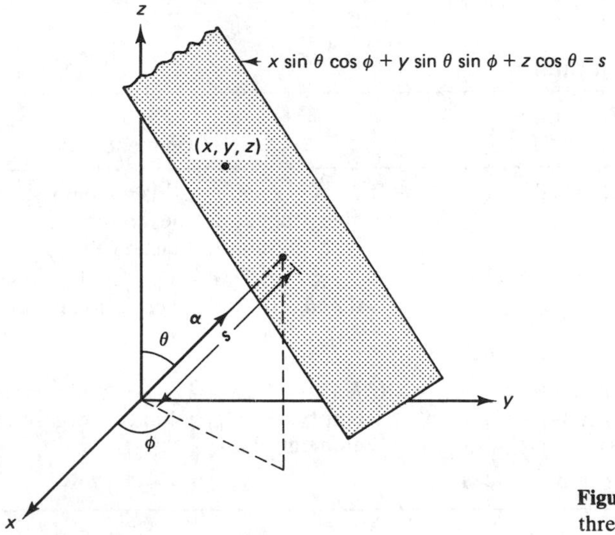

Figure 10.19 Projection geometry for three-dimensional objects.

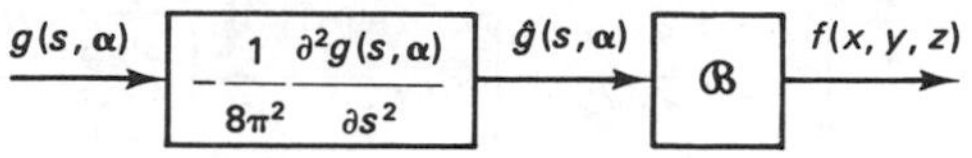

Figure 10.20 Three-dimensional inverse radon transform.

where

$$F(\xi_1, \xi_2, \xi_3) = F(\boldsymbol{\xi}) \triangleq \iiint_{-\infty}^{\infty} f(\mathbf{x}) e^{-j2\pi(\mathbf{x}^T\boldsymbol{\xi})}\, d\mathbf{x}, \qquad \boldsymbol{\xi} \triangleq [\xi_1, \xi_2, \xi_3]^T \tag{10.93}$$

Three-Dimensional Inverse Radon Transform Theorem. The inverse of the three-dimensional Radon transform of (10.90) is given by (Fig. 10.20)

$$f(\mathbf{x}) = \mathscr{B}\hat{g} \triangleq \int_0^{\pi}\int_0^{2\pi} \hat{g}(\mathbf{x}^T\boldsymbol{\alpha}, \boldsymbol{\alpha}) \sin\theta\, d\phi\, d\theta \tag{10.94}$$

where

$$\hat{g}(s, \boldsymbol{\alpha}) = -\frac{1}{8\pi^2}\frac{\partial^2 g(s, \boldsymbol{\alpha})}{\partial s^2} = \mathscr{F}_1^{-1}\left\{G(\xi, \boldsymbol{\alpha})\frac{\xi^2}{2}\right\} \tag{10.95}$$

Proofs and extensions of these theorems may be found in [13].

Three-Dimensional Reconstruction Algorithms

The preceding theorem gives a direct three-dimensional digital reconstruction algorithm. If $g(s_m, \boldsymbol{\alpha}_n)$ are the sampled values of the projections given for $s_m = md$, $-M/2 \le m \le M/2 - 1$, $0 \le n \le N - 1$, for instance, then we can approximate the second partial derivative by a three-point digital filter to give

$$\hat{g}(s_m, \boldsymbol{\alpha}_n) \simeq \frac{1}{8\pi^2 d^2}[2g(md, \boldsymbol{\alpha}_n) - g(\overline{m-1}d, \boldsymbol{\alpha}_n) - g(\overline{m+1}d, \boldsymbol{\alpha}_n)] \tag{10.96}$$

In order to approximate the back-projection integral by a sum, we have to sample ϕ and θ, where (ϕ_k, θ_j) define the direction $\boldsymbol{\alpha}_n$. A suitable arrangement for the projection angles (ϕ_k, θ_j) is one that gives a uniform distribution of the projections over the surface of a sphere. Note that if θ and ϕ are sampled uniformly, then the projections will have higher concentration near the poles compared to the equator. If θ is sampled uniformly with $\Delta\theta = \pi/J$, we will have $\theta_j = (j + \frac{1}{2})\Delta\theta$, $j = 0, 1, \ldots, J - 1$. The density of projections at elevation angles θ_j will be proportional to $1/\sin\theta_j$. Therefore, for uniform distribution of projections, we should increment ϕ_k in proportion to $1/\sin\theta_j$, that is, $\Delta\phi = c/\sin\theta_j$, $\phi_k = k\Delta\phi$, $k = 0, \ldots, K - 1$, where c is a proportionality constant. Since ϕ_K must equal 2π, we obtain

$$K = K_j = \frac{2\pi}{\Delta\phi} = \frac{2\pi \sin\theta_j}{c}, \qquad \phi_k = \frac{2\pi k}{K_j}, \qquad k = 0, 1, \ldots, K_j - 1 \tag{10.97}$$

This relation can be used to estimate K_j to the nearest integer value for different θ_j.

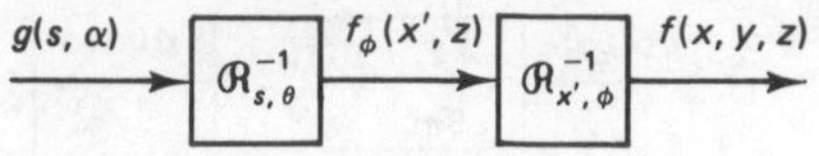

Figure 10.21 Two-stage reconstruction.

The back-projection integral is approximated as

$$\mathcal{B}\hat{g} \simeq \hat{f}(x, y, z) \triangleq \frac{2\pi^2}{J} \sum_{j=0}^{J-1} \sin\theta_j \frac{1}{K_j} \sum_{k=0}^{K_j-1} \cdot \qquad (10.98)$$
$$\cdot\, \hat{g}(x \sin\theta_j \cos\phi_k + y \sin\theta_j \sin\phi_k + z \cos\theta_j, \boldsymbol{\alpha}_n)$$

where (ϕ_k, θ_j) define the direction $\boldsymbol{\alpha}_n$, $n = 0, \ldots, N-1$, where

$$N = J \sum_{j=0}^{J-1} K_j \qquad (10.99)$$

There is an alternate reconstruction algorithm (Fig. 10.21), which contains two stages of two-dimensional inverse Radon transforms. In the first stage, ϕ is held constant (by $\mathcal{R}^{-1}_{s,\theta}$) and in the second stage, z is held constant (by $\mathcal{R}^{-1}_{x',\phi}$). Details are given in Problem 10.17. Compared to the direct method, where an arbitrary section of the object can be reconstructed, the two-stage algorithm necessitates reconstruction of parallel cross sections of the object. The advantage is that only two-dimensional reconstruction algorithms are needed.

10.13 SUMMARY

In this chapter we have studied the fundamentals and algorithms for reconstruction of objects from their projections. The projection theorem is a fundamental result of Fourier theory, which leads to useful algorithms for inverting the Radon transform. Among the various algorithms discussed here, the convolution back projection is the most widely used. Among the various filters, the stochastic reconstruction filter seems to give the best results, especially for noisy projections. For low levels of noise, the modified Shepp-Logan filter performs equally well.

The Radon transform theory itself is very useful for filtering and representation of multidimensional signals. Most of the results discussed here can also be extended to multidimensions.

PROBLEMS

10.1 Prove the properties of the Radon transform listed in Table 10.1.

10.2 Derive the expression for the Radon transform of the ellipse shown in Figure 10.4b.

10.3 Express the Radon transform of an object $f_p(r, \phi)$ given in polar coordinates.

10.4 Find the Radon transform of

a. $\exp[-\pi(x^2 + y^2)], \quad \forall x, y$

b. $\exp\left[\frac{j2\pi}{L}(kx + ly)\right], \; -\frac{L}{2} \le x, y \le \frac{L}{2}$

c. $\cos\dfrac{\pi kx}{L}\cos\dfrac{\pi ly}{L}, \ -\dfrac{L}{2} \le x, y \le \dfrac{L}{2}$

d. $\cos 2\pi(\alpha x + \beta y), \sqrt{x^2 + y^2} \le a$

Assume the given functions are zero outside the region of support defined.

10.5 Find the impulse response of the following linear systems.

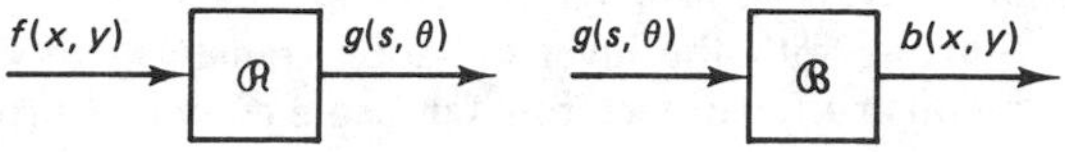

Figure P10.5

10.6 In order to prove (10.19), use the definitions of $\mathcal{B}$ and $\mathcal{R}$ to show that

$$\tilde{f}(x, y) = \iint_{-\infty}^{\infty} f(x', y')\left[\int_0^{\pi} \delta((x' - x)\cos\theta + (y' - y)\sin\theta)\, d\theta\right] dx'\, dy'$$

Now using the identity (10.9), prove (10.19). Transform to polar coordinates and prove (10.20).

10.7 Show that the back-projection operator $\mathcal{B}$ is the adjoint of $\mathcal{R}$. This means for any $a(x, y) \in \mathcal{U}$ and $b(s, \theta) \in \mathcal{V}$, show that $\langle \mathcal{R}a, b\rangle_v = \langle a, \mathcal{B}b\rangle_u$.

10.8 Find the Radon transforms of the functions defined in Problem 10.4 by applying the projection theorem.

10.9 Apply the projection theorem to the function $f(x, y) \overset{\Delta}{=} f_1(x, y) \circledast f_2(x, y)$ and show $\mathcal{F}_1[\mathcal{R}f] = G_1(\xi, \theta)G_2(\xi, \theta)$. From this, prove the convolution-projection theorem.

10.10 Using the formulas (10.26) or (10.27) verify the inverse Radon transforms of the results obtained in Problem 10.8.

10.11 If an object is space limited by a circle of diameter D and if ξ_0 is the largest spatial frequency of interest in the polar coordinates of the Fourier domain, show the number of projections required to avoid aliasing effects due to angular sampling in the transform domain must be $N > \pi D \xi_0$.

10.12* Compute the frequency responses of the linearly interpolated digital filter responses shown in Figure 10.10. Plot and compare these with $H(\xi)$.

10.13 **a.** *(Proof of Theorem 10.1)* Using the fact that $|\xi|^{1/2}$ is real and symmetric, first show that $\mathcal{H}^{1/2}[\mathcal{H}^{1/2}]^+ g = (1/2\pi)\tilde{\mathcal{H}}\mathcal{D}g$, where $\tilde{\mathcal{H}}$ and $\mathcal{D}$ are defined in (10.32) and (10.33). Then show that $\mathcal{R}^+\mathcal{R} = \mathcal{B}\mathcal{H}^{1/2}\mathcal{H}^{1/2}\mathcal{R} = \mathcal{B}\mathcal{H}\mathcal{R} = \mathcal{R}^{-1}\mathcal{R} = \mathcal{I}$ (identity). Now observe that $\langle \tilde{g}, \tilde{g}\rangle_v = \langle g, \hat{g}\rangle_v = \langle \mathcal{R}f, \hat{g}\rangle_v$ and use Problem 10.7 to show $\langle \mathcal{R}f, \hat{g}\rangle_v = \langle f, \mathcal{B}\hat{g}\rangle_u = \langle f, f\rangle_u$.

b. *(Proof of Theorem 10.3)* From Fig. 10.14 obscrvc that $\tilde{g} = \tilde{\mathcal{R}}f = \mathcal{R}\tilde{f}$. Using this show that [22]

(i) $E[\tilde{f}(x, y)\tilde{g}(s, \theta)] = \tilde{r}_p(s - x\cos\theta - y\sin\theta), \tilde{r}_p(s, \theta) \overset{\Delta}{=} \mathcal{F}_1^{-1}\{|\xi| S_p(\xi, \theta)\}$

(ii) $E[\tilde{g}(s, \theta)\tilde{g}(s', \theta')] = \displaystyle\int_{-\infty}^{\infty} |\xi| S_p(\xi, \theta)\exp(j2\pi\xi s')\alpha(s, \theta; \xi, \theta')\, d\xi$

where $\alpha(s, \theta; \xi, \theta')$ is the Radon transform of the plane wave $\exp\{-j2\pi\xi(x\cos\theta' + y\sin\theta')\}$ and equals $\exp(-j2\pi\xi s)\delta(\theta - \theta')/|\xi|$. Simplify this and obtain (10.56a). Combine (10.56b) and (10.58) to prove (10.60).

10.14 Write the observation model of (10.68) as $v(x, y) \overset{\Delta}{=} \mathcal{R}^{-1}w = h(x, y) \circledast f(x, y) + \eta(x, y)$ where $h(x, y) \overset{\mathcal{F}_2}{\longleftrightarrow} H(\xi_1, \xi_2) = H_p(\xi, \theta) \overset{\mathcal{F}_1}{\longleftrightarrow} h_p(s, \theta)$ and $\eta \overset{\Delta}{=} \mathcal{R}^{-1}\nu$, whose power spectrum density is given by (10.67). Show that the frequency response of the two-dimensional Wiener filter for $f(x, y)$, written in polar coordinates, is $\tilde{A}(\xi, \theta) = H_p^* S_p[|H_p|^2 S_p + |\xi| S_\nu^{-1}]^{-1}$. Implement this filter in the Radon transform domain, as shown in Fig. 10.9, to arrive at the filter $A_p = |\xi| \tilde{A}_p$.

10.15 Compare the operation counts of the Fourier method with the convolution/filter back-projection methods. Assume $N \times N$ image size with αN projections, $\alpha \overset{\Delta}{=}$ constant.

10.16 *(Radon inversion formula for divergent rays)*

a. Starting with the inverse Radon transform in polar coordinates, show that the reconstructed object from fan-beam geometry projections $b(\sigma, \beta)$ can be written as

$$f_p(r, \phi) = \frac{1}{4\pi^2} \int_0^{2\pi} \int_{-\gamma}^{\gamma} \frac{\left[\dfrac{\partial b(\sigma, \beta)}{\partial \sigma} - \dfrac{\partial b(\sigma, \beta)}{\partial \beta}\right]}{r \cos(\sigma + \beta - \phi) - R \sin \sigma} \, d\sigma \, d\beta$$

where $|\sigma| \le \gamma$.

b. Rewrite the preceding result as a generalized convolution back-projection result, called the *Radon inversion formula for divergent rays,* as

$$f_p(r, \phi) = \frac{1}{4\pi^2} \int_0^{2\pi} \int_{-\infty}^{\infty} \frac{\psi(\sigma, \beta, \sigma')}{\sigma' - \sigma} \, d\sigma \, d\beta$$

where

$$\psi(\sigma, \beta, \sigma') \overset{\Delta}{=} \begin{cases} \dfrac{\sigma' - \sigma}{\sin(\sigma' - \sigma)} \left[\dfrac{1}{\rho} \dfrac{\partial b(\sigma, \beta)}{\partial \sigma} - \dfrac{1}{\rho} \dfrac{\partial b(\sigma, \beta)}{\partial \beta}\right], & |\sigma| \le \gamma \\ 0, & |\sigma| > \gamma \end{cases}$$

$$\sigma' \overset{\Delta}{=} \tan^{-1} \frac{r \cos(\beta - \phi)}{R + r \sin(\beta - \phi)}, \qquad \rho \overset{\Delta}{=} \{[r \cos(\beta - \phi)]^2 + [R + r \sin(\beta - \phi)]^2\}^{1/2} > 0$$

Show that σ' and ρ correspond to a ray (σ', β) that goes through the object at location (r, ϕ) and ρ is the distance between the source and (r, ϕ). The inner integral in the above Radon inversion formula is the Hilbert transform of $\psi(\sigma, \cdot, \cdot)$ and the outer integral is analogous to back projection.

c. Develop a practical reconstruction algorithm by replacing the Hilbert transform by a bandlimited filter, as in the case of parallel beam geometry.

10.17 *(Two-stage reconstruction in three dimensions)*

a. Referring to Fig. 10.19, rotate the x- and y-axes by an angle ϕ, that is, let $x' = x \cos \phi + y \sin \phi$, $y' = -x \sin \phi + y \cos \phi$, and obtain

$$g(s, \phi, \theta) = g(s, \alpha) = \iint_{-\infty}^{\infty} f_\phi(x', z) \delta(x' \sin \theta + z \cos \theta - s) \, dx' \, dz$$

where f_ϕ and g are the two-dimensional Radon transforms of f (with z constant) and f_ϕ (with ϕ constant), respectively, that is,

$$f_\phi = \mathcal{R}_{x', \phi} f, \qquad g = \mathcal{R}_{s, \theta} f_\phi.$$

b. Develop the block diagram for a digital implementation of the two-stage reconstruction algorithm.

BIBLIOGRAPHY

Section 10.1

For image formation models of CT, PET, MRI and overview of computerized tomography:

1. *IEEE Trans. Nucl. Sci.* Special Issue on topics related to image reconstruction. NS-21, no. 3 (1974); NS-26, no. 2 (April 1979); NS-27, no. 3 (June 1980).
2. *IEEE Trans. Biomed. Engineering.* Special Issue on computerized medical imaging. BME-28, no. 2 (February 1981).
3. *Proc. IEEE.* Special Issue on Computerized Tomography. 71, no. 3 (March 1983).
4. A. C. Kak. "Image Reconstruction from Projections," in M. P. Ekstrom (ed.). *Digital Image Processing Techniques.* New York: Academic Press, 1984, pp. 111–171.
5. G. T. Herman (ed.). *Image Reconstruction from Projections. Topics in Applied Physics,* vol. 32. New York: Springer-Verlag, 1979.
6. G. T. Herman. *Image Reconstruction from Projections—The Fundamentals of Computerized Tomography.* New York: Academic Press, 1980.
7. H. J. Scudder. "Introduction to Computer Aided Tomography." *Proc. IEEE* 66, no. 6 (June 1978).
8. Z. H. Cho, H. S. Kim, H. B. Song, and J. Cumming. "Fourier Transform Nuclear Magnetic Resonance Tomographic Imaging." *Proc. IEEE,* 70, no. 10 (October 1982): 1152–1173.
9. W. S. Hinshaw and A. H. Lent. "An Introduction to NMR Imaging," in [3].
10. D. C. Munson, Jr., J. O'Brien, K. W. Jenkins. "A Tomographic Formulation of Spotlight Mode Synthetic Aperture Radar." *Proc. IEEE,* 71, (August 1983): 917–925.

Literature on image reconstruction also appears in other journals such as: *J. Comput. Asst. Tomo., Science, Brit. J. Radiol., J. Magn. Reson. Medicine, Comput. Biol. Med.,* and *Medical Physics.*

11. J. L. C. Sanz, E. B. Hinkle, A. K. Jain. *Radon and Projection Transform-Based Machine Vision: Algorithms, A Pipeline Architecture, and Industrial Applications,* Berlin: Springer-Verlag, (1988). Also see, *Journal of Parallel and Distributed Computing,* vol. 4, no. 1 (Feb. 1987): 45–78.

Sections 10.2–10.5

Fundamentals of Radon transform theory appear in several of the above references, such as [4–7], and:

12. J. Radon. "Über die Bestimmung von Funktionen durch ihre Integralwerte Tangs gewisser Mannigfaltigkeiten" (On the determination of functions from their integrals along certain manifolds). Bertichte Saechsiche Akad. Wissenschaften (Leipzig), *Math. Phys. Klass* 69, (1917): 262–277.
13. D. Ludwig. "The Radon Transform on Euclidean Space." *Commun. Pure Appl. Math.* 19, (1966): 49–81.

14. D. E. Kuhl and R. Q. Edwards. "Image Separation Isotope Scanning," *Radiology* 80, no. 4 (1963): 653–662.
15. P. F. C. Gilbert. "The Reconstruction of a Three-Dimensional Structure from Projections and its Application to Electron Microscopy: II. Direct Methods." *Proc. Roy. Soc. London* Ser. B, vol. 182, (1972): 89–102.
16. P. R. Smith, T. M. Peters, and R. H. T. Bates. "Image Reconstruction from Finite Number of Projections." *J. Phys. A: Math Nucl. Gen.* 6, (1973): 361–382. Also see *New Zealand J. Sci.*, 14, (1971): 883–896.
17. S. R. Deans. *The Radon Transform and Some of Its Applications*. New York: Wiley, 1983.

Section 10.6

For convolution/filter back-projection algorithms, simulations, and related details:

18. S. W. Rowland, in [5], pp. 9–79.
19. G. N. Ramachandran and A. V. Lakshminarayanan. "Three-Dimensional Reconstruction from Radiographs and Electron Micrographs: II. Application of convolutions instead of Fourier Transforms." *Proc. Nat. Acad. Sci.*, 68 (1971): 2236–2240. Also see *Indian J. Pure Appl. Phys.* 9 (1971): 997–1003.
20. R. N. Bracewell and A. C. Riddle. "Inversion of Fan-Beam Scans in Radio Astronomy." *Astrophys. J.* 150 (1967): 427–437.
21. L. A. Shepp and B. F. Logan. "The Fourier Reconstruction of a Head Section." *IEEE Trans. Nucl. Sci.* NS-21, no. 3 (1974): 21–43.

Sections 10.7–10.8

Results on Radon transform of random fields were introduced in

22. A. K. Jain and S. Ansari. "Radon Transform Theory for Random Fields and Image Reconstruction From Noisy Projections." *Proc. ICASSP,* San Diego, 1984.
23. A. K. Jain. "Digital Image Processing: Problems and Methods," in T. Kailath (ed.), *Modern Signal Processing*. Washington: Hemisphere Publishing Corp., 1985.

For reconstruction from noisy projections see the above references and:

24. Z. Cho and J. Burger. "Construction, Restoration, and Enhancement of 2- and 3-Dimensional Images," *IEEE Trans. Nucl. Sci.* NS-24, no. 2 (April 1977): 886–895.
25. E. T. Tsui and T. F. Budinger. "A Stochastic Filter for Transverse Section Reconstruction." *IEEE Trans. Nucl. Sci.* NS-26, no. 2 (April 1979): 2687–2690.

Section 10.9

26. R. N. Bracewell. "Strip Integration in Radio Astronomy." *Aust. J. Phys.* 9 (1956): 198–217.
27. R. A. Crowther, D. J. Derosier, and A. Klug. "The Reconstruction of a Three-Dimensional Structure from Projections and Its Application to Electron Microscopy."

Proc. Roy. Soc. London Ser. A, vol. 317 (1970): 319–340. Also see *Nature* (London) 217 (1968): 130–134.

28. G. N. Ramachandran. "Reconstruction of Substance from Shadow: I. Mathematical Theory with Application to Three Dimensional Radiology and Electron Microscopy." *Proc. Indian Acad. Sci.* 74 (1971): 14–24.
29. R. M. Merseau and A. V. Oppenheim. "Digital Reconstruction of Multidimensional Signals From Their Projections." *Proc. IEEE* 62, (1974): 1319–1332.
30. R. F. King and P. R. Moran. "Unified Description of NMR Imaging Data Collection Strategies and Reconstruction." *Medical Physics* 11, no. 1 (1984): 1–14.

Sections 10.10–10.13

For fan-beam reconstruction theory, see [6, 7] and Horn in [1(iii), pp. 1616–1623]. For algebraic techniques and ART algorithms, see [5, 6] and:

31. S. Kaczmarz. "Angenäherte Auflösung von Systemen Linearer Gleichungen." *Bull. Acad. Polon. Sci. Lett. A.* 35, (1937): pp. 355–357.
32. R. Gordon, "A Tutorial on ART (Algebraic Reconstruction Techniques)." *IEEE Trans. Nucl. Sci.* NS-21, (1974): 78.
33. P. F. C. Gilbert. "Iterative Methods for the Reconstruction of Three-Dimensional Objects from Projections." *J. Theor. Biol.* 36 (1972): 105–117.
34. G. T. Herman, A. Lent, and S. W. Rowland. "ART: Mathematics and Applications," *J. Theor. Biol.* 42 (1973): 1–32.
35. A. M. Cormack. "Representation of a Function by its Line Integrals with Some Radiological Applications." *J. Appl. Phys.* 34 (1963): 2722–2727. Also see Part II, *J. Appl. Phys.* 35 (1964): 2908–2913.

For other applications of Radon transform and its extensions:

36. M. Bernfield. "CHIRP Doppler Radar." *Proc. IEEE,* vol. 72, no. 4 (April 1984): 540–541.
37. J. Raviv, J. F. Greenleaf, and G. T. Herman (eds.). *Computer Aided Tomography and Ultrasonics in Medicine.* Amsterdam: North-Holland, 1979.

11

Image Data Compression

11.1 INTRODUCTION

Image data compression is concerned with minimizing the number of bits required to represent an image. Perhaps the simplest and most dramatic form of data compression is the sampling of bandlimited images, where an infinite number of pixels per unit area is reduced to one sample without any loss of information (assuming an ideal low-pass filter is available). Consequently, the number of samples per unit area is infinitely reduced.

Applications of data compression are primarily in transmission and storage of information. Image transmission applications are in broadcast television, remote sensing via satellite, military communications via aircraft, radar and sonar, teleconferencing, computer communications, facsimile transmission, and the like. Image storage is required for educational and business documents, medical images that arise in computer tomography (CT), magnetic resonance imaging (MRI) and digital radiology, motion pictures, satellite images, weather maps, geological surveys, and so on. Application of data compression is also possible in the development of fast algorithms where the number of operations required to implement an algorithm is reduced by working with the compressed data.

Image Raw Data Rates

Typical television images have spatial resolution of approximately 512×512 pixels per frame. At 8 bits per pixel per color channel and 30 frames per second, this translates into a rate of nearly 180×10^6 bits/s. Depending on the application, digital image raw data rates can vary from 10^5 bits per frame to 10^8 bits per frame or higher. The large-channel capacity and memory requirements (see Table 1.1b) for digital image transmission and storage make it desirable to consider data compression techniques.

Data Compression versus Bandwidth Compression

The mere process of converting an analog video signal into a digital signal results in increased bandwidth requirements for transmission. For example, a 4-MHz television signal sampled at Nyquist rate with 8 bits per sample would require a bandwidth of 32 MHz when transmitted using a digital modulation scheme, such as phase shift keying (PSK), which requires 1 Hz per 2 bits. Thus, although digitized information has advantages over its analog form in terms of processing flexibility, random access in storage, higher signal to noise ratio for transmission with the possibility of errrorless communication, and so on, one has to pay the price in terms of this eightfold increase in bandwidth. Data compression techniques seek to minimize this cost and sometimes try to reduce the bandwidth of the digital signal below its analog bandwidth requirements.

Image data compression methods fall into two common categories. In the first category, called predictive coding, are methods that exploit *redundancy* in the data. Redundancy is a characteristic related to such factors as *predictability, randomness,* and *smoothness* in the data. For example, an image of constant gray levels is fully predictable once the gray level of the first pixel is known. On the other hand, a white noise random field is totally unpredictable and every pixel has to be stored to reproduce the image. Techniques such as delta modulation and *differential pulse code modulation* fall into this category. In the second category, called *transform coding,* compression is achieved by transforming the given image into another array such that a large amount of information is *packed* into a small number of samples. Other image data compression algorithms exist that are generalizations or combinations of these two methods. The compression process inevitably results in some distortion because of accompanying A to D conversion as well as rejection of some relatively insignificant information. Efficient compression techniques tend to minimize this distortion. For digitized data, distortionless compression techiques are possible. Figure 11.1 gives a summary classification of various data compression techniques.

Information Rates

Raw image data rate does not necessarily represent its average information rate, which for a source with L possible independent symbols with probabilities p_i,

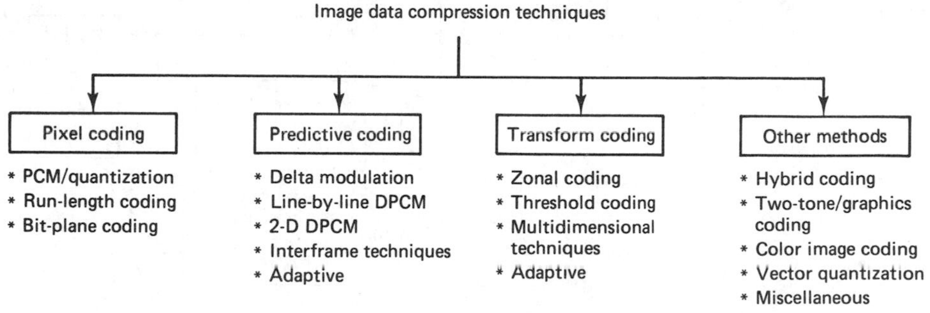

Figure 11.1 Image data compression techniques.

$i = 0, \ldots, L-1$, is given by the entropy

$$H = -\sum_{i=0}^{L-1} p_i \log_2 p_i \quad \text{bits per symbol} \tag{11.1}$$

According to Shannon's *noiseless coding theorem* (Section 2.13) it is possible to code, without distortion, a source of entropy H bits per symbol using $H + \varepsilon$ bits per symbol, where ε is an arbitrarily small positive quantity. Then the maximum achievable compression C, defined by

$$C = \frac{\text{average bit rate of the original raw data } (B)}{\text{average bit rate of the encoded data } (H + \varepsilon)} \tag{11.2}$$

is $B/(H + \varepsilon) \simeq B/H$. Computation of such a compression ratio for images is impractical, if not impossible. For example an $N \times M$ digital image with B bits per pixel is one of $L = 2^{BNM}$ possible image patterns that could occur. Thus if p_i, the probability of the ith image pattern, were known, one could compute the entropy, that is, the information rate for B bits per pixel $N \times M$ images. Then one could store all the L possible image patterns and encode the image by its address—using a suitable encoding method, which will require approximately H bits per image or H/NM bits per pixel.

Such a method of coding is called *vector quantization,* or *block coding* [12]. The main difficulty with this method is that even for small values of N and M, L can be prohibitively large; for example, for $B = 8$, $N = M = 16$ and $L = 2^{2048} \simeq 10^{614}$. Figure 11.2 shows a practical adaptation of this idea for vector quantization of 4×4 image blocks with $B = 6$. Each block is normalized to have zero mean and unity variance. Using a few prototype training images, the most probable subset containing $L' \ll L$ images is stored. If the input block is one of these L' blocks, it is coded by the address of the block; otherwise it is replaced by its mean value.

The entropy of an image can also be estimated from its conditional entropy. For a block of N pixels $u_0, u_1, \ldots, u_{N-1}$, with B bits per pixel and arranged in an arbitrary order, the Nth-order conditional entropy is defined as

$$H_N \triangleq -\sum_{u_0} \cdots \sum_{u_{N-1}} p(u_0, u_1, \ldots, u_{N-1}) \log_2 p(u_0 | u_1, \ldots, u_{N-1}) \tag{11.3}$$

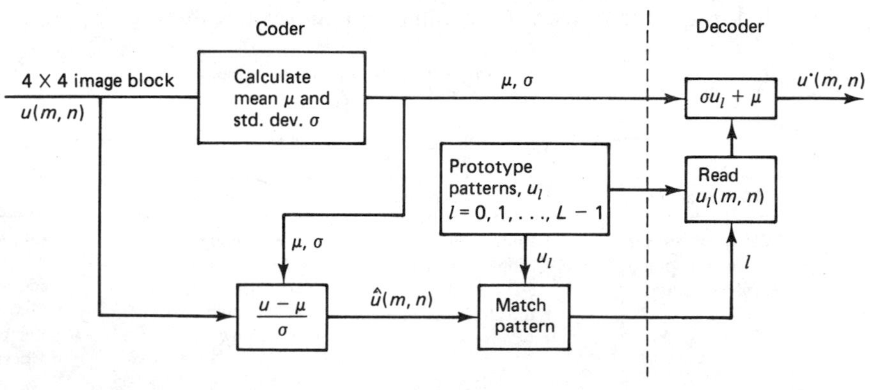

Figure 11.2 Vector quantization of images.

where each u_i, $i = 0, \ldots, N-1$, takes 2^B values, and $p(\cdot,\cdot,\ldots)$ represent the relevant probabilities. For 8-bit monochrome television-quality images, the zero- to second-order entropies (with nearest-neighbor ordering) generally lie in the range of 2 to 6 bits/pixel. Theoretically, for ergodic sequences, as $N \to \infty$, H_N converges to H, the per-pixel entropy. Shannon's theory tells us that the bit rate of any exact coding method can never be below the entropy H.

Subsampling, Coarse Quantization, Frame Repetition, and Interlacing

One obvious method of data compression would be to reduce the sampling rate, the number of quantization levels, and the refresh rate (number of frames per second) down to the limits of aliasing, contouring, and flickering phenomena, respectively. The distortions introduced by subsampling and coarse quantization for a given level of compression are generally much larger than the more sophisticated methods available for data compression. To avoid flicker in motion images, successive frames have to be refreshed above the critical fusion frequency (CFF), which is 50 to 60 pictures per second (Section 3.12). Typically, to capture motion a refresh rate of 25 to 30 frames per second is generally sufficient. Thus, a compression of 2 to 1 could be achieved by transmitting (or storing) only 30 frames per second but refreshing at 60 frames per second by repeating each frame. This requires a frame storage, but an image breakup or jump effect (not flicker) is often observed. Note that the frame repetition rate is chosen at 60 per second rather than 55 per second, for instance, to avoid any interference with the line frequency of 60 Hz (in the United States).

Instead of frame skipping and repetition, line interlacing is found to give better visual rendition. Each frame is divided into an *odd field* containing the odd line addresses and an *even field* containing the even line addresses; frames are transmitted alternately. Each field is displayed at half the refresh rate in frames per second. Although the jump or image breakup effect is significantly reduced by line interlacing, spatial frequency resolution is somewhat degraded because each field is a subsampled image. An appropriate increase in the scan rate (that is, lines per frame) with line interlacing gives an actual compression of about 37% for the same subjective quality at the 60 frames per second refresh rate without repetition. The success of this method rests on the fact that the human visual system has poor response for simultaneously occurring high spatial and temporal frequencies. Other interlacing techniques, such as vertical line interlacing in each field (Fig. 4.9), can reduce the data rate further without introducing aliasing if the spatial frequency spectrum does not contain simultaneously horizontal and vertical high frequencies (such as diagonal edges). Interlacing techniques are unsuitable for the display of high resolution graphics and other computer generated images that contain sharp edges and transitions. Such images are commonly displayed on a large raster (e.g., 1024×1024) refreshed at 60 Hz.

11.2 PIXEL CODING

In these techniques each pixel is processed independently, ignoring the inter pixel dependencies.

PCM

In PCM the incoming video signal is sampled, quantized, and coded by a suitable code word (before feeding it to a digital modulator for transmission) (Fig. 11.3). The quantizer output is generally coded by a fixed-length binary code word having B bits. Commonly, 8 bits are sufficient for monochrome broadcast or video-conferencing quality images, whereas medical images or color video signals may require 10 to 12 bits per pixel.

The number of quantizing bits needed for visual display of images can be reduced to 4 to 8 bits per pixel by using companding, contrast quantization, or dithering techniques discussed in Chapter 4. Halftone techniques reduce the quantizer output to 1 bit per pixel, but usually the input sampling rate must be increased by a factor of 2 to 16. The compression achieved by these techniques is generally less than 2 : 1.

In terms of a mean square distortion, the minimum achievable rate by PCM is given by the rate-distortion formula

$$R_{PCM} = \frac{1}{2}\log_2 \frac{\sigma_u^2}{\sigma_q^2}, \qquad \sigma_q^2 < \sigma_u^2 \tag{11.4}$$

where σ_u^2 is the variance of the quantizer input and σ_q^2 is the quantizer mean square distortion.

Entropy Coding

If the quantized pixels are not uniformly distributed, then their entropy will be less than B, and there exists a code that uses less than B bits per pixel. In entropy coding the goal is to encode a block of M pixels containing MB bits with probabilities $p_i, i = 0, 1, \ldots, L-1, L = 2^{MB}$, by $-\log_2 p_i$ bits, so that the average bit rate is

$$\sum_i p_i(-\log_2 p_i) = H$$

This gives a variable-length code for each block, where highly probable blocks (or symbols) are represented by small-length codes, and vice versa. If $-\log_2 p_i$ is not an integer, the achieved rate exceeds H but approaches it asymptotically with increasing block size. For a given block size, a technique called *Huffman coding* is the most efficient fixed to variable length encoding method.

The Huffman Coding Algorithm

1. Arrange the symbol probabilities p_i in a decreasing order and consider them as leaf nodes of a tree.

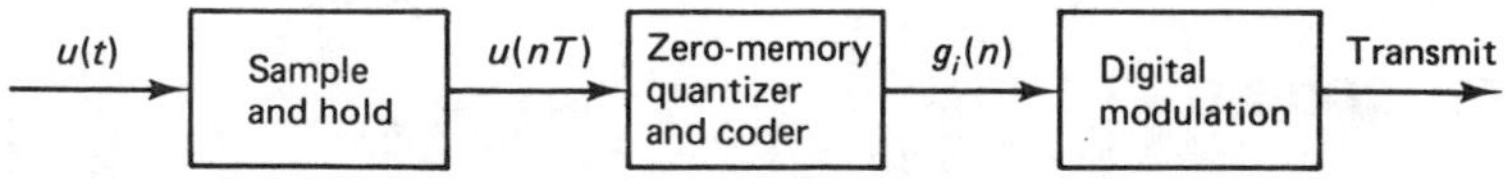

Figure 11.3 Pulse code modulation (PCM).

2. While there is more than one node:
   - Merge the two nodes with smallest probability to form a new node whose probability is the sum of the two merged nodes.
   - Arbitrarily assign 1 and 0 to each pair of branches merging into a node.
3. Read sequentially from the root node to the leaf node where the symbol is located.

The preceding algorithm gives the *Huffman code book* for any given set of probabilities. Coding and decoding is done simply by looking up values in a table. Since the code words have variable length, a buffer is needed if, as is usually the case, information is to be transmitted over a constant-rate channel. The size of the code book is L and the longest code word can have as many as L bits. These parameters become prohibitively large as L increases. A practical version of Huffman code is called the *truncated Huffman code*. Here, for a suitably selected $L_1 < L$, the first L_1 symbols are Huffman coded and the remaining symbols are coded by a prefix code followed by a suitable fixed-length code.

Another alternative is called the *modified Huffman code*, where the integer i is represented as

$$i = qL_1 + j, \qquad 0 \le q \le \text{Int}\left[\frac{(L-1)}{L_1}\right], 0 \le j \le L_1 - 1 \tag{11.5}$$

The first L_1 symbols are Huffman coded. The remaining symbols are coded by a prefix code representing the quotient q, followed by a *terminator code*, which is the same as the Huffman code for the remainder j, $0 \le j \le L_1 - 1$.

The long-term histogram for television images is approximately uniform, although the short-term statistics are highly nonstationary. Consequently entropy coding is not very practical for raw image data. However, it is quite useful in predictive and transform coding algorithms and also for coding of binary data such as graphics and facsimile images.

Example 11.1

Figure 11.4 shows an example of the tree structure and the Huffman codes. The algorithm gives code words that can be uniquely decoded. This is because no code word can be a prefix of any larger-length code word. For example, if the Huffman coded bit stream is

0 0 0 1 0 1 1 0 1 0 1 1 . . .

then the symbol sequence is $s_0 s_2 s_5 s_3 \ldots$. A prefix code (circled elements) is obtained by reading the code of the leaves that lead up to the first node that serves as a root for the truncated symbols. In this example there are two prefix codes (Fig. 11.4). For the truncated Huffman code the symbols $s_4, \ldots, s_7$ are coded by a 2-bit binary code word. This code happens to be less efficient than the simple fixed length binary code in this example. But this is not typical of the truncated Huffman code.

Run-Length Coding

Consider a binary source whose output is coded as the number of 0s between two successive 1s, that is, the length of the runs of 0s are coded. This is called *run-length*

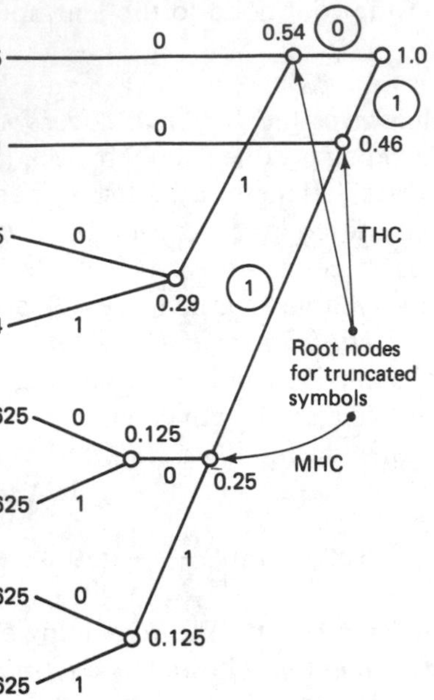

Symbol	Binary code	p_i	Huffman code (HC)	Truncated Huffman code, $L_1 = 2$ (THC)	Modified Huffman code, $L_1 = 4$ (MHC)
s_0	0 0 0	0.25	0 0	0 0	0 0
s_1	0 0 1	0.21	1 0	1 0	1 0
s_2	0 1 0	0.15	0 1 0	x': 0 1, y: 0 0	0 1 0
s_3	0 1 1	0.14	0 1 1	0 1 1 0	0 1 1
s_4	1 0 0	0.0625	1 1 0 0	x: 1 1, y: 0 0	x'': 1 1, z: 0 0
s_5	1 0 1	0.0625	1 1 0 1	1 1 0 1	1 1 1 0
s_6	1 1 0	0.0625	1 1 1 0	1 1 1 0	1 1 0 1 0
s_7	1 1 1	0.0625	1 1 1 1	1 1 1 1	1 1 0 1 1
Average code length	3.0	2.781 (entropy)	2.79	3.08	2.915
Code efficiency H/B_a	92.7%		99.7%	90.3%	95.4%

Figure 11.4 Huffman coding example. x, x', x'' = prefix codes, y = fixed length code, z = terminator code. In general x, x', x'' can be different.

coding (RLC). It is useful whenever large runs of 0s are expected. Such a situation occurs in printed documents, graphics, weather maps, and so on, where p, the probability of a 0 (representing a white pixel) is close to unity. (See Section 11.9.)

Suppose the runs are coded in maximum lengths of M and, for simplicity, let $M = 2^m - 1$. Then it will take m bits to code each run by a fixed-length code. If the successive 0s occur independently, then the probability distribution of the run lengths turns out to be the geometric distribution

$$g(l) = \begin{cases} p^l(1-p), & 0 \le l \le M-1 \\ p^M, & l = M \end{cases} \tag{11.6}$$

Since a run length of $l \le M - 1$ implies a sequence of l 0s followed by a 1, that is,

$(l + 1)$ symbols, the average number of symbols per run will be

$$\mu = \sum_{l=0}^{M-1} (l+1)p^l(1-p) + Mp^M = \frac{(1-p^M)}{(1-p)} \tag{11.7}$$

Thus it takes m bits to establish a run-length code for a sequence of μ binary symbols, on the average. The compression achieved is, therefore,

$$C = \frac{\mu}{m} = \frac{(1-p^M)}{m(1-p)} \tag{11.8}$$

For $p = 0.9$ and $M = 15$ we obtain $m = 4$, $\mu = 7.94$, and $C = 1.985$. The achieved average rate is $B_a = m/\mu = 0.516$ bit per pixel and the code efficiency, defined as H/B_a, is $0.469/0.516 = 91\%$. For a given value of p, the optimum value of M can be determined to give the highest efficiency. RLC efficiency can be improved further by going to a variable length coding method such as Huffman coding for the blocks of length m. Another alternative is to use *arithmetic coding* [10] instead of the RLC.

Bit-Plane Encoding [11]

A 256 gray-level image can be considered as a set of eight *1-bit planes,* each of which can be run-length encoded. For 8-bit monochrome images, compression ratios of 1.5 to 2 can be achieved. This method becomes very sensitive to channel errors unless the significant bit planes are carefully protected.

11.3 PREDICTIVE TECHNIQUES

Basic Principle

The philosophy underlying predictive techniques is to remove mutual redundancy between successive pixels and encode only the new information. Consider a sampled sequence $u(m)$, which has been coded up to $m = n - 1$ and let $u^\cdot(n-1)$, $u^\cdot(n-2), \ldots$ be the values of the reproduced (decoded) sequence. At $m = n$, when $u(n)$ arrives, a quantity $\bar{u}^\cdot(n)$, an estimate of $u(n)$, is predicted from the previously decoded samples $u^\cdot(n-1), u^\cdot(n-2)\ldots$, that is,

$$\bar{u}^\cdot(n) = \psi(u^\cdot(n-1), u^\cdot(n-2), \ldots) \tag{11.9}$$

where $\psi(\cdot,\cdot,\ldots)$ denotes the prediction rule. Now it is sufficient to code the *prediction error*

$$e(n) \triangleq u(n) - \bar{u}^\cdot(n) \tag{11.10}$$

If $e^\cdot(n)$ is the quantized value of $e(n)$, then the reproduced value of $u(n)$ is taken as

$$u^\cdot(n) = \bar{u}^\cdot(n) + e^\cdot(n) \tag{11.11}$$

The coding process continues *recursively* in this manner. This method is called *differential pulse code modulation* (DPCM) or *differential* PCM. Figure 11.5 shows

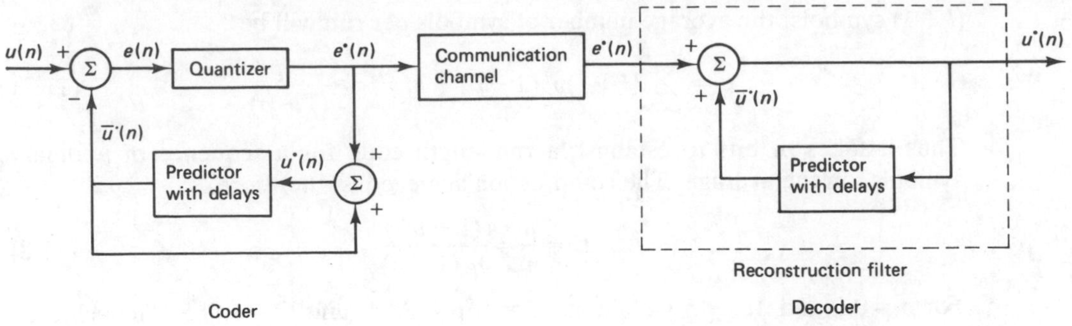

Figure 11.5 Differential pulse code modulation (DPCM) CODEC.

the DPCM *codec* (coder-decoder). Note that the coder has to calculate the reproduced sequence $u^{\cdot}(n)$. The decoder is simply the predictor loop of the coder. Rewriting (11.10) as

$$u(n) = \bar{u}^{\cdot}(n) + e(n) \tag{11.12}$$

and subtracting (11.11) from (11.12), we obtain

$$\delta u(n) \triangleq u(n) - u^{\cdot}(n) = e(n) - e^{\cdot}(n) = q(n) \tag{11.13}$$

Thus, the pointwise coding error in the input sequence is exactly equal to $q(n)$, the quantization error in $e(n)$. With a reasonable predictor the mean square value of the differential signal $e(n)$ is much smaller than that of $u(n)$. This means, for the same mean square quantization error, $e(n)$ requires fewer quantization bits than $u(n)$.

Feedback Versus Feedforward Prediction

An important aspect of DPCM is (11.9) which says the prediction is based on the output—the quantized samples—rather than the input—the unquantized samples. This results in the predictor being in the feedback loop around the quantizer, so that the quantizer error at a given step is fed back to the quantizer input at the next step. This has a stabilizing effect that prevents dc drift and accumulation of error in the reconstructed signal $u^{\cdot}(n)$.

On the other hand, if the prediction rule is based on the *past inputs* (Fig. 11.6a), the signal reconstruction error would depend on all the past and present

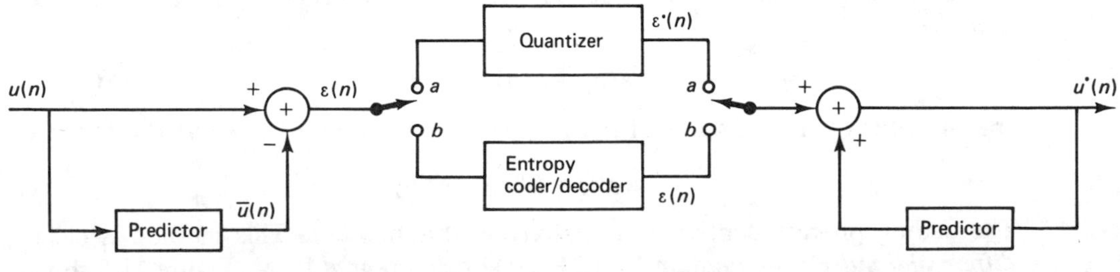

Figure 11.6 Feedforward coding (a) with distortion; (b) distortionless.

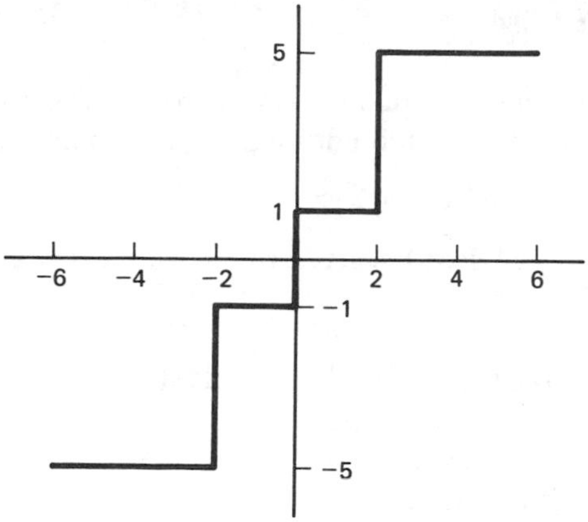

Figure 11.7 Quantizer used in Example 11.2.

quantization errors in the feedforward prediction-error sequence $\varepsilon(n)$. Generally, the mean square value of this reconstruction error will be greater than that in DPCM, as illustrated by the following example (also see Problem 11.3).

Example 11.2

The sequence 100, 102, 120, 120, 120, 118, 116 is to be predictively coded using the previous element prediction rule, $\bar{u}^{\cdot}(n) = u^{\cdot}(n-1)$ for DPCM and $\bar{u}(n) = u(n-1)$ for the feedforward predictive coder. Assume a 2-bit quantizer shown in Fig. 11.7 is used, except the first sample is quantized separately by a 7-bit uniform quantizer, giving $u^{\cdot}(0) = u(0) = 100$. The following table shows how reconstruction error builds up with a feedforward predictive coder, whereas it tends to stabilize with the feedback system of DPCM.

Input			DPCM					Feedforward Predictive Coder				
	n	$u(n)$	$\bar{u}^{\cdot}(n)$	$e(n)$	$e^{\cdot}(n)$	$u^{\cdot}(n)$	$\delta u(n)$	$\bar{u}(n)$	$\varepsilon(n)$	$\varepsilon^{\cdot}(n)$	$u^{\cdot}(n)$	$\delta u(n)$
	0	100	—	—	—	100	0	—	—	—	100	0
	1	102	100	2	1	101	1	100	2	1	101	1
Edge →	2	120	101	19	5	106	14	102	18	5	106	14
	3	120	106	14	5	111	9	120	0	−1	105	15
	4	120	111	9	5	116	4	120	0	−1	104	16
	5	118	116	2	1	117	1	120	−2	−5	99	19

Distortionless Predictive Coding

In digital processing the input sequence $u(n)$ is generally digitized at the source itself by a sufficient number of bits (typically 8 for images). Then, $u(n)$ may be considered as an integer sequence. By requiring the predictor outputs to be integer values, the prediction error sequence will also take integer values and can be entropy coded without distortion. This gives a distortionless predictive codec (Fig. 11.6b), whose minimum achievable rate would be equal to the entropy of the prediction-error sequence $\varepsilon(n)$.

Performance Analysis of DPCM

Denoting the mean square values of quantization error $q(n)$ and the prediction error $e(n)$ by σ_q^2 and σ_e^2, respectively, and noting that (11.13) implies

$$E[(\delta u(n))^2] = \sigma_q^2 \tag{11.14}$$

the minimum achievable rate by DPCM is given by the rate-distortion formula [see (2.116)]

$$R_{DPCM} = \tfrac{1}{2}\log_2\left(\frac{\sigma_e^2}{\sigma_q^2}\right) \quad \text{bits/pixel} \tag{11.15}$$

In deducing this relationship, we have used the fact that common zero memory quantizers (for arbitrary distributions) do not achieve a rate lower than the Shannon quantizer for Gaussian distributions (see Section 4.9). For the same distortion $\sigma_q^2 \le \sigma_e^2$, the reduction in DPCM rate compared to PCM is [see (11.4)]

$$R_{PCM} - R_{DPCM} = \tfrac{1}{2}\log_2\left(\frac{\sigma_u^2}{\sigma_e^2}\right) \simeq \frac{1}{0.6}\log_{10}\left(\frac{\sigma_u^2}{\sigma_e^2}\right) \quad \text{bits/pixel} \tag{11.16}$$

This shows the achieved compression depends on the reduction of the variance ratio (σ_u^2/σ_e^2), that is, on the ability to predict $u(n)$ and, therefore, on the intersample redundancy in the sequence. Also, the recursive nature of DPCM requires that the predictor be causal. For minimum prediction-error variance, the optimum predictor is the conditional mean $E[u(n)|u^{\cdot}(m), m \le n-1]$. Because of the quantizer, this is a nonlinear function and is difficult to determine even when $u(n)$ is a stationary Gaussian sequence. The optimum feedforward predictor is linear and shift invariant for such sequences, that is,

$$\bar{u}(n) = \phi(u(n-1), \ldots) = \sum_k a(k)u(n-k) \tag{11.17}$$

If the feedforward prediction error $\varepsilon(n)$ has variance β^2, then

$$\beta^2 \le \sigma_e^2 \tag{11.18}$$

This is true because $\bar{u}^{\cdot}(n)$ is based on the quantization noise containing samples $\{u^{\cdot}(m), m \le n\}$ and could never be better than $\bar{u}(n)$. As the number of quantization levels is increased to infinity, σ_e^2 will approach β^2. Hence, a lower bound on the rate achievable by DPCM is

$$R_{\min} = \tfrac{1}{2}\log_2\frac{\beta^2}{\sigma_q^2} < R_{DPCM} \tag{11.19}$$

When the quantization error is small, R_{DPCM} approaches $R_{\min}$. This expression is useful because it is much easier to evaluate β^2 than σ_e^2 in (11.15), and it can be used to estimate the achievable compression. The SNR corresponding to σ_q^2 is given by

$$(\text{SNR})_{DPCM} = 10\log_{10}\frac{\sigma_u^2}{\sigma_q^2} = 10\log_{10}\frac{\sigma_u^2}{\sigma_e^2 f(B)} \le 10\log_{10}\frac{\sigma_u^2}{\beta^2 f(B)} \tag{11.20}$$

where $f(B)$ is the quantizer mean square distortion function for a unit variance input and B quantization bits. For equal number of bits used, the gain in SNR of DPCM over PCM is

$$(\text{SNR})_{DPCM} - (\text{SNR})_{PCM} = 10 \log_{10}\left(\frac{\sigma_u^2}{\sigma_e^2}\right) \le 10 \log_{10}\left[\frac{\sigma_u^2}{\beta^2}\right] \text{dB} \tag{11.21}$$

which is proportional to the log of the variance ratio (σ_u^2/β^2). Using (11.16) we note that the increase in SNR is approximately $6\,(R_{PCM} - R_{DPCM})$ dB, that is, 6 dB per bit of available compression.

From these measures we see that the performance of predictive coders depends on the design of the predictor and the quantizer. For simplicity, the predictor is designed without considering the quantizer effects. This means the prediction rule deemed optimum for $\bar{u}(n)$ is applied to estimate $\bar{u}^{\cdot}(n)$. For example, if $\bar{u}(n)$ is given by (11.17) then the DPCM predictor is designed as

$$\bar{u}^{\cdot}(n) = \phi(u^{\cdot}(n-1), u^{\cdot}(n-2), \ldots) = \sum_k a(k)u^{\cdot}(n-k) \tag{11.22}$$

In two (or higher) dimensions this approach requires finding the optimum causal prediction rules. Under the mean square criterion the minimum variance causal representations can be used directly. Note that the DPCM coder remains *nonlinear* even with the linear predictor of (11.22). However, the decoder will now be a linear filter. The quantizer is generally designed using the statistical properties of the innovations sequence $\varepsilon(n)$, which can be estimated from the predictor design. Figure 11.8 shows a typical prediction error signal histogram. Note the prediction

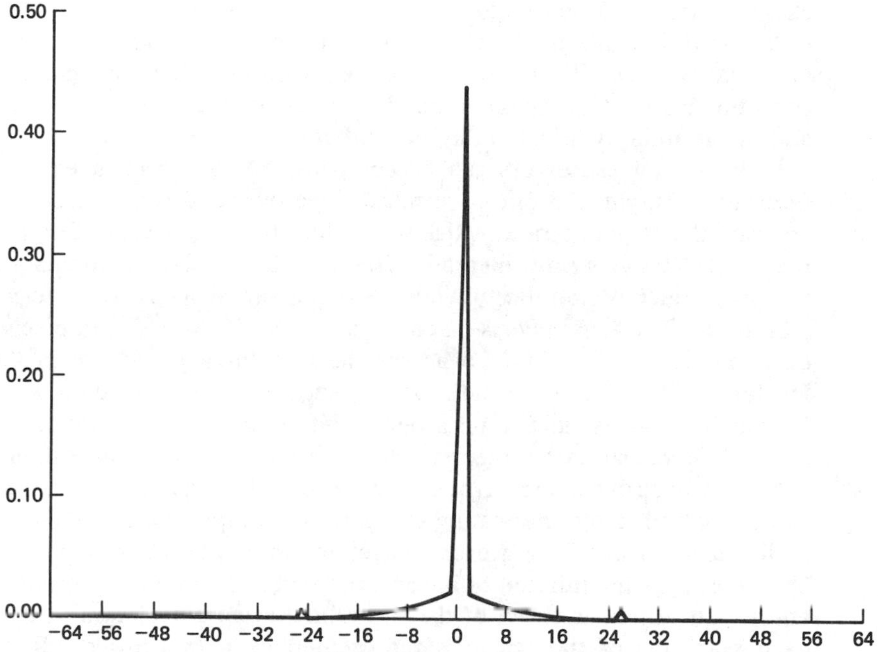

Figure 11.8 Predictions = error histogram.

error takes large values near the edges. Often, the prediction error is modeled as a zero mean uncorrelated sequence with a Laplacian probability distribution, that is,

$$p(\varepsilon) = \frac{1}{\beta^2} \exp\left(\frac{-\sqrt{2}}{\beta}|\varepsilon|\right) \tag{11.23}$$

where β^2 is its variance. The quantizer is generally chosen to be either the Lloyd-Max (for a constant bit rate at the output) or the optimum uniform quantizer (followed by an entropy coder to minimize the average rate). Practical predictive codecs differ with respect to realizations and the choices of predictors and quantizers. Some of the common classes of predictive codecs for images are described next.

Delta Modulation

Delta modulation (DM) is the simplest of the predictive coders. It uses a one-step delay function as a predictor and a 1-bit quantizer, giving a 1-bit representation of the signal. Thus

$$\bar{u}^{\cdot}(n) = u^{\cdot}(n-1), \qquad e(n) = u(n) - u^{\cdot}(n-1) \tag{11.24}$$

A practical DM system, which does not require sampling of the input signal, is shown in Fig. 11.9a. The predictor integrates the quantizer output, which is a sequence of binary pulses. The receiver is a simple integrator. Figure 11.9b shows typical input-output signals of a delta modulator. Primary limitations of delta modulation are (1) slope overload, (2) granularity noise, and (3) instability to channel errors. Slope overload occurs whenever there is a large jump or discontinuity in the signal, to which the quantizer can respond only in several delta steps. Granularity noise is the steplike nature of the output when the input signal is almost constant. Figure 11.10b shows the blurring effect of slope overload near the edges and the granularity effect in the constant gray-level background.

Both of these errors can be compensated to a certain extent by low-pass filtering the input and output signals. Slope overload can also be reduced by increasing the sampling rate, which will reduce the interpixel differences. However, the higher sampling rate will tend to lower the achievable compression. An alternative for reducing granularity while retaining simplicity is to go to a *tristate delta modulator*. The advantage is that a large number (65 to 85%) of pixels are found to be in the *level*, or 0, state, whereas the remaining pixels are in the ± 1 states. Huffman coding the three states or run-length coding the 0 states with a 2-bit code for the other states yields rates around 1 bit per pixel for different images [14].

The reconstruction filter, which is a simple integrator, is unstable. Therefore, in the presence of channel errors, the receiver output can accumulate large errors. It can be stabilized by attenuating the predictor output by a positive constant $\rho < 1$ (called *leak*). This will, however, not retain the simple realization of Fig. 11.9a.

For delta modulation of images, the signal is generally presented line by line and no advantage is taken of the two-dimensional correlation in the data. When each scan line of the image is represented by a first-order AR process (after subtracting the mean),

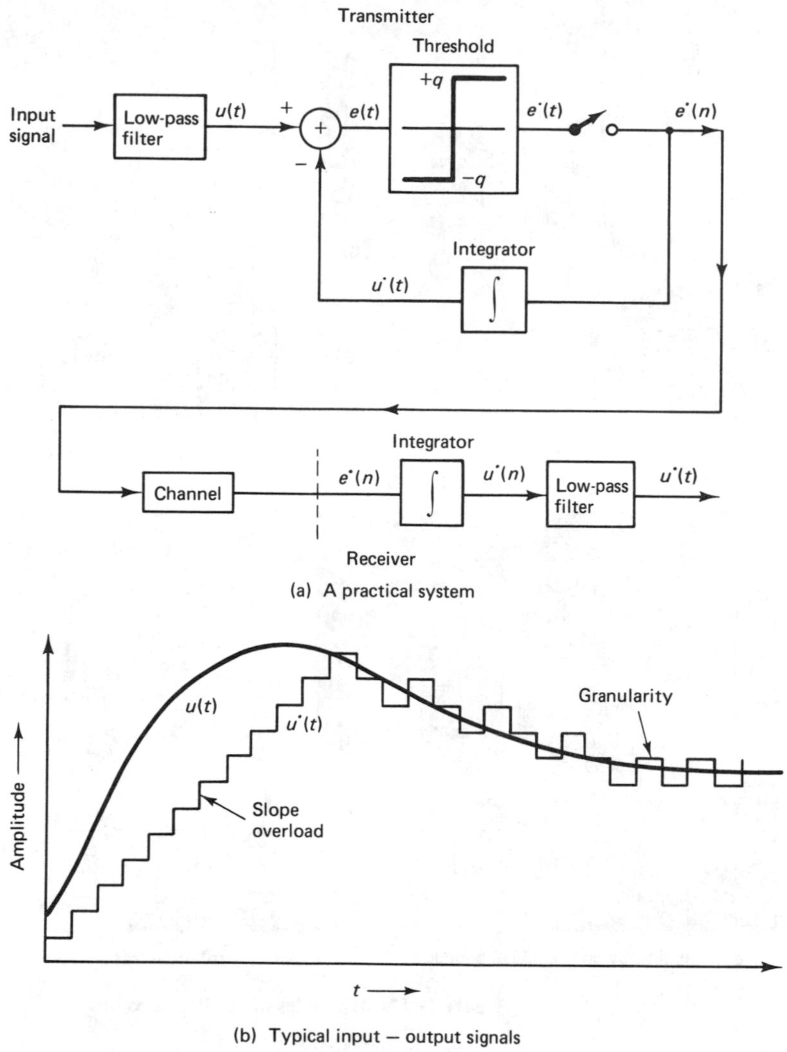

Figure 11.9 Delta modulation.

$$u(n) = \rho u(n-1) + \varepsilon(n), \qquad E[\varepsilon(n)] = 0, \qquad E[\varepsilon(n)\varepsilon(m)] = (1-\rho^2)\sigma_u^2\,\delta(m-n) \tag{11.25}$$

the SNR of the reconstructed signal is given, approximately, by (see Problem 11.4)

$$(\mathrm{SNR})_{DM} = 10 \log_{10} \frac{\{1-(2\rho-1)f(1)\}}{\{2(1-\rho)f(1)\}} \quad \mathrm{dB} \tag{11.26}$$

Assuming the prediction error to be Gaussian and quantized by its Lloyd-Max quantizer and $\rho = 0.95$, the SNR is 12.8 dB, which is an 8.4-dB improvement over PCM at 1 bit per pixel. This amounts to a compression of 2.5, or a savings of about 1.5 bits per pixel. Equations (11.25) and (11.26) indicate the SNR of delta modu-

(a) Original; (b) delta modulation, leak = 0.9;

(c) line-by-line DPCM, 3 bits/pixel; (d) two-dimensional DPCM, 3 bits/pixel.

Figure 11.10 Examples of predictive coding.

lation can be improved by increasing ρ, which can be done by increasing the sampling rate of the quantizer output. For example, by doubling the sampling rate in this example, ρ will be increased to 0.975, and the SNR will increase by 3 dB. At the same time, however, the data rate is also doubled. Better performance can be obtained by going to adaptive techniques or increasing the number of quantizer bits, which leads to DPCM. In fact, a large number of the ills of delta modulation can be cured by DPCM, thereby making it a more attractive alternative for data compression.

Line-by-Line DPCM

In this method each scan line of the image is coded independently by the DPCM technique. Generally, a suitable AR representation is used for designing the predictor. Thus if we have a pth-order, stationary AP sequence (see Section 6.2)

$$u(n) - \sum_{k=1}^{p} a(k)u(n-k) = \varepsilon(n), \qquad E[(\varepsilon(n)^2] = \beta^2 \tag{11.27}$$

the DPCM system equations are

$$\text{Predictor:} \quad \bar{u}^{\cdot}(n) = \sum_{k=1}^{p} a(k)u^{\cdot}(n-k) \tag{11.28a}$$

$$\text{Quantizer input:} \quad e(n) = u(n) - \bar{u}^{\cdot}(n), \qquad \text{quantizer output} = e^{\cdot}(n) \tag{11.28b}$$

Reconstruction filter: $u^{\cdot}(n) = \bar{u}^{\cdot}(n) + e^{\cdot}(n)$
(reproduced output)

$$= \sum_{k=1}^{p} a(k)u^{\cdot}(n-k) + e^{\cdot}(n) \tag{11.28c}$$

For the first-order AR model of (11.25), the SNR of a B-bit DPCM system output can be estimated as (Problem 11.6)

$$(\text{SNR})_{DPCM} = 10 \log_{10}\left[\frac{(1-\rho^2 f(B))}{(1-\rho^2)f(B)}\right] \quad \text{dB} \tag{11.29}$$

For $\rho = 0.95$ and a Laplacian density-based quantizer, roughly 8-dB to 10-dB SNR improvement over PCM can be expected at rates of 1 to 3 bits per pixel. Alternatively, for small distortion levels ($f(B) \simeq 0$), the rate reduction over PCM is [see (11.16)]

$$R_{PCM} - R_{DPCM} = \tfrac{1}{2} \log_2 \frac{1}{(1-\rho^2)} \quad \text{bits/pixel} \tag{11.30}$$

This means, for example, the SNR of 6-bit PCM can be achieved by 4-bit line-by-line DPCM for $\rho = 0.97$. Figure 11.10c shows a line-by-line DPCM coded image at 3 bits per pixel.

Two-Dimensional DPCM

The foregoing ideas can be extended to two dimensions by using the causal MVRs discussed in chapter 6 (Section 6.6), which define a predictor of the form

$$\bar{u}(m,n) = \sum\sum_{(k,l) \in \hat{W}_1} a(k,l)u(m-k, n-l) \tag{11.31}$$

where $\hat{W}_1$ is a causal prediction window. The coefficients $a(k, l)$ are determined by solving (6.66) for $x = 1$, which minimizes the variance of the prediction error in the image. For common images it has been found that increasing size of $\hat{W}_1$ beyond the four nearest (causal) neighbors (Fig. 11.11) does not give any appreciable reduction in prediction error variance. Thus for row-by-row scanned images, it is sufficient to consider predictors of the form

$$\bar{u}(m,n) = a_1 u(m-1,n) + a_2 u(m, n-1) + a_3 u(m-1, n-1) + a_4 u(m-1, n+1) \tag{11.32}$$

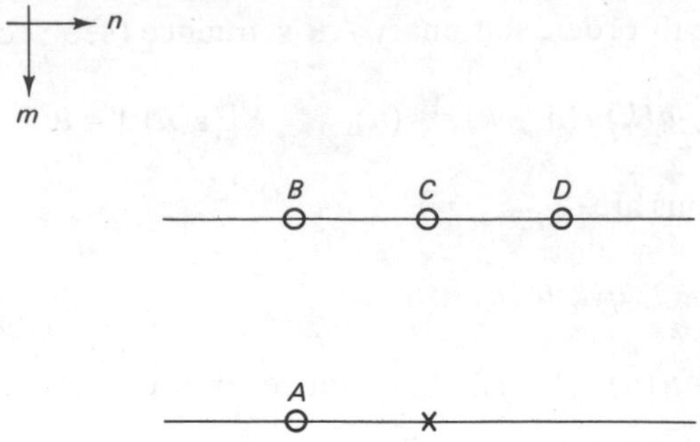

Figure 11.11 Pixels (A, B, C, D) used in two-dimensional prediction.

Here a_1, a_2, a_3, a_4, and β^2 are obtained by solving the linear equations

$$\begin{aligned} r(1,0) &= a_1 r(0,0) + a_2 r(1,-1) + a_3 r(0,1) + a_4 r(0,1) \\ r(0,1) &= a_1 r(1,-1) + a_2 r(0,0) + a_3 r(1,0) + a_4 r(1,-2) \\ r(1,1) &= a_1 r(0,1) + a_2 r(1,0) + a_3 r(0,0) + a_4 r(0,2) \\ r(1,-1) &= a_1 r(0,1) + a_2 r(1,-2) + a_3 r(0,2) + a_4 r(0,0) \\ \beta^2 &= E[\varepsilon^2(m,n)] \\ &= r(0,0) - a_1 r(1,0) - a_2 r(0,1) - a_3 r(1,1) - a_4 r(1,-1) \end{aligned} \tag{11.33}$$

where $r(k, l)$ is the covariance function of $u(m, n)$. In the special case of the separable covariance function of (2.84), we obtain

$$\begin{aligned} &a_1 = \rho_1, \qquad a_2 = \rho_2, \qquad a_3 = -\rho_1 \rho_2, \qquad a_4 = 0, \\ &\beta^2 = \sigma^2 (1 - \rho_1^2)(1 - \rho_2^2) \end{aligned} \tag{11.34}$$

Recall from Chapter 6 that unlike the one-dimension case, this solution of (11.33) can give rise to an unstable causal model. This means while the prediction error variance will be minimized (ignoring the quantization effects), the reconstruction filter could be unstable causing any channel error to be amplified greatly at the receiver. Therefore, the predictor has to be tested for stability and, if not stable, it has to be modified (at the cost of either increasing the prediction error variance or increasing the predictor order). Fortunately, for common monochrome image data (such as television images), this problem is rarely encountered.

Given the predictor as just described, the equations for a two-dimensional DPCM system become

$$\text{Predictor:} \quad \bar{u}^{\cdot}(m,n) = a_1 u^{\cdot}(m-1,n) + a_2 u^{\cdot}(m,n-1) + a_3 u^{\cdot}(m-1,n-1) + a_4 u^{\cdot}(m-1,n+1) \tag{11.35a}$$

$$\text{Quantizer input:} \quad e(m,n) = u(m,n) - \bar{u}^{\cdot}(m,n) \tag{11.35b}$$

$$\text{Reconstruction filter:} \quad u^{\cdot}(m,n) = \bar{u}^{\cdot}(m,n) + e^{\cdot}(m,n) \tag{11.35c}$$

The performance bounds of this method can be evaluated via (11.19) and (11.20). An example of a two-dimensional DPCM coding at 3 bits per pixel is shown in Fig. 11.10d.

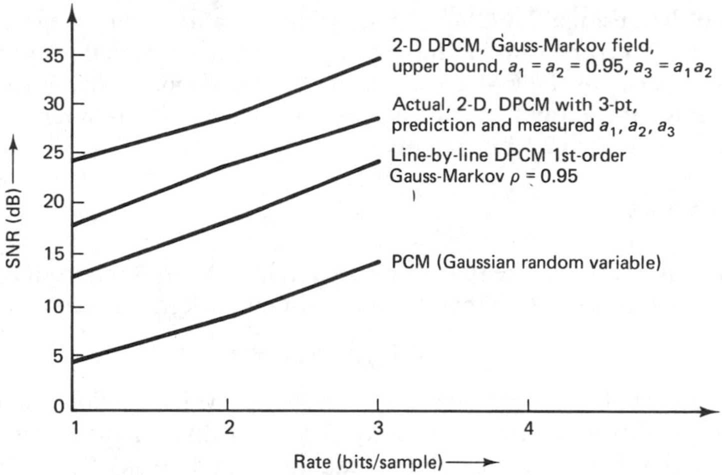

Figure 11.12 Performance of predictive codes.

Performance Comparisons

Figure 11.12 shows the theoretical SNR versus bit rate of two-dimensional DPCM of images modeled by (11.34) and (11.35) with $a_4 = 0$. Comparison with one-dimensional line-by-line DPCM and PCM is also shown. Note that delta modulation is the same as 1-bit DPCM in these curves. In practice, two-dimensional DPCM does not achieve quite as much as a 20-dB improvement over PCM, as expected for random fields with parameters of (11.34). This is because the two-dimensional separable covariance model is overly optimistic about the variance of the prediction error. Figure 11.13 compares the coding-error images in one- and

Figure 11.13 One- and two-dimensional DPCM images coded at 1 bit (upper images) and 3 bits (lower images) and their errors in reproduction.

two-dimensional DPCM. The subjective quality of an image and its tolerance to channel errors can be improved by two-dimensional predictors. Generally a 3-bit-per-pixel DPCM coder can give very good quality images. With Huffman coding, the output rate of a 3-bit quantizer in two-dimensional DPCM can be reduced to 2 to 2.5 bits per pixel average.

Remarks

Strictly speaking, the predictors used in DPCM are for zero mean data (that is, the dc value is zero). Otherwise, for a constant background μ, the predicted value

$$\bar{u}^{\cdot}(m, n) = (a_1 + a_2 + a_3 + a_4)\mu \tag{11.36}$$

would yield a bias of $(1 - a_1 - a_2 - a_3 - a_4)\mu$, which would be zero only if the sum of the predictor coefficients is unity. Theoretically, this will yield an unstable reconstruction filter (e.g., in delta modulation with no leak). This bias can be minimized by (1) choosing the predictors coefficients whose sum is close to but less than unity, (2) designing the quantizer reconstruction level to be zero for inputs near zero, and (3) tracking the mean of the quantizer output and feeding the bias correction to the predictor.

The quantizer should be designed to limit the three types of degradations, *granularity, slope overload,* and *edge-busyness.* Coarsely placed inner levels of the quantizer cause granularity in the flat regions of the image. Slope overload occurs at high-contrast edges where the prediction error exceeds the extreme levels of the quantizer, resulting in blurred edges. Edge-busyness is caused at less sharp edges, where the reproduced pixels on adjacent scan lines have different quantization levels. In the region of edges the optimum mean square quantizer based on Laplacian density for the prediction error sequence turns out to be too companded; that is, the inner quantization steps are too small, whereas the outer levels are too coarse, resulting in edge-busyness. A solution for minimizing these effects is to increase the number of quantizer levels and use an entropy coder for its outputs. This increases the dynamic range and the resolution of the quantizer. The average coder rate will now depend on the relative occurrences of the edges. Another alternative is to incorporate visual properties in the quantizer design using the visibility function [18]. In practice, standard quantizers are optimized iteratively to achieve appropriate subjective picture quality.

In hardware implementations of two-dimensional DPCM, the predictor is often simplified to minimize the number of multiplications per step. With reference to Fig. 11.11, some simplified prediction rules are discussed in Table 11.2.

The choice of prediction rule is also influenced by the response of the reconstruction filter to channel errors. See Section 11.8 for details.

For interlaced image frames, the foregoing design principles are applied to each field rather than each frame. This is because successive fields are $\frac{1}{60}$ s apart and the intrafield correlations are expected to be higher (in the presence of motion) than the pixel correlations in the de-interlaced adjacent lines.

Overall, DPCM is simple and well suited for real-time (video rate) hardware implementation. The major drawbacks are its sensitivity to variations in image statistics and to channel errors. Adaptive techniques can be used to improve the compression performance of DPCM. (Channel-error effects are discussed in Section 11.8.)

Adaptive Techniques

The performance of DPCM can be improved by adapting the quantizer and predictor characteristics to variations in the local statistics of the image data. Adaptive techniques use a range of quantizing characteristics and/or predictors from which a "current optimum" is selected according to local image properties. To eliminate the overhead due to the adaptation procedure, previously coded pixels are used to determine the mode of operation of the adaptive coder. In the absence of transmission errors, this allows the receiver to follow the same sequence of decisions made at the transmitter. Adaptive predictors are generally designed to improve the subjective image quality, especially at the edges. A popular technique is to use several predictors, each of which performs well if the image is highly correlated in a certain direction. The direction of maximum correlation is computed from previously coded pixels and the corresponding predictor is chosen.

Adaptive quantization schemes are based on two approaches, as discussed next.

Adaptive gain control. For a fixed predictor, the variance of the prediction error will fluctuate with changes in spatial details of the image. A simple adaptive quantizer updates the variance of the prediction error at each step and adjusts the spacing of the quantizer levels accordingly. This can be done by normalizing the prediction error by its updated standard deviation and designing the quantizer levels for unit variance inputs (Fig. 11.14a).

Let $\sigma_e^2(j)$ and $\dot{\sigma}_e^2(j)$ denote the variances of the quantizer input and output, respectively, at step j of a DPCM loop. (For a two-dimensional system, this means we are mapping (m, n) into j.) Since $e^{\cdot}(j)$ is available at the transmitter as well as

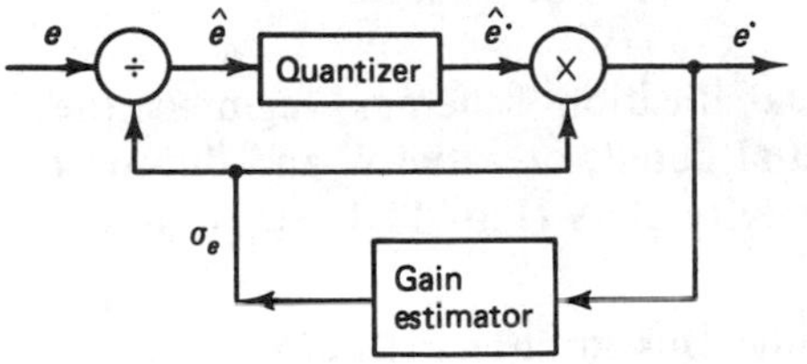

(a) Adaptive gain control

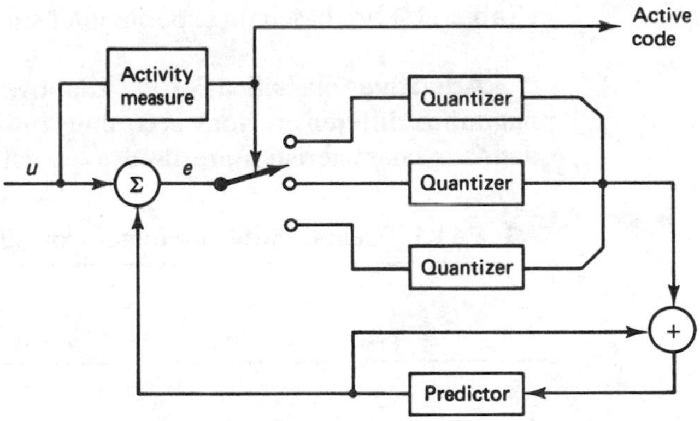

(b) Adaptive classification

Figure 11.14 Adaptive quantization.

the receiver, it is easy to estimate $\dot{\sigma}_e^2(j)$. A simple estimate, called the *exponential average variance estimator*, is of the form

$$\dot{\sigma}_e^2(j+1) = (1-\gamma)[e^\cdot(j)]^2 + \gamma\dot{\sigma}_e^2(j), \quad \dot{\sigma}_e^2(0) = (e^\cdot(0))^2, \quad j = 0, 1, \dots \tag{11.37}$$

where $0 \le \gamma \le 1$. For small quantization errors, we may use $\dot{\sigma}_e(j)$ as an estimate of $\sigma_e(j)$. For Lloyd-Max quantizers, since the variance of the input equals the sum of the variances of the output and the quantization error [see (4.47)], we can obtain the recursion for $\sigma_e^2(j)$ as

$$\sigma_e^2(j+1) = \frac{1-\gamma}{1-f(B)}[e^\cdot(j)]^2 + \gamma\sigma_e^2(j), \qquad j = 0, 1, \dots \tag{11.38}$$

In practice the estimate of (11.37) may be replaced by

$$\dot{\sigma}_e(j) \simeq \gamma \sum_{m=1}^{N} |e^\cdot(j-m)| \tag{11.39}$$

where γ is a constant determined experimentally so that the mean square error is minimized. The above two estimates become poor at low rates, for example, when $B = 1$. An alternative, originally suggested for adaptive delta modulation [7], is to define a gain $\sigma_e = g(m, n)$, which is recursively updated as

$$g(m, n) = \sum\sum_{(k,l) \in W} \alpha_{k,l}\, g(m-k, n-l) M(|q_{m-k,n-l}|), \qquad g_{\min} \le g(m,n) \le g_{\max} \tag{11.40}$$

where $M(|q_i|)$ is a multiplier factor that depends on the quantizer levels q_i and $\alpha_{k,l}$ are weights which sum up to unity. Often $\alpha_{k,l} = 1/N_w$, where N_w is the number of pixels in the causal window W. For example (see Table 11.1), for a three-level quantizer ($L = 3$) using the predictor neighbors of Fig. 11.11 and the gain-control formula

$$g(m, n) = \tfrac{1}{2}[g(m-1, n)M(|q_{m-1,n}|) + g(m, n-1)M(|q_{m,n-1}|)] \tag{11.41}$$

the multiplier factor $M(|q|)$ takes the values $M(0) = 0.7, M(\pm q_1) = 1.7$. The values in Table 11.1 are based on experimental studies [19] on 8-bit images.

Adaptive classification. Adaptive classification schemes segment the image into different regions according to spatial detail, or activity, and different quantizer characteristics are used for each activity class (Fig. 11.14b). A simple

TABLE 11.1 Gain-Control Parameters for Adaptive Quantization in DPCM

			Multipliers $M(q)$			
L	$g_{\min}$	$g_{\max}$	$q = 0$	$\pm q_1$	$\pm q_2$	$\pm q_3$
3	5	55	0.7	1.7		
5	5	40	0.8	1.0	2.6	
7	4	32	0.6	1.0	1.5	4.0

measure of activity is the variance of the pixels in the neighborhood of the pixel to be predicted. The flat regions are quantized more finely than edges or detailed areas. This scheme takes advantage of the fact that noise visibility decreases with increased activity. Typically, up to four activity classes are sufficient. An example would be to divide the image into 16×16 blocks and classify each block into one of four classes. This requires only a small overhead of 2 bits per block of 256 pixels.

Other Methods [17, 20]

At low bit rates ($B = 1$) the performance of DPCM deteriorates rapidly. One reason is that the predictor and the quantizer, which were designed independently, no longer operate at near-optimum levels. Thus the successive inputs to the quan-

TABLE 11.2 Summary of Predictive Coding

Design Parameter	Comments
Predictor	Predictors of orders 3 to 4 are adequate.
Linear mean square	Determined from image correlations. Performs very well as long as image class does not change very much.
Previous element γA	Sharp vertical or diagonal edges are blurred and exhibit edge-busyness. Channel error manifests itself as a horizontal streak.
Averaged prediction a. $\gamma\left(\frac{A+D}{2}\right)$ b. $\gamma\left(\frac{A+C}{2}\right)$ c. $\gamma\left(\frac{A+(C+D)/2}{2}\right)$	Significant improvement over previous element prediction for vertical and most sloping edges. Horizontal and gradual rising edges blurred. The two predictors using pixel D perform equally well but better than $(A + C)/2$ on gradual rising edges. Edge-busyness and sensitivity to channel errors much reduced (Fig. 11.38).
Planar prediction a. $\gamma(A + (C - B))$ b. $\gamma\left(A + \frac{(D-B)}{2}\right)$	Better than previous element prediction but worse than averaged prediction with respect to edge busyness and channel errors (Fig. 11.38).
Leak (γ)	$0 < \gamma < 1$. As the leak is increased, transmission errors become less visible, but granularity and contouring become more visible.
Quantizer	
a. Optimum mean square (Lloyd-Max)	Recommended when the compression ratio is not too high (≤ 3) and a fixed length code is used. Prediction error may be modeled by Laplacian or Gaussian probability densities.
b. Visual	Difficult to design. One alternative is to perturb the levels of the max quantizer to obtain an increased subjective quality.
c. Uniform	Useful in high-compression schemes (>3) where the quantizer output is entropy coded.

tizer may have significant correlation, and the predictor may not be good enough. Two methods that can improve the performance are

1. Delayed predictive coding
2. Predictive vector quantization

In the first method [17], a tree code is generated by the prediction filter excited by different quantization levels. As successive pixels are coded, the predictor selects a *path* in the tree (rather than a branch value, as in DPCM) such that the mean square error is minimized. Delays are introduced in the predictor to enable development of a tree with sufficient look-ahead paths.

In the second method [20], the successive inputs to the quantizer are entered in a shift register, whose state is used to define the quantizer output value. Thus the quantizer current output depends on its previous outputs.

11.4 TRANSFORM CODING THEORY

The Optimum Transform Coder

Transform coding, also called *block quantization,* is an alternative to predictive coding. A block of data is unitarily transformed so that a large fraction of its total energy is packed in relatively few transform coefficients, which are quantized independently. The optimum transform coder is defined as the one that minimizes the mean square distortion of the reproduced data for a given number of total bits. This turns out to be the KL transform.

Suppose an $N \times 1$ random vector $\mathbf{u}$ with zero mean and covariance $\mathbf{R}$ is linearly transformed by an $N \times N$ (complex) matrix $\mathbf{A}$, not necessarily unitary, to produce a (complex) vector $\mathbf{v}$ such that its components $v(k)$ are mutually uncorrelated (Fig. 11.15). After quantizing each component $v(k)$ independently, the output vector $\mathbf{v}^{\bullet}$ is linearly transformed by a matrix $\mathbf{B}$ to yield a vector $\mathbf{u}^{\bullet}$. The problem is to find the optimum matrices $\mathbf{A}$ and $\mathbf{B}$ and the optimum quantizers such that the overall average mean square distortion

$$D = \frac{1}{N} E\left[\sum_{n=1}^{N} (u(n) - u^{\bullet}(n))^2\right] = \frac{1}{N} E[(\mathbf{u} - \mathbf{u}^{\bullet})^T (\mathbf{u} - \mathbf{u}^{\bullet})] \qquad (11.42)$$

is minimized. The solution of this problem is summarized as follows:

1. For an arbitrary quantizer the optimal reconstruction matrix $\mathbf{B}$ is given by

$$\mathbf{B} = \mathbf{A}^{-1}\mathbf{\Gamma} \qquad (11.43)$$

where $\mathbf{\Gamma}$ is a diagonal matrix of elements γ_k defined as

$$\gamma_k \triangleq \frac{\tilde{\lambda}_k}{\lambda_k^{\bullet}} \qquad (11.44a)$$

$$\tilde{\lambda}_k \triangleq E[v(k)v^{\bullet *}(k)], \qquad \lambda_k^{\bullet} \triangleq E[|v^{\bullet}(k)|^2] \qquad (11.44b)$$

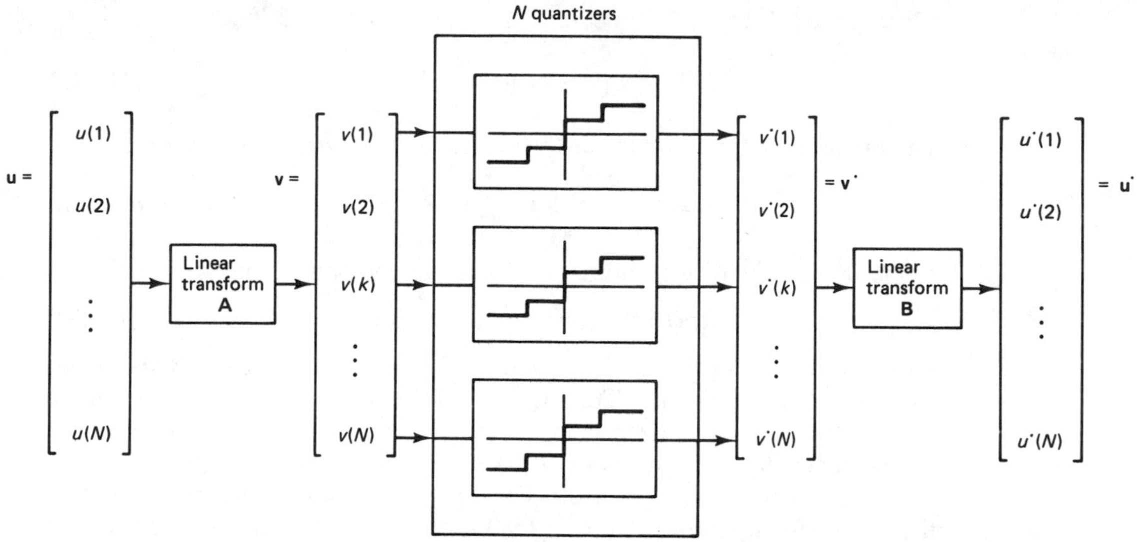

Figure 11.15 One-dimensional transform coding.

2. The Lloyd-Max quantizer for each $v(k)$ minimizes the overall mean square error giving

$$\mathbf{\Gamma} = \mathbf{I} \tag{11.45}$$

3. The optimal decorrelating matrix $\mathbf{A}$ is the KL transform of $\mathbf{u}$, that is, the rows of $\mathbf{A}$ are the orthonormalized eigenvectors of the autocovariance matrix $\mathbf{R}$. This gives

$$\mathbf{B} = \mathbf{A}^{-1} = \mathbf{A}^{*T} \tag{11.46}$$

Proofs

1. In terms of the transformed vectors $\mathbf{v}$ and $\mathbf{v}^{\cdot}$, the distortion can be written as

$$D = \frac{1}{N} E\{Tr[\mathbf{A}^{-1}\mathbf{v} - \mathbf{B}\mathbf{v}^{\cdot}][\mathbf{A}^{-1}\mathbf{v} - \mathbf{B}\mathbf{v}^{\cdot}]^{*T}\} \tag{11.47a}$$

$$= \frac{1}{N} \mathrm{Tr}[\mathbf{A}^{-1}\mathbf{\Lambda}(\mathbf{A}^{-1})^{*T} + \mathbf{B}\mathbf{\Lambda}^{\cdot}\mathbf{B}^{*T} - \mathbf{A}^{-1}\tilde{\mathbf{\Lambda}}\mathbf{B}^{*T} - \mathbf{B}\tilde{\mathbf{\Lambda}}^{*T}(\mathbf{A}^{-1})^{*T}] \tag{11.47b}$$

where

$$\mathbf{\Lambda} \triangleq E[\mathbf{v}\mathbf{v}^{*T}], \qquad \mathbf{\Lambda}^{\cdot} \triangleq E[\mathbf{v}^{\cdot}(\mathbf{v}^{\cdot})^{*T}], \qquad \tilde{\mathbf{\Lambda}} \triangleq E[\mathbf{v}(\mathbf{v}^{\cdot})^{*T}] \tag{11.47c}$$

Since $v(k), k = 0, 1, \ldots, N-1$ are uncorrelated and are quantized independently, the matrices $\mathbf{\Lambda}$, $\mathbf{\Lambda}^{\cdot}$, and $\tilde{\mathbf{\Lambda}}$ are diagonal with λ_k, $\lambda_k^{\cdot}$, and $\tilde{\lambda}_k$ as their respective diagonal elements. Minimizing D by differentiating it with respect to $\mathbf{B}^{*T}$ (or $\mathbf{B}$), we obtain (see Problem 2.15)

$$\mathbf{B}\mathbf{\Lambda}^{\cdot} - \mathbf{A}^{-1}\tilde{\mathbf{\Lambda}} = 0 \;\Rightarrow\; \mathbf{B} = \mathbf{A}^{-1}\tilde{\mathbf{\Lambda}}(\mathbf{\Lambda}^{\cdot})^{-1} \tag{11.48}$$

which gives (11.43) and

$$D = \frac{1}{N} \operatorname{Tr}[\mathbf{A}^{-1} E[\mathbf{v} - \mathbf{\Gamma v}^\cdot][\mathbf{v} - \mathbf{\Gamma v}^\cdot]^{*T} (\mathbf{A}^{-1})^{*T}] \tag{11.49a}$$

2. The preceding expression for distortion can also be written in the form

$$D = \frac{1}{N} \sum_{i=0}^{N-1} \sum_{k=0}^{N-1} |[\mathbf{A}^{-1}]_{i,k}|^2 \hat{\sigma}_q^2(k) \lambda_k \tag{11.49b}$$

where $\hat{\sigma}_q^2(k)$ is the distortion as if $v(k)$ had unity variance, that is

$$\hat{\sigma}_q^2(k) \triangleq E[|v(k) - \gamma_k v^\cdot(k)|^2]/\lambda_k \tag{11.49c}$$

From this it follows that $\hat{\sigma}_q^2(k)$ should be minimized for each k by the quantizer no matter what **A** is.† This means we have to minimize the mean square error between the quantizer input $v(k)$ and its scaled output $\gamma_k v^\cdot(k)$. Without loss of generality, we can absorb γ_k inside the quantizer and require it to be a minimum mean square quantizer. For a given number of bits, this would be the Lloyd-Max quantizer. Note that γ_k becomes unity for any quantizer whose output levels minimize the mean square quantization error regardless of its decision levels. For the Lloyd-Max quantizer, it is not only that γ_k equals unity, but also its decision levels are such that the mean square quantization error is minimum. Thus (11.45) is true and we get $\mathbf{B} = \mathbf{A}^{-1}$. This gives

$$\sigma_q^2(k) = \lambda_k \hat{\sigma}_q^2(k) = E[|v(k) - v^\cdot(k)|^2] = \lambda_k f(n_k) \tag{11.50}$$

where $f(x)$ is the distortion-rate function of an x-bit Lloyd-Max quantizer for unity variance inputs (Table 4.4). Substituting (11.50) in (11.49b), we obtain

$$D = \frac{1}{N} \operatorname{Tr}[\mathbf{A}^{-1} \mathbf{F} \mathbf{\Lambda} (\mathbf{A}^{-1})^{*T}], \qquad \mathbf{F} \triangleq \operatorname{Diag}\{f(n_k)\} \tag{11.51}$$

Since **v** equals **Au**, its covariance is given by the diagonal matrix

$$E[\mathbf{v}\mathbf{v}^{*T}] \triangleq \mathbf{\Lambda} = \mathbf{A R A}^{*T} \tag{11.52}$$

Substitution for $\mathbf{\Lambda}$ in (11.51) gives

$$D = \frac{1}{N} \operatorname{Tr}[\mathbf{A}^{-1} \mathbf{F A R}] \tag{11.53}$$

where **F** and **R** do not depend on **A.** Minimizing D with respect to **A,** we obtain (see Problem 2.15)

$$\mathbf{0} = \frac{\partial D}{\partial \mathbf{A}} = -\frac{1}{N} [\mathbf{A}^{-1} \mathbf{F A R A}^{-1}]^T + \frac{1}{N} [\mathbf{R A}^{-1} \mathbf{F}]^T$$
$$\Rightarrow \quad \mathbf{F}(\mathbf{A R A}^{-1}) = (\mathbf{A R A}^{-1})\mathbf{F} \tag{11.54}$$

Thus, **F** and $\mathbf{ARA}^{-1}$ commute. Because **F** is diagonal, $\mathbf{ARA}^{-1}$ must also be diagonal. But $\mathbf{ARA}^{*T}$ is also diagonal. Therefore, these two matrices must be related by a diagonal matrix **G,** as

† Note that $\hat{\sigma}_q^2(k)$ is independent of the transform **A**.

$$\mathbf{ARA}^{*T} = (\mathbf{ARA}^{-1})\mathbf{G} \tag{11.55}$$

This implies $\mathbf{AA}^{*T} = \mathbf{G}$, so the columns of $\mathbf{A}$ must be orthogonal. If $\mathbf{A}$ is replaced by $G^{1/2}\mathbf{A}$, the overall result of transform coding will remain unchanged because $\mathbf{B} = \mathbf{A}^{-1}$. Therefore, $\mathbf{A}$ can be taken as a unitary matrix, which proves (11.46). This result and (11.52) imply that $\mathbf{A}$ is the KL transform of $\mathbf{u}$ (see Sections 2.9 and 5.11).

Remarks

Not being a fast transform in general, the KL transform can be replaced either by a fast unitary transform, such as the cosine, sine, DFT, Hadamard, or Slant, which is not a perfect decorrelator, or by a fast decorrelating transform, which is not unitary. In practice, the former choice gives better performance (Problem 11.9).

The foregoing result establishes the optimality of the KL transform among all decorrelating transformations. It can be shown that it is also optimal among all the unitary transforms (see Problem 11.8) and also performs better than DPCM (which can be viewed as a nonlinar transform; Problem 11.10).

Bit Allocation and Rate-Distortion Characteristics

The transform coefficient variances are generally unequal, and therefore each requires a different number of quantizing bits. To complete the transform coder design we have to allocate a given number of total bits among all the transform coefficients so that the overall distortion is minimum. Referring to Fig. 11.15, for any unitary transform $\mathbf{A}$, arbitrary quantizers, and $\mathbf{B} = \mathbf{A}^{-1} = \mathbf{A}^{*T}$; the distortion becomes

$$D = \frac{1}{N}\sum_{k=0}^{N-1} E[|v(k) - v^{\cdot}(k)|^2] = \frac{1}{N}\sum_{k=0}^{N-1} \sigma_k^2 f(n_k) \tag{11.56}$$

where σ_k^2 is the variance of the transform coefficient $v(k)$, which is allocated n_k bits, and $f(\cdot)$, the quantizer distortion function, is monotone convex with $f(0) = 1$ and $f(\infty) = 0$. We are given a desired average bit rate per sample, B; then the rate for the $\mathbf{A}$-transform coder is

$$R_{\mathbf{A}} \triangleq \frac{1}{N}\sum_{k=0}^{N-1} n_k = B \tag{11.57}$$

The bit allocation problem is to find $n_k \geq 0$ that minimize the distortion D, subject to (11.57). Its solution is given by the following algorithm.

Bit Allocation Algorithm

Step 1. Define the inverse function of $f'(x) \triangleq df(x)/dx$ as $h(x) \triangleq f'^{-1}(x)$, or $h(f'(x)) = x$. Find θ, the root of the nonlinear equation

$$R_{\mathbf{A}} \triangleq \frac{1}{N}\sum_{k:\sigma_k^2 > \theta} h\left(\frac{(\theta f'(0))}{\sigma_k^2}\right) = B \tag{11.58}$$

The solution may be obtained by an iterative technique such as the Newton method. The parameter θ is a threshold that controls which transform coefficients are to be coded for transmission.

Step 2. The number of bits allocated to the kth transform coefficient are given by

$$n_k = \begin{cases} 0, & \sigma_k^2 < \theta \\ h(\theta f'(0)/\sigma_k^2), & \sigma_k^2 \geq \theta \end{cases} \tag{11.59}$$

Note that the coefficients whose mean square value falls below θ are not coded at all.

Step 3. The minimum achievable distortion is then

$$D = \frac{1}{N}\left[\sum_{k:\sigma_k^2 \geq \theta} \sigma_k^2 f(n_k) + \sum_{k:\sigma_k^2 < \theta} \sigma_k^2\right] \tag{11.60}$$

Sometimes we specify the average distortion $D = d$ rather than the average rate B. In that case (11.60) is first solved for θ. Then (11.59) and (11.58) give the bit allocation and the minimum achievable rate. Given n_k, the number of quantizer levels can be approximated as $\text{Int}[2^{n_k}]$. Note that n_k is not necessarily an integer. This algorithm is also useful for calculating the rate versus distortion characteristics of a transform coder based on a given transform $\mathbf{A}$ and a quantizer with distortion function $f(x)$.

In the special case of the Shannon quantizer, we have $f(x) = 2^{-2x}$, which gives

$$f'(x) = -(2\log_e 2)2^{-2x} \Rightarrow h(x) = -\tfrac{1}{2}\log_2\left(\frac{-x}{2\log_e 2}\right)$$

$$n_k = \max\left\{0, \tfrac{1}{2}\log_2\left(\frac{\sigma_k^2}{\theta}\right)\right\} \tag{11.61}$$

$$D = \frac{1}{N}\left[\sum_{\sigma_k^2 \leq \theta} \sigma_k^2 + \sum_{\sigma_k^2 > \theta} \theta\right] = \frac{1}{N}\sum_{k=0}^{N-1} \min(\theta, \sigma_k^2) \tag{11.62}$$

$$\begin{aligned} R_{\mathbf{A}} &= \frac{1}{N}\left[\sum_{\sigma_k^2 \leq \theta} 0 + \sum_{\sigma_k^2 > \theta} \tfrac{1}{2}\log_2\left(\frac{\sigma_k^2}{\theta}\right)\right] \\ &= \frac{1}{N}\sum_{k=0}^{N-1} \max\left(0, \tfrac{1}{2}\log_2\frac{\sigma_k^2}{\theta}\right) \end{aligned} \tag{11.63}$$

More generally, when $f(x)$ is modeled by piecewise exponentials as in Table 4.4, we can similarly obtain the bit allocation formulas [32]. Equations (11.62) and (11.63) give the *rate-distortion bound* for transform coding of an $N \times 1$ Gaussian random vector $\mathbf{u}$ by a unitary transform $\mathbf{A}$. This means for a fixed distortion D, the rate $R_{\mathbf{A}}$ will be lower than the rate achieved by using any practical quantizer. When D is small enough so that $0 < \theta < \min_k \{\sigma_k^2\}$, we get $\theta = D$, and

$$R_{\mathbf{A}} = \frac{1}{N}\sum_{k=0}^{N-1} \tfrac{1}{2}\log_2\frac{\sigma_k^2}{D} = \frac{1}{2N}\log_2\left(\prod_{k=0}^{N-1}\sigma_k^2\right) - \tfrac{1}{2}\log_2 D \tag{11.64}$$

In the case of the KL transform, $\sigma_k^2 = \lambda_k$ and $\prod_k \lambda_k = |\mathbf{R}|$, which gives

$$R_{KL} = \frac{1}{2N} \log_2 |\mathbf{R}| - \tfrac{1}{2} \log_2 D, \qquad D < \min_k \{\lambda_k\} \tag{11.65}$$

For small but equal distortion levels,

$$R_{\mathbf{A}} - R_{KL} = \frac{1}{2N} \log_2 \left[\left(\prod_{k=0}^{N-1} \sigma_k^2 \right) \Big/ |\mathbf{R}| \right] \geq 0 \tag{11.66}$$

where we have used (2.43) to give

$$|\mathbf{R}| = |\mathbf{ARA}^{*T}| \leq \prod_{k=0}^{N-1} [\mathbf{ARA}^{*T}]_{k,k} = \prod_{k=0}^{N-1} \sigma_k^2$$

For PCM coding, it is equivalent to assuming $\mathbf{A} = \mathbf{I}$, so that

$$R_{PCM} - R_{KL} = -\frac{1}{2N} \log_2 |\hat{\mathbf{R}}| \geq 0 \tag{11.67}$$

where $\hat{\mathbf{R}} = \{r(m,n)/\sigma_m^2\}$ is the correlation matrix of $\mathbf{u}$, and σ_m^2 are the variances of its elements.

Example 11.3

The determinant of the covariance matrix $\mathbf{R} = \{\rho^{|m-n|}\}$ of a Markov sequence of length N is $|R| = (1-\rho^2)^{N-1}$. This gives

$$R_{KL} = \frac{N-1}{2N} \log_2 (1-\rho^2) - \tfrac{1}{2} \log_2 D, \qquad D < \min\{\lambda_k\} \tag{11.68}$$

For $N = 16$ and $\rho = 0.95$, the value of $\min\{\lambda_k\}$ is 0.026 (see Table 5.2). So for $D = 0.01$, we get $R_{KL} = 1.81$ bits per sample. Rearranging (11.68) we can write

$$R_{KL} = \tfrac{1}{2} \log_2 \frac{(1-\rho^2)}{D} - \frac{1}{2N} \log_2 (1-\rho^2) \tag{11.69}$$

As $N \to \infty$, the rate R_{KL} goes down to a lower bound $R_{KL}(\infty) = \frac{1}{2} \log_2 (1-\rho^2)/D$, and $R_{PCM} - R_{KL}(\infty) = -\frac{1}{2} \log_2 (1-\rho^2) = 1.6$ bits per sample. Also, as $N \to \infty$, the eigenvalues of $\mathbf{R}$ follow the distribution $\lambda(\omega) = (1-\rho^2)/(1+\rho^2+2\rho\cos\omega)$, which gives $\min\{\lambda_k\} = (1-\rho^2)/(1+\rho)^2 = (1-\rho)/(1+\rho)$. For $\rho = 0.95$, $D = 0.01$ we obtain $R_{KL}(\infty) = 1.6$ bits per sample.

Integer Bit Allocation Algorithm. The number of quantizing bits n_k are often specified as integers. Then the solution of the bit allocation problem is obtained by applying a theory of marginal analysis [6, 21], which yields the following simple algorithm.

Step 1. Start with the allocation $n_k^0 = 0, 0 \leq k \leq N-1$. Set $j = 1$.
Step 2. $n_k^j = n_k^{j-1} + \delta(k-i)$, where i is any index for which

$$\Delta_k \triangleq \sigma_k^2 [f(n_k^{j-1}) - f(n_k^{j-1}+1)]$$

is maximum. Δ_k is the reduction in distortion if the jth bit is allocated to the kth quantizer.

Step 3. If $\sum_k n_k^j \geq NB$, stop; otherwise $j \rightarrow j+1$ and go to Step 2.

If ties occur for the maximizing index, the procedure is successively initiated with the allocation $n_k^j = n_k^{j-1} + \delta(i-k)$ for each i. This algorithm simply means that the marginal returns

$$\Delta_{k,j} \triangleq \sigma_k^2[f(n_k^j) - f(n_k^j + 1)], \qquad k = 0, \ldots, N-1, j = 1, \ldots, NB \quad (11.70)$$

are arranged in a decreasing order and bits are assigned one by one according to this order. For an average bit rate of B, we have to search N marginal returns NB times. This algorithm can be speeded up whenever the distortion function is of the form $f(x) = a2^{-bx}$. Then $\Delta_{k,j} = (1-2^{-b})\sigma_k^2 f(n_k^{j-1})$, which means the quantizer having the largest distortion, at any step j, is allocated the next bit. Thus, as we allocate a bit, we update the quantizer distortion and the step 2 of the algorithm becomes:

Step 2: Find the index i such that

$$D_i = \max_k [\sigma_k^2 f(n_k^{j-1})] \text{ is maximum}$$

Then

$$n_k^j = n_k^{j-1} + \delta(k-i)$$

$$D_i = 2^{-b} D_i$$

The piecewise exponential models of Table 4.4 can be used to implement this step.

11.5 TRANSFORM CODING OF IMAGES

The foregoing one-dimensional transform coding theory can be easily generalized to two dimensions by simply mapping a given $N \times M$ image $u(m, n)$ to a one-dimensional $NM \times 1$ vector $\mathbf{u}$. The KL transform of $\mathbf{u}$ would be a matrix of size $NM \times NM$. In practice, this transform is replaced by a separable fast transform such as the cosine, sine, Fourier, Slant, or Hadamard; these, as we saw in chapter 5, pack a considerable amount of the image energy in a small number of coefficients.

To make transform coding practical, a given image is divided into small rectangular blocks, and each block is transform coded independently. For an $N \times M$ image divided into NM/pq blocks, each of size $p \times q$, the main storage requirements for implementing the transform are reduced by a factor of NM/pq. The computational load is reduced by a factor of $\log_2 MN/\log_2 pq$ for a fast transform requiring $\alpha N \log_2 N$ operations to transform an $N \times 1$ vector. For 512×512 images divided into 16×16 blocks, these factors are 1024 and 2.25, respectively. Although the operation count is not greatly reduced, the complexity of the hardware for implementing small-size transforms is reduced significantly. However, smaller block sizes yield lower compression, as shown by Fig. 11.16. Typically, a block size of 16×16 is used.

Two-Dimensional Transform Coding Algorithm. We now state a practical transform coding algorithm for images (Fig. 11.17).

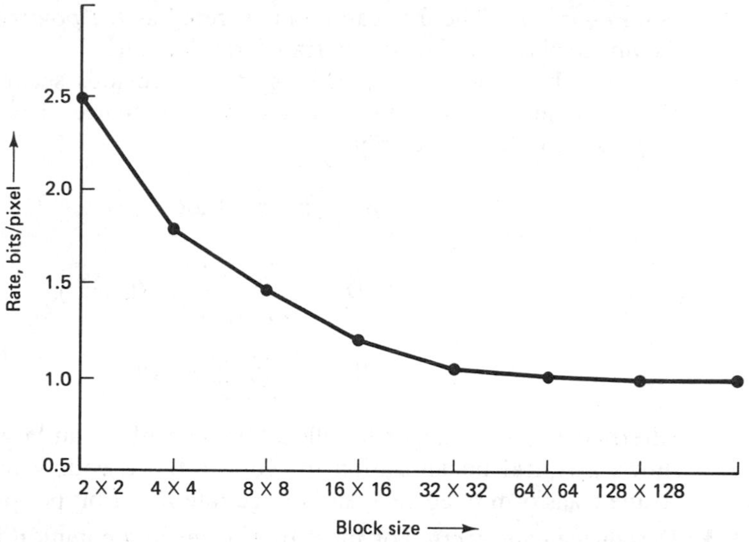

Figure 11.16 Rate achievable by block KL transform coders for Gaussian random fields with separable covariance function, $\rho = \rho_2 = 0.95$, at distortion $D = 0.25\%$.

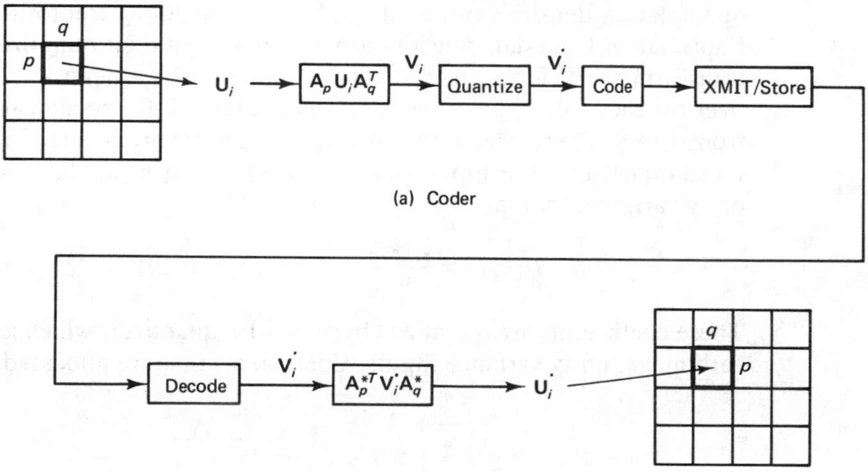

Figure 11.17 Two-dimensional transform coding.

1. *Divide the given image.* Divide the image into small rectangular blocks of size $p \times q$ and transform each block to obtain $\mathbf{V}_i, i = 0, \ldots, I-1, I \triangleq NM/pq$.
2. *Determine the bit allocation.* Calculate the transform coefficient variances $\sigma^2_{k,l}$ via (5.36) or Problem 5.29b if the image covariance function is given. Alternatively, estimate the variances $\hat{\sigma}^2_{k,l}$ from the ensemble of coefficients $v_i(k, l)$, $i = 0, \ldots, I-1$, obtained from a given prototype image normalized to have unity variance. From this, the $\sigma^2_{k,l}$ for the image with variance σ^2 are estimated

as $\sigma^2_{k,l} = \hat{\sigma}^2_{k,l}\sigma^2$. The $\hat{\sigma}^2_{k,l}$ can be interpreted as the *power spectral density* of the image blocks in the chosen transform domain.

The bit allocation algorithms of the previous section can be applied after mapping $\sigma^2_{k,l}$ into a one-dimensional sequence. The ideal case, where $f(x) = 2^{-2x}$, yields the formulas

$$n_{k,l} = \max\left(0, \tfrac{1}{2}\log_2 \frac{\sigma^2_{k,l}}{\theta}\right),$$

$$D = \frac{1}{pq}\sum_{k=0}^{p-1}\sum_{l=0}^{q-1} \min(\theta, \sigma^2_{k,l}), \tag{11.71}$$

$$R_A = \frac{1}{pq}\sum_{k=0}^{p-1}\sum_{l=0}^{q-1} n_{k,l}(\theta)$$

Alternatively, the integer bit allocation algorithm can be used. Figure 11.18 shows a typical bit allocation for 16×16 block coding of an image by the cosine transform to achieve an average rate $B = 1$ bit per pixel.

3. *Design the quantizers.* For most transforms and common images (which are nonnegative) the dc coefficient $v_i(0,0)$ is nonnegative, and the remaining coefficients have zero mean values. The dc coefficient distribution is modeled by the Rayleigh density (see Problem 4.15). Alternatively, one-sided Gaussian or Laplacian densities can be used. For the remaining tranform coefficients, Laplacian or Gaussian densities are used to design their quantizers. Since the transform coefficients are allocated unequal bits, we need a different quantizer for each value of $n_{k,l}$. For example, in Fig. 11.18 the allocated bits range from 1 to 8. Therefore, eight different quantizers are needed. To implement these quantizers, the input sample $v_i(k, l)$ is first normalized so that it has unity variance, that is,

$$\hat{v}_i(k,l) \triangleq \frac{v_i(k,l)}{\sigma_{k,l}}, \qquad (k,l) \neq (0,0) \tag{11.72}$$

These coefficients are quantized by an $n_{k,l}$-bit quantizer, which is designed for zero mean, unity variance inputs. Coefficients that are allocated zero bits are

j →

i ↓

8	7	6	5	3	3	2	2	2	1	1	1	1	1	0	0
7	6	5	4	3	3	2	2	1	1	1	1	1	0	0	0
6	5	4	3	3	2	2	2	1	1	1	1	1	0	0	0
5	4	3	3	3	2	2	2	1	1	1	1	1	0	0	0
3	3	3	3	2	2	2	1	1	1	1	1	0	0	0	0
3	3	2	2	2	2	2	1	1	1	1	1	0	0	0	0
2	2	2	2	2	2	1	1	1	1	1	0	0	0	0	0
2	2	2	2	1	1	1	1	1	1	1	0	0	0	0	0
2	1	1	1	1	1	1	1	1	1	0	0	0	0	0	0
1	1	1	1	1	1	1	1	1	0	0	0	0	0	0	0
1	1	1	1	1	1	1	1	0	0	0	0	0	0	0	0
1	1	1	1	1	1	0	0	0	0	0	0	0	0	0	0
1	1	1	1	0	0	0	0	0	0	0	0	0	0	0	0
1	0	0	0	0	0	0	0	0	0	0	0	0	0	0	0
0	0	0	0	0	0	0	0	0	0	0	0	0	0	0	0
0	0	0	0	0	0	0	0	0	0	0	0	0	0	0	0

Figure 11.18 Bit allocation for 16×16 block cosine transform coding of images modeled by isotropic covariance function with $\rho = 0.95$. Average rate = 1 bit per pixel.

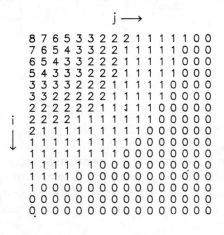

not processed at all. At the decoder, which knows the bit allocation table in advance, the unprocessed coefficients are replaced by zeros (that is, their mean values).

4. *Code the quantizer output.* Code the output into code words and transmit or store.
5. *Reproduce the coefficients.* Assuming a noiseless channel, reproduce the coefficients at the decoder as

$$v_i^{\cdot}(k,l) = \begin{cases} v_i^{\cdot}(k,l)\sigma_{k,l}, & (k,l) \in I_t \\ 0, & \text{otherwise} \end{cases} \tag{11.73}$$

where I_t denotes the set of transmitted coefficients. The inverse transformation $\mathbf{U}_i^{\cdot} = \mathbf{A}^{*T}\mathbf{V}_i^{\cdot}\mathbf{A}^*$ gives the reproduced image blocks.

Once a bit assignment for transform coefficients has been determined, the performance of the coder can be estimated by the relations

$$D = \frac{1}{pq}\sum_{k=0}^{p-1}\sum_{l=0}^{q-1}\sigma_{k,l}^2 f(n_{k,l}), \qquad R_{\mathbf{A}} = \frac{1}{pq}\sum_{k=0}^{p-1}\sum_{l=0}^{q-1} n_{k,l} \tag{11.74}$$

Transform Coding Performance Trade-Offs and Examples

Example 11.4 Choice of transform

Figure 11.19 compares the performance of different transforms for 16×16 block coding of a random field. Table 11.3 shows examples of SNR values at different rates. The cosine transform performance is superior to the other fast transforms and is almost indistinguishable from the KL transform. Recall from Section 5.12 that the cosine transform has near-optimal performance for first-order stationary Markov sequences with $\rho > 0.5$. Considerations for the choice of other transforms are summarized in Table 11.4.

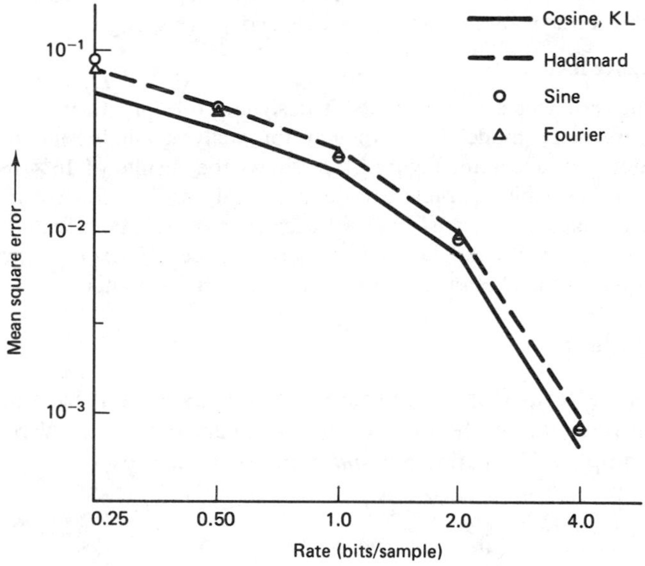

Figure 11.19 Distortion versus rate characteristics for different transforms for a two-dimensional isotropic random field.

TABLE 11.3 SNR Comparisons of Various Transform Coders for Random Fields with Isotropic Covariance Function, $\rho = 0.95$

Block size	Rate bits/pixel	SNR (dB)				
		KL	Cosine	Sine	Fourier	Hadamard
8×8	0.25	11.74	11.66	9.08	10.15	10.79
	0.50	13.82	13.76	11.69	12.27	12.65
	1.00	16.24	16.19	14.82	14.99	15.17
	2.00	20.95	20.89	19.53	19.73	19.86
	4.00	31.61	31.54	30.17	30.44	30.49
16×16	0.25		12.35	10.37	10.77	10.99
	0.50		14.25	12.82	12.87	12.78
	1.00		16.58	15.65	15.52	15.27
	2.00		21.26	20.37	20.24	20.01
	4.00		31.90	31.00	30.88	30.69

Note: The KL transform would be nonseparable here.

Example 11.5 Choice of block size

The effect of block size on coder performance can easily be analyzed for the case of separable covariance random fields, that is, $r(m, n) = \rho^{|m| + |n|}$. For a block size of $N \times N$, the covariance matrix can be written as

$$\mathcal{R} = \mathbf{R} \otimes \mathbf{R} \quad \Rightarrow \quad |\mathcal{R}| = |\mathbf{R}|^{2N} = (1 - \rho^2)^{2N(N-1)}$$

where $\mathbf{R}$ is given by (2.68). Applying (11.69) the rate achievable by an $N \times N$ block KL transform coder is

$$R_{KL}(N) = \tfrac{1}{2} \log_2 \frac{(1-\rho^2)^2}{D} - \frac{1}{2N} \log_2 (1-\rho^2)^2, \qquad D \le \left(\frac{1-\rho}{1+\rho}\right)^2 \tag{11.75}$$

When plotted as a function of N (Fig. 11.16), this shows the block size of 16×16 is suitable for $\rho = 0.95$. For higher values of the correlation parameter, ρ, the block size should be increased. Figure 11.20 shows some 16×16 block coded results.

Example 11.6 Choice of covariance model

The transform coefficient variances are important for designing the quantizers. Although the separable covariance model is convenient for analysis and design of transform coders, it is not very accurate. Figure 11.20 shows the results of 16×16 cosine transform coders based on the separable covariance model, the isotropic covariance model, and the actual measured transform coefficient variances. As expected, the actual measured variances yield the best coder performance. Generally, the isotropic covariance model performs better than the separable covariance model.

Zonal Versus Threshold Coding

Examination of bit allocation patterns (Fig. 11.18) reveals that only a small *zone* of transformed image is transmitted (unless the average rate was very high). Let N_t be the number of transmitted samples. We define a *zonal mask* as the array

$$m(k, l) = \begin{cases} 1, & k, l \in I_t \\ 0, & \text{otherwise} \end{cases} \tag{11.76}$$

(a) separable, SNR′ = 37.5 dB, the right side shows error images

(b) isotropic, SNR′ = 37.8 dB

(c) measured covariance, SNR′ = 40.3 dB

Figure 11.20 Two-dimensional 16 × 16 block cosine transform coding at 1 bit/pixel rate using different covariance models. The right half shows error images. SNR′ is defined by eq. (3.13).

which takes the unity value in the zone of largest N_t variances of the transformed samples. Figure 11.21a shows a typical zonal mask. If we apply a zonal mask to the transformed blocks and encode only the nonzero elements, then the method is called *zonal coding*.

In *threshold coding* we encode the N_t coefficients of largest amplitude rather than the N_t coefficients having the largest variances, as in zonal coding. The address set of transmitted samples is now

$$I_t' = \{k, l; |v(k, l)| > \eta\} \tag{11.77}$$

where η is a suitably chosen threshold that controls the achievable average bit rate. For a given ensemble of images, since the transform coefficient variances are fixed,

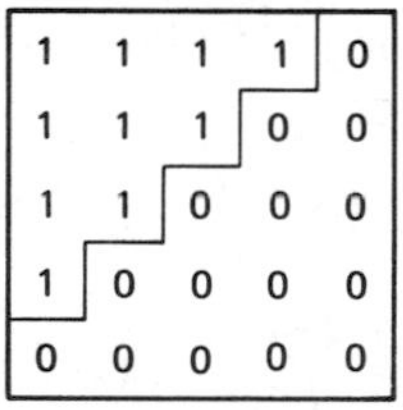

(a) Zonal mask

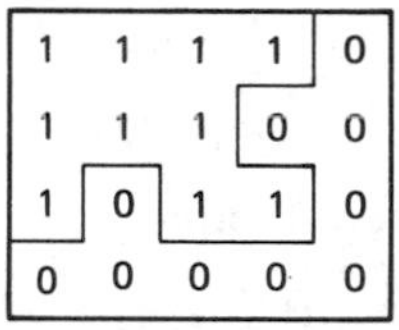

(b) Threshold mask

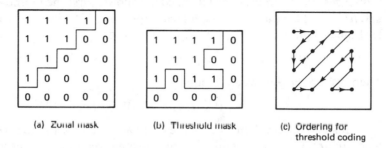

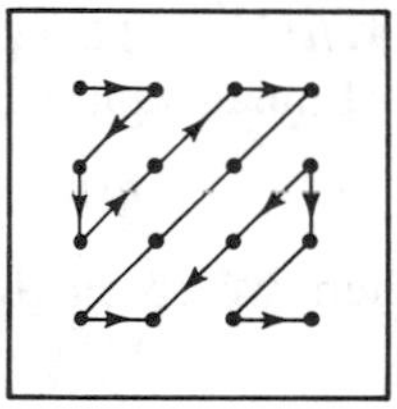

(c) Ordering for threshold coding

Figure 11.21 Zonal and threshold masks.

the zonal mask remains unchanged from one block to the next for a fixed bit rate. However, the *threshold mask* m_η, (Fig. 11.21b) defined as

$$m_\eta(k,l) = \begin{cases} 1, & k,l \in I_t' \\ 0, & \text{otherwise} \end{cases} \tag{11.78}$$

can change because I_t', the set of largest amplitude coefficients, need not be the same for different blocks. The samples retained are quantized by a suitable uniform quantizer followed by an entropy coder.

For the same number of transmitted samples (or quantizing bits), the threshold mask gives a better choice of transmission samples (that is, lower distortion). However, it also results in an increased rate because the addresses of the transmitted samples, that is, the boundary of the threshold mask, has to be coded for every image block. One method is to run-length code the transition boundaries in the threshold mask line by line. Alternatively, the two-dimensional transform coefficients are mapped into a one-dimensional sequence arranged in a predetermined order, such as in Fig. 11.21c. The thresholded sequence transitions are then run-length coded. Threshold coding is adaptive in nature and is useful for achieving high compression ratios when the image contents change considerably from block to block so that a fixed zonal mask would be inefficient.

Fast KL Transform Coding

For first-order AR sequences and for certain random fields represented by low-order noncausal models, fast KL transform coding approaches or *exceeds* the data compression efficiency of block KL transform coders. Recall from Section 6.5 that an $N \times 1$ vector $\mathbf{u}$ whose elements $u(n), 1 \le n \le N$, come from a first-order, stationary, AR sequence with zero mean and correlation ρ has the decomposition

$$\mathbf{u} = \mathbf{u}^0 + \mathbf{u}^b \tag{11.79}$$

where $\mathbf{u}^b$ is completely determined by the boundary variables $u(0)$ and $u(N+1)$ (see Fig. 6.8) and $\mathbf{u}^0$ and $\mathbf{u}^b$ are mutually orthogonal random vectors. The KL transform of the sequence $\{u^0(n), 1 \le n \le N\}$ is the sine transform, which is a fast transform. Thus (11.79) expresses the $N \times 1$ segment of a stationary Markov process as a two-source model. The first source has a fast KL transform, and the second source has only two degrees of freedom (that is, it is determined by two variables).

Suppose we are given the $N+2$ elements $u(n), 0 \le n \le N+1$. Then the $N \times 1$ sequences $u^0(n)$ and $u^b(n)$ are realized as follows. First the boundary variables $u(0)$ and $u(N+1)$ are passed through an interpolating FIR filter, which gives $u^b(n)$, the best mean square estimate of $u(n), 1 \le n \le N$, as

$$u^b(n) = \alpha[\mathbf{Q}^{-1}]_{n,1}\, u(0) + \alpha[\mathbf{Q}^{-1}]_{n,N}\, u(N+1), \qquad 1 \le n \le N \tag{11.80}$$

Then, we obtain the residual sequence

$$u^0(n) \triangleq u(n) - u^b(n), \qquad i \le n \le N \tag{11.81}$$

Instead of transform coding the original $(N+2) \times 1$ sequence by its KL transform, $\mathbf{u}^0$ and $\mathbf{u}_b$ can be coded separately using three different methods [6, 27]. One of these methods, called *recursive block coding*, is discussed here.

Recursive block coding. In the conventional block transform coding methods, the successive blocks of data are processed independently. The block size should be large enough so that interblock redundancy is minimal. But large block sizes spell large hardware complexity for the transformer. Also, at low bit rates (less than 1 bit per pixel) the block boundaries start becoming visibly objectionable.

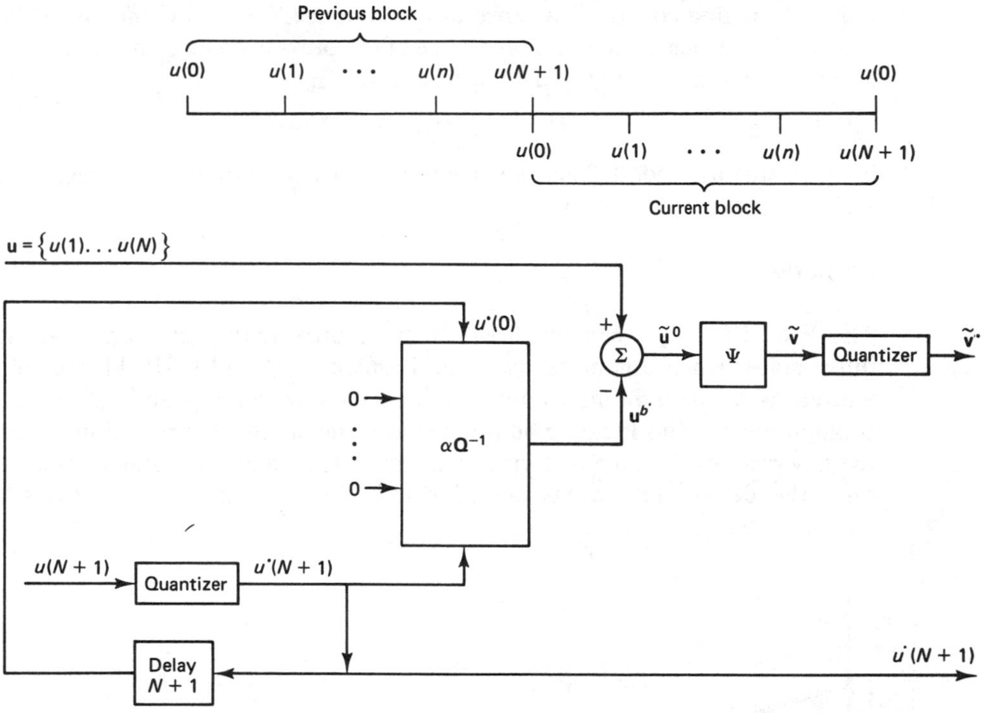

(a) Coder;

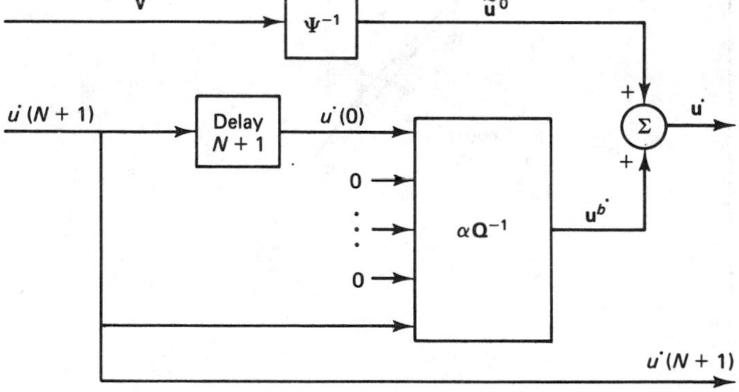

(b) Decoder.

Figure 11.22 Fast KL transform coding (recursive block coding). Each successive block brings $(N + 1)$ new pixels.

In *recursive block coding* (RBC), the correlation between successive blocks of data is exploited through the use of block boundaries. This yields additional compression and allows the use of smaller block sizes (8×8 or less) without sacrificing performance. Moreover, this method significantly reduces the block boundary effects.

Figure 11.22 shows this method. For each block the boundary variable $u(N+1)$ is first coded. The reproduced value $u^{\cdot}(N+1)$ together with the initial value $u^{\cdot}(0)$ (which is the boundary value of the previous block) are used to generate $u^{b^{\cdot}}(n)$, an estimate of $u^{b}(n)$. The difference

$$\tilde{u}^0(n) \triangleq u(n) - u^{b^{\cdot}}(n)$$

is sine transform coded. This yields better performance than the conventional block KLT coding (see Fig. 11.23).

Remarks

The FIR filter $\alpha\mathbf{Q}^{-1}$ can be shown to be approximately the simple straight-line interpolator when ρ is close to 1 (see Problem 6.13 and [27]). Hence, $\mathbf{u}^b$ can be viewed as a *low-resolution copy* obtained by subsampling and interpolating the original image. This fact can be utilized in image archival applications, where only the low resolution image is retrieved in *search mode* and the residual image is called once the desired image has been located. This way the search process can be speeded up.

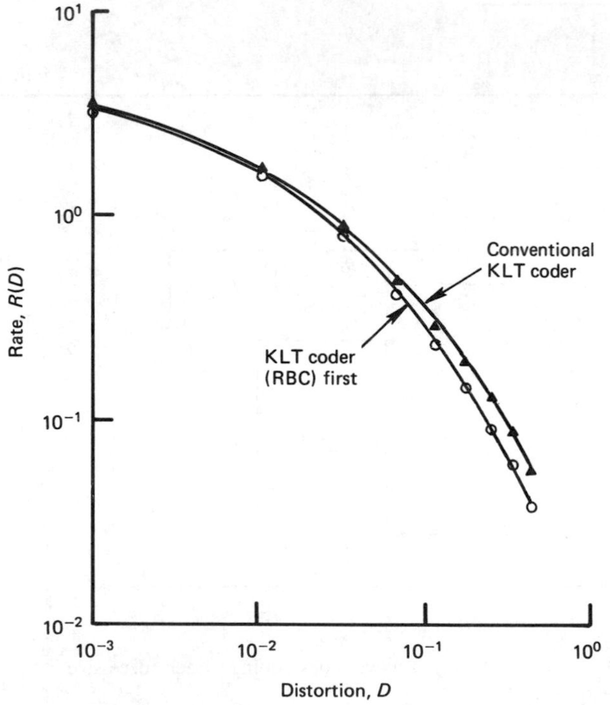

Figure 11.23 Rate versus distortion characters.

The foregoing theory can also be extended to second-order and higher AR models [27]. In these cases the KL transform of the residuals $u^0(n)$ is no longer a fast transform. This may not be a disadvantage for the recursive block coding algorithm because the transform size can now be quite small, so that a fast transform is not necessary.

In two dimensions many noncausal random field models yield fast KLT decompositions (see Example 6.17). Two-dimensional fast KLT coding algorithms similar to the ones just discussed can be designed using these decompositions. Figure 11.24 shows a two-dimensional recursive block coder. Figure 11.25 compares results of recursive block coding with cosine transform coding. The error images show the reduction in the block effects.

Two-Source Coding

The decomposition of (11.79) represents a *stationary source* by two sources, which can be realized from the image (block) and its boundary values. We could extend

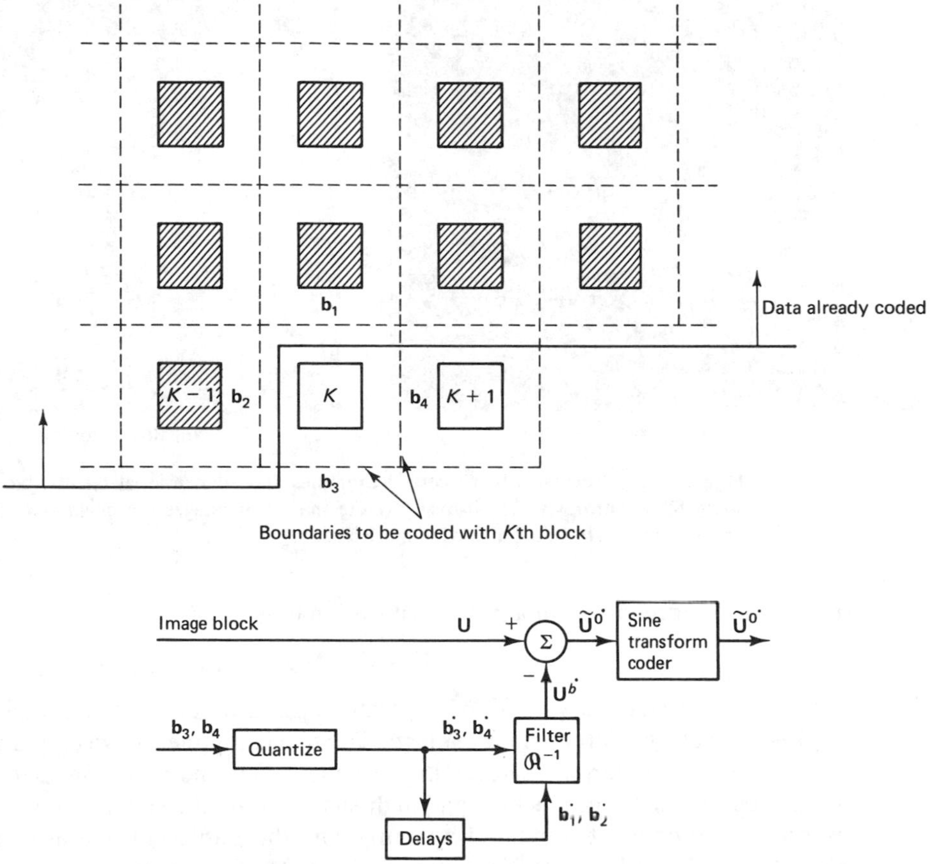

Figure 11.24 Two-dimensional recursive block coding.

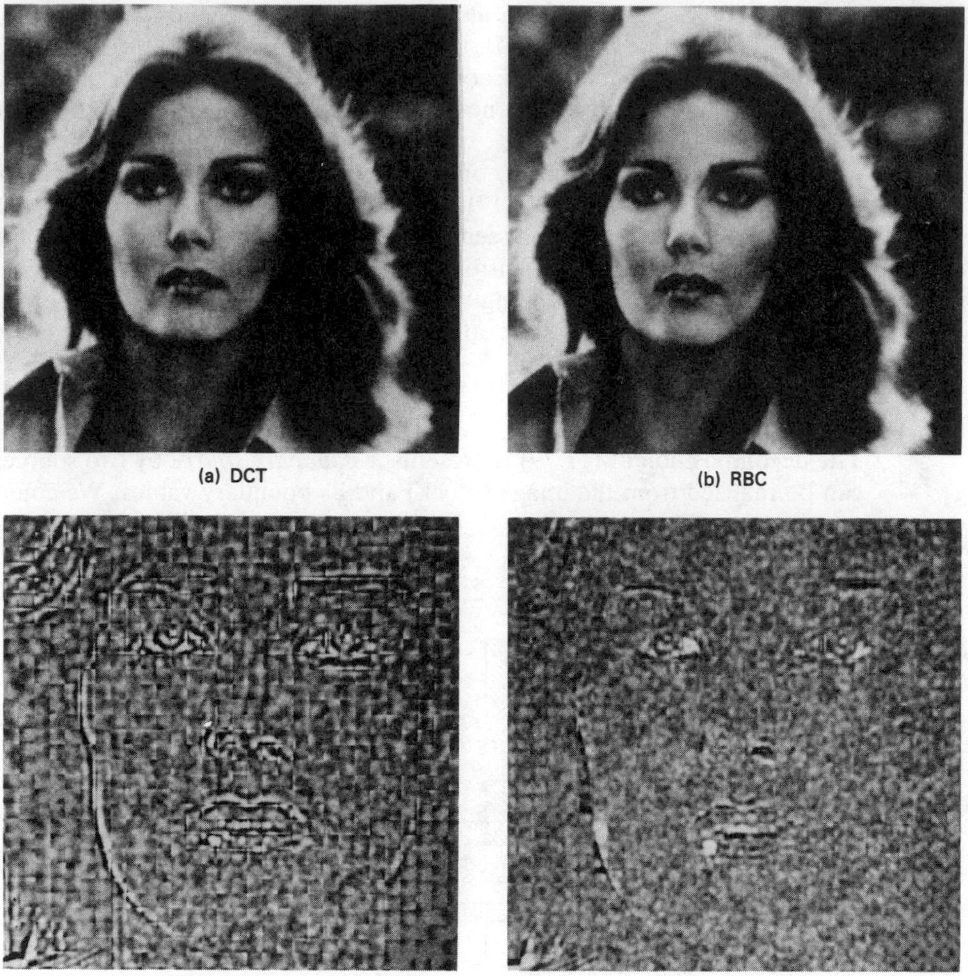

(a) DCT (b) RBC

(c) DCT coder distortion (d) RBC distortion

Figure 11.25 Recursive block coding examples. Two-dimensional results, block size = 8 × 8, entropy = .24 bits/pixel. (Note that error images are magnified and enhanced to show coding distortion more clearly.)

this idea to represent an image as a nonstationary source

$$\mathbf{U} = \mathbf{U}_s + \mathbf{U}_f \tag{11.82}$$

where $\mathbf{U}_s$ is a stationary random field and $\mathbf{U}_f$ is a deterministic component that represents certain features in the image. The two components are coded separately to preserve the different features in the image. One method, considered in [28], separates the image into its low- and high-spatial-frequency components. The high-frequency component, obtained by employing the Laplacian operator, is used to detect the edges whose locations are encoded. The low-frequency component can easily be encoded by transform or DPCM techniques. At the receiver a quantity

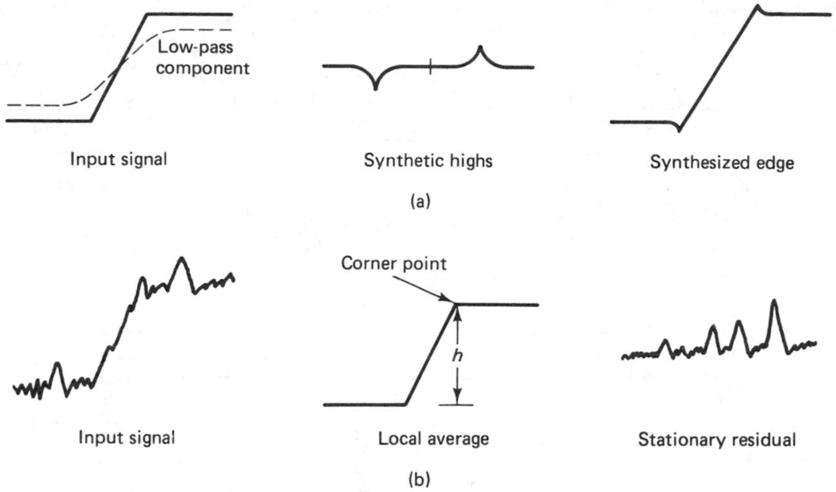

Figure 11.26 Two-source coding via (a) synthetic heights; (b) detrending.

proportional to the Laplacian of a ramp function, called *synthetic highs*, is generated at the location of the edges. The reproduced image is the sum of the low-frequency component and the synthetic highs (Fig. 11.26a).

In other two source-coding schemes (Yan and Sakrison in [1c]), the stationary source is found by subtracting from the image a local average found by piecewise fitting planes or low-order surfaces through boundary or corner points. The corner points and changes in amplitude are coded (Fig. 11.26b). The residual signal, which is a much better candidate for a stationary random field model, is transform coded.

Transform Coding Under Visual Criteria [30]

From Section 3.3 we know that a weighted mean square criterion can be useful for visual evaluation of images. An FFT coder that incorporates this criterion (Fig. 11.27) quantizes the transform coefficients of the image contrast field weighted by $H(k, l)$, the sampled frequency response function of the visual system. Inverse weighting followed by the inverse FFT gives the reconstructed contrast field. To apply this method to block image coding, using arbitrary transforms, the image contrast field should first be convolved with $h(m, n)$, the sampled Fourier inverse of $H(\xi_1, \xi_2)$. The resulting field can then be coded by any desired method. At the receiver, the decoded field must then be convolved with the inverse filter whose frequency response is $1/H(\xi_1, \xi_2)$.

Adaptive Transform Coding

There are essentially three types of adaptation for transform coding:

1. Adaptation of transform
2. Adaptation of bit allocation
3. Adaptation of quantizer levels

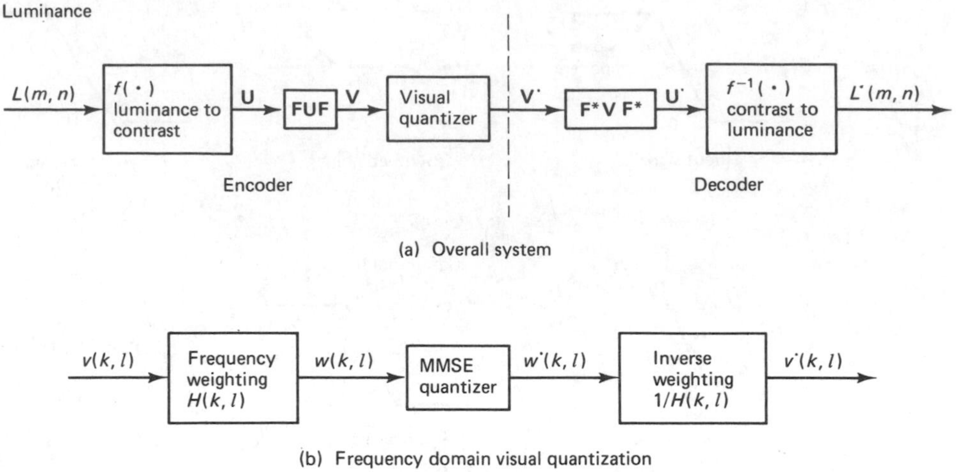

Figure 11.27 Transform coding under a visual criterion; $\mathbf{F}$ = unitary DFT.

Adaptation of the transform basis vectors is most expensive because a new set of KL basis vectors is required whenever any change occurs in the statistical parameters. A more practical method is to adapt the bit assignment of an image block, classified into one of several predetermined categories, according to the spatial activity (for instance, the variance of the data) in that block [1(c), p. 1285]. This results in a variable average rate from block to block but gives a better utilization of the total bits over the entire ensemble of image blocks. Another adaptive scheme is to allocate bits to image blocks so that each block has the same distortion [29]. This results in a uniform degradation of the image and appears less objectionable to the eye.

In adaptive quantization schemes, the bit allocation is kept constant but the quantizer levels are adjusted according to changes in the variances of the transform coefficients. Transform domain variances may be estimated either by updating the statistical parameters of the covariance model or by local averaging of the squared magnitude of the transform domain samples. Examples of adaptive transform coding are given in Section 11.7, where we consider interframe transform coding.

Summary of Transform Coding

In summary, transform coding achieves relatively larger compression than predictive methods. Any distortion due to quantization and channel errors gets distributed, during inverse transformation, over the entire image (Fig. 11.40). Visually, this is less objectionable than predictive coding errors, which appear locally at the source. Although transform and predictive coding schemes are theoretically close in performance at low distortion levels for one-dimensional Markov sequences, their performance difference in two dimensions is substantial. This is because of two reasons. First, predictive coding is quite sensitive to changes in the statistics of the data. Therefore, in practice, only adaptive predictive coding schemes achieve the efficiency of (nonadaptive) transform coding methods. Second, in two dimensions, finite-order causal predictors may never achieve compression ability close to transform coding because a finite-order causal representation of a two-dimensional random field may not exist. From an implementation point of view, predictive coding has much lower complexity both in terms of memory requirements and the number of operations to be performed. However, with the rapidly decreasing cost of digital hardware and computer memory, the hardware complexity of transform coders will not remain a disadvantage for very long. Table 11.4 summarizes the

TABLE 11.4 Practical Considerations in Designing Transform Coders

Design variables	Comments
1. Covariance model	
Separable see (2.84)	Convenient to analyze. Actual performance lower compared to nonseparable exponential.
Nonseparable see (2.85)	Works well when σ^2, α_1, α_2 are matched to the image class.
Noncausal NC2 (see Section 6.9)	Useful in designing 2-D fast KLT coders [6].
2. Block size (N) 16×16 to 64×64	Choice depends on available memory size and the value of the one-step correlation ρ. For $\rho \leq 0.9$, 16×16 size is adequate. A rule of thumb is to pick N such that $\rho^N \ll 1$ (say 0.2). For smaller block size, recursive block coding is helpful.
3. Transform	This choice is important if block size is small, say $N \leq 64$.
Cosine	Performs best for highly correlated data ($\rho \geq 0.5$)
Sine	For fast KL or recursive block coders.
DFT	Requires working with complex variables. Recommended if use of frequency domain is mandatory, such as in visual coding, and in CT, MRI image data, where source data has to pass through the Fourier domain.
Hadamard	Useful for small block sizes ($\simeq 4 \times 4$). Implementation is much simpler than sinusoidal fast transforms.
Haar	Useful if higher spatial frequencies are to be emphasized. Poor compression on mean square basis.
KL	Optimum on mean square basis. Difficult to implement. Cosine or other sinusoidal transforms are preferable.
Slant	Best performance among nonsinusoidal fast transforms.
4. Quantizer	
Lloyd-Max	Either Laplacian or Gaussian density may be used. For dc coefficient a Rayleigh density may be used.
Optimum uniform	Useful if the output is entropy coded or if the number of quantization levels is very large.

TABLE 11.5 Overall Performance of Transform Coding Methods

Method	Typical compressions ratios for images
One-dimensional	2–4
Two-dimensional	4–8
Two-dimensional adaptive	8–16
Three-dimensional	8–16
Three-dimensional adaptive	16–32

practical considerations in developing a design for transform coders. Table 11.5 compares the various transform coding schemes in terms of their compression ability. Here the compression ratio is the ratio of the number of bits per pixel in the original digital image (typically, 8) and the average number of bits per pixel required in encoding. Compression ratio values listed are to achieve SNR's in the 30- to 36-dB range.

11.6 HYBRID CODING AND VECTOR DPCM

Basic Idea

The predictive coding techniques of Section 11.3 are based on raster scanning and scalar recursive prediction rules. If the image is vector scanned, for instance, a column at a time, then it is possible to generalize the DPCM techniques by considering vector recursive predictors. Hybrid coding is a method of implementing an $N \times 1$ vector DPCM coder by N decoupled scalar DPCM coders. This is achieved by combining transform and predictive coding techniques. Typically, the image is unitarily transformed in one of its dimensions to decorrelate the pixels in that direction. Each transform coefficient is then sequentially coded in the other direction by one-dimensional DPCM (Fig. 11.28). This technique combines the advantages of hardware simplicity of DPCM and the robust performance of transform coding. The hardware complexity of this method is that of a one-dimensional transform coder and at most N DPCM channels. In practice the number of DPCM

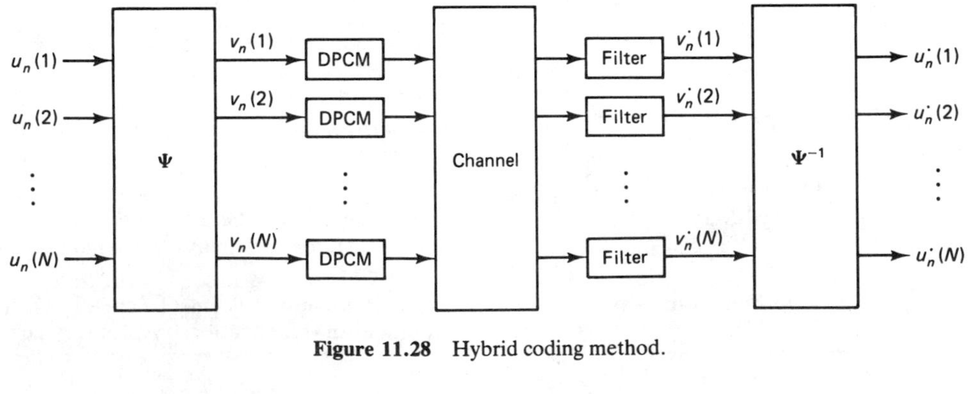

Figure 11.28 Hybrid coding method.

channels is significantly less than N because many elements of the transformed vector are allocated zero bits and are therefore not coded at all.

Hybrid Coding Algorithm. Let $\mathbf{u}_n, n = 0, 1, \ldots,$ denote $N \times 1$ columns of an image, which are transformed as

$$\mathbf{v}_n = \mathbf{\Psi u}_n, \qquad n = 0, 1, 2, \ldots \tag{11.83}$$

For each k, the sequence $v_n(k)$ is usually modeled by a first-order AR process [32], as

$$\begin{aligned} v_n(k) &= a(k)v_{n-1}(k) + b(k)e_n(k), \qquad 1 \le k \le N, n \ge 0 \\ E[e_n(k)e_{n'}(k')] &= \sigma_e^2(k)\delta(k-k')\delta(n-n') \end{aligned} \tag{11.84}$$

The parameters of this model can be identified from the covariances of $v_n(k)$, $n = 0, 1, \ldots,$ for each k (see Section 6.4). Some semicausal representations of images can also be reduced to such models (see Section 6.9). The DPCM equations for the kth channel can now be written as

$$\textit{Predictor:} \quad \overline{v}_n^{\bullet}(k) = a(k)v_{n-1}^{\bullet}(k) \tag{11.85a}$$

$$\textit{Quantizer input:} \quad \tilde{e}_n(k) \triangleq \frac{v_n(k) - \overline{v}_n^{\bullet}(k)}{b(k)}$$

$$\textit{Quantizer output:} \quad \tilde{e}_n^{\bullet}(k) \tag{11.85b}$$

$$\textit{Filter:} \quad v_n^{\bullet}(k) = \overline{v}_n^{\bullet}(k) + b(k)\tilde{e}_n^{\bullet}(k) \tag{11.85c}$$

The receiver simply reconstructs the transformed vectors according to (11.85c) and performs the inverse transformation $\mathbf{\Psi}^{-1}$. Ideally, the transform $\mathbf{\Psi}$ should be the KL transform of $\mathbf{u}_n$. In practice, a fast sinusoidal transform such as the cosine or sine is used.

To complete the design we now need to specify the quantizer in each DPCM channel. Let B denote the average desired bit rate in bits per pixel, n_k be the number of bits allocated to the kth DPCM channel, and $\sigma_e^2(k)$ be the quantizer mean square error in the kth channel, that is,

$$B = \frac{1}{N}\sum_{k=1}^{N} n_k, \qquad n_k \ge 0 \tag{11.86}$$

Assuming that all the DPCM channels are in their steady state, the average mean square distortion in the coding of any vector (for noiseless channels) is simply the average of distortions in the various DPCM channels, that is,

$$D = \frac{1}{N}\sum_{k=1}^{N} g_k(n_k)\sigma_e^2(k), \qquad g_k(x) \triangleq \frac{f(x)}{1 - |a(k)|^2 f(x)} \tag{11.87}$$

where $f(x)$ and $g_k(x)$ are the distortion-rate functions of the quantizer and the kth DPCM channel, respectively, for unity variance prediction error (see Problem 11.6). The bit allocation problem for hybrid coding is to minimize (11.87) subject to (11.86). This is now in the framework of the problem defined in Section 11.4, and the algorithms given there can be applied.

Example 11.7

Suppose the semicausal model (see section 6.9, and Eq. (6.106))

$$u(m, n) = \alpha[u(m-1, n) + u(m+1, n)] + \gamma u(m, n-1) + \varepsilon(m, n),$$
$$u(m, 0) = 0, \qquad \forall m$$
$$E[\varepsilon(m, n)] = 0, \qquad E[\varepsilon(m, n)\varepsilon(i, j)] = \beta^2 \delta(m-i, n-j)$$

is used to represent an $N \times M$ image with high interpixel correlation. At the boundaries we can assume $u(0, n) = u(1, n)$ and $u(N, n) = u(N+1, n)$. With these boundary conditions, this model has the realization of (11.84) for cosine transformed columns of the image with $a(k) \triangleq \gamma/\lambda(k)$, $b(k) \triangleq 1/\lambda(k)$, $\sigma_\epsilon^2(k) = \beta^2$, $\lambda(k) \triangleq 1 - 2\alpha \cos(k-1)\pi/N$, $1 \le k \le N$. In an actual experiment, a 256×256 image was coded in blocks of 16×256. The integer bit allocation for $B = 1$ was obtained as

$$3, 3, 3, 2, 2, 1, 1, 1, 0, 0, 0, 0, 0, 0, 0, 0$$

Thus, only the first eight cosine transform coefficients of each 16×1 column are used for DPCM coding. Figure 11.29a shows the result of hybrid coding using this model.

Adaptive Hybrid Coding

By updating the AR model parameters with variations in image statistics, adaptive hybrid coding algorithms can be designed [32]. One simple method is to apply the *adaptive variance estimation* algorithm [see Fig. 11.14a and eq. (11.38)] to each DPCM channel. The *adaptive classification method* discussed earlier gives better results especially at low rates. Each image column is classified into one of I (which is typically 4) predetermined classes that are fixed according to the variance distribution of the columns. Bits are allocated among different classes so that columns of high dynamic activity are assigned more bits than those of low activity. The class

(a) (b)

Figure 11.29 Hybrid encoded images at 1 bit/pixel. (a) Nonadaptive; (b) adaptive classification.

membership information requires additional overhead of $\log_2 I$ bits per column or $(1/N) \log_2 I$ bits per pixel. Figure 11.29 shows the result of applying adaptive algorithms to each DPCM channel of Example 11.7. Generally, the adaptive hybrid coding schemes can improve upon the nonadaptive techniques by about 3 dB, which is significant at low rates.

Hybrid Coding Conclusions

Hybrid coders combine the advantages of simple hardware complexity of DPCM coders and the high performance of transform coders, particularly at moderate bit rates (around 1 bit per pixel). Its performance lies between transform coding and DPCM. It is easily adaptable to noisy images [6, 32] and to changes in image statistics. It is particularly useful for interframe image data compression of motion images, as we shall see in Section 11.7. It is less sensitive to channel errors than DPCM but is not as robust as transform coding. Hybrid coders have been implemented for real-time data compression of images acquired by remotely piloted vehicles (RPV) [33].

11.7 INTERFRAME CODING

Teleconferencing, broadcast, and many medical images are received as sequences of two-dimensional image frames. Interframe coding techniques exploit the redundancy between the successive frames. The differences between successive frames are due to object motion or camera motion, panning, zooming, and the like.

Frame Repetition

Beyond the horizontal and/or vertical line interlace methods discussed in Section 11.1, a simple method of interframe compression is to subsample and frame-repeat interlaced pictures. This, however, does not produce good-quality moving images. An alternative is selective replenishment, where the frames are transmitted at a reduced rate according to a fixed, predetermined updating algorithm. At the receiver, any nonupdated data is refreshed from the previous frame stored in the frame memory. This method is reasonable for slow-moving areas only.

Resolution Exchange

The response of the human visual system is poor for dynamic scenes that simultaneously contain high spatial and temporal frequencies. Thus, rapidly changing areas of a scene can be represented with reduced amplitude and spatial resolution when compared with the stationary areas. This allows exchange of spatial resolution with temporal resolution and can be used to produce good-quality images at data rates of 2–2.5 bits per pixel. One such method segments the image into stationary and moving areas by thresholding the value of the frame-difference signal. In stationary areas frame differences are transmitted for every other pixel and the remaining pixels are repeated from the previous frame. In moving areas 2 : 1 horizontal subsampling is used, with intervening elements restored by interpolation along the scan lines. Using 5-bits-per-pixel frame-differential coding, a channel rate of 2.5 bits per pixel can be achieved. The main distortion occurs at sharp edges moving with moderate speed.

Conditional Replenishment

This technique is based on detection and coding of the moving areas, which are replenished from frame to frame. Let $u(m, n, i)$ denote the pixel at location (m, n) in frame i. The interframe difference signal is

$$e(m, n, i) = u(m, n, i) - u^{\cdot}(m, n, i-1) \tag{11.88}$$

where $u^{\cdot}(m, n, i-1)$ is the reproduced value of $u(m, n, i-1)$ in the $(i-1)$st frame. Whenever the magnitude of $e(m, n, i)$ exceeds a threshold η, it is quantized and coded for transmission. At the receiver, a pixel is reconstructed either by repeating the value of that pixel location from the previous frame if it came from a stationary area, or it is replenished by the decoded difference signal if it came from a moving area, giving

$$u^{\cdot}(m, n, i) = \begin{cases} u^{\cdot}(m, n, i-1) + e^{\cdot}(m, n, i), & \text{if } |e(m, n, i)| > \eta \\ u^{\cdot}(m, n, i-1), & \text{otherwise} \end{cases} \tag{11.89}$$

For transmission, code words representing the quantized values and their addresses are generated. Isolated points or very small clusters of moving areas are ignored to make the address coding scheme efficient. A reasonable-size buffer with appropriate buffer-control strategy is necessary to achieve a steady bit rate. With insufficient buffer size, its control can require extreme action (such as stopping the coder temporarily), which can cause jerky reproduction of motion (Fig. 11.30a). Simulation studies [6, 39] have shown that with a suitably large buffer a 1-bit-per-pixel rate can be achieved conveniently with an average SNR′ of about 34 dB (39 dB in stationary areas and 30 dB in moving areas). Figure 11.30b shows an encoded image and the encoding error magnitudes for a typical frame.

Adaptive Predictive Coding

Adaptations to motion characteristics can yield considerable gains in performance of interframe predictive coding methods. Figure 11.31 shows one such method, where the incoming pixel is classified as belonging to an area of stationary (C_S), moderate/slow (C_M), or rapid (C_R) motion. Classification is based on an activity index $\alpha(m, n, i)$, which is the absolute sum of interframe differences of a neighborhood $\mathcal{N}$ (Fig. 11.11) of previously encoded pixels, that is,

$$\alpha(m, n, i) = \sum_{(x, y) \in \mathcal{N}} |u^{\cdot}(m + x, n + y, i) - u^{\cdot}(m + x, n + y, i - 1)| \tag{11.90}$$

$$\mathcal{N} \triangleq \{(0, -s), (-1, -1), (-1, 0), (-1, 1)\} \tag{11.91}$$

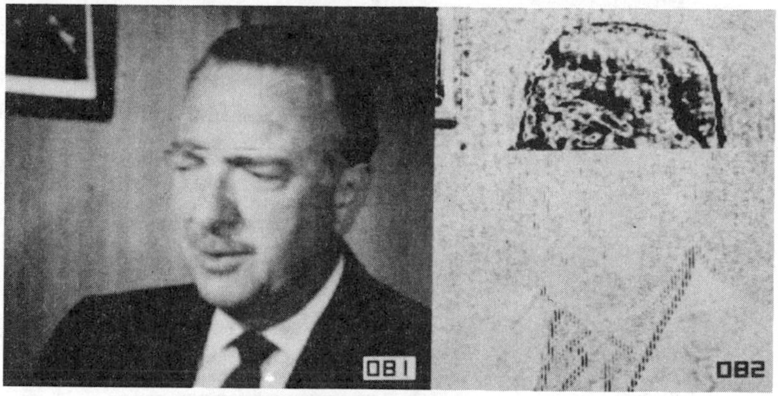

(a) Frame Replenishment Coding.
Bit-rate = .5 bit/pixel, SNR′ = 28.23 dB.

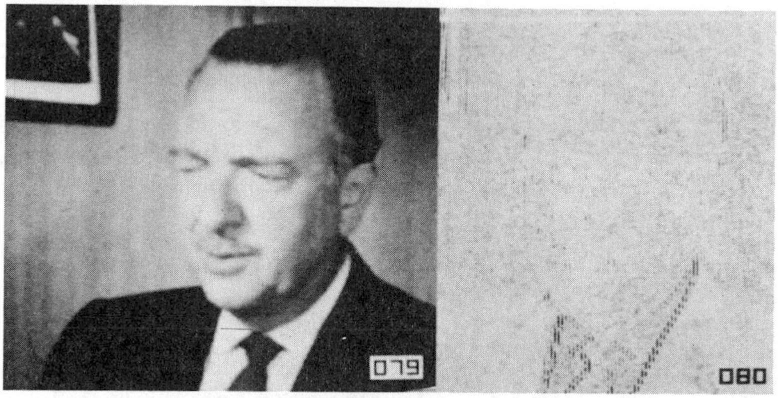

(b) Frame Replenishment Coding.
Bit-rate = 1 bit/pixel, SNR′ = 34.19 dB.

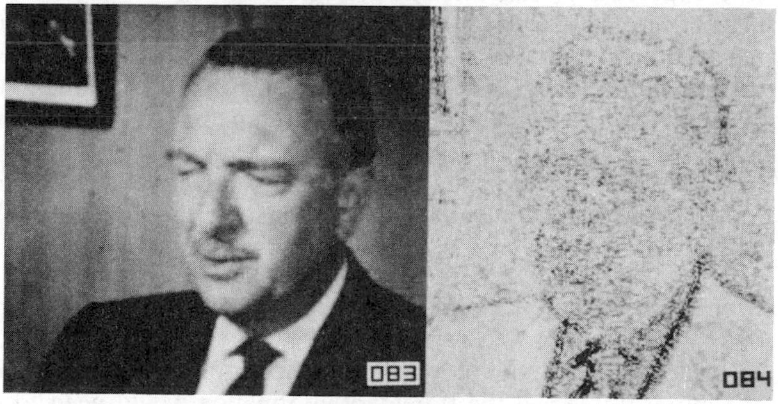

(c) Adaptive Classification Prediction Coding.
Bit-rate = .5 bit/pixel, SNR′ = 34.35 dB.

Figure 11.30 Results of interframe predictive schemes.

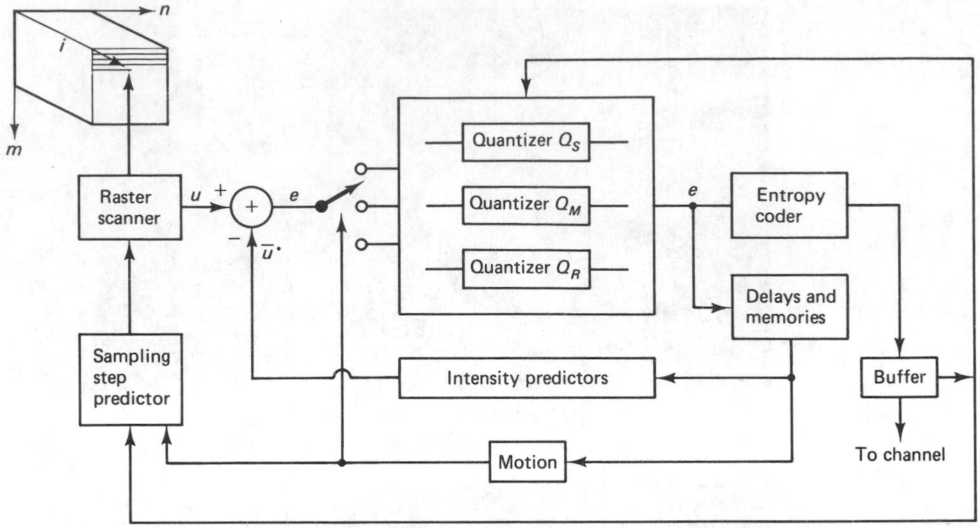

Figure 11.31 Interframe adaptive predictive coding.

where $s = 2$ in 2:1 subsampling mode and is unity otherwise. A large value of $\alpha(m, n, i)$ indicates large motion in the neighborhood of the pixel. The predicted value of the current pixel, $u(m, n, i)$, is

$$\bar{u}^{\cdot}(m, n, i) = \begin{cases} u^{\cdot}(m, n, i-1), & (m, n) \in C_S \\ u^{\cdot}(m-p, n-q, i-1), & (m, n) \in C_M \\ \rho_1^s u^{\cdot}(m, n-1, i) + \rho_2 u^{\cdot}(m-1, n, i) \\ \qquad - \rho_1\rho_2 u^{\cdot}(m-1, n-1, i), & (m, n) \in C_R \end{cases} \tag{11.92}$$

where ρ_1 and ρ_2 are the one-pixel correlations coefficients along m and n, respectively. Displacements p and q are chosen by estimating the average displacement of the neighborhood $\mathcal{N}$ that gives the minimum activity. Observe that for the case of rapid motion, the two-dimensional predictor of (11.35a) is used. This is because temporal prediction would be difficult for rapidly changing areas. The number of quantizer levels used for each class is proportional to its activity. This method achieves additional compression by a factor of two over the conditional replenishment while maintaining approximately the same SNR (Fig. 11.30c). For greater details on coder design, see [39].

Predictive Coding with Motion Compensation

In principle, if the motion trajectory of each pixel could be measured, then only the initial frame and the trajectory information would need to be coded. To reproduce the images we could simply propagate each pixel along its trajectory. In practice, the motion of objects in the scene can be approximated by piecewise displacements from frame to frame. The displacement vector is used to direct the motion-compensated interframe predictor. The success of a motion-compensated coder depends on accuracy, speed, and robustness (with respect to noise) of the displacement estimator.

Displacement Estimation Algorithms

1. *Search techniques.* Search techniques look for a displacement vector $\mathbf{d} \stackrel{\Delta}{=} [p, q]^T$ such that a distortion function $D(p, q)$ between a reference frame and the current frame is minimized. Examples are template matching, logarithmic search, hierarchical search, conjugate direction, and gradient search techniques, which are discussed in Section 9.12. In these techniques the log search converges most rapidly when the search area is large. For small search areas, the conjugate direction search method is simpler. These techniques are quite robust and are useful especially when the displacement is constant for a block of pixels. For interframe motion estimation the search can be usually limited to a window of 5×5 pixels.
2. *Recursive displacement estimation.* To understand this algorithm it is convenient to consider the continuous function $u(x, y, t)$ representing the image frame at time t. Given a displacement error measure $f(x)$, one possibility is to update $\mathbf{d}$ recursively via the gradient algorithm [37]

$$\begin{aligned} \mathbf{d}_k &= \mathbf{d}_{k-1} - \varepsilon \nabla_{\mathbf{d}} f_{k-1} \\ f_k &\stackrel{\Delta}{=} f[u(x, y, t) - u(x - p_k, y - q_k, t - \tau)] \end{aligned} \tag{11.93}$$

where $\mathbf{d}_k$ is the displacement estimate at pixel k for the adopted scanning sequence, $\nabla_{\mathbf{d}}$ is the gradient of f with respect to $\mathbf{d}$, and τ is the interframe interval. The ε is a small positive quantity that controls the correction at each recursion. For $f(x) = x^2/2$, (11.93) reduces to

$$\begin{aligned} \mathbf{d}_k = \mathbf{d}_{k-1} - \varepsilon[u(x, y, t) - u(x - p_{k-1}, y - q_{k-1}, t - \tau)] \cdot \\ \cdot \begin{bmatrix} \dfrac{\partial u}{\partial x}(x - p_{k-1}, y - q_{k-1}, t - \tau) \\ \dfrac{\partial u}{\partial y}(x - p_{k-1}, y - q_{k-1}, t - \tau) \end{bmatrix} \end{aligned} \tag{11.94}$$

Since $u(x, y, t)$ is available only at sampling locations, interpolation is required to evaluate u, $\partial u/\partial x$, and $\partial u/\partial y$ at $(x - p_k, y - q_k, t - \tau)$. The advantage of this algorithm is that it estimates displacement values for each pixel. However, it lacks the robustness of block search techniques.
3. *Differential techniques.* These algorithms are based on the fact that the gray-level value of a pixel remains constant along its path of motion, that is,

$$u(x(t), y(t), t) = \text{constant} \tag{11.95}$$

where $x(t), y(t)$ is the motion trajectory. Differentiating both sides with respect to t, we obtain

$$\frac{\partial u}{\partial t} + v_1 \frac{\partial u}{\partial x} + v_2 \frac{\partial u}{\partial y} = 0, \qquad (x, y) \in \mathcal{R} \tag{11.96}$$

where $v_1 \triangleq dx(t)/dt$, $v_2 \triangleq dy(t)/dt$, are the two velocity components and $\mathcal{R}$ is the region of moving pixels having the same motion trajectory. The displacement vector $\mathbf{d}$ can be estimated as

$$\mathbf{d} = \int_{t_0}^{t_0+\tau} \mathbf{v}\,dt \simeq \mathbf{v}\tau \tag{11.97}$$

assuming the velocity remains constant during the frame intervals. The velocity vector can be estimated from the interframe data after it has been segmented into stationary and moving areas [41] by minimizing the function

$$J \triangleq \iint_{\mathcal{R}} \left[\frac{\partial u}{\partial t} + v_1 \frac{\partial u}{\partial x} + v_2 \frac{\partial u}{\partial y}\right]^2 dx\,dy \tag{11.98}$$

Setting $\nabla_{\mathbf{v}} J = \mathbf{0}$ gives the solution as

$$\begin{bmatrix} v_1 \\ v_2 \end{bmatrix} = -\begin{bmatrix} c_{xx} & c_{xy} \\ c_{yx} & c_{yy} \end{bmatrix}^{-1} \begin{bmatrix} c_{xt} \\ c_{yt} \end{bmatrix} \tag{11.99}$$

where $c_{\alpha\beta}$ denotes the correlation between $\partial u/\partial\alpha$ and $\partial u/\partial\beta$, that is,

$$c_{\alpha\beta} \triangleq \iint_{\mathcal{R}} \frac{\partial u}{\partial \alpha}\frac{\partial u}{\partial \beta}\,dx\,dy \qquad \text{for } \alpha, \beta = x, y, t \tag{11.100}$$

This calculation can be speeded up by estimating the correlations as [40]

$$c_{\alpha\beta} = \iint_{\mathcal{R}} \frac{\partial u}{\partial \alpha}\,\text{sgn}\left(\frac{\partial u}{\partial \beta}\right) dx\,dy \tag{11.101}$$

Thus $\mathbf{v}$ can be calculated from the partial differentials of $u(x, y, t)$, which can be approximated from the given interframe sampled data. Like the block search algorithms, the differential techniques also give the displacement vector for a block or a region. These methods are faster, although still not as robust as the block search algorithms.

Having estimated the motion, compression is achieved by skipping image frames and reproducing the missing frames at the receiver either by frame repetition or by interpolation along the motion trajectory. For example, if the alternate frames, say $u(m, n, 2i)$, $i = 1, 2, \ldots$, have been skipped, then with motion compensation we have

$$\text{Frame repetition:} \quad u^{\cdot}(m, n, 2i) = u^{\cdot}(m-p, n-q, 2i-1) \tag{11.102}$$

$$\text{Frame interpolation:} \quad u^{\cdot}(m, n, 2i) = \tfrac{1}{2}[u^{\cdot}(m-p, n-q, 2i-1) + u^{\cdot}(m-p', n-q', 2i+1)] \tag{11.103}$$

where (p, q) and (p', q') are the displacement vectors relative to the preceding and following frames, respectively. Without motion compensation, we would set $p = q = p' = q' = 0$. Figure 11.32 shows the advantage of motion compensation in frame skipping. The improvement due to motion compensation, roughly 10 dB, is quite significant.

Interframe Hybrid Coding

Hybrid coding is particularly useful for interframe image data compression of motion images. A two-dimensional $M \times N$ block of the ith frame, denoted by $\mathbf{U}_i$, is first transformed to give $\mathbf{V}_i$. For each (k, l) the sequence $v_i(k, l)$, $i = 1, 2, \dots,$ is considered a one-dimensional random process and is coded independently by a suitable one-dimensional DPCM method. The receiver reconstructs $v_i^{\bullet}(k, l)$ and

Along temporal axis, SNR′ = 16.90 dB

Along motion trajectory, SNR′ = 26.69 dB

(a) Frame repetition (or interframe prediction) based on the preceding frame

Figure 11.32 Effects of motion compensation on interframe prediction and interpolation.

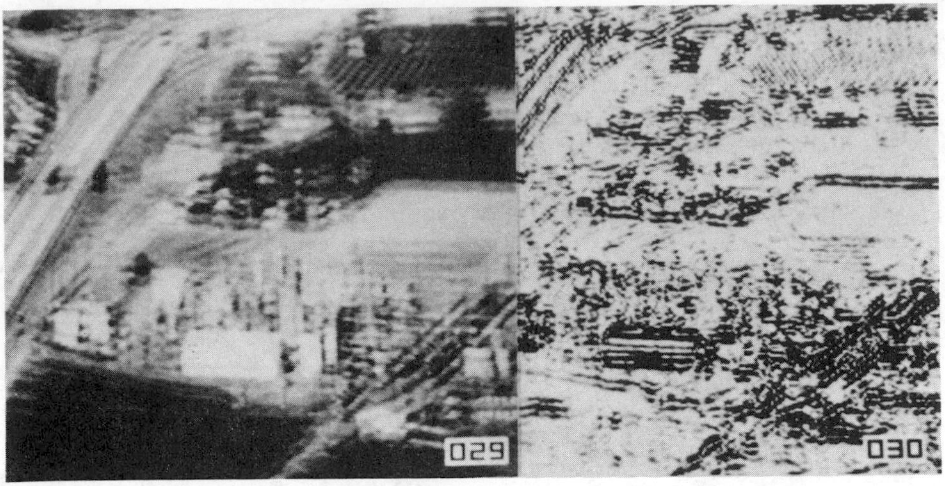

Along temporal axis, SNR′ = 19.34 dB

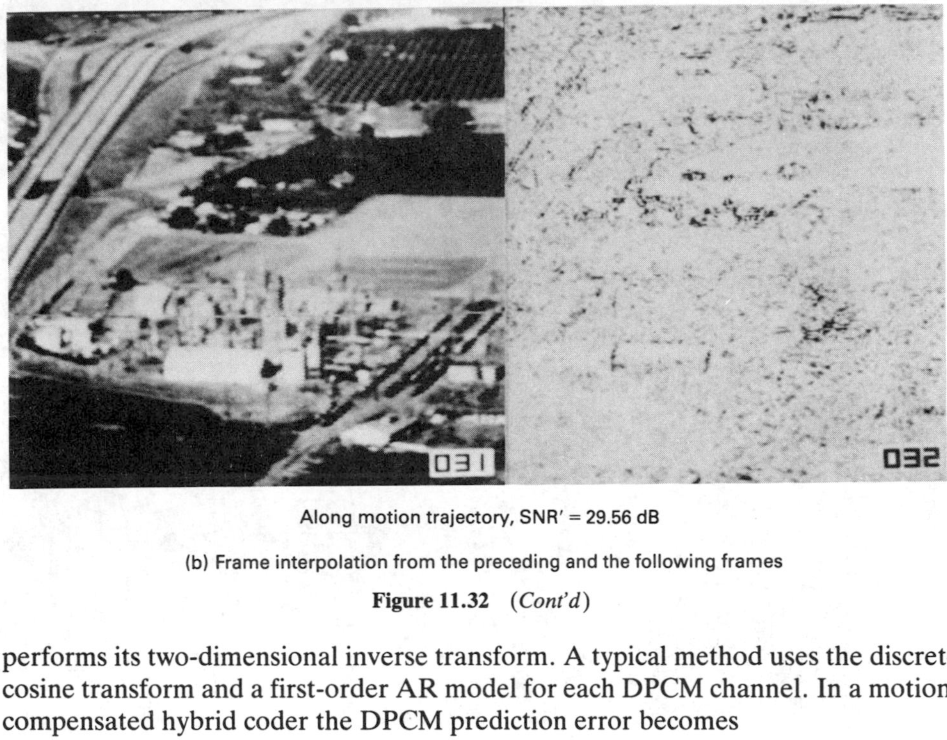

Along motion trajectory, SNR′ = 29.56 dB

(b) Frame interpolation from the preceding and the following frames

Figure 11.32 (*Cont'd*)

performs its two-dimensional inverse transform. A typical method uses the discrete cosine transform and a first-order AR model for each DPCM channel. In a motion-compensated hybrid coder the DPCM prediction error becomes

$$e_i(k, l) = v_i(k, l) - \alpha \tilde{v}'_{i-1}(k, l) \tag{11.104}$$

where $\tilde{v}'_{i-1}(k, l)$ are obtained by transforming the motion-compensated sequence $u'_{i-1}(m-p, n-q)$. If α, the prediction coefficient, is constant for each channel (k, l), then $e_i(k, l)$ would be the same as the transform of $u_i(m, n) - \alpha u'_{i-1}(m-p, n-q)$. This yields a motion-compensated hybrid coder, as shown in

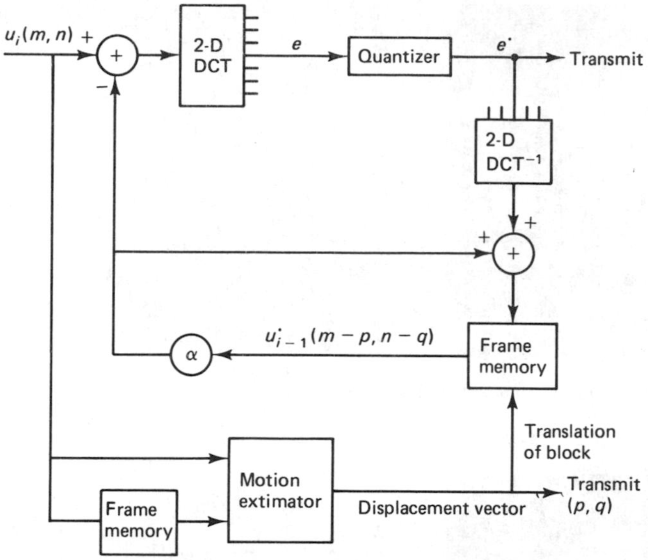

Figure 11.33 Interframe hybrid coding with motion compensation.

Fig. 11.33. Results of different interframe hybrid coding methods are shown in Fig. 11.34. These and other results [6, 36] show that with motion compensation, the adaptive hybrid coding method performs better than adaptive predictive coding and adaptive three-dimensional transform coding. However, the coder now requires two sets of two-dimensional transformations.

Three-Dimensional Transform Coding

In many applications, (for example, in multispectral imaging, interframe video imaging, medical cineangiography, CT scanning, and so on), we have to work with three- (or higher-) dimensional data. Transform coding schemes are possible for compression of such data by extending the basic ideas of Section 11.5. A three-dimensional (separable) transform of a $M \times N \times I$ sequence $u(m, n, i)$ is defined as

$$v(k, l, j) \triangleq \sum_{m=0}^{M-1} \sum_{n=0}^{N-1} \sum_{i=0}^{I-1} u(m, n, i) a_M(k, m) a_N(l, n) a_I(j, i) \qquad (11.105)$$

where $0 \le (k, m) \le M-1$, $0 \le (l, n) \le N-1$, $0 \le (j, i) \le I-1$, and $\{a_M(k, m)\}$ are the elements of an $M \times M$ unitary matrix $\mathbf{A}_M$, and so on. The transform coefficients given by (11.105) are simply the result of taking the **A**-transform with respect to each index and will require $MNI \log_2(MNI)$ operations for a fast transform. The storage requirement for the data is MNI. As before, the practical approach is to partition the data into small blocks (such as $16 \times 16 \times 16$) and process each block independently. The coding algorithm after transformation is the same as before except that we are working with triple indexed variables. Figure 11.35 shows results for one frame of a sequence of cosine transform coded images. The result of Fig. 11.35a corresponds to the use of the three-dimensional separable covariance model

$$r(m, n, i) = \sigma^2 \rho_1^{|m|} \rho_2^{|n|} \rho_3^{|i|}$$

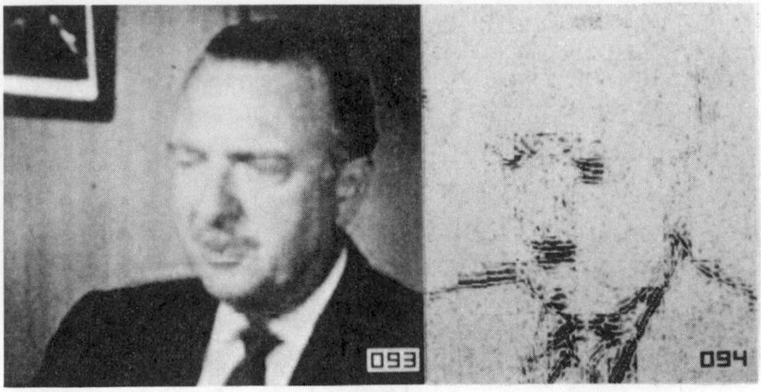

(a) 0.5 bit/pixel, nonadaptive, SNR′ = 34 dB;

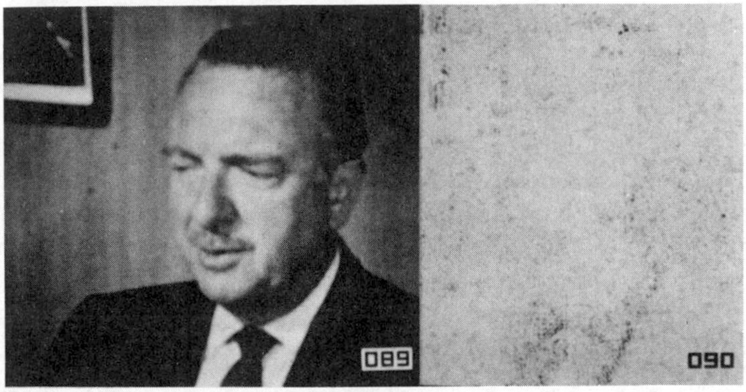

(b) 0.5 bit/pixel, adaptive, SNR′ = 40.3 db;

(c) 0.125 bit/pixel, adaptive with motion compensation, SNR′ = 36.7 dB

Figure 11.34 Interframe hybrid coding

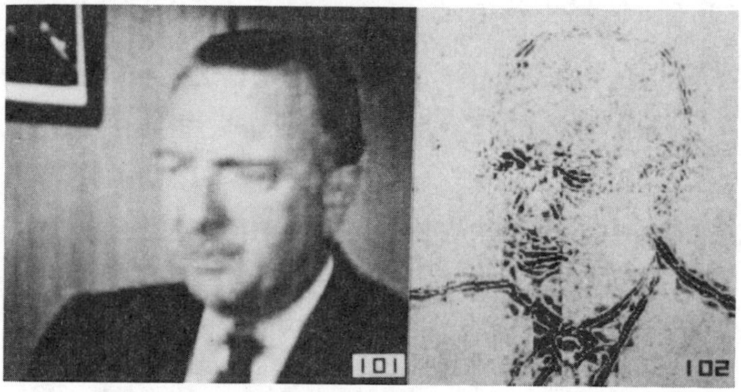

(a) 0.5 bit/pixel, separable covariance model, SNR′ = 32.1 dB;

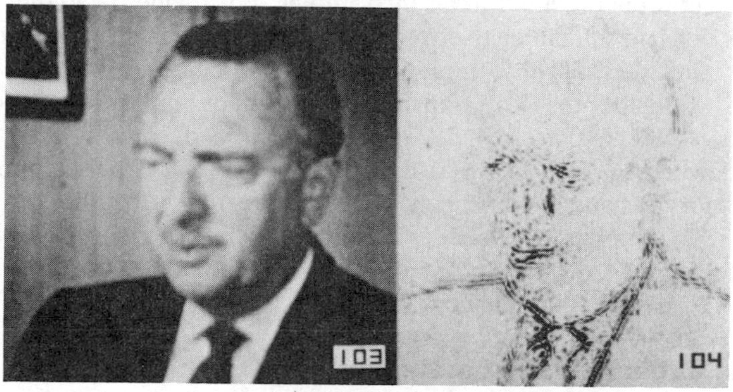

(b) 0.5 bit/pixel, measured covariances, SNR′ = 36.8 dB;

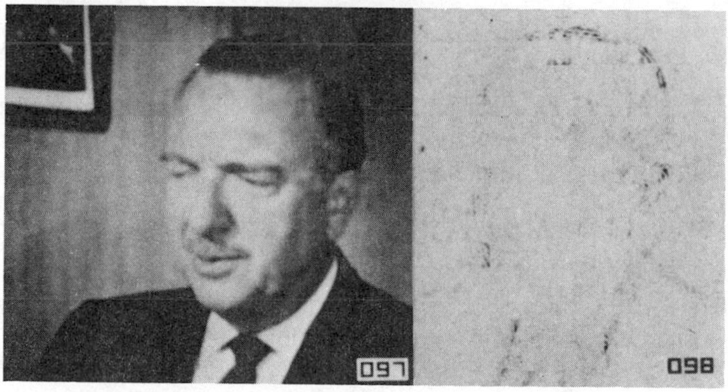

(c) 0.5 bit/pixel, measured covariances, adaptive, SNR′ = 41.2 dB.

Figure 11.35 Interframe transform coding

which, as expected, performs poorly. Also, the adaptive hybrid coding with motion compensation performs better than three-dimensional transform coding. This is because incorporating motion information in a three-dimensional transform coder requires selecting spatial blocks along the motion trajectory, which is not a very attractive alternative.

11.8 IMAGE CODING IN THE PRESENCE OF CHANNEL ERRORS

So far we have assumed the channel between the coder and the decoder to be noiseless. To account for channel errors, we have to add redundancy to the input by appending error correcting bits. Thus a proper trade-off between source coding (redundancy removal) and channel coding (redundancy injection) has to be achieved in the design of data compression systems. Often, the error-correcting codes are designed to reduce the probability of bit errors, and for simplicity, equal protection is provided to all the samples. For image data compression algorithms, this does not minimize the overall error. In this section we consider source-channel-encoding methods that minimize the overall mean square error.

Consider the PCM transmission system of Fig. 11.36, where a quantizer generates k-bit outputs $x_i \in S$, which are mapped, one-to-one, into n-bit $(n \geq k)$ codewords $\mathbf{g}_i \in C$. Let $\beta(\cdot)$ denote this mapping. The channel is assumed to be memoryless and binary symmetric with bit error probability p_e. It maps the set C of $K = 2^k$ possible n-bit code words into a set V of 2^n possible n-bit words. At the receiver, $\lambda(\cdot)$ denotes the mapping of elements of V into the elements on the real line $\mathbb{R}$. The identity element of V is the vector $\mathbf{0} \triangleq [0, 0, \ldots, 0]$.

The Optimum Mean Square Decoder

The mean square error between the decoder output and the encoder input is given by

$$\sigma_c^2 = \sigma_c^2(\beta, \lambda) \triangleq E[(y-x)^2] = \sum_{\mathbf{v} \in V} p(\mathbf{v}) \left\{ \sum_{x \in S} (\lambda(\mathbf{v}) - x)^2 p(x|\mathbf{v}) \right\} \tag{11.106}$$

and depends on the mappings $\beta(\cdot)$ and $\lambda(\cdot)$. From estimation theory (see Section 2.12) we know that given the encoding rule β, the decoder that minimizes this error is given by the conditional mean of x, that is,

$$y = \lambda(\mathbf{v}) = \sum_{x \in S} x p(x|\mathbf{v}) = E[x|\mathbf{v}] \tag{11.107}$$

where $p(x|\mathbf{v})$ is the conditional density of x given the channel output $\mathbf{v}$. The function $\lambda(\mathbf{v})$ need not map the channel output into the set S even if $n = k$.

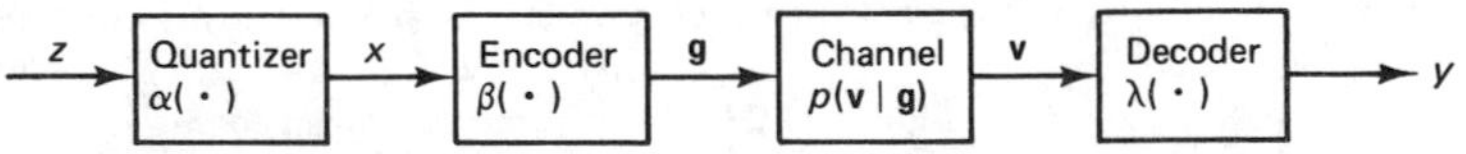

Figure 11.36 Channel coding for PCM transmission.

The Optimum Encoding Rule

The optimum encoding rule $\beta(\cdot)$ that minimizes (11.106) requires an exhaustive search for the optimum subspace C over all subspaces of V. Practical solutions are found by restricting the search to a particular set of subspaces [39, 42]. Table 11.6 shows one set of practical basis vectors for uniformly distributed sources, from

TABLE 11.6 Basic Vectors $\{\phi_i, i = 1, \ldots, k\}$ for (n, k) Group Codes

i	k \ $n-k$	0	1	2	3	4
1		10000000	100000001	1000000011	10000000110	
2		01000000	010000000	0100000001	01000000101	
3		00100000	001000000	0010000000	00100000011	
4	8	00010000	000100000	0001000000	00010000111	
5		00001000	000010000	0000100000	00001000000	
6		00000100	000001000	0000010000	00000100000	
7		00000010	000000100	0000001000	00000010000	
8		00000001	000000010	0000000100	00000001000	
1		1000000	10000001	100000011	1000000110	10000001110
2		0100000	01000000	010000001	0100000101	01000001010
3		0010000	00100000	001000000	0010000011	00100000101
4	7	0001000	00010000	000100000	0001000111	00010000011
5		0000100	00001000	000010000	0000100000	00001000000
6		0000010	00000100	000001000	0000010000	00000100000
7		0000001	00000010	000000100	0000001000	00000010000
1		100000	1000001	10000011	100000110	1000001110
2		010000	0100000	01000001	010000101	0100001010
3	6	001000	0010000	00100000	001000011	0010000101
4		000100	0001000	00010000	000100111	0001000011
5		000010	0000100	00001000	000010000	0000100000
6		000001	0000010	00000100	000001000	0000010000
1		10000	100001	1000011	10000110	100001110
2		01000	010000	0100001	01000101	010001010
3	5	00100	001000	0010000	00100011	001000101
4		00010	000100	0001000	00010111	000100011
5		00001	000010	0000100	00001000	000010000
1		1000	10001	100011	1000110	10001110
2	4	0100	01000	010001	0100101	01001010
3		0010	00100	001000	0010011	00100101
4		0001	00010	000100	0001111	00010011
1		100	1001	10011	100110	1001110
2	3	010	0100	01001	010101	0101010
3		001	0010	00100	001011	0010101
1	2	10	101	1011	10110	010111
2		01	010	0101	01101	101110
1	1	1	11	111	1111	11111

$n - 11, k - 6$

$i \rightarrow$	1	2	3	4	5	6
$g_i \rightarrow$	10000011101	01000010100	00100001011	000100001100	00001000001	00000100000

which $\beta(\cdot)$ is obtained as follows. Let $\mathbf{b} \overset{\Delta}{=} [b(1), b(2), \ldots, b(k)]$ be the binary representation of an element of S; then

$$\mathbf{g} = \beta(\mathbf{b}) = \sum_{i=1}^{k} \oplus b(i) \cdot \boldsymbol{\phi}_i \tag{11.108}$$

where $\Sigma \oplus$ denotes exclusive-OR summation and $\cdot$ denotes the binary product. The codes generated by this method are called the (n, k) *group codes.*

Example 11.8

Let $n = 4$ and $k = 2$, so that $n - k = 2$. Then $\boldsymbol{\phi}_1 = [1 \quad 0 \quad 1 \quad 1]$, $\boldsymbol{\phi}_2 = [0 \quad 1 \quad 0 \quad 1]$, and $\beta(\cdot)$ is given as follows

x	$\mathbf{b}$	$\mathbf{g} = \beta(\mathbf{b})$
0	0 0	$0\ 0\ 0\ 0 = 0 \cdot \boldsymbol{\phi}_1 \oplus 0 \cdot \boldsymbol{\phi}_2$
1	0 1	$0\ 1\ 0\ 1 = 0 \cdot \boldsymbol{\phi}_1 \oplus 1 \cdot \boldsymbol{\phi}_2$
2	1 0	$1\ 0\ 1\ 1 = 1 \cdot \boldsymbol{\phi}_1 \oplus 0 \cdot \boldsymbol{\phi}_2$
3	1 1	$1\ 1\ 1\ 0 = \boldsymbol{\phi}_1 \oplus \boldsymbol{\phi}_2$

In general the basis vectors $\boldsymbol{\phi}_i$ depend on the bit error probability p_e and the source probability distribution. For other distributions, Table 11.5 is found to lower the channel coding performance only slightly for $p_e \ll 1$ [39]. Therefore, these group codes are recommended for all mean square channel coding applications.

Optimization of PCM Transmission

If η_c and η_q denote the channel and the quantizer errors, we can write (from Fig. 11.36) the input sample as

$$z = x + \eta_q = y + \eta_c + \eta_q \tag{11.109}$$

This gives the total mean square error as

$$\sigma_t^2 = E[(z - y)^2] = E[(\eta_c + \eta_q)^2] \tag{11.110}$$

For a fixed channel coder $\beta(\cdot)$, this error is minimum when [6, 39] (1) η_c and η_q are orthogonal and (2) $\sigma_c^2 \overset{\Delta}{=} E[\eta_c^2]$ and $\sigma_q^2 \overset{\Delta}{=} E[\eta_q^2]$ are minimum. This requires

$$y \overset{\Delta}{=} \lambda(\mathbf{v}) = E[x|\mathbf{v}] \tag{11.111}$$

$$x = \alpha(z) = E[z|z \in \mathscr{T}_i] \tag{11.112}$$

where $\mathscr{T}_i, i = 1, \ldots, 2^k$ denotes the ith quantization interval of the quantizer.

This result says that the optimum decoder is independent of the optimum quantizer, which is the Lloyd-Max quantizer. Thus the overall optimal design can be accomplished by optimizing the quantizer and the decoder individually. This gives

$$\sigma_t^2 = \sigma_c^2 + \sigma_q^2 \tag{11.113}$$

Let $f(k)$ and $c(n, k)$ denote the mean square distortions due to the k-bit Lloyd-Max quantizer and the channel, respectively, when the quantizer input is a unit variance random variable (Tables 11.7 and 11.8). Then we can write the total

TABLE 11.7 Quantizer Distortion Function $f(k)$ for Unity Variance Inputs

Density \ $k \rightarrow$	1	2	3	4	5	6	7	8
Gaussian	0.3634	0.1175	0.0346	0.0095	2.5×10^{-3}	6.4×10^{-4}	1.6×10^{-4}	4×10^{-5}
Laplacian	0.3634	0.1762	0.0545	0.0154	4.1×10^{-3}	1.06×10^{-3}	2.7×10^{-4}	7×10^{-5}
Uniform	0.2500	0.0625	0.0156	0.0039	9.77×10^{-4}	2.44×10^{-4}	6.1×10^{-5}	1.52×10^{-5}

TABLE 11.8 Channel Distortion $c(n, k)$ for (n, k) Block Coding of Outputs of a Quantizer with Unity Variance Input

Input Density		$p_e = 0.01$					$p_e = 0.001$				
	k \ $n-k$	0	1	2	3	4	0	1	2	3	4
	1	0.0252	0.0129	0.00075	0.0004	0.00003	0.00254	0.0013	0.8×10^{-5}	0.4×10^{-6}	$<10^{-6}$
	2	0.0483	0.0280	0.0066	0.0018	0.0010	0.0050	0.0028	0.0005	0.00002	0.00001
	3	0.0656	0.0372	0.0115	0.0033	0.0025	0.0069	0.0038	0.0010	0.00004	0.00003
	4	0.0765	0.0423	0.0138	0.0056	0.0032	0.0083	0.0044	0.0012	0.00006	0.00003
	5	0.0821	0.0450	0.0149	0.0062	0.0036	0.0093	0.0047	0.0013	0.0001	0.00006
	6	0.0856	0.0463	0.0154	0.0064	0.0037	0.0101	0.0049	0.0014	0.0001	0.00008
	7	0.0923	0.0477	0.0156	0.0068	0.0037	0.0112	0.0051	0.0014	0.00012	0.00008
	8	0.1050	0.0508	0.0169	0.0076		0.0143	0.0056	0.0015	0.00015	
	1	0.0193	0.0098	0.0006	0.0003	$<10^{-4}$	0.0020	0.0010	$<10^{-4}$	$<10^{-4}$	$<10^{-5}$
	2	0.0554	0.0312	0.0075	0.0022	0.0001	0.0058	0.0032	0.0006	$<10^{-4}$	$<10^{-4}$
	3	0.0934	0.0480	0.0151	0.0042	0.0029	0.0102	0.0049	0.0013	$<10^{-4}$	$<10^{-4}$
	4	0.1230	0.0601	0.0195	0.0083	0.0040	0.0148	0.0063	0.0017	$<10^{-4}$	$<10^{-4}$
	5	0.1381	0.0681	0.0219	0.0091	0.0048	0.0241	0.0075	0.0020	0.0002	0.0001
	6	0.1404	0.0719	0.0231	0.0098	0.0050	0.0210	0.0083	0.0021	0.0002	0.0001
	7	0.1406	0.0733	0.0234	0.0099	0.0051	0.0217	0.0088	0.0022	0.0002	0.0001
	8	0.14065	0.07395	0.02385	0.00995		0.02185	0.00905	0.00235	0.00025	
	1	0.0297	0.0151	0.0009	0.0005	$<10^{-4}$	0.0030	0.0015	$<10^{-4}$	$<10^{-4}$	$<10^{-4}$
	2	0.0371	0.0226	0.0053	0.0015	0.0010	0.0038	0.0023	0.0004	$<10^{-4}$	$<10^{-4}$
	3	0.0390	0.0245	0.0072	0.0026	0.0021	0.0040	0.0025	0.0006	$<10^{-4}$	$<10^{-4}$
	4	0.0391	0.0249	0.0077	0.0034	0.0024	0.0040	0.0025	0.0006	$<10^{-4}$	$<10^{-4}$
	5	0.0396	0.0250	0.0077	0.0035	0.0025	0.0040	0.0025	0.0006	$<10^{-4}$	$<10^{-4}$
	6	0.0396	0.0250	0.0078	0.0035	0.0025	0.0040	0.0025	0.00066	0.00006	0.00004
	7	0.0396	0.0250	0.0078	0.0035	0.0025	0.0040	0.0025	0.0006	0.00005	0.00004
	8	0.0396	0.0251	0.0078	0.0035		0.0040	0.0025	0.00065	0.00005	

mean square error for the input of z of variance σ_z^2 as

$$\left.\begin{aligned} \sigma_t^2 &= \sigma_z^2 \hat{\sigma}_t^2, \qquad & \hat{\sigma}_t^2 &\triangleq [f(k) + c(n, k)] \\ \sigma_q^2 &\triangleq \sigma_z^2 f(k), \qquad & \sigma_c^2 &\triangleq \sigma_z^2 c(n, k) \end{aligned}\right\} \tag{11.114}$$

For a fixed n and $k \le n$, $f(k)$ is a monotonically decreasing function of k, whereas $c(n, k)$ is a monotonically increasing function of k. Hence for every n there is an optimum value of $k = k_0(n)$ for which $\hat{\sigma}_t^2$ is minimized. Let $d(n)$ denote the minimum value of $\hat{\sigma}_t^2$ with respect to k, that is,

$$d(n) = \min_k \{\hat{\sigma}_t^2(n, k)\} \triangleq \hat{\sigma}_t^2(n, k_0(n)) \tag{11.115}$$

Figure 11.37 shows the plot of the distortions $\hat{\sigma}_t^2(n, n)$ and $d(n)$ versus the rate for n-bit PCM transmission of a Gaussian random variable when $p_e = 0.01$. The quantity $\hat{\sigma}_t^2(n, n)$ represents the distortion of the PCM system if no channel error protection is provided and all the bits are used for quantization. It shows, for example, that optimum combination of error protection and quantization could improve the system performance by about 11 dB for an 8-bit transmission.

Channel Error Effects in DPCM

In DPCM the total error in the reconstructed image can be written as

$$\delta u(i, j) = \eta_q(i, j) + \sum_{i'} \sum_{j'} h(i - i', j - j')\eta_c(i', j') \tag{11.116}$$

where η_q is now the DPCM quantizer noise and $h(i, j)$ is the impulse response of the reconstruction filter.

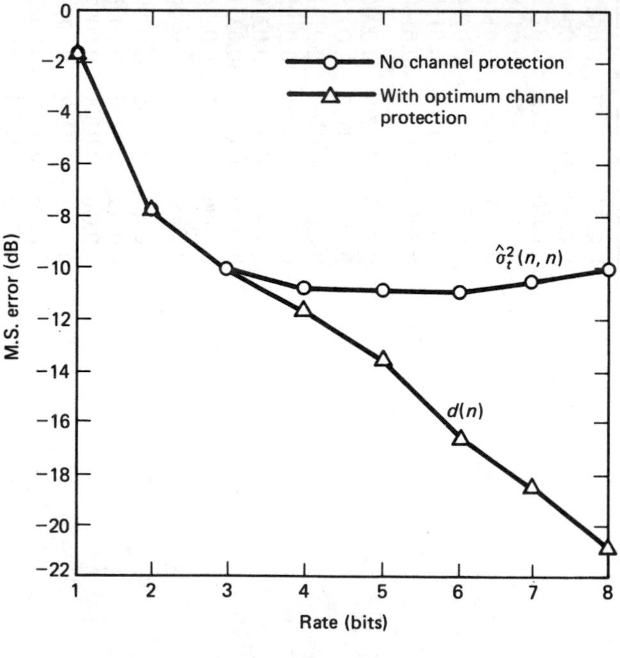

Figure 11.37 Distortion versus rate characteristics of PCM transmission over a binary symmetric channel.

It is essential that the reconstruction filter be stable to prevent the channel errors from accumulating to arbitrarily large values. Even when the predictor models are stable, the channel mean square error gets amplified by a factor σ_u^2/β^2 by the reconstruction filter where β^2 is the theoretical prediction error variance (without quantizer) (Problem 11.5). For the optimum mean square channel decoder the total mean square error in the reconstructed pixel at (i, j) can be written as

$$\sigma_t^2 = \sigma_q^2 + \frac{\sigma_c^2\sigma_u^2}{\beta^2} = \sigma_e^2\left[f(k) + \frac{1}{\hat{\beta}^2}c(n, k)\right] \tag{11.117}$$

where $\hat{\beta}^2 = \beta^2/\sigma_u^2$, and σ_e^2 is the variance of the actual prediction error in the DPCM loop. Recall that high compression is achieved for small values of $\hat{\beta}^2$. Equation (11.117) shows that the higher the compression, the larger is the channel error amplification. Visually, channel noise in DPCM tends to create two-dimensional patterns that originate at the channel error locations and propagate until the reconstruction filter impulse response decays to zero (see Fig. 11.38). In line-by-line DPCM, streaks of erroneous lines appear. In such cases, the erroneous line can be replaced by the previous line or by an average of neighboring lines. A median filter operating orthogonally to the scanning direction can also be effective.

To minimize channel error effects, σ_t^2 given by (11.117) must be minimized to find the quantizer optimum allocation $k = k(n)$ for a given overall rate of n bits per pixel.

Example 11.9

A predictor with $a_1 = 0.848, a_2 = 0.755, a_3 = -0.608, a_4 = 0$ in (11.35a) and $\hat{\beta}^2 = 0.019$ is used for DPCM of images. Assuming a Gaussian distribution for the quantizer input, the optimum pairs $[n, k(n)]$ are found to be:

TABLE 11.9 Optimum Pairs $[n, k(n)]$ for DPCM Transmission

	$p_e = 0.01$					$p_e = 0.001$				
n	1	2	3	4	5	1	2	3	4	5
$k(n)$	1	1	1	1	2	1	2	2	2	3

This shows that if the error rate is high ($p_e = 0.01$), it is better to protect against channel errors than to worry about the quantizer errors. To obtain an optimum pair, we evaluate σ_t^2/σ_e^2 via (11.117) and Tables 11.6, 11.7, and 11.8 for different values of k (for each n). Then the values of k for which this quantity is minimum is found. For a given choice of (n, k), the basis vectors from Table 11.6 can be used to generate the transmission code words.

Optimization of Transform Coding

Suppose a channel error causes a distortion $\delta v(k, l)$ of the (k, l)th transform coefficient. This error manifests itself by spreading in the reconstructed image in proportion to the (k, l)th basis image, as

$$\delta u(i, j) = \delta v(k, l)\phi_k(i)\phi_l(j) \tag{11.118}$$

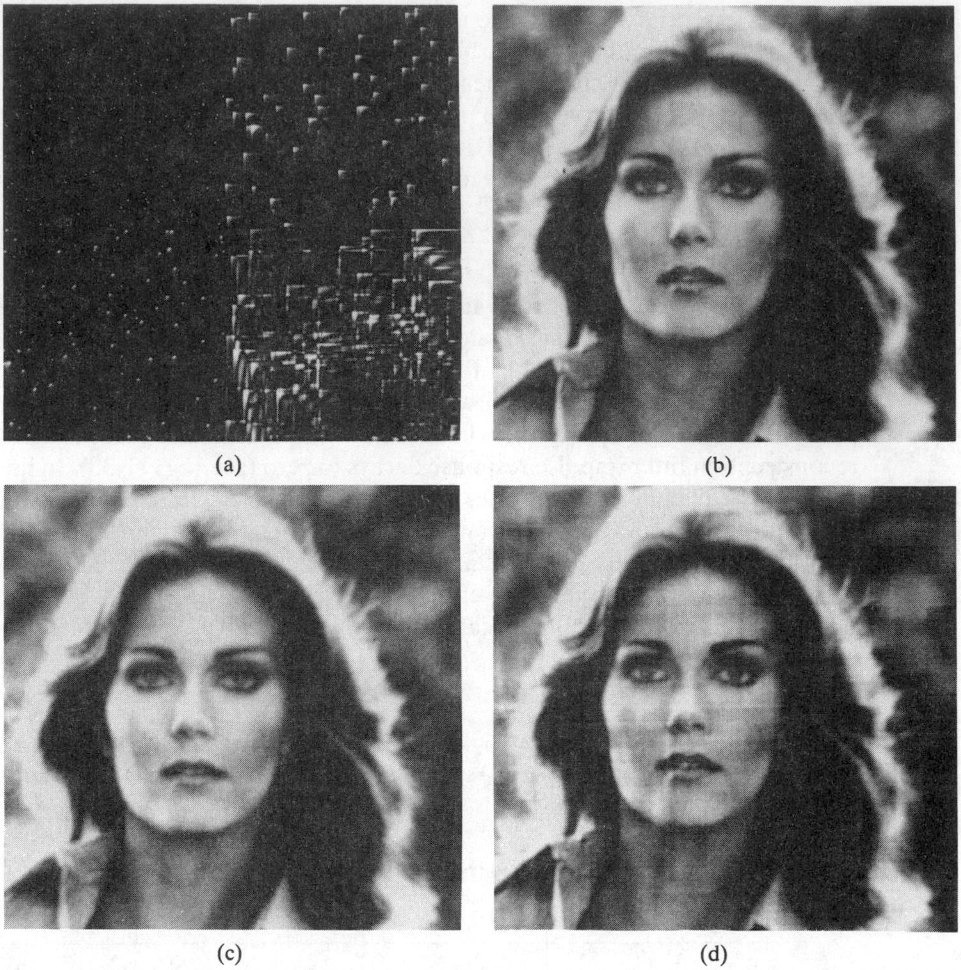

Figure 11.38 Two bits/pixel DPCM coding in the presence of a transmission error rate of 10^{-3}. (a) Propagation of transmission errors for different predictors. Clockwise from top left, error location: optimum, three point, and two point predictors. (b) Optimum linear predictor. (c) Two point predictor $\gamma\frac{(A+D)}{2}$. (d) Three-point predictor $\gamma(A+C-B)$.

This is actually an advantage of transform coding over DPCM because, for the same mean square value, localized errors tend to be more objectionable than distributed errors. The foregoing results can be applied for designing transform coders that protect against channel errors. A transform coder contains several PCM channels, each operating on one transform coefficient. If we represent z_j as the jth transform coefficient with variance σ_j^2, then the average mean square distortion of a transform coding scheme in the presence of channel errors becomes

$$D = \sum_j \sigma_j^2 d(n_j) \tag{11.119}$$

where n_j is the number of bits allocated to the jth PCM channel. The bit allocation algorithm for a transform coder will now use the function $d(n)$, which can be j ⟶ (columns), i ↓ (rows)

6	6	5	4	3	3	2	2	1	1	1	1	1	0	0	0
6	5	4	3	3	2	2	2	1	1	1	1	1	0	0	0
5	4	4	3	3	2	2	2	1	1	1	1	0	0	0	0
4	3	3	3	2	2	2	1	1	1	1	1	0	0	0	0
3	3	3	2	2	2	2	1	1	1	1	1	0	0	0	0
2	2	2	2	2	2	1	1	1	1	1	0	0	0	0	0
2	2	2	2	2	1	1	1	1	1	1	0	0	0	0	0
2	2	2	1	1	1	1	1	1	1	0	0	0	0	0	0
1	1	1	1	1	1	1	1	1	0	0	0	0	0	0	0
1	1	1	1	1	1	1	1	0	0	0	0	0	0	0	0
1	1	1	1	1	1	1	0	0	0	0	0	0	0	0	0
1	1	1	1	1	0	0	0	0	0	0	0	0	0	0	0
1	1	0	0	0	0	0	0	0	0	0	0	0	0	0	0
0	0	0	0	0	0	0	0	0	0	0	0	0	0	0	0
0	0	0	0	0	0	0	0	0	0	0	0	0	0	0	0
0	0	0	0	0	0	0	0	0	0	0	0	0	0	0	0

(a) Bit allocated to transform quantizers, $k_0(i, j)$

j ⟶ (columns), i ↓ (rows)

5	4	3	2	0	0	0	0	0	0	0	0	0	0	0	0
4	3	2	2	0	0	0	0	0	0	0	0	0	0	0	0
3	2	2	0	0	0	0	0	0	0	0	0	0	0	0	0
2	2	0	0	0	0	0	0	0	0	0	0	0	0	0	0
0	0	0	0	0	0	0	0	0	0	0	0	0	0	0	0
0	0	0	0	0	0	0	0	0	0	0	0	0	0	0	0
0	0	0	0	0	0	0	0	0	0	0	0	0	0	0	0
0	0	0	0	0	0	0	0	0	0	0	0	0	0	0	0
0	0	0	0	0	0	0	0	0	0	0	0	0	0	0	0
0	0	0	0	0	0	0	0	0	0	0	0	0	0	0	0
0	0	0	0	0	0	0	0	0	0	0	0	0	0	0	0
0	0	0	0	0	0	0	0	0	0	0	0	0	0	0	0
0	0	0	0	0	0	0	0	0	0	0	0	0	0	0	0
0	0	0	0	0	0	0	0	0	0	0	0	0	0	0	0
0	0	0	0	0	0	0	0	0	0	0	0	0	0	0	0
0	0	0	0	0	0	0	0	0	0	0	0	0	0	0	0

(b) Channel protection bits allocated to quantizer output; $n(i, j) - k_0(i, j)$

Figure 11.39 Bit allocations for quantizers and channel protection in 16×16 block, cosine transform coding of images modeled by the isotropic covariance function. Average bit rate is 1 bit/pixel, and channel error is 1%.

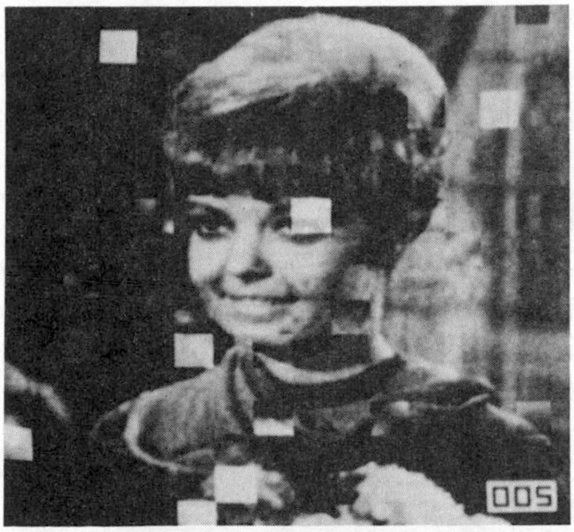

(a) Original image, 8 bits/pixel,

(b) 1 bit/pixel, $p = 0$,

(c) 1 bit/pixel, $p = 10^{-2}$, without channel error protection

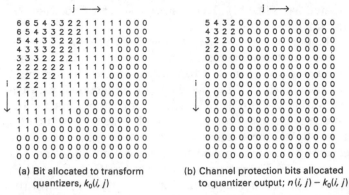

(d) 1 bit/pixel, $p = 10^{-2}$, with channel error protection

Figure 11.40 Transform coding in the presence of channel errors.

evaluated via (11.115). Knowing n_j, we can find $k_j = k(n_j)$, the corresponding optimum number of quantizer bits.

Figure 11.39 shows the bit allocation pattern $k_0(i, j)$ for the quantizers and the allocation of channel protection bits $(n(i, j) - k_0(i, j))$ at an overall average bit rate of 1 bit per pixel for 16×16 block coding of images modeled by the isotropic covariance function. As expected, more protection is provided to samples that have larger variances (and are, therefore, more important for transmission). The overhead due to channel protection, even for the large value of $p_e = 0.01$, is only 15%. For $p_e = 0.001$, the overhead is about 4%. Figure 11.40 shows the results of the preceding technique applied for transform coding of an image in the presence of channel errors. The improvement in SNR is 10 dB at $p = 0.01$ and is also significant visually. This scheme has been found to be quite robust with respect to fluctuations in the channel error rates [6, 39].

11.9 CODING OF TWO-TONE IMAGES

The need for electronic storage and transmission of graphics and two-tone images such as line drawings, letters, newsprint, maps, and other documents has been increasing rapidly, especially with the advent of personal computers and modern telecommunications. Commercial products for document transmission over telephone lines and data lines already exist. The CCITT† has recommended a set of eight documents (Fig. 11.41) for comparison and evaluation of different binary image coding algorithms. The CCITT standard sampling rates for typical A4 ($8\frac{1}{2}$-in. by 11-in.) documents for transmission over the so-called Group 3 digital facsimile apparatus are 3.85 lines per millimeter at normal resolution and 7.7 lines per millimeter at high resolution in the vertical direction. The horizontal sampling rate standard is 1728 pixels per line, which corresponds to 7.7 lines per millimeter resolution or 200 *points per inch* (ppi). For newspaper pages and other documents that contain text as well as halftone images, sampling rates of 400 to 1000 ppi are used. Thus, for the standard $8\frac{1}{2}$-in. by 11-in. page, 1.87×10^6 bits will be required at 200 ppi × 100 lpi sampling density. Transmitting this information over a 4800-bit/s telephone line will take over 6 min. Compression by a factor of, say, 5 can reduce the transmission time to about 1.3 minutes.

Many compression algorithms for binary images exploit the facts that (1) most pixels are white and (2) the black pixels occur with a regularity that manifests itself in the form of characters, symbols, or connected boundaries. There are three basic concepts of coding such images: (1) coding only transition points between black and white, (2) skipping white, and (3) pattern recognition. Figure 11.42 shows a convenient classification of algorithms based on these concepts.

Run-Length Coding

In run-length coding (RLC) the lengths of black and white runs on the scan lines are coded. Since white (1s) and black (0s) runs alternate, the color of the run need not

† Comité Consultatif International de Téléphonie et Télégraphie.

THE SLEREXE COMPANY LIMITED

SAPORS LANE . BOOLE . DORSET . BH 25 8 ER

TELEPHONE BOOLE (945 13) 51617 . TELEX 123456

Our Ref. 350/PJC/EAC 18th January, 1972.

Dr. P.N. Cundall,
Mining Surveys Ltd.,
Holroyd Road,
Reading,
Berks.

Dear Pete,

Permit me to introduce you to the facility of facsimile transmission.

In facsimile a photocell is caused to perform a raster scan over the subject copy. The variations of print density on the document cause the photocell to generate an analogous electrical video signal. This signal is used to modulate a carrier, which is transmitted to a remote destination over a radio or cable communications link.

At the remote terminal, demodulation reconstructs the video signal, which is used to modulate the density of print produced by a printing device. This device is scanning in a raster scan synchronised with that at the transmitting terminal. As a result, a facsimile copy of the subject document is produced.

Probably you have uses for this facility in your organisation.

Yours sincerely,

Phil.

P.J. CROSS
Group Leader - Facsimile Research

Registered in England No. [illegible]
Registered Office [illegible] Vicars Lane, Ilford. Essex.

Document 1

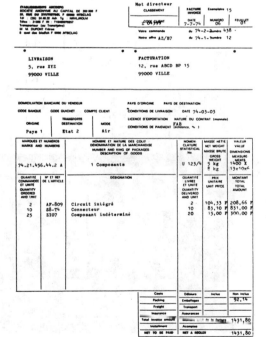

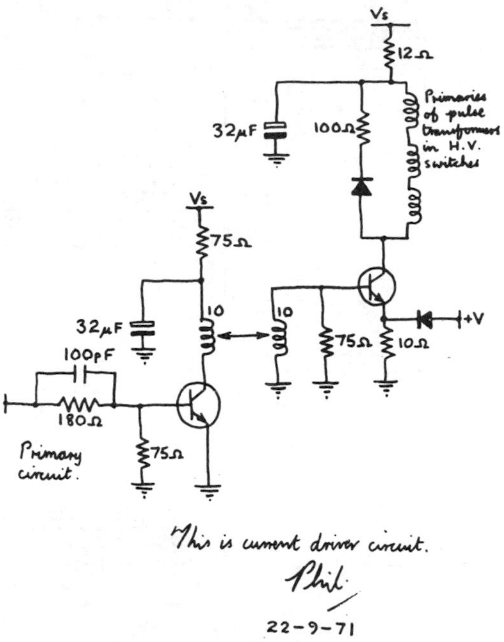

Document 2

ETABLISSEMENTS ABCDEFG
SOCIETE ANONYME AU CAPITAL DE 200 000 F
[illegible]
Tél : (35) 24.40.22 Adr. Tg : [illegible]
Télex : 31586 F [illegible]
Transporteur (ou Transitaire)
M. M. DUPONT Frères
8 quai des [illegible] F 0000 [illegible]

Mot directeur

CLASSEMENT	FACTURE INVOICE	Exemplaire 15	
CODE CLIENT Z 04399	DATE 7-7-74	NUMERO 06	FEUILLET 01

Votre commande du 74-2-2 numéro 438
Notre offre AZ/B7 du 74-1- numéro 12

LIVRAISON
5, rue XYZ
99000 VILLE

FACTURATION
12, rue ABCD BP 15
99000 VILLE

DOMICILIATION BANCAIRE DU VENDEUR
CODE BANQUE CODE GUICHET COMPTE CLIENT

ORIGINE	TRANSPORTS DESTINATION	MODE
Paye 1	Etat 2	Air

PAYS D'ORIGINE PAYS DE DESTINATION
CONDITIONS DE LIVRAISON DATE 74-03-03
LICENCE D'EXPORTATION NATURE DU CONTRAT (monnaie) FAB
CONDITIONS DE PAIEMENT (échéance, %)

MARQUES ET NUMEROS MARKS AND NUMBERS		NOMBRE ET NATURE DES COLIS DÉNOMINATION DE LA MARCHANDISE NUMBER AND KING OF PACKAGES DESCRIPTION OF GOODS	NOMENCLATURE STATISTICAL No	MASSE NETTE NET WEIGHT MASSE BRUTE GROSS WEIGHT	VALEUR VALUE DIMENSIONS MEASUREMENTS
74.21.456.44.2 A		1 Composants	U 123/4	5 kg 8 kg	1400 X 13x10x6
QUANTITÉ COMMANDÉE ET UNITÉ QUANTITY ORDERED AND UNIT	N° ET REF DE L'ARTICLE	DÉSIGNATION	QUANTITÉ LIVRÉE ET UNITÉ QUANTITY DELIVERED AND UNIT	PRIX UNITAIRE UNIT PRICE	MONTANT TOTAL TOTAL AMOUNT
2	AF-809	Circuit intégré	2	104,33 F	208,66 F
10	S8-T4	Connecteur	10	83,10 F	831,00 F
25	ZI07	Composant indéterminé	20	15,00 F	300,00 F

Costs	Débours	Inclus	Non inclus
Packing	Emballages		92,14
Freight	Transport		
Insurance	Assurances		
Total invoice amount	Montant [illegible] de la facture		1431,80
Installment	Acomptes		
NET TO BE PAID	NET A RÉGLER		1431,80

Document 3

\- 34 -

L'ordre de lancement et de réalisation des applications fait l'objet de décisions au plus haut niveau de la Direction Générale des Télécommunications. Il n'est certes pas question de construire ce système intégré "en bloc" mais bien au contraire de procéder par étapes, par paliers successifs. Certaines applications, dont la rentabilité ne pourra être assurée, ne seront pas entreprises. Actuellement, sur trente applications qui ont pu être globalement définies, six en sont au stade de l'exploitation, six autres se sont vu donner la priorité pour leur réalisation.

Chaque application est confiée à un "chef de projet", responsable successivement de sa conception, de son analyse-programmation et de sa mise en oeuvre dans une région-pilote. La généralisation ultérieure de l'application réalisée dans cette région-pilote dépend des résultats obtenus et fait l'objet d'une décision de la Direction Générale. Néanmoins, le chef de projet doit dès le départ considérer que son activité a une vocation nationale donc refuser tout particularisme régional. Il est aidé d'une équipe d'analystes-programmeurs et entouré d'un "groupe de conception" chargé de rédiger le document de "définition des objectifs globaux" puis le "cahier des charges" de l'application, qui sont adressés pour avis à tous les services utilisateurs potentiels et aux chefs de projet des autres applications. Le groupe de conception comprend 6 à 10 personnes représentant les services les plus divers concernés par le projet, et comporte obligatoirement un bon analyste attaché à l'application.

II - L'IMPLANTATION GEOGRAPHIQUE D'UN RESEAU INFORMATIQUE PERFORMANT

L'organisation de l'entreprise française des télécommunications repose sur l'existence de 20 régions. Des calculateurs ont été implantés dans le passé au moins dans toutes les plus importantes. On trouve ainsi des machines Bull Gamma 30 à Lyon et Marseille, des GE 425 à Lille, Bordeaux, Toulouse et Montpellier, un GE 437 à Massy, enfin quelques machines Bull 300 TI à programmes câblés étaient récemment ou sont encore en service dans les régions de Nancy, Nantes, Limoges, Poitiers et Rouen ; ce parc est essentiellement utilisé pour la comptabilité téléphonique.

A l'avenir, si la plupart des fichiers nécessaires aux applications décrites plus haut peuvent être gérés en temps différé, un certain nombre d'entre eux devront nécessairement être accessibles, voire mis à jour en temps réel : parmi ces derniers le fichier commercial des abonnés, le fichier des renseignements, le fichier des circuits, le fichier technique des abonnés contiendront des quantités considérables d'informations.

Le volume total de caractères à gérer en phase finale sur un ordinateur ayant en charge quelques 500 000 abonnés a été estimé à un milliard de caractères au moins. Au moins le tiers des données seront concernées par des traitements en temps réel.

Aucun des calculateurs énumérés plus haut ne permettait d'envisager de tels traitements.

L'intégration progressive de toutes les applications suppose la création d'un support commun pour toutes les informations, une véritable "Banque de données", répartie sur des moyens de traitement nationaux et régionaux, et qui devra rester alimentée, mise à jour en permanence, à partir de la base de l'entreprise, c'est-à-dire les chantiers, les magasins, les guichets des services d'abonnement, les services de personnel etc.

L'étude des différents fichiers à constituer a donc permis de définir les principales caractéristiques du réseau d'ordinateurs nouveaux à mettre en place pour aborder la réalisation du système informatif. L'obligation de faire appel à des ordinateurs de troisième génération, très puissants et dotés de volumineuses mémoires de masse, a conduit à en réduire substantiellement le nombre.

L'implantation de sept centres de calcul interrégionaux constituera un compromis entre d'une part le désir de réduire le coût économique de l'ensemble, de faciliter la coordination des équipes d'informaticiens; et d'autre part le refus de créer des centres trop importants difficiles à gérer et à diriger, et posant des problèmes délicats de sécurité. Le regroupement des traitements relatifs à plusieurs régions sur chacun de ces sept centres permettra de leur donner une taille relativement homogène. Chaque centre "gérera" environ un million d'abonnés à la fin du VIème Plan.

La mise en place de ces centres a débuté au début de l'année 1971 : un ordinateur IRIS 50 de la Compagnie Internationale pour l'Informatique a été installé à Toulouse en février ; la même machine vient d'être mise en service au centre de calcul interrégional de Bordeaux.

Photo n° 1 - Document très dense lettre 1,5mm de haut -
Restitution photo n° 9

Document 4

Figure 11.41 CCITT test documents.

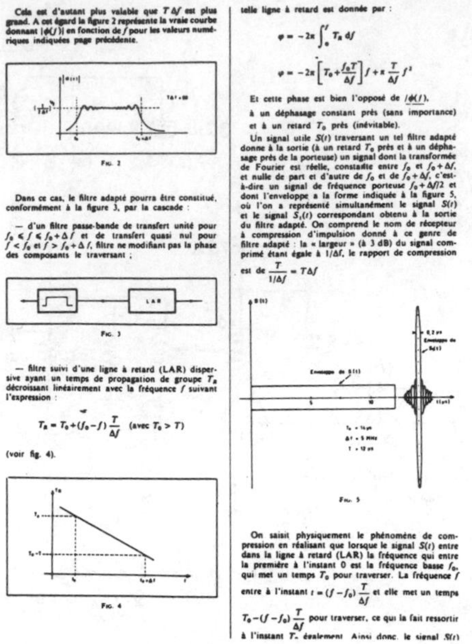

Cela est d'autant plus valable que $T\Delta f$ est plus grand. A cet égard la figure 2 représente la vraie courbe donnant $|\phi(f)|$ en fonction de f pour les valeurs numériques indiquées page précédente.

Fig. 2

Dans ce cas, le filtre adapté pourra être constitué, conformément à la figure 3, par la cascade :

— d'un filtre passe-bande de transfert unité pour $f_0 \leq f \leq f_0 + \Delta f$ et de transfert quasi nul pour $f < f_0$ et $f > f_0 + \Delta f$, filtre ne modifiant pas la phase des composants le traversant ;

Fig. 3

— filtre suivi d'une ligne à retard (LAR) dispersive ayant un temps de propagation de groupe T_R décroissant linéairement avec la fréquence f suivant l'expression :

$$T_R = T_0 + (f_0 - f)\frac{T}{\Delta f} \quad (\text{avec } T_0 > T)$$

(voir fig. 4).

Fig. 4

telle ligne à retard est donnée par :

$$\varphi = -2\pi \int_0^f T_R \, df$$

$$\varphi = -2\pi \left[T_0 + \frac{f_0 T}{\Delta f}\right] f + \pi \frac{T}{\Delta f} f^2$$

Et cette phase est bien l'opposé de $/\phi(f)$,

à un déphasage constant près (sans importance) et à un retard T_0 près (inévitable).

Un signal utile $S(t)$ traversant un tel filtre adapté donne à la sortie (à un retard T_0 près et à un déphasage près de la porteuse) un signal dont la transformée de Fourier est réelle, constante entre f_0 et $f_0 + \Delta f$, et nulle de part et d'autre de f_0 et de $f_0 + \Delta f$, c'est-à-dire un signal de fréquence porteuse $f_0 + \Delta f/2$ et dont l'enveloppe a la forme indiquée à la figure 5, où l'on a représenté simultanément le signal $S(t)$ et le signal $S_1(t)$ correspondant obtenu à la sortie du filtre adapté. On comprend le nom de récepteur à compression d'impulsion donné à ce genre de filtre adapté : la « largeur » (à 3 dB) du signal comprimé étant égale à $1/\Delta f$, le rapport de compression est de $\frac{T}{1/\Delta f} = T\Delta f$

Fig. 5

On saisit physiquement le phénomène de compression en réalisant que lorsque le signal $S(t)$ entre dans la ligne à retard (LAR) la fréquence qui entre la première à l'instant 0 est la fréquence basse f_0, qui met un temps T_0 pour traverser. La fréquence f entre à l'instant $t = (f - f_0)\frac{T}{\Delta f}$ et elle met un temps $T_0 - (f - f_0)\frac{T}{\Delta f}$ pour traverser, ce qui la fait ressortir à l'instant T_0 également. Ainsi donc, le signal $S(t)$

Document 5

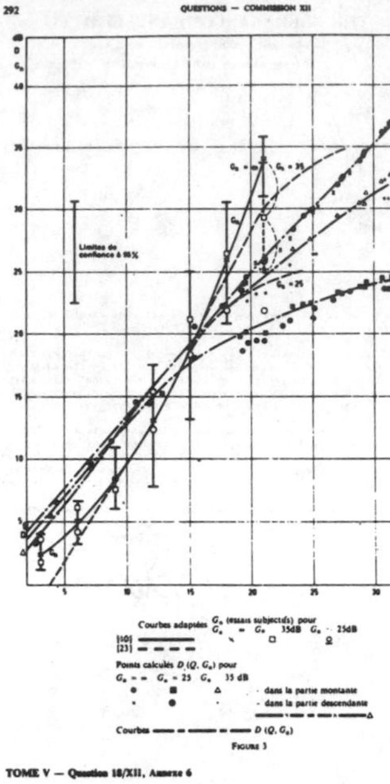

Document 6

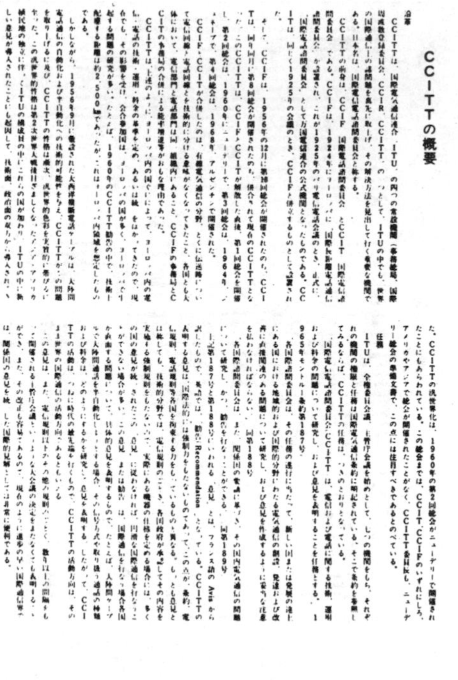

Document 7

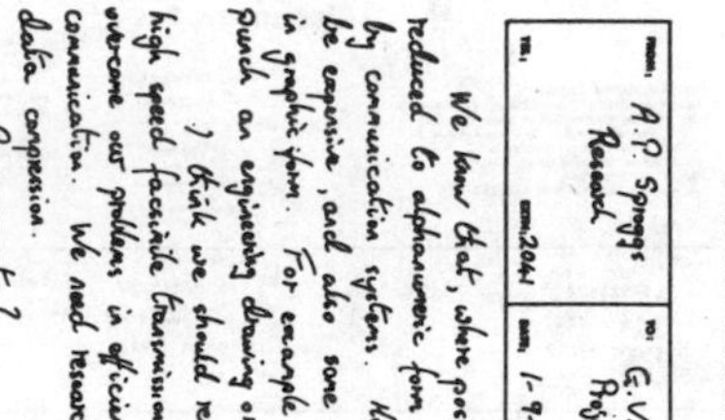

Document 8

Figure 11.41 (*Cont'd*)

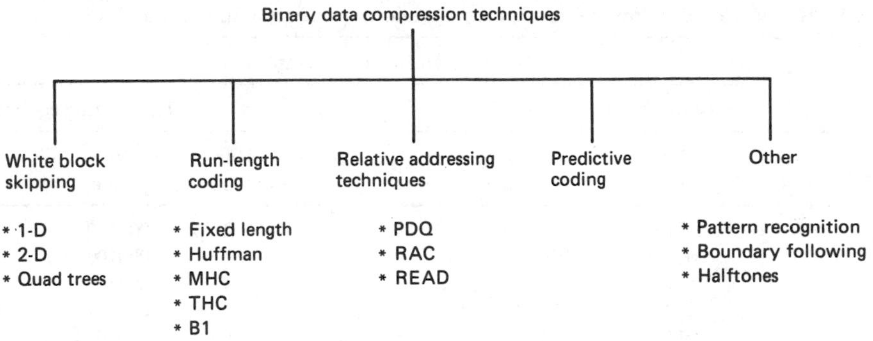

Figure 11.42 Binary data compression techniques.

be coded (Fig. 11.43). The first run is always a white run with length zero, if necessary. The run lengths can be coded by fixed-length m-bit code words, each representing a block of maximum run length $M - 1$, $M \triangleq 2^m$, where M can be optimized to maximize compression (see Section 11.2 and Problem 11.17).

A more-efficient technique is to use Huffman coding. To avoid a large code book, the truncated and the modified Huffman codes are used. The truncated Huffman code assigns separate Huffman code words for white and black runs up to lengths L_w and L_b, respectively. Typical values of these runs have been found to be $L_w = 47$, $L_b = 15$ [1c, p. 1426]. Longer runs, which have lower probabilities, are assigned by a fixed-length code word, which consists of a prefix code plus an 11-bit binary code of the run length.

The modified Huffman code, which has been recommended by the CCITT as a one-dimensional standard code for Group 3 facsimile transmission, uses $L_w = L_b = 63$. Run lengths smaller than 64 are Huffman coded to give the *terminator* code. The remaining runs are assigned two code words, consisting of a *make-up* code and a terminator code. Table 11.10 gives the codes for the Group 3 standard. It also gives the *end-of-line* (EOL) code and an extended code table for larger paper widths up to A3 in size, which require up to 2560 pixels per line.

Other forms of variable-length coding that simplify the coding-decoding procedures are *algorithm based.* Noteworthy among these codes are the A_N and B_N codes [1c, p. 1406]. The A_N codes, also called L_N codes, are multiple fixed-length codes that are nearly optimal for exponentially distributed run lengths. They belong to a class of linear codes whose length increases approximately linearly with the number of messages. If l_k, $k = 1, 2, 3, \ldots$, are the run lengths, then the A_N code of block size N is obtained by writing $k = q(2^N - 1) + r$, where $1 \le r \le 2^N - 1$ and q is a

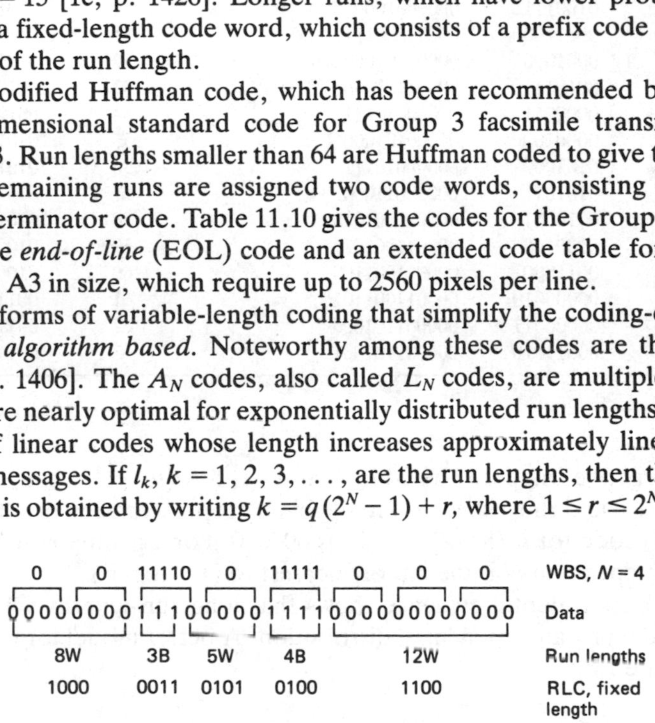

Figure 11.43 Run-length coding and white block skipping.

TABLE 11.10 Modified Huffman Code Tables for One-dimensional Run-length Coding

Modified Huffman code table					
Terminating code words			Terminating code words		
Run length	White runs	Black runs	Run length	White runs	Black runs
0	00110101	0000110111	32	00011011	000001101010
1	000111	010	33	00010010	000001101011
2	0111	11	34	00010011	000011010010
3	1000	10	35	00010100	000011010011
4	1011	011	36	00010101	000011010100
5	1100	0011	37	00010110	000011010101
6	1110	0010	38	00010111	000011010110
7	1111	00011	39	00101000	000011010111
8	10011	000101	40	00101001	000001101100
9	10100	000100	41	00101010	000001101101
10	00111	0000100	42	00101011	000011011010
11	01000	0000101	43	00101100	000011011011
12	001000	0000111	44	00101101	000001010100
13	000011	00000100	45	00000100	000001010101
14	110100	00000111	46	00000101	000001010110
15	110101	000011000	47	00001010	000001010111
16	101010	0000010111	48	00001011	000001100100
17	101011	0000011000	49	01010010	000001100101
18	0100111	0000001000	50	01010011	000001010010
19	0001100	00001100111	51	01010100	000001010011
20	0001000	00001101000	52	01010101	000000100100
21	0010111	00001101100	53	00100100	000000110111
22	0000011	00000110111	54	00100101	000000111000
23	0000100	00000101000	55	01011000	000000100111
24	0101000	00000010111	56	01011001	000000101000
25	0101011	00000011000	57	01011010	000001011000
26	0010011	000011001010	58	01011011	000001011001
27	0100100	000011001011	59	01001010	000000101011
28	0011000	000011001100	60	01001011	000000101100
29	00000010	000011001101	61	00110010	000001011010
30	00000011	000001101000	62	00110011	000001100110
31	00011010	000001101001	63	00110100	000001100111

nonnegative integer. The codeword for l_k has $(q+1)N$ bits, of which the first qN bits are 0 and the last N bits are the binary representation of r. For example if $N = 2$, the A_2 code for l_8 $(8 = 2 \times 3 + 2)$ is 000010. For a geometric distribution with mean μ, the optimum N is the integer nearest to $(1 + \log_2 \mu)$.

Experimental evidence shows that long run lengths are more common than predicted by an exponential distribution. A better model for run-length distribution is of the form

$$P(l) = \frac{c}{l^{\alpha}}, \qquad \alpha > 0, c = \text{constant} \tag{11.120}$$

TABLE 11.10 (Continued)

Modified Huffman code table

Make-up code words

Run length	White runs	Black runs	Run length	White runs	Black runs
64	11011	0000001111	960	011010100	0000001110011
128	10010	000011001000	1024	011010101	0000001110100
192	010111	000011001001	1088	011010110	0000001110101
256	0110111	000001011011	1152	011010111	0000001110110
320	00110110	000000110011	1216	011011000	0000001110111
384	00110111	000000110100	1280	011011001	0000001010010
448	01100100	000000110101	1344	011011010	0000001010011
512	01100101	0000001101100	1408	011011011	0000001010100
576	01101000	0000001101101	1472	010011000	0000001010101
640	01100111	0000001001010	1536	010011001	0000001011010
704	011001100	0000001001011	1600	010011010	0000001011011
768	011001101	0000001001100	1664	011000	0000001100100
832	011010010	0000001001101	1728	010011011	0000001100101
896	011010011	0000001110010	EOL	000000000001	000000000001

Extended Modified Huffman code table

Run length (Black or White)	Make-up code word
1792	00000001000
1856	00000001100
1920	00000001101
1984	000000010010
2048	000000010011
2112	000000010100
2176	000000010101
2240	000000010110
2304	000000010111
2368	000000011100
2432	000000011101
2496	000000011110
2560	000000011111

which decreases less rapidly with l than the exponential distribution.

The B_N codes, also called H_N codes, are also multiples of fixed-length codes. Word length of B_N increases roughly as the logarithm of N. The fixed block length for a B_N code is $N + 1$ bits. It is constructed by listing all possible N-bit words, followed by all possible $2N$-bit words, and then $3N$-bit words, and so on. An additional bit is inserted after every block of N bits. The inserted bit is 0 except for the bit inserted after the last block, which is 1.

Table 11.11 shows the construction of the B_1 code, which has been found useful for RLC.

TABLE 11.11 Construction of B_1 Code. Inserted bits are underscored.

Run length	kN-bit words	B_1 Code
1	0	$0\underline{1}$
2	1	$1\underline{1}$
3	0 0	$0\underline{0}0\underline{1}$
4	0 1	$0\underline{0}1\underline{1}$
5	1 0	$1\underline{0}0\underline{1}$
6	1 1	$1\underline{0}1\underline{1}$
7	000	$0\underline{0}0\underline{0}0\underline{1}$
⋮	⋮	⋮

Example 11.10

Table 11.12 lists the averages μ_w and μ_b and the entropies H_w and H_b of white and black run lengths for CCITT documents. From this data an upper bound on achievable compression can be obtained as

$$C_{\max} = \frac{\mu_w + \mu_b}{H_w + H_b} \tag{11.121}$$

which is also listed in the table. These results show compression factors of 5 to 20 are achievable by RLC techniques.

White Block Skipping [1c, p. 1406, 44]

White block skipping (WBS) is a very simple but effective compression algorithm. Each scan line is divided into blocks of N pixels. If the block contains all white pixels, it is coded by a 0. Otherwise, the code word has $N + 1$ bits, the first bit being 1, followed by the binary pattern of the block (Fig. 11.43). The bit rate for this method is

$$\begin{aligned} R_N &= \frac{(1 - p_N)(N + 1) + p_N}{N} \\ &= \left(1 - p_N + \frac{1}{N}\right) \quad \text{bits/pixel} \end{aligned} \tag{11.122}$$

TABLE 11.12 Run-Length Measurements for CCITT Documents

Document number	μ_w	μ_b	H_w	H_b	$C_{\max}$
1	156.3	6.8	5.5	3.6	18.0
2	257.1	14.3	8.2	4.5	21.4
3	89.8	8.5	5.7	3.6	10.6
4	39.0	5.7	4.7	3.1	5.7
5	79.2	7.0	5.7	3.3	9.5
6	138.5	8.0	6.2	3.6	14.9
7	45.3	4.4	5.9	3.1	5.6
8	85.7	70.9	6.9	5.8	12.4

where p_N is the probability that a block contains all white pixels. This rate depends on the block size N. The value $N \simeq 10$ has been found to be suitable for a large range of images. Note that WBS is a simple form of the truncated Huffman code and should work well especially when an image contains large white areas.

An adaptive WBS scheme improves the performance significantly by coding all white scan lines separately. A 0 is assigned to an all-white scan line. If a line contains at least one black pixel, a 1 precedes the regular WBS code for that line. The WBS method can also be extended to two dimensions by considering $M \times N$ blocks of pixels. An all-white block is coded by a 0. Other blocks are coded by $(MN + 1)$ bits, whose first bit is 1 followed by the block bit pattern. An adaptive WBS scheme that uses variable block size proceeds as follows. If the initial block contains all white pixels, it is represented by 0. Otherwise, a prefix of 1 is assigned and the block is subdivided into several subblocks, each of which is then treated similarly. The process continues until an elementary block is reached, which is coded by the regular WBS method. Note that this method is very similar to generating the quad-tree code for a region (see Section 9.7).

Prediction Differential Quantization [1c, p. 1418]

Prediction differential quantization (PDQ) is an extension of run-length coding, where the correlation between scan lines is exploited. The method basically encodes the overlap information of a black run in successive scan lines. This is done by coding differences Δ' and Δ'' in black runs from line to line together with messages, new start (NS) when black runs start, and merge (M) when there is no further overlap of that run. A new start is coded by a special code word, whereas a merge is represented by coding white and black runlengths r_w, r_b, as shown in Fig. 11.44.

Relative Address Coding [1f, p. 834]

Relative address coding (RAC) uses the same principle as the PDQ method and computes run-length differences by tracking either the last transition on the same line or the nearest transition on the previous line. For example, the transition pixel Q (Fig. 11.45) is encoded by the shortest distance PQ or $Q'Q$, where P is the preceding transition element on the current line and Q' is the nearest transition element to the right of P on the previous line, whose direction of transition is the same as that of Q. If P does not exist, then it is considered to be the imaginary pixel to the right of the last pixel on the preceding line. The distance QQ' is coded as $+N$

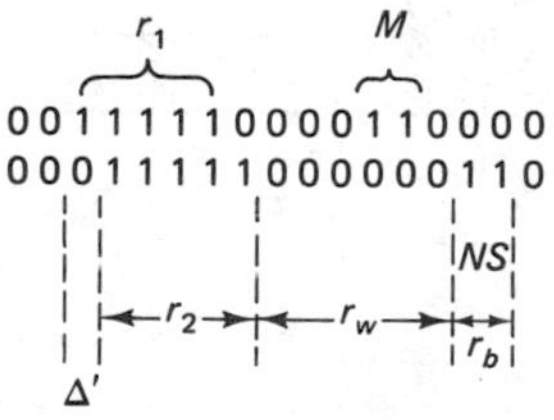

Figure 11.44 The PDQ method: $\Delta'' = r_2 - r_1$.

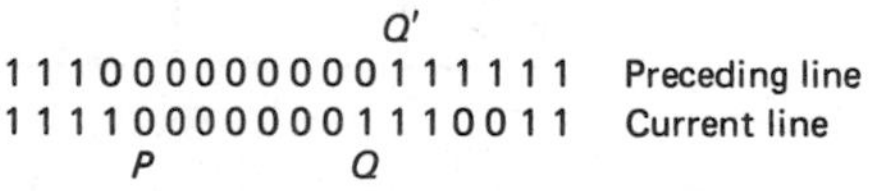

Figure 11.45 RAC method. $PQ = 1$, $QQ' = -1$, RAC distance $= -1$

TABLE 11.13 Relative Address Codes. $x \ldots x$ = binary representation of N.

Distance	Code	N	$F(N)$
+0	0	1–4	0xx
+1	100	5–20	10$xxxx$
−1	101	21–84	110$xxxxxx$
$N\ (N>1)$	111 $F(N)$	85–340	1110$xxxxxxxx$
$+N\ (N>2)$	1100 $F(N)$	341–1364	11110$xxxxxxxxxx$
$-N\ (N>2)$	1101 $F(N)$	1365–5460	111110$xxxxxxxxxxxx$
		⋮	⋮

if Q' is $N(\geq 0)$ pixels to the left or $-N$ if Q' is $N(\geq 1)$ pixels to the right of Q on the preceding line. Distance PQ is coded as $N(\geq 1)$ if it is N pixels away. The RAC distances are coded by a code similar to the B_1 code, except for the choice of the reference line and for very short distances, +0, +1, −1 (see Table 11.13).

CCITT Modified Relative Element Address Designate Coding

The modified relative element address designate (READ) algorithm has been recommended by CCITT for two-dimensional coding of documents. It is a modification of the RAC and other similar codes [1f, p. 854]. Referring to Fig. 11.46 we define a_0 as the reference transition element whose position is defined by the previous coding mode (to be discussed shortly). Initially, a_0 is taken to be the imaginary white transition pixel situated to the left of the first pixel on the coding

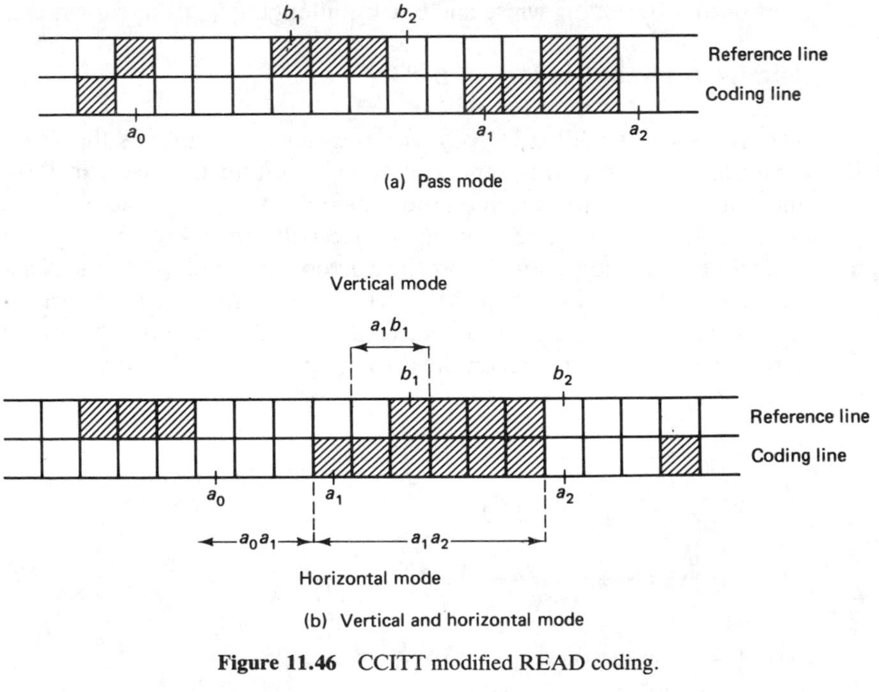

Figure 11.46 CCITT modified READ coding.

line. The next pair of transition pixels to the right of a_0 are labeled a_1 and a_2 on the coding line and b_1 and b_2 on the reference line and have alternating colors to a_0. Any of the elements a_1, a_2, b_1, b_2 not detected for a particular coding line is taken as the imaginary pixel to the right of the last element of its respective scan line. Pixel a_1 represents the next transition element to be coded. The algorithm has three modes of coding, as follows.

Pass mode. b_2 is to the left of a_1 (Fig. 11.46a). This identifies the white or black runs on the reference line that do not overlap with the corresponding white or black runs on the coding line. The reference element a_0 is set below b_2 in preparation for the next coding.

Vertical mode. a_1 is coded relative to b_1 by the distance $a_1 b_1$, which is allowed to take the values $0, 1, 2, 3$ to the right or left of b_1. These are represented by $V(0)$, $V_R(x)$, $V_L(x)$, $x = 1, 2, 3$ (Table 11.14). In this mode a_0 is set at a_1 in preparation for the next coding.

Horizontal mode. If $|a_1 b_1| > 3$, the vertical mode is not used and the run lengths $a_0 a_1$ and $a_1 a_2$ are coded using the modified Huffman codes of Table 11.10. After coding, the new position of a_0 is set at a_2. If this mode is needed for the first element on the coding line, then the value $a_0 a_1 - 1$ rather than $a_0 a_1$ is coded. Thus if the first element is black, then a run length of zero is coded.

TABLE 11.14 CCITT Modified READ Code Table [1f, p. 865]

Mode	Elements to be coded		Notation	Code Word
Pass	b_1, b_2		P	0001
Horizontal	a_0a_1, a_1a_2		H	$001 + M(a_0a_1) + M(a_1a_2)$
Vertical	a_1 just under b_1	$a_1b_1 = 0$	$V(0)$	1
	a_1 to the right of b_1	$a_1b_1 = 1$	$V_R(1)$	011
		$a_1b_1 = 2$	$V_R(2)$	000011
		$a_1b_1 = 3$	$V_R(3)$	0000011
	a_1 to the left of b_1	$a_1b_1 = 1$	$V_L(1)$	010
		$a_1b_1 = 2$	$V_L(2)$	000010
		$a_1b_1 = 3$	$V_L(3)$	0000010
	2-D extensions			0000001*xxx*
	1-D extensions			000000001*xxx*
	End-of-line (EOL) code word			000000000001
	1-D coding of next line			EOL + '1'
	2-D coding of next line			EOL + '0'

$M(a_0 a_1)$ and $M(a_1 a_2)$ are code words taken from the modified Huffman code tables given in Table 11.10. The bit assignment for the *xxx* bits is 111 for the uncompressed mode.

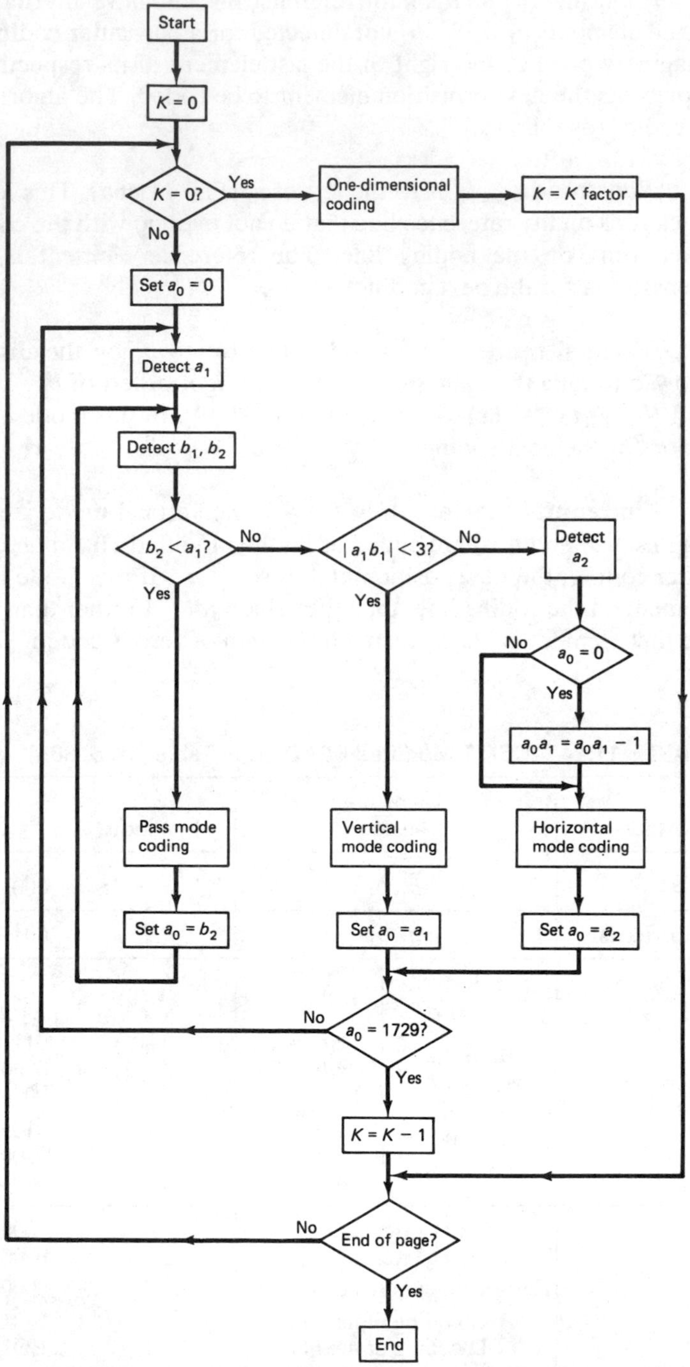

Figure 11.47 CCITT modified READ coding algorithm.

The coding procedure along a line continues until the imaginary transition element to the right of the last actual element on the line has been detected. In this way exactly 1728 pixels are coded on each line. Figure 11.47 shows the flow diagram for the algorithm. Here K is called the K-factor, which means that after a one-dimensionally coded line, no more than $K-1$ successive lines are two-dimensionally coded. CCITT recommended values for K are 2 and 4 for documents scanned at normal resolution and high resolution, respectively. The K-factor is used to minimize the effect of channel noise on decoded images. The one-dimensional and two-dimensional extension code words listed in Table 11.14, with *xxx* equal to 111, are used to allow the coder to enter *the uncompressed mode,* which may be desired when the run lengths are very small or random, such as in areas of halftone images or cross hatchings present in some business forms.

Predictive Coding

The principles of predictive coding can be easily applied to binary images. The main difference is that the prediction error is also a binary variable, so that a quantizer is not needed. If the original data has redundancy, then the prediction error sequence will have large runs of 0s (or 1s). For a binary image $u(m, n)$, let $\bar{u}(m, n)$ denote its predicted value based on the values of pixels in a prediction window W, which contains some of the previously coded pixels. The prediction error is defined as

$$\begin{aligned} e(m, n) &= \begin{cases} 1, & \bar{u}(m, n) \neq u(m, n) \\ 0, & \bar{u}(m, n) = u(m, n) \end{cases} \\ &= u(m, n) \oplus \bar{u}(m, n) \end{aligned} \tag{11.123}$$

The sequence $e(m, n)$ can be coded by a run-length or entropy coding method. The image is reconstructed from $e(m, n)$ simply as

$$u(m, n) = \bar{u}(m, n) \oplus e(m, n) \tag{11.124}$$

Note that this is an errorless predictive coding method. An example of a prediction window W for a raster scanned image is shown in Fig. 11.48.

A reasonable prediction criterion is to minimize the prediction error probability. For an N-element prediction window, there are 2^N different states. Let S_k, $k = 1, 2, \dots, 2^N$ denote the kth state of W with probability p_k and define

$$q_k = \text{Prob}[u(m, n) = 1 | S_k] \tag{11.125}$$

Then the optimum prediction rule having minimum prediction error probability is

$$\bar{u}(m, n) = \begin{cases} 1, & \text{if } q_k \geq 0.5 \\ 0, & \text{if } q_k < 0.5 \end{cases} \tag{11.126}$$

If the random sequence $u(m, n)$ is strict-sense stationary, then the various probabilities will remain constant at every (m, n), and therefore the prediction rule stays the same. In practice a suitable choice of N has to be made to achieve a trade-off between prediction error probability and the complexity of the predictor

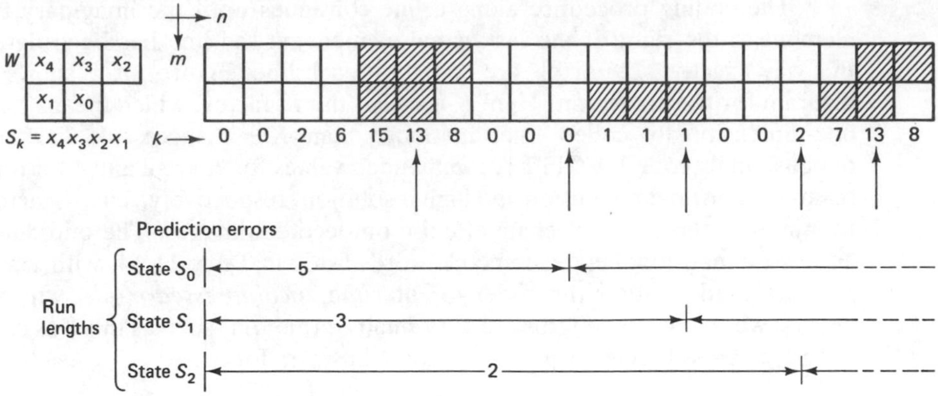

Figure 11.48 TUH method of predictive coding. The run length l_k of state S_k means a prediction error has occurred after state S_k has repeated l_k times.

due to large values of N. Experimentally, 4 to 7 pixel predictors have been found to be adequate. Corresponding to the prediction rule of (11.126), the minimized prediction error is

$$p_e = \sum_{k=1}^{2^N} p_k \min(q_k, 1 - q_k) \tag{11.127}$$

If the random sequence $u(m, n)$ is Markovian with respect to the prediction window W, then the run lengths for each state S_k are independent. Hence, the prediction-error run lengths for each state (Fig. 11.47) can be coded by the truncated Huffman code, for example. This method has been called the Technical University of Hannover (TUH) code [1c, p. 1425].

Adaptive Predictors

Adaptive predictors are useful in practice because the image data is generally nonstationary. In general, any pattern classifier or a discriminant function could be used as a predictor. A simple classifier is a *linear learning machine* or *adaptive threshold logic unit* (TLU), which calculates the threshold q_k as a linear functional of the states of the pixels in the prediction window. Another type of pattern classifier is a network of TLUs called layered machines and includes piecewise linear discriminant functions and the so-called α-perceptron. A practical adaptive predictor uses a counter C_k of L bits for each state [43]. The counter runs from 0 to $2^L - 1$. The adaptive prediction rule is

$$\bar{u}(m, n) = \begin{cases} 1, & \text{if } C_k \geq 2^{L-1} \\ 0, & \text{if } C_k < 2^{L-1} \end{cases} \tag{11.128}$$

After prediction of a pixel has been performed, the counter is updated as

$$C_k = \begin{cases} \min(C_k + 1, 2^L - 1), & \text{if } u(m, n) = 1 \\ \max(C_k - 1, 0), & \text{otherwise} \end{cases} \tag{11.129}$$

TABLE 11.15 Compression Ratios of Different Binary Coding Algorithms

	One-dimensional codes			Two-dimensional codes					
	Normal or high resolution			Normal resolution			High resolution		
Document	B_1-code	Truncated Huffman code	Modified Huffman code	RAC code $K = 4$	TUH code $K = 4$	CCITT READ code, $K = 2$	RAC code $K = 4$	TUH code $K = 4$	CCITT READ code, $K = 4$
1	13.62	17.28	16.53	16.67	20.32	15.71	19.68	24.66	19.77
2	14.45	15.05	16.34	24.66	24.94	19.21	28.67	28.99	26.12
3	8.00	8.88	9.42	10.42	11.57	9.89	12.06	14.96	12.58
4	4.81	5.74	5.76	4.52	5.94	5.02	5.42	7.52	6.27
5	7.67	8.63	9.15	9.39	11.45	9.07	10.97	13.82	11.63
6	9.78	10.14	10.98	15.41	17.26	13.63	17.59	19.96	18.18
7	4.60	4.69	5.20	4.56	5.28	5.10	5.36	6.62	6.30
8	8.54	7.28	8.70	12.85	13.10	11.13	14.73	15.03	15.55
Average	8.93	9.71	10.26	12.31	13.71	11.10	14.31	16.45	14.55

The value $L = 3$ has been found to yield minimum prediction error for a typical printed page.

Comparison of Algorithms

Table 11.15 shows a comparison of the compression ratios achievable by different algorithms. The compression ratios for one-dimensional codes are independent of the vertical resolution. At normal resolution the two-dimensional codes improve the compression by only 10 to 30% over the modified Huffman code. At high resolution the improvements are 40 to 60% and are significant enough to warrant the use of these algorithms. Among the two-dimensional codes, the TUH predictive code is superior to the relative address techniques, especially for text information. However, the latter are simpler to code. The CCITT READ code, which is a modification of the RAC, performs somewhat better.

Other Methods

Algorithms that utilize higher-level information, such as whether the image contains a known type (or font) of characters or graphics, line drawings, and the like, can be designed to obtain very high-compression ratios. For example, in the case of printed text limited to the 128 ASCII characters, for instance, each character can be coded by 7 bits. The coding technique would require a character recognition algorithm. Likewise, line drawings can be efficiently coded by boundary-following algorithms, such as chain codes, line segments, or splines. Algorithms discussed here are not directly useful for halftone images because the image area has been modulated by pseudorandom noise and thresholded thereafter. In all these cases special preprocessing and segmentation is required to code the data efficiently.

11.10 COLOR AND MULTISPECTRAL IMAGE CODING

Data compression techniques discussed so far can be generalized to color and multispectral images, as shown in Fig. 11.49. Each pixel is represented by a $p \times 1$

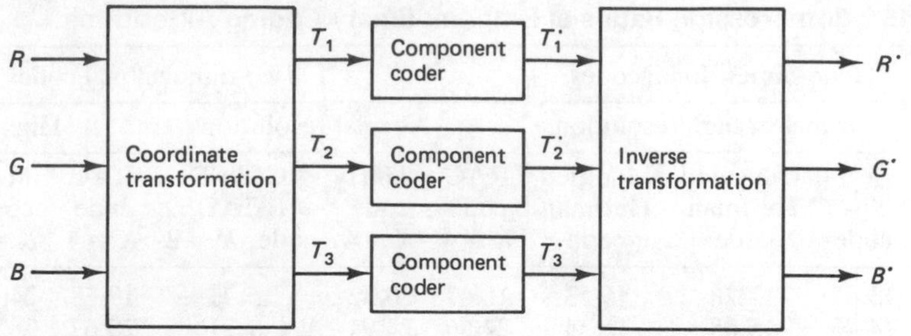

Figure 11.49 Component coding of color images. For multispectral images the input vector has a dimension greater than or equal to 2.

vector. For example, in the case of color, the input is a 3×1 vector containing the R, G, B components. This vector is transformed to another coordinate system, where each component can be processed by an independent spatial coder.

In coding color images, consideration should be given to the facts that (1) the luminance component (Y) has higher bandwidth than the chrominance components (I, Q) or (U, V) and (2) the color-difference metric is non-Euclidean in these coordinates, that is, equal noise power in different color components is perceived differently. In practical image coding schemes, the lower-bandwidth chrominance signals are sampled at correspondingly lower rates. Typically, the I and Q signals are sampled at one-third and one-sixth of the sampling rate of the luminance signal. Use of color-distance metric(s) is possible but has not been used in practical systems primarily because of the complexity of the color vision model (see Chapter 3).

An alternate method of coding color images is by processing the composite color signal. This is useful in broadcast applications, where it is desired to manage only one signal. However, since the luminance and color signals are not in the same frequency band, the foregoing monochrome image coding techniques are not very efficient if applied directly. Typically, the composite signal is sampled at $3f_{sc}$ (the lowest integer multiple of subcarrier frequency above the Nyquist rate) or $4f_{sc}$, and

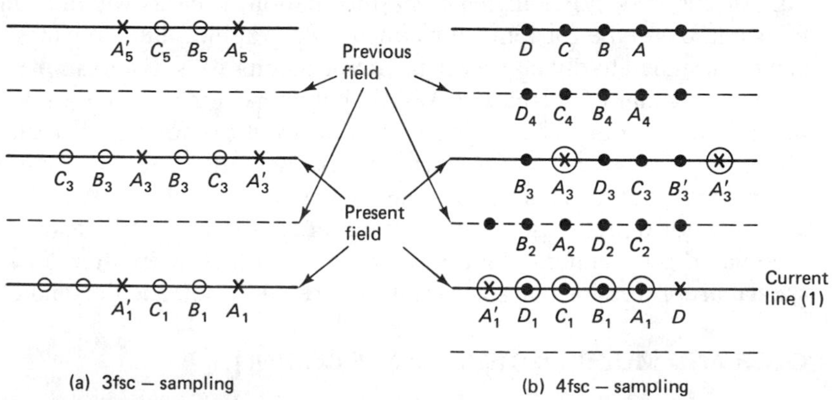

Figure 11.50 Subcarrier phase relationships in sampled NTSC signal. Pixels having the same labels have the same subcarrier phase.

the designs of predictors (for DPCM coding) or the block size (for transform coding) take into account the relative phases of the pixels. For example, at $3f_{sc}$ sampling, the adjacent samples of subcarrier have 120° phase difference and at $4f_{sc}$ sampling, the phase difference is 90°. Figure 11.50 shows the subcarrier phase relationships of neighboring pixels in the sampled NTSC signal. Due to the presence of the modulated subcarrier, higher-order predictors are required for DPCM of composite signals. Table 11.16 gives examples of predictors for predictive coding of the NTSC signal.

Table 11.17 lists the practical bit rates achievable by different coding algorithms for broadcast quality reproduced images. These results show that the chrominance components can be coded by as few as $\frac{1}{2}$ bit per pixel (via adaptive transform coding) to 1 bit per pixel (via DPCM).

Due to the flexibility in the design of coders, component coding performs somewhat better than composite coding and may well become the preferred choice with the advent of digital television.

For multispectral images, the input data is generally KL transformed in the temporal direction to obtain the principal components (see Section 7.6). Each

TABLE 11.16 Predictors for DPCM of Composite NTSC Signal. $z^{-1} = 1$ pixel delay, $z^{-N} = 1$ line delay, $z^{-262N} = 1$ field delay, ρ (leak) ≤ 1.

Sampling rate		Predictor, P(z)
$2f_{sc}$	1-D	$\frac{1}{2}z^{-1} + \rho z^{-2}(1 - \frac{1}{2}z^{-1})$
$3f_{sc}$	1-D	$\frac{1}{2}z^{-1} + \rho z^{-3}(1 - \frac{1}{2}z^{-1})$
	2-D	ρz^{-2N}
	3-D	$\frac{1}{2}z^{-1} - \rho z^{-262N}(1 - \frac{1}{2}z^{-1})$
$4f_{sc}$	1-D	$\frac{3}{4}(z^{-1} - z^{-2} + z^{-3}) + \frac{1}{4}z^{-4}$
	2-D	$\frac{1}{2}(z^{-1} - z^{-2} + z^{-3}) + \frac{1}{6}[z^{-4} + z^{-N}(z^{-2} + z^{2})]$
	2-D	$\frac{1}{2}z^{-1} + \rho z^{-2N}(1 - \frac{1}{2}z^{-1})$
	3-D	$\frac{1}{2}z^{-1} + \rho z^{-262N}(1 - \frac{1}{2}z^{-1})$

TABLE 11.17 Typical Performance of Component Coding Algorithms on Color Images

Method	Components coded	Description	Rate per component bits/component/pixel	Average rate bits/pixel
PCM	R, G, B	Raw data	8	24
	U^*, V^*, W^*	Color space quantizer	1024 color cells	10
	Y, I, Q	I, Q subsampled	8	12
DPCM One-step predictor	Y, I, Q	I, Q subsampled	2 to 3	3 to 4.5
Transform (cosine, slant)	Y, I, Q	No subsampling	Y (1.75 to 2) I, Q (0.75 to 1)	2.5 to 3
	Y, I, Q Y, U, V	Same as above with adaptive classification	Variable	1 to 2

TABLE 11.18 Summary of Image Data Compression Methods

Method	Typical average rates bits/pixel	Comments
Zero-memory methods		Simple to implement.
PCM	6–8	
Contrast quantization	4–5	
Pseudorandom noise—quantization	4–5	
Line interlace	4	
Dot interlace	2–4	
Predictive coding		
Delta modulation	1	Performance poorer than DPCM, oversample data for improvement.
Intraframe DPCM	2–3	Predictive methods are generally simple to implement, but sensitive to data statistics. Adaptive techniques improve performance substantially. Channel error effects are cumulative and visibly degrade image quality.
Intraframe adaptive DPCM	1–2	
Interframe conditional—replenishment	1–2	
Interframe DPCM	1–1.5	
Interframe adaptive DPCM	0.5–1	
Transform coding		
Intraframe	1–1.5	Achieve high performance, small sensitivity to fluctuation in data statistics, channel and quantization errors distributed over the image block. Easy to provide channel protection. Hardware complexity is high.
Intraframe adaptive	0.5–1	
Interframe	0.5–1	
Interframe adaptive	0.1–0.5	
Hybrid coding		
Intraframe	1–2	Achieve performance close to transform coding at moderate rates (0.5 to 1 bit/pixel). Complexity lies midway between transform coding and DPCM.
Intraframe adaptive	0.5–1.5	
Interframe	0.5–1	
Interframe adaptive	0.25–0.5	
Color image coding		
Intraframe	1–3	The above techniques are applicable.
Interframe	0.25–1	
Two-tone image coding		
1-D Methods	0.06–0.2	Distortionless coding. Higher compression achievable by pattern recognition techniques.
2-D Methods	0.03–0.2	

component is independently coded by a two-dimensional coder. An alternative is to identify a finite number of clusters in a suitable feature space. Each multispectral pixel is represented by the information pertaining to (e.g., the centroid of) the cluster to which it belongs. Usually the fidelity criterion is the classification (rather than mean square) accuracy of the encoded data [45, 46].

11.11 SUMMARY

Image data compression techniques are of significant practical interest. We have considered a large number of compression algorithms that are available for implementation. We conclude this chapter by providing a summary of these methods in Table 11.18.

PROBLEMS

11.1* For an 8-bit integer image of your choice, determine the Nth-order prediction error field $\varepsilon_N(m, n) \triangleq u(m, n) - \bar{u}_N(m, n)$ where $\bar{u}_N(m, n)$ is the best mean square causal predictor based on the N nearest neighbors of $u(m, n)$. Truncate $\bar{u}_N(m, n)$ to the nearest integer and calculate the entropy of $\varepsilon_N(m, n)$ from its histogram for $N = 0, 1, 2, 3, 4, 5$. Using these as estimates for the Nth-order entropies, calculate the achievable compression.

11.2 The output of a binary source is to be coded in blocks of M samples. If the successive outputs are independent and identically distributed with $p = 0.95$ (for a 0), find the Huffman codes for $M = 1, 2, 3, 4$ and calculate their efficiencies.

11.3 For the AR sequence of (11.25), the predictor for feedforward predictive coding (Fig 11.6) is chosen as $\bar{u}(n) \triangleq \rho u(n-1)$. The prediction error sequence $\varepsilon(n) \triangleq u(n) - \bar{u}(n)$ is quantized using B bits. Show that in the steady state,

$$E[|\delta u(n)|^2] = \frac{\sigma_q^2}{1-\rho^2} = \sigma_u^2 f(B)$$

where $\sigma_q^2 = \sigma_u^2(1-\rho^2)f(B)$ is the mean square quantization error of $\varepsilon(n)$. Hence the feedforward predictive coder cannot perform better than DPCM because the preceding result shows its mean square error is precisely the same as in PCM. This result happens to be true for arbitrary stationary sequences utilizing arbitrary linear predictors. A possible instance where the feedforward predictive coder may be preferred over DPCM is in the distortionless case, where the quantizer is replaced by an entropy coder. The two coders will perform identically, but the feedforward predictive coder will have a somewhat simpler hardware implementation.

11.4 (*Delta modulation analysis*) For delta modulation of the AR sequence of (11.25), write the prediction error as $e(n) = \varepsilon(n) - (1-\rho)u(n-1) + \delta e(n-1)$. Assuming a 1-bit Lloyd Max quantizer and $\delta e(n)$ to be an uncorrelated sequence, show that

$$\sigma_e^2(n) = 2(1-\rho)\sigma_u^2 + (2\rho - 1)\sigma_e^2(n-1)f(1)$$

from which (11.26) follows after finding the steady-state value of $\sigma_u^2/\sigma_e^2(n)$.

11.5 Consider images with power spectral density function

$$S(z_1, z_2) = \frac{\beta^2}{A(z_1, z_2)A(z_1^{-1}, z_2^{-1})}$$

$$A(z_1, z_2) \triangleq 1 - P(z_1, z_2)$$

where $A(z_1, z_2)$ is the minimum-variance, causal, prediction-error filter and β^2 is the variance of the prediction error. Show that the DPCM algorithm discussed in the text takes the form shown in Fig. P11.5, where $H(z_1, z_2) = 1/A(z_1, z_2)$. If the channel adds independent white noise of variance σ_η^2, show that the total noise in the reconstructed output would be $\sigma_t^2 = \sigma_q^2 + (\sigma_\eta^2 \sigma_u^2)/\beta^2$, where $\sigma_u^2 \triangleq E[(u(m, n))^2]$.

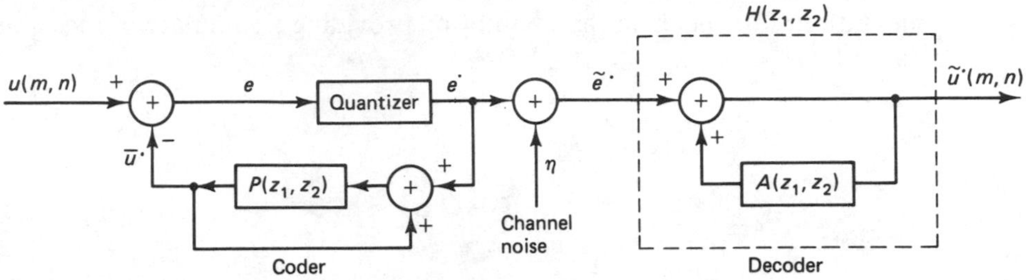

Figure P11.5

11.6 (*DPCM analysis*) For DPCM of the AR sequence of (11.25), write the prediction error as $e(n) = \varepsilon(n) + \rho\delta e(n-1)$. Assuming $\delta e(n)$ to be an uncorrelated sequence, show that the steady-state distortion due to DPCM is

$$D \triangleq E[\delta e^2(n)] = \sigma_e^2(n) f(B) = \frac{\sigma_u^2(1-\rho^2) f(B)}{1 - \rho^2 f(B)}$$

For $\rho = 0.95$, plot the normalized distortion D/σ_u^2 as a function of bit rate B for $B = 1, 2, 3, 4$ for a Laplacian density quantizer and compare it with PCM.

11.7* For a 512×512 image of your choice, design DPCM coders using mean square predictors of orders up to four. Implement the coders for $B = 3$ and compare the reconstructed images visually as well as on the basis of their mean square errors and entropies.

11.8 **a.** Using the transform coefficient variances given in Table 5.2 and the Shannon quantizer based rate distortion formulas (11.61) to (11.63), compare the distortion versus rate curves for the various transforms. (*Hint:* An easy way is to arrange σ_j^2 in decreasing order and let $\theta = \sigma_j^2$, $j = 0, \ldots, N-1$ and plot $D_j \triangleq 1/N \sum_{k=j}^{N-1} \sigma_k^2$ versus $R_j \triangleq 1/2N \sum_{k=0}^{j} \log_2 \sigma_k^2/\sigma_j^2$).

b. Compare the cosine transform R versus D function when the bit allocation is determined first by truncating the real numbers obtained via (11.61) to nearest integers and second, using the integer bit allocation algorithm.

11.9 (*Whitening transform* versus *unitary transform*) An $N \times 1$ vector $\mathbf{u}$ with covariance $\mathbf{R} = \{\rho^{|m-n|}\}$ is transformed as $\mathbf{v} = \mathbf{L}\mathbf{u}$, where $\mathbf{L}$ is a lower triangular (nonunitary) matrix whose elements are

$$l_{i,j} = \begin{cases} 1, & i = j \\ -\rho, & i - j = 1 \\ 0, & \text{otherwise} \end{cases}$$

a. Show that $\mathbf{v}$ is a vector of uncorrelated elements with $\sigma_v^2(0) = 1$ and $\sigma_v^2(k) = 1 - \rho^2$, $1 \le k \le N - 1$.

b. If $v(k)$ is quantized using n_k bits, show the reproduced vector $\mathbf{u}^\cdot \triangleq \mathbf{L}^{-1}\mathbf{v}^\cdot$ has the mean square distortion

$$D = \frac{1}{N}\sum_{k=0}^{N-1} w_k f(n_k), \qquad w_0 \triangleq \frac{(1-\rho^{2N})}{1-\rho^2}, \qquad w_k \triangleq 1 - \rho^{2(N-k)}$$

c. For $N = 15$, $\rho = 0.95$, and $f(x) = 2^{-2x}$, find the optimum rate versus distortion function of this coder and compare it with that of the cosine transform coder.

From these results conclude that it is more advantageous to replace the (usually slow) KL transform by a fast unitary transform rather than a fast decorrelating transform.

11.10 (*Transform coding* versus *DPCM*) Suppose a sequence of length N has a causal representation

$$u(n) = \bar{u}(n) + \varepsilon(n) \triangleq \sum_{k=0}^{n-1} a(n,k)u(k) + \varepsilon(n), \qquad 0 \le n \le N-1$$

where $\bar{u}(n)$ is the optimum linear predictor of $u(n)$ and $\varepsilon(n)$ is an uncorrelated sequence of variance β_n^2. Writing this in vector notation as $\mathbf{Lu} = \boldsymbol{\varepsilon}$, $\varepsilon(0) \triangleq u(0)$, where $\mathbf{L}$ is an $N \times N$ unit lower triangular matrix, it can be shown that $\mathbf{R}$, the covariance matrix of $\mathbf{u}$, satisfies $|\mathbf{R}| = \prod_{k=0}^{N-1} \beta_k^2$. If $u(n)$ is DPCM coded, then for small levels of distortion, the average minimum achievable rate is

$$R_{DPCM} = \frac{1}{2N}\sum_{k=0}^{N-1} \log_2 \frac{\sigma_e^2(k)}{D}, \qquad D < \min_k \{\beta_k^2\}$$

a. Show that at small distortion levels, the above results imply

$$R_{DPCM} > \frac{1}{2N}\sum_{k=0}^{N-1} \log_2 \frac{\beta_k^2}{D} = \frac{1}{2N}\log_2 |R| - \frac{1}{2}\log_2 D = R_{KL}$$

that is, the average minimum rate achievable by KL transform coding is lower than that of DPCM.

b. Suppose the Markov sequence of (11.25) with $\sigma_u^2 = 1$, $\beta_0^2 = 1$, $\beta_n^2 = (1-\rho^2)$ for $1 \le n \le N-1$, is DPCM coded and the steady state exists for $n \ge 1$. Using the results of Problem 11.6, with $f(B) = 2^{-2B}$, show that the number of bits B required to achieve a distortion level D is given by

$$B = \tfrac{1}{2}\log_2\left(\rho^2 + \frac{(1-\rho^2)}{D}\right), \qquad 1 \le n \le N-1.$$

For the initial sample $\varepsilon(0)$, the same distortion level is achieved by using $\frac{1}{2}\log_2(1/D)$ bits. From these and (11.69) show that

$$R_{DPCM} = \frac{1}{2N}\log_2\frac{1}{D} + \frac{(N-1)}{2N}\log_2\left(\rho^2 + \frac{(1-\rho)}{D}\right)$$

$$R_{DPCM} - R_{KL} = \frac{(N-1)}{2N}\log_2\left(1 + \frac{\rho^2 D}{(1-\rho^2)}\right), \qquad D < \frac{(1-\rho)}{(1+\rho)}$$

Calculate this difference for $N = 16$, $\rho = 0.95$, and $D = 0.01$, and conclude that at low levels of distortion the performance of KL transform and DPCM coders is close for Markov sequences. This is a useful result, which can be generalized for AR sequences. For ARMA sequences, bandlimited sequences, and two-dimensional random fields, this difference can be more significant.

11.11 For the separable covariance model used in Example 11.5, with $\rho = 0.95$, plot and compare the R versus D performances of (a) various transform coders for 16×16 size block utilizing Shannon quantizers (*Hint:* Use the data of Table 5.2.) and (b) $N \times N$ block cosine transform coders with $N = 2^n$, $n = 1, 2, \ldots, 8$. (*Hint:* Use eq. (P5.28–2).]

11.12 Plot and compare the R versus D curves for 16×16 block transform coding of images modeled by the nonseparable exponential covariance function $0.95^{\sqrt{m^2+n^2}}$ using the discrete, Fourier, cosine, sine, Hadamard, slant, and Haar transforms. (*Hint:* Use results of Problem P5.29 to calculate transform domain variances.)

11.13* Implement the zonal transform coding algorithm of Section 11.5 on 16×16 blocks of an image of your choice. Compare your results for average rates of 0.5, 1.0, and 2.0 bits per pixel using the cosine transform or any other transform of your choice.

11.14* Develop a chart of adaptive transform coding algorithms containing details of the algorithms and their relative merits and complexities. Implement your favorite of these and compare it with the 16×16 block cosine transform coding algorithm.

11.15 The motivation for hybrid coding comes from the following example. Suppose an $N \times N$ image $u(m, n)$ has the autocorrelation function $r(k, l) = \rho^{|k| + |l|}$.

a. If each column of the image transformed as $\mathbf{v}_n = \boldsymbol{\Phi}\mathbf{u}_n$, where $\boldsymbol{\Phi}$ is the KLT of $\mathbf{u}_n$, then show that the autocorrelation of $\mathbf{v}_n$, that is, $E[v_n(k)v_{n'}(k')] = \lambda_k \rho^{|n-n'|}\delta(k-k')$. What are $\boldsymbol{\Phi}$ and λ_k?

b. This means the transformed image is uncorrelated across the rows and show that the pixels along each row can be modeled by the first order AR process of (11.84) with $a(k) = \rho$, $b(k) = 1$, and $\sigma_e^2(k) = (1 - \rho^2)\lambda_k$.

11.16 For images having separable covariance function with $\rho = 0.95$, find the optimum pairs n, $k(n)$ for DPCM transmission over a noisy channel with $p_e = 0.001$ employing the optimum mean square predictor. (Hint: $\hat{\beta}^2 = (1 - \rho^2)^2$.)

11.17 Let the transition probabilities $q_0 = p(0|1)$ and $q_1 = p(1|0)$ be given. Assuming all the runs to be independent, their probabilities can be written as

$$g_i(l) = q_i(1 - q_i)^{l-1}, \qquad l \geq 1, i = 0(\text{white}), 1(\text{black})$$

a. Show that the average run lengths and entropies of white and black runs are $\mu_i = 1/q_i$ and $H_i = (-1/q_i)[q_i \log_2 q_i + (1 - q_i)\log_2(1 - q_i)]$. Hence the achievable compression ratio is $(H_0 P_0/\mu_0 + H_1 P_1/\mu_1)$, where $P_i = q_i/(q_0 + q_1)$, $i = 0, 1$ are the a priori probabilities of white and black pixels.

b. Suppose each run length is coded in blocks of m-bit words, each word representing the $M - 1$ run lengths in the interval $[kM, (k+1)M - 1]$, $M = 2^m$, $k = 0, 1, \ldots$, and a block terminator code. Hence the average number of bits used for white and black runs will be $m \sum_{k=0}^{\infty}(k+1)P[kM \leq l_i \leq (k+1)M - 1]$, $i = 0, 1$. What is the compression achieved? Show how to select M to maximize it.

BIBLIOGRAPHY

Section 11.1

Data compression has been a topic of immense interest in digital image processing. Several special issues and review papers have been devoted to this. For details and extended bibliographies:

1. Special Issues (a) *Proc. IEEE* 55, no. 3 (March 1967), (b) *IEEE Commun. Tech.* COM-19, no. 6, part I (December 1971), (c) *IEEE Trans. Commun.* COM-25, no. 11 (November 1977), (d) *Proc. IEEE* 68, no. 7 (July 1980), (e) *IEEE Trans. Commun.* COM-29 (December 1981), (f) *Proc. IEEE* 73, no. 2 (February 1985).
2. T. S. Huang and O. J. Tretiak (eds.). *Picture Bandwidth Compression.* New York: Gordon and Breach, 1972.
3. L. D. Davisson and R. M. Gray (eds.). *Data Compression.* Benchmark Papers in Electrical Engineering and Computer Science, Stroudsberg, Penn.: Dowden Hunchinson & Ross, Inc., 1976.
4. W. K. Pratt (ed.). *Image Transmission Techniques.* New York: Academic Press, 1979.
5. A. N. Netravali and J. O. Limb. "Picture Coding: A Review." *Proc. IEEE* 68, no. 3 (March 1980): 366–406.
6. A. K. Jain. "Image Data Compression: A Review." *Proc. IEEE* 69, no. 3 (March 1981): 349–389.
7. N. S. Jayant and P. Noll. *Digital Coding of Waveforms.* Englewood Cliffs, N.J.: Prentice-Hall, 1984.
8. A. K. Jain, P. M. Farrelle, and V. R. Algazi. "Image Data Compression." In *Digital Image Processing Techniques,* M. P. Ekstrom, ed. New York: Academic Press, 1984.
9. E. Dubois, B. Prasada, and M. S. Sabri. "Image Sequence Coding." In *Image Sequence Analysis,* T. S. Huang, (ed.) New York: Springer-Verlag, 1981, pp. 229–288.

Section 11.2

For entropy coding, Huffman coding, run-length coding, arithmetic coding, vector quantization, and related results of this section see papers in [3] and:

10. J. Rissanen and G. Langdon. "Arithmetic Codic." *IBM J. Res. Develop.* 23, (March 1979): 149–162. Also see *IEEE Trans. Comm.* COM-29, no. 6 (June 1981): 858–867.
11. J. W. Schwartz and R. C. Barker. "Bit-Plane Encoding: A Technique for Source Encoding." *IEEE Trans. Aerospace Electron. Syst.* AES-2, no. 4 (July 1966): 385–392.
12. A. Gersho. "On the Structure of Vector Quantizers." *IEEE Trans. Inform. Theory* IT-28 (March 1982): 157–165. Also see vol. IT-25 (July 1979): 373–380.

Section 11.3

For some early work on predictive coding, Delta modulation and DPCM see Oliver, Harrison, O'Neal, and others in *Bell Systems Technical Journal* issues of July 1952, May–June 1966, and December 1972. For more recent work:

13. R. Steele. *Delta Modulation Systems.* New York: John Wiley, 1975.
14. I. M. Paz, G.C. Collins and B. H. Batson. "A Tri-State Delta Modulator for Run-Length Encoding of Video." *Proc. National Telecomm. Conf.* Dallas, Texas, vol. I (November 1976): 6.3-1–6.3-6.

For adaptive delta modulation algorithms and applications to image transmission, see [7], Cutler (pp. 898–906), Song, et al. (pp. 1033–1044) in [1b], Lei et al. in [1c]. For more recent work on DPCM of two-dimensional images, see Musmann in [4], Habibi (pp. 948–956) in [1b], Sharma et al. in [1c],

15. J. B. O'Neal, Jr. "Differential Pulse-Code Modulation (DPCM) with Entropy Coding." *IEEE Trans. Inform. Theory* IT-21, no. 2 (March 1976): 169–174. Also see vol. IT-23 (November 1977): 697–707.
16. V. R. Algazi and J. T. DeWitte. "Theoretical Performance of Entropy Coded DPCM." *IEEE Trans. Commun.* COM-30, no. 5 (May 1982): 1088–1095.
17. J. W. Modestino and V. Bhaskaran. "Robust Two-Dimensional Tree Encoding of Images." *IEEE Trans. Commun.* COM-29, no. 12 (December 1981): 1786–1798.
18. A. N. Netravali. "On Quantizers for DPCM Coding of Picture Signals." *IEEE Trans. Inform. Theory* IT-23 (May 1977): 360–370. Also see *Proc. IEEE* 65 (April 1977): 536–548.

For adaptive DPCM, see Zschunk (pp. 1295–1302) and Habibi (pp. 1275–1284) in [1c], Jain and Wang [32], and:

19. L. H. Zetterberg, S. Ericsson, C. Couturier. "DPCM Picture Coding with Two-Dimensional Control of Adaptive Quantization." *IEEE Trans. Commun.* COM-32, no. 4 (April 1984): 457–462.
20. H. M. Hang and J. W. Woods. "Predictive Vector Quantization of Images," *IEEE Trans. Commun.*, (1985).

Section 11.4

For results related to the optimality of the KL transform, see Chapter 5 and the bibliography of that chapter. For the optimality of KL transform coding and bit allocations, see [6], [32], and:

21. A. Segall. "Bit Allocation and Encoding for Vector Sources." *IEEE Trans. Inform. Theory* IT-22, no. 2 (March 1976): 162–169.

Section 11.5

For early work on transform coding and subsequent developments and examples of different transforms and algorithms, see Pratt and Andrews (pp. 515–554), Woods and Huang (pp. 555–573) in [2], and:

22. A. Habibi and P. A. Wintz. "Image Coding by Linear Transformation and Block Quantization." *IEEE Trans. Commun. Tech.* COM-19, no. 1 (February 1971): 50–63.

23. P. A. Wintz. "Transform Picture Coding." *Proc. IEEE* 60, no. 7 (July 1972): 809–823.
24. W. K. Pratt, W. H. Chen, and L. R. Welch. "Slant Transform Image Coding." *IEEE Trans. Commun.* COM-22, no. 8 (August 1974): 1075–1093.
25. K. R. Rao, M. A. Narasimhan, and K. Revuluri. "Image Data Processing by Hadamard-Haar Transforms." *IEEE Trans. Computers* C-23, no. 9 (September 1975): 888–896.

The concepts of fast KL transform and recursive block coding were introduced in [26 and Ref 17, Ch 5]. For details and extensions see [6], Meiri et al. (pp. 1728–1735) in [1e], Jain et al. in [8], and:

26. A. K. Jain. "A Fast Karhunen-Loève Transform for a Class of Random Processes." *IEEE Trans. Commun.* COM-24 (September 1976): 1023–1029.
27. A. K. Jain and P. M. Farrelle. "Recursive Block Coding." Sixteenth Annual Asilomar Conference on Circuits, Systems, and Computers, November 1982. Also see P. M. Farrelle and A. K. Jain. *IEEE Trans. Commun.* (February 1986), and P. M. Farrelle, Ph.D. Dissertation, U.C. Davis, 1988.

For results on two-source coding, adaptive transform coding, and the like, see Yan and Sakrison (pp. 1315–1322) in [1c], Jain and Wang [32], Tasto and Wintz (pp. 956–972) in [1b], Graham (pp. 336–346) in [1a], Chen and Smith in [1c], and:

28. W. F. Schreiber, C. F. Knapp, and N. D. Kay. "Synthetic Highs: An Experimental TV Bandwidth Reduction System." *J. Soc. Motion Picture and Television Engineers* 68 (August 1959): 525–537.
29. V. R. Algazi and D. J. Sakrison. "Encoding of a Counting Rate Source with Orthogonal Functions." *Computer Processing in Communications,* N.Y.: Polytechnic Institute of Brooklyn, 1969, pp. 85–100.
30. J. L. Mannos and D. J. Sakrison. "The Effects of a Visual Fidelity Criterion on the Encoding of Images." *IEEE Trans. Inform. Theory* IT-20 (July 1974): 525–536.

Section 11.6

Hybrid coding principle, its analysis and relationship with semicausal models, and its applications can be found in [6, 8], Jones (Chapter 5) in [4], and:

31. A. Habibi. "Hybrid Coding of Pictorial Data." *IEEE Trans. Commun.* COM-22 (May 1974): 614–626.
32. A. K. Jain and S. H. Wang. "Stochastic Image Models and Hybrid Coding," *Final Report,* NOSC contract N00953-77-C-003MJE, Department of Electrical Engineering, SUNY Buffalo, New York, October 1977. Also see Technical Report #SIPL-79-6, Signal and Image Processing Laboratory, ECE Dept., University of California at Davis, September 1979.
33. R. W. Means, E. H. Wrench and H. J. Whitehouse. "Image Transmission via Spread Spectrum Techniques." *ARPA Quarterly Technical Reports* ARPA-QR6, QR8 Naval Ocean Systems Center, San Diego, Calif., January–December 1975.
34. A. K. Jain. "Advances in Mathematical Models for Image Processing." *Proc. IEEE* 69 (March 1981): 502–528.

Section 11.7

For interframe predictive coding, see [5, 6, 8], Haskell et al. (pp. 1339–1348) in [1c], Haskell (Chapter 6) in [4], and:

35. J. C. Candy, et al. "Transmitting Television as Clusters of Frame-to-Frame Differences." *Bell Syst. Tech. J.* 50 (August 1971): 1889–1917.

36. J. R. Jain and A. K. Jain. "Displacement Measurement and Its Application in Interframe Image Coding." *IEEE Trans. Comm.* COM-29 (December 1981): 1799–1808.

37. A. N. Netravali and J. D. Robbins. "Motion Compensated Television Coding—Part I," *Bell Syst. Tech. J.* (March 1979): 631–670.

Interframe hybrid and transform coding techniques are discussed in [5, 6, 8, 9], Roese et al. (pp. 1329–1338), Natrajan and Ahmed (pp. 1323–1329) in [1c], and:

38. J. A. Stuller and A. N. Netravali. "Transform Domain Motion Estimation," *Bell Syst. Tech. J.* (September 1979): 1623–1702. Also see pages 1703–1718 of the same issue for application to coding.

39. J. R. Jain and A. K. Jain. "Interframe Adaptive Data Compression Techniques for Images." *Tech. Rept., Signal and Image Processing Laboratory,* ECE Department, University of California at Davis, August 1979.

40. J. O. Limb and J. A. Murphy. "Measuring the Speed of Moving Objects from Television Signals," *IEEE Trans. Commun.* COM-23 (April 1975): 474–478.

41. C. Cafforio and F. Rocca. "Methods for Measuring Small Displacements of Television Images." *IEEE Trans. Inform. Theory* IT-22 (September 1976): 573–579.

Section 11.8

For the optimal mean square encoding and decoding results and their application, we follow [6, 39] and:

42. G. A. Wolf. "The Optimum Mean Square Estimate for Decoding Binary Block Codes." Ph.D. Thesis, University of Wisconsin at Madison, 1973. Also see G. A. Wolf and R. Redinbo. *IEEE Trans. Inform. Theory* IT-20 (May 1974): 344–351.

For extended bibliography, see [6, 8].

Section 11.9

[1f] is devoted to coding of two tone images. Details of CCITT standards and various algorithms are available here. Some other useful references are Arps (pp. 222–276) in [4], Huang in [1c, 2], Musmann and Preuss in [1c], and:

43. H. Kobayashi and L. R. Bahl. "Image Data Compression by Predictive Coding I: Prediction Algorithms." and "II: Encoding Algorithms." *IBM J. Res. Dev.* 18, no. 2 (March 1974): 164–179.

44. T. S. Huang and A. B. S. Hussain. "Facsimile Coding by Skipping White." *IEEE Trans. Commun.* COM-23, no. 12 (December 1975): 1452–1466.

Section 11.10

Color and multispectral coding techniques discussed here have been discussed by Limb et al. in [1c], Pratt in [1b], and:

45. F. E. Hilbert. "Cluster Compression Algorithm, A Joint Clustering/Data Compression Concept." JPL Publication 77-43, Jet Propulsion Laboratory, Pasadena, Calif., December 1, 1977.

46. J. N. Gupta and P. A. Wintz. "A Boundary Finding Algorithm and its Applications." *IEEE Trans. Circuits and Systems* CAS-22 (April 1976): 351–362.

Section 11.11

Several techniques not discussed in this chapter include nonuniform sampling techniques combined with interpolation (such as using splines), use of singular value decompositions, autoregressive (AR) model synthesis, and the like. Summary discussion and the relevant sources of these and other useful methods are given in [6, 8].

Index

Aberrations, 269
Adaptive prediction, 495, 520, 522, 552
Adaptive transform coding, 516
Affine transformation, 321
Algebraic methods, 464
Aliasing, 87, 89
All-pole models, 192
All-zero models, 195
Alternating projection method, 325
Analytic continuation, 323
Anticausal, 21
Area, 391
Area correlation, 400, 402
Arithmetric coding, 483
Atomic regions, 412
Autocorrelation function (ACF), 395
Autoregressive (AR) models, 190, 193, 198, 374
Autoregressive moving average (ARMA) representations, 195, 207
Average energy, 344
Average least squares, mean square, 59, 292

Back-projection, 439
Band pass filtering, 250
Bandlimited, 84
Basis functions, 5, 132, 135, 148
Basis restriction error, 166
Bayesian methods, classifier, 319, 416
Bending energy, 392
Best linear estimate, 277
Best-fit ellipse, 394
Bipolar cells, 50
Bit allocation algorithm, 501, 503
Bit extraction, 239
Bit-plane encoding, 483
Blending function, 364
Blind deconvolution, 322, 323
Bloch's law, 75
Block coding, 478
Block matrices, 28
Block quantization, 498
Boundary, 357, 362, 371, 387, 411
Boundary response, 168
Bounding rectangle, 393
Brightness, 49, 51, 60
Bubble sort, 58

Causal systems, prediction, 5, 21, 190, 205, 213, 219
Center of mass, 392
Cepstrum, 219, 259, 292
Chain codes, 363
Channel error effects, 532, 536
Chromaticity diagram, 65
Chrominance, 82
Circulant, 25, 28, 150
Classification techniques, 313, 414, 415, 417
Clipping, 235
Closing, 387
Clustering, 412, 418–21
Coarse quantization, 479
Coarse-fine search, 407
Coding of two tone images, 540
Color, 60, 73
Color and multi-spectral image coding, 553
Color gamut, 66
Color image enhancement, 262
Color, vision model, 73
Colorimetry, 65
Commuting distance, 176
Compactness, 392
Compandor, 113
Compass operators, 348, 350
Completeness, 134
Component labeling, 409
Computerized tomography, 431
Concurrence matrix, 346
Condition number, 301
Conditional replenishment, 522
Cones, 49, 62
Connectivity, 357
Constained least square restoration, 297
Contour following, 358
Contouring, 119, 120

Contrast, 49, 51, 120, 235
Control points, 364, 366
Covariance generating function (CGF), 37
Convolution, 14, 18, 144, 148, 444
Convolution back-projection, 446, 448, 450
Corners, 392
Cosine transform, 150
Covariance models, 36, 189
Crack-following algorithm, 358
Critical fusion frequency (CFF), 75
Cross relation, 396

Data compression, 9, 132, 138, 476–567
Decision levels, 99
Decision tree classifier, 415
Decorrelation, 140
Delta modulation, 488
Detector models, 273
Diagonal forms, 27
Dictionary ordering, 23
Differential pulse coding modulation (DPCM), 100, 114, 204, 477, 483–95
Digital image processing, 1
Digital negative, 238
Dilation, 384
Directional smoothing, 245
Discrete cosine transform (DCT), 150
Discrete fourier transform (DFT), 141, 145, 147
Discrete prolate spheroidal sequences (DPSS), 327
Discriminant function, 414
Dispersion, 344
Displacement estimation algorithms, 404, 525
Distance transform, 382
Dither, 120
Dynamic programming, 359
Dynamic range, 102

Eccentricity, 394
Edge detection, linking, density, 160, 335, 347, 358, 396
Edge-busyness, 494
Emission tomography, 433
Energy conservation compaction, 138, 139
Entropy, 42
Entropy coding, 480
Erosion, 384
Estimation theory, 39
Euler number, 392
Even field, 479
Extrapolation of signals, 322, 324

Fan-bean reconstruction, 464
Fast Fourier transform (FFT), 142
Fast KL transform, 168, 202, 223, 510
Fast transform, 137, 138, 148
Feature extraction, 342, 343
Filter back-projection, 446, 448, 451
Filtering using image transforms, 292
Finite impulse response, (FIR), 13, 284
Fitting line segments, 364
Fixed-interval smoother (Wiener filter), 306
Flat field, 99
Fold-over frequencies, 87
Fourier descriptors (FDS), 370
Fourier reconstruction method, 462
Fourier transform, 15, 18
Fovea, 50
Frame, 81, 479, 521
Frequency response, 17, 19, 281

Gamma of the film, 273
Ganglion cells, 50
Gaussian distribution, 32, 258
Generalized inverse, 276, 299, 300
Generalized linear filtering, 256, 293
Generalized Wiener filtering, 293
Geodesic, 72
Geometric correction, 320
Geometric mean filter, 291
Geometry features, 391
Gradient methods, 301, 302, 330
Gradient operators, 348
Granularity, 494
Grassman's laws, 65

Haar Functions, 160
Hadamard transform, order, 156, 157
Halftone, 84, 121
Hankel transform, 18
Hexagonal sampling, 92
High pass, 250
Hilbert transform, 445
Histogram modeling, equalization, features, 241–243, 344, 396
Hit-miss transform, 387
Homomorphic filtering, 259, 315, 316
Hough transform, 362
Hue, 60
Huffman coding, 480
Hybrid algorithms, 204, 226, 518, 519

Idempotent, 326
Image analysis, 7, 342
Image data compression, 9, 476–567
Image enhancement, 6, 233, 260
Image filtering, 431
Image perception, 49
Image reconstruction, 8, 431, 452, 458, 469
Image representation and modeling, 4, 189, 268, 269
Image restoration, 7, 267–335
Image scanning, 80, 196
Image subtraction and change detection, 240, 400
Image transforms, 5, 132, 134, 135, 396
Image understanding, 421
Impulse response, 13, 14
Infinite impulse response (IIR), 13
Information, 41–43, 477
Inner product, 118, 135
Innovations process, 191, 305
Intensity level slicing, 238
Intensity ratios, 260
Interframe coding, 521, 527
Interpolation, 204, 253, 297
Inverse contrast ratio, 252
Inverse filtering, 275
Iterative methods, 299, 327, 420

Just noticeably different (jnd), 68, 72

Kalman filtering, 304
KL transform, 34, 92, 163, 176
Knots, 364
Kronecker products, 28, 31, 137
Kurtosis, 344

Lagrange interpolation, 98
Laplace operator, 351
Lateral inhibition, 54
Laurent series, 20
Leak, 488
Leakage, 98
Least square filters, 297
Levinson algorithm, 198
Lexicographic ordering, 23
Light, 49
Line detection, 356
Line of purples, 66
Line quincunx, 92
Linear learning machine, 552
Linear prediction, 204
Linear predictive coding (LPC), 194
Linear systems, 13, 189
Linearly separable, 415
Lloyd Max quantizer, 101
Log-ratios, 261
Luminance, 49, 50
Luminosity coefficient, 64

Macadam ellipses, 68
Mach bands, 53
Magnetic resonance imaging, 434
Magnification and interpolation (zooming), 253
Markov process, random field, 33, 196, 210
Masking function, 55
Masks, 204, 348
Matched filtering, 403
Matrix theory results, 22–31
Maximum a posteriori (MAP), 319
Maximum entropy, 193, 316, 318
Maximum likelihood (ML), 319
Mean square, error, estimate, extrapolation, 40, 59, 60, 319, 328, 330, 344
Medial axis, 381, 382
Median, 344
Median filtering, 246, 249
Medical imaging, 432
Mesotropic, 50
Minimum norm least squares (MNLS) solution, 300, 326
Minimum variance representation (MVR), 191, 206–8
Minimum-phase filter, 197
Mode, 344
Modulation transfer function (MTF), 21, 54
Moire effect, 99
Moments, invariants, matching, features, 377–81, 392
Monochrome vision model, 56
Morphological processing, 384, 387
Mosaic models, 399
Moving average (MA) representations, 194
Multispectral image enhancement, 260
Munssel book of color, 73

Neural, 54
Nodes, 365
Noise models, 273
Noiseless coding theorem, 478
Noncausal systems, 21, 200–204, 206, 212, 224
Non-destructive testing, 432
Nonlinear filters, 291
Nonparametric techniques, 417
Nonsymmetric half-plane (NSHP), 205
Normal distribution, 32
Normal equations, 192
NTSC, 11, 82
Nyquist frequency interval, 87

Odd field, 479
One-dimensional system, 29
Online filter, 305
Open, opening, 358, 387
Optic nerve, 50
Optical transfer function (OTF), 21
Optimal sampling, 92
Optimum mean square quantizer, filter, decoder, 40, 101, 115, 458, 532
Optimum transform, 138, 498
Orientation, 392
Orthogonality, 26, 34, 35, 324
Orthogonal decomposition, 203
Orthogonal gradient, 351
Orthogonal moments, 379
Orthonormality, 134, 324
Outer product expansion, 176, 177

Perimeter, 391
Periodic random fields, 223
Phase alternating line (PAL) system, 82
Photopic, 50
Picture element (PEL), 4
Pixel coding, 479
Pixel labeling, 409
Point operations, 235
Point spread function (PSF), 7, 13
Positron emission tomography (pet), 433
Power spectral density, 506
Prediction, predictor, 191, 208, 305, 477, 483, 484
Prediction differential quantization, 547
Predictive coding, 485, 526, 551
Primary sources, 62
Primitives, 389
Principal components, 261
Principal of optimality, 359
Profile spread, 396
Projection geometry, operators, theorems, 376, 434, 442, 457–69
Pruning, 387
Pseudocolor, 262
Pseudoinverse, filters, 276, 299, 327
Pulse code modulator (PCM), 100, 480, 534

Quad-trees, 375, 412
Quantization, 80, 99–131

Radar cross section, 344
Radii, 391
Radon transform, 434, 444, 446, 447, 452
Ram-Lak filter, 448
Random variable, signals, image field, 31–35, 189
Range compression, 240
Rate-distortion function, 43, 168, 502
Rational SDF, 197
Ray-sum, 435
Realization of MVRS, 198, 215, 216
Receiver primary system, 69
Reconstruction (*see* Image reconstruction.)
Reconstruction levels, 99
Reconstruction from samples, 85
Recursive block coding, 510, 511, 512
Recursive filtering, 304–12
Reflection, 380
Reflection coefficients, 199
Reflection tomography, 432, 433
Region growing, 412
Region of support, 13

Region-based approaches, 375, 412
Relative address and READ coding, 548
Relative luminous efficiency function, 50
Resolution exchange, 521
Riccati equation, 305
Ripple, 98
Rods, 49
Root filtering, 258, 291
Root law, 52
Rotation, 380, 392
Roundness, 392
Row and column ordering, 23
Run-length coding, 375, 481, 541
Run length connectivity analysis, 409

Sampling, 80–99, 147, 275, 448
Saturation, 60
Scaling, 380
Scene matching, 400
Scotopic, 49
Search algorithms, 358, 407, 525
Second-degree spread, 396
Segmentation, 343, 407, 413
Semicausal systems, 206, 213, 220, 222, 225, 311
Separability, 12, 16, 31, 36, 134, 214, 303
Sequency, 138, 156
Sequential Coleur a Memoire (SECAM), 82
Sequential probability ration test (SPRT), 417
Shadowgram, 431
Shannon Quantizer, bound, 117
Shape features, 390
Shift invariant, 13, 14
Signal-to-noise ratio (SNR), 59
Similarity measure, 419
Singular value decomposition (SVD), 113, 176, 177, 299
Sinusoidal family of transforms, 175, 203
Skeleton, 382, 387, 389
Skewness, 344
Slant transform, 161
Slope overload, 494
Spatial averaging, 244, 315
Spatial correlation, 18
Spatial frequency, 16, 138
Spatial low-pass filtering, 244, 250
Spatially invariant, 14, 35
Spatially varying, 14, 287
Spatial masks, 244
Speckle, 274, 313
Spectral density function (sdf), 37, 38
Spectral estimation, 213
Spectral factorization, 190, 196, 198, 213, 219
Spectral matching curves, 63
Spectral representation, 177
Spectral responses, 61
Splines, B-, cubic, interpolating, 296, 364
Spot detection, 356
Stability, 21, 190, 212
State variable models, 195
Stationary processes, 32, 310, 513
Statistical scaling, 252
Steepest descent, 301
Stochastic decoupling, 223
Stochastic gradients, 353
Structural approaches, 398
Structure, 381
Structuring element, 384
Summation algorithm, 441
Super-resolution, 323
Supervised learning, 414, 421
Syntactic representations, 389
Synthetic highs, 515

Television standards, 81
Template matching, 400, 404, 413
Temporal properties of vision, 75
Texture, 394, 397, 413
Thickening, 387
Thinning, 382, 387
Threshold representation, coding, 168
Thresholding, 235, 407, 509
Toeplitz, 25, 28
Transfer function, 20, 21
Transform coding, 477, 498, 504, 507, 515, 516, 529, 537
Transform coefficients, features, 134, 138, 139, 256, 346
Transition levels, 99
Translation invariant, 35
Transmission tomography, 431, 432
Tristimulus values, space, 63, 72
Two-dimensional logarithmic search, 404
Two-dimensional spectral factorization problem, 213
Two-dimensional systems, 11
Two-source coding, 513

Uniform chromaticity scale (UCS), 68
Unitary matrices, transforms, 26, 134, 138, 141, 145, 177
Unsharp masking, 249

Van Cittert filter, 301
Vector quantization, 478
Visibility function, 55, 56
Visual quantization, 119

Walsh functions, 156
Wave number, 138
Weber's law, 51
White block skipping, 546
White noise drive representation (WNDR), 207
White noise field, 35
Whitening filter, 191
Whitening transform, 456
Wide-sense stationary, 33, 35
Wiener filter, 7, 276–92, 306
Wiener-Doob homomorphic transformation, 219
Window slicing, 407

Z-transform, 20
Zernike moments, polynomials, 381
Zero crossing, 351
Zonal coding, 508, 509
Zonal mask, 173, 508